FREDERICK A. BETTELHEIM
Adelphi University

WILLIAM H. BROWN
Beloit College

JERRY MARCH
Adelphi University

INTRODUCTION TO
GENERAL, ORGANIC & BIOCHEMISTRY

sixth edition

HARCOURT COLLEGE PUBLISHERS

Fort Worth Philadelphia San Diego New York Orlando Austin San Antonio Toronto Montreal London Sydney Tokyo

*This text is dedicated to
the living memory of Jerry March
(1929 – 1997)*

Publisher: Emily Barrosse
Publisher/Acquisitions Editor: John Vondeling
Marketing Strategist: Pauline Mula
Associate Editor: Marc Sherman
Project Editor: Frank Messina
Production Manager: Charlene Catlett Squibb
Art Director: Lisa Adamitis
Text Designer: Caroline McGowan

Cover Credit: Monarch butterfly. *(© Photodisc)*

INTRODUCTION TO GENERAL, ORGANIC & BIOCHEMISTRY, Sixth Edition,
ISBN: 0-03-025547-3
Library of Congress Catalog Card Number: 00-103641

Address for domestic orders:
Harcourt College Publishers, 6277 Sea Harbor Drive, Orlando, FL 32887-6777
1-800-782-4479
e-mail   collegesales@harcourt.com

Address for international orders:
International Customer Service, Harcourt, Inc.
6277 Sea Harbor Drive, Orlando, FL 32887-6777
(407) 345-3800
Fax   (407) 345-4060
e-mail   hbintl@harcourt.com

Address for editorial correspondence:
Harcourt College Publishers, Public Ledger Building, Suite 1250,
150 S. Independence Mall West, Philadelphia, PA 19106-3412

Web Site Address
http://www.harcourtcollege.com

Printed in the United States of America

0 1 2 3 4 5 6 7 8 9   032   10 9 8 7 6 5 4 3 2 1

Here's What Your Colleagues Are Saying About Bettelheim, Brown & March...

The main reason I continue to choose Bettelheim for my course is the readability of the text. The large number of medical application boxes is also extremely important in showing the relevancy of the material to [the students'] future careers in allied health. Another strong feature of the text is the large number of worked example problems and an extremely good collection of end-of-chapter problems.

— *Eric Johnson, Ball State University*

The key strength of this text is its readability. Students actually read this textbook! The writing style and level of presentation are appropriate for the audience that this text is aimed at. The boxes, I believe, are another strength of the book. The Test Bank is very good, as are the illustrations.

— *Steve Socol, Southern Utah University*

Breaking up the Organic chapters into more and shorter chapters is an excellent idea . . . . The separate chapter on Chirality is also an excellent idea . . . the applications throughout the text are very good and bring a practical aspect to the subject.

— *Jack Hefley, Blinn College*

This text has one of the lowest incidences of factual and typographical errors that I have seen. Students have no complaints about being able to understand the authors' meaning and message. Text is also well-illustrated; color is used effectively . . . . I like the new Organic presentation very much, especially since it does allow the individual instructor some flexibility without drastically altering the flow of the text.

— *Larry McGahey, The College of St. Scholastica*

The clarity of the writing and the style of presentation are the real strengths of this textbook. Students find it easy to read and to understand.

— *James Yuan, Old Dominion University*

Setting off the subject of Chirality is an excellent decision . . . many students get the impression that chirality is strongly associated with carbohydrates and only marginally transferable to other biomolecules.

— *William Scovell, Bowling Green State University*

The authors do a good job of addressing the chemistry of the various functional groups that are important in biochemistry . . . . The excellent discussions of reactions in the organic section can only make it easier for students to understand the metabolic reactions they encounter later.

— *David Reinhold, Western Michigan University*

The chapters dealing with DNA and genetic expression are excellent. They were written in general terms, not too descriptive, with an appropriate level of detail. The end-of-chapter questions were excellent. These chapters are among the best I have seen describing this material.

— *Bobby Stanton, University of Georgia*

I particularly like the placement and coverage of the chapter on Nuclear Chemistry. Other authors seem inclined to include it up front, combined with chapters on atomic structure.

— *Richard Hoffman, Illinois Central College*

The concepts are well explained, clear and concise. There are enough homework problems to help the students. Some other textbooks have been watered down so much that it is difficult for students to fully understand concepts. Other textbooks don't have enough problems to help the students learn the concepts thoroughly.

— *Jennifer Tan, Ohlone College*

ABOUT THE AUTHORS

Frederick A. Bettelheim

Frederick A. Bettelheim is a Distinguished University Research Professor at Adelphi University, and a Visiting Scientist at the National Eye Institute. He has coauthored every edition of *Introduction to General, Organic & Biochemistry* and several laboratory manuals, including *Laboratory Experiments for General, Organic & Biochemistry* and *Experiments for Introduction to Organic Chemistry*. He is the author of *Experimental Physical Chemistry* and coauthor of numerous monographs and research articles. Professor Bettelheim received his Ph.D. from the University of California, Davis, and his areas of specialization have included the biochemistry of proteins and carbohydrates and the physical chemistry of polymers. He was Fulbright Professor at the Weizman Institute, Israel, and Visiting Professor at the University of Uppsala, Sweden, at Technion, Israel, and at the University of Florida. He is keynote lecturer at the 16th International Conference on Chemical Education, August 2000, in Budapest, Hungary, where his lecture topic is Modern Trends in Teaching Chemistry to Nurses and Other Health-Related Professionals.

William H. Brown

William H. Brown, a new coauthor for the sixth edition of *Introduction to General, Organic & Biochemistry,* is Professor of Chemistry at Beloit College, where he has twice been named Teacher of the Year. He is also the author of two best-selling undergraduate texts, *Introduction to Organic Chemistry,* Second Edition, and *Organic Chemistry,* Second Edition. His regular teaching responsibilities include organic chemistry, advanced organic chemistry, and special topics in pharmacology and drug synthesis. He is the leader of a team revising the organic chemistry material appearing in the upcoming *Encyclopaedia Britannica CD-ROM*. Professor Brown received his Ph.D. from Columbia University under the direction of Gilbert Stork and did postdoctoral work at the California Institute of Technology and the University of Arizona.

Jerry March

Jerry March (d. 1997) was the coauthor of *Introduction to General, Organic & Biochemistry* for its first five editions. He was the sole author of the best-selling *Advanced Organic Chemistry* text, now in its fifth edition, which has been translated into many languages including Russian and Japanese. He was a member of the Physical Organic Chemistry Commission of the International Union of Pure and Applied Chemistry, and as such was instrumental in the development of several nomenclature systems. Professor March received his Ph.D. at Pennsylvania State University and was Professor of Chemistry at Adelphi University, specializing in organic chemistry.

CONTENTS OVERVIEW

(James Balog / Tony Stone Images)

(NASA)

CONTENTS

(*Charles D. Winters*)

(Christy Carter / Grant Heilman Photography, Inc.)

PART THREE BIOCHEMISTRY

CHAPTER 19
Carbohydrates 440

(© Meckes / Ottawa / Photo Researchers, Inc.)

CHAPTER 20
Lipids 470

CHAPTER 24
Nucleotides, Nucleic Acids, and Heredity   566

CHAPTER 25
Gene Expression and Protein Synthesis   586

CHAPTER 26
Bioenergetics. How the Body Converts Food to Energy   607

CHAPTER 27
Specific Catabolic Pathways: Carbohydrate, Lipid, and Protein Metabolism   628

CHAPTER 28
Biosynthetic Pathways 651

CHAPTER 29
Nutrition and Digestion 664

CHAPTER 30
Immunochemistry 681

CHAPTER 31
Body Fluids 697

(Charles D. Winters)

In me are hidden constellations.

> Once I managed to sight one
> through the lens of equations
> that could be solved only
> approximately. Still
> with that imperfect rule
> I taught others the electrons
> lobed motions. . . .
> I work this wild chemical
> garden with one old tool.

Let me show others new ways to see.

Roald Hoffmann: Gaps and Verges

It is the dream of every educator to convey the subject for all to see, to elicit excitement in learning, to discover in wonderment, and through the discovery to let the self be discovered. In writing this Preface for the sixth edition of our textbook, it is tempting to recall the excitement and elation that greeted the first edition 16 years ago. We had a book that was novel in its coverage and its pedagogy that made concepts relevant by frequent examples of their applications. Throughout the years, the cutting edge of this presentation was acknowledged and rewarded by the adoption of this book by our colleagues for their courses. The ideas originated and developed in our book became the standard and were incorporated into many textbooks.

Now it is time to provide a new look: we brighten an old gem, polish its surface, turn its edges toward the Sun, and let it glitter. We hope that in this new edition we have managed to achieve the most comprehensive treatment of the subject of chemistry in a clearcut presentation for the edification and enjoyment of our students.

This sixth edition intends to be even more readable and understandable than earlier editions, and we have forged a greater unity of the three domains of the text: general, organic, and biochemistry. Chemistry, especially biochemistry, is a fast-developing discipline, and we include new, relevant material in the text. We have done this not only by upgrading information, but also by enlarging the scope of the book in the text and in the boxes containing medical and other applications of chemical principles. At the same time, we are aware of the need to keep the book to a manageable size and proportion. Approximately 25 percent of the problems are new, and we have increased the number of more challenging, thought-provoking problems (marked by asterisks).

Audience

(Charles D. Winters)

As were the previous editions, this book is intended for nonchemistry majors, mainly those entering health science and related fields (such as nursing, medical technology, physical therapy, and nutrition). It also can be used by students in environmental studies. In its entirety, it can be used for a one-year (two-semester or three-quarter) course in chemistry, or parts of the book can be used in a one-term chemistry course.

We assume that the students using the book have little or no background in chemistry. Therefore, we introduce the basic concepts slowly at the beginning and increase the tempo and the level of sophistication as we go on. We progress from the basic tenets of general chemistry to organic and, finally, to biochemistry. We consider this progression an ascent in terms of both practical importance and sophistication. However, we integrate the three parts by keeping a unified view of chemistry. We do not consider the general chemistry sections to be the exclusive domain of inorganic compounds, so we frequently use organic and biological substances to illustrate general principles.

While teaching the chemistry of the human body is our ultimate goal, we try to show that each subsection of chemistry is important in its own right, besides being required for future understanding.

Boxes (Medical and Other Applications of Chemical Principles)

The boxes contain applications of the principles discussed in the text. Comments from users of the earlier editions indicate that these have been especially well-received, providing a much requested relevance to the text. The large number of boxes deal mainly with health-related applications, including ones related to the environment. Some boxes from the fifth edition have been dropped, and a few others have been incorporated into the text or other boxes.

Numerous new boxes dealing with diverse topics such as the human isotopic abundance and astrochemistry, using coral to repair broken bones, reverse osmosis and desalinization, chiral drugs, toxicity and drug dosage, transport across cell membranes, protein conformation–dependent diseases, and tumor suppressor genes, among others, have been added. Many boxes have been enlarged and updated. For example, boxes on nitric oxide, anti-inflammatory drugs, Alzheimer's disease, laser surgery, and AIDS now contain recent information.

The presence of boxes allows a considerable degree of flexibility. If an instructor wants to assign only the main text, the boxes do not interrupt continuity, and the essential material will be covered. However, most instructors will probably wish to assign at least some of the boxes, since they enhance the core material. In our experience, students are eager to read the relevant boxes without assignments and they do so with discrimination. From such a large number of boxes, the instructor can select those that best fit the particular needs of the course and of the students. **Problems are provided at the end of each chapter for nearly all of the boxes.**

Metabolism: Color Code

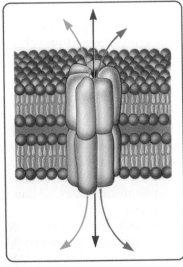

The biological functions of chemical compounds are explained in each of the biochemistry chapters and in many of the organic chapters. Emphasis is placed on chemistry rather than on physiology. We have received much positive feedback regarding the way in which we have organized the topic of metabolism (Chapters 26, 27, and 28). We have maintained this organization.

First we introduce the common metabolic pathway through which all food will be utilized (citric acid cycle; oxidative phosphorylation), and only after that do we discuss the specific pathways leading to the common pathway. We find this a useful pedagogic device, and it enables us to sum up the caloric values of each type of food because their utilization through the common pathway has already been learned. Finally, we separate the catabolic pathways from the anabolic pathways by treating them in different chapters, emphasizing the different ways the body breaks down and builds up different molecules.

The topic of metabolism is a difficult one for most students. We have tried to explain it as clearly as possible. As in the previous edition, we enhance the clarity of presentation by the use of a color code for the most important biological compounds discussed in Chapters 26, 27, and 28. Each type of compound is screened in a specific color, which remains the same throughout the three chapters. These colors are as follows:

ATP and other nucleoside triphosphates

ADP and other nucleoside diphosphates

The oxidized coenzymes NAD^+ and FAD

The reduced coenzymes NADH and $FADH_2$

Acetyl coenzyme A

The circled numbers in the figures showing the steps involved in the various metabolism pathway are always in yellow.

In addition to this main use of a color code, other figures in various parts of the book are color-coded, so that the same color is used for the same entity throughout. For example, in Chapter 22, enzymes are always shown in blue and substrates in orange in all the figures that show enzyme-substrate interactions.

Features

One of the main features of this book, as in earlier editions, is the **number of applications of chemical concepts presented in the boxes.** Another important feature is the **Glossary,** which is now separate from the Index. The definition of each term is given along with the number of the section in which the term appears. Another feature is the list of **Key Terms** at the end of each chapter, with notation of the section number in which the term is introduced. Some chapters list **Key Reactions** after the Key Terms. Many students find these lists to be helpful study tools.

Other features are the **Summary** at the end of each chapter and the substantial number of **Margin Notes.** In this sixth edition, we have added **Margin Definitions** to complement nearby text. Other new features include **Conceptual Problems** at the end of every chapter, and a new type of problem called **Reading the Labels** in Chapter 3, in order to provide a balance of conceptual understanding and reinforcement of critical skills. Here is a summary of the new features:

(James King–Holmes / Science Photo Library / Photo Researchers, Inc.)

New Feature	Location	Benefit
Margin definitions	Throughout text	Helps student learn terminology; avoids cluttering narrative with definitions
Molecular artwork	Throughout text	Helps student visualize molecular properties and reactions
Conceptual Problems	At end of each chapter	Helps student frame the problems in the proper conceptual framework; avoids rote problem-solving
Key Reactions	At end of most Organic chapters	Provides convenient reference
Reading the Labels	At end of Chapter 3	Provides important practice in understanding product labels
Glossary separate from Index; terms identified by section number	At end of text, precedes Index	Allows for more terms to be defined and thus improves student comprehension of terminology

Style

Feedback from colleagues and students alike indicates that the style of the book, which addresses the students directly in simple and clear phrasing, is one of its major assets. We continue to make special efforts to provide clear and concise writing. Our hope is that this facilitates the understanding and absorption of difficult concepts.

Problems

About 25 percent of the problems in this edition are new. The number of starred problems, which represent the more challenging, thought-provoking questions, has been increased. The end-of-chapter problems are grouped and given subheads in order of topic coverage. In the last group, headed "Additional Problems," problems are not arranged in any specific order. The answers to all the in-text problems and to the odd-numbered end-of-chapter problems are given at the end of the book. Answers to the even-numbered problems are included in the Instructor's Manual.

Ancillaries

(Charles D. Winters)

An Introduction to Organic & Biochemistry, a modified version of this text, is available to those instructors who omit the general chemistry chapters. Both the full version and the modified texts are accompanied by a number of ancillary publications.

- **Student CD-ROM** New to this edition, this high-quality interactive CD-ROM, created by William Vining at the University of Massachusetts, features tutorials in general chemistry, practice in visualizing organic molecules, and some biochemistry animations.

- **Web Site** An interactive Web site containing, among other things, extra practice exercises, teaching and learning tips correlated to each chapter, and PowerPoint™ presentations.

- **Study Guide** by William Scovell, Bowling Green State University. Includes reviews of chapter objectives, important terms and comparisons, focused reviews of concepts, and self-tests.

- **Instructor's Manual** by text authors contains suggested course outlines and answers to the even-numbered problems.

- **Test Bank** by Peter Krieger of Palm Beach Community College and Shawn Farrell of Colorado State University contains more than 1600 multiple-choice questions—more than twice as many questions as in the last edition.

- **ExaMaster™ Computerized Test Bank** is the software version of the printed test bank. Instructors can create thousands of questions in the multiple-choice format. A command reformats a multiple-choice question into a short-answer question. New problems can be added and existing problems modified, and graphics can be incorporated. ExaMaster™ has grade-book capabilities for recording and graphing students' grades.

- Approximately 150 **Overhead Transparencies** in full color are available. Figures and tables are taken from the text.

- **Laboratory Experiments for General, Organic & Biochemistry, 4/e,** by Frederick A. Bettelheim and Joseph M. Landesberg.

Fifty-two experiments—including a new one on analysis of drinking water and one on the law of definite proportions—illustrate important concepts and principles in general, organic, and biochemistry. Many experiments have been revised to miniscale the use of chemicals for environmental concerns and economic reasons. The large number of experiments allows sufficient flexibility for the instructor.

- **Instructor's Manual** to accompany Laboratory Experiments. This manual will help instructors in grading the answers to the questions and in assessing the range of experimental results obtained by students.

- **Chemistry 2001 Instructor CD-ROM** provides imagery from the text and other quality Saunders College Publishing chemistry texts. Available as a presentation tool, this CD-ROM can be used in conjunction with commercial presentation packages, such as PowerPoint™, Persuasion™, and Podium™, as well as the Saunders LectureActive™ presentation software. Available in both Macintosh® and Windows™ platforms.

- **Flash Cards** by H. Akers, Lamar University, comprise 200 bi-directional flash cards that offer drills on important reactions, terms, structures, and classifications.

Acknowledgments

The publication of a book such as this requires the efforts of many more people than merely the authors. We would like to thank the following professors who offered many valuable suggestions for the new edition:

David Ball	Cleveland State University
Steven E. Bottle	Queensland University of Technology
Julie Brady	Delaware Technical & Community College
Robert Bruner	University of California—Berkeley
Timothy Burch	Milwaukee Area Technical College
Shawn Farrell	Colorado State University
Roger Frampton	Tidewater Community College
Robert Gooden	Southern University—Baton Rouge
Lee Harris	University of Arizona
Donald Harriss	University of Minnesota—Duluth
Jack Hefley	Blinn College
Richard Hoffman	Illinois Central College
Larry Jackson	Montana State University
Eric Johnson	Ball State University
Jesse Jones	Baylor University

(Charles D. Winters)

Josephine Kohn	Olympic College
Peter Krieger	Palm Beach Community College
Krishnan Madappat	San Antonio College
Larry McGahey	The College of St. Scholastica
Barbara Mowery	Thomas Nelson Community College
Howard Ono	California State University—Fresno
Jennifer Powers	Kennesaw State University
David Reinhold	Western Michigan University
David Rislove	Winona State University
William Scovell	Bowling Green State University
Margareta Séquin	San Francisco State University
Robert Smith	Longview Community College
Steve Socol	Southern Utah University
Bobby Stanton	The University of Georgia
Jennifer Tan	Ohlone College
James Yuan	Old Dominion University

We also wish to thank several of our colleagues at Adelphi University for their useful advice. These include Stephen Goldberg, Joseph Landesberg, Sung Moon, Donald Opalecky, Reuben Rudman, Charles Shopsis, Kevin Terrance, and Stanley Windwer. We thank Larry Fishel for accuracy checking all of the answers. We are grateful for the support of John Vondeling, Vice President/Publisher at Harcourt College Publishing. We thank Marc Sherman, Senior Associate Editor, and Frank Messina, Senior Project Editor, for their congenial steady assistance. We would like to express our appreciation to Charlene Squibb for supervising the production of this edition and to Lisa Adamitis for supervising the art program. We also thank Pauline Mula, Marketing Strategist, for the successful marketing of our textbook. J/B Woolsey Associates transformed our drawings into pieces of fine art. Last but not least, we want to thank Beverly March and Charles D. Winters for their many excellent photographs, as well as Jane Sanders for conducting the photo research.

Frederick A. Bettelheim
William H. Brown
June 2000

REVISION SUMMARY

The major organizational changes are in the organic and biochemistry chapters. The order of chapters has been changed to:

1. Provide more logical order in pedagogy.

2. Emphasize the shift in importance of biochemical topics and the tremendous developments in some fields of biochemistry.

3. Help solidify the connection between organic and biochemistry. The logical order emphasizes the presentation of a topic (e.g., proteins), first emphasizing the structure, the chemistry, and biological properties, before discussion of interactions.

In general, every chapter has been updated to reflect advances in the disciplines and professions involving chemistry. Every chapter has about four new **Conceptual Problems** and 25 percent new problems overall. Every chapter has **Margin Definitions** of those terms that were not defined in the text. Throughout the text, particularly in the organic chapters, we have added ball-and-stick and space-filling **molecular models** to encourage students to think of molecules as three-dimensional objects.

Below is a short outline of the overall features of the presentation. *Please read the rest of the Preface for more details.*

(Charles D. Winters)

Feature	Benefit
Greater integration of general, organic, and biochemistry	Student acquires a unified view of chemistry.
Logical presentation of structures first, followed by metabolism	All factors in metabolism are established before their interactions are treated.
Ample illustrations of principles in boxes, more than any competing text	Generates interest when students see applications to every principle.
Unique, extensive coverage of the discipline, such as chemical communication and immunochemistry	Students advancing to health professions will apply this knowledge in various procedures and administering of drugs.
Challenging problems identified by asterisks	Gives instructor flexibility in assigning problems and avoids rote problem-solving.
Many new photographs and molecular models	Helps students visualize difficult topics.

General Chemistry

The general chemistry chapters retain their organization from previous editions. Significant changes include the following.

We placed the extensive, operational treatment of exponential notation in the Appendix together with that of significant figures. An introduction to these concepts was kept in Section 1.3 of **Chapter 1, Matter, Energy, and Measurements.** In **Chapter 2, Atoms,** we extended the discussion on manmade new elements and introduced Lewis structures immediately after the discussion of the Periodic Table. There is a new box on Isotopic Abundance and Astrochemistry. In **Chapter 3, Covalent Bonding,** we added more exercises in drawing Lewis structures; increased emphasis on

(Charles D. Winters)

the difference between polar and nonpolar molecules; and added new material on bond angles and shapes of molecules. A new box is Coral Chemistry and Broken Bones.

In **Chapter 4, Chemical Reactions,** we emphasize the concept and use of limiting reagent. A new box is on Solubility and Tooth Decay. In **Chapter 5, Gases, Liquids, and Solids,** significant new topic coverage includes nanotubes as a carbon isoform, surface tension, and phase diagram of water. New information on digital blood pressure monitors is presented in Box 5C. In **Chapter 6, Solutions and Colloids,** we emphasize the nanoscale revolution in material science and add various medical applications. There is a new box on Reverse Osmosis and Desalinization.

Rate constants are introduced and calculated in **Chapter 7, Reaction Rates and Equilibrium.** A working concept of stronger acid + stronger base → weaker acid + weaker base is developed in **Chapter 8, Acids and Bases,** and working with pOH is discussed in greater detail. **Chapter 9, Nuclear Chemistry,** contains extended discussions and illustrations of imaging in medical diagnosis and of the use of iodine-125 in treating prostate cancer. We have updated the coverage of atomic waste disposal and radiation damage by repeated exposure.

Organic Chemistry

(Connie Brown)

Reviewers have commented that the reorganization of the organic chemistry material allows for **greater flexibility** in presentation. Also, the separation of some topics makes each less intimidating to the student. These chapters do retain their concentration on the structure, properties, and reactions of only those **functional groups** necessary to lay a foundation for the coverage of biochemistry that follows. We consider the progress from general to organic to biochemistry to be an ascent and, in the organic chapters, we concentrate on the classes of organic compounds and reactions that have physiological activity of one sort or another and have **biological importance.**

The organic section begins with **Chapter 10, Organic Chemistry.** This short chapter concentrates on the structure of organic compounds and an introduction to the concept of functional groups: the hydroxyl group of alcohols, the amino group of amines, the carbonyl group of aldehydes and ketones, and the carboxyl group of carboxylic acids.

Chapter 11, Alkanes and Cycloalkanes, concentrates on the structure of these two classes of saturated hydrocarbons and further develops an appreciation of molecular shapes through the study of conformations of alkanes and cycloalkanes, and cis-trans isomerism in cycloalkanes.

Chapter 12, Alkenes and Alkynes, introduces students to one aspect of the molecular logic of living systems—namely, that in building large molecules (in this case, terpenes), small subunits (in this case, isoprene units) are strung together by a series of enzyme-catalyzed reactions. Chapter 12 also introduces students to organic reactions, with concentration on the acid-catalyzed hydration of alkenes. It is only in this chapter that reaction mechanisms are discussed—namely, carbocation intermediates in alkene addition reactions. **We limit our discussion of mechanisms deliberately;** in the relatively brief portion of the text devoted to organic chemistry, students have little time to learn anything substantial about them. Chapter 12 closes with a brief overview of polymerization of ethylene and substituted ethylenes, including the structure and properties of both low-density and high-density polyethylenes.

Chapter 13 is devoted to the structure, nomenclature, and important reactions of **Alcohols, Ethers, and Thiols.** Then follows **Chapter 14,**

which concentrates on the structure of **Benzene and Its Derivatives,** as well as the acidity of phenols.

Chapter 15, unique to this text, is an early introduction to **Chirality** and the significance of chirality in the biological world. It has been common in the past to introduce chirality within the context of carbohydrates or amino acids, but the subject of chirality is broader than this. Furthermore, chirality is an additional demonstration of the importance of viewing molecules as three-dimensional objects. The R,S system for assigning configuration to a stereocenter is developed in a simple and direct manner. It is our belief that a greater understanding of molecular shape facilitates the mastery of molecular behavior, knowledge that these students will apply both in this course and in their profession. *Note:* This chapter can be moved up closer to Chapter 19, Carbohydrates, if desired.

Chapter 16 concentrates on the most important chemical property of **Amines**—namely, their basicity. **Chapters 17 and 18** are devoted to the chemistry of **Aldehydes and Ketones** and **Carboxylic Acids and Their Derivatives**. Included in Chapter 18 is an overview of step-growth polymerization and the structure of representative polyamides, polyesters, and polycarbonates.

Bioorganic chemistry is emphasized throughout the organic chapters in 30 boxes and in problems. There are numerous references to *The Merck Index* (Susan Budavari, Editor, 12th Edition, Merck Research Laboratories, 1996). At no point are students required to access this valuable resource. Rather, these new references are given for those who wish to learn more about the chemistry and biochemistry of organic compounds mentioned in the text.

New to these chapters are a large number of ball-and-stick and space-filling **molecular models.** Their purpose is to assist students in visualizing the three-dimensional nature of organic molecules. All models have been prepared using CambridgeSoft Corporation ChemDraw and Chem3D software.

An end-of-chapter summary of **Key Reactions** highlights each new reaction presented in the organic chapters and keys each to the section where it is discussed.

Biochemistry

The logical order emphasizes the presentation of a topic first in separate chapters—(e.g., Carbohydrates and Lipids), discussing the structure, chemistry, and biological properties—and the interrelationship with other groups of compounds in later chapters on metabolism, nutrition, immunochemistry, and so on.

The Nucleic Acids and Protein Synthesis chapter from the fifth edition has been separated into two chapters to properly represent the exploding information available in this field. A new chapter on Immunochemistry has been created to pull together the essential information in this rapidly developing field, which is so important in the health sciences. The Nutrition chapter has been moved to follow immediately after Metabolism (Chapters 26 to 28) in order to emphasize the connection between the two topics. More details follow:

In **Chapter 19, Carbohydrates,** we make the distinction between the R,S system and the D,L system of configuration, offer extra treatment of chair conformations, and add new boxes on Glucose Assay and Blood Types.

In **Chapter 20, Lipids,** we have added topics, including phosphatidyl inositol as a signaling agent, COX enzymes in prostaglandin production, and thromboxanes. There is a new box on transport across cell membranes and new material in Box 20J, Action of Anti-Inflammatory Drugs.

(© David Scharf/Peter Arnold, Inc.)

There is an abundance of new topics in **Chapter 21, Proteins,** including fetal hemoglobin; classification of glycoproteins on the basis of the carbohydrate-protein linkage; hydroxyurea, a new treatment of sickle cell disease, protein conformation–dependent diseases (mad cow, Jacob–Kreutzfeld, prion); the power of quaternary structure in mechanical stress and strain, and laser surgery (denaturation by physical means).

In **Chapter 22, Enzymes,** we examine enzymes of thermophilic bacteria and PCR technique, the pH environment of *Helicobacter,* a detailed 3D map of the active site of pyruvate kinase, and protein modification (phosphorylation) and enzyme activity.

A major rewriting of **Chapter 23, Chemical Communication: Neurotransmitters and Hormones** combines discussion of neurotransmitters and hormones and treats them from the point of view of the chemical nature of the messenger. New emphasis is placed on the trinity of mode of transmission, receptor/ligand/secondary messengers, and the fact that most drugs in use affect chemical communications in one way or another and that drugs may act as (a) agonists or (b) antagonists of receptors, (c) influencing the release of chemical messengers, (d) influencing the decomposition of chemical messengers. Other covered topics include ligand-gated ion channel, signal transduction-G-protein-adenylate cyclase cascade, amino acid neurotransmitters and NMDA receptor, P-protein, and steroid hormone action. New boxes on calcium as secondary messenger and breast cancer and tamoxifen have been added, and the box on nitric oxide has been completely revised.

In **Chapter 24, Nucleic Acids and Heredity,** we have updated our treatment of polymerase factories and PCR techniques and added new boxes on telomerase and immortality and on apoptosis (programmed cell death). In **Chapter 25, Transfer of Information: Protein Synthesis,** we introduce ribozymes, update the discussion of signaling for transcription, present the 3D structure of ribosomes, and add a new box on tumor suppressor genes.

In **Chapter 26, Bioenergetics,** we have revised the discussion of oxidative phosphorylation, present structural details of proton translocating ATPase, and examine superoxide dismutase. In **Chapter 27, Specific Catabolic Pathways,** we detail the pentose phosphate pathway and examine ubiquitin and cystic fibrosis. The role of RubisCo enzyme in photosynthesis is included in **Chapter 28, Biosynthetic Pathways.**

In **Chapter 29, Nutrition and Digestion,** we have condensed the presentation of RDA and other information on vitamins and minerals in table form, cited the new Dietary Reference Intakes (a set of dietary recommendations new to the field of nutrition), and added a new box on dieting. **Chapter 30, Immunochemistry** is an entirely **new chapter** that will help prepare students for further work and study in health fields. Topics include location of the immune system, antigens and antigen presentation, immunoglobulins, T cells and their responses, control of immune response, and cytokines. There are boxes on myasthenia gravis, antibodies and cancer therapies, immunization, and mobilization of leukocytes. The final chapter, **Chapter 31, Body Fluids,** presents new problems, and updated information in all boxes.

A final note on organization and coverage:
This new edition has a total of five more chapters than the fifth edition, yet the amount of material is basically the same in order to ease the burden of teaching. The new organization of the organic and biochemistry chapters has been carefully planned to permit greater flexibility in coverage.

HEALTH-RELATED TOPICS

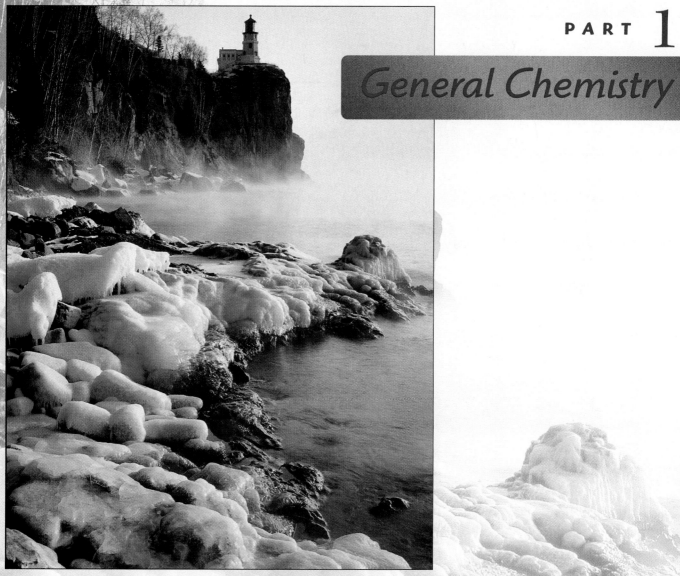

PART 1

General Chemistry

Water in three phases: Solid, liquid, and gas. *(© Tom Till / Tony Stone Images)*

Above: Woman climbing a frozen waterfall in British Columbia. *(James Balog/Tony Stone Images)*

Matter, Energy, and Measurement

1.1 Introduction

There was a time—only a few hundred years ago—when physicians were powerless to treat many diseases. Cancer, tuberculosis, smallpox, typhus, plague, and many other sicknesses struck people seemingly at random, and doctors, who had no idea how any of these diseases were caused, could do little or nothing about them (Figure 1.1). Doctors treated them with magic as well as with such measures as bleeding, laxatives, hot plasters, and pills made from powdered stag's horn, saffron, or gold. None of these treatments was effective, and the doctors, because they came into direct contact with highly contagious diseases, died at a much higher rate than the general public.

Reportable Diseases in the United States[a]				
Disease	**1930**	**1950**	**1970**	**1997**
AIDS	—	—	—	21.9
Diphtheria	54.1	3.8	0.2	—
Measles	340.8	210.1	23.2	0.1
Syphilis	185.4	154.2	44.8	17.4
Tuberculosis	101.5	80.4	18.3	7.4
Whooping cough	135.6	80.1	2.1	2.5

Source: U.S. Census Bureau.
[a] Rates per 100,000 population.

FIGURE 1.1 This painting shows plague victims at a hospital in Hamburg. *(© Archiv/Photo Researchers, Inc.)*

Medicine has made great progress since those times. Although people still get sick, they do on the average live much longer, and many once-feared diseases have been essentially eliminated or made easily curable. Smallpox has been completely eradicated, and polio, typhus, bubonic plague, diphtheria, and other diseases that once killed millions no longer pose a serious problem, at least not in the developed countries.

How has this medical progress come about? Because diseases could not be cured until they were understood, this understanding has come about through a knowledge of how the body functions. Progress in our understanding of the principles of biology, chemistry, and physics has led to these advances in medicine.

Because so much of modern medicine depends on chemistry, it is essential that students who intend to enter the health professions have some understanding of basic chemistry. This book was written to help you achieve that goal. Even if you choose a different profession, you will find that the chemistry you learn in this course will enrich your life.

Chemistry As the Study of Matter

The universe consists of matter, energy, and empty space. **Matter is anything that has mass and takes up space. Chemistry is the science that deals with matter:** the structure and properties of matter and the transformations from one form of matter to another. We discuss energy in Section 1.8.

It has long been known that matter can change, or be made to change, from one form to another. In a **chemical change** or, as it is more commonly called, a **chemical reaction,** substances are used up (disappear), and others are formed to take their place. An example is the burning of propane ("bottled gas"). When this chemical change takes place, propane and oxygen from the air are converted to carbon dioxide and water. Another chemical change is shown in Figure 1.2.

Matter also undergoes other kinds of changes, called **physical changes.** These changes differ from chemical reactions in that substances do not change their identity. Most physical changes are changes of state—for example, the melting of solids and the boiling of liquids. Water remains water whether in the liquid state or in the form of ice or steam. Conversion from one state to another is a physical, not a chemical, change. Another important type of physical change involves making or separating mixtures. Dissolving sugar in water is a physical change.

When we talk about the **chemical properties** of a substance, we mean the chemical reactions it undergoes. **Physical properties** are all properties that do not involve chemical reactions; for example, density, color, melting point, and physical state (liquid, solid, gas) are all physical properties.

Digestion of food also involves chemical changes.

The mixing of vinegar, salad oil, and eggs to make mayonnaise is a physical change.

(a)

(b)

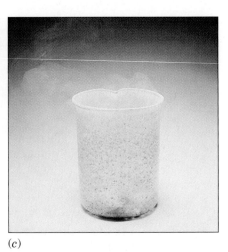
(c)

FIGURE 1.2   A chemical reaction. (*a*) Bromine, an orange-brown liquid, and aluminum, a silver-colored metal. (*b*) These two substances react so vigorously that the aluminum becomes molten and glows white hot on the bottom of the beaker. The yellowish vapor consists of vaporized bromine and some of the product of the reaction, white aluminum bromide. (*c*) Once the reaction is completed, the beaker is coated with aluminum bromide and the products of its reaction with atmospheric moisture. (Note: This reaction is dangerous! Under no circumstances should it be done except under properly supervised conditions.) *(Charles D. Winters)*

1.2   The Scientific Method

Before about 1600, science as such did not exist except as a branch of philosophy.

Scientists acquire knowledge by using a tool called the **scientific method.** The heart of the scientific method is the testing of theories. It was not always so. Before about 1600, philosophers believed statements because they sounded right. For example, the great philosopher Aristotle (384– 322 BCE) believed that if you took the gold out of a mine, it would grow back. He believed this because it fitted in with a more general picture that he had about the workings of nature. In ancient times, this was the way most thinkers behaved. If a statement sounded right, they believed it without testing it.

About 1600 the scientific method came into use. Let us look at an example to see how the scientific method operates. The Greek physician Galen (130–200 CE) recognized that the blood on the left side of the heart somehow gets to the right side. This is a fact. A **fact** is a statement that is obvious to anyone who cares to look, that is, a statement based on direct experience. Having observed this fact, Galen proposed a hypothesis to explain it. A **hypothesis** is a statement that is not obvious but is offered to explain the facts. Because Galen could not actually see how the blood got from the left side to the right side of the heart, he came up with the hypothesis that there must be tiny holes present in the muscular wall that separates the two halves.

For example, it is a fact that humans must breathe in order to live.

Up to this point in time, modern scientists and ancient philosophers behaved the same way. Each offered a hypothesis to explain the facts. From this point on, however, their methods differ. To Galen, his explanation sounded right, and that was enough to make him believe it, even though he couldn't see any holes. His hypothesis was, in fact, believed by virtually all physicians for more than 1000 years. When we use the scientific method, however, we do not believe a hypothesis just because it sounds right. We test it, using the most rigorous testing we can think of.

Around 1600, William Harvey (1578–1657) tested Galen's hypothesis by dissecting human and animal hearts and blood vessels. He discovered that there are one-way valves separating the two chambers of the heart. He also discovered that the heart is a pump that, by contracting and expanding, pushes the blood out. Harvey's teacher Fabricius (1537–1619) had previously observed that there are one-way valves in the veins so that blood in the veins can travel only toward the heart and not in the other direction.

Harvey put all these facts together to come up with a new hypothesis: Blood is pumped by the heart and circulates throughout the body. This was a better hypothesis than Galen's because it fitted the facts more closely. Even so, it was still a hypothesis and, according to the scientific method, had to be tested further. One important test took place in 1661, four years after Harvey died. Harvey predicted that, because there had to be a way for the blood to get from the arteries to the veins, there must be tiny blood vessels connecting them. In 1661, the Italian anatomist Malpighi (1628–1694), using the newly invented microscope, found these tiny vessels, which are now called capillaries.

Malpighi's discovery supported the blood circulation hypothesis by fulfilling Harvey's prediction. When a hypothesis passes the tests, we have more confidence in it and call it a **theory.** A theory, in this sense, is the same as a hypothesis except that we have a stronger belief in it because it is supported by more evidence. No matter how much confidence we have in a theory, however, if new facts that conflict with it are discovered or if it does not pass newly devised tests, the theory must be altered or rejected. In the history of science, many firmly established theories have eventually been thrown out because they could not pass new tests.

The scientific method is thus very simple. We don't accept a hypothesis or a theory just because it sounds right. We devise tests, and only if the hypothesis or theory passes the tests do we accept it. The enormous progress made since 1600 in chemistry, biology, and the other sciences is a testimony to the value of the scientific method.

You may get the impression from the previous discussion that science progresses in one direction: facts first, hypothesis second, and theory last. Nothing is that simple, however. Hypotheses and theories may attract the attention of scientists who then look for and discover new facts. An example of this scenario is the discovery of the element germanium. In 1871, Mendeleev's Periodic Table predicted the existence of a new element whose properties would be similar to that of silicon. He called it eka-silicon. In 1886, this new element was discovered in Germany (hence the name), and its properties were truly similar to those predicted by theory.

On the other hand, many scientific discoveries are the result of **serendipity,** or chance observation. An example of serendipity occurred in 1926 when James Sumner of Cornell University left an enzyme preparation of jack bean urease in a refrigerator over the weekend. Upon his return, he found that his solution contained crystals that turned out to be a protein. This chance discovery led to the hypothesis that all enzymes are proteins. But serendipity is not enough. Scientists must have the creativity and insight to recognize the significance of their observations. Sumner fought for more than 15 years for his hypothesis to gain acceptance because people believed that only small molecules can form crystals. Eventually he won, and he was awarded a Nobel Prize in Chemistry in 1946.

Galen. *(Courtesy of the National Library of Medicine)*

Theory A model that explains many interrelated facts and can be used to make predictions about natural phenomena. Examples are Newton's theory of gravitation and the kinetic–molecular theory of gases, which we will meet in Section 5.2. This type of theory is also subject to testing and will be discarded or modified if it is contradicted by new facts.

1.3 Exponential Notation and Significant Figures

Scientists often deal with numbers that are very large or very small. For example, a penny (before 1982, when pennies in the United States were still made of copper) contains approximately

29 500 000 000 000 000 000 000 atoms of copper

and a single copper atom weighs

0.000 000 000 000 000 000 000 000 23 pound

Exponential notation is also called scientific notation.

Many years ago, an easy way to handle such large and small numbers was devised. The method is called **exponential notation,** and it is based on powers of 10. In exponential notation, the number of copper atoms in a penny is written

For example, 10^6 means a one followed by six zeros, or 1 000 000, and 10^2 means 100.

$$2.95 \times 10^{22} \text{ atoms}$$

and the weight of a single copper atom is written

$$2.3 \times 10^{-25} \text{ pound}$$

Problems 1.16 through 1.23 deal with exponential notation, and calculations using exponential notation can be found in Appendix I.

When it comes to measurements, not all the numbers you can generate in your calculator or computer are of equal importance. Only the number of digits that are known with certainty are significant. Let us say that you measured the weight of an object as 3.4 g on a balance that you can read to the nearest 0.1 g. You can report the weight as 3.4 g but not 3.40 or 3.400 g because you do not know the added zeros with certainty. A detailed account of using **significant figures** is presented in Appendix II. Discussion on accuracy, precision, and significant figures can also be found in your Bettelheim and Landesberg *Laboratory Manual, fourth edition* (Experiment 2).

1.4 Measurements

In our daily lives we are constantly making measurements. We measure ingredients for recipes, distances when driving, gallons of gasoline, weights of fruits and vegetables, and the timing of TV programs. Doctors and nurses measure pulse rates, blood pressures, temperatures, and drug dosages. Chemistry, like other sciences, is based on measurements.

A measurement consists of two parts: a number and a unit. A number without a unit is usually meaningless. If you were told that a person's weight is 57, the information would be of very little use. Is it 57 lb, which would indicate that the person is very likely a child, or 57 kg, which is the weight of an average woman or a small man? Or is it perhaps some other unit? Because there are so many units, a number by itself is not enough; the unit must also be stated.

In the United States, most measurements in everyday life are made with the English system of units: pounds, miles, gallons, and so on. In Germany, France, Russia, Japan, and most other countries, however, almost

FIGURE 1.3 A highway sign in Missouri. *(Beverly March)*

FIGURE 1.4 Soft drink bottles in metric sizes (here, 1-and 2-L sizes). *(George Semple)*

nobody could describe a pound or a gallon. These countries use the **metric system,** a system that began in France about 1800 and has since spread throughout the world. Even in the United States, metric measurements are slowly being introduced (Figure 1.3). For example, many soft drinks and most alcoholic beverages now come in metric sizes (Figure 1.4). *Scientists in the United States have been using metric units all along.*

Around 1960, international scientific organizations adopted another system, called the **International System of Units** (abbreviated **SI**). The SI is based on the metric system and uses some metric units. The main difference is that the SI is more restrictive: It discourages the use of certain metric units and favors others. Although the SI has advantages over the older metric system, it also has significant disadvantages. For this reason, U.S. chemists have been very slow to adopt it. At this time, approximately 40 years after its introduction, not many U.S. chemists use the entire SI, although some of its preferred units are gaining ground.

In this book we shall use the metric system (Table 1.1). Occasionally we shall mention the preferred SI unit.

> **Metric system** A system of units of measurement in which the divisions to subunits are made by a power of ten

"SI" is actually an abbreviation of the French title *Système Internationale.*

Length

The key to the metric system (and the SI) is that there is one base unit for each kind of measurement and that other units are related to the base unit only by powers of ten. For example, let us look at measurements of length. In the English system, we have the inch, the foot, the yard, and the mile (not to mention such older units as the league, furlong, ell, and rod). If you want to convert one to another, you must memorize or look up these conversion factors:

$$12 \text{ inches} = 1 \text{ foot}$$

$$3 \text{ feet} = 1 \text{ yard}$$

$$1760 \text{ yards} = 1 \text{ mile}$$

All this is unnecessary in the metric system (and the SI). In both systems, the base unit of length is the **meter (m).** To convert to larger or smaller units, we do not use arbitrary numbers like 12, 3, and 1760 but

TABLE 1.1 Base Units in the Metric System

Dimension	Unit
Length	meter
Volume	liter
Mass	gram
Time	second
Temperature	kelvin
Pressure	mm Hg
Energy	calorie
Amount of substance	mole

TABLE 1.2 The Most Common Metric Prefixes

Prefix	Symbol	Value
giga	G	$10^9 = 1\ 000\ 000\ 000$ (one billion)
mega	M	$10^6 = 1\ 000\ 000$ (one million)
kilo	k	$10^3 = 1000$ (one thousand)
deci	d	$10^{-1} = 0.1$ (one tenth)
centi	c	$10^{-2} = 0.01$ (one hundredth)
milli	m	$10^{-3} = 0.001$ (one thousandth)
micro	μ	$10^{-6} = 0.000\ 001$ (one millionth)
nano	n	$10^{-9} = 0.000\ 000\ 001$ (one billionth)

The official definition of the meter is the distance that light travels in $\dfrac{1}{299\ 792\ 458}$ of a second.

only 10, 100, $\frac{1}{10}$, $\frac{1}{100}$, or other powers of ten. This means that **to convert from one metric or SI unit to another, we only have to move the decimal point.** Furthermore, the other units are named by putting prefixes in front of "meter," and **these prefixes are the same throughout the metric system and the SI.** The most important of these prefixes are shown in Table 1.2. If we put some of these prefixes in front of "meter," we have

$$1 \text{ kilometer (km)} = 1000 \text{ meters (m)}$$

$$1 \text{ centimeter (cm)} = 0.01 \text{ m}$$

$$1 \text{ nanometer (nm)} = 10^{-9} \text{ m}$$

For people who have grown up using English units, it is helpful to have some idea of the size of metric units, and so some conversion factors are shown in Table 1.3.

Volume

Volume is space. The volume of a liquid, solid, or gas is the space occupied by that substance. The base unit of volume in the metric system is the **liter (L).** This unit is a little larger than a quart (Table 1.3). The only other common metric unit for volume is the milliliter (mL), which is equal to 10^{-3} L:

$$1000 \text{ mL} = 1 \text{ L}$$

One milliliter is exactly equal to one cubic centimeter (cc or cm^3):

$$1 \text{ mL} = 1 \text{ cc}$$

which means that there are 1000 cc in 1 L.

Hypodermic syringe

Volume measurements are important in chemistry and medicine. In a hypodermic syringe, the device for injecting the drug (the needle) is combined with the means for measuring the volume injected (the barrel), usually in cubic centimeters.

TABLE 1.3 Some Conversion Factors Between the English and Metric Systems

Length	Mass	Volume
1 in. = 2.54 cm	1 oz (avdp) = 28.35 g	1 qt = 0.946 L
1 m = 39.37 in.	1 lb (avdp) = 453.6 g	1 gal = 3.785 L
1 mile = 1.609 km	1 kg = 2.205 lb	1 L = 33.81 fl oz
	1 g = 15.43 grains	1 fl oz = 29.57 mL
		1 L = 1.057 qt

FIGURE 1.5 Two laboratory balances. *(Courtesy of Brinkman Instruments Co.)*

Mass

Mass is the quantity of matter in an object. The base unit of mass in the metric system is the **gram (g).** As always in the metric system, larger and smaller units are indicated by prefixes. The ones in common use are

$$1 \text{ kilogram (kg)} = 1000 \text{ g}$$

$$1 \text{ milligram (mg)} = 0.001 \text{ g}$$

The gram is a small unit; there are 453.6 g in 1 lb (Table 1.3).

In chemistry, we use a device called a balance to measure mass. Two types of laboratory balances are shown in Figure 1.5.

There is a fundamental difference between mass and weight. Mass is independent of location. The mass of a stone, for example, is the same whether we measure it at sea level, on top of a mountain, or in the depths of a mine. In contrast, weight is not independent of location. **Weight is the force a mass experiences under the pull of gravity.** This was dramatically demonstrated when astronauts walked on the surface of the moon. The moon, being a smaller body than the Earth, exerts a weaker gravitational pull. Consequently, even though the astronauts wore space suits and equipment that would be heavy on Earth, they felt lighter on the moon and could execute great leaps and bounces during their walks.

Although mass and weight are different concepts, they are related to each other by the force of gravity. We frequently use the words interchangeably because we weigh objects by comparing their masses to standard reference masses on a balance, and the gravitational pull is the same on the unknown object and on the standard masses. Because the force of gravity is essentially constant, mass is always directly proportional to weight.

The official kilogram is defined by the mass of a particular platinum-iridium bar kept at the International Bureau of Standards in Paris.

Both mass and weight are independent of temperature.

BOX 1A

Drug Dosage and Body Mass

In many cases, drug dosages are prescribed on the basis of body mass. For example, the recommended dosage of a drug may be 3 mg of drug for each kilogram of body weight. In this case, a 50-kg (110-lb) woman would receive 150 mg, and an 82-kg (180-lb) man would get 246 mg. This is especially important for children because a dose suitable for an adult will generally be too much for a child, who has much less body mass. For this reason, manufacturers package and sell smaller doses of certain drugs, such as acetaminophen, for children.

Drug dosage may also vary with age. Occasionally, when an elderly patient has impaired kidney or liver function, the clearance of a drug from the body is delayed, and the drug may stay in the body longer than is normal. This can cause dizziness, vertigo, and migraine-like headaches, resulting in falls and broken bones. Such delayed clearance must be monitored so that the drug dosage can be adjusted accordingly.

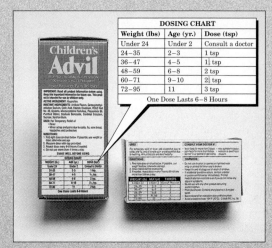

DOSING CHART		
Weight (lbs)	**Age (yr.)**	**Dose (tsp)**
Under 24	Under 2	Consult a doctor
24–35	2–3	1 tsp
36–47	4–5	1½ tsp
48–59	6–8	2 tsp
60–71	9–10	2½ tsp
72–95	11	3 tsp
One Dose Lasts 6–8 Hours		

■ These packages of Advil and Tylenol have charts showing the proper dose for children of a given weight.   *(Beverly March)*

Time

Time is the one quantity for which the units are the same in all systems: English, metric, and SI. The base unit is the **second (s),** and

$$60 \text{ s} = 1 \text{ min}$$
$$60 \text{ min} = 1 \text{ h}$$

Temperature

Most people in the United States are familiar with the Fahrenheit scale of temperature. In the metric system, the **Celsius** scale is used. In this scale, the boiling point of water is set at 100°C, and the freezing point is 0°C. We can convert from one scale to the other by using the formulas

$$°F = \tfrac{9}{5} °C + 32$$
$$°C = \tfrac{5}{9} (°F - 32)$$

The 32 in these equations is a defined number and is, therefore, treated as if it had an infinite number of zeros following the decimal point. (See Appendix A.)

EXAMPLE 1.1

Normal body temperature of 98.6°F. Convert this to °C.

Solution

$$°C = \tfrac{5}{9} (98.6 - 32) = \tfrac{5}{9} (66.6) = 37.0°C$$

Problem 1.1

Convert: (a) 64.0°C to °F (b) 47°F to °C. ■

The relationship between the Fahrenheit and Celsius scales is shown in Figure 1.6.

There is another temperature scale, called the **Kelvin (K)** scale or the absolute scale. The size of the Kelvin degree is the same as that of the Celsius degree. The only difference between these two scales is the zero point. The temperature $-273°C$ is taken as the zero point on the Kelvin scale. This makes conversions between Kelvin and Celsius very easy. To go from Celsius to Kelvin, just *add* 273; to go from Kelvin to Celsius, *subtract* 273:

$$K = °C + 273$$

$$°C = K - 273$$

The relationship between the Kelvin and Celsius scales is shown in Figure 1.6. Note that we don't use the degree symbol in the Kelvin scale: $100°C$ equals 373 K, not $373°K$.

Why was $-273°C$ chosen as the zero point on the Kelvin scale? The reason is that **$-273°C$, or 0 K, is the lowest possible temperature.** Because of this, 0 K is called **absolute zero.** Temperature depends on how fast molecules move. The more slowly they move, the colder it gets. At absolute zero, molecules stop moving altogether, and it is not possible for them to move more slowly than that. Therefore, the temperature cannot get any lower. For some purposes, it is convenient to have a scale that begins at the lowest possible temperature; the Kelvin scale fulfills this need.

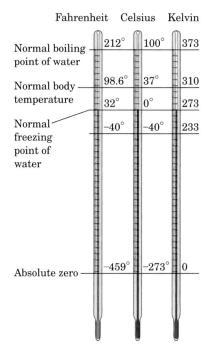

FIGURE 1.6 Three temperature scales. The kelvin is the SI unit.

Scientists in laboratories have been able to attain temperatures as low as 0.001 K, but a temperature of 0 K is impossible.

> **Factor-label method** A procedure in which the equations are set up so that all the unwanted units cancel and only the desired units remain

> **Conversion factor** A ratio of two different units

1.5 Unit Conversions. The Factor-Label Method

We frequently need to convert a measurement from one unit to another. The best and most foolproof way to do this is the **factor-label method.** In this method, we follow the rule that *when multiplying numbers we also multiply units, and when dividing numbers we also divide units.*

For conversions between one unit and another, it is always possible to set up two fractions, called **conversion factors.** Suppose that we wish to convert 381 g to pounds. We can see in Table 1.3 that there are 453.6 g in 1 lb. The conversion factors between grams and pounds, therefore, are

$$\frac{1 \text{ lb}}{453.6 \text{ g}} \quad \text{and} \quad \frac{453.6 \text{ g}}{1 \text{ lb}}$$

To convert 381 g to pounds, all we need to do is multiply by the proper conversion factor. But which one? Let us try both and see what happens. First, let us multiply by 1 lb/453.6 g:

$$381 \text{ g} \times \frac{1 \text{ lb}}{453.6 \text{ g}} = 0.840 \text{ lb}$$

Following the procedure of multiplying and dividing units when we multiply and divide numbers, we find that dividing grams by grams cancels out the grams, and we are left with pounds, which is the answer we want. Thus 1 lb/453.6 g is the correct conversion factor because it converts grams to pounds. Suppose that we had done it the other way, multiplying by 453.6 g/1 lb:

$$381 \text{ g} \times \frac{453.6 \text{ g}}{1 \text{ lb}} = 173\,000 \, \frac{\text{g}^2}{\text{lb}}$$

When we multiply grams by grams, we get g² (grams squared). Dividing by pounds gives g²/lb. Because this is not the unit we want, this must be the incorrect method.

The advantage of the factor-label method is that it lets us know when we have made a wrong calculation. *If the units of the answer are not the ones we are looking for, the calculation must be wrong.* Incidentally, this principle works not only in unit conversions but in all problems where we make calculations using measured numbers.

The factor-label method gives the correct mathematical solution for a problem. However, it is a mechanical technique and does not require you to think through the problem. Thus, it may not provide a deeper understanding. For this reason and also to check your answer (because it is easy to make mistakes in arithmetic, for example, by punching wrong numbers into a calculator), it is beneficial to ask yourself if the answer you have obtained is reasonable. For example, the question might ask for the mass of a single oxygen atom. If your answer comes out 8.5×10^6 g, it is not reasonable. A single atom cannot weigh more than you do! In such a case, you have obviously made a mistake and should go back and take another look to see where you went wrong. Everybody makes mistakes at times, but if you check you can at least determine whether your answer is reasonable. If it is not, you will immediately know that there is a mistake and will be able to correct it.

Checking whether an answer is reasonable gives you a deeper understanding of the problem because it forces you to think through the relationship between the question and the answer. We will now give a few examples of unit conversions and then test the answers to see if they are reasonable. To save space, we will do this only in this chapter, but you should use a similar approach in later chapters. In unit conversion problems, you should always check two things. First, the numeric factor you multiply by tells you whether the answer is going to be larger or smaller than the number being converted. Second, the factor tells you how much larger or smaller than the number you start with your answer should be. For example, if you start with a number 96 to be converted and your conversion factor tells you that your answer should be about twice as large, then an answer about 200 is reasonable but an answer of 0.2 or 2000 is not reasonable.

EXAMPLE 1.2

The distance between Rome and Milan (the two largest cities in Italy) is 358 miles. How many kilometers if this?

Solution
We want to convert miles to km. Table 1.3 shows that 1 mile = 1.609 km. From this we get two conversion factors:

$$\frac{1 \text{ mile}}{1.609 \text{ km}} \quad \text{and} \quad \frac{1.609 \text{ km}}{1 \text{ mile}}$$

Which should we use? We use the one that gives the answer in km:

$$358 \text{ miles} \times \text{Conversion factor} = ? \text{ km}$$

This means that the miles must cancel, so the conversion factor 1.609 km/1 mile is appropriate.

$$358 \text{ miles} \times \frac{1.609 \text{ km}}{1 \text{ mile}} = 576 \text{ km}$$

Is it reasonable? We want to convert miles to kilometers. The conversion factor in Table 1.3 tells us that, in a given distance, the number of kilometers is larger than the number of miles. How much larger? The exact number is 1.609, which is approximately 1.5 times larger. Thus, we expect that the answer in kilometers will be about 1.5 times greater than the number given in miles. The number given in miles is 358, which, *for the purpose of checking whether or not our answer is reasonable,* we can round off to 400. Multiplying this by 1.5 gives an approximate answer of 600 kilometers. Since our actual answer, 576 km, is of the same order of magnitude as the estimated answer, we can say that it is reasonable. If the estimated answer had been 6 km or 60 km or 6000 km, we would suspect that we had made a mistake in calculating the actual answer.

Problem 1.2

How many kg are in 241 lb? Check your answer to see if it is reasonable. ■

EXAMPLE 1.3

The label on a container of olive oil says 1.844 gal. How many mL is this?

Solution

Table 1.3 shows no factor for converting gallons to milliliters, but it does show that 1 gal = 3.785 L. Since we know that 1000 mL = 1 L, we can solve this problem by multiplying by two conversion factors, making certain that all units cancel except milliliters:

$$1.844 \text{ gal} \times \frac{3.785 \text{ L}}{1 \text{ gal}} \times \frac{1000 \text{ mL}}{1 \text{ L}} = 6980 \text{ mL}$$

Is it reasonable? The conversion factor in Table 1.3 tells us that there are more liters in a given volume than gallons. How much more? Approximately four times more. We also know that any volume in milliliters is one thousand times larger than the same volume in liters. Thus, we expect that the volume expressed in milliliters will be 4 × 1000 or 4000 times more than in gallons. The estimated volume in milliliters will be approximately 1.8 × 4000 or 7000 mL. But we also expect that the actual answer should be somewhat less than the estimated figure because we overestimated the conversion factor (4 rather than 3.785). Thus the answer, 6980 mL, is quite reasonable.

Problem 1.3

Calculate the number of km in 8.55 miles. Check your answer to see if it is reasonable. ■

EXAMPLE 1.4

The maximum speed limit on many roads in the United States is 65 mph. How many m/s is this?

Solution

Here we have a double conversion problem. We must convert miles to meters and hours to seconds. We use as many conversion factors as necessary, always making sure that we use them in such a way that the proper units cancel:

$$65 \ \frac{\text{miles}}{\text{h}} \times \frac{1.609 \ \text{km}}{1 \ \text{mile}} \times \frac{1000 \ \text{m}}{1 \ \text{km}} \times \frac{1 \ \text{h}}{60 \ \text{min}} \times \frac{1 \ \text{min}}{60 \ \text{s}} = 29 \ \frac{\text{m}}{\text{s}}$$

Is it reasonable? To estimate the 65 mph speed in meters per second, we must first establish the relationship between miles and meters. As in Example 1.2, we know that there are more kilometers than miles in a given distance. How much more? Since there are approximately 1.5 km in a mile, there must be approximately 1500 times more meters. We also know that in 1 hour there are $60 \times 60 = 3600$ seconds. There are approximately half as many more meters in a mile than seconds in an hour: 1500/3600. Therefore, we estimate that the speed in meters per second will be about one half of that in miles per hour, which comes out to 32 m/s. Once again, the actual answer, 29 m/s, is not far from the estimate of 32 m/s, so the answer is reasonable.

> Estimating the answer is a good thing to do when working any mathematical problem, not just unit conversions.

As shown in these examples, we do not cancel the numbers when canceling units. The numbers are multiplied and divided in the ordinary way.

Problem 1.4 ■■■■■

Convert the speed of sound, 332 m/s, to mph. Check your answer to see if it is reasonable. ■

1.6 The States of Matter

> The most familiar gas is the air around us, which we depend on for life itself.

Matter can exist in three states: gas, liquid, and solid. *Gases* have no definite shape or volume. They expand to fill whatever container they are put into. On the other hand, they are highly compressible and can be forced into small containers. *Liquids* also have no definite shape. However, they do have a definite volume that remains the same when they are poured from one container to another. Liquids are only slightly compressible. *Solids* have definite shapes and definite volumes. They are essentially incompressible.

Whether a substance is a gas, a liquid, or a solid depends on its temperature. On a cold winter day, a puddle of liquid water turns to ice; it becomes a solid. If we boil water, the liquid becomes a gas—we call it steam. Most substances can exist in the three states: they are gases at high temperature, liquefy at a lower temperature, and solidify when their tem-

(a)   (b)   (c)

FIGURE 1.7 The three states of matter for bromine: (*a*) solid, (*b*) liquid, and (*c*) gas. (*Charles D. Winters*)

perature becomes low enough. Figure 1.7 shows a single substance in the three states.

The chemical identity of a substance does not change when it is converted from one state to another. Water is still water whether in the form of ice, steam, or liquid water. We discuss the three states of matter, and the changes between one state and another, at greater length in Chapter 5.

1.7 Density and Specific Gravity

One of the many pollution problems that the world faces is the spillage of petroleum in the oceans from oil tankers or from offshore drilling. When oil spills into the ocean, it floats on top of the water. The oil doesn't sink because water has a higher density than oil. When two liquids are mixed (assuming that one does not dissolve in the other), the one of lower density floats on top (Figure 1.8).

The **density** of any substance is defined as its **mass per unit volume.** Not only do all liquids have a density, but so do all solids and gases, too. Density is calculated by dividing the mass of a substance by its volume:

$$d = \frac{m}{V} \qquad d = \text{density}, \; m = \text{mass}, \; V = \text{volume}$$

Water

Carbon tetrachloride

Diethyl ether

Water

FIGURE 1.8 Two separatory funnels containing water and another liquid. In each case the liquid with the lower density is on top.

EXAMPLE 1.5

If 73.2 mL of a liquid has a mass of 61.5 g, what is its density in g/mL?

Solution

$$d = \frac{m}{V} = \frac{61.5 \text{ g}}{73.2 \text{ mL}} = 0.840 \; \frac{\text{g}}{\text{mL}}$$

The spillage of more than 10 million gallons of petroleum in Prince William Sound, Alaska, in March 1989 caused a great deal of environmental damage. *(Ken Graham / Tony Stone Images)*

EXAMPLE 1.6

The density of iron is 7.86 g/mL. What is the volume, in mL, of an irregularly shaped piece of iron that has a mass of 524 g?

Solution

Here we are given the mass and the density. In this type of problem, it is useful to derive a conversion factor from the density. We know that the density is 7.86 g/mL. This means that 1 mL has a mass of 7.86 g. From this we can get two conversion factors:

$$\frac{1 \text{ mL}}{7.86 \text{ g}} \quad \text{and} \quad \frac{7.86 \text{ g}}{1 \text{ mL}}$$

As usual, we multiply the mass by whichever conversion factor results in the cancellation of all but the correct unit:

$$524 \text{ g} \times \frac{1 \text{ mL}}{7.86 \text{ g}} = 66.7 \text{ mL}$$

Is it reasonable? The density 7.86 g/mL tells us that the volume in mL of any piece of iron is always less than its mass in grams. How much less? Approximately eight times less. Thus, we expect the volume to be approximately 500/8 = 63 mL. Because the actual answer is 66.7 mL, it is reasonable.

Problem 1.5 ▬▬▬▬▬▬▬▬

The density of titanium is 4.54 g/mL. What is the mass, in g, of 17.3 mL of titanium? Check your answer to see if it is reasonable. ▪

Problem 1.6 ▬▬▬▬▬▬▬▬

An unknown substance has a mass of 56.8 g and occupies a volume of 23.4 mL. What is its density in g/mL? Check your answer to see if it is reasonable. ▪

In old movies, villains frequently used a chloroform-wetted handkerchief to subdue their victims.

The density of any liquid or solid is a physical property that is constant, which means that it always has the same value at a given temperature. We use physical properties to help identify a substance. For example, the density of chloroform (a liquid formerly used as an anesthetic) is 1.483 g/mL at 20°C. If we want to find out if an unknown liquid is chloroform, we might measure its density at 20°C. If the density is, say, 1.355 g/mL, we know the liquid isn't chloroform. If the density comes out to be 1.483 g/mL, we cannot be sure that the liquid is chloroform because other liquids might also have this density, but we can then measure other physical properties (the boiling point, for example). If all the physical properties we measure match those of chloroform, we can be reasonably sure that the liquid is chloroform.

We have said that the density of a liquid or solid is a constant at a given temperature. Density changes when the temperature changes. Density almost always decreases with increasing temperature. This is true because mass does not change when a substance is heated, but volume almost always increases because atoms and molecules tend to get farther apart as the temperature increases. Since $d = m/V$, if m stays the same and V gets larger, d must get smaller.

The most common liquid, water, provides a partial exception to this rule. As the temperature increases from 4°C to 100°C, the density of water does decrease, but from 0°C to 4°C, the density increases. In other words, water has its maximum density at 4°C.

Because density is equal to mass divided by volume, it always has units, most commonly g/mL or g/cc (g/L for gases). **Specific gravity** is numerically the same as density, but it has no units (it is dimensionless). The reason is that specific gravity is defined as a comparison of the density of a substance with that of water, which is taken as a standard. For example, the density of copper at 20°C is 8.92 g/mL. The density of water at the same temperature is 1.00 g/mL. Therefore, copper is 8.92 times as dense as water, and its specific gravity at 20°C is 8.92. Because water is taken as the standard and because the density of water is 1.00 g/mL at 20°C, the specific gravity of any substance is always numerically equal to its density, provided that the density is measured in g/mL or g/cc.

EXAMPLE 1.7

The density of ethanol at 20°C is 0.789 g/mL. What is its specific gravity?

Solution

$$\text{Specific gravity} = \frac{0.789 \ \cancel{\text{g/mL}}}{1.00 \ \cancel{\text{g/mL}}} = 0.789$$

Problem 1.7
The specific gravity of a urine sample is 1.016. What is its density in g/mL? ∎

1.8 Energy

Energy is defined as the capacity to do work. Energy can be described as being either kinetic energy or potential energy.

Kinetic energy is the energy of motion. Any object that is moving possesses kinetic energy. We can calculate how much kinetic energy it has by using the formula $KE = \frac{1}{2}mv^2$, where m is the mass of the object and v is its velocity. This means that kinetic energy increases (1) when an object moves faster and (2) when a heavier object is moving. When a truck and a bicycle are moving at the same velocity, the truck has more kinetic energy.

Potential energy is stored energy. The potential energy possessed by an object arises from its capacity to move or to cause motion. For example, a book balanced at the edge of a table contains potential energy—it is capable of doing work. If given a slight push, it will fall. The potential energy the book has while it is on the table is converted to the kinetic energy it has while it is falling. Another way in which potential energy is converted to kinetic energy is shown in Figure 1.9.

An important principle in nature is that things have a tendency to seek their lowest possible potential energy. We all know that water always flows downhill and not uphill.

There are several forms of energy, of which the most important are: (a) mechanical energy, light, heat, and electrical energy, which are all examples of kinetic energy possessed by all moving objects, whether elephants or molecules or electrons; and (b) chemical energy and nuclear energy, which are examples of potential energy or stored energy. In chemistry, the most important form of potential energy is chemical energy. This

Potential energy is stored in this drawn bow and becomes kinetic energy in the arrow when released. *(Eric Lars Baleke/Black Star)*

An example of energy conversion: Light energy from the sun is converted to electrical energy by solar cells. The electricity runs the refrigerator on the back of the camel, keeping the vaccines inside cool so that they can be delivered to remote locations. *(Courtesy of Siemens Solar Industries, Camarillo, CA)*

Hydroelectric plants use the kinetic energy of falling water.

Potential energy ready to do work   Kinetic energy doing work

FIGURE 1.9   The water held back by the dam possesses potential energy, which will be converted to kinetic energy when the water is released.

is the energy stored within chemical substances. The energy is given off when they take part in a chemical reaction. For example, a piece of paper possesses chemical energy. When the paper is ignited, the chemical energy (potential) is released in the form of heat and light.

The various forms of energy can be converted from one to another. In fact, we do this all the time. A power plant operates either on the chemical energy derived from burning fuel or on nuclear energy. This energy is converted to heat, which is converted to the electricity that is sent over transmission wires into houses and factories. Here we convert the electricity to light, heat (in an electrical heater, for example), or mechanical energy (in the motors of refrigerators, vacuum cleaners, etc.).

Although one form of energy can be converted to another, the *total amount* of energy in any system does not change. **Energy can neither be created nor destroyed.** This statement is called the **law of conservation of energy.***

1.9   Heat

Heat and Temperature

One form of energy that is particularly important in chemistry is **heat.** This is the form of energy that most frequently accompanies chemical reactions. Heat is not the same as temperature. Heat is a form of energy, but temperature is not.

The difference between heat and temperature can be seen in the following example. If we have two beakers, one containing 100 mL of water and the other 1 L of water at the same temperature, the heat content of the water in the larger beaker is ten times that of the water in the smaller beaker, even though the temperature is the same in both. If you were to dip your hand accidentally into a liter of boiling water, you would be much more severely burned than if only one drop fell on your hand. Even though the water is at the same temperature in both cases, the liter of boiling water has much more heat.

---

* This statement is not completely true. As discussed in Box 9F, it is possible to convert matter to energy, and vice versa. Therefore, a more correct statement would be **matter-energy can neither be created nor destroyed.** However, the law of conservation of energy is valid for most purposes and is highly useful.

Hypothermia and Hyperthermia

The human body cannot tolerate temperatures that are too low. A person who is outside in very cold weather (say, −20°F, or −29°C) and is not protected by heavy clothing will eventually freeze to death because the body loses heat. Normal body temperature is 37°C, and when the outside temperature is lower than that, heat flows out of the body. When the air temperature is moderate (10 to 25°C), this poses no problem and is, in fact, necessary because the body produces more heat than it needs and must lose some. At extremely low temperatures, however, too much heat is lost,

and body temperature drops; this is a condition called **hypothermia.** A drop in body temperature of 1 or 2°C causes shivering, which is the body's attempt to increase its temperature by the heat generated through muscular action. A greater drop than that results in unconsciousness and eventually death.

The opposite condition is **hyperthermia.** This can be caused either by high outside temperatures or by the body itself when an individual develops a high fever. A sustained body temperature higher than 41.7°C (107°F) is usually fatal.

As we saw in Section 1.4, temperature is measured in degrees. Heat can be measured in various units, the most common of which is the **calorie,** which is defined as **the amount of heat necessary to raise the temperature of 1 g of liquid water by 1°C.** This is a small unit, and chemists more often use the kilocalorie (kcal):

$$1 \text{ kcal} = 1000 \text{ cal}$$

Nutritionists use the word "Calorie" (with a capital C) to mean the same thing as "kilocalorie"; that is, 1 Cal = 1000 cal = 1 kcal. The calorie is not part of the SI. The official SI unit for heat is the **joule (J),** which is about one fourth as big as the calorie:

$$1 \text{ cal} = 4.184 \text{ J}$$

Calories and joules are units not only of heat but of other forms of energy as well.

Specific Heat

As we noted, it takes 1 cal of heat to raise the temperature of 1 g of liquid water by 1°C. **Specific heat is the amount of heat (in calories) necessary to raise the temperature of 1 g of any substance by 1°C.** Each substance has its own specific heat, which is a physical property of that substance, like density or melting point. For example, the specific heat of copper is 0.092 cal/g·°C. Therefore, if we had 1 g of copper at 20°C, it would require only 0.092 cal to increase the temperature to 21°C. Specific heats for a few common substances are shown in Table 1.4. Note that ice

Specific heat is sometimes called specific heat capacity.

Water has the highest specific heat of any common substance.

TABLE 1.4 Specific Heat Values for Some Common Substances

Substance	Specific Heat (cal/g·°C)	Substance	Specific Heat (cal/g·°C)
Water	1.00	Wood (typical)	0.42
Ice	0.48	Glass (typical)	0.12
Steam	0.48	Rock (typical)	0.20
Iron	0.11	Ethanol	0.59
Aluminum	0.22	Methanol	0.61
Copper	0.092	Ether	0.56
Lead	0.038	Carbon tetrachloride	0.21

and steam do not have the same specific heat as liquid water. It is easy to make calculations involving specific heats. The equation for the amount of heat absorbed or released is

$$\text{Amount of heat} = \text{Specific heat} \times \text{Mass} \times \text{Change in temperature}$$
$$= SH \times m \times (T_2 - T_1)$$

where T_2 is the final and T_1 is the initial temperature in °C.

EXAMPLE 1.8

How many calories are required to heat 352 g of water from 23°C to 95°C?

Solution

$$\text{Amount of heat} = SH \times m \times (T_2 - T_1)$$

$$= \frac{1.0 \text{ cal}}{g \cdot °C} \times 352 \text{ g} \times (95 - 23) °C$$

$$= 2.5 \times 10^4 \text{ cal} = 25 \text{ kcal}$$

Is it reasonable? Each gram of water requires 1 cal to raise its temperature by one degree. We have approximately 350 g of water. To raise its temperature by one degree would, therefore, require approximately 350 cal. But we are raising the temperature not by one degree but by approximately 70 degrees (from 23 to 95). Thus the total number of calories will be approximately $70 \times 350 = 24\,500$ cal or 25 kcal. In this case, the estimate comes out exactly the same as the calculated answer.

Problem 1.8 ▬▬▬
How many calories are required to heat 731 g of water from 8°C to 74°C? Check your answer to see if it is reasonable. ■

EXAMPLE 1.9

If we add 450 cal of heat to 37 g of ethanol at 20°C, what is the final temperature?

Solution
The specific heat of ethanol is 0.59 cal/g · °C (Table 1.4).

$$\text{Amount of heat} = SH \times m \times (T_2 - T_1)$$

$$450 \text{ cal} = \frac{0.59 \text{ cal}}{g \cdot °C} \times 37 \text{ g} \times (T_2 - T_1)$$

$$(T_2 - T_1) = \frac{450 \text{ cal}}{0.59 \text{ cal/}°C \times 37} = 21°C$$

Since the starting temperature is 20°C, the final temperature is 41°C.

Is it reasonable? The specific heat of ethanol is 0.59 cal/g · °C. This is close to 0.5, meaning that about half a calorie will raise the temperature of 1 g by 1°C. A mass of 37 g of ethanol needs approximately 40 times as many calories for a 1°C rise, and $40 \times \frac{1}{2} = 20$ cal. But we are adding 450 cal, which is about 20 times as much. Thus, we expect the temperature to rise by about 20°C, from 20°C to 40°C. The actual answer, 41°C, is, therefore, quite reasonable.

Problem 1.9 ▬▬▬
A 100 g piece of iron at 25°C is heated by adding 230 cal. What will be the final temperature? Check your answer to see if it is reasonable. ■

EXAMPLE 1.10

We heat 50.0 g of an unknown substance by adding 205 cal, and its temperature rises by 7.0°C. What is its specific heat?

Solution
To find the specific heat,

$$SH = \frac{\text{Amount of heat}}{m \times (T_2 - T_1)}$$

$$SH = \frac{205 \text{ cal}}{50.0 \text{ g} \times 7.0°C} = 0.59 \text{ cal/g} \cdot °C$$

Is it reasonable? If we would have water instead of an unknown substance with $SH = 1$ cal/g · °C, raising the temperature of 50.0 g by 7.0°C would require $50 \times 7.0 = 350$ cal. But we added only approximately 200 cal. Therefore, the specific heat of the unknown substance must be less than 1.0. How much less? Approximately 200/350 = 0.6. The actual answer, 0.59 cal/g · °C is, therefore, quite reasonable.

Problem 1.10 ▬▬▬
It required 88.2 cal to heat 13.4 g of an unknown substance from 23°C to 176°C. What is the specific heat of the unknown substance? Check your answer to see if it is reasonable. ■

BOX 1C

Cold Compresses

The high specific heat of water is useful in making cold compresses stay cold a long time. For example, consider two patients who are using cold compresses; one compress is made by soaking a towel with water, and the other is made with a towel soaked in ethanol, both at 0°C. Each gram of water in the water compress requires 25 cal to make the temperature of the compress rise to 25°C (after which it must be changed).

Because the specific heat of ethanol is 0.59 cal/g · °C (Table 1.4), each gram of the ethanol requires only 15 cal to reach 25°C. If the two patients give off heat at the same rate, the ethanol compress will be less effective because it will reach 25°C a good deal sooner than the water compress and will need to be changed sooner.

SUMMARY

To understand modern medicine, we must know some chemistry. **Chemistry** is the science that deals with the structure of matter and the changes it can undergo. In chemical changes (chemical reactions), substances are used up, and others are formed. A **chemical change** or **chemical reaction** produces new arrangements of atoms, the basic particles of every matter, without a gain or loss in the number of atoms. In **physical changes,** substances do not change their identity.

The **scientific method** is a tool used in science and medicine. The heart of the scientific method is the testing of **hypotheses** and **theories.**

In chemistry, we use the **metric system** for measurements. The base units are the meter for length, the liter for volume, and the gram for mass. Other units are indicated by prefixes that represent powers of ten. Temperature is measured in degrees Celsius or in kelvins. Conversions from one unit to another are best done by the **factor-label method,** in which units are multiplied and divided.

There are three states of matter: **gas, liquid,** and **solid. Density** is mass per unit volume. **Specific gravity** is density relative to water and, thus, has no units. Density usually decreases with increasing temperature.

Kinetic energy is energy of motion; **potential energy** is stored energy. Energy can be neither created nor destroyed, but it can be converted from one form to another.

Heat is a form of energy and is measured in calories. A calorie is the amount of heat necessary to raise the temperature of 1 g of liquid water by 1°C. Every substance has a **specific heat,** which is a physical constant. The specific heat is the number of calories required to raise the temperature of 1 g of the substance by 1°C.

KEY TERMS

Chemical change (Section 1.1)
Chemical property (Section 1.1)
Chemical reaction (Section 1.1)
Chemistry (Section 1.1)
Conversion factor (Section 1.5)
Density (Section 1.7)
Energy (Section 1.8)
Exponential notation (Section 1.3;
   Appendix I)
Fact (Section 1.2)
Factor-label method (Section 1.5)
Gas (Section 1.6)

Heat (Section 1.9)
Hyperthermia (Box 1B)
Hypothermia (Box 1B)
Hypothesis (Section 1.2)
Kinetic energy (Section 1.8)
Liquid (Section 1.6)
Mass (Section 1.4)
Matter (Section 1.1)
Metric system (Section 1.4)
Physical change (Section 1.1)
Physical property (Section 1.1)
Potential energy (Section 1.8)

Scientific method (Section 1.2)
Serendipity (Section 1.2)
SI (Section 1.4)
Significant figures (Section 1.3;
   Appendix II)
Solid (Section 1.6)
Specific gravity (Section 1.7)
Specific heat (Section 1.9)
Temperature (Sections 1.4, 1.9)
Theory (Section 1.2)
Volume (Section 1.4)
Weight (Section 1.4)

CONCEPTUAL PROBLEMS

1.A The life expectancy of a citizen in the United States is 76 years. Eighty years ago it was 56 years. In your opinion, what was the major contribution to this spectacular increase in life expectancy? Explain your answer.

1.B You read in a newspaper that Dr. X claims to have found a new remedy to cure diabetes. The remedy is an extract of carrots. Would you classify this claim as (a) fact, (b) theory, (c) hypothesis, or (d) hoax? Explain your answer.

1.C You drive in Canada, where distances are marked in km. The sign says you are 80 km from Ottawa. You are traveling at 75 mph. Will you reach Ottawa in less than an hour, in exactly one hour, or in more than an hour?

1.D While you drive your car, the battery is charged. How would you describe this process in terms of kinetic and potential energy?

PROBLEMS

Difficult problems are designated by an asterisk.

Chemistry and the Scientific Method _____

1.11 Why is the study of chemistry necessary for students planning a career in the health sciences?

1.12 Define: (a) matter   (b) chemistry.

1.13 What is the difference between a hypothesis and a fact?

1.14 Are new discoveries always driven by theories or hypotheses?

1.15 Classify each as a chemical or physical change:
(a) Burning gasoline
(b) Making ice cubes
(c) Boiling oil
(d) Melting lead
(e) Rusting iron
(f) Making ammonia from nitrogen and hydrogen
(g) Digesting food

Exponential Notation _____

1.16 Write in exponential notation:
(a) 37.5   (b) 62 900   (c) 0.091   (d) 0.000 000 28

1.17 Write out in full:
(a) 4.03×10^5 (b) 3.2×10^3
(c) 7.13×10^{-5} (d) 5.55×10^{-10}

1.18 Multiply:
(a) $(4.73 \times 10^5)(1.37 \times 10^2)$
(b) $(2.7 \times 10^{-4})(5.9 \times 10^8)$
(c) $(6.49 \times 10^7)(7.22 \times 10^{-3})$
(d) $(3.4 \times 10^{-5})(8.2 \times 10^{-11})$

1.19 Divide:
(a) $\dfrac{6.02 \times 10^{23}}{3.01 \times 10^5}$ (b) $\dfrac{3.14}{2.30 \times 10^{-5}}$
(c) $\dfrac{7.08 \times 10^{-8}}{300}$ (d) $\dfrac{5.8 \times 10^{-6}}{6.6 \times 10^{-8}}$
(e) $\dfrac{7.05 \times 10^{-3}}{4.51 \times 10^5}$

1.20 Add:
(a) $(7.9 \times 10^4) + (5.2 \times 10^4)$
(b) $(8.73 \times 10^4) + (6.7 \times 10^3)$
(c) $(3.61 \times 10^{-4}) + (4.776 \times 10^{-3})$

1.21 Subtract:
(a) $(8.50 \times 10^3) - (7.61 \times 10^2)$
(b) $(9.120 \times 10^{-2}) - (3.12 \times 10^{-3})$
(c) $(1.3045 \times 10^2) - (2.3 \times 10^{-1})$

*1.22 Solve: $\dfrac{(3.14 \times 10^3) \times (7.80 \times 10^5)}{(5.50 \times 10^2)}$

*1.23 Solve: $\dfrac{(9.52 \times 10^4) \times (2.77 \times 10^{-5})}{(1.39 \times 10^7) \times (5.83 \times 10^{-3})}$

Significant Figures _____

1.24 How many significant figures are there in
(a) 3216   (b) 0.007   (c) 0.0910   (d) 7.3004
(e) 16.104   (f) 0.0001   (g) 0.0000900?

1.25 How many significant figures are there in
(a) 5.71×10^{13} (b) 4.4×10^5 (c) 3×10^{-6}
(d) 4.000×10^{-11} (e) 5.5550×10^{-3}?

1.26 Round off to two significant figures:
(a) 91.621   (b) 7.329   (c) 0.677
(d) 0.003249   (e) 5.88

1.27 Multiply these numbers, using the correct number of significant figures in your answer:
(a) 3630.15×6.8 (b) 512×0.0081
(c) $5.79 \times 1.85825 \times 1.4381$

1.28 Divide these numbers, using the correct number of significant figures in your answer:
(a) $\dfrac{3.185}{2.08}$ (b) $\dfrac{6.5}{3.0012}$ (c) $\dfrac{0.0035}{7.348}$

1.29 Add these groups of measured numbers:

(a)	(b)	(c)
37.4083	84	51.51
5.404	8.215	100.27
10 916.3	0.01	16.878
3.94	151.7	3.6817
0.0006		

Measurements _____

1.30 In the metric system, what is the base unit of
(a) mass (b) length (c) volume?

1.31 How many grams are there in (a) 1 kg (b) 1 mg?

1.32 Estimate without actually calculating which one is the shorter distance.
(a) 20 mm or 0.3 m
(b) 1 inch or 30 mm
(c) 2000 m or 1 mile

1.33 Which answer is most reasonable?
(a) A baseball bat has a length of 100 mm or 100 cm or 100 m
(b) A glass of milk holds 23 cc or 230 mL or 23 L
(c) A man weighs 75 mg or 75 g or 75 kg
(d) A tablespoon contains 15 mL or 150 mL or 1.5 L
(e) A paper clip weighs 50 mg or 50 g or 50 kg
(f) Your hand has a width of 100 mm or 100 cm or 100 m
(g) An audiocassette weighs 40 mg or 40 g or 40 kg

1.34 What is the difference between mass and weight?

1.35 Convert to degrees Celsius and to kelvins:
(a) 320°F   (b) 212°F   (c) 0°F   (d) −250°F

1.36 Convert to degrees Fahrenheit and to kelvins:
(a) 25°C   (b) 40°C   (c) 250°C   (d) −273°C

Unit Conversions _____

1.37 Make the following conversions (conversion factors are given in Table 1.3):
(a) 42.6 kg to lb   (b) 1.62 lb to g
(c) 34 in. to cm   (d) 37.2 km to miles
(e) 2.73 gal to L   (f) 62 g to oz (avdp)
(g) 33.61 qt to L   (h) 43.7 L to gal
(i) 1.1 miles to km   (j) 34.9 mL to fl oz

1.38 Make the following metric conversions:
  (a) 96.4 mL to L (b) 275 mm to cm
  (c) 45.7 kg to g (d) 475 cm to m
  (e) 21.64 cc to mL (f) 3.29 L to cc
  (g) 0.044 L to mL (h) 711 g to kg
  (i) 63.7 mL to cc (j) 0.073 kg to mg
  (k) 83.4 m to mm (l) 361 mg to g

1.39 What is your weight in kg? Your height in m?

1.40 The speed limit in some European cities in 80 km/h. How many mph is this?

1.41 Your car gets 25.00 miles on a gallon of gas. What would be your car efficiency in km/L?

Density and Specific Gravity

1.42 The volume of a rock weighing 1.075 kg is 334.5 mL. What is the density of the rock in g/mL? Express your answer to three significant figures.

1.43 The density of manganese is 7.21 g/mL, that of calcium chloride is 2.15 g/mL, and that of sodium acetate is 1.528 g/mL. If you place these three solids in a liquid that has a density of 2.15 g/mL, which will sink to the bottom, which will stay on the top, and which will stay in the middle of the liquid?

1.44 The density of titanium is 4.54 g/mL. What is the volume, in mL, of 163 g of titanium?

1.45 A 335.0-cc sample of urine has a mass of 342.6 g. What is the density, in g/mL, to three decimal places?

1.46 The density of methanol at 20°C is 0.791 g/mL. What is the mass, in g, of a 280-mL sample?

1.47 The density of dichloromethane, a liquid insoluble in water, is 1.33 g/cc. If dichloromethane and water are placed in a separatory funnel, which will be the upper layer?

*1.48 A sample of 10.00 g of oxygen has a volume of 6702 mL. The same weight of carbon dioxide occupies 5058 mL.
  (a) What are the densities in g/L?
  (b) Carbon dioxide is used as a fire extinguisher to cut off the fire's supply of oxygen. Do the densities of these two gases explain the fire-extinguishing ability of carbon dioxide?

*1.49 Two 1-quart containers of butter are for sale. Butter A has a density of 1.2 g/mL and costs $5.00; butter B has a density of 0.96 g/mL and costs $4.00. Which is a better buy?

Energy

1.50 What is the difference between kinetic and potential energy?

Heat

1.51 The specific heat of steam is 0.48 cal/g · °C. How many kcal are needed to raise the temperature of 10.5 kg of steam from 120°C to 150°C?

1.52 How many calories are required to heat the following (specific heats are given in Table 1.4)?
  (a) 52.7 g of aluminum from 100°C to 285°C
  (b) 93.6 g of methanol from −35°C to 55°C

  (c) 3.4 kg of lead from −33°C to 730°C
  (d) 71.4 g of ice from −77°C to −5°C

*1.53 If 168 g of an unknown liquid requires 2750 cal of heat to raise its temperature from 26°C to 74°C, what is the specific heat of the liquid?

Boxes

1.54 (Box 1A) The recommended dose of a medication is 5 mg/kg body weight. You have a patient whose weight is 125 lb. The pharmacy offers the medication in three different pills containing (a) 500 mg, (b) 250 mg, and (c) 100 mg. Which pill would you administer to the patient?

1.55 (Box 1A) If the recommended dose of a drug is 445 mg for a 180-lb man, what would be a suitable dose for a 135-lb man?

1.56 (Box 1A) The average lethal dose of heroin is 150 mg/kg of body weight. Estimate how many g of heroin would be lethal for a 200-lb man.

1.57 (Box 1B) How does the body react to hypothermia?

1.58 (Box 1B) Low temperatures often cause people to shiver. What is the function of this involuntary body action?

1.59 (Box 1C) Which would make a more efficient cold compress, ethanol or methanol? (Refer to Table 1.4.)

Additional Problems

1.60 The meter is a measure of length. Tell what each of the following units measures:
  (a) cm^3 (b) mL (c) kg (d) cal
  (e) g/cc (f) joule (g) °C (h) cm/s

1.61 A brain weighing 1 lb occupied a volume of 620 mL. What is the specific gravity of the brain?

*1.62 If the density of air is 1.25×10^{-3} g/cc, what is the mass in kg of the air in a room that is 5.3 m long, 4.2 m wide, and 2.0 m high?

1.63 Is it possible to convert 83 mL to g?

*1.64 The kinetic energy possessed by an object with a mass of 1 g moving with a velocity of 1 cm/s is called 1 erg. What is the kinetic energy, in ergs, of an athlete with a mass of 127 lb running at a velocity of 14.7 mph?

1.65 A European car advertises an efficiency of 22 km/L, whereas that of an American car claims an economy of 30 miles per gallon. Which car is more efficient?

1.66 Classify these as kinetic or potential energy:
  (a) Water held by a dam
  (b) A speeding train
  (c) A book on its edge before falling
  (d) A falling book
  (e) Electric current in a light bulb

1.67 Shivering is the body's response to increase the body temperature. What kind of energy is generated by shivering?

1.68 When the astronauts walked on the moon, they could make giant leaps despite their heavy gear.
  (a) Why were their weights on the moon so low?
  (b) Were their masses different on the moon than on the Earth?

Astronaut Edwin E. Aldrin, Jr., walking on the Moon after the *Apollo 11* lunar landing. *(NASA)*

1.69 Which of the following is the largest mass and which is the smallest?
(a) 41 g   (b) 3×10^3 mg   (c) 8.2×10^6 μg
(d) 4.1×10^{-8} kg

1.70 What is the difference between heat and temperature?

1.71 In Japan, high-speed bullet trains move with an average speed of 220 km/h. If Dallas and Los Angeles were connected by such a train, how long would it take to travel nonstop between these cities (1490 miles)?

*1.72 The specific heats of some elements at 25°C are: aluminum = 0.215 cal/g·°C; carbon (graphite) = 0.170 cal/g·°C; iron = 0.107 cal/g·°C; mercury = 0.0331 cal/g·°C.

(a) Which element would require the smallest amount of heat to raise the temperature of 100 g of the element by 10°C?
(b) If the same amount of heat needed to raise the temperature of 1 g of aluminum by 25°C were applied to 1 g of mercury, by how many degrees would its temperature be raised?
(c) If a certain amount of heat is used to raise the temperature of 1.6 g of iron by 10°C, the temperature of 1 g of which element would also be raised by 10°C, using the same amount of heat?

1.73 The density of gasoline is 0.71 g/mL. What is this in lb/gal?

1.74 One quart of milk costs 80 cents, and 1 L costs 86 cents. Which is the better buy?

*1.75 Consider butter, density 0.860 g/mL, and sand, density 2.28 g/mL.
(a) If 1.00 mL of butter is thoroughly mixed with 1.00 mL of sand, what is the density of the mixture?
(b) What would the density of the mixture be if 1.00 g of the same butter were mixed with 1.00 g of the same sand?

1.76 Which speed is the fastest?
(a) 70 mph   (b) 140 km/h   (c) 4.5 km/s
(d) 48 miles/min

1.77 In calculating the specific heat of a substance, the following data are used: Mass = 92.15 g; heat = 3.200 kcal; rise in temperature = 45°C. How many significant figures should you report in calculating the specific heat?

*1.78 A solar cell generates 500 kJ of energy per hour. In order to keep a refrigerator at 4°C, one needs 250 kcal/h. Can the solar cell supply sufficient energy per hour to maintain the temperature of the refrigerator?

Some ancient Greeks said that water is a liquid because its atoms are smooth and round and can easily glide over each other, whereas atoms of iron are rough and hard.

Above: Silicon atoms viewed through a scanning tunneling microscope. (© Science VU/BMRL/Visuals Unlimited)

Atoms

2.1 Introduction

What is matter made of? This question was discussed for thousands of years, long before human beings had any reasonable way of getting an answer. In ancient Greece, two schools of thought tried to answer this question. One group, led by a scholar named Democritus (about 470–380 BCE), believed that all matter is made of very small particles—much too small to see. Democritus called these particles atoms. Some of his followers developed the idea that there were different kinds of atoms, with different properties, and that the properties of the atoms caused ordinary matter to have the properties we all know.

But these ideas were not accepted by all ancient thinkers. Another group, led by Zeno of Elea (born about 450 BCE), did not believe in atoms at all. They insisted that matter is infinitely divisible; that is, if you took any object such as a coin or a piece of wood, or a liquid such as water or olive oil, you could cut it or otherwise divide it into two parts, divide each of these into two more parts, and continue the process forever. According to Zeno and his followers, you would never reach a particle of matter that could no longer be divided.

Today we know that Democritus was right and Zeno was wrong. **Atoms are the basic units of matter.** But there is a great difference in the way we look at this question. Today our ideas are based on *evidence*. Democritus had no evidence to prove that matter cannot be divided an infinite number of times, just as Zeno had no evidence for his claim that matter can be divided infinitely. Both claims were based not on evidence but on visionary belief: one in unity, the other in diversity.

BOX 2A

Elements Necessary for Human Life

Only a few elements are necessary for human life. The four most important of these—carbon, hydrogen, oxygen, and nitrogen—make up the subjects of organic chemistry and biochemistry (Chapters 10–31), so we need not consider them here. Seven other elements are also quite important, and our bodies use at least nine additional ones (trace elements) in very small quantities. Table 2A shows these 20 elements and their functions in the human organism. Many of these elements are more fully discussed later in the book. For the average daily requirements, sources in foods, and symptoms of deficiencies, see Chapter 29.

TABLE 2A   Elements and Their Functions in the Human Body

Element	Function
The Big Four	
Carbon (C) Hydrogen (H) Nitrogen (N) Oxygen (O)	Discussed in Chapters 10–31
The Next Seven	
Calcium (Ca)	Strengthens bones and teeth; aids in blood clotting
Phosphorus (P)	Present in phosphates of bones and involved in energy transfer
Potassium (K)	Helps regulate electrical balance of body fluids; essential for nerve conduction
Sulfur (S)	An essential component of proteins; important in protein structure
Chlorine (Cl)	Helps regulate electrical balance of body fluids
Sodium (Na)	Helps regulate electrical balance of body fluids
Magnesium (Mg)	Helps nerve and muscle action; present in bones
The Trace Elements	
Chromium (Cr)	Increases effectiveness of insulin
Cobalt (Co)	A part of vitamin B_{12}
Copper (Cu)	Strengthens bones; assists in enzyme action
Fluorine (F)	Reduces incidence of dental cavities
Iodine (I)	An essential part of thyroid hormones
Iron (Fe)	An essential part of some proteins, such as hemoglobin, myoglobin, cytochromes, and FeS protein
Magnesium (Mg)	Present in bone-forming enzymes; aids in fat and carbohydrate metabolism
Molybdenum (Mo)	Necessary for action of certain enzymes
Zinc (Zn)	Necessary for normal growth

In Section 2.3 we will discuss the evidence for the existence of atoms, but first we need to look at the diverse forms of matter.

2.2 Classification of Matter

Matter can be divided into three classes: elements, compounds, and mixtures.

Elements

The elements technetium, promethium, astatine, francium, and all the other elements in the Periodic Table higher than uranium do not occur naturally on Earth but have been synthesized by chemists and physicists.

The symbol for tungsten, W, comes from the German *wolfram*. The symbol for iron, Fe, comes from the Latin *ferrum*.

An element is a pure substance that consists of one type of atom. *An element cannot be broken down into simpler substances by chemical reactions.* At this time, 114 elements are known. Of these, 88 occur in nature; the others have been made by chemists and physicists. The elements are the building blocks of all matter. *A list of the known elements is given on the inside back cover of this book, along with their symbols.* These symbols, consisting of one, two, or three letters, are the same in all countries. Most of the symbols correspond directly to the name in English (for example, C for carbon, H for hydrogen, Li for lithium), but a few come from the Latin or German names. Photographs of some elements are shown in Figure 2.1. Some of the more important elements, their symbols, and their functions in the human body are listed in Table 2A.

BOX 2B

Should We Be Afraid of Chemicals?

Are chemicals dangerous? Many people seeing something like "one tenth of 1 percent sodium sulfite," "carrageenan," or "carboxymethylcellulose" on a package of food will automatically react negatively, as if any chemical with a hard-to-pronounce name must be harmful. This reaction, while understandable in people who know little about chemistry, is quite mistaken. All the food we eat is made entirely of chemicals, and so are the trees and grass, the ground itself, all animals, and even our bodies. All matter is chemical.

To answer our opening question, some chemicals are dangerous; others are not; still others are essential to life (oxygen is a chemical). In many cases, a small amount of a chemical is essential, while a larger amount of the same chemical may well be toxic.

The important thing is not to be frightened by words just because they sound unfamiliar. The names of chemicals are often long and, to the student, hard to pronounce. The reason is that millions of chemicals are known, and each must be given a different name, with the name in most cases carrying enough information for chemists to be able to tell the composition from the name. An example is the chemical 1,3,3-trimethyl-2-(9-hydroxy-3,7-dimethylnona-1,3,5,7-tetraenyl) cyclohexene. Far from being harmful, this chemical, also

■ All our food contains chemicals. *(Charles D. Winters)*

known as vitamin A, is essential to human life (although excessive amounts can indeed be harmful; see Box 29B).

The intelligent approach to chemicals is to obtain an understanding of what the substances are and what they can do. In this book we will be doing just that.

Phosphorus (P)

Magnesium (Mg) (left), calcium (Ca) (right)

FIGURE 2.1 Some elements.
(Charles D. Winters)

Clockwise from the top: lead (Pb), tin (Sn), carbon (C), silicon (Si)

Compounds

A compound is a pure substance made up of two or more elements in a fixed proportion by weight. For example, water is a compound made up of hydrogen and oxygen, and table salt is a compound made up of sodium and chlorine. There are millions of known compounds, many of which we will meet in this book. A compound is characterized by its formula. The formula tells us the ratios of its constituent elements and identifies each element by its atomic symbol. For example, in table salt the combining ratio of sodium and chlorine is 1:1. Given that Na is the symbol of sodium and Cl the symbol of chlorine, the formula of table salt is NaCl. In water, the combining ratio is two hydrogens to one oxygen. The symbol of hydrogen is H, and that of oxygen is O. The formula of water is H_2O. The subscripts following the atomic symbols indicate the ratio of the combining elements. The number one in these ratios is omitted from the subscript. It is understood that NaCl means a ratio of 1:1 and that H_2O represents a ratio of 2:1. You will find more about the nature of combining elements in a compound and about their names and formulas in Chapter 3.

EXAMPLE 2.1

(a) In a compound, magnesium, atomic symbol Mg, and fluorine, atomic symbol F, combine in a ratio of 1:2. What is the formula of the compound?

(b) A compound's formula is $HClO_4$. What are the combining ratios of the elements in this compound?

Solution

(a) The formula is MgF_2. We do not write a subscript of one after Mg.

(b) Both H and Cl have no subscripts. This means that hydrogen and chlorine have a combining ratio of 1:1. The subscript on oxygen is four. Therefore, the combining ratios in $HClO_4$ are 1:1:4.

Problem 2.1 ▬▬

Write the formulas of compounds in which the combining ratios are as follows:

(a) Sodium:chlorine:oxygen   1:1:3

(b) Aluminum, atomic symbol Al, and fluorine, atomic symbol F   1:3 ■

Mixtures

Elements and compounds are pure substances. The third category of matter consists of **mixtures** of pure substances. In our daily life most of the matter we meet (including our own bodies) consists of mixtures rather than pure substances. For example, blood, butter, gasoline, soap, the metal in a wedding ring, the air we breathe, and the earth we walk on are all mixtures of pure substances.

The difference between mixtures and compounds can be illustrated by a simple experiment. The elements zinc and sulfur are shown in Figure 2.2(a) and (b). If we mix them, Figure 2.2(c), we have a mixture of zinc and sulfur. We

Mixture Matter made of more than one kind of pure substance in any proportion

(a)

(b)

(c)

(d)

FIGURE 2.2 Mixtures and compounds. *(a)* Zinc. *(b)* Sulfur. *(c)* A mixture of zinc and sulfur. Both yellow sulfur and gray zinc particles are present. *(d)* The compound zinc sulfide. The compound is white. Neither gray nor yellow particles are present. *(Beverly March)*

can put the mixture in a blender, so as to make it a very thorough mixture, with the particles evenly distributed, but it still consists of tiny pieces of sulfur and tiny pieces of zinc. On the other hand, if we heat the mixture of zinc and sulfur, we get the compound zinc sulfide, ZnS, shown in Figure 2.2(d). This compound is an entirely different substance, with properties that are completely different from those of either zinc or sulfur. An important difference between a compound and a mixture is that the proportions by weight of the elements in a compound are fixed (Section 2.3), whereas, for example, **in a mixture** of zinc and sulfur, the two **elements can be mixed in any proportion.**

2.3 Dalton's Atomic Theory

Around 1805 an English chemist, John Dalton (1766–1844), put forth a scientific atomic theory. The big difference between Dalton's theory and that of Democritus (Section 2.1) was that Dalton based his theory on evidence. First, let us state the theory and then see what kind of evidence supported it.

1. All matter is made up of very tiny indivisible particles, which Dalton called **atoms,** using the same name that Democritus had used.

2. All atoms of the same element are identical.

3. Atoms of any one element are different from those of any other clement.

4. Atoms combine to form molecules. A **molecule** is a **tightly bound group of atoms that acts as a unit.**

> **Atom** The smallest particle of an element that shows the chemical behavior of that element

The Evidence for Dalton's Atomic Theory

The Law of Conservation of Mass

The great French chemist Antoine Laurent Lavoisier (1743–1794) discovered the *law of conservation of mass,* which states that **matter can be neither created nor destroyed.** Lavoisier proved this by many experiments in which he showed that, no matter what chemical changes took place, the total mass of the matter at the end of the experiment was exactly the same as at the beginning. Dalton's theory explained this fact very nicely. If all matter is made up of indestructible atoms, then any chemical reaction just changes the attachments between atoms but does not destroy the atoms themselves. For example, carbon monoxide reacts with lead oxide to give carbon dioxide and lead. This is how Dalton might have explained what happens:

$$CO + PbO \longrightarrow CO_2 + Pb$$
Carbon monoxide   Lead oxide   Carbon dioxide   Lead

In this illustration, the carbon atom is shown as a black ball, the lead atom as a gray ball, and the oxygen atoms as red balls. All the original atoms are still there in the end; they have only changed partners. This explains why the total mass at the end is the same as at the beginning.

> The real atoms, of course, do not have colors.

The Law of Constant Composition

Before Dalton presented his theory, another French chemist, Joseph Proust (1754–1826), demonstrated the *law of constant composition,* which states that **any compound is always made up of elements in the same proportion by mass.** For example, the compound water contains 89 percent oxygen and 11 percent hydrogen by mass. Proust said that all pure water contains these two elements in these exact proportions, whether the water comes from the Atlantic Ocean or the Seine River or is collected as rain, squeezed out of a watermelon, or distilled from urine. Water is never formed in any other proportions of oxygen and hydrogen. This fact was also evidence for Dalton's theory. If a water molecule is made up of two atoms of hydrogen and one of oxygen, and if an oxygen atom has a mass 16 times that of a hydrogen atom (as today we know it has), water must always contain 89 percent oxygen and 11 percent hydrogen and can never be found in any other proportions.

Monatomic, Diatomic, and Polyatomic Elements

Some elements—for example, helium and neon—consist of single atoms that are not connected to each other. They are monatomic. In contrast, oxygen, in its most common form, contains two atoms in each molecule, connected to each other by a chemical bond. We write this as O_2, the subscript showing the number of atoms in the molecule. Other elements that occur as diatomic molecules are hydrogen (H_2), nitrogen (N_2), fluorine (F_2), chlorine (Cl_2), bromine (Br_2), and iodine (I_2). It is important to understand that under normal conditions the free atoms O, H, N, F, Cl, Br, and I do not exist. These elements occur only as diatomic molecules.

Three representations of a water molecule

H_2O

| Molecular formula | Space-filling model | Ball-and-stick bond model |

Three representations of a S_8 molecule

| Molecular formula | Space-filling model viewed from above | Ball-and-stick bond model viewed along its edge |

Some elements have even more atoms in each molecule. Ozone has three oxygen atoms (O_3) and in one form of phosphorus, each molecule has four atoms (P_4). One form of sulfur forms S_8 molecules. There are even elements whose molecules are much larger than this. For example, diamond has millions of carbon atoms all bonded together in a gigantic cluster. Just the same, all these are elements because they all contain only one kind of atom. They are referred to as polyatomic elements.

2.4 Inside an Atom

Three Elementary Particles

Today we know that matter is more complex than Dalton believed. A wealth of evidence obtained over the last 100 years or so has convinced us that atoms are not indivisible but are made up of even smaller particles

TABLE 2.1 Properties and Location of the Elementary Particles

Elementary Particle	Mass (g)	Mass (amu)	Charge	Location
Proton	1.672×10^{-24}	1	+1	In the nucleus
Electron	9.110×10^{-28}	1/1837	−1	Outside the nucleus
Neutron	1.675×10^{-24}	1	0	In the nucleus

called elementary particles. **Three elementary particles make up all atoms: protons, electrons, and neutrons.** The main things we know about these three particles are their charge and their mass (Table 2.1).

A **proton** has a positive charge. By convention we say that the magnitude of the charge is 1. Thus, the proton has a charge of +1 (two protons have a total charge of +2, and so forth). The mass of a proton is 1.672×10^{-24} g, but this number is so small that it is more convenient to use another unit, called the atomic mass unit (amu):

$$1 \text{ amu} = 1.661 \times 10^{-24} \text{ g}$$

The proton has a mass of approximately 1 amu.

An **electron** has a charge of −1, equal in magnitude to the charge on a proton but opposite in sign. The mass of an electron is approximately 1/1837 that of the proton. Thus, it takes approximately 1837 electrons to equal the mass of one proton.

Like charges repel, and unlike charges attract. Two protons repel each other, but a proton and an electron attract each other.

Neutrons have no charge at all. Therefore, they do not attract or repel each other or any other particle. The mass of a neutron is slightly greater than that of a proton: 1.675×10^{-24} g. This mass is so close to that of a proton, however, that for most purposes we say that these two particles have the same mass, namely 1 amu.

These are the three particles that make up atoms, but where are they found? Protons and neutrons are found in a tight cluster in the center of an atom (Figure 2.3). This part of an atom is called the **nucleus,** and both protons and neutrons are called **nucleons.** We will discuss the nucleus in greater detail in Chapter 9. Electrons are found as a diffuse cloud outside the nucleus.

Mass Number

Each atom has a fixed number of protons, electrons, and neutrons. **The total number of protons and neutrons (nucleons) in any atom is called the mass number.** For example, an atom with 5 protons, 5 electrons, and 6 neutrons has a mass number of 11 (electrons are not counted in determining mass number because their mass is negligible compared to that of the protons and neutrons, Table 2.1).

EXAMPLE 2.2

What is the mass number of an atom containing:
(a) 58 protons, 58 electrons, and 78 neutrons?
(b) 17 protons, 17 electrons, and 20 neutrons?

Solution
(a) The mass number is 58 + 78 = 136.
(b) The mass number is 17 + 20 = 37.
In each case, we add the number of protons and neutrons and ignore the electrons.

There are many other elementary particles, but we do not deal with them in this book.

To be more exact, the mass of the proton is 1.0073 amu.

The exact mass of the neutron is 1.0087 amu.

Repel   Repel   Attract

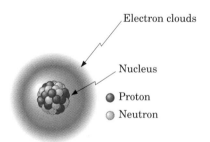

FIGURE 2.3 A typical atom. Protons and neutrons make up the nucleus. The electrons are found as a cloud outside the nucleus. This figure is not drawn to scale.

Problem 2.2 ▬

What is the mass number of an atom containing:
(a) 15 protons, 15 electrons, and 16 neutrons?
(b) 86 protons, 86 electrons, and 136 neutrons? ■

Atomic Number

Atomic numbers for all the elements are given in the atomic weight table on the inside back cover.

The **atomic number** of an element is **the number of protons in its nucleus.** At the present time, 114 elements have been discovered; these elements have atomic numbers from 1 to 114. A few atoms with 116 protons and few atoms containing 118 protons were created in 1999 in Berkeley, California, but their existence still awaits confirmation. The smallest atomic number belongs to the element hydrogen, which has only one proton, and the largest (so far) to the yet unnamed heaviest known element, which contains 118 protons.

If you know the atomic number and the mass number of an element, you can properly identify it. For example, the element with 6 protons, 6 electrons, and 6 neutrons has an atomic number of 6 and a mass number of 12. The element with atomic number 6 is called carbon. Since its mass number is 12, we call this atomic nucleus carbon-12. The symbol of carbon is C. We can write this atomic nucleus with the symbol of

$$^{12}_{6}C$$

In this form of writing, the atomic mass is always in the upper left corner (superscript) and the atomic number in the lower left corner (subscript) of the atomic symbol.

EXAMPLE 2.3

If you know the name of an element, you can look up its atomic number from the atomic weight table on the inside back cover. Conversely, if you know the atomic number of an element you can look up its symbol from the Periodic Table on the inside front cover. The atomic number of each element is in red in the Periodic Table.

Name the elements mentioned in Example 2.2. Write the symbols of their atomic nuclei.

Solution
(a) Since there are 58 protons, we learn from the Periodic Table on the inside front cover that its symbol is Ce. We find in the atomic weight table on the inside back cover that the name of this atom with 58 protons is cerium. This element, having a mass number of 136, is then cerium-136. Its symbol is $^{136}_{58}Ce$.
(b) This atom has 17 protons making it a chlorine (Cl) atom. Since its atomic mass number is 37, we call the element chlorine-37. Its symbol is $^{37}_{17}Cl$.

Problem 2.3 ▬

Name the elements mentioned in Problem 2.2. Write the symbols of their atomic nuclei. ■

EXAMPLE 2.4

A number of elements have an equal number of protons and neutrons in their nuclei. Among these are oxygen, nitrogen, and neon. What are the atomic numbers of these elements? How many protons and neutrons does an atom of each have? Write the name and the symbol of each of these atomic nuclei.

Solution

Atomic numbers are found for these elements in the list of elements on the inside back cover. This table shows that oxygen (O) has atomic number 8, nitrogen 7, and neon 10. This means that the oxygen has 8 protons and 8 neutrons. Its name is oxygen-16, and its symbol is $^{16}_{8}O$. The nitrogen (N) in this problem has 7 protons and 7 neutrons. Its name is nitrogen-14 and its symbol is $^{14}_{7}N$. The neon (Ne) has 10 protons and 10 neutrons. Its name is neon-20 and its symbol is $^{20}_{10}Ne$.

Problem 2.4 ▬▬▬
(a) What are the atomic numbers of mercury (Hg) and lead (Pb)?
(b) How many protons does an atom of each have?
(c) If both Hg and Pb have 120 neutrons in their nuclei, what is the mass number of each?
(d) Write the name and the symbol of each. ∎

Ions

Atoms are electrically neutral; that is, they have no charge. This means that the number of electrons in any atom must be equal to the number of protons. The mercury atom, for example, with 80 protons, also has 80 electrons. So why don't we say that the number of electrons establishes the identity of the atom? The reason is that atoms can lose or gain electrons but do not thereby change to other elements. If a mercury atom, with 80 protons and 80 electrons, loses two electrons, it becomes a mercury ion, with 80 protons and 78 electrons. We no longer call it an atom, but it is still mercury because it still has 80 protons. We call it an **ion,** which is defined as **a particle with unequal number of protons and electrons.** Atoms do not normally lose or gain protons or neutrons. However, they can gain or lose electrons easily. The mercury ion we just mentioned has 78 electrons and 80 protons. The total charge of the 78 electrons (-78) is canceled by the charge on 78 of the protons ($+78$), but there are no electrons present to cancel the other two positive charges. Therefore, this particular mercury ion has a charge of $+2$, and we write it as Hg^{2+}.

If in an ion there are more protons than electrons, the ion has a positive charge. If there are more electrons than protons, the ion has a negative charge. We will discuss ions again in Sections 3.2 and 3.3.

A sulfide ion, S^{2-}, has 16 protons and 18 electrons.

Isotopes

Although we can say that a mercury atom or ion always has 80 protons and that a mercury atom has 80 electrons, we cannot say that a mercury atom (or ion) must have any particular number of neutrons. Some of the mercury atoms found in nature have 122 neutrons; the mass number of these atoms is 202, and they are designated as either ^{202}Hg or mercury-202. Other mercury atoms have only 120 neutrons (and therefore a mass number of 200). Still others have 118, 119, or 121 neutrons, and others have as many as 124. **Atoms with the same number of protons but different numbers of neutrons** are called **isotopes.** All isotopes of mercury contain 80 protons (or they wouldn't be mercury). If they are neutral atoms, they all contain 80 electrons. Each isotope contains a different number of neutrons and, therefore, has a different mass number.

The fact that isotopes exist means that the second statement of Dalton's theory (Section 2.3) is not correct.

EXAMPLE 2.5

How many neutrons are in each isotope of oxygen?
(a) Oxygen-16   (b) Oxygen-17   (c) Oxygen-18

Solution
Each oxygen atom has 8 protons. The difference between the mass number and the number of protons gives the number of neutrons.
(a) $16 - 8 = 8$ neutrons
(b) $17 - 8 = 9$ neutrons
(c) $18 - 8 = 10$ neutrons

Problem 2.5 ■
Two iodine isotopes are used in medical treatments: iodine-125 and iodine-131. How many neutrons has each iodine? Write their symbols. ■

The properties of isotopes of the same element are almost identical, and for most purposes we regard them as identical. They differ, however, in radioactivity properties, which we will discuss in Chapter 9.

Atomic Weight

Most elements are found on Earth as mixtures of isotopes, in a more or less constant ratio. For example, any naturally occurring sample of the element chlorine contains 75.5 percent chlorine-35 (18 neutrons) and 24.5 percent chlorine-37 (20 neutrons). Silicon is composed of a fixed ratio of three isotopes, with 14, 15, and 16 neutrons. For some elements, these ratios may vary slightly, but for most purposes the slight variations can be ignored. The **atomic weight** of an element is a weighted average of the masses of the isotopes (the mass of an isotope is approximately the same as its mass number). Thus, the atomic weight of chlorine is 35.5 amu, which is a weighted average of the masses of the chlorine atoms, 75.5 percent of which have a mass of 35 amu and 24.5 percent of which have a mass of 37 amu:

$$\left(\frac{75.5}{100} \times 35.0 \text{ amu}\right) + \left(\frac{24.5}{100} \times 37.0 \text{ amu}\right) = 35.5 \text{ amu}$$

Some elements—for example, gold, fluorine, and aluminum—occur naturally as only one isotope. The atomic weights of these elements are close to whole numbers (gold, 196.97 amu; fluorine, 18.998 amu; aluminum, 26.98 amu).

A table of atomic weights is found on the inside back cover of this book.

> **Atomic weight** The ratio of the weighted average mass of the atom compared with the mass of the carbon isotope whose mass number is 12 and whose mass is taken to be 12.000 0 amu. Thus, the amu is defined as one twelfth the mass of a ^{12}C atom

This illustrates how to calculate a weighted average.

Some biochemists use **daltons** as a unit of atomic weight.

EXAMPLE 2.6

Magnesium has three stable isotopes, with abundances as follows: ^{24}Mg—78.6 percent, ^{25}Mg—10.1 percent, and ^{26}Mg—11.3 percent. Calculate the atomic weight of magnesium.

Solution
To calculate the weighted average of the masses of the isotopes, we must multiply each atomic mass by its abundance and add these together:

$$\left(\frac{78.6}{100} \times 24.0 \text{ amu}\right) + \left(\frac{10.1}{100} \times 25.0 \text{ amu}\right) + \left(\frac{11.3}{100} \times 26.0 \text{ amu}\right)$$

$$= 24.3 \text{ amu}$$

BOX 2C

Isotopic Abundance and Astrochemistry

When we speak of **isotopic abundance** and the atomic weights derived from the abundance, we refer to the conditions that exist on Earth. With the advance of space exploration, it became evident that other parts of the solar system: Sun, Moon, Mars, Jupiter, asteroids and comets, and the like, as well as the stars in galaxies and intergalactic gases may have different isotopic abundance. For example, the deuterium/hydrogen ratio $^2H/^1H$ on Mars is five times larger than on Earth. The $^{17}O/^{18}O$ ratio is also higher on Mars than on Earth. Such differences in isotopic abundance are used to establish theories on the origins and history of the solar system. Furthermore, a comparison of isotopic abundance in certain meteorites found on Earth enables us to guess their origin. For example, it is now believed that certain meteorites that landed on Earth came from Mars. Scientists speculate that they were ejected from its surface when the Red Planet collided with some other large body, perhaps an asteroid.

■ Mars. *(NASA)*

Problem 2.6 ■▬▬
The atomic weight of lithium is 6.941 amu. Lithium has only two naturally occurring isotopes, lithium-6 and lithium-7. Estimate which one is in greater abundance. ■

2.5 The Periodic Table

Origin of the Periodic Table

In the 1860s, a number of scientists, most notably the Russian Dmitri Mendeleev (1834–1907), produced periodic tables. In its modern form (see the inside front cover), the **Periodic Table** consists of all the elements arranged in order of atomic number. When this is done, we find that the elements fall into rows, in such a way that elements in the same vertical column (**group**) have similar properties. This is best understood by looking at specific examples in the table. The elements fluorine (atomic number 9), chlorine (17), bromine (35), iodine (53), and astatine (85) are all in the same column in the table. These elements are called *halogens*. They are all colored substances, and the colors deepen as we go down the table (Figure 2.4). All form compounds with sodium that have the formula NaX (for example, NaCl, NaBr), and not NaX_2, Na_2X, Na_3X, or anything else. Elements not in this column do not share these properties.

At this point, we must say a word about the numbering of the columns of the Periodic Table. Mendeleev gave them numerals and added the letter A for some columns and B for others. A newer way is to use a different set of numbers without added letters, 1 to 18, going from left to right. Thus, by Mendeleev's numbering, the halogens are in Group 7A, but, in the new numbering, they are in Group 17. Although in this book we use the traditional numbering, both kinds are shown on the Periodic Table on the inside front cover.

> **Period** One of the horizontal rows of elements in the Periodic Table

> **Group** One of the vertical columns of elements in the Periodic Table

"X" is a general symbol for halogens.

Note that hydrogen (H) appears in columns 1A and 7A but is neither a halogen nor an alkali metal.

BOX 2D

Strontium-90

Elements in the same column of the Periodic Table show similar properties, which can have biological consequences. One important example is the similarity of strontium to calcium (strontium is just above calcium on the Periodic Table). Calcium is an important element for humans because our bones and teeth consist largely of calcium compounds. We need some of this mineral in our diet every day, and we get it mostly from milk, cheese, and other dairy products, where it is present in the form of calcium ions.

One of the products released by nuclear explosions is the isotope strontium-90. This isotope is radioactive, with a half-life of 28.1 years. (Half-life is discussed in Section 9.3). Strontium-90, present in the fallout from aboveground nuclear test

explosions, was carried all over the Earth and slowly settled to the ground, where it was eaten by cows and other animals. Strontium-90 got into milk and eventually into our bodies. If it were not similar to calcium, our bodies would eliminate all of it within a few days, but because it is similar to calcium, some of the strontium-90 ions were deposited in bones and teeth (especially of children), subjecting all of us to a small amount of radioactivity for long periods of time. Aboveground nuclear testing was banned by a 1963 treaty between the United States and the former Soviet Union. Although a few other countries still conduct occasional tests, there is reason to hope that such tests will be completely halted in the near future.

Dimitri Mendeleev. *(Courtesy of E. F. Smith Memorial Collection, Special Collection Department, Van Pelt Library, University of Pennsylvania)*

What Is Periodic About the Periodic Table?

Not only do the elements in any one column have similar properties, but the properties also vary in some fairly regular way as we go up or down a column. For instance, as shown in Table 2.2, the melting and boiling points of the halogens regularly increase as we go down a column. Another example is found in Group 1A. Once again, the elements here (called *alkali met-*

FIGURE 2.4   Three halogens: chlorine, a gas; bromine, a liquid; and iodine, a solid. *(Larry Cameron)*

TABLE 2.2	Melting and Boiling Points of Group 7A Elements	
Element	Melting Point (°C)	Boiling Point (°C)
Fluorine	−220	−188
Chlorine	−101	−35
Bromine	−7	59
Iodine	114	184
Astatine	302	337

als) all share similar properties. All are metals with low melting points and are soft enough to be cut with a knife. All react with water (Figure 2.5), and all form compounds with fluorine, whose formula is MF, where "M" stands for an alkali metal. Also, as we go down the column, the melting and boiling points of the alkali metals change in a regular way, the violence of the reaction with water increases, as does the softness of the metal. The elements in the last column (Group 8A) provide still another example. All are gases at room temperature and form either no compounds at all or very few. For that reason, they are often called *noble* or *inert gases*.

The same regularity is found in all the other columns of the Periodic Table. The table is so useful that it hangs in nearly all chemistry classrooms and in most chemical laboratories throughout the world. What makes it so useful is that it correlates a vast amount of data about the elements and their compounds and allows us to make many predictions about physical as well as chemical properties. For example, if you were told that the boiling point of germane (GeH_4) is −88°C and that of methane (CH_4) is −164°C, could you predict the boiling point of silane (SiH_4)? The position of silicon in the table, between germanium and carbon, might lead you to a prediction of about −120°C. The actual boiling point of silane is −112°C, not far from this prediction.

If we look at the Periodic Table on the inside front cover of this book, we see that columns 1A and 2A extend above the rest, and so do columns 3A to 8A. The elements in these eight columns are called **representative elements.** The elements in the B columns and those in Group 8 (Groups 3 to 12 in the new numbering) are called **transition elements.** Notice that elements 58 to 71 and 90 to 103 are not included in the main body of the table but are shown separately at the bottom. These elements (called **inner transition elements**) actually belong in the main body, between columns 3B and 4B (between La and Hf). As is customary, we put them outside the main body only in order to make a more compact presentation. So, if you will, you may mentally take a pair of scissors, cut through the heavy line between columns 3B and 4B, move them apart, and insert the inner transition elements. You will now have a table with 32 columns.

Classification of Elements

There are three classes of elements. **Metals** belong to the first class of elements. **Metals are shiny substances whose atoms tend to give up electrons.** They are usually good conductors of electricity and heat and are malleable, which means that they can be hammered into flat sheets, and ductile, which means that they can be drawn into thin wires. Most elements are metals, including all the transition and inner transition ele-

FIGURE 2.5 A piece of metallic sodium reacting with water. *(Charles D. Winters)*

BOX 2E

The Use of Metals As Historical Landmarks

The malleability of metals played an important role in the development of human society. In the early days, tools were made from stone, which has no malleability. This period is called the Stone Age. Then about 9000 or 10 000 years ago it was discovered that copper found on the surface of the Earth could be hammered into sheets, which made it suitable for vessels, utensils, and religious or artistic objects. This period is sometimes called the Copper Age. Pure copper on the surface of the Earth is scarce, however, around 5000 BCE humans found a way to obtain copper from its ore by putting the green copper-containing stone, malachite, into a fire. Malachite yielded pure copper at the relatively low temperature of 200°C. Copper is a soft metal made of layers of copper crystals. It can easily be drawn into wires as the layers slip past each other. When hammered, the large crystals break into smaller ones with rough edges. The layers cannot slip, so hammered copper sheets were harder than the draw copper. As a result, the ancient profession of coppersmith was born, and beautiful plates, pots, and ornaments were produced.

Around 4000 BCE, it was discovered that by mixing molten copper with tin a greater hardness could be achieved. Copper, even when hammered, does not take an edge, but a copper-tin mixture (an alloy called bronze) does take an edge. As a result, knives and swords could be manufactured. Thus, the Bronze Age was born somewhere in the Middle East and quickly spread to China and all over the world.

An even harder metal was soon to come. The first raw iron was found in meteorites. (The ancient Sumerian name of iron

■ King Solomon's copper mines, Timna, Israel.
(Courtesy of Holy Views, Ltd., Jerusalem)

is "metal from heaven.") Later, around 2500 BCE, it was smelted from its ores, and the Iron Age began. Higher technology was needed because iron melts at a high temperature (about 1500°C). Thus, it took a longer time to perfect this smelting process and to learn how to manufacture steel (about 90 to 95 percent iron and 5 to 10 percent carbon). Steel objects appeared first in India around 100 BCE.

Modern historians look back at ancient cultures and use the discovery of a new metal as a landmark for that age.

---

ments. In addition, the elements in columns 1A and 2A are all metals, as are those in columns 3A to 6A below the heavy step-like line.

A second class of elements are the **nonmetals,** which do not conduct electricity, and their atoms do not give up electrons but rather tend to accept them. The nonmetals are the ones above the heavy step-like line in the Periodic Table.

Some of the elements right next to the heavy line have both metal and nonmetal properties and are called **metalloids.** The following elements are classified as metalloids: B, Si, Ge, As, Sb, Te, Po, and At.

In general, all representative elements behave more like metals in going down a column of the Periodic Table, and in going from right to left.

Noble (or inert) gases are the six elements in Group 8A. They have properties different from those of the other nonmetals. They neither give nor take electrons. Apart from them, the nonmetals are H, C, N, O, F, P, S, Cl, Se, Br, and I. Although there are only 11 nonmetals, they make up nearly all the compounds we shall study when we get to organic chemistry and biochemistry later in this book. Figure 2.6 shows the classification of the elements in the Periodic Table.

H																	H	He

Metals   Metalloids
Nonmetals   Noble gases

Li	Be												B	C	N	O	F	Ne
Na	Mg												Al	Si	P	S	Cl	Ar
K	Ca	Sc	Ti	V	Cr	Mn	Fe	Co	Ni	Cu	Zn	Ga	Ge	As	Se	Br	Kr
Rb	Sr	Y	Zr	Nb	Mo	Tc	Ru	Rh	Pd	Ag	Cd	In	Sn	Sb	Te	I	Xe
Cs	Ba	La	Hf	Ta	W	Re	Os	Ir	Pt	Au	Hg	Tl	Pb	Bi	Po	At	Rn
Fr	Ra	Ac	Rf	Db	Sg	Bh	Hs	Mt	Uun	Uuu	Uub						

Ce	Pr	Nd	Pm	Sm	Eu	Gd	Tb	Dy	Ho	Er	Tm	Yb	Lu
Th	Pa	U	Np	Pu	Am	Cm	Bk	Cf	Es	Fm	Md	No	Lr

FIGURE 2.6 Classification of the elements.

2.6 The Electron Structure of Atoms

In discussing the electron structure of atoms, we begin with hydrogen because it is the simplest, with only one electron. Before we do that, however, it is necessary to mention the work of the Danish physicist Niels Bohr (1885–1962). In 1913 Bohr found that the electron in a hydrogen atom could have one of a number of energies. An electron is always moving around the nucleus and so possesses kinetic energy. What Bohr proposed is that certain values were possible for this energy, but not others. This was a very surprising proposal. If you were told that you could drive your car at 23.4 mph or 28.9 mph or 34.2 mph, but never at any speed in between these values, you wouldn't believe it. Yet this is just what Bohr said about electrons in atoms. There is a lowest possible energy level; we call it the **ground state** (electrons in atoms cannot stop moving, so the ground-state energy is not zero). If an electron is to have more energy than in the ground state, however, only certain values are allowed; values in between are not permitted. We can liken the situation to walking up a flight of stairs (Figure 2.7). You can put your foot on any step, but you cannot stand anyplace between two steps. Bohr was unable to explain why these levels exist, but the accumulated evidence forced him to the conclusion that they do. We say that the energy of the electron is **quantized.**

Electrons in the lowest energy state of atoms—the ground state—occupy a spherical cloud surrounding the nucleus (Figure 2.8). Clouds of this type are called **orbitals.** They are designated by numbers and letters. This one is called the 1s orbital. There are also orbitals of higher energy. Some of them (2s and 2p) are shown in Figure 2.8. There are three 2p orbitals (designated $2p_x$, $2p_y$, and $2p_z$), which have equal energy. These three orbitals look like dumbbells (but are still clouds) and are at right angles to each other.

There are also orbitals of still higher energy. After 2p come 3s, 3p, 4s, and others even higher. All s orbitals are spherical; all p orbitals look like dumbbells and come in sets of three. Besides s and p orbitals, there are two

Niels Bohr of Denmark was awarded the Nobel Prize for physics in 1922. *(AIP Niels Bohr Library)*

FIGURE 2.7   The energy stairway. Unlike our stairways, the spaces between levels are not equal; they get smaller as they go up.

The d and f orbitals are less important to us, and we shall not discuss their shapes.

There are no $1p$, $1d$, or $1f$ orbitals.

other kinds, called d and f. All d orbitals come in sets of five, and f orbitals come in sets of seven.

The number used in the designation of orbitals is called the **principal energy level.** The first principal energy level (the 1 level) contains only one orbital, the $1s$ orbital. The second principal energy level (the 2 level) contains four orbitals (one $2s$ and three $2p$). The number of orbitals in each level increases as the principal energy level increases.

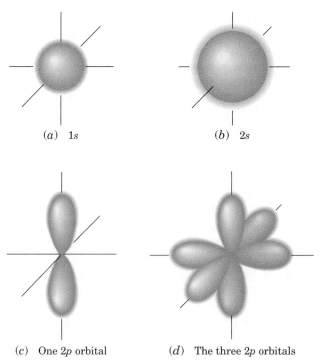

FIGURE 2.8   The $1s$, $2s$, and $2p$ orbitals.

All the orbitals in any principal energy level constitute a **shell.** Successive shells are at increasingly greater distances from the nucleus. The 1 shell, the innermost shell, contains only the 1s orbital. The 2 shell contains one 2s and three 2p orbitals. A **subshell** consists of all the orbitals with the same number and letter. For example, there are two subshells in the 2 shell: One of them consists of the three 2p orbitals, and the other contains only the single 2s orbital.

Note that the orbital designations show both the energy aspect and the spatial aspect of each orbital. The number tells the energy level and the size of the orbital; the letter tells its shape.

> A subshell can also be called a **sublevel.**

The Hydrogen Atom

We now have a good picture of the electron structure of a hydrogen atom. In its ground state (1s), an atom looks like the picture shown in Figure 2.8(a), with the nucleus at the center. This is the ground state, but it is not the only possible state. The electron can also exist in any one of the higher energy levels, called **excited states.** This is how it works: A hydrogen atom is in its ground state, looking like Figure 2.9(a). It then receives additional energy from some outside source, usually in the form of heat or light. This additional energy, when added to the energy the electron already has, equals the amount of energy the electron needs to reach some higher orbital—the $2p_x$, for example. In any individual hydrogen atom, only one orbital is occupied at any time. At room temperature, this orbital is overwhelmingly the 1s.

Rules for Other Atoms

Now that we have seen the electron structure of hydrogen, what about other atoms? The situation is similar except that these atoms have more than one electron. The orbitals available to all atoms are the same as in hydrogen: 1s, 2s, 2p, and so on. In the ground state of each atom, only the orbitals of lowest energy are occupied; all higher ones are empty. The electrons fill in according to the following rules:

1. Each orbital can hold two electrons, but no more. Furthermore, if there are two electrons in an orbital, they must have opposite *spins* clockwise and counterclockwise. This is a statement of the **Pauli exclusion principle.** When there is only one electron in an orbital, we call the electron unpaired and say that the orbital is singly occupied.

2. When we have a set of orbitals of equal energy (such as the $2p_x$, $2p_y$, and $2p_z$ orbitals), each orbital becomes half filled before any is completely filled. This is a statement of **Hund's rule.**

3. The electrons in any atom go into whichever unoccupied orbitals have the lowest energies, until there are no more electrons left. If we want to know which orbitals are occupied, we have to know the order of energies. This is shown in Figure 2.9. Note that the lowest p orbitals are 2p, the lowest d orbitals are 3d, and the lowest f orbitals are 4f. Within any shell, the energy levels increase in the order s, p, d, f. That is, s is the lowest, and f is the highest.

> **Hund's rule** Each orbital of a given subshell is occupied by a single electron before pairing begins

FIGURE 2.9 Energy levels of atomic orbitals up to the 4f level.

The Electron Buildup

Using the rules just given, we now determine the ground-state electron configuration for the first 36 elements. (We use boxes to designate orbitals.)

1. Hydrogen (H) As we have already seen, the single hydrogen electron is in the 1*s* orbital in the ground state. The arrow in the box indicates the direction of spin, either clockwise or counterclockwise.

This digit tells us that there is one electron in this orbital.

H configuration: $1s^1$

2. Helium (He) There are two electrons in this atom. Since any orbital can hold two electrons, both are able to go into the lowest energy orbital, the 1*s*. The arrows indicate that the two electrons have opposite spins.

The 2 tells us that there are two electrons in this orbital.

He configuration: $1s^2$

All electrons at the 1 level (the first principal energy level) constitute the first shell. This shell can hold only two electrons, so helium has a complete shell.

3. Lithium (Li) The first two of lithium's three electrons fill the 1*s* orbital. The third must go into the orbital of next lowest energy, the 2*s*.

Li configuration: $1s^2 2s^1$

In lithium the first shell (the 1 shell) is complete, and there is one electron in the 2 shell. The 1 shell is the inner shell, and the 2 shell is the incomplete outer shell.

4. Beryllium (Be) The first two electrons fill the 1*s* orbital (and the 1 shell). The remaining two fill the 2*s* orbital.

Be configuration: $1s^2 2s^2$

5. Boron (B) Now the 1*s* and 2*s* orbitals are filled, and there is one electron left over, which must go into one of the three 2*p* orbitals. It doesn't matter which one; we shall say $2p_x$.

B configuration: $1s^2 2s^2 2p^1$

6. Carbon (C) We now come to our first application of Hund's rule. The sixth electron of carbon must go into one of the 2*p* orbitals, but which one? Hund's rule says that it cannot fill the $2p_x$ orbital while either of the other two 2*p* orbitals is vacant, and so it goes into the $2p_y$

or $2p_z$ (it doesn't matter which one; we shall say $2p_y$). A carbon atom thus has two unpaired electrons.

1s   2s $2p_x$ $2p_y$ $2p_z$

C configuration: $1s^2 2s^2 2p^2$

7. Nitrogen (N) Hund's rule puts the seventh electron into the $2p_z$ orbital. A nitrogen atom has three unpaired electrons.

1s   2s $2p_x$ $2p_y$ $2p_z$

N configuration: $1s^2 2s^2 2p^3$

There is experimental proof for Hund's rule. Magnetic measurements show that a nitrogen atom has three unpaired electrons.

8–10. Oxygen, Fluorine, Neon (O, F, Ne) Now that all three $2p$ orbitals are half filled; the next three elements fill these orbitals, one at a time.

8	O	$1s^2 2s^2 2p^4$
9	F	$1s^2 2s^2 2p^5$
10	Ne	$1s^2 2s^2 2p^6$

1s   2s $2p_x$ $2p_y$ $2p_z$

Oxygen has two unpaired electrons, and fluorine has one. The tenth electron of neon, in the $2p_z$ orbital, completes the entire 2 shell. Neon is the second element with all its electrons in completely filled shells (helium was the first).

11. Sodium (Na) Since the 1 and 2 shells are now complete, the eleventh electron of sodium goes into the $3s$ orbital and begins the 3 shell. The 2 shell is now an inner shell, and the 3 shell becomes the outer shell.

1s   2s $2p_x$ $2p_y$ $2p_z$ 3s $3p_x$ $3p_y$ $3p_z$

Na configuration: $1s^2 2s^2 2p^6 3s^1$ or $[\text{Ne}]3s^1$

Since the configuration of the two inner shells is the same as that of the neon atom, we use the abbreviation [Ne] to show the configuration of the two inner shells.

12–18. The next seven elements follow the same pattern as elements 4 through 10 (Be through Ne), but at the 3 rather than the 2 level.

12	Mg	$[\text{Ne}]3s^2$
13	Al	$[\text{Ne}]3s^2 3p^1$
14	Si	$[\text{Ne}]3s^2 3p^2$
15	P	$[\text{Ne}]3s^2 3p^3$
16	S	$[\text{Ne}]3s^2 3p^4$
17	Cl	$[\text{Ne}]3s^2 3p^5$
18	Ar	$[\text{Ne}]3s^2 3p^6$

Argon is the third element to have all its electrons in complete shells. This is true, even though the 3*d* orbitals are vacant. We say this because the behavior of argon is very similar to that of helium and neon (none of the three readily form any compounds). For this reason, we regard argon as an element with no incomplete shells.

19–20. Potassium, Calcium (K, Ca) The next two electrons go into the 4*s* orbital (Figure 2.9).

$$19 \quad K \quad [Ar]4s^1$$

$$20 \quad Ca \quad [Ar]4s^2$$

21. Scandium (Sc) If scandium followed the previous pattern, the twenty-first electron would go into the 4*p* orbital. It does not do so, however, because the 3*d* orbital has a lower energy (Figure 2.9).

$$21 \quad Sc \quad [Ar]4s^2 3d^1$$

22–30. The next nine elements fill the five 3*d* orbitals; the last of these is **zinc (Zn).**

$$30 \quad Zn \quad [Ar]4s^2 3d^{10}$$

Zinc does not have a complete outer shell. The 3 shell is complete, but the 3 shell is an inner shell here. The occupied shell with the highest number is always the outer shell. In the case of zinc, the occupied shell with the highest number is the 4 shell, and that one is not complete because the 4*p*, 4*d*, and 4*f* orbitals are vacant.

31–36. The next six elements fill the 4*p* orbitals.

$$36 \quad Kr \quad [Ar]4s^2 3d^{10} 4p^6$$

Krypton is the fourth element with a complete outer shell (even though the 4*d* and 4*f* orbitals are vacant).

EXAMPLE 2.7

Show how to find the electron configuration of these elements.
(a) Iron (26)   (b) Potassium (19)   (c) Potassium ion (K^+)

Solution
(a) Iron (26). The closest preceding element with a complete outer shell is argon (18), so we know that the first 18 electrons of iron will fill the 1*s*, 2*s*, 2*p*, 3*s*, and 3*p* orbitals, for a total of 18 electrons. Since iron has 26 electrons, there are eight left over. Figure 2.9 shows that the subshell of next higher energy is the 4*s* followed, not by the 4*p*, but by the 3*d*. Therefore, two of the eight electrons go into the 4*s*, and the remaining six go into the five 3*d* orbitals. Hund's rule tells us that one of these will get two electrons, and the other will get four, one each.

Fe configuration: $1s^2 2s^2 2p^6 3s^2 3p^6 4s^2 3d^6$

(b) Potassium (19). The previous element with a complete outer shell is argon (18). The first 18 electrons of potassium, therefore, must have the same configuration as argon. The extra electron is in the orbital of next highest energy, the 4s.

1s 2s 2p 3s 3p 4s

K configuration: $1s^2 2s^2 2p^6 3s^2 3p^6 4s^1$

(c) Potassium ion (K$^+$). Potassium ion is formed by removing one electron from the outer shell. Hence, the configuration is the same as that of argon.

1s 2s 2p 3s 3p

K$^+$ configuration: $1s^2 2s^2 2p^6 3s^2 3p^6$

TABLE 2.3 Electron Configurations of the First 36 Elements

1	H	$1s^1$	
2	He	$1s^2$	
3	Li	$1s^2 2s^1$	or [He]$2s^1$
4	Be	$1s^2 2s^2$	or [He]$2s^2$
5	B	$1s^2 2s^2 2p^1$	or [He]$2s^2 2p^1$
6	C	$1s^2 2s^2 2p^2$	or [He]$2s^2 2p^2$
7	N	$1s^2 2s^2 2p^3$	or [He]$2s^2 2p^3$
8	O	$1s^2 2s^2 2p^4$	or [He]$2s^2 2p^4$
9	F	$1s^2 2s^2 2p^5$	or [He]$2s^2 2p^5$
10	Ne	$1s^2 2s^2 2p^6$	or [He]$2s^2 2p^6$
11	Na	$1s^2 2s^2 2p^6 3s^1$	or [Ne]$3s^1$
12	Mg	$1s^2 2s^2 2p^6 3s^2$	or [Ne]$3s^2$
13	Al	$1s^2 2s^2 2p^6 3s^2 3p^1$	or [Ne]$3s^2 3p^1$
14	Si	$1s^2 2s^2 2p^6 3s^2 3p^2$	or [Ne]$3s^2 3p^2$
15	P	$1s^2 2s^2 2p^6 3s^2 3p^3$	or [Ne]$3s^2 3p^3$
16	S	$1s^2 2s^2 2p^6 3s^2 3p^4$	or [Ne]$3s^2 3p^4$
17	Cl	$1s^2 2s^2 2p^6 3s^2 3p^5$	or [Ne]$3s^2 3p^5$
18	Ar	$1s^2 2s^2 2p^6 3s^2 3p^6$	
19	K	$1s^2 2s^2 2p^6 3s^2 3p^6 4s^1$	or [Ar]$4s^1$
20	Ca	$1s^2 2s^2 2p^6 3s^2 3p^6 4s^2$	or [Ar]$4s^2$
21	Sc	$1s^2 2s^2 2p^6 3s^2 3p^6 4s^2 3d^1$	or [Ar]$4s^2 3d^1$
22	Ti	$1s^2 2s^2 2p^6 3s^2 3p^6 4s^2 3d^2$	or [Ar]$4s^2 3d^2$
23	V	$1s^2 2s^2 2p^6 3s^2 3p^6 4s^2 3d^3$	or [Ar]$4s^2 3d^3$
24	Cr	$1s^2 2s^2 2p^6 3s^2 3p^6 4s^1 3d^{5\text{a}}$	or [Ar]$4s^1 3d^5$
25	Mn	$1s^2 2s^2 2p^6 3s^2 3p^6 4s^2 3d^5$	or [Ar]$4s^2 3d^5$
26	Fe	$1s^2 2s^2 2p^6 3s^2 3p^6 4s^2 3d^6$	or [Ar]$4s^2 3d^6$
27	Co	$1s^2 2s^2 2p^6 3s^2 3p^6 4s^2 3d^7$	or [Ar]$4s^2 3d^7$
28	Ni	$1s^2 2s^2 2p^6 3s^2 3p^6 4s^2 3d^8$	or [Ar]$4s^2 3d^8$
29	Cu	$1s^2 2s^2 2p^6 3s^2 3p^6 4s^1 3d^{10\text{a}}$	or [Ar]$4s^1 3d^{10}$
30	Zn	$1s^2 2s^2 2p^6 3s^2 3p^6 4s^2 3d^{10}$	or [Ar]$4s^2 3d^{10}$
31	Ga	$1s^2 2s^2 2p^6 3s^2 3p^6 4s^2 3d^{10} 4p^1$	or [Ar]$4s^2 3d^{10} 4p^1$
32	Ge	$1s^2 2s^2 2p^6 3s^2 3p^6 4s^2 3d^{10} 4p^2$	or [Ar]$4s^2 3d^{10} 4p^2$
33	As	$1s^2 2s^2 2p^6 3s^2 3p^6 4s^2 3d^{10} 4p^3$	or [Ar]$4s^2 3d^{10} 4p^3$
34	Se	$1s^2 2s^2 2p^6 3s^2 3p^6 4s^2 3d^{10} 4p^4$	or [Ar]$4s^2 3d^{10} 4p^4$
35	Br	$1s^2 2s^2 2p^6 3s^2 3p^6 4s^2 3d^{10} 4p^5$	or [Ar]$4s^2 3d^{10} 4p^5$
36	Kr	$1s^2 2s^2 2p^6 3s^2 3p^6 4s^2 3d^{10} 4p^6$	

[a] The electron configurations of chromium and copper are slightly irregular because a filled (10 electrons) or half-filled (5 electrons) d subshell has a special stability.

Problem 2.7 ▬▬▬▬

Show how to find the electron configuration of (a) Aluminum (Al) and
(b) Al^{3+}. ■

We have now described the electron configuration of the first 36 elements (they are summarized in Table 2.3). The remaining elements fill up the next orbitals (5s, 4d, and so on) in a similar manner, in the order shown in Figure 2.9.

2.7 Electron Configuration of Atoms and the Periodic Table

Outer-Shell Electron Configuration

When Mendeleev published his first periodic table in 1869, he could not explain why it worked, and indeed nobody else had any good explanation either. It was not until after the discovery of electron configuration that chemists finally understood why the table works. The answer is very simple. *The Periodic Table works because elements in the same column have the same configuration of electrons in the outer shell.* For example, let us look at the elements in column 1A. We already know the configuration for lithium, sodium, and potassium. We now add rubidium and cesium:

Li $[He]2s^1$

Na $[Ne]3s^1$

K $[Ar]4s^1$

Rb $[Kr]5s^1$

Cs $[Xe]6s^1$

All these atoms have complete inner shells and an outer shell consisting of one s electron. The properties of elements largely depend on the electron configuration of their outer shell. This being so, it is not surprising that these elements, all of which have the same outer-shell configuration, should have such similar properties.

We could not expect *identical* properties because the number of protons in the nucleus is different, and so is the number of inner shells. These numbers also have an effect on the properties and allow us to explain why properties change regularly as we move down a column of the Periodic Table. For example, all the elements in Group 1A, with the exception of hydrogen, are metals, which means that they tend to give up electrons. In this case, they give up only the single electron in the outer shell, but cesium gives it up most easily, rubidium next, and so on up to lithium. We explain this by saying that the 2s electron is closer to the positive nucleus in lithium than is the 3s electron in sodium, and so the 2s electron is more attracted to its nucleus. Furthermore, sodium has one more filled inner shell than lithium, and this filled inner shell (made up of negatively charged electrons) "shields" the 3s electron from the nucleus more effectively than the fewer inner electrons of lithium. Because atoms get bigger as we go down the table, these effects explain why loss of the outer electron becomes easier as we go down a column.

Because the highest energy electrons of the elements in Groups 1A and 2A are in s orbitals, these elements are often called **s-block elements.** Similarly, the **p-block elements** are those in columns 3A to 8A (except hydrogen), the **d-block elements** are the transition elements (the eight B columns), and the **f-block** elements are the inner transition elements.

TABLE 2.4	Lewis Structures for Elements 1 – 18 of the Periodic Table						

1A	2A	3A	4A	5A	6A	7A	8A
H·							He :
Li·	Be :	Ḃ:	·Ċ:	·N̈:	:Ö:	:F̈:	:N̈e :
Na·	Mg :	Ȧl:	·Si:	·P̈:	:S̈:	:C̈l:	:Ȧr:

Lewis Structures of Atoms

The electrons in the outermost shell of an atom are called **valence electrons.** They are involved in chemical bond formation (Sections 3.2 to 3.4). The energy level in which the valence electrons are found is called **valence shell.** To show the outermost electrons of an atom, we commonly use the representation called a Lewis structure, after the American chemist Gilbert N. Lewis (1875–1946) who devised this notation. **Lewis structures** show the symbol of the element surrounded by a number of dots equal to the number of valence electrons of an atom of that element. In Lewis structures, the atomic symbol represents the nucleus and all the filled inner shells. Table 2.4 shows Lewis structures for the first 18 elements of the Periodic Table. Noble gases (Group 8A) have filled valence shells. The valence shell of all other elements in Table 2.4 contain fewer than eight electrons. The Lewis structure of carbon (C) in Table 2.4 shows four valence electrons. These are the two paired $2s$ electrons and the single $2p_x$ and $2p_y$ electrons shown in Table 2.3.

Ionization Energy

Ionization energy is the energy necessary to remove one electron from an atom in the gaseous state. We can measure its numerical value for each atom. Table 2.5 shows the ionization energy for atoms of some of the representative elements. We can see from Table 2.5 that ionization energy decreases as we go down any column. We have already seen why this is so: Electrons are easier to remove the farther they get from the nucleus and the greater the number of inner electrons shielding them from the attraction of the positive charge of the nucleus. Table 2.5 also shows that, with a few exceptions, ionization energy increases regularly as we go from left to right across any row. This trend is consistent with the change in metallic character mentioned in Section 2.5. As we go from left to right in the table,

Ionization energy The amount of energy required to remove the most loosely held electron of an isolated gaseous atom or ion

In all cases these regularities exist because elements in the same column have the same outer-shell electron configuration.

TABLE 2.5	Ionization Energy in Kilocalories Per Mole of Some Representative Elements						

1A	2A	3A	4A	5A	6A	7A	8A
Li	Be	B	C	N	O	F	Ne
124	215	191	260	335	314	402	497
Na	Mg	Al	Si	P	S	Cl	Ar
119	176	138	188	242	239	299	363
K	Ca	Ga	Ge	As	Se	Br	Kr
100	141	138	182	226	225	272	323

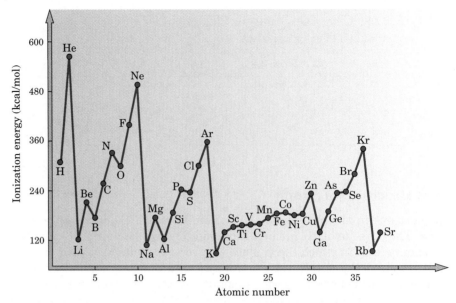

FIGURE 2.10 A plot of ionization energy versus atomic number for the first 38 elements. Many other properties, such as atomic radii, give similar plots.

elements become less metallic. Metals tend to give up electrons more readily than nonmetals, and the information in Table 2.5 largely bears this out. Figure 2.10 shows graphically how ionization energy varies with increasing atomic number.

Electron configuration also explains the presence of the transition elements. Let us look at the electron structure of iron, a typical transition element:

Two incomplete shells

$$26 \qquad Fe \qquad [Ar]4s^2 3d^6$$

In this atom, the outer shell—the 4 shell—contains two electrons and is incomplete because the $4p$ orbitals, which can hold six more electrons, are empty. The 3 shell, an inner shell, is also incomplete. There are six electrons in the five $3d$ orbitals, and these orbitals can hold a total of ten electrons. So iron has not only an incomplete outer shell but also an incomplete inner shell. We define a **transition element as an element whose atoms have one or more incomplete inner shells.** (As you might have guessed, inner transition elements have two incomplete inner shells.)

The presence of unfilled inner shells gives transition elements special properties. For example, most of them form more than one kind of ion.

2.8 How Small Are Atoms?

In the previous section we examined the electron structure of atoms. It is useful to get some idea of just how small atoms and their components really are.

The Mass of an Atom

A typical heavy atom (although not the heaviest) is lead-208, a lead atom with 82 protons, 82 electrons, and 126 neutrons. It has a mass of 3.5×10^{-22} g. You would need 1.3×10^{24} lead-208 atoms to make 1 lb of lead. This is a very large number. There are about six billion people on

BOX 2F

Abundance of Elements on Earth and in People

Table 2F shows the abundance of the elements present in the human body. Oxygen is the most abundant element by weight, followed by carbon, hydrogen, and nitrogen. If we go by number of atoms instead of by weight, hydrogen is even more abundant than oxygen.

Table 2F also shows the abundance of elements in the Earth's crust. Although 88 elements are found in the Earth's crust (we know very little about the interior of the Earth because we have not been able to penetrate very far), they are not present in anything close to equal amounts. In the Earth's crust as well as in the human body, the most abundant element is oxygen, but there the similarity ends. Silicon, aluminum, and iron, which are not major elements in the body (although iron is an important trace element, as mentioned in Box 2A), are the three next most abundant in the Earth's crust, and only six other elements (Ca, Na, K, Mg, H, and Ti)

■ Of the four principal elements—H, O, C, and N—only oxygen is a principal element in the sand, gravel, water, and water buffaloes. *(Beverly March)*

make up even as much as one half of 1 percent by weight of the Earth's crust.

TABLE 2F	**Relative Abundance of Elements in the Human Body and in the Earth's Crust, Including Atmosphere and Oceans**		
	Percent in Human Body		**Percent in Earth's Crust, by Weight**
Element	**By Number of Atoms**	**By Weight**	
H	63.0	10.0	0.9
O	25.4	64.8	49.3
C	9.4	18.0	0.08
N	1.4	3.1	0.03
Ca	0.31	1.8	3.4
P	0.22	1.4	0.12
K	0.06	0.4	2.4
S	0.05	0.3	0.06
Cl	0.03	0.2	0.2
Na	0.03	0.1	2.7
Mg	0.01	0.04	1.9
Si	—	—	25.8
Al	—	—	7.6
Fe	—	—	4.7
Others	0.01	—	—

Earth right now. If you divided 1 lb of these atoms among all the people on Earth, each person would get about 2.2×10^{14} atoms. This is still a large number. If these atoms had a value of 1 cent per 1000 atoms, every person on Earth could get more than $2 billion by selling his or her share.

The Size of an Atom

An atom of lead has a diameter of about 3.5×10^{-8} cm. If you could line them up with the atoms just touching, it would take 73 million lead atoms to make a line 1 inch long. Despite their tiny size, we can actually see

FIGURE 2.11 The surface of graphite is revealed with a scanning tunneling microscope. The contours represent the arrangement of individual carbon atoms on a crystal surface. *(Courtesy of Roger Freedman and Paul K. Hansma, University of California—Santa Barbara)*

atoms, in certain cases, by the use of special microscopic techniques. An example is shown in Figure 2.11.

The Mass of an Electron and of a Nucleus

As shown in Table 2.1, the mass of an electron, 9.1×10^{-28} g, is very much smaller than that of a proton or neutron. Since protons and neutrons are in the nucleus and electrons are outside, this means that almost all the mass of an atom is in the nucleus. For example, the total mass of the 82 electrons in a lead-208 atom is 7.5×10^{-26} g, and the mass of the whole atom is 3.5×10^{-22} g.

The Size and Density of the Nucleus

Although the mass of the nucleus is very great compared to the mass of the electrons in an atom, the size of the nucleus is very small. The nucleus of a lead-208 atom has a diameter of about 1.6×10^{-12} cm. When we compare this with the diameter of the whole atom, which is 3.5×10^{-8} cm, we see that the nucleus occupies only a tiny fraction of the volume of the atom. If the nucleus of a lead-208 atom were the size of a baseball, then the whole atom would be much larger than a baseball stadium. In fact, it would be a sphere about one mile in diameter. With such a large mass in such a small volume, it is obvious that the nucleus must be very dense. For example, the density of the lead-208 nucleus is extremely high, 1.8×10^{14} g/cc. Nothing in our daily life has a density anywhere near as high. If a paper clip had this density, it would weigh about 10 million (10^7) tons.

SUMMARY

Matter can be classified as **elements** (substances all of whose atoms are the same), **compounds** (substances made up of at least two elements), or **mixtures.** Dalton's atomic theory says that all atoms of a given element are identical and that the atoms of any one element are different from those of any other element. The theory is based on the law of conservation of mass (matter cannot be created or destroyed) and on the law of constant composition (any compound is always made up of elements in the same proportion by mass). An **atom** is the smallest unit of an element that shows the chemical behavior of the element. A **molecule** is a cluster of two or more atoms that act as a unit.

Atoms consist of **protons** and **neutrons** inside the nucleus and electrons outside it. **Electrons** are very light and have a charge of -1. Protons and neutrons are much heavier; their masses are about equal. Protons have a charge of $+1$; neutrons have no charge. The sum of the number of protons and neutrons is the **mass number.** The **atomic number** of an element is the number of protons; this determines the identity of the element. Atoms have an equal number of electrons and protons, but **ions,** which are atoms that have lost or gained electrons, have an

unequal number of electrons and protons and therefore are charged. **Isotopes** are atoms with the same number of protons but different numbers of neutrons. The **atomic weight** of an element is a weighted average of the masses of the isotopes as they occur in nature.

The **Periodic Table** arranges elements into columns of elements with similar properties, which gradually change as we move down a column. Metallic elements, which tend to lose electrons, are below the heavy line in the Periodic Table. Nonmetals are above the line.

Electrons in atoms can exist in a number of **energy levels.** The energy levels are clouds called **orbitals.** All the orbitals in a principal energy level constitute a shell that contains subshells. All s orbitals are spherical, and all p orbitals are shaped like dumbbells. Each orbital can hold a maximum of two electrons of opposite spin. Electrons in the **ground state** (lowest energy state) of atoms fill up the available orbitals of lowest energy.

Electrons in the outermost shell of an atom are called **valence electrons.** The valence shell contains all the valence electrons. The **Lewis structure** is a representation of an atom in which the symbol of the element is surrounded by a number of dots equal to the number of its valence electrons.

The Periodic Table works because elements in the same column have the same outer-shell electron configuration. **Ionization energy,** which is the energy necessary to remove an electron from an atom, decreases as we go down the Periodic Table and increases as we go from left to right. Transition elements have one or more incomplete inner shells.

Atoms are very tiny, with a very small mass, almost all of which is in the nucleus. The nucleus is extremely tiny, with a high density.

KEY TERMS

Alkali metal (Section 2.5)
Atom (Section 2.3)
Atomic number (Section 2.4)
Atomic weight (Section 2.4)
Compound (Section 2.2)
Electric charge (Section 2.4)
Electron (Section 2.4)
Element (Section 2.2)
Excited state (Section 2.6)
Ground state (Section 2.6)
Halogens (Section 2.5)
Hund's rule (Section 2.6)
Inner transition element
  (Sections 2.5, 2.7)
Ion (Section 2.4)

Ionization energy (Section 2.7)
Isotope (Section 2.4)
Isotopic abundance (Box 2C)
Lewis structure (Section 2.7)
Mass number (Section 2.4)
Metal (Section 2.5)
Metalloid (Section 2.5)
Mixture (Section 2.2)
Molecule (Section 2.3)
Neutron (Section 2.4)
Noble (or inert) gas (Section 2.5)
Nonmetal (Section 2.5)
Nucleon (Section 2.4)
Nucleus (Section 2.4)

Orbital (Section 2.6)
Outer shell (Section 2.6)
Pauli exclusion principle
  (Section 2.6)
Periodic Table (Section 2.5)
Principal energy level (Section 2.6)
Proton (Section 2.4)
Quantized (Section 2.6)
Representative element (Section 2.5)
Shell (Section 2.6)
Subshell (Section 2.6)
Transition element (Sections 2.5, 2.7)
Valence electrons (Section 2.7)
Valence shell (Section 2.7)

CONCEPTUAL PROBLEMS

2.A Suppose that you face a problem similar to Mendeleev: to predict the properties of an element not yet discovered. What will element 118 be like if and when enough of element 118 has been made to study its physical and chemical properties?

2.B Silicon, element 14, is in the 4A group. How many subshells are occupied by the valence electrons of Si in its ground state?

2.C You are presented with a Lewis structure of element X as X:. To which group in the Periodic Table does this element belong?

2.D Based on the data presented in Figure 2.10, which of the following atoms would have the highest ionization energy: I, Cs, Sn, or Xe?

PROBLEMS

Difficult problems are designated by an asterisk.

Classifications of Matter

2.8 Classify each of the following as an element, a compound, or a mixture.
  (a) Oxygen   (b) Table salt   (c) Sea water
  (d) Wine   (e) Air   (f) Silver
  (g) Diamond   (h) A pebble   (i) Gasoline
  (j) Milk   (k) Carbon dioxide

2.9 A jeweler can make a ring in which gold and silver are combined in any proportion. Is such a ring made of a compound or of a mixture?

2.10 Name these elements (try not to look at a table).
  (a) O   (b) Pb   (c) Ca   (d) Na   (e) C
  (f) S   (g) Fe   (h) H   (i) K

2.11 Write the formulas of compounds in which the combining ratios are as follows:
  (a) potassium : oxygen, 2 : 1
  (b) sodium : phosphorus : oxygen, 3 : 1 : 4
  (c) lithium : nitrogen : oxygen, 1 : 1 : 2

Dalton's Atomic Theory

2.12 How does Dalton's atomic theory explain (a) the law of conservation of mass (b) the law of constant composition?

2.13 When 2.16 g of mercury oxide is decomposed to yield 2.00 g of mercury and 0.16 g of oxygen, which law is supported by the experiment?

2.14 The compound carbon monoxide contains 42.9 percent carbon and 57.1 percent oxygen. The compound carbon dioxide contains 27.2 percent carbon and 72.7 percent oxygen. Does this disprove Proust's law of constant composition?

Inside the Atom

2.15 Where in the atom are these particles located?
(a) Protons   (b) Electrons   (c) Neutrons

2.16 What is the mass number of an atom or ion with
(a) 22 protons, 18 electrons, and 26 neutrons?
(b) 76 protons, 76 electrons, and 114 neutrons?
(c) 34 protons, 36 electrons, and 45 neutrons?
(d) 94 protons, 94 electrons, and 150 neutrons?

2.17 Name each element in Problem 2.16 and indicate whether it is an atom or ion.

2.18 Given the mass number and the number of neutrons, what is the name of each of these elements?
(a) mass number 45; 24 neutrons
(b) mass number 48; 26 neutrons
(c) mass number 107; 60 neutrons
(d) mass number 246; 156 neutrons
(e) mass number 36; 18 neutrons

2.19 If each of the atoms in Problem 2.18 acquired two more neutrons, what would their names be?

2.20 How many neutrons are there in:
(a) A carbon atom of mass number 13?
(b) A germanium atom of mass number 73?
(c) An osmium atom of mass number 188?
(d) An yttrium atom of mass number 89?

2.21 How many protons and how many neutrons does each of these isotopes of radon contain: (a) Rn-210 (b) Rn-218 (c) Rn-222?

2.22 How many nucleons are there in each isotope?
(a) ^{22}Ne (b) ^{104}Pd (c) ^{35}Cl
(d) Tellurium-128   (e) Lithium-7   (f) Uranium-238

2.23 State the number of protons and neutrons in each isotope in Problem 2.22.

2.24 Write the symbols of the elements with the following makeup.
(a) 17p, 18e, 18n   (b) 82p, 80e, 124n

2.25 Tin-118 is one of the isotopes of tin. Name the isotopes of tin that contain two, three, and six more neutrons than tin-118.

2.26 Define: (a) ion (b) isotope.

2.27 Write the symbol for each atom or ion, given the number of protons (p) and electrons (e) (example: 3p, 2e = Li^+):
(a) 9p, 10e   (b) 19p, 18e   (c) 18p, 18e   (d) 1p, 0e
(e) 11p, 10e   (f) 16p, 18e   (g) 35p, 36e   (h) 25p, 23e

2.28 For each atom or ion, tell how many electrons, protons, and neutrons are present.
(a) Cl with a mass number of 37
(b) K^+ with a mass number of 39
(c) Pb^{2+} with a mass number of 207
(d) I^- with a mass number of 127
(e) Pd^{2+} with a mass number of 106
(f) W with a mass number of 186

*2.29 Lithium has two major naturally occurring isotopes: lithium-6 and lithium-7. The abundance of the former is 7.4 percent and of the latter is 92.6 percent. What is the atomic weight of lithium? Compare the value you calculated to that reported in the Periodic Table. Estimate whether the minor isotopes not included in your calculation are heavier or lighter than lithium-7.

2.30 If the three isotopes of hydrogen occurred in an extraterrestrial sample as follows: 98.00 percent ^1H, 1.800 percent ^2H, and 0.200 0 percent ^3H, what would be the atomic weight of hydrogen?

*2.31 If there are only two naturally occurring isotopes of antimony, ^{121}Sb and ^{123}Sb, and the atomic weight of antimony is 121.75, what are the abundances of the two isotopes?

The Periodic Table

2.32 Which of the following elements would you expect to have fairly similar properties (look at the Periodic Table): As, I, Ne, F, Mg, K, Ca, Ba, Li, He, N, P.

2.33 Which are transition elements?
(a) Pd   (b) K   (c) Co
(d) Ce   (e) Br   (f) Cr

2.34 Which element in each pair is more metallic?
(a) Silicon or aluminum
(b) Arsenic or phosphorus
(c) Gallium or germanium
(d) Gallium or aluminum

2.35 Classify these elements as metal, nonmetal, or metalloid:
(a) Argon   (b) Boron   (c) Lead
(d) Arsenic   (e) Potassium   (f) Silicon
(g) Iodine   (h) Antimony   (i) Vanadium
(j) Sulfur   (k) Nitrogen

Electron Structure of Atoms

2.36 What is the maximum number of electrons that can go into the 3 shell?

2.37 Write ground-state electron configurations for these elements.
(a) Li   (b) Ne   (c) Be   (d) C
(e) He   (f) Mn   (g) Cl   (h) P

2.38 What is common and what is different in the electron configuration of (a) Na and Cs (b) O and Te (c) C and Ge?

2.39 Positive ions are obtained from neutral atoms by removing the number of electrons indicated by the charge on the ion. Give the ground-state electron configuration of (a) Mg and Mg^{2+} (b) Na and Na^+ (c) Al and Al^{3+}.

2.40 Negative ions are obtained from neutral atoms by adding the number of electrons indicated by the charge on the ion. Give the ground-state electron configuration of (a) F and F^- (b) S and S^{2-} (c) P and P^{3-}.

2.41 The electron configurations for the elements with atomic numbers higher than 36 follow the same rules as given in the text for the first 36. Write the ground-state electron configuration for (a) Y (b) Ba (c) I.

*2.42 The element radon (atomic number 86) has the electron configuration $[Xe]6s^25d^{10}4f^{14}6p^6$. The element francium (atomic number 87) has the same configuration plus one more electron. Which orbital does this extra electron occupy?

2.43 What is the total number of electrons that can fit into the 6s, 6p, 6d, and 6f orbitals?

Electron Configuration and the Periodic Table ___

2.44 Assume that a new element has been discovered with atomic number 117. Its chemical property should be similar to At. Predict whether its ionization energy will be greater than, the same as, or smaller than that of (a) At (b) Uub?

2.45 Why do the elements in column 1A of the Periodic Table (the alkali metals) have similar but not identical properties?

*2.46 List in order of increasing ionization energy: Al, Li, N, Ar, B, O, K, Cl, C. Draw the Lewis structure of K and Kr.

Boxes ___

2.47 (Box 2A) Why does the body need sulfur, calcium, and iron?

2.48 (Box 2B) Are all chemicals poisonous? Explain.

2.49 (Box 2C) Consider the isotopic abundance of hydrogen on Mars. Would the atomic weight of hydrogen on Mars be greater than, the same as, or smaller than that on Earth?

2.50 (Box 2D) Why is strontium-90 more dangerous to human beings than most other radioactive isotopes that are present in fallout?

2.51 (Box 2E) Bronze is an alloy of which two metals?

2.52 (Box 2E) Copper is a soft metal. How can one make it harder?

2.53 (Box 2F) Which are the two most abundant elements, by weight, in (a) the Earth's crust (b) the human body?

Additional Problems ___

2.54 Give the designation of all subshells in the (a) 1 shell (b) 2 shell (c) 3 shell (d) 4 shell.

2.55 Tell whether metals or nonmetals are more likely to have each of the following characteristics:
(a) Conduct electricity and heat
(b) Accept electrons
(c) Be malleable
(d) Be a gas at room temperature
(e) Be a transition element

2.56 What is the outer-shell electron configuration of the elements in (a) Group 3A (b) Group 7A (c) Group 5A?

2.57 Determine the number of protons, electrons, and neutrons present in:
(a) ^{35}Cl (b) ^{98}Mo (c) $^{44}Ca^{2+}$
(d) $^{127}I^{-}$ (e) ^{158}Gd (f) $^{212}Bi^{3+}$

2.58 Write the abbreviated electron configuration of Sr^{2+}.

2.59 Do isotopes of the heavy elements (for example from 37 to 53) contain more, the same, or fewer neutrons than protons?

2.60 What is the symbol for the following elements? (Try not to look at a table.)
(a) Phosphorus   (b) Potassium   (c) Sodium
(d) Nitrogen   (e) Bromine   (f) Silver
(g) Calcium   (h) Carbon   (i) Tin
(j) Zinc

2.61 This state of C

$$1s^2 2s^2 2p^1 3s^1$$

represents (a) a C^+ ion (b) a C^- ion (c) the ground state of a C atom (d) an excited state of a C atom.

*2.62 Write the electron structure of an excited state of the Be atom.

*2.63 The natural abundance of titanium isotopes is as follows: titanium-46 = 7.95 percent, titanium-47 = 7.75 percent, titanium-48 = 73.45 percent, titanium-49 = 5.51 percent, and titanium-50 = 5.34 percent. Calculate the atomic weight of titanium.

2.64 Write the symbol for each atom or ion, given the number of protons (p) and electrons (e) (example: 3p, 2e = Li^+).
(a) 10p, 10e   (b) 19p, 18e   (c) 35p, 36e
(d) 56p, 54e   (e) 1p, 2e   (f) 21p, 18e

2.65 How many electrons are there in the outer shells of the following elements?
(a) Si   (b) Br   (c) P   (d) K   (e) He
(f) Ca   (g) Kr   (h) Pb   (i) Se   (j) O

*2.66 What percent of the mass of radon-222 is contributed by neutrons?

*2.67 After the s, p, d, and f orbitals, the next higher series is the g series.
(a) How many electrons does each g subshell hold?
(b) What is the lowest principal energy level in which g orbitals can exist?

2.68 The mass of a proton is 1.67×10^{-24} g. The mass of a grain of salt is 1.0×10^{-2} g. How many protons would it take to have the same mass as a grain of salt?

2.69 (a) What are the charges of an electron, a proton, and a neutron?
(b) What are the masses (in amu) of an electron, a proton, and a neutron?

2.70 Define the excited state of an atom. Can an atom be excited into more than one unique excited state?

*2.71 Rubidium has two natural isotopes: rubidium-85 and rubidium-87. What is the natural abundance of each of these isotopes if the atomic weight of rubidium is 85.47?

2.72 What is the name of this element and how many protons and neutrons does it have in its nucleus, $^{131}_{54}Xe$?

Above: Sodium chloride crystal.
(Charles D. Winters)

Chemical Bonds

3.1 Introduction

In Chapter 2 we stated that molecules are tightly bound groups of atoms. In this chapter, we will see that atoms in these groups are held together by powerful forces of attraction called chemical bonds. There are two main types: ionic bonds and covalent bonds. In order to talk about ionic bonds, we must first discuss ions.

3.2 Lewis Model of Bonding

Formation of Ions

In 1916, Gilbert N. Lewis (Section 2.7) devised a beautifully simple model that unified many of the observations about chemical bonding and chemical reactions. He pointed out that the lack of chemical reactivity of the noble gases indicates a high degree of stability of their electron configurations: helium with a valence shell of two electrons ($1s^2$), neon with a valence shell of eight electrons ($2s^2 2p^6$), and argon with a valence shell of eight electrons ($3s^2 3p^6$). The tendency to react in ways that achieve an outer shell of eight valence electrons is particularly common among the main-group elements (Groups 1A–7A) and is given the special name **octet**

rule. An atom that loses electrons from its valence shell becomes a positively charged ion called a **cation.** An atom that gains electrons in its valence shell becomes a negatively charged ion called an **anion.**

EXAMPLE 3.1

Show how the following chemical changes obey the octet rule:
(a) Sodium loses an electron to form Na^+.

$$Na \longrightarrow Na^+ + \boxed{e^-}$$

(b) Chorine gains an electron to form Cl^-.

$$Cl + \boxed{e^-} \longrightarrow Cl^-$$

Solution

To see how each chemical change follows the octet rule, first write the ground-state electron configuration of the atom involved in the chemical change and the ion it forms, and then compare them.
(a) The ground-state electron configurations for Na and Na^+ are:

$$Na \ (11 \ electrons): \qquad 1s^2 2s^2 2p^6 \boxed{3s^1}$$
$$Na^+ \ (10 \ electrons): \qquad 1s^2 2s^2 2p^6$$

A sodium atom has one electron in its valence shell. Loss of this one valence electron changes the sodium atom to a sodium cation, Na^+, which has a complete octet of electrons in its valence shell and has the same electron configuration as neon, the noble gas nearest it in atomic number. We can write this chemical change using Lewis structures:

$$Na \cdot \longrightarrow Na^+ + \boxed{e^-}$$

(b) The ground-state electron configurations for Cl and Cl^- are:

$$Cl \ (17 \ electrons): \qquad 1s^2 2s^2 2p^6 3s^2 \boxed{3p^5}$$
$$Cl^- \ (18 \ electrons): \qquad 1s^2 2s^2 2p^6 3s^2 \boxed{3p^6}$$

A chlorine atom has seven electrons in its valence shell. Gain of one electron changes the chlorine atom to a chloride anion, Cl^-, which has a complete octet of electrons in its valence shell and has the same electron configuration as argon, the noble gas nearest it in atomic number. We can write this chemical change using Lewis structures:

$$:\overset{\cdot\cdot}{\underset{\cdot\cdot}{Cl}}\cdot + \boxed{e^-} \longrightarrow :\overset{\cdot\cdot}{\underset{\cdot\cdot}{Cl}}:^-$$

Problem 3.1 ▬▬▬
Show how the following chemical changes obey the octet rule.
(a) Magnesium forms Mg^{2+}.
(b) Sulfur forms S^{2-}. ∎

Table 3.1 gives the names and symbols for some common cations and anions formed from single atoms. Cations formed from a single atom are named by giving the name of the atom followed by the word "ion." Anions

Octet rule When undergoing chemical reaction, atoms of main-group elements gain, lose, or share electrons to achieve a configuration having eight valence electrons

Cation An ion with a positive electric charge

Anion An ion with a negative electric charge

G. N. Lewis. *(Francis Simon / AIP Niels Bohr Library)*

TABLE 3.1 Some Common Cations and Anions Formed from Single Atoms

Group	Ion Name	Ion Symbol	Group	Ion Name	Ion Symbol
	Hydrogen	H^+	6A	Oxide	O^{2-}
	Hydride	H^-		Sulfide	S^{2-}
1A	Lithium	Li^+	7A	Fluoride	F^-
	Sodium	Na^+		Chloride	Cl^-
	Potassium	K^+		Bromide	Br^-
				Iodide	I^-
2A	Magnesium	Mg^{2+}			
	Calcium	Ca^{2+}	Transition elements	Copper(I)	Cu^+
	Strontium	Sr^{2+}		Copper(II)	Cu^{2+}
	Barium	Ba^{2+}		Iron(II)	Fe^{2+}
				Iron(III)	Fe^{3+}
3A	Aluminum	Al^{3+}		Mercury(I)	Hg^+
				Mercury(II)	Hg^{2+}
4A	Lead	Pb^{2+}		Silver	Ag^+

formed from a single atom are named by replacing the ending in the name of the atom with "-ide" and adding the word "ion." For example, sodium loses one electron and forms sodium ion; fluorine gains one electron and forms fluoride ion.

The octet rule gives us a good way to understand why Groups 1A–7A elements form the ions they do. However, it is not perfect for two reasons.

1. Concentrated charges on small ions are unstable. Boron, for example, has three valence electrons. If it lost these three, it would become B^{3+} and have a complete outer shell like that of helium. It seems, however, that this is too concentrated a charge for such a small ion, and this ion is not found in stable compounds. By the same reasoning, carbon does not lose its four valence electrons to become C^{4+}, nor does it gain four valence electrons to become C^{4-}. To do so would place too concentrated a charge on too small an ion.

2. The octet rule does not apply to transition metals, the Groups 1B–7B elements, most of which form ions with different positive charges. We use Roman numerals to show the number of positive charges on a transition metal ion. Copper, for example, may lose one valence electron to form Cu^+, named copper(I) ion; alternatively, it may lose two valence electrons to form Cu^{2+}, named copper(II) ion. Note that even though silver is a transition metal, it forms only Ag^+; therefore, there is no need to use a Roman numeral to show its charge.

The transition elements mostly form ions with two or more positive charges.

Atoms and Their Ions

It is important to understand that there are enormous differences between the properties of atoms and the properties of their ions. Atoms and their ions are completely different chemical species and have completely different chemical and physical properties. Consider, for example, sodium and chlorine. Sodium is a soft metal made of sodium atoms and reacts violently

Salt, sodium chloride. This chemical compound (*right*) is composed of the elements sodium (*left*) and chlorine (*center*) in chemical combination. Salt is very different in appearance and properties from the elements that constitute it. *(Charles D. Winters)*

with water. Chlorine atoms are very unstable and even more reactive than sodium atoms. Both sodium and chlorine are poisonous. NaCl, common table salt, is made up of sodium ions and chloride ions. These two ions are quite stable and unreactive. Neither sodium ions nor chloride ions react with water at all.

Because atoms and their ions are different chemical species, we must be careful to distinguish one from the other. An example is the drug commonly known as "lithium," which is used to treat manic-depression. The element lithium, like sodium, is a soft metal that reacts with water. The drug used to treat manic-depression is not lithium (Li) at all but lithium ions (Li^+), usually given in the form of lithium carbonate (Li_2CO_3). Another example comes from the fluoridation of drinking water and toothpastes and dental gels. The element fluorine (F_2) is an extremely poisonous and corrosive gas and is not what is used for this purpose. Fluoridation is done using fluoride ions (F^-) in the form of sodium fluoride (NaF), a compound that is unreactive and nonpoisonous in the concentrations used.

Lithium carbonate is categorized as an antimanic. It is prescribed under at least 18 different proprietary names.

Formation of Chemical Bonds

According to the Lewis model of chemical bonding, atoms bond together in such a way that each atom participating in a bond acquires a valence-shell electron configuration the same as that of the noble gas nearest it in atomic number. Atoms acquire completed valence shells in two ways.

1. An atom may lose or gain enough electrons to acquire a filled valence shell and becomes an ion as it does so. The attraction between positive and negative ions is called an **ionic bond.**

2. An atom may share electrons with one or more other atoms to acquire a filled valence shell. The attraction between atoms resulting from sharing electrons is called a **covalent bond.**

Ionic bond The attraction between a positive and a negative ion

Covalent bond The interatomic attraction resulting from the sharing of electrons between atoms

TABLE 3.2   Electronegativity Values for Some Atoms (Pauling Scale)

							H 2.1										

1A 2A   **3A 4A 5A 6A 7A**

Li 1.0	Be 1.5						8B			1B	2B	B 2.0	C 2.5	N 3.0	O 3.5	F 4.0
Na 0.9	Mg 1.2	**3B**	**4B**	**5B**	**6B**	**7B**						Al 1.5	Si 1.8	P 2.1	S 2.5	Cl 3.0
K 0.8	Ca 1.0	Sc 1.3	Ti 1.5	V 1.6	Cr 1.6	Mn 1.5	Fe 1.8	Co 1.8	Ni 1.8	Cu 1.9	Zn 1.6	Ga 1.6	Ge 1.8	As 2.0	Se 2.4	Br 2.8
Rb 0.8	Sr 1.0	Y 1.2	Zr 1.4	Nb 1.6	Mo 1.8	Tc 1.9	Ru 2.2	Rh 2.2	Pd 2.2	Ag 1.9	Cd 1.7	In 1.7	Sn 1.8	Sb 1.9	Te 2.1	I 2.5
Cs 0.7	Ba 0.9	La 1.1	Hf 1.3	Ta 1.5	W 1.7	Re 1.9	Os 2.2	Ir 2.2	Pt 2.2	Au 2.4	Hg 1.9	Tl 1.8	Pb 1.8	Bi 1.9	Po 2.0	At 2.2

◻ <1.0   ◻ 1.5 – 1.9   ◻ 2.5 – 2.9
◻ 1.0 – 1.4   ◻ 2.0 – 2.4   ◻ 3.0 – 4.0

Electronegativity A measure of an atom's attraction for the electrons it shares in a chemical bond with another atom

Linus Pauling (1901–1994) was awarded the 1954 Nobel Prize in chemistry for his pioneering studies of chemical bonding. Shortly after World War II, Pauling and his wife began a crusade to limit nuclear testing, a crusade that came to fruition in the Limited Test Ban Treaty of 1963. For this effort, Pauling was awarded the 1963 Nobel Prize for peace. Never before had any person received two unshared Nobel Prizes.

Electronegativity and Chemical Bonds

We now ask how to determine whether two atoms bond by an ionic bond or a covalent bond. One way to answer this question is to compare the electronegativities of the atoms involved. **Electronegativity** is a measure of an atom's attraction for the electrons it shares in a chemical bond with another atom. The most widely used scale of electronegativities (Table 3.2) was devised by Linus Pauling in the 1930s.

On the Pauling scale, fluorine, the most electronegative element, is assigned an electronegativity of 4.0, and all other elements are assigned values in relation to fluorine. As you study the electronegativity values in this table, note that they generally increase from left to right within a row of the Periodic Table and decrease from top to bottom within a column. Values increase from left to right because of the increasing positive charge on the nucleus, which leads to a stronger attraction for electrons in the valence shell. Values decrease from top to bottom because of the increasing distance of the valence electrons from the nucleus, which leads to a weaker attraction.

EXAMPLE 3.2

Judging from their relative positions in the Periodic Table, which element in each set has the larger electronegativity?
(a) Lithium or carbon   (b) Nitrogen or oxygen
(c) Carbon or oxygen

Solution
The elements in each set are in the second period of the Periodic Table. Within a period, electronegativity increases from left to right.
(a) C > Li   (b) O > N   (c) O > C

Problem 3.2 ▬▬▬
Judging from their relative positions in the Periodic Table, which element in each set has the larger electronegativity?
(a) Lithium or potassium   (b) Nitrogen or phosphorus
(c) Carbon or silicon ■

3.3 Ionic Bonds

Formation of Ionic Bonds

According to the Lewis model of bonding, an ionic bond is formed by the transfer of valence-shell electrons from an atom of lower electronegativity to the valence shell of an atom of higher electronegativity. The more electronegative atom gains one or more valence electrons and becomes an anion; the less electronegative atom loses one or more valence electrons and becomes a cation. We say that a bond is ionic if the difference in electronegativity between the bonded atoms is greater than 1.9. The compound formed by the combination of positive and negative ions is called an **ionic compound.**

> **Ionic compound** A compound formed by the combination of positive and negative ions

An example of an ionic bond is that formed between sodium (electronegativity 0.9) and chlorine (electronegativity 3.5). The difference in electronegativity between these two elements is 2.6. In forming the ionic compound Na^+Cl^-, the single $3s$ valence electron of sodium is transferred to the partially filled valence shell of chlorine.

$$Na(1s^22s^22p^63s^1) + Cl(1s^22s^22p^63s^23p^5) \longrightarrow Na^+(1s^22s^22p^6) + Cl^-(1s^22s^22p^63s^23p^6)$$

In the following equation, we use a single-headed curved arrow to show this transfer of one electron

$$Na \cdot + \cdot \ddot{C}l : \longrightarrow Na^+ \; : \ddot{C}l :^-$$

The ionic bond in solid sodium chloride is the result of the force of attraction between positive sodium ions and negative chloride ions. In its solid (crystalline) form, sodium chloride consists of a three-dimensional array of Na^+ and Cl^- ions arranged as shown in Figure 3.1.

In a crystal of sodium chloride, each sodium ion is surrounded by six chloride ions, and each chloride ion is surrounded by six sodium ions (Figure 3.2). Although ionic compounds do not consist of molecules, they do have a definite ratio of one kind of ion to another; this ratio is shown in their formulas. For example, NaCl represents the simplest ratio of sodium ions to chloride ions, namely 1:1.

FIGURE 3.1 The structure of a sodium chloride crystal.

Predicting Formulas of Ionic Compounds

Ions are charged particles, but the matter we see all around us and deal with every day is electrically neutral (uncharged). If there are ions present in any sample of matter, the total number of positive charges must equal the total number of negative charges. Therefore, we cannot have a sample

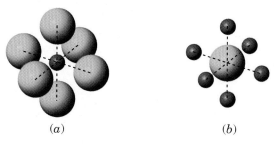

(a)   (b)

FIGURE 3.2 The environment around (a) an Na^+ ion and (b) a Cl^- ion in solid NaCl.

BOX 3A

Ionic Compounds in Medicine

There are many ionic compounds that have medical uses, some of which are shown here.

Formula	Name	Medical Use
$AgNO_3$	Silver nitrate	Astringent (external)
$BaSO_4$	Barium sulfate	Radiopaque medium for x-ray work
$CaSO_4$	Calcium sulfate	Plaster casts
$FeSO_4$	Iron(II) sulfate	For iron deficiency
$KMnO_4$	Potassium permanganate	Anti-infective (external)
KNO_3	Potassium nitrate (saltpeter)	Diuretic
Li_2CO_3	Lithium carbonate	Antimanic
$MgSO_4$	Magnesium sulfate (epsom salts)	Cathartic
$NaHCO_3$	Sodium bicarbonate (baking soda)	Antacid
NaI	Sodium iodide	Source of iodide for thyroid
NH_4Cl	Ammonium chloride	To acidify the digestive system
$(NH_4)_2CO_3$	Ammonium carbonate	Expectorant
SnF_2	Tin(II) fluoride	To strengthen teeth (external)
ZnO	Zinc oxide	Astringent (external)

■ Drinking a "barium cocktail" containing barium sulfate makes the intestinal tract visible on an x-ray. *(Susan Leavines/Science Source/Photo Researchers, Inc.)*

■ A plaster cast being applied to a patient's arm. *(Charles D. Winters)*

containing only Na^+ ions. Any sample that contains Na^+ ions must also contain negative ions, such as Cl^-, Br^-, or S^{2-}, and the sum of the positive charges must equal the sum of the negative charges.

EXAMPLE 3.3

Write the formula for the ionic compound formed from:
(a) Lithium ion and bromide ion
(b) Barium ion and iodide ion
(c) Aluminum ion and sulfide ion

Solution
(a) We see from Table 3.1 that the charge on lithium ion is $+1$, and the charge on bromide ion is -1. The formula for lithium bromide is, therefore, LiBr.

(b) The charge on barium ion is $+2$, and the charge on iodide ion is -1. Two I^- ions are required to neutralize one Ba^{2+} ion. The formula for barium iodide is, therefore, BaI_2.

(c) The charge on aluminum ion is $+3$; the charge on sulfide ion is -2. It takes three sulfide ions to neutralize the positive charge of two aluminum ions. The formula for aluminum sulfide is, therefore, Al_2S_3.

Problem 3.3 ▬▬

Write the formula for the ionic compound formed from:
(a) Potassium ion and chloride ion
(b) Calcium ion and fluoride ion
(c) Iron(III) ion and oxide ion ■

Remember that the subscripts in the formulas for ionic compounds represent the ratio of the ions. Thus, a crystal of BaI_2 has twice as many iodide ions as barium ions. For ionic compounds, when both charges are 2, as in the compound formed from Pb^{2+} and O^{2-}, we must "reduce to lowest terms." That is, lead oxide is PbO, and not Pb_2O_2. The reason is that we are looking at ratios only, and the ratio of ions in lead oxide is $1:1$.

3.4 Covalent Bonds

Formation of a Covalent Bond

A covalent bond is formed when electron pairs are shared between two atoms whose difference in electronegativity is 1.9 or less. According to the Lewis model, a pair of electrons in a covalent bond functions in two ways simultaneously; it is shared by two atoms and at the same time fills the valence shell of each. The simplest example of a covalent bond is that in the hydrogen molecule, H_2. When two hydrogen atoms bond, the single electrons from each combine to form an electron pair. A bond formed by sharing a pair of electrons is called a **single bond** and is represented by a single line between the two atoms. The electron pair shared between the two hydrogen atoms in H_2 completes the valence shell of each hydrogen. Thus, we say that in H_2, each hydrogen has two electrons in its valence shell and an electron configuration like that of helium, the noble gas nearest it in atomic number.

$$H\cdot \ + \ \cdot H \longrightarrow H—H$$

> **Single bond** A bond formed by sharing one pair of electrons; represented by a single line between two bonded atoms

Although all covalent bonds involve the sharing of electrons, they differ widely in the degree of sharing. We divide covalent bonds into two categories, **nonpolar covalent** and **polar covalent,** depending on the difference in electronegativity between the bonded atoms. There is no sharp dividing line between these two categories, nor, for that matter, is there a sharp dividing line between polar covalent bonds and ionic bonds. Nonetheless, the rule-of-thumb guidelines given in Table 3.3 will help you decide whether a given bond is more likely to be nonpolar covalent, polar covalent, or ionic.

A covalent bond between carbon and hydrogen, for example, is classified as nonpolar covalent because the difference in electronegativity between these two atoms is $2.5 - 2.1 = 0.4$. An example of a polar covalent bond is that of H—Cl, in which the difference in electronegativity between the bonded atoms is $3.0 - 2.1 = 0.9$.

> **Nonpolar covalent bond** A covalent bond between atoms whose difference in electronegativity is less than 0.5

> **Polar covalent bond** A covalent bond between atoms whose difference in electronegativity is between 0.5 and 1.9

TABLE 3.3 Classification of Chemical Bonds

Difference in Electronegativity Between Bonded Atoms	Type of Bond
less than 0.5	nonpolar covalent
0.5 to 1.9	polar covalent
greater than 1.9	ionic

EXAMPLE 3.4

Classify each bond as nonpolar covalent, polar covalent, or ionic.
(a) O—H   (b) N—H   (c) Na—F   (d) C—Mg   (e) C—S

Solution

Bond	Difference in Electronegativity	Type of Bond
(a) O—H	$3.5 - 2.1 = 1.4$	polar covalent
(b) N—H	$3.0 - 2.1 = 0.9$	polar covalent
(c) Na—F	$4.0 - 0.9 = 3.1$	ionic
(d) C—Mg	$2.5 - 1.2 = 1.3$	polar covalent
(e) C—S	$2.5 - 2.5 = 0.0$	nonpolar covalent

Problem 3.4 ▬▬▬
Classify each bond as nonpolar covalent, polar covalent, or ionic.
(a) S—H   (b) P—H   (c) C—F   (d) C—Cl ■

An important consequence of the unequal sharing of electrons in a polar covalent bond is that the more electronegative atom gains a greater fraction of the shared electrons and acquires a partial negative charge, indicated by the symbol δ^- (read delta minus). The less electronegative atom has a lesser fraction of the shared electrons and acquires a partial positive charge, indicated by the symbol δ^+ (read delta plus).

EXAMPLE 3.5

Using the symbols δ^- and δ^+, indicate the polarity in each polar covalent bond.
(a) C—O   (b) N—H   (c) C—Mg

Solution
Electronegativity is given beneath each atom. The more electronegative atom has the partial negative charge, and the less electronegative atom has the partial positive charge.

$$
\begin{array}{ccc}
\overset{\delta^+}{\text{(a) C}}-\overset{\delta^-}{\text{O}} & \overset{\delta^-}{\text{(b) N}}-\overset{\delta^+}{\text{H}} & \overset{\delta^-}{\text{(c) C}}-\overset{\delta^+}{\text{Mg}} \\
\phantom{\text{(a) }}2.5\quad 3.5 & \phantom{\text{(b) }}3.0\quad 2.1 & \phantom{\text{(c) }}2.5\quad 1.2
\end{array}
$$

TABLE 3.4 Lewis Structures for Several Small Molecules

$H—\overset{..}{\underset{..}{O}}—H$ $H—\overset{\overset{H}{|}}{\underset{\underset{H}{|}}{\overset{..}{N}}}—H$

H—Ö—H   H—N̈—H   H—C—H   H—C̈l:
                     | |
                     H   H
                                 (with H on top of C)

H_2O (8) NH_3 (8) CH_4 (8)   HCl (8)
Water   Ammonia   Methane   Hydrogen chloride

$$\underset{\underset{H}{|}}{\overset{\overset{H}{|}}{C}}=\underset{\underset{H}{|}}{\overset{\overset{H}{|}}{C}}$$

H—C≡C—H

$$\underset{\underset{H}{|}}{\overset{\overset{H}{|}}{C}}=\overset{..}{\underset{..}{O}}:$$

$$H—\overset{..}{\underset{..}{O}}\diagdown\underset{\underset{\overset{..}{O}:}{||}}{C}\diagup\overset{..}{\underset{..}{O}}—H$$

C_2H_4 (12) C_2H_2 (10) CH_2O (12) H_2CO_3 (24)
Ethylene   Acetylene   Formaldehyde   Carbonic acid

Problem 3.5 ▬▬▬
Using the symbols δ^- and δ^+, indicate the polarity in each polar covalent bond.
(a) C—N   (b) N—O   (c) C—Cl ■

Drawing Lewis Structures

The ability to write **Lewis structures** for molecules is a fundamental skill for the study of chemistry. The following guidelines will help you to do this. As you study these guidelines, look at the examples in Table 3.4.

1. Determine the number of valence electrons in the molecule. To do this, add the number of valence electrons contributed by each atom. For example, the Lewis structure of a water molecule, H_2O, must show eight valence electrons: one from each hydrogen and six from oxygen.

2. Determine the arrangement of atoms in the molecule. Except for the simplest molecules, this arrangement must be determined experimentally. For some molecules given as examples in the text, we ask you to propose an arrangement of atoms. For most, however, we give you the experimentally determined arrangement.

3. Connect the atoms with single bonds. Then, arrange the remaining electrons in pairs so that each atom in the molecule has a complete outer shell. Each hydrogen atom must be surrounded by two electrons. Each carbon, oxygen, nitrogen, and halogen atom must be surrounded by eight electrons (per the octet rule).

4. A pair of electrons involved in a covalent bond (**bonding electrons**) is shown as a single line; an unshared pair of electrons (**nonbonding electrons**) is shown as a pair of dots.

5. In a **single bond,** two atoms share one pair of electrons. In a **double bond,** they share two pairs of electrons; a double bond is shown by two lines between the bonded atoms. In a **triple bond,** two atoms share three pairs of electrons; a triple bond is shown by three lines between the bonded atoms.

Table 3.4 shows Lewis structures and names for several small molecules. The number of valence electrons each molecule contains is shown in

Bonding electrons Valence electrons involved in forming a covalent bond; that is, shared electrons

Nonbonding electrons Valence electrons not involved in forming covalent bonds; that is, unshared electrons

Double bond A bond formed by sharing two pairs of electrons; represented by two lines between the two bonded atoms

Triple bond A bond formed by sharing three pairs of electrons, represented by three lines between the two bonded atoms parentheses. Notice that, in these molecules, each hydrogen is surrounded by two valence electrons, and each carbon, nitrogen, oxygen, and chlorine is surrounded by eight valence electrons. Furthermore, each carbon has four bonds, each nitrogen has three bonds and one unshared pair of electrons, each oxygen has two bonds and two unshared pairs of electrons, and chlorine (and other halogens as well) has one bond and three unshared pairs of electrons.

EXAMPLE 3.6

Draw a Lewis structure for each molecule.
(a) Hydrogen peroxide, H_2O_2
(b) Methanol, CH_4O
(c) Acetic acid, CH_3COOH

Solution
(a) A Lewis structure for hydrogen peroxide, H_2O_2, must contain six valence electrons from each oxygen and one from each hydrogen for a total of $12 + 2 = 14$ valence electrons. We know that hydrogen forms only one covalent bond, so the order of attachment of atoms must be

$$H—O—O—H$$

The three single bonds account for six valence electrons. The remaining eight valence electrons must be placed on the oxygen atoms to give each a complete octet:

$$H—\overset{\cdot\cdot}{\underset{\cdot\cdot}{O}}—\overset{\cdot\cdot}{\underset{\cdot\cdot}{O}}—H$$

(b) A Lewis structure for methanol, CH_4O, must contain four valence electrons from carbon, one from each hydrogen, and six from oxygen for a total of $4 + 4 + 6 = 14$ valence electrons. The order of attachment of atoms in methanol is given below on the left. This partial structure shows five single bonds and accounts for ten valence electrons. The remaining four valence electrons must be placed on oxygen to give it a complete octet.

$$\begin{array}{ccc} & H & & & H \\ & | & & & | \\ H—\!\!\!&C&\!\!\!—O—H & \quad H—\!\!\!&C&\!\!\!—\overset{}{\underset{\cdot\cdot}{O}}—H \\ & | & & & | \\ & H & & & H \end{array}$$

The order of attachment   Lewis structure
of atoms   for methanol

(c) A molecule of acetic acid, CH_3COOH, must contain four valence electrons from each carbon, six from each oxygen, and one from each hydrogen for a total of $8 + 12 + 4 = 24$ valence electrons. The order of attachment of atoms, shown on the left, contains seven single bonds, which accounts for 14 valence electrons. The remaining ten electrons must be added in such a way that each carbon and oxygen atom has a complete outer shell of eight electrons. This can be done in only one way, which creates a double bond between carbon and one of the oxygens:

The order of attachment of atoms

Lewis structure for acetic acid

In this Lewis structure, each carbon has four bonds: One has four single bonds, and the other has two single bonds and one double bond. Further, each oxygen has two bonds and two unshared pairs of electrons; one oxygen has one double bond and two unshared pairs of electrons, and the other oxygen has two single bonds and two unshared pairs of electrons.

Problem 3.6 ━━━
Draw a Lewis structure for each molecule. Each has only one possible order of attachment of its atoms, which is left for you to determine.
(a) Ethane, C_2H_6 (b) Chloromethane, CH_3Cl
(c) Hydrogen cyanide, HCN ■

Exceptions to the Octet Rule

The Lewis model of covalent bonding focuses on valence electrons and the necessity for each atom other than H to have a completed valence shell of eight electrons. Although most molecules formed by main-group elements (Groups 1A–7A) have structures that satisfy the octet rule, there are important exceptions. One exception is molecules that contain an atom with more than eight electrons in its valence shell. Atoms of second-period elements use $2s$ and $2p$ orbitals for bonding, and these orbitals can contain only eight valence electrons, hence the octet rule. Atoms of third-period elements, however, also have five $3d$ orbitals and can accommodate more than eight electrons in their valence shells. Following are Lewis structures for phosphine, phosphorus pentachloride, and phosphoric acid. The first compound has eight electrons in the valence shell of phosphorus; the second and third compounds have ten electrons in the valence shell of phosphorus.

Phosphine   Phosphorus pentachloride   Phosphoric acid

Sulfur, another third-period element, forms compounds in which its valence shell contains 8, 10, or 12 electrons.

Hydrogen sulfide   Dimethyl sulfoxide   Sulfuric acid

3.5 Bond Angles and Shapes of Molecules

> **Bond angle** The angle between two atoms bonded to a central atom

In Section 3.4, we used a shared pair of electrons as the fundamental unit of a covalent bond and drew Lewis structures for several small molecules containing various combinations of single, double, and triple bonds (see, for example, Table 3.4). We can predict **bond angles** in these and other molecules using the **valence-shell electron-pair repulsion (VSEPR) model.** According to this model, the valence electrons of an atom may be involved in the formation of single, double, or triple bonds, or they may be unshared. Each combination creates a negatively charged region of electron density, and, because like charges repel each other, the various regions of electron density around an atom spread so that each is as far away from the others as possible.

We use the VSEPR model in the following way to predict the shape of a molecule of methane, CH_4. The Lewis structure for CH_4 shows a carbon atom surrounded by four regions of electron density, each containing a pair of electrons forming a covalent bond to a hydrogen atom. According to the VSEPR model, the four regions radiate from carbon so that they are as far away from one another as possible. This occurs when the angle between any two regions of electron density is 109.5°. Therefore, we predict all H—C—H bond angles to be 109.5°, and the shape of the molecule to be **tetrahedral** (Figure 3.3). The H—C—H bond angles in methane have been measured experimentally and found to be 109.5°. Thus, the bond angles and shape of methane predicted by the VSEPR model are identical to those observed.

A regular tetrahedron is the geometric figure made up of four equilateral triangles. The central atom is inside the tetrahedron, and the four atoms or electron pairs point to the four corners. Another way to think of it is as a triangle-based pyramid.

We predict the shape of an ammonia molecule, NH_3, in the same way. The Lewis structure of NH_3 shows nitrogen surrounded by four regions of electron density. Three regions contain single pairs of electrons forming covalent bonds with hydrogen atoms. The fourth region contains an unshared pair of electrons [Figure 3.4(a)]. Using the VSEPR model, we predict that the four regions are arranged in a tetrahedral manner and that the three H—N—H bond angles in this molecule are 109.5°. The observed bond angles are 107.3°. This small difference between the predicted and observed angles can be explained by proposing that the unshared pair of electrons on nitrogen repels adjacent electron pairs more strongly than bonding pairs repel one another.

Figure 3.5 shows a Lewis structure and ball-and-stick model of a water molecule. In H_2O, oxygen is surrounded by four regions of electron density. Two of these regions contain pairs of electrons used to form covalent bonds to hydrogens; the remaining two contain unshared electron pairs. Using the VSEPR model, we predict that the four regions of electron density around oxygen are arranged in a tetrahedral manner and that the H—O—H bond angle is 109.5°. Experimental measurements show that the actual H—O—H bond angle in a water molecule is 104.5°, a value smaller than that predicted. This difference between the predicted and observed bond angle can be explained by proposing, as we did for NH_3, that unshared pairs of electrons repel adjacent pairs more strongly than do bonding pairs. Note that the distortion from 109.5° is greater in H_2O, which has two unshared pairs of electrons, than it is in NH_3, which has only one unshared pair.

A general prediction emerges from this discussion. If a Lewis structure shows four regions of electron density around an atom, the VSEPR model predicts a tetrahedral distribution of electron density and bond angles of approximately 109.5°.

Ammonia gas is "drilled" into the soil of a farm field. Most of the ammonia manufactured in the world is used as fertilizer because ammonia supplies the nitrogen needed by green plants. *(Arthur C. Smith III, from Grant Heilman Photography)*

FIGURE 3.3 The shape of a methane molecule, CH_4. (*a*) Lewis structure and (*b*) ball-and-stick model. The hydrogens occupy the four corners of a regular tetrahedron, and all H—C—H bond angles are 109.5°.

FIGURE 3.4 The shape of an ammonia molecule, NH_3. (*a*) Lewis structure and (*b*) ball-and-stick model. The H—N—H bond angles are 107.3°, slightly smaller than the H—C—H bond angles of methane.

FIGURE 3.5 The shape of a water molecule, H_2O. (*a*) Lewis structure, (*b*) ball-and-stick model.

In many of the molecules we encounter, an atom is surrounded by three regions of electron density. Figure 3.6 shows Lewis structures and ball-and-stick models for formaldehyde, CH_2O, and ethylene, C_2H_4.

In the VSEPR model, a double bond is treated as a single region of electron density. In formaldehyde, carbon is surrounded by three regions of electron density: Two regions contain single pairs of electrons forming sin-

The geometry about an atom surrounded by three regions of electrons density is said to be trigonal planar.

FIGURE 3.6 Shapes of formaldehyde, CH_2O, and ethylene, C_2H_4.

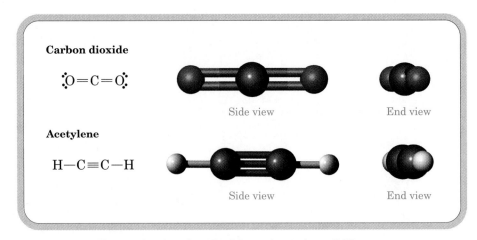

FIGURE 3.7 Shapes of carbon dioxide, CO_2, and acetylene, C_2H_2.

gle bonds to hydrogen atoms; the third region contains two pairs of electrons forming a double bond to oxygen. In ethylene, each carbon atom is also surrounded by three regions of electron density; two contain single pairs of electrons, and the third contains two pairs of electrons.

Three regions of electron density about an atom are farthest apart when they lie in a plane and make angles of 120° with one another. Thus, the predicted H—C—H and H—C—O bond angles in formaldehyde and the H—C—H and H—C—C bond angles in ethylene are all 120°. Further, all bonds in each molecule lie in a plane. Both formaldehyde and ethylene are planar molecules.

In still other types of molecules, a central atom is surrounded by two regions of electron density. Figure 3.7 shows Lewis structures and ball-and-stick models of carbon dioxide, CO_2, and acetylene, C_2H_2. In carbon dioxide, carbon is surrounded by two regions of electron density: Each contains two pairs of electrons and forms a double bond to an oxygen atom. In acetylene, each carbon is also surrounded by two regions of electron density: One contains a single pair of electrons and forms a single bond to a hydrogen atom, and the other contains three pairs of electrons and forms a triple bond to a carbon atom. In each case, the two regions of electron density are farthest apart if they form a straight line through the central atom and create an angle of 180°. Both carbon dioxide and acetylene are linear molecules.

Predictions of the VSEPR model are summarized in Table 3.5. In this table, three-dimensional shapes are shown using a solid wedge to represent a bond coming toward you, out of the plane of the paper. A broken wedge represents a bond going away from you, behind the plane of the paper, and a line represents a bond in the plane of the paper.

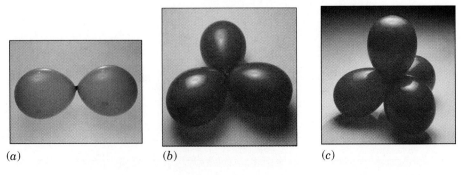

(a)　　(b)　　(c)

When balloons are tied together, they naturally assume these shapes because that is how they can keep farthest apart. (a) Two balloons are linear. (b) Three balloons are trigonal. (c) Four balloons are tetrahedral. (*Charles D. Winters*)

TABLE 3.5 Predicted Molecular Shapes (VSEPR Model)

Regions of Electron Density Around Central Atom	Predicted Distribution of Electron Density	Predicted Bond Angles	Examples		
4	tetrahedral	109.5°	Methane	Ammonia	Water
3	trigonal planar	120°	Ethylene	Formaldehyde	
2	linear	180°	Carbon dioxide	Acetylene	

EXAMPLE 3.7

Predict all bond angles and the shape of each molecule.
(a) CH_3Cl (b) $CH_2=CHCl$

Solution

(a) The Lewis structure for CH_3Cl shows carbon surrounded by four regions of electron density. Therefore, we predict that the distribution of electron pairs about carbon is tetrahedral, all bond angles are 109.5°, and the shape of CH_3Cl is tetrahedral.

(b) In the Lewis structure for $CH_2=CHCl$, each carbon is surrounded by three regions of electron density. Therefore, we predict that all bond angles are 120° and that the molecule is planar.

Top view   Viewed along
the C=C bond

Problem 3.7 ■
Predict all bond angles for these molecules.
(a) CH_3OH (b) CH_2Cl_2 (c) H_2CO_3 (carbonic acid) ■

3.6 Polar and Nonpolar Molecules

In Section 3.4, we used the term "polar" to describe a covalent bond in which one atom bears a partial positive charge and the other bears a partial negative charge. Further, we saw that we can use the difference in electronegativity between bonded atoms to determine the relative polarities of covalent bonds. We can now combine our understanding of bond polarity and molecular geometry (Section 3.5) to predict the polarity of molecules.

A molecule is polar if (1) it has polar bonds and (2) its center of partial positive charge lies at a different place within the molecule than its center of partial negative charge. Consider first carbon dioxide, CO_2, a molecule with two polar carbon-oxygen double bonds. The oxygen on the left pulls electrons of the $O{=}C$ bond toward it, giving it a partial negative charge. Similarly, the oxygen on the right pulls electrons of the $C{=}O$ bond toward it by the same amount, giving this oxygen the same partial negative charge as the oxygen on the left. Carbon bears a partial positive charge. Because carbon dioxide is a linear molecule, its centers of negative and positive partial charge coincide; therefore, it is a nonpolar molecule.

$$\overset{\delta^-}{\ddot{O}}{=}\overset{\delta^+}{C}{=}\overset{\delta^-}{\ddot{O}}$$

Carbon dioxide
(a nonpolar molecule)

In a water molecule, each O—H bond is polar, with oxygen, the more electronegative atom, bearing a partial negative charge and each hydrogen bearing a partial positive charge. The center of partial positive charge in a water molecule is halfway between the two hydrogen atoms, and the center of partial negative charge is on oxygen. Thus, a water molecule has polar bonds and, because of its geometry, is a polar molecule. Ammonia has three polar N—H bonds, and, because of its geometry, the centers of partial positive and negative charges are at different places within the molecule. Thus, ammonia has polar bonds and, because of its geometry, is a polar molecule.

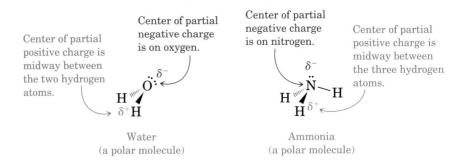

Center of partial positive charge is midway between the two hydrogen atoms.

Center of partial negative charge is on oxygen.

Center of partial negative charge is on nitrogen.

Center of partial positive charge is midway between the three hydrogen atoms.

Water
(a polar molecule)

Ammonia
(a polar molecule)

EXAMPLE 3.8

Which of these molecules are polar?
(a) CH_3Cl (b) CH_2O (c) C_2H_2

Solution
Both chloromethane, CH_3Cl, and formaldehyde, CH_2O, have polar bonds and, because of their geometry, are polar molecules. Because acetylene, C_2H_2, contains no polar bonds, it is a nonpolar molecule.

(a) Chloromethane (a polar molecule)
(b) Formaldehyde (a polar molecule)
(c) Acetylene (a nonpolar molecule)

Problem 3.8 ▬▬▬▬
Which of these molecules are polar?
(a) CH_2Cl_2 (b) HCN (c) C_2H_6 ■

3.7 Polyatomic Ions

Formulas for Polyatomic Ions

A **polyatomic ion** is an ion that contains more than one atom. An example is the hydroxide ion, OH^-. An important compound containing this ion is sodium hydroxide, NaOH, which contains positive Na^+ ions and the negative OH^- ions. The structure of solid NaOH is quite similar to that of NaCl (Figure 3.1). The difference is that Cl^- is a monatomic ion, and OH^- is a polyatomic ion. Two polyatomic ions especially important in biochemistry are the phosphate ion, PO_4^{3-}, and the bicarbonate ion, HCO_3^-. A list of these and other important polyatomic ions is given in Table 3.6.

Polyatomic ion An ion that contains more than one atom

TABLE 3.6 Some Important Polyatomic Ions

Polyatomic Ion	Name	Polyatomic Ion	Name
NH_4^+	Ammonium	HCO_3^-	Bicarbonate
OH^-	Hydroxide	SO_3^{2-}	Sulfite
NO_2^-	Nitrite	HSO_3^-	Bisulfite
NO_3^-	Nitrate	SO_4^{2-}	Sulfate
CH_3COO^-	Acetate	HSO_4^-	Bisulfate
CN^-	Cyanide	PO_4^{3-}	Phosphate
MnO_4^-	Permanganate	HPO_3^{2-}	Hydrogen phosphate
CrO_4^{2-}	Chromate	$H_2PO_4^-$	Dihydrogen phosphate
CO_3^{2-}	Carbonate		

3.8 Naming Simple Inorganic Compounds

It is relatively easy to name simple inorganic compounds just by looking at their formulas; conversely, it is relatively easy to write their formulas by looking at their names.

Binary Ionic Compounds

A **binary compound** contains only two elements, and, in a **binary ionic compound,** both elements are present as ions. The name of the compound is the name of the metal from which the positive ion was formed followed

Binary ionic compound A compound containing two elements, each present as ions by the name of the negative ion. The name of the negative ion derived from a single element ends in "-ide" (Section 3.2). Subscripts are ignored in naming binary ionic compounds. For example, $AlCl_3$ is named aluminum chloride. You know there must be three chloride ions in this compound because positive and negative charges must match; one Al^{3+} ion must combine with three Cl^- ions to give a neutral compound.

EXAMPLE 3.9

Name these binary ionic compounds.
(a) LiBr   (b) Ag_2S (c) NaBr

Solution
(a) Lithium bromide   (b) Silver sulfide   (c) Sodium bromide

Problem 3.9 ■■■■
Name these binary ionic compounds.
(a) MgO   (b) BaI_2 (c) KCl ■

EXAMPLE 3.10

Write formulas for these binary ionic compounds.
(a) Barium hydride   (b) Sodium fluoride   (c) Calcium oxide

Solution
(a) BaH_2 (b) NaF   (c) CaO

Problem 3.10 ■■■■
Write formulas for these binary ionic compounds.
(a) Magnesium chloride   (b) Aluminum oxide   (c) Lithium iodide ■

Binary Ionic Compounds of Metals That Form Different Positive Ions

We saw in Table 3.1 that some metals form different positive ions, as for example copper, which can form both Cu^+ and Cu^{2+}. In these cases, we use Roman numerals in the name to show the charge. The Cu^+ ion, for example, is named Cu(I), and the Cu^{2+} ion is named Cu(II). The symbol for copper, Cu, is derived from its Latin name *cuprum*.

Copper(I) oxide (*left*) and copper(II) oxide. The different copper ion charges result in different colors. (*Charles D. Winters*)

EXAMPLE 3.11

Name these binary ionic compounds.
(a) CuO   (b) Cu_2O

Solution
Part (a) is a compound of copper(II). For part (b), remember that subscripts are ignored in naming binary ionic compounds. Therefore, the 2 in Cu_2O is not indicated. You know that there are two copper(I) ions because there must be two positive charges to balance the two negative charges on O^{2-} ion.
(a) Copper(II) oxide   (b) Copper(I) oxide

Problem 3.11 ■
Name these binary ionic compounds.
(a) FeO (b) $HgCl_2$ ■

An alternative way to distinguish between two different positive ions a metal may form is to use the suffix "-ous" to show the lower positive charge and "-ic" to show the higher positive charge. By this method, Cu_2O is named cuprous oxide, and CuO is named cupric oxide.

EXAMPLE 3.12

Name these binary ionic compounds by two different methods.
(a) FeO (b) Fe_2O_3

Solution
(a) Iron(II) oxide and ferrous oxide
(b) Iron(III) oxide and ferric oxide

Problem 3.12 ■
Name these binary ionic compounds by two different methods.
(a) CuBr (b) $FeBr_3$ ■

Some common ionic compounds. Clockwise from the top they are: A box of salt (sodium chloride, NaCl), a clear crystal of calcite (calcium carbonate, $CaCO_3$), a pile of $CoCl_2 \cdot 6H_2O$ [cobalt(II) chloride hexahydrate], and an octahedral crystal of fluorite (calcium fluoride, CaF_2). *(Charles D. Winters)*

Ionic Compounds That Contain Polyatomic Ions

To name ionic compounds containing polyatomic ions, name the positive ion first followed by the name of the negative ion.

EXAMPLE 3.13

Name these ionic compounds, each of which contains a polyatomic ion.
(a) $NaNO_3$ (b) $CaCO_3$ (c) $(NH_4)_2SO_3$ (d) NaH_2PO_4

Solution
Note that the name of the $H_2PO_4^-$ ion is dihydrogen phosphate.
(a) Sodium nitrate (b) Calcium carbonate
(b) Ammonium sulfite (d) Sodium dihydrogen phosphate

Problem 3.13 ■
Name these ionic compounds, each of which contains a polyatomic ion.
(a) K_2HPO_4 (b) $Al_2(SO_4)_3$ (c) $FeCO_3$ ■

The White Cliffs of Dover, England, are composed mainly of calcium carbonate ($CaCO_3$). *(Harold L. Levin)*

Binary Molecular Compounds

A **binary molecular compound** is a binary compound that does not contain ions; that is, it is one in which all bonds are covalent. In naming them, follow these two rules:

1. The less electronegative element (see Table 3.2) is given first, in both the formula and the name.

2. Prefixes "di-," "tri-," "tetra-," and so on, are used to show the number of atoms of each element. The prefix "mono-," however, is commonly

Binary molecular compound A compound containing two elements joined by a covalent bond

Coral Chemistry and Broken Bones

Bone is a highly structured matrix consisting of both inorganic and organic materials. The inorganic material is chiefly hydroxyapatite, $Ca_5(PO_4)_3OH$, which makes up about 70 percent of bone by dry weight. By comparison, the enamel of teeth is almost entirely hydroxyapatite. Chief among the organic components of bone are collagen fibers (proteins, Section 21.9), which thread through the inorganic matrix, providing extra strength and allowing bones to flex under stress. Also threading through the hydroxyapatite-collagen framework are blood vessels, which supply nutrients.

A problem faced by orthopedic surgeons is how to repair bone damage. For a minor fracture, usually a few weeks in a cast are sufficient for the normal process of bone growth to repair the damaged area. For severe fractures, especially where there is bone loss, a bone graft may be needed. An alternative to bone graft is an implant of synthetic bone material. One such material, called Pro Osteon, has been developed by Interpore International of Irvine, California. Their bone graft material is derived by heating coral with ammonium hydrogen phosphate to form a hydroxyapatite similar to that of bone. Throughout the heating process, the porous structure of the coral, which resembles that of bone, is retained.

■ A wrist fracture repaired with bone cement (white area). *(Norian Corp., Cupertino, CA 95014)*

$$5CaCO_3 + 3(NH_4)_2HPO_4 \xrightarrow[24-60 \text{ hours}]{200°C}$$

Coral

$$Ca_5(PO_4)_3OH + 3(NH_4)_2CO_3 + 2H_2CO_3$$

Hydroxyapatite

The surgeon can then shape a piece of this material to match the bone void, implant it, stabilize the area by metal plates and/or screws, and let new bone tissue grow into the pores of the implant.

An alternative process has been developed by Norian Corporation of Cupertino, California. First, a dry mixture of calcium dihydrogen phosphate monohydrate, $Ca(H_2PO_4)_2 \cdot H_2O$; calcium phosphate, $Ca_3(PO_4)_2$; and calcium carbonate, $CaCO_3$, is prepared. Just before the surgical implant is to be carried out, these chemicals are mixed with a solution of sodium phosphate to form a paste, which is then injected into the bony area to be repaired. The fractured bony area is thus held in place while the natural process of bone rebuilding replaces the implant with living bone tissue.

omitted from names. One exception to this rule, however, is CO, which is named carbon monoxide.

EXAMPLE 3.14

Name these binary molecular compounds.
(a) NO   (b) SF_2 (c) P_2O_3

Solution
(a) Nitrogen monoxide (more commonly called nitric oxide)
(b) Sulfur difluoride
(c) Diphosphorus trioxide

Problem 3.14 ■
Name these binary molecular compounds.
(a) NO_2 (b) PBr_3 (c) SCl_2 ■

BOX 3C

Nitric Oxide—Air Pollutant and Biological Messenger

Nitric oxide is a colorless gas whose importance in the environment has been known for several decades, but whose biological importance is only now being fully recognized. This molecule has 11 valence electrons. Because its number of electrons is odd, it is not possible to draw a structure for it that obeys the octet rule; there must be one unpaired electron, here shown on the less electronegative nitrogen atom.

An unpaired electron $:N=\overset{\cdot}{O}:$
Nitric oxide

The importance of NO in the environment arises from the fact that it is formed as a byproduct of the combustion of fossil fuels. Under the temperature conditions of the internal combustion engine and other combustion sources, nitrogen and oxygen of the air react to form small quantities of NO. When inhaled, NO passes from the lungs into the blood stream and interacts with the iron in hemoglobin, decreasing its ability to carry oxygen.

What makes nitric oxide so hazardous in the environment is that it reacts almost immediately with oxygen to form NO_2. When dissolved in water, NO_2 forms nitric acid, which is a major acidifying component of acid rain.

$$2NO + O_2 \longrightarrow 2NO_2$$
Nitric oxide   Nitrogen dioxide

$$NO_2 + H_2O \longrightarrow HNO_3$$
Nitrogen dioxide   Nitric acid

■ Colorless nitric oxide gas, coming from the tank, bubbles through the water. When it reaches the air it is oxidized to brown NO_2. *(Photograph by Charles D. Winters.)*

Imagine scientists' surprise when they discovered within the last two decades that this highly reactive, seemingly hazardous compound is synthesized in humans and plays a vital role as a signaling molecule in the cardiovascular system (Box 23F).

SUMMARY

According to the **Lewis model of bonding,** atoms bond together in such a way that each atom participating in the bond acquires a completed valence-shell electron configuration resembling that of the noble gas nearest it in atomic number. Atoms that lose electrons from their valence shell become **cations;** those that gain electrons in their valence shell become **anions.**

Electronegativity is a measure of the force of attraction an atom exerts on the electrons it shares in a chemical bond. Electronegativity increases from left to right in a row and from bottom to top in a column of the Periodic Table.

An **ionic bond** is formed by the transfer of valence-shell electrons from an atom of lower elec-

tronegativity to the valence shell of an atom of higher electronegativity. A chemical bond is considered to be ionic if the difference in electronegativity between the bonded atoms is greater than 1.9. In an ionic compound, the total number of positive charges must equal the total number of negative charges.

A **covalent bond** is a chemical bond formed by the sharing of electron pairs between atoms. The tendency of elements of Groups 1A–7A (the main-group elements) to achieve an outer shell of eight valence electrons is called the **octet rule.** A **nonpolar covalent bond** is a covalent bond in which the difference in electronegativity between bonded atoms is small, usually less than 0.5. A **polar covalent bond** is a covalent bond in which the difference in electronegativity between bonded atoms is between 0.5 and 1.9. In a polar covalent bond, the more electronegative atom bears a partial negative charge (δ^-) and the less electronegative atom bears a partial positive charge (δ^+).

A **Lewis structure** for a molecule or ion must show (1) the correct arrangement of atoms, (2) the correct number of valence electrons, (3) no more than two electrons in the outer shell of hydrogen, and (4) no more than eight electrons in the outer shell of any second-period element. An exception to the octet rule is compounds of third-period elements, such as phosphorus and sulfur, which may have as many as 10 and 12 electrons in their valence shells.

Bond angles of molecules and polyatomic ions can be predicted using Lewis structures and the **valence-shell electron-pair repulsion (VSEPR) model.** For atoms surrounded by four regions of electron density, predict bond angles of 109.5°; by

three regions, predict bond angles of 120°; and by two regions, predict bond angles of 180°.

A molecule is polar if it has one or more polar bonds and its centers of partial positive and negative charge are at different places within the molecule.

A **polyatomic ion** is an ion that contains more than one type of atom.

To name a **binary ionic compound,** name the positive ion first, followed by the negative ion. Where a metal ion may form different cations, use a Roman numeral to show its positive charge. Alternatively, use the suffix "-ous" to show the lower positive charge, and "-ic" to show the higher positive charge. To name an ionic compound that contains polyatomic ions, name the positive ion first followed by the name of the negative ion. To name a **binary molecular compound,** use the prefixes "di-", "tri", "tetra-", and so on to show the presence of two or more atoms of the same kind.

KEY TERMS

Anion (Section 3.2)
Binary compound (Section 3.8)
Binary ionic compound (Section 3.8)
Binary molecular compound
   (Section 3.8)
Bond angle (Section 3.5)
Bonding electrons (Section 3.4)
Cation (Section 3.2)
Covalent bond (Section 3.4)
Double bond (Section 3.4)

Electronegativity (Section 3.2)
Exceptions to the octet rule
   (Section 3.4)
Formation of ions (Section 3.2)
Ionic bond (Section 3.3)
Ionic compound (Section 3.3)
Lewis model of bonding (Section 3.2)
Lewis structure (Section 3.4)
Nonbonding electrons (Section 3.4)

Nonpolar covalent bond (Section 3.4)
Octet rule (Section 3.2)
Polar covalent bond (Section 3.4)
Polar molecule (Section 3.6)
Polyatomic ion (Section 3.7)
Single bond (Section 3.4)
Tetrahedral (Section 3.5)
Triple bond (Section 3.4)
VSEPR model (Section 3.5)

CONCEPTUAL PROBLEMS

Difficult problems are designated by an asterisk.

3.A Without looking at the Periodic Table, develop an argument to predict how the electronegativities of elements change based on the composition of their atoms.

*3.B Show how to use the valence-shell electron-pair repulsion model to predict the shape of a molecule in which a central atom is surrounded by five regions of electron density, as for example phosphorus pentafluoride, PF_5. Hint: Use molecular models, or if you do not have a set handy, use a marshmallow or gum drop and toothpicks.

*3.C Show how to use the valence-shell electron-pair repulsion model to predict the shape of a molecule in which a central atom is surrounded by six regions of electron density, as for example sulfur hexafluoride, SF_6.

3.D Knowing what you do about covalent bonding in compounds of carbon, nitrogen, and oxygen and given the fact that silicon is just below carbon in the Periodic Table, phosphorus is just below nitrogen, and sulfur is just below oxygen, predict the molecular formula for the compound formed by (a) silicon and chlorine, (b) phosphorus and hydrogen, and (c) sulfur and hydrogen.

PROBLEMS

Difficult problems are designated by an asterisk.

Ions and Ionic Bonding

3.15 Calcium is a soft metal that reacts with water. When a physician prescribes calcium to treat osteoporosis, what species of calcium is prescribed?

3.16 How many electrons must each atom gain or lose to acquire an electron configuration identical to that of the noble gas nearest it in atomic number?
(a) Li (b) Cl (c) P (d) Al
(e) Sr (f) S (g) Si (h) O

3.17 Write the formula for the most stable ion formed by each element.
(a) Mg (b) F (c) Al (d) S (e) K (f) Br

3.18 Predict which ions are stable.
(a) I^- (b) Se^{2+} (c) Na^+ (d) S^{2-}
(e) Li^{2+} (f) Ba^{3+} (g) Br^{2-} (h) C^{4-}
(i) Ca^+ (j) Ar^+ (k) Na^- (l) Cs^+

3.19 Why is Li^- not a stable ion?

3.20 Why are carbon and silicon reluctant to form ionic bonds?

*3.21 Table 3.1 shows the following ions of copper: Cu^+ and Cu^{2+}. Do these violate the octet rule? Explain.

3.22 Describe the structure of sodium chloride in the solid state.

3.23 Because sodium sulfate, Na_2SO_4, is an ionic compound, there are no discrete Na_2SO_4 molecules. What does the 2 in Na_2SO_4 mean?

3.24 Complete the chart by writing formulas for the compounds formed.

	Br^-	ClO_4^-	O^{2-}	NO_3^-	SO_4^{2-}	PO_4^{3-}	OH^-
Li^+							
Ca^{2+}							
Co^{3+}							
K^+							
Cu^{2+}							

*3.25 Write a formula for the ionic compound formed from each pair of elements.
(a) Sodium and bromine
(b) Sodium and oxygen
(c) Aluminum and chlorine
(d) Barium and chlorine
(e) Magnesium and oxygen

3.26 Write the formulas for the ions present in each compound.
(a) NaBr (b) $FeSO_3$ (c) $Mg_3(PO_4)_2$
(d) KH_2PO_4 (e) $NaHCO_3$ (f) $Ba(NO_3)_2$

3.27 What is the charge on each ion in these compounds?
(a) CaS (b) MgF_2 (c) Cs_2O
(d) $ScCl_3$ (e) Al_2S_3

Covalent Bonding

3.28 Explain how a covalent bond is formed.

3.29 How many covalent bonds are normally formed by each element?
(a) N (b) F (c) C (d) Br (e) O

3.30 What is (a) a single bond (b) a double bond (c) a triple bond?

*3.31 Draw a structural formula for each covalent compound.
(a) CH_4 (b) C_2H_2 (c) C_2H_4
(d) BF_3 (e) CH_2O (f) C_3Cl_8

*3.32 Some of the following structural formulas are incorrect because they contain one or more atoms that do not have their normal number of covalent bonds. Which structural formulas are incorrect, and which atom or atoms in each have the incorrect number of bonds?

Lewis Structures

3.33 What is the difference between a molecular formula, a structural formula, and a Lewis structure?

3.34 Draw the Lewis structure for each atom.
(a) K (b) Se (c) N (d) I
(e) Ar (f) Be (g) Cl

3.35 Write the total number of valence electrons in each molecule or ion.
(a) NH_3 (b) C_3H_6 (c) $C_2H_4O_2$ (d) C_2H_6O
(e) CCl_4 (f) NO_3^- (g) PO_4^{3-} (h) O_2

3.36 Draw a Lewis structure for each molecule or ion. In each case, there is only one way the atoms can be connected.
(a) Br_2 (b) H_2S (c) N_2H_4 (d) N_2H_2
(e) CN^- (f) NH_4^+ (g) N_2 (h) O_2

*3.37 Complete the Lewis structure for each molecule or ion. In each case, the order of attachment of atoms is given after the molecular formula.
(a) HCN H—C—N
(b) HNO_2 H—O—N—O

(c) H_2SO_4

$$H-O-\overset{\displaystyle O}{\underset{\displaystyle O}{\overset{|}{\underset{|}{S}}}}-O-H$$

(d) $C_2H_3O_2{}^-$

$$H-\overset{\displaystyle H}{\underset{\displaystyle H}{\overset{|}{\underset{|}{C}}}}-\overset{\displaystyle O}{\overset{\|}{C}}-O$$

(e) $CO_3{}^{2-}$

$$O-\overset{\displaystyle O}{\overset{\|}{C}}-O$$

3.38 What is the difference between (a) a bromine atom, (b) a bromine molecule, and (c) a bromide ion? Write the Lewis structure for each.

3.39 (a) Acetylene (C_2H_2), (b) hydrogen cyanide (HCN), and (c) nitrogen (N_2) each contain a triple bond. Draw a Lewis structure for each molecule.

The Shapes of Molecules

3.40 State the shape of a molecule whose central atom is surrounded by:
(a) Two regions of electron density
(b) Three regions of electron density
(c) Four regions of electron density

*3.41 Predict the shape of each molecule or ion.
(a) CH_4 (b) PH_3 (c) CHF_3 (d) SO_2
(e) SO_3 (f) CCl_2F_2 (g) $NO_2{}^-$ (h) NH_3
(i) $NH_4{}^+$ (j) PCl_3

Electronegativity and Dipoles

3.42 Toward which atom are the electrons shifted in a bond between:
(a) H and Cl   (b) N and O   (c) C and O
(d) Cl and Br   (e) C and S   (f) P and S
(g) H and O

3.43 Which of these bonds is the most polar? The least polar?
(a) C—N   (b) C—C   (c) C—O

*3.44 Like ethylene, the chlorinated ethylenes are planar (flat). Which, if any, of these molecules are polar?

*3.45 Both CO_2 and SO_2 have polar bonds. Account for the fact that CO_2 is nonpolar and that SO_2 is polar.

Kinds of Bonds

3.46 In each case, tell whether the bond is ionic, polar covalent, or nonpolar covalent:
(a) Br_2 (b) BrCl   (c) HCl   (d) SrF_2
(e) SiH_4 (f) CO   (g) N_2 (h) CsCl

*3.47 Predict whether a bond will form between the elements in each pair and, if so, whether it will be ionic or covalent:
(a) Cl and I   (b) Li and K   (c) O and K
(d) Al and N   (e) Na and Cr   (f) Ne and S
(g) Na and Ca   (h) Li and He

Polyatomic Ions

3.48 Name each polyatomic ion.
(a) $HCO_3{}^-$ (b) $NO_2{}^-$ (c) $SO_4{}^{2-}$
(d) $HSO_4{}^-$ (e) $H_2PO_4{}^-$

3.49 Name the polyatomic ion in each compound.
(a) Na_2SO_3 (b) KNO_3 (c) Cs_2CO_3
(d) NH_4OH (e) K_2HPO_4

Naming Inorganic Compounds

3.50 Write formulas for the following inorganic compounds (see Tables 3.1 and 3.6):
(a) Potassium bromide   (b) Calcium oxide
(c) Lead(II) hydroxide   (d) Copper(II) phosphate
(e) Lithium sulfate   (f) Iron(III) sulfide
(g) Ammonium bisulfite   (h) Magnesium acetate
(i) Strontium dihydrogen phosphate
(j) Silver carbonate   (k) Strontium chloride
(l) Barium permanganate
(m) Mercury(II) oxide

3.51 Name these inorganic compounds.
(a) NaF   (b) MgS
(c) Al_2O_3 (d) $BaCl_2$
(e) $Ca(HSO_3)_2$ (f) KI
(g) $Sr_3(PO_4)_2$ (h) $Fe(OH)_2$
(i) NaH_2PO_4 (j) $Pb(CH_3COO)_2$
(k) BaH_2 (l) $(NH_4)_2HPO_4$

Reading the Labels

3.52 Name and write the formula for the fluorine-containing compound present in fluoride dental gels. (You will find it is a compound of tin, Sn, the symbol for which is derived from its Latin name, *stannum*. Tin forms both Sn^{2+} and Sn^{4+} ions.)

3.53 If you read the labels of sun-blocking lotions, you will find that a common UV-blocking agent is a compound containing titanium. Name and write the formula of this titanium-containing compound.

3.54 On packaged table salt, it is common to see a label stating that the salt "supplies iodide, a necessary nutrient." Name and write the formula of the iodine-containing compound present in iodized salt.

3.55 We are constantly warned about the dangers of "lead-based" paints. Name and write the formula for a lead-containing compound present in lead-based paints.

3.56 If you read the labels of several liquid and tablet antacid preparations, you will find that in many of them the active ingredients are two compounds containing hydroxide ions. Name and write formulas for these two hydroxide ion-containing compounds.

3.57 Iron forms Fe^{2+} and Fe^{3+} ions. Which ion is present in the over-the-counter preparations intended to treat "iron-poor blood"?

*3.58 Read the labels of several multivitamin/multimineral formulations. Among their components you will find a number of so-called trace elements, elements required in the diet of a healthy adult in amounts less than 100 mg per day, or present in the body in amounts less than 0.01 percent of body weight. Following are 18 trace elements. Name and write a structural formula for at least one form of each trace element present in one or more multivitamin formulations. Note: Some formulations do not contain all 18 of these trace elements.

(a) Phosphorus   (b) Magnesium
(c) Potassium   (d) Iron
(e) Calcium   (f) Zinc
(g) Manganese   (h) Titanium
(i) Silicon   (j) Copper
(k) Boron   (l) Molybdenum
(m) Chromium   (n) Iodine
(o) Selenium   (p) Vanadium
(q) Nickel   (r) Tin

3.59 Write formulas for these compounds.
(a) Calcium sulfite, which is used in preserving cider and other fruit juices
(b) Calcium bisulfite, which is used in an aqueous solution for washing casks in brewing to prevent souring and cloudiness of beer and to prevent secondary fermentation
(c) Calcium hydroxide, which is used in mortar, plaster, cement, and other building and paving materials
(d) Calcium hydrogen phosphate, which is used in animal feeds and as a mineral supplement in cereals and other foods

3.60 Many paint pigments contain transition metal compounds. Name the compounds in these pigments using a Roman numeral to show the charge on the metal ion.
(a) Yellow, CdS   (b) Green, Cr_2O_3
(c) White, TiO_2 (d) Purple, $Mn_3(PO_4)_2$
(e) Blue, Co_2O_3 (f) Ochre, Fe_2O_3

Boxes

3.61 (Box 3A) Why is sodium iodide often present in the table salt we buy at the grocery store?

3.62 (Box 3A) What is a medical use of barium sulfate?

3.63 (Box 3A) What is a medical use of potassium permanganate?

3.64 (Box 3B) What is the main metal ion present in bone and tooth enamel?

3.65 (Box 3B) What are the three main inorganic components of one dry mixture currently used to create synthetic bone?

3.66 (Box 3C) In what way does the gas nitric oxide, NO, contribute to the acidity of acid rain?

Additional Problems

3.67 The ion O^{6+} has a complete outer shell. Why is this ion never found in normal matter?

3.68 Is there such a thing as a KF molecule? Explain.

3.69 Draw a structural formula for a molecule in which a carbon atom is bonded by a double bond to (a) another carbon atom (b) an oxygen atom (c) a nitrogen atom.

*3.70 Why can't a carbon atom be bonded to a chlorine atom by a double bond?

3.71 Explain why argon does not form (a) ionic bonds (b) covalent bonds.

3.72 What is the difference between (a) a chlorine atom, (b) a chlorine molecule, and (c) a chlorine ion? Write the Lewis structure for each.

*3.73 Which of these have an atom that does not obey the octet rule (not all these are stable molecules)?
(a) BF_3 (b) CF_4 (c) BeF_2 (d) C_2H_4
(e) CH_3 (f) N_2 (g) NO

*3.74 Account for the fact that chloromethane, CH_3Cl, which has only one polar C—Cl bond, is a polar molecule, but carbon tetrachloride, CCl_4, which has four polar C—Cl bonds, is a nonpolar molecule.

3.75 Write molecular formulas for these molecular compounds.
(a) Carbon monoxide
(b) Bromine chloride
(c) Phosphorus pentachloride
(d) Silicon tetrabromide
(e) Dinitrogen trioxide
(f) Sulfur hexafluoride

3.76 Name these molecular compounds.
(a) H_2S (b) IF   (c) IF_7 (d) NO
(e) N_2O_5 (f) SO_2 (g) SeO_2

Chapter 4

Above: Fireworks are a spectacular display of chemical reactions. *(Charles Steele)*

Chemical Reactions

4.1 Introduction

In Chapter 1, we learned that chemistry is mainly concerned with two things: the structure of matter and the transformations of one form of matter to another. In Chapters 2 and 3, we discussed the first of these topics, and now we are ready to turn our attention to the second. In a chemical change, also called a chemical reaction, one or more original substances (called the **starting materials** or **reactants**) are converted to one or more new substances (called the **products**). Chemical reactions are all around us. They fuel and keep alive the cells of living tissues; they occur when we light a match, cook dinner, start a car, listen to a portable radio, or watch television (Figure 4.1). Most of the world's manufacturing processes involve chemical reactions (Figure 4.2); they include petroleum refining, metal-smelting operations, and food processing, as well as the manufacture of drugs, plastics, artificial fibers, explosives, and many other materials.

In this chapter we discuss three aspects of chemical reactions: (a) mass relationships, (b) types of reactions, and (c) heat gains and losses.

(a)

(b)

FIGURE 4.1 Examples of chemical reactions. (*a*) Baking. *(Beverly March)*
(*b*) Rocket blast-off. *(Courtesy of NASA)*

4.2 Formula Weight

We begin our study of mass relationships with formula weight. In Section 2.4, we learned that every atom has an atomic weight. The **formula weight** of any substance is the **sum of the atomic weights of all the atoms in the molecular formula.**

Molecular formulas mean different things for molecular compounds (those that contain only covalent bonds) and ionic compounds. Molecular compounds are made of molecules, and the molecular formula tells how many atoms of each element are present in each molecule. For example, a CCl_4 molecule contains one C atom and four Cl atoms. Since there are no molecules in an ionic compound, however, its "molecular" formula represents the ratio of ions. For example, Na_2SO_4 means that two Na^+ ions are present for every one SO_4^{2-} ion.

The term "formula weight" can be used for both ionic and molecular compounds and tells nothing about whether the compound is ionic or mo-

A table of atomic weights is given on the inside back cover. Atomic weights can also be found in the Periodic Table on the inside front cover.

FIGURE 4.2 A chemical factory that makes isopropyl alcohol, as well as other compounds. *(Beverly March)*

Many people use "molecular weight" for both molecular and ionic compounds.

lecular. There is an older term, **molecular weight,** that is strictly correct only when used for molecular compounds. In this book, we use it only for them. Strictly speaking, a more accurate term than "weight" would be "mass" (Section 1.4), but the expressions "molecular weight" and "formula weight" are more commonly used.

EXAMPLE 4.1

In this book, we take formula weights to one decimal place.

What is the formula weight of (a) CO_2 (b) $(NH_2)_2CO$ (urea)?

Solution

(a)

	AW (amu)		Total weight
C	12.0×1	=	12.0
O	16.0×2	=	32.0
		CO_2 =	44.0 amu

(b)

	AW (amu)		Total weight
N	14.0×2	=	28.0
H	1.0×4	=	4.0
C	12.0×1	=	12.0
O	16.0×1	=	16.0
		$(NH_2)_2CO$ =	60.0 amu

Note that in (b) the $(NH_2)_2$ tells us that there are two nitrogen atoms and four hydrogen atoms.

Problem 4.1 ▰
What is the formula weight of (a) aspirin, $C_9H_8O_4$, (b) barium phosphate, $Ba_3(PO_4)_2$? ▪

4.3 The Mole

Avogadro was the first to propose a relationship between the mass and the number of molecules of a compound. The actual number of molecules in a mole was determined much later. One mole of pennies placed side by side would stretch for more than 1 million lightyears, a distance far outside our solar system and even our own galaxy. Six moles of freshman chemistry books would weigh as much as the Earth.

Atoms and molecules are so tiny (Section 2.8) that chemists are seldom able to deal with them one at a time (very special and expensive apparatus is necessary for that). When a chemist weighs even a very small quantity of a molecular compound, huge numbers of molecules (perhaps 10^{19}) are present. To overcome this problem, chemists long ago defined a unit called the mole. A **mole** is the amount of substance that contains as many entities as there are atoms in exactly 12 g of carbon-12. This happens to be 6.02×10^{23} atoms. Therefore, **one mole of anything contains 6.02×10^{23} particles** of anything.

A mole of hydrogen atoms is 6.02×10^{23} hydrogen atoms, a mole of sugar molecules is 6.02×10^{23} sugar molecules, and a mole of apples is 6.02×10^{23} apples. As you can see, there is nothing mysterious about the mole. Just as we call 12 of anything a dozen, 20 a score, and 144 a gross, we call 6.02×10^{23} of anything a mole. The number 6.02×10^{23} is called **Avogadro's number,** after the Italian physicist Amedeo Avogadro (1776–1856).

One mole of any substance is its formula weight expressed in g. For instance, the formula weight of CO_2, as we learned in our last example, is 44.0 amu; therefore, 44.0 g of CO_2 is 1 mole of CO_2. Likewise, the formula weight of $(NH_2)_2CO$ is 60.0 amu, and so 60.0 g of $(NH_2)_2CO$ is 1 mole of this compound. For atoms, 1 mole is the atomic weight expressed in g: 12 g of carbon is 1 mole of carbon; 32 g of sulfur is 1 mole of sulfur. One mole quantities of several compounds are shown in Figure 4.3.

Let us see why the fact that 1 mole always contains 6.02×10^{23} molecules is very useful information. An atomic weight table tells us that the mass of a silicon atom is 28 amu and that the mass of a nitrogen atom is 14 amu; that is, a silicon atom has twice the mass of a nitrogen atom (to the nearest integer). Suppose we had ten of each. Ten Si atoms have a mass of 280 amu, and ten N atoms have a mass of 140 amu. The ratio is still $2:1$. In fact, the mass ratio is $2:1$ no matter how many atoms we have, as long as we have the same number of N and Si atoms. Now suppose that we have a mole of each. One mole containing Avogadro's number of N atoms has a mass of 14 g. Therefore, the same number of Si atoms must have a mass of 28 g.

The same argument can be made for any other two atoms or molecules. In 1 mole each of any two substances, the mass ratios are the same as they are for the individual atoms or molecules. This fact allows chemists to measure out a given *number* of atoms or molecules by *weighing out* a sample. It is impossible for a chemist to weigh out one molecule or even 100 molecules, but it is very easy to weigh out 1 mole of molecules, or 0.5 mole, or 2.7 moles, or any convenient fraction or multiple desired.

Even 100 molecules represent such a small mass that no balance is sensitive enough to detect it.

FIGURE 4.3   One-mole quantities (clockwise from upper left) $CuSO_4 \cdot 5H_2O$ (249.7 g), $K_2Cr_2O_7$ (294.2 g), $(NH_2)_2CO$ (urea, 60.0 g), $KMnO_4$ (158.0 g), $C_2H_5NO_2$ (glycine, 75.1 g). *(Beverly March)*

EXAMPLE 4.2

We have 27.5 g of stannous fluoride, SnF_2. How many moles is this?

Solution

The formula weight of $SnF_2 = 118.7 + 2(19.0) = 156.7$ amu. This means that each mole of SnF_2 has a mass of 156.7 g, allowing us to use the conversion factor 1 mole $SnF_2 = 156.7$ g SnF_2:

$$27.5 \text{ g } \cancel{SnF_2} \times \frac{1 \text{ mole } SnF_2}{156.7 \text{ g } \cancel{SnF_2}} = 0.175 \text{ mole } SnF_2$$

Note that such calculations can be performed for ionic compounds, such as SnF_2, as well as molecular compounds, such as CO_2 and urea.

Problem 4.2
A man drinks 1500 g of water per day. How many moles is this? ■

EXAMPLE 4.3

We wish to weigh out 3.41 moles of ethanol, C_2H_5OH. How many g is this?

Solution

The formula weight of C_2H_5OH is $2(12.0) + 6(1.0) + 16.0 = 46.0$ amu, and so the conversion factor is 1 mole $C_2H_5OH = 46.0$ g C_2H_5OH. In order for the proper units to cancel, we multiply as follows:

$$3.41 \text{ moles } \cancel{C_2H_5OH} \times \frac{46.0 \text{ g } C_2H_5OH}{1 \text{ mole } \cancel{C_2H_5OH}} = 157 \text{ g } C_2H_5OH$$

Problem 4.3
We wish to weigh out 2.84 moles of sodium sulfide, Na_2S. How many g is this? ■

TNT (trinitrotoluene) is a powerful explosive.

EXAMPLE 4.4

How many moles of nitrogen atoms and oxygen atoms are there in 21.4 moles of TNT, $C_7H_5(NO_2)_3$?

Solution

The formula $C_7H_5(NO_2)_3$ tells us that each molecule of TNT contains three nitrogen atoms. Therefore, each mole of TNT must contain 3 moles of nitrogen atoms. The number of moles of N atoms in 21.4 moles of TNT is, therefore,

$$21.4 \text{ moles TNT} \times \frac{3 \text{ moles N}}{1 \text{ mole TNT}} = 64.2 \text{ moles N}$$

Similarly, each mole of TNT contains 6 moles of oxygen atoms so that the total number of moles of oxygen atoms is

$$21.4 \text{ moles TNT} \times \frac{6 \text{ moles O}}{1 \text{ mole TNT}} = 128.4 \text{ moles O}$$

Problem 4.4

How many moles of C atoms, H atoms, and O atoms are there in 2.5 moles of glucose, $C_6H_{12}O_6$? ■

EXAMPLE 4.5

How many moles of sodium ions, Na^+, are there in 5.63 g of Na_2SO_4?

Solution

First, we find out how many moles of Na_2SO_4 we have. As in the SnF_2 example (Example 4.2), we obtain the conversion factor by calculating the formula weight: $2(23.0) + 32.1 + 4(16.0) = 142.1$ amu. Then,

$$5.63 \text{ g Na}_2SO_4 \times \frac{1 \text{ mole Na}_2SO_4}{142.1 \text{ g Na}_2SO_4} = 0.0396 \text{ mole Na}_2SO_4$$

The formula Na_2SO_4 tells us that there are two Na^+ ions in each mole of Na_2SO_4. The number of moles of Na^+ ions in 5.63 g of Na_2SO_4 is, therefore,

$$0.0396 \text{ mole Na}_2SO_4 \times \frac{2 \text{ mole Na}^+}{1 \text{ mole Na}_2SO_4} = 0.0792 \text{ mole Na}^+$$

Problem 4.5

How many moles of copper(I) ions, Cu^+, are there in 0.062 g of copper(I) nitrate, $CuNO_3$? ■

EXAMPLE 4.6

A typical aspirin tablet, $C_9H_8O_4$, contains 0.36 g of aspirin. (The rest of it is starch or similar other fillers.) How many aspirin molecules are present?

Solution

The formula weight of aspirin is $9(12.0) + 8(1.0) + 4(16.0) = 180.0$ amu. First we find out how many moles of aspirin there are in 0.36 g:

$$0.36 \text{ g aspirin} \times \frac{1 \text{ mole aspirin}}{180.0 \text{ g aspirin}} = 0.0020 \text{ mole aspirin}$$

Each mole contains 6.02×10^{23} molecules, and so the number of molecules is

$$(2.0 \times 10^{-3} \text{ mole}) \left(6.02 \times 10^{23} \frac{\text{molecules}}{\text{mole}} \right) = 1.2 \times 10^{21} \text{ molecules}$$

A single aspirin tablet contains 1.2×10^{21} molecules of $C_9H_8O_4$.

Chemists seldom need to find out the number of molecules in anything because they usually work with grams and moles. If they ever do need this information, however, it is very easy to determine, as Example 4.6 shows.

Problem 4.6 ▬

How many molecules of water, H_2O, are there in a glass of water (235 g)? ∎

4.4 Chemical Equations

Writing a Chemical Equation

A typical chemical reaction is the burning of propane. In this reaction, propane reacts with the oxygen in the air; both of these reactants are converted to the products: carbon dioxide and water. We could write this

> propane plus oxygen gives carbon dioxide and water

If we didn't know the molecular formulas, we would in fact be forced to write it this way. We do know the formulas, however, so a better way to write it is the following, where the arrow indicates the direction in which the reaction proceeds.

$$\underset{\text{Propane}}{C_3H_8} + \underset{\text{Oxygen}}{O_2} \longrightarrow \underset{\substack{\text{Carbon} \\ \text{dioxide}}}{CO_2} + \underset{\text{Water}}{H_2O}$$

This kind of expression is called a **chemical equation.**

The equation we have written is incomplete, however. Even though it does tell us the formulas for all the starting materials and all the products (which every chemical equation must do), it does not give the amounts correctly. That is, it is qualitatively correct but quantitatively incorrect. It is *not balanced,* which means that the number of atoms to the left of the arrow is not the same as the number of atoms to the right of the arrow. From the law of conservation of mass, we know that, in chemical reactions, atoms are never destroyed or created; they are only shifted from one substance to another. This means that all the atoms present at the start (on the left side of the arrow) must still be there at the end (on the right side), as seen in Figure 4.4. In the equation we have written, there are three car-

Propane burning in air. *(Charles D. Winters)*

The arrow indicates that the starting materials (on the left) are converted to the products (on the right).

Chemical equation An expression in which molecular formulas are used to describe a chemical reaction. The reactants are on the left side of the equation, and the products are on the right.

bon atoms on the left and only one on the right. This means that the equation cannot be correct. In order to make it correct, we must balance it.

Balancing Chemical Equations

Most equations are easily balanced. All we need to do is place numbers in front of the formulas until the equation is balanced. These numbers are called **coefficients** and represent numbers of molecules. As an example, let us balance our propane equation:

$$C_3H_8 + O_2 \longrightarrow CO_2 + H_2O$$

How do we choose our coefficients? Let us begin with carbon. There are three carbon atoms on the left and one on the right. If we put a 3 in front of the CO_2 (indicating that three CO_2 molecules are formed), there will be three carbons on each side, and the carbons will be balanced:

$$C_3H_8 + O_2 \longrightarrow 3CO_2 + H_2O$$

Next we look at the hydrogens. There are eight on the left and two on the right. We therefore put a 4 in front of the H_2O to make a total of eight hydrogens on the right (4×2). We now have

$$C_3H_8 + O_2 \longrightarrow 3CO_2 + 4H_2O$$

Coefficient The number in front of a chemical symbol in a balanced chemical equation of a compound indicating the number of molecules or moles of the compound

The only thing still unbalanced is oxygen. There are two oxygen atoms on the left and ten on the right (make sure you can find ten). If we put the coefficient 5 in front of the oxygen on the left, we balance the oxygens and the whole equation:

$$C_3H_8 + 5O_2 \longrightarrow 3CO_2 + 4H_2O \qquad \text{Balanced equation}$$

At this point, the equation ought to be balanced, but we should always check, just to make sure. In a balanced equation there must be the same number of atoms of each element on both sides. A check of our work shows three C, ten O, and eight H atoms on each side. The equation is indeed balanced.

Note that the numbers we put in to balance an equation are always coefficients. We are not permitted to change formulas. For example, when we started to balance $C_3H_8 + O_2 \rightarrow CO_2 + H_2O$, we saw that we needed three C atoms on the right. Why not change CO_2 to C_3O_2? We cannot do this because the product of the reaction is carbon dioxide and the formula for carbon dioxide is CO_2, not C_3O_2. In balancing equations, we are permitted to say that three CO_2 molecules are formed (or whatever number is necessary for a balanced equation), but we are not permitted to write incorrect formulas. Here are three other examples.

EXAMPLE 4.7

Balance this equation:

$$Ca(OH)_2 + HCl \longrightarrow CaCl_2 + H_2O$$

Solution

The calcium is already balanced: there is one on each side. There is one Cl on the left and two on the right. In order to balance them, we add the coefficient 2 to HCl. The equation now is

$$Ca(OH)_2 + 2HCl \longrightarrow CaCl_2 + H_2O$$

In the next step, we note that there are four hydrogens on the left but only two on the right. Placing the coefficient 2 in front of H_2O remedies the situation:

$$Ca(OH)_2 + 2HCl \longrightarrow CaCl_2 + 2H_2O \qquad \text{Balanced equation}$$

Note that the last step also balanced the oxygen atoms.

Problem 4.7 ▬▬▬▬▬▬▬▬

Balance this equation:

$$P_4 + Cl_2 \longrightarrow PCl_3 \ \blacksquare$$

EXAMPLE 4.8

Balance this equation:

$$C_3H_6 + O_2 \longrightarrow H_2O + CO_2$$

Solution

This equation, which represents the combustion of propene, is very similar to the first one we did. As before, we put a 3 in front of the CO_2, but this time we put a 3 in front of the H_2O (since there are six hydrogens on the left):

$$C_3H_6 + O_2 \longrightarrow 3H_2O + 3CO_2$$

When we count the oxygens, we find 2 on the left and 9 on the right. The only way to balance the equation is to put a $4\frac{1}{2}$ in front of the O_2:

$$C_3H_6 + 4\tfrac{1}{2}O_2 \longrightarrow 3H_2O + 3CO_2$$

Although there are times when chemists have good reason to write equations with fractional coefficients, it is common practice to use only whole-number coefficients. We accomplish that by multiplying everything by 2.

$$2C_3H_6 + 9O_2 \longrightarrow 6H_2O + 6CO_2 \qquad \text{Balanced equation}$$

A check will show that the equation is balanced.

Problem 4.8 ▬▬▬▬▬▬▬▬

Balance this equation:

$$C_6H_{14} + O_2 \longrightarrow H_2O + CO_2 \ \blacksquare$$

EXAMPLE 4.9

Balance this equation:

$$Na_2SO_3 + H_3PO_4 \longrightarrow H_2SO_3 + Na_3PO_4$$

Solution

The key to balancing equations like this one is to realize that polyatomic ions like SO_3^{2-} and PO_4^{3-} usually remain intact on both sides of the equation. We can begin by balancing the sodiums, which is done by putting a 3 in front of Na_2SO_3 and a 2 in front of Na_3PO_4, thus giving us six sodiums on each side:

$$3Na_2SO_3 + H_3PO_4 \longrightarrow H_2SO_3 + 2Na_3PO_4$$

There are three SO_3 units on the left and only one on the right, so we put a 3 in front of H_2SO_3 and now have

$$3Na_2SO_3 + H_3PO_4 \longrightarrow 3H_2SO_3 + 2Na_3PO_4$$

Since there are two PO_4 units on the right, but only one on the left, we put a 2 in front of H_3PO_4:

$$3Na_2SO_3 + 2H_3PO_4 \longrightarrow 3H_2SO_3 + 2Na_3PO_4 \qquad \text{Balanced equation}$$

In doing so, we balanced not only the PO_4 units but also the hydrogens. Finally, we check to make sure the equation is balanced.

Problem 4.9 ▬▬▬

Balance this equation:

$$K_2C_2O_4 + Ca_3(AsO_4)_2 \longrightarrow K_3AsO_4 + CaC_2O_4 \quad ■$$

One more thing. If

$$C_3H_8 + 5O_2 \longrightarrow 3CO_2 + 4H_2O$$

is a correctly balanced equation (and indeed it is), would it be correct if we doubled all the coefficients?

$$2C_3H_8 + 10O_2 \longrightarrow 6CO_2 + 8H_2O$$

The answer is that it is mathematically and scientifically correct, but chemists do not normally write equations with coefficients that are all divisible by a common number. A correctly balanced equation is almost always written with the coefficients expressed "in lowest terms."

4.5   Mass Relationships in Chemical Reactions

As we saw in Section 4.4, a balanced chemical equation tells us not only which substances react and which are formed but also in what proportions. This is very useful information when we want to carry out a chemical reaction. For example, we can easily calculate how much starting materials to weigh out if we want to produce a particular mass of a product. **The study of mass relationships in chemical reactions** is called **stoichiometry.**

"Stoichiometry" comes from the Greek *stoicheion*, element, and *metron*, measure.

Let us look once again at the balanced equation for the burning of propane:

$$C_3H_8 + 5O_2 \longrightarrow 3CO_2 + 4H_2O$$

This equation tells us not only that propane and oxygen are converted to carbon dioxide and water, but also that 1 mole of propane combines with 5 moles of oxygen to produce 3 moles of carbon dioxide and 4 moles of water; that is, we know the quantities involved. The same is true for any other balanced equation. This fact allows us to answer questions such as:

1. How many moles of any particular product are produced if we start with a given mass of any starting material?

2. How many grams (or moles) of one starting material are necessary to react completely with a given number of grams (or moles) of another starting material?

3. How many grams (or moles) of starting material are needed if we want to produce a certain number of grams (or moles) of a certain product?

4. How many grams (or moles) of another product is obtained when a certain amount of main product is produced?

It sounds as if we have four different types of problems here, and yet they are all solved by the same very simple procedure, which involves three steps.

In Section 4.4, we saw that the coefficients in an equation represent numbers of molecules. Since moles are proportional to molecules (Section 4.3), the coefficients in an equation also represent numbers of moles.

EXAMPLE 4.10

Consider the reaction

$$N_2 + 3H_2 \longrightarrow 2NH_3$$

How many g of N_2 are necessary to produce 7.5 g of NH_3?

Solution

Step 1 The coefficients in an equation refer to moles, not g. Therefore, we must first find out how many moles of NH_3 there are in 7.5 g. As before, we multiply the 7.5 g by the proper conversion factor (1 mole NH_3 = 17.0 g NH_3). At this stage, we don't do the actual multiplication, but just leave it set up:

$$7.5 \text{ g } NH_3 \times \frac{1 \text{ mole } NH_3}{17.0 \text{ g } NH_3} = \text{ number of moles of } NH_3$$

Step 2 Next we turn to the balanced equation, which tells us that 2 moles of NH_3 are produced from 1 mole of N_2. We multiply the number of moles of NH_3 by this fraction:

$$7.5 \text{ g } NH_3 \times \frac{1 \text{ mole } NH_3}{17.0 \text{ g } NH_3} \times \frac{1 \text{ mole } N_2}{2 \text{ moles } NH_3} = \text{ number of moles of } N_2$$

This gives us the number of moles of N_2 that are necessary.

Step 3 We now find the number of grams of N_2, as usual, by multiplying by the proper conversion factor (28.0 g N_2 = 1 mole N_2):

$$7.5 \text{ g } NH_3 \times \frac{1 \text{ mole } NH_3}{17.0 \text{ g } NH_3} \times \frac{1 \text{ mole } N_2}{2 \text{ moles } NH_3} \times \frac{28.0 \text{ g } N_2}{1 \text{ mole } N_2} = 6.2 \text{ g } N_2$$

At this point, we do the arithmetic to get our answer, 6.2 g N_2.

Ammonia, NH_3, which is used to make fertilizer, is produced by this reaction (the Haber process).

The fraction obtained from the balanced equation is shown in color.

In all such problems, we are given a mass (or number of moles) of one compound and asked to find the mass (or number of moles) of another compound. The two compounds can be on the left or the right, on the same side of the equation, or on opposite sides. All such problems can be done by the three simple steps we have just used:

1. If you are given a mass, convert to moles by multiplying by the proper conversion factor. If you are given the number of moles, just write down the value.

Note that we look at the coefficients of the balanced equation only in step 2. These coefficients play no part in steps 1 and 3.

2. Multiply by a fraction consisting of the coefficient of the compound whose mass is unknown divided by the coefficient of the compound whose mass is known. Because this information is obtained from the balanced equation, we cannot do such problems without a balanced equation.

3. If you are asked for the mass of a compound, multiply by the appropriate conversion factor. If you are asked for moles, this step is unnecessary, and the answer is already present at the end of step 2.

The simplicity of the method is shown in the following examples.

EXAMPLE 4.11

Silicon to be used in computer chips is manufactured by a process represented by the reaction

$$SiCl_4 + 2Mg \longrightarrow Si + 2MgCl_2$$

A sample of 225 g of silicon tetrachloride, $SiCl_4$, is reacted with an excess (more than necessary) amount of Mg. How many moles of Si are produced?

Excess means more than is necessary for complete reaction.

Solution

Step 1   1 mole $SiCl_4$ = 170 g $SiCl_4$;

225 g $SiCl_4 \times$ (1 $SiCl_4$ mole/170 g $SiCl_4$) = 1.32 moles $SiCl_4$

Step 2   Multiply by the fraction obtained from the equation:

(1 mole Si/1mole $SiCl_4$) $\times$ 1.32 moles $SiCl_4$ = 1.32 moles Si

Step 3   This step is unnecessary because we were asked for moles.

Problem 4.10 ◼━━━━━━━━
In the reaction

$$CH_4 + 2O_2 \longrightarrow CO_2 + 2H_2O$$

how many moles of CH_4 are required to produce 16.6 moles of H_2O? ◼

EXAMPLE 4.12

When urea, $(NH_2)_2CO$, is acted on by the enzyme urease in the presence of water, ammonia and carbon dioxide are produced:

$$(NH_2)_2CO + H_2O \longrightarrow 2NH_3 + CO_2$$

If excess water is present (more than necessary for the reaction), how many g each of CO_2 and NH_3 are produced from 0.83 mole of urea?

Solution
For CO_2:

Step 1   Once again, we merely write down the number of moles: 0.83 mole urea:

Step 2   Multiply by the fraction obtained from the equation:

$$0.83 \text{ mole urea} \times \frac{1 \text{ mole } CO_2}{1 \text{ mole urea}}$$

Step 3   Multiply by the conversion factor (44.0 g CO_2 = 1 mole CO_2):

$$0.83 \text{ mole urea} \times \frac{1 \text{ mole } CO_2}{1 \text{ mole urea}} \times \frac{44.0 \text{ g } CO_2}{1 \text{ mole } CO_2} = 37 \text{ g } CO_2$$

For NH_3:
Step 1 is the same, but in step 2 the equation gives a different fraction so that, when we complete steps 2 and 3, we have:

$$0.83 \text{ mole urea} \times \frac{2 \text{ mole } NH_3}{1 \text{ mole urea}} \times \frac{17.0 \text{ g } NH_3}{1 \text{ mole } NH_3} = 28 \text{ g } NH_3$$

Problem 4.11 ▬▬▬
In the reaction

$$C_2H_2 + 2Br_2 \longrightarrow C_2H_2Br_4$$

how many g of $C_2H_2Br_4$ are produced from 7.24 moles of Br_2 (assume that there is excess C_2H_2)? ■

Frequently reactants are mixed in nonstoichiometric proportions, which are different from those that appear in a balanced equation. For example, if 5 moles of N_2 are mixed with 1 mole of O_2 in the reaction

$$N_2 + O_2 \longrightarrow 2NO$$

only 1 mole of N_2 will react consuming the 1 mole of O_2. The oxygen is used up completely; it disappears first. The **limiting reagent is the reagent that is used up first.** In the preceding reaction, oxygen is the limiting reagent. It will govern the extent of the reaction. The other reagent (N_2) is in excess.

EXAMPLE 4.13

In the reaction

$$C + O_2 \longrightarrow CO_2$$

12 g of C is mixed with 64 g of O_2.
(a) Which is the limiting reagent and which is in excess?
(b) How many g of CO_2 will be formed?

Solution
(a) The formula weight of C is 12 amu, and that of O_2 is 32 amu. Step 1 gives us

$$12 \text{ g } \cancel{C} \times \frac{1 \text{ mole C}}{12 \text{ g } \cancel{C}} = 1 \text{ mole C}$$

$$64 \text{ g } \cancel{O_2} \times \frac{1 \text{ mole } O_2}{32 \text{ g } \cancel{O_2}} = 2 \text{ moles } O_2$$

According to the balanced equation, 1 mole of C needs 1 mole of O_2. Therefore, C is the limiting reagent, and O_2 is in excess.
(b) Steps 1, 2, and 3 give us

$$12 \text{ g } \cancel{C} \times \frac{1 \text{ mole } \cancel{C}}{12 \text{ g } \cancel{C}} \times \frac{1 \text{ mole } \cancel{CO_2}}{1 \text{ mole } \cancel{C}} \times \frac{44 \text{ g } CO_2}{1 \text{ mole } \cancel{CO_2}} = 44 \text{ g } CO_2$$

Problem 4.12
In the reaction

$$C + 2H_2 \longrightarrow CH_4$$

6.0 g of C and 2.1 g H_2 are mixed.
(a) Which is the limiting reagent and which is in excess?
(b) How many g of CH_4 are produced in the reaction? ∎

4.6 Percent Yield

When carrying out chemical reactions in the laboratory, we often get less of a product than expected from the type of calculation we performed in Section 4.5. For example, if we start with 1 mole (32.0 g) of CH_3OH and add excess PCl_5 in the reaction

$$2CH_3OH + PCl_5 \longrightarrow 2CH_3Cl + POCl_3 + H_2O$$

we should get 1 mole (50.5 g) of CH_3Cl. However, we often get less. Does this mean that the law of conservation of mass is being violated? No, it does not. We get less than 50.5 g of CH_3Cl either because some of the CH_3OH does not react or because some of it reacts in some other way, or perhaps because our laboratory technique is not perfect and we lose a little in transferring from one container to another.

Reactions that do not give the main product are called *side reactions*.

Actual yield is the **mass of product we get in the laboratory. Theoretical yield** is the mass of the same product calculated by the method of Section 4.5. **Percent yield** is **actual yield divided by theoretical yield multiplied by 100:**

$$\% \text{ yield} = \frac{\text{Actual yield}}{\text{Theoretical yield}} \times 100$$

Theoretical yield The expected mass of a product based on calculations using a balanced chemical equation

EXAMPLE 4.14

In an experiment using the reaction just described, the actual yield of CH_3Cl was 46.8 g (theoretical yield is 50.5 g). What is the percent yield?

Solution

$$\frac{46.8 \text{ g}}{50.5 \text{ g}} \times 100 = 92.7\%$$

Most chemists would consider this a pretty good percent yield for this reaction.

Occasionally the percent yield is larger than 100 percent. For example, if a chemist uses sloppy technique and fails to dry the product completely before weighing it, the product weighs more than it should because it also contains water. In such cases, the actual yield is larger than it should be and the percent yield may be greater than 100 percent.

Problem 4.13 ■■■■

In one method for making aspirin, the theoretical yield is 153.7 g. If the actual yield obtained by a student is 124 g, what is the percent yield? ■

4.7 Reactions Between Ions in Aqueous Solution

Aqueous solution A solution in which the solvent is water

Many ionic compounds are soluble in water. As we saw in Section 3.3, ionic compounds always consist of both positive and negative ions. When they dissolve in water, the positive and negative ions are separated from each other by the water molecules. We call such separation a **dissociation.** For example,

$$NaCl(s) \xrightarrow{H_2O} Na^+(aq) + Cl^-(aq)$$

H_2O above the arrow means that the reaction takes place in water.

The (s) stands for solid and the (aq) for aqueous solution.

What happens when we mix two different ionic solutions? Does a reaction take place between the ions? The answer depends on the ions. If any of the negative and positive ions come together to form a compound, then a reaction takes place; otherwise, there is no reaction. For example, if we mix a solution of NaCl with one of silver nitrate, $AgNO_3$, four ions are present in the solution: Ag^+, Na^+, Cl^-, and NO_3^-. It so happens that two of these, Ag^+ and Cl^- form the compound AgCl (silver chloride), which is insoluble in water. A reaction therefore takes place

$$Ag^+(aq) + Cl^-(aq) \longrightarrow AgCl(s)$$

forming a white precipitate that slowly sinks to the bottom of the container. What about the Na^+ and NO_3^- ions? They do nothing at all and merely remain dissolved in the water. Ions like these, which do not participate in a reaction, are called **spectator ions,** certainly an appropriate name.

Spectator ion An ion that appears unchanged on both sides of the chemical equation

In general, ions in solution react with each other only when one of these things can happen:

1. Two of them form a solid that is insoluble in water (AgCl is one example; Figure 4.5).

2. Two of them form a gas. In this case, the reaction takes place because the gas is not very soluble in water. An example (Figure 4.6) is

The (g) stands for gas and the (l) for liquid.

$$HCO_3^-(aq) + H^+(aq) \longrightarrow CO_2(g) + H_2O(l)$$
$$\text{Bicarbonate}$$
$$\text{ion}$$

3. An acid neutralizes a base. Acid-base reactions are so important that we devote a whole chapter to them (Chapter 8).

4. One of the ions can oxidize another. This type of reaction is also important enough to be discussed separately (Section 4.8).

In many cases, no reaction takes place when we mix solutions of ions because none of these situations holds. For example, if we mix a solution of copper nitrate, $Cu(NO_3)_2$, and potassium sulfate, K_2SO_4, we merely have a mixture containing Cu^{2+}, K^+, NO_3^-, and SO_4^{2-} ions dissolved in water. None of these react with each other; therefore, we see nothing happening (Figure 4.7).

The equations we write for ions in solution are called ionic equations. Like all other chemical equations, they must be balanced. We balance them the same way as any other equation, except that we must make sure that charges balance as well as atoms. For example, the precipitation of arsenic sulfide takes place according to this equation:

$$2As^{3+}(aq) + 3S^{2-}(aq) \longrightarrow As_2S_3(s)$$

Not only are these two arsenic and three sulfur atoms on each side, but the total charge on the left side is the same as the total charge on the right side; they are both zero. **Net ionic equations show only the ions that react.** Spectator ions are not shown.

FIGURE 4.5 Adding Cl$^-$ ions to a solution of Ag$^+$ ions produces a white precipitate of silver chloride. *(Charles Steele)*

(a)   (b)

FIGURE 4.6 (a) The left beaker contains a solution of HCO$_3^-$ ions; the right beaker contains a solution of H$^+$ ions. (b) When the contents of the beakers are mixed, the reaction between HCO$_3^-$ ions and H$^+$ ions produces the gas CO$_2$, which can be seen as bubbles. *(Beverly March)*

(a)   (b)

FIGURE 4.7 (a) The beakers contain copper nitrate (blue) and potassium sulfate (colorless). (b) When they are mixed, no reaction takes place. The blue color gets lighter because the copper nitrate is less concentrated, but nothing else happens. (Beverly March)

EXAMPLE 4.15

When a solution of barium nitrate, $Ba(NO_3)_2$, is added to a solution of potassium sulfate, K_2SO_4, a white precipitate of barium sulfate, $BaSO_4$, forms. Write the net ionic equation for this reaction.

Solution

Because they are ionic compounds, both barium nitrate and potassium sulfate exist in water as the dissociated ions:

$$Ba^{2+}(aq) + 2NO_3^-(aq) + 2K^+(aq) + SO_4^{2-}(aq)$$

We are told that a precipitate of barium sulfate forms:

$$Ba^{2+}(aq) + 2NO_3^-(aq) + 2K^+(aq) + SO_4^{2-}(aq) \longrightarrow$$
$$BaSO_4(s) + 2K^+(aq) + 2NO_3^-(aq)$$

Since K^+ and NO_3^- appear on both sides (nothing happens to them), we cancel them (they are spectator ions) and are left with

$$Ba^{2+}(aq) + SO_4^{2-}(aq) \longrightarrow BaSO_4(s) \qquad \text{Net ionic equation}$$

Problem 4.14 ■

When a solution of copper(II) chloride, $CuCl_2$, is added to a solution of potassium sulfide, K_2S, a black precipitate of copper(II) sulfide, CuS, forms. Write the net ionic equation for the reaction. ■

Of the four ways for ions to react in water, one of the most common is the formation of an insoluble compound. We can predict when this will happen if we know the solubilities of the ionic compounds. There is no simple way to remember which ionic solids are soluble in water and which are not, but there are some useful generalizations.

1. All compounds containing Na^+, K^+, or NH_4^+ ions are soluble in water.

2. All nitrates (NO_3^-) and acetates (CH_3COO^-) are soluble in water.

The solubilities of compounds are found in handbooks.

(a)   (b)   (c)

FIGURE 4.8   Some insoluble compounds: (a) barium sulfate, (b) ferric hydroxide, and (c) copper carbonate. *(Beverly March)*

FIGURE 4.9   Sea animals of the mollusk family often use insoluble $CaCO_3$ to construct their shells. *(Charles D. Winters)*

3. Most chlorides (Cl^-) and sulfates (SO_4^{2-}) are soluble in water. Some important exceptions are silver chloride (AgCl), barium sulfate ($BaSO_4$), and lead sulfate ($PbSO_4$), which are insoluble in water.

4. Most carbonates (CO_3^{2-}), phosphates (PO_4^{3-}), sulfides (S^{2-}), and hydroxides (OH^-) are insoluble in water (Figures 4.8 and 4.9). Important exceptions are those of Na^+, K^+, and NH_4^+ salts, as well as barium hydroxide, $Ba(OH)_2$.

4.8 Oxidation–Reduction

One of the most important, as well as one of the most common, types of reaction is oxidation-reduction. We define **oxidation** as **the loss of electrons. Reduction** is defined as **the gain of electrons.** An oxidation-reduction reaction (often called a **redox reaction**) is the transfer of electrons from one species to another. An example is the oxidation of zinc by copper ions:

There must also be negative ions present, such as Cl^- or NO_3^-, for example, but we do not write them in the equation because they are spectator ions.

$$Zn(s) + Cu^{2+}(aq) \longrightarrow Zn^{2+}(aq) + Cu(s)$$

BOX 4A

Solubility and Tooth Decay

The outermost protective layer of a tooth is the enamel. Hydroxyapatite, $Ca_{10}(PO_4)_6(OH)_2$, is its main component (95 percent), and the remaining 5 percent is collagen (see also Box 21F). As with most other phosphates and hydroxides, hydroxyapatite is insoluble in water. This property protects the tooth. However, in acidic media, hydroxyapatite dissolves to a slight extent, yielding Ca^{2+}, PO_4^{3-}, and OH^- ions. This loss of enamel creates pits and cavities in the tooth.

What causes acidity in the mouth? Remnants of food, especially carbohydrates, are fermented by bacteria, producing acid. Once pits and cavities form in the enamel, the bacteria can hide there and cause further damage in the underlying softer material called dentine. The fluoridation of water brings F^- ions to the hydroxyapatite. It exchanges the OH^- ions for F^- ions. A less-soluble fluoroapatite, $Ca_{10}(PO_4)_6F_2$, is formed. Fluoride-containing toothpaste enhances this exchange process and provides further protection.

Blue solution
of Cu^{2+} ions

Zinc bar

Coating of copper
plating out, and
zinc bar partially
dissolved

Colorless solution
of Zn^{2+} ions

FIGURE 4.10 When a piece of zinc is added to a solution containing Cu^{2+} ions, the zinc is oxidized by the Cu^{2+} ions. *(Charles D. Winters)*

If you put a piece of zinc metal into a beaker containing copper ions in aqueous solution, you will see, after a short time, three things happening (Figure 4.10):

1. Some of the zinc metal dissolves and goes into solution as Zn^{2+}.

2. Copper metal deposits on the zinc metal.

3. The blue color of Cu^{2+} ions gradually disappears.

The zinc metal loses electrons to the copper ions. The zinc is being oxidized:

$$Zn(s) \longrightarrow Zn^{2+}(aq) + 2e^- \quad \text{Oxidation of Zn}$$

At the same time, the Cu^{2+} ions gain electrons from the zinc. The copper ions are being reduced:

$$Cu^{2+}(aq) + 2e^- \longrightarrow Cu(s) \quad \text{Reduction of } Cu^{2+}$$

It is evident that oxidation and reduction are not independent reactions. A species cannot gain electrons from nowhere; a species cannot lose electrons to nothing. *There can be no oxidation without an accompanying reduction, and vice versa.* In this reaction, Cu^{2+} is oxidizing Zn. Thus, we call Cu^{2+} an **oxidizing agent.** Similarly, Zn is reducing Cu^{2+}, and so we call Zn a **reducing agent.**

Although the definitions we have given for oxidation (loss of electrons) and reduction (gain of electrons) are easy to apply in many redox reactions, they are not so easy to apply in others. For example, another redox reaction is the combustion (burning) of methane, CH_4, in which the CH_4 is oxidized to CO_2 while O_2 is reduced to CO_2 and H_2O:

$$CH_4(g) + 2O_2(g) \longrightarrow CO_2(g) + 2H_2O(l)$$

It is not easy to see the electron loss and gain in cases like this, and so chemists have another definition of oxidation and reduction, one that is easier to apply in many cases, especially where organic compounds are involved:

The oxidizing agent is the species that gets reduced. The reducing agent is the species that gets oxidized.

> Oxidation is the gain of oxygen or the loss of hydrogen.
> Reduction is the loss of oxygen or the gain of hydrogen.

BOX 4B

Voltaic Cells

In Figure 4.10, we see that when a piece of zinc metal is put in a solution containing Cu^{2+} ions, the zinc atoms give electrons to the Cu^{2+} ions. If we change the experiment by putting the zinc metal in one beaker and the Cu^{2+} ions in another and connect the two beakers by a piece of wire and a salt bridge (Figure 4B), the reaction still takes place; that is, zinc atoms still give electrons to the Cu^{2+} ions, but now the electrons must flow through the wire to get from the Zn to the Cu^{2+}. This flow of electrons is an electric current, and the electrons keep flowing until either the zinc or the Cu^{2+} is used up. This apparatus is, therefore, a device that generates an electric current by using a redox reaction; in other words, it is a **voltaic cell** or, as we commonly say, a battery.

The electrons produced at the zinc carry negative charges. This end of the battery is the negative electrode (called the anode). The electrons released at the anode go through an outside circuit and in doing so produce the electric current of the battery. At the other end of the battery, at the positively charged electrode (called the cathode), reduction takes place, consuming electrons by changing Cu^{2+} ions to copper metal.

To see why a salt bridge is necessary, we must look at the Cu^{2+} solution. Since we cannot have positive charges in any place without an equivalent number of negative charges, there must be negative ions in the beaker also, which may be sulfate, SO_4^{2-}, nitrate, NO_3^-, or some other ion. When electrons come over the wire, the Cu^{2+} is converted to Cu:

$$Cu^{2+}(aq) + 2e^- \longrightarrow Cu(s)$$

■ Figure 4B   A voltaic cell. The electron flow, over the wire from Zn to Cu^{2+}, is an electric current that causes the light bulb to glow.

This reaction diminishes the number of Cu^{2+} ions, but the number of negative ions is unchanged. The salt bridge is necessary to carry some of these negative ions to the other beaker, where they are needed to balance the Zn^{2+} ions being produced by the reaction

$$Zn(s) \longrightarrow Zn^{2+}(aq) + 2e^-$$

In fact, this second definition is much older than the one involving electron transfer; it is the definition given by Lavoisier when he first discovered oxidation and reduction more than 200 years ago. Note that here, too, oxidation and reduction must occur together. Note also that we could not apply this definition to our zinc-copper example.

EXAMPLE 4.16

In each equation, identify the substance that is oxidized, the one that is reduced, and the oxidizing and reducing agents.
(a) $Al(s) + Fe^{3+}(aq) \rightarrow Al^{3+}(aq) + Fe(s)$
(b) $CH_3OH(l) + O_2(g) \rightarrow HCOOH(l) + H_2O(l)$
    Methanol   Formic acid

Solution
(a) It is evident that aluminum has lost three electrons, since it went from an aluminum atom, which has no charge, to an aluminum ion, with a charge of +3. Since loss of electrons is oxidation, the Al has been oxidized. On the other hand, the iron, which went from Fe^{3+} to Fe, must have gained three electrons and has, therefore, been reduced. In every redox reaction, one substance is

reduced and another oxidized. The substance that has been oxidized (Al in this case) is the reducing agent, and Fe^{3+} is the oxidizing agent.

(b) It is not easy to see electron loss or gain here, so we apply the second definitions. The methanol, which has one oxygen and four hydrogens, is converted to a compound that has two oxygens and two hydrogens. It has, therefore, both gained oxygen and lost hydrogen, meeting the definition of a compound that has been oxidized. The O_2 molecule has been converted to H_2O, losing one oxygen and gaining two hydrogens in the process. Again, the substance that has been oxidized (methanol) is the reducing agent, and O_2 is the oxidizing agent.

Problem 4.15 ▬▬▬

In each equation, identify the substance that is oxidized, the one that is reduced, and the oxidizing and reducing agents:

(a) $Ni^{2+}(aq) + Cr(s) \rightarrow Cr^{2+}(aq) + Ni(s)$

(b) $H_2CO(g) + H_2(g) \rightarrow CH_3OH(l)$ ▪

We have said that redox reactions are extremely common. Here are some important categories.

1. Combustion All combustion (burning) reactions are redox reactions in which the compounds or mixtures that are burned are oxidized by the oxygen in the air. This includes the burning of gasoline, diesel oil, fuel oil, natural gas, coal, wood, and paper. All these materials contain carbon, and all except coal also contain hydrogen. If the combustion is complete, the carbon is oxidized to carbon dioxide and the hydrogen to water. In an incomplete combustion, these elements are oxidized to other compounds, many of which cause air pollution. It is unfortunate that much of the combustion that takes place in gasoline engines and in furnaces is incomplete and so contributes to air pollution.

2. Respiration Humans and animals get their energy by respiration. The oxygen in the air we breathe oxidizes carbon-containing compounds in our cells to produce CO_2 and H_2O. In fact, this oxidation reaction is really the same as combustion, although it takes place more slowly and at a much lower temperature. We discuss these reactions more fully in Chapter 26. The important product of these reactions is not the CO_2 (which the body gets rid of anyhow) or the H_2O, but the energy.

3. Rusting We all know that when iron or steel objects are left out in the open air, they eventually rust (steel is mostly iron but contains certain other elements also). In rusting (Figure 4.11), iron is oxidized to a mixture of oxides. We can represent the main reaction by the equation

$$4Fe(s) + 3O_2(g) \longrightarrow 2Fe_2O_3(s)$$

4. Bleaching Most bleaching is oxidation, and common bleaches are oxidizing agents. The colored compounds being bleached are usually organic compounds; oxidation converts them to colorless compounds.

5. Batteries A voltaic cell (Box 4B) is a device in which electricity is generated from a chemical reaction. Such cells are often called batteries. We are all familiar with them in our cars and in such portable devices as radios and flashlights (Figure 4.12). In all cases, the reaction taking place in the battery is a redox reaction.

Air pollution is caused by incomplete fuel combustion. *(© Russell D. Curtis / Photo Researchers, Inc.)*

FIGURE 4.11 The rusting of iron and steel can be a serious problem in an industrial society. *(M. Ippolito)*

Anode

Cathode

Negative plates:
lead grills filled
with spongy
lead

Positive plates:
lead grills filled
with PbO$_2$

(a)

(b)

FIGURE 4.12 (a) Dry cell batteries are used in a wide variety of applications. *(Charles D. Winters)* (b) A lead storage battery.

4.9 Heat of Reaction

The heat given off increases the temperature of the reaction mixture.

In almost all chemical reactions, not only are starting materials converted to products, but there is also a gain or loss of heat. For example, when 1 mole of carbon is oxidized by oxygen to give CO_2, 94 000 cal of heat is given off:

$$C(s) + O_2(g) \longrightarrow CO_2(g) + 94\ 000\ cal$$

BOX 4C

Oxidizing Antiseptics

An **antiseptic** is a compound that kills bacteria. Antiseptics are used to treat wounds—not to heal them any faster but to prevent them from becoming infected by bacteria. Some antiseptics operate by oxidizing (hence destroying) compounds essential to the normal functioning of the bacteria. One example is iodine, I_2, which was used as a household antiseptic for minor cuts and bruises for many years, not in the pure form but as a dilute solution in ethanol, called a tincture. Pure I_2 is a steel-gray solid that gives a purple vapor when heated. The tincture is a brown liquid. Other examples of oxidizing antiseptics are dilute solutions of hydrogen peroxide, H_2O_2, and potassium permanganate, $KMnO_4$, both of which oxidize a large number of organic compounds. Sodium hypochlorite (a bleaching agent), $NaOCl$, was also once used as an antiseptic. It was one of the main ingredients of Dakin's solution, which was used to treat the wounded in World War I.

Oxidizing antiseptics are often regarded as too harsh, however. They not only kill bacteria but also harm skin and other normal tissues. For this reason, they have largely been replaced by phenolic antiseptics (Section 14.5). **Disinfectants** are also used to kill bacteria, but they are used on inanimate objects rather than on living tissues. Many disinfectants are oxidizing agents. Two important examples are chlorine, Cl_2, a pale green gas, and ozone, O_3, a colorless gas.

■ These tablets, which produce chlorine when mixed with water, are added to a swimming pool filter. *(© Yoav Levy/Phototake NYC)*

Both gases are added in small quantities to municipal water supplies to kill any harmful bacteria that may be present. Both gases must be handled carefully because they are very poisonous.

Photography

Like many other processes that make up a part of our modern civilization, photography is a chemical process. It was invented early in the nineteenth century when it was discovered that light has the ability to reduce silver salts to metallic silver. The basic redox reaction is

$$2Ag^+ + 2Br^- + light \longrightarrow 2Ag + Br_2$$

Photographic film consists of tiny crystals of silver bromide, AgBr, embedded in a gelatin emulsion. When light hits the emulsion, the silver ions are reduced to metallic silver, turning the film black in just those places where the light hits but not where the light does not hit. The extent of reduction is proportional to the intensity and duration of the light exposure.

The reduction is further aided by organic reducing agents in the developing process. The photographic image is provided by the dark silver metal atoms still embedded in the emulsion. In the final step, called fixation, the unreacted AgBr crystals are dissolved and washed away together with the gelatin emulsion. Both the development and the fixation must be done in a dark room. The negative, thus obtained, can be converted to the final photograph by the following process: The negative is used as a screen through which the light passes before it hits a piece of photographic paper on which AgBr crystals are embedded in a gelatin emulsion layer. The processing of the positive photograph again involves development and fixation. What is obtained is a black-and-white photo. The description of color photography is more complicated, but it too involves chemical reactions.

(a)   (b)

■ The principle on which conventional photography is based. (*a*) Silver chloride crystals are exposed to light, which will darken them. (*b*) When the paper clip is removed, the nondarkened area stands out vividly. Silver chloride and silver bromide undergo the same redox reaction. (*© 1995 Richard Megna/Fundamental Photographs*)

The heat lost or gained in a reaction is called the **heat of reaction** (Figure 4.13). A reaction that gives off heat is said to be **exothermic;** a reaction that gains heat is **endothermic.** The amount of heat gained or lost is proportional to the amount of material. For example, when 2 moles of pure carbon are oxidized, $2 \times 94\ 000 = 188\ 000$ cal of heat is lost.

The energy changes accompanying a chemical reaction are not limited to heat. In some reactions, as in voltaic cells (Box 4B), the energy output is in the form of electricity. In other reactions, such as photosynthesis (the reaction whereby plants convert water and carbon dioxide to sugar and oxygen), the energy absorbed is in the form of light. Thus, a more general expression is in use with the terms **exergonic** and **endergonic** (Figure 14.4).

An example of an endothermic (endergonic) reaction is the decomposition of mercury(II) oxide:

$$2HgO(s) + 43\ 400\ cal \longrightarrow 2Hg(l) + O_2(g)$$

> An exergonic reaction is one that loses energy in any form, not necessarily in the form of heat. An endergonic reaction is one that gains energy in any form. In most reactions the energy change is in the form of heat.

(a)   (b)

FIGURE 4.13   The thermite reaction, which gives off a large quantity of heat.
(a) After the reaction is started with a fuse of burning magnesium wire, iron(III)
oxide reacts with aluminum powder. (b) The reaction generates so much heat that the
product, iron, is produced in the molten state.   *(Charles D. Winters)*

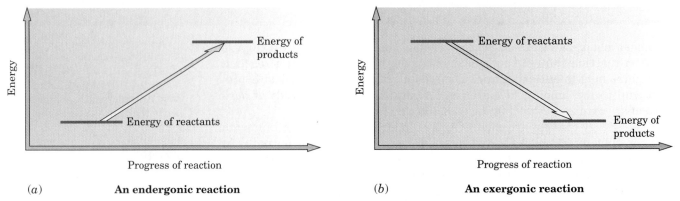

(a)   **An endergonic reaction**   (b)   **An exergonic reaction**

FIGURE 4.14   (a) In an endergonic reaction, the products have more energy than the
starting compounds. Energy has been absorbed. (b) In an exergonic reaction, the products have less energy than the starting compounds. Energy has been lost (given off).

BOX 4E

The Caloric Value of Foods

All of us have seen the food package labels that say, for example, that one serving has 50 calories (see Section 29.3). What does this mean? We pointed out that combustion is essentially the same reaction as the metabolism of food. In both cases, the starting materials are converted to CO_2 and H_2O, although metabolism takes place at 37°C and by a series of reactions, whereas combustion is a high-temperature process that takes place directly. Nevertheless, the heat of reaction depends only on the starting compounds and the final products, not on the pathway. (This is called **Hess's law.**) This means that if we take a given weight of any food (such as sugar) and burn it in order to measure the amount of heat given off, then we know that the same amount of heat

is given off in the body when that weight of that food is metabolized.

By this simple method, we have determined the energy content of many foods (Table 4E). Of course, we all know that it works two ways. If we don't take in enough calories in our food to replace the heat we lose or use for energy, our bodies begin to "burn" some of our normal tissues, and we lose weight. On the other hand, if we eat food with more than enough calories to match our losses, our bodies turn that extra energy to fatty tissue, and we gain weight. That is why many people carefully count the number of calories they eat. Remember (Section 1.9) that a nutritionist's calorie is equal to 1000 chemist's calories.

TABLE 4E Caloric Values of Some Foods

Food	Calories	Food	Calories
Milk, 8 oz	165	Apple, 5 oz	70
Swiss cheese, 1 oz	105	Banana, 5 oz	85
Egg, boiled, large	80	Grapefruit, $\frac{1}{2}$ medium	50
Hamburger patty, 3 oz	245	Orange juice, 8 oz	100
Beef, lean, 2 oz	115	Bread, rye, 1 slice	55
Lamb, shoulder, lean 2 oz	125	Bread, white, 1 slice	60
		Fudge, 1 oz	115
Haddock, fried, 3 oz	135	Honey, 1 tablespoon	60
Shrimp, 3 oz	110	Cookie, 3-in.	110
Peanuts, roasted, 2 oz	210	Cornflakes, 1 oz	110
Carrot, raw, 8 oz	45	Noodles, 8 oz	200
Corn, 5-in. ear	65	Butter, 1 tablespoon	100
Potato, 5 oz	90	Margarine, 1 tablespoon	100
Tomato, 5 oz	30		

This equation tells us that, if we want to break down 2 moles of mercury(II) oxide into the elements Hg and O_2, we must add 43 400 cal of energy to HgO.

Incidentally, the law of conservation of energy tells us that the reverse reaction, the oxidation of mercury, must give off exactly the same amount of heat:

$$2Hg(l) + O_2(g) \longrightarrow 2HgO(s) + 43\ 400\ cal$$

Especially important are heats of reaction for combustion reactions. As we saw in the previous section, combustion reactions are the most important heat-producing reactions, since most of our energy required for modern society is derived from them. All combustions are exothermic. The heat given off in a combustion reaction is called the **heat of combustion.** Accurate heat of combustion values are available for thousands of compounds.

Mercury(II) oxide, a red compound, decomposes when heated into two elements: mercury (a metal) and oxygen (a nonmetal). Mercury vapor condenses on the cooler upper portion of the test tube. *(Charles D. Winters)*

SUMMARY

In a chemical reaction, the starting materials are called **reactants,** and the new materials to which they are converted are called **products.**

The **formula weight** of a compound is the sum of the atomic weights of all atoms in the molecular formula of the compound. A **mole** of a substance is defined as **Avogadro's number** (6.02×10^{23}) of particles of that substance. The mass of this number of molecules of any compound is numerically equal to the formula weight expressed in grams.

A **chemical equation** is an expression telling which reactants are converted to which products. All equations must be balanced. A balanced equation tells how many moles of each starting material are converted to how many moles of each product. The reagent that is used up first in a reaction is called the limiting reagent. A reactant is in excess when its amount is more than necessary for the reaction. **Percent yield** equals actual yield divided by **theoretical yield** multiplied by 100.

When ions are mixed in aqueous solution, they react with each other only if (a) an insoluble solid precipitate is formed, (b) a gas is formed, (c) an acid neutralizes a base, or (d) an oxidation–reduction takes place. Ions that do not react are called **spectator ions. Net ionic equations** show only the ions that react. In a net ionic equation, the charges as well as the atoms must be balanced.

In **oxidation,** electrons are lost; in **reduction,** electrons are gained. These two processes must take place together. The joint process is often called a **redox reaction.** Oxidation can also be defined as the gain of oxygen or loss of hydrogen; reduction is the reverse. Almost all reactions are accompanied by either a loss or a gain of heat. This heat is called the **heat of reaction.** Reactions that give off heat are **exothermic;** those that absorb heat are **endothermic.** The heat of a combustion reaction is called the **heat of combustion.** Reactions that give off energy in whatever form are called **exergonic,** and those that absorb energy in whatever form are called **endergonic.**

KEY TERMS

Actual yield (Section 4.6)
Antiseptic (Box 4C)
Avogadro's number (Section 4.3)
Chemical equation (Section 4.4)
Coefficients (Section 4.4)
Disinfectant (Box 4C)
Dissociation (Section 4.7)
Endergonic (Section 4.9)
Endothermic (Section 4.9)
Exergonic (Section 4.9)
Exothermic (Section 4.9)

Formula weight (Section 4.2)
Heat of combustion (Section 4.9)
Heat of reaction (Section 4.9)
Hess's law (Box 4E)
Limiting reagent (Section 4.5)
Mole (Section 4.3)
Molecular weight (Section 4.2)
Net ionic equation (Section 4.7)
Oxidation (Section 4.8)
Oxidizing agent (Section 4.8)
Percent yield (Section 4.6)

Product (Section 4.1)
Reactant (Section 4.1)
Redox reaction (Section 4.8)
Reducing agent (Section 4.8)
Reduction (Section 4.8)
Spectator ion (Section 4.7)
Starting material (Section 4.1)
Stoichiometry (Section 4.5)
Theoretical yield (Section 4.6)
Voltaic cell (Box 4B)

CONCEPTUAL PROBLEMS

4.A The structure of methanol, wood alcohol, can be represented by the ball-and-stick molecular model. The black ball represents C, the red represents O, and the white balls represent H. Write the formula of methanol and calculate its formula weight.

Methanol

4.B Draw a box containing six red balls and three blue diatomic balls. To the right of this box, draw another one that contains six red-blue diatomic balls. Which reaction is described by your drawings? The reactions are given in their unbalanced forms.
(a) $Na + F_2 \longrightarrow NaF$
(b) $H_2 + Cl_2 \longrightarrow HCl$
(c) $Al + Cl_2 \longrightarrow AlCl_3$

4.C When 4 moles of H_2 react with 2 moles of C to form C_2H_6, ethane, which is the limiting reagent?

PROBLEMS

Difficult problems are designated by an asterisk.

Formula Weight and Moles

4.16 Calculate the formula weight of:
(a) Cl_2 (b) Ar   (c) P_4 (d) N_2 (e) He

4.17 Calculate the formula weight of:
(a) KCl   (b) Na_3PO_4 (c) $Fe(OH)_2$
(d) sucrose, $C_{12}H_{22}O_{11}$ (e) $NaAl(SO_3)_2$
(f) glycine, $C_2H_7O_2N$ (g) $Al_2(SO_4)_3$
(h) $(NH_4)_2CO_3$ (i) DDT, $CCl_3CH(C_6H_4Cl)_2$

4.18 How many
(a) Moles of CH_4 are 32 g of CH_4?
(b) Moles of NO are in 345.6 g of NO?
(c) Moles of cyanuric chloride, $C_3N_3Cl_3$, are in 184.43 g of cyanuric chloride?

4.19 How many
(a) Grams of nitrogen dioxide, NO_2, are there in 1.77 moles of NO_2?
(b) Grams of 2-propanol, C_3H_8O (rubbing alcohol), are there in 0.84 mole of 2-propanol?
(c) Grams of tungsten pentachloride WCl_5 are there in 3.69 moles of WCl_5?
(d) Grams of galactose, $C_6H_{12}O_6$, are there in 0.348 mole of galactose?
(e) Grams of vitamin C, $C_6H_8O_6$, are there in 4.9×10^{-2} moles of vitamin C?

4.20 How many
(a) Moles of O atoms are there in 18.1 moles of formaldehyde, CH_2O?
(b) Moles of Br atoms are there in 0.41 mole of bromoform, $CHBr_3$?
(c) Moles of O atoms are there in 3.5×10^3 moles of $Al_2(SO_4)_3$?
(d) Moles of Hg atoms are there in 87 g of HgO?

4.21 How many
(a) Moles of S^{2-} ions are in 6.56 moles of Na_2S?
(b) Moles of Mg^{2+} ions are in 8.320 moles of $Mg_3(PO_4)_2$?
(c) Moles of acetate ion, CH_3COO^-, are in 0.43 mole of $Ca(CH_3COO)_2$?

4.22 A single atom of cerium weighs just about twice as much as a single atom of gallium. What is the weight ratio of 25 atoms of cerium to 25 atoms of gallium?

4.23 What is the mass in grams of (a) 100 (b) 3000 (c) 5.0×10^6 (d) 2.0×10^{24} molecules of formaldehyde, CH_2O?

4.24 How many molecules are in 3.1×10^{-1} g of aspirin, $C_9H_8O_4$?

4.25 How many
(a) Molecules of TNT, $C_7H_5N_3O_6$, are there in 2.9 moles of TNT?
(b) Molecules are in one drop (0.050 0 g) of water?

4.26 A typical cholesterol deposit in an artery might have a mass of 3.9 mg of cholesterol ($C_{27}H_{46}O$). How many molecules of cholesterol are in this mass?

4.27 Protein molecules are very large. The molecular weight of hemoglobin is about 68 000. What is the mass in g of a single molecule of hemoglobin?

Balancing Equations

4.28 Balance each equation.
(a) $H_2 + I_2 \longrightarrow HI$
(b) $Al + O_2 \longrightarrow Al_2O_3$
(c) $Na + Cl_2 \longrightarrow NaCl$
(d) $CaCO_3 \longrightarrow CO_2 + CaO$

4.29 Balance each equation.
(a) $HI + NaOH \longrightarrow NaI + H_2O$
(b) $Ba(NO_3)_2 + H_2S \longrightarrow BaS + HNO_3$
(c) $CH_4 + O_2 \longrightarrow CO_2 + H_2O$
(d) $C_4H_{10} + O_2 \longrightarrow CO_2 + H_2O$
(e) $Fe + CO_2 \longrightarrow Fe_2O_3 + CO$
(f) $Al + HBr \longrightarrow AlBr_3 + H_2$
(g) $P_4 + O_2 \longrightarrow P_4O_{10}$

Mass Relationships in Chemical Reactions

4.30 Define (a) stoichiometry (b) aqueous solution.

4.31 For the reaction

$$2N_2(g) + 3O_2(g) \longrightarrow 2N_2O_3(g)$$

(a) How many moles of N_2 are required to react completely with 1 mole of O_2?
(b) How many moles of N_2O_3 are produced from the complete reaction of 1 mole of O_2?
(c) How many moles of O_2 are required to produce 8 moles of N_2O_3?

4.32 For the reaction

$$Mg + H_2SO_4 \longrightarrow MgSO_4 + H_2$$

how many moles of H_2 are produced by the complete reaction of 0.023 g of Mg?

4.33 For the reaction

$$CH_4(g) + 3Br_2(l) \longrightarrow CHBr_3(l) + 3HBr(g)$$
Bromoform

how many g of Br_2 are needed to produce 1.50 moles of bromoform?

4.34 Acetaldehyde, C_2H_4O, can be produced from acetylene by the reaction

$$C_2H_2(g) + H_2O(l) \longrightarrow C_2H_4O(g)$$

How many grams of C_2H_4O can be produced from 81.7 g of C_2H_2, if we assume enough water is present?

4.35 Ethanol, C_2H_6O, can be added to gasoline to produce "gasohol," a fuel for automobile engines. The equation for the combustion of ethanol is

$$C_2H_6O(l) + 3O_2(g) \longrightarrow 2CO_2(g) + 3H_2O(g)$$

(© John Sohlden / Visuals Unlimited)

How many g of O_2 are required for the combustion of 421 g of C_2H_6O?

4.36 In the process called photosynthesis, plants convert CO_2 and H_2O to glucose, $C_6H_{12}O_6$, by the reaction

$$6CO_2(g) + 6H_2O(l) \longrightarrow C_6H_{12}O_6(s) + 6O_2(g)$$

How many g of CO_2 are required to produce 5.1 g of glucose?

4.37 Iron ore is converted to iron by a process that can be represented as

$$2Fe_2O_3(s) + 6C(s) + 3O_2(g) \longrightarrow 4Fe(s) + 6CO_2(g)$$

If the process is run until 3940 g of Fe is produced, how many grams of CO_2 will also be produced?

4.38 Given the reaction in Problem 4.37, how many g of C are necessary to react completely with 0.58 g of Fe_2O_3?

4.39 Aspirin can be made in the laboratory by the reaction

$$\underset{\substack{\text{Salicylic} \\ \text{acid}}}{C_7H_6O_3(s)} + \underset{\substack{\text{Acetic} \\ \text{anhydride}}}{C_2H_6O_4(l)} \longrightarrow \underset{\text{Aspirin}}{C_9H_8O_4(s)} + CH_3COOH(l)$$

How many g of aspirin are produced if 85.0 g of salicylic acid is treated with excess acetic anhydride?

4.40 Benzene reacts with bromine as follows:

$$C_6H_6 + Br_2 \longrightarrow C_6H_5Br + HBr$$

If 60.0 g of benzene is mixed with 135 g of bromine,
(a) Which is the limiting reagent?
(b) How many g of bromobenzene is formed in the reaction?

*4.41 If 29.7 g of N_2 is added to 3.31 g of H_2 in the reaction

$$N_2(g) + 3H_2(g) \longrightarrow 2NH_3(g)$$

(a) Which reactant is completely used up?
(b) How many g of the other reactant is left over?
(c) How many g of NH_3 is formed if the reaction goes to completion?

Percent Yield

4.42 A student reported an actual yield of quinine preparation as 45.2 g. The theoretical yield was calculated to be 39.2 g. What was the percentage yield? Was anything unusual in this percentage yield?

*4.43 Diethyl ether, $(C_2H_5)_2O$, can be made by the reaction

$$2C_2H_5OH(l) \longrightarrow (C_2H_5)_2O(l) + H_2O(l)$$

If a chemist using 517 g of ethanol, C_2H_5OH, produces 391 g of diethyl ether, what is the percent yield?

4.44 A chemist ran this reaction

$$CH_3CH_2Cl(l) + KOH(aq) \longrightarrow CH_3CH_2OH(l) + KCl(aq)$$

by treating 47.2 g of CH_3CH_2Cl with excess KOH and obtained a yield of 28.7 g of CH_3CH_2OH. What was the percent yield?

Reactions Between Ions in Aqueous Solution

4.45 In the equation

$$Na^+(aq) + CO_3{}^{2-}(aq) + Sr^{2+}(aq) + Cl^-(aq) \longrightarrow$$
$$Na^+(aq) + SrCO_3(s) + Cl^-(aq)$$

(a) Identify the spectator ions.
(b) Write the balanced net ionic equation.

4.46 Balance these net ionic equations.
(a) $Ag^+(aq) + Br^-(aq) \longrightarrow AgBr(s)$
(b) $Cd^{2+}(aq) + S^{2-}(aq) \longrightarrow CdS(s)$
(c) $Sc^{3+}(aq) + SO_4{}^{2-}(aq) \longrightarrow Sc_2(SO_4)_3(s)$
*(d) $Sn^{2+}(aq) + Fe^{2+}(aq) \longrightarrow Fe^{3+}(aq) + Sn(s)$
(e) $K(s) + H_2O(l) \longrightarrow K^+(aq) + OH^-(aq) + H_2(g)$

4.47 When a solution of ammonium chloride, NH_4Cl, is added to a solution of lead(II) nitrate, $Pb(NO_3)_2$, a white precipitate lead(II) chloride, $PbCl_2$, is formed. Write a balanced net ionic equation for this reaction. Both ammonium chloride and lead nitrate exist as dissociated ions in aqueous solution.

4.48 When a solution of iron(II) chloride, $FeCl_2$, is added to a solution of potassium hydroxide, KOH, iron(II) hydroxide, $Fe(OH)_2$, comes out of solution as a pale green precipitate. Write a balanced net ionic equation for this reaction. Both $FeCl_2$ and KOH exist as dissociated ions in aqueous solutions.

4.49 When a solution of hydrochloric acid, HCl, is added to a solution of sodium sulfite, Na_2SO_3, sulfur dioxide gas, SO_2, is released from the solution. Write a net ionic equation for this reaction. An aqueous solution of HCl contains H^+ and Cl^- ions, and Na_2SO_3 exists as dissociated ions in aqueous solution.

4.50 When a solution of sodium hydroxide, NaOH, is added to a solution of ammonium carbonate, $(NH_4)_2CO_3$, ammonia gas, NH_3, is released when the solution is heated. Write a net ionic equation for this reaction. Both NaOH and $(NH_4)_2CO_3$ exist as dissociated ions in aqueous solution.

4.51 Using the solubility generalizations give in Section 4.7, predict which of these ionic compounds are soluble in water.
(a) KCl   (b) NaOH   (c) $BaSO_4$
(d) Na_2SO_4 (e) Na_2CO_3 (f) $Fe(OH)_2$

4.52 Using the solubility generalizations given in Section 4.7, predict which of these ionic compounds are soluble in water.
(a) $MgCl_2$ (b) $CaCO_3$ (c) Na_2SO_3
(d) NH_4NO_3 (e) $Pb(OH)_2$

4.53 Predict whether or not a precipitate will form when aqueous solutions of the following compounds are mixed. If a precipitate will form, write its formula.
(a) $CaCl_2 + K_3PO_4 \longrightarrow$
(b) $KCl + Na_2SO_4 \longrightarrow$
(c) $(NH_4)_2CO_3 + Ba(NO_3)_2 \longrightarrow$
(d) $FeCl_2 + KOH \longrightarrow$
(e) $Ba(NO_3)_2 + NaOH \longrightarrow$
(f) $Na_2S + SbCl_3 \longrightarrow$
(g) $Pb(NO_3)_2 + K_2SO_4 \longrightarrow$

Oxidation–Reduction

4.54 Give two definitions of oxidation and two definitions of reduction.

4.55 Can a reduction take place without an oxidation? Explain.

4.56 In the reaction

$$Pb(s) + 2Ag^+(aq) \longrightarrow 2Ag(s) + Pb^{2+}(aq)$$

(a) Which species gets oxidized and which gets reduced?
(b) Which is the oxidizing agent and which is the reducing agent?

4.57 In the reaction

$$C_7H_{12}(l) + 10O_2(g) \longrightarrow 7CO_2(g) + 6H_2O(g)$$

(a) Which species gets oxidized and which gets reduced?
(b) Which is the oxidizing agent and which is the reducing agent?

Heat of Reaction

4.58 What is the difference between endothermic and endergonic?

4.59 Which of these reactions are endothermic and which are exothermic?
(a) $2NH_3(g) + 22\ 000\ cal \longrightarrow N_2(g) + 3H_2(g)$
(b) $H_2(g) + F_2(g) \longrightarrow 2HF(g) + 124\ 000\ cal$
(c) $C(s) + O_2(g) \longrightarrow CO_2(g) + 94\ 000\ cal$
(d) $H_2(g) + CO_2(g) + 9800\ cal \longrightarrow H_2O(g) + CO(g)$
(e) $C_3H_8(g) + 5O_2(g) \longrightarrow$
$$3CO_2(g) + 4H_2O(g) + 531\ 000\ cal$$

4.60 In the reaction

$$H_2(g) + CO_2(g) + 9800\ cal \longrightarrow H_2O(l) + CO(g)$$

9800 cal are used up. How much heat would be given off (or used up) if 2 moles of water reacted with 2 moles of carbon monoxide?

4.61 The equation for the combustion of acetone is

$$\underset{\text{Acetone}}{2C_3H_6O(l)} + 8O_2(g) \longrightarrow 6H_2O(g) + 6CO_2(g) + 853\ 600\ cal$$

How much heat will be given off if 0.37 mole of acetone is completely burned?

4.62 To convert 1 mole of iron(III) oxide to its elements requires 196 500 cal:

$$Fe_2O_3(s) + 196\ 500\ cal \longrightarrow 2Fe(s) + 3O_2(g)$$

How many g of iron can be produced if 156 000 calories of heat are absorbed by a large-enough sample of iron(III) oxide?

Boxes

4.63 (Box 4A) How does fluoride ion protect the enamel against tooth decay?

4.64 (Box 4A) What kind of ions is hydroxyapatite made of?

4.65 (Box 4B) A voltaic cell is represented by the following equation:

$$Fe(s) + Zn^{2+}(aq) \longrightarrow Zn(s) + Fe^{2+}(aq)$$

Which electrode is the anode and which is the cathode?

4.66 (Box 4C) Hydrogen peroxide is not only an antiseptic but also an oxidizing agent. A typical equation is

$$4H_2O_2(l) + CH_4(g) \longrightarrow CO_2(g) + 6H_2O(l)$$

In this reaction, what is oxidized and what is reduced? What will become of the compound that is being reduced?

4.67 (Box 4D) Explain how an image is formed on a photographic film when light hits it in a camera.

4.68 (Box 4E) How many ounces of milk are calorically equivalent to one slice of rye bread?

Additional Problems

4.69 In a certain reaction, Cu^+ was converted to Cu^{2+}. What happened to the Cu^+ ion?

4.70 In the equation

$$Fe_2O_3(s) + 3CO(g) \longrightarrow 2Fe(s) + 3CO_2(g)$$

(a) How many moles of Fe_2O_3 are required to produce 38.4 moles of Fe?
(b) How many g of CO are required to produce 38.4 moles of Fe?

4.71 Explain this statement: Our modern civilization depends on the heat obtained from chemical reactions.

4.72 An aqueous solution of NH_4Cl reacts with a solution of $Cd(NO_3)_2$. A precipitate is formed. Write a net ionic equation and identify the spectator ions.

4.73 The active ingredient in one analgesic pill is 488 mg of aspirin, $C_9H_8O_4$. How many moles of aspirin does the pill contain? How many molecules?

*4.74 Chlorophyll, the compound responsible for the green color of leaves and grass, contains one atom of magnesium in each molecule. If the percentage by weight of magnesium in chlorophyll is 2.72 percent, what is the molecular weight of chlorophyll?

*4.75 The 1992 Clean Air Act mandated that an additive such as MTBE, methyl *tert*-butyl ether, $CH_3OC(CH_3)_3$, be added to gasoline. Although MTBE is now phased out in many states because of ground water contamination, you still find this additive in your gasoline.
(a) How many moles of CO_2 are produced in the complete combustion of 100 g of methyl *tert*-butyl ether?
(b) To the extent that this ether is used in gasoline, it replaces, weight-for-weight, the alkanes normally present. A typical alkane is heptane, C_7H_{16}. How many moles of CO_2 are produced in the complete combustion of 100 g of heptane?

*4.76 If 7 g of N_2 is added to 11 g of H_2 in the Haber reaction,

$$N_2(g) + 3H_2(g) \longrightarrow 2NH_3(g)$$

which reactant is in excess?

*4.77 In the reaction

$$3Pb(NO_3)_2 + 2AlCl_3 \longrightarrow 3PbCl_2 + 2Al(NO_3)_3$$

8.00 g of lead nitrate interact with 2.67 g of aluminum chloride. The reaction yields 5.55 g of lead chloride.
(a) Which reagent is in excess?
(b) What is the percentage yield?

4.78 Define (a) exergonic (b) endergonic.

*4.79 A test tube contains some calcium oxide, CaO. The total weight of the test tube and the CaO is 10.860 g. The CaO reacts with water vapor in the air to give calcium hydroxide, by this equation:

$$CaO(s) + H_2O(g) \longrightarrow Ca(OH)_2(s)$$

After completely reacting with the water vapor, the test tube and its contents weigh 11.149 g. What is the weight of the empty test tube?

4.80 Define spectator ion.

4.81 Assume that the average red blood cell has a mass of 2×10^{-8} g and that 20 percent of its content is hemoglobin (a protein whose molecular weight is 68 000). How many molecules of hemoglobin are there in one red blood cell?

4.82 (a) Balance:

$$C_4H_8 + O_2 \longrightarrow CO_2 + H_2O$$

(b) In this reaction, what is oxidized and what is reduced?
(c) What is the oxidizing agent and what is the reducing agent?

Gases, Liquids, and Solids

5.1 Introduction

Various forces hold matter together and cause it to take different forms. In the atomic nucleus, very strong forces keep the protons and neutrons together (Chapter 9). In the atom itself, there are attractions between the positive nucleus and the negative electrons. Within molecules, atoms are attracted to each other by covalent bonds, the arrangement of which causes the molecules to assume a particular shape. Within ionic crystals, three-dimensional shapes are brought about by electrostatic attractions between ions.

In addition to all these forces, there are also **attractive forces between molecules,** and these **intermolecular forces** are the subject of this chapter. These forces are weaker than any already mentioned; nevertheless, they help to determine whether a particular compound or element is a solid, a liquid, or a gas at room temperature.

Above: Two forms of carbon. (diamond, Charles D. Winters; nanotubes, Z.F. Ren et al., SUNY–Buffalo)

Entropy, a Measure of Disorder

Section 5.1 states that the molecules of a gas move randomly and that the higher the temperature, the faster they move. In general, random motion means disorder. When the temperature is lowered, the molecules slow down and more and more order is apparent. When all the molecules of a system become motionless and line up perfectly, we achieve the greatest possible order. (Of course, at this point the substance is a solid—it can no longer be a gas, or even a liquid.) The measure of such order is called **entropy.** When the order is perfect, the entropy of the system is zero. If the system has the slightest impurity, the perfect order is broken and the entropy increases. When molecules rotate or move from one place to another, the disorder increases and so does the entropy. Thus when a crystal melts, there is an increase in entropy; when a liquid vaporizes, the disorder, and hence the entropy, in-

creases. When we combine two pure substances and a mixture is formed, disorder increases and so does the entropy. When the temperature increases, the entropy of any substance or mixture of substances increases—not just that of gases but also of liquids and solids—because an increase in temperature means increased molecular motions. Conversely, when a system is cooled, its entropy decreases.

We learned about the absolute, or Kelvin, temperature scale in Section 1.4. Absolute zero or −273 °C, is the lowest possible temperature. At this temperature, all molecular motions cease, and perfect order reigns. No such perfect order can be reached in reality, although scientists have been able to produce temperatures within a few hundreths of a degree of absolute zero. Of course, any substance at a temperature this low must have extremely little entropy.

All these attractive forces provide the potential energy that holds matter together and keeps it from falling apart by counteracting another form of energy—kinetic energy—which tends to disorganize matter. In the absence of attractive forces, the kinetic energy that particles possess keeps them constantly moving, mostly in a random, disorganized way. This kinetic energy increases with increasing temperature. Therefore, *the higher the temperature, the greater the tendency of particles to fly randomly in every direction.*

The physical state of matter thus depends on a balance between the kinetic energy of particles and the attractive forces between them (Figure 5.1). At high temperatures, molecules possess a large amount of kinetic energy and move so fast that intermolecular attractive forces are too weak to hold them together. This is the **gaseous state.** At lower temperatures, the molecules slow down. When the temperature is sufficiently reduced, the

Gas

- Molecules far apart and disordered
- Negligible interactions between molecules

Liquid

- Intermediate situation

Solid

- Molecules close together and ordered
- Strong interactions between molecules

FIGURE 5.1   Three states of matter. A gas has no definite shape, and its volume is the volume of the container. A liquid has a definite volume but no definite shape. A solid has a definite shape and a definite volume.

gas condenses to form the **liquid state.** The molecules are still moving past each other, but they are traveling much more slowly than in the gaseous state. The intermolecular attractive forces now become important. When the temperature goes even lower, the molecules no longer have enough velocity to move past each other. This is the **solid state.** In this state each molecule has a certain number of nearest neighbors, and these neighbors do not change.

The attractive forces between molecules are the same in all three states. The difference is that, in the gaseous state (and to a lesser degree in the liquid), the kinetic energy of the molecules is great enough to overcome the attractive forces.

Most substances can exist in any of the three states. Typically a solid, when heated to a sufficiently high temperature, melts and becomes a liquid. The temperature at which this change takes place is called the **melting point.** Further heating causes the temperature to rise to the point at which the liquid boils and becomes a gas. This temperature is the **boiling point** of the liquid. Not all substances, however, can exist in all three states. For example, wood and paper cannot be melted. On heating, they either decompose or burn (depending on whether air is present), but they do not melt. Another example is sugar, which does not melt when heated but rather forms a dark substance called caramel.

> **Melting point** A temperature at which solid and liquid are in equilibrium

> **Boiling point** The temperature at which a liquid's vapor pressure is equal to the pressure of the atmosphere in contact with its surface

5.2 Gases

Of the three states of matter, the gaseous state is best understood. The most important characteristic of a gas is that it completely fills whatever container it is in. In doing so, the gas exerts a pressure on the walls of the container. This behavior, and all other properties of gases, can be explained by the **kinetic molecular theory,** which assumes that

> The kinetic molecular theory explains the behavior of ideal gases.

1. Gases consist of molecules that are constantly moving through space in straight lines, randomly, and with various speeds.

2. The average kinetic energy of the gas molecules is proportional to the absolute (Kelvin) temperature. This means that, as the temperature increases, the gas molecules move faster.

3. The gas molecules collide with each other, as billiard balls do, bouncing off each other and changing direction. Each time they collide, they may exchange kinetic energies (one moves faster than before, the other slower), but the total kinetic energy is conserved (assuming that the molecules do not react with each other).

4. The molecules of a gas are very tiny compared to the distances between the molecules. They are so tiny that the kinetic molecular theory assumes that the molecules have no volume at all. The **volume of a gas** is thus the **volume of the container** it is in.

5. Gas molecules have no attractions between them; therefore, they do not stick together after collision.

6. The molecules collide with the walls of the container, and these collisions constitute the pressure of the gas. The greater the number of collisions per unit time, the greater the pressure.

These six assumptions of the kinetic molecular theory give an idealized picture of what is going on. In real gases there are intermolecular attractions, and the molecules really do occupy some volume. Because of this, a gas described by these six assumptions is called an **ideal gas.** There cannot actually be an ideal gas; all gases are real. Under normal environmen-

> **Ideal gas** A gas that obeys the assumptions of the kinetic molecular theory

Mercury As Poison

Mercury is used in most manometers and barometers because it is a liquid over a wide temperature range and so can be used for pressure measurements between −30 and 300 °C. However, mercury is poisonous. Mercury spills should be collected and stored under water. Frequent and prolonged contact with mercury should be avoided (use gloves). Prolonged inhalation of mercury vapor can cause teeth to fall out; later, disorders of the nervous system can develop.

Beginning in the middle of the seventeenth century, industrial mercury poisoning became prevalent among people who made felt hats. Felt hats are made of the fine hair of rabbits and hares. The workers treated the hairs with mercury salts, which twisted them and made them limp, aiding in the felting process. This contaminated the air with mercury vapors, which the hat makers breathed in. As a consequence, they developed twitches, a symptom of nervous disorders.

■ Mercury has many industrial uses because of its wide temperature range as a liquid, but exposure should be minimized. *(Charles D. Winters)*

The term "mad hatter" (as in *Alice in Wonderland*) originates from this industrial disease.

tal conditions, however, real gases behave in much the same way that an ideal gas would, so we can safely use these assumptions.

Pressure The force per unit area exerted against a surface

5.3 Pressure

The barometer was invented in 1643 by Evangelista Torricelli, an Italian physicist and mathematician (1608–1647).

1 atm = 760 torr = 760 mm Hg.

On the earth we live under a blanket of air that presses down on us, on the ground, and on everything else around us. As we all know from weather reports, the pressure of the atmosphere varies from day to day. A device called a **barometer** (Figure 5.2) is used to measure atmospheric pressure. In this instrument, there is a vacuum inside the tube, and the atmosphere presses down on the liquid in the dish. The atmospheric pressure is measured as the difference between the height of the column and that of the open dish. This difference in height is used directly as a unit of pressure. **Pressure** is measured in **millimeters of mercury** (mm Hg). (The weather bureau usually gives values of pressure in inches of mercury.) In honor of Torricelli, we

Vacuum

Height (mm)

Mercury surface

Atmospheric pressure

FIGURE 5.2 The mercury barometer. A long glass tube completely filled with mercury is inverted into a pool of mercury in a dish. There is no air at the top of the mercury column in the tube because there is no way air could get in. Thus, there is no gas pressure exerted on the mercury column. However, on the open dish of mercury, the whole atmosphere exerts its pressure. The difference in the heights of the two mercury levels is the atmospheric pressure.

define another unit, the **torr,** as being equal to 1 mm Hg. At sea level the average pressure of the atmosphere is 760 mm Hg, and we use this number to define still another unit of pressure, the **atmosphere** (atm). There are several other units to measure pressure. The SI unit is the pascal (1 atm = 101 325 pascals). In this book we use only mm Hg and atm.

A barometer is fine for measuring the pressure of the atmosphere, but to measure the pressure of a gas in a container, we use a simpler instrument called a **manometer.** One form of manometer consists of a U-shaped tube containing mercury (Figure 5.3) in which one arm (A) has been evacuated and sealed. This arm has zero pressure. The other arm (B) is connected to the container in which the gas sample is enclosed. The pressure of the gas depresses the level of the mercury in the B arm of the U-shaped tube, and the difference between the two mercury levels gives the pressure directly in mm Hg. If more gas is added to the sample container in Figure 5.3, the mercury level in B will be pushed down and that in A will rise because the pressure in the bulb increased.

FIGURE 5.3 A manometer.

5.4 Gas Laws

By observing the behavior of gases, scientists have established a number of relationships.

1. **Boyle's law**   At constant temperature, the volume of any sample of gas decreases as its pressure increases, and vice versa. This means that, for a fixed mass of gas at a constant temperature, the volume is inversely proportional to the pressure. This is what happens in the cylinder of a car engine; as the pistons move up, the volume of gas decreases and the pressure increases (Figure 5.4).

2. **Charles's law**   At constant pressure, the volume of any sample of gas increases as its temperature increases, and vice versa. Putting it more precisely, the volume of a fixed mass of gas at a constant pressure is directly proportional to the absolute temperature. This is the basis of the hot-air balloon (Figure 5.5). The air in the balloon expands as it is heated from below; this makes the air inside the balloon less dense than the surrounding air, providing the lift. (Charles was one of the first balloonists.)

These gas laws hold not only for pure gases but also for any mixture of gases.

Robert Boyle (1627–1691), author of *The Sceptical Chymist* (1661), studied the expansion and compression of gases.

Although Jacques Charles (1746–1823) had earlier formulated a crude version of his law, it was first stated mathematically by Joseph Gay-Lussac (1778–1850) in 1802.

FIGURE 5.4 Boyle's law illustrated in an automobile cylinder. When the piston moves up, the volume occupied by the gas (a mixture of air and gasoline vapor) decreases. When the temperature is constant, the pressure must increase. When the gas is compressed, the spark plug ignites it.

BOX 5C

Blood Pressure Measurement

Blood pressure is the result of the blood's pushing against the walls of the blood vessels. When the heart ventricles contract, pushing blood out into the arteries, the blood pressure is high (systolic blood pressure); when the ventricles relax, the blood pressure is low (diastolic pressure).

Blood pressure varies with age. In young adults the normal range is 100 to 120 mm Hg systolic and 60 to 80 mm Hg diastolic. In older people the corresponding normal pressures are 115 to 135 and 75 to 85. Blood pressure is usually expressed as a fraction showing systolic over diastolic pressure, for instance, 120/80.

■ A sphygmomanometer.

FIGURE 5.5 Charles's law illustrated in a hot-air balloon. Because the balloon can stretch, the pressure inside it is constant. When the air in the balloon is heated, its volume increases, expanding the balloon. Its density, however, decreases because the volume but not the mass of the air has increased. The air inside the balloon, being less dense than the air outside, provides the lift so that the balloon can ascend.

Normal blood pressure is lower for women than for men. Blood pressures higher than normal may lead to heart attack, stroke, or kidney failure. Early diagnosis is important because blood pressure can be reduced by proper diet (low in sodium ions) and drugs. Nurses and physicians routinely measure blood pressure.

The sphygmomanometer consists of a bulb, a cuff, a manometer, and a stethoscope. The cuff is wrapped around a bare arm, and the bulb is repeatedly squeezed. This increases the amount of air in the cuff and thus the pressure. This increased pressure is transmitted to the upper arm. The pressure applied can be read on the manometer. At the beginning, the cuff is inflated to a pressure of 200 mm Hg. Then a release valve is opened slowly, and the pressure is allowed to drop 2 or 3 mm Hg/s. While the pressure is dropping, a stethoscope is placed over the brachial artery above the elbow, and the nurse or physician listens for a clear tapping sound. At the start no sound is heard because, in most cases, 200 mm Hg applied pressure is greater than the blood pressure and so no pulsating blood flows under the stethoscope. When the first faint tapping sound is heard, the pressure on the manometer is read. This is the systolic pressure. The applied pressure just matches the blood pressure when the ventricle contracts, allowing pulsating blood to flow into the lower arm. The clear tapping sound becomes louder as the applied cuff pressure drops and eventually fades away. When the last tapping sound is heard, the manometer is read again. This is the diastolic pressure. The applied cuff pressure now matches the blood pressure when the ventricle is relaxed, allowing continuous

■ A man monitors his own blood pressure using a digital sphygmomanometer.
(Saturn Stills/SPL/Photo Researchers, Inc.)

blood flow into the lower arm. Repeated blood pressure measurements can be taken, but the cuff should be deflated completely for 2 or 3 min before the next reading is attempted.

Digital blood pressure monitors are available for home or office use. In these instruments the stethoscope and manometer are combined in a sensory device that reads out the systolic and diastolic blood pressures together with the pulse rate. The cuff and the inflation bulb are used the same way as in traditional sphygmomanometers.

3. Gay-Lussac's law At constant volume, as the temperature increases, the pressure of any sample of gas increases, and vice versa. For a fixed mass of gas at constant volume, the pressure is directly proportional to the absolute temperature. This is what happens inside an autoclave (Figure 5.6). Steam generated inside the autoclave at 1 atm pressure has a temperature of 100°C. When this steam is then heated further at constant volume, the increased temperature of the steam increases its pressure. A valve controls the pressure within the autoclave; if the pressure exceeds the designated maximum, the valve opens, releasing the steam. At maximum pressure, the temperature may be as high as 120 to 150°C. All microorganisms in the equipment are destroyed at these high temperatures.

FIGURE 5.6 An autoclave used to sterilize hospital equipment. *(Custom Medical Stock Photo)*

Mathematical expressions of these three laws are shown in Table 5.1. The three gas laws can be combined and expressed by a mathematical equation called the **combined gas law:**

$$\frac{P_1 V_1}{T_1} = \frac{P_2 V_2}{T_2}$$

In this equation, P_1, V_1, and T_1 are the pressure, volume, and absolute (Kelvin) temperature of any sample of gas, respectively. When any change

TABLE 5.1 Mathematical Expressions of the Three Gas Laws for a Fixed Mass of Gas

Name	Expression	Constant
Boyle's law	$P_1V_1 = P_2V_2$	T
Charles's law	$\dfrac{V_1}{T_1} = \dfrac{V_2}{T_2}$	P
Gay-Lussac's law	$\dfrac{P_1}{T_1} = \dfrac{P_2}{T_2}$	V

is made to the sample, P_2, V_2, and T_2 are the new pressure, volume, and temperature.

EXAMPLE 5.1

A gas occupies 3.00 L at 2.00 atm pressure. First, estimate and then calculate its volume when we increase the pressure to 10.15 atm at the same temperature.

Solution

(a) Estimate: At constant temperature, the volume is inversely proportional to pressure. Because we increased the pressure approximately fivefold, the volume should decrease fivefold: 3/5 = 0.6 L.

(b) Calculate: First we identify the known quantities:

$$P_1 = 2.00 \text{ atm}; \quad V_1 = 3.00 \text{ L}; \quad P_2 = 10.15 \text{ atm}; \quad T_1 = T_2$$

$$\frac{P_1V_1}{T_1} = \frac{P_2V_2}{T_2}$$

We want the new volume, so we solve our equation for V_2. Because T_1 and T_2 are the same in this example and consequently cancel each other, we don't even need to know the temperature. Thus,

$$V_2 = \frac{P_1V_1\cancel{T_2}}{\cancel{T_1}P_2} = \frac{(2.00 \cancel{\text{ atm}})(3.00 \text{ L})}{10.15 \cancel{\text{ atm}}} = 0.591 \text{ L}$$

The new volume is 0.591 L, or 591 mL. The estimate of 0.6 L is reasonable.

Problem 5.1 ▬

A gas occupies 3.8 L at 0.70 atm pressure. If we expand the volume at constant temperature to 6.5 L, what is the new pressure? ■

EXAMPLE 5.2

In an autoclave, steam at 100°C is generated at 1.00 atm. After the autoclave is closed, the steam is heated at constant volume until the pressure gauge indicates 1.13 atm. What is the new temperature in the autoclave?

Solution

All temperatures in gas law calculations are in kelvins; we consequently, first convert the Celsius temperature to Kelvin temperature:

$$100°\text{C} = 100 + 273 = 373 \text{ K}$$

In this type of problem, we first identify the quantities that remain constant. Then we solve the equation for the property (P, V, or T) we are interested in and substitute the numbers.

BOX 5D

Breathing and Boyle's Law

We breathe about 12 times a minute, each time taking in and exhaling about 500 mL of air. When we breathe, we lower the diaphragm or raise the rib cage (Figure 5D). Either of these motions increases the volume of the chest cavity. In accord with Boyle's law, the pressure in the chest cavity is therefore decreased and becomes lower than the outside pressure. Air thus flows from the higher-pressure area outside the body into the lungs. The difference is only about 3 mm Hg, but this is enough to make the air go in. In breathing out we reverse the process: We raise the di-aphragm or lower the rib cage. The resulting decrease in volume increases the pressure inside the chest cavity, causing air to flow out.

In certain diseases the chest becomes paralyzed, and the patient cannot move the diaphragm or the rib cage. In such cases artificial respirators are used, pushing down on the chest and forcing the air inside the patient to go out the only way possible, through the lungs. The pressure in the respirator is then lowered below atmospheric pressure, causing the patient to breathe in, and the cycle is constantly repeated.

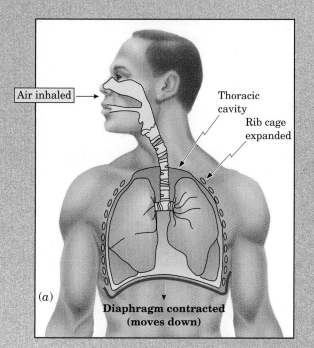

(a) **Diaphragm contracted (moves down)**

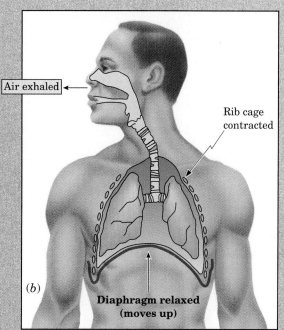

(b) **Diaphragm relaxed (moves up)**

■ **FIGURE 5.D** Schematic drawings of the chest cavity. In (a) the lungs fill with air. In (b) air empties from the lungs.

We then identify the known quantities:

$$P_1 = 1.00 \text{ atm}; \quad T_1 = 373 \text{ K}; \quad P_2 = 1.13 \text{ atm}; \quad V_1 = V_2$$

$$\frac{P_1 V_1}{T_1} = \frac{P_2 V_2}{T_2}$$

Since we want a new temperature, we solve this equation for T_2:

$$T_2 = \frac{P_2 V_2 T_1}{P_1 V_1}$$

$$T_2 = \frac{(1.13 \text{ atm})(373 \text{ K})}{(1.00 \text{ atm})} = 421 \text{ K}$$

The new temperature is 421 K, or $421 - 273 = 148°\text{C}$. In this example the two volumes cancel out.

BOX 5E

Hyperbaric Medicine

Ordinary air contains 20.9% oxygen. Under certain conditions, the cells of tissues are starved for oxygen (hypoxia), and there is a need for quick oxygen delivery. Increasing the percentage of oxygen in the air supplied to the patient is one way to remedy the situation, but sometimes even breathing pure (100%) oxygen may not be enough. For example, in carbon monoxide poisoning, the hemoglobin in the blood, which usually carries most of the O_2 from the lungs to the tissues, binds CO and cannot take up any O_2 in the lungs. Without any help, the tissues would soon be starved for oxygen and the patient would die. If oxygen is administered, especially under high pressure (2 or 3 atm), it dissolves in the plasma to such a degree that the tissues receive enough to recover. Many large hospitals now have hyperbaric units where pure oxygen under pressure is given to patients. The entire unit is under high pressure, and the patient and the accompanying medical staff must undergo gradual compression and, at the end of the treatment, decompression.

Besides treating carbon monoxide poisoning, hyperbaric units are also used along with other treatment. For example, in radiation necrosis, when tissues are damaged by x-rays during cancer treatment or accidents, hyperbaric medicine is beneficial in 80 to 100% of the cases. Chronic refractory osteomyelitis, a bone infection that does not heal, also responds well to hyperbaric medicine. Other conditions for which hyperbaric medicine is used are gas gangrene, smoke inhalation, cyanide poisoning, skin grafts, and thermal burns.

■ Hyperbaric oxygen chamber at Medical City Dallas Hospital. *(© Gregory G. Dimijian/Photo Researchers, Inc.)*

Breathing pure oxygen for prolonged periods, however, is toxic. For example, if O_2 is administered at 2 atm for more than 6 h, the lung tissues begin to be destroyed, and the central nervous system can be damaged. Therefore, recommended exposures of O_2 are 2 h at 2 atm and 90 min at 3 atm.

The benefits of hyperbaric medicine must be carefully weighed against the contraindications. Reports have been published showing that this treatment can cause nuclear cataract formation, thus necessitating postrecovery eye surgery.

Problem 5.2

A constant volume of oxygen gas is heated from 120°C to 212°C. The final pressure is 20.3 atm. What was the initial pressure? ■

EXAMPLE 5.3

A gas is under 1.0 atm pressure in a flexible container with a volume of 0.50 L at a temperature of 393 K. When the gas is heated to 500 K and the volume is expanded to 3.0 L, what will the new pressure be?

Solution

The known quantities are

$$P_1 = 1.0 \text{ atm}; \quad V_1 = 0.50 \text{ L}; \quad T_1 = 393 \text{ K}; \quad V_2 = 3.0 \text{ L}; \quad T_2 = 500 \text{ K}$$

$$\frac{P_1V_1}{T_1} = \frac{P_2V_2}{T_2}$$

Solving for P_2, we get

$$P_2 = \frac{P_1V_1T_2}{T_1V_2} = \frac{(1.0 \text{ atm})(0.50 \text{ L})(500 \text{ K})}{(3.0 \text{ L})(393 \text{ K})} = 0.21 \text{ atm}$$

The new pressure will be 0.21 atm, or

$$0.21 \; \cancel{atm} \times \frac{760 \text{ mm Hg}}{1 \; \cancel{atm}} = 160 \text{ mm Hg}$$

Problem 5.3 ▬▬▬▬▬
A gas is expanded from an initial volume of 20.5 L at 0.92 atm at room temperature (23.0°C) to a volume of 340.6 L. During the expansion the gas temperature cooled to 12.0°C. What is the new pressure? ■

5.5 Avogadro's Law and the Ideal Gas Law

The combined gas law is valid only for a given constant amount of gas. That is, it does not work if any more gas is added or any taken away. It is often necessary, however, to work with changing amounts of gas. The relationship between the amount of gas present and its volume is described by **Avogadro's law,** which states that **equal volumes of gases at the same temperature and pressure contain equal numbers of molecules.** For example, Figure 5.7 shows two tanks of equal volume, one containing hydrogen gas and the other, carbon dioxide gas. Avogadro's law tells us that, if the temperature and pressure are the same, the two tanks contain the same number of molecules, regardless of what the gases are. Avogadro's law is valid no matter what the gases are, as long as they don't become liquids. The actual temperature and pressure don't matter. It is convenient to select one temperature and one pressure as standard, however, and chemists have chosen 1 atm as the standard pressure and 0°C (273 K) as the standard temperature. These conditions are called **STP (standard temperature and pressure).**

Just as at any other temperature and pressure, all gases at STP contain the same number of molecules in any given volume. But how many is that? In Chapter 4 we saw that a mole (which contains 6.02×10^{23} particles) is a convenient quantity of matter. What volume of gas at STP contains 1 mole of molecules? This quantity, when measured, turns out to be 22.4 L.

> 1 mole of any gas at STP occupies a volume of 22.4 L.

Avogadro's law allows us to write a gas law that is valid not only for any pressure, volume, and temperature, but also for any quantity of gas. This law, called the **ideal gas law,** is

$$PV = nRT$$

where P = pressure of the gas in atmospheres
  V = volume of the gas in liters
  n = amount of the gas in moles
  T = temperature of the gas in kelvins
  R = a constant for all gases, for all conditions of $n, P, V,$ and T.

We can find the value of the constant, R, by using the fact that 1 mole of any gas at STP occupies 22.4 L:

$$PV = nRT$$

$$R = \frac{PV}{nT} = \frac{(1.00 \text{ atm})(22.4 \text{ L})}{(1.00 \text{ mole})(273 \text{ K})} = 0.0821 \; \frac{\text{L} \cdot \text{atm}}{\text{mole} \cdot \text{K}}$$

Thus R, called the **universal gas constant,** is equal to 0.0821 L·atm/ mole·K.

Avogadro's law holds only for gases, not liquids or solids.

H$_2$ CO$_2$

FIGURE 5.7 Avogadro's law: Two tanks of gas of equal volume at the same temperature and pressure contain the same number of molecules.

Real gases behave most like ideal gases at low pressures (1 atm or less) and high temperatures (300 K or higher).

The ideal gas law, $PV = nRT$, is a universal law. It holds for all gases, including all mixtures of gases, at any temperature, pressure, volume, or amount, as long as they are ideal gases. As we mentioned before, real gases, although not exactly ideal, in most cases behave as ideal gases, and so we can use this equation with little trouble. Using $PV = nRT$, we can calculate any one quantity—P, V, T, or n—if we know the other three. Furthermore, since the number of moles of any substance is its mass, m, in grams divided by its molecular weight (Section 4.3),

MW stands for molecular weight.

$$n = \frac{m}{MW}$$

we can replace n in the ideal gas equation to give

$$PV = \frac{mRT}{MW}$$

This equation allows us to calculate the molecular weight of a gas if we know the other quantities.

EXAMPLE 5.4

One (1.00) mole of CH_4 gas occupies 20.0 L at 1.00 atm pressure. What is the temperature of the gas in kelvins?

Solution
Solve the ideal gas law for T:

$$T = \frac{PV}{nR}$$

$$T = \frac{(1.00 \text{ atm})(20.0 \text{ L})}{(1.00 \text{ mole})(0.082\ 1 \text{ L} \cdot \text{atm/mole} \cdot \text{K})} = 244 \text{ K}$$

Note that the answer would be the same if the gas were CO_2, N_2, NH_3, or any other gas.

Problem 5.4 ▬▬▬
If 2.00 moles of NO gas occupy 10.0 L at 295 K, what is the pressure in atm? ∎

EXAMPLE 5.5

If there is 5.0 g of CO_2 gas in a 10-L vessel at 350 K, what is the pressure?

Solution
Here we are given a mass, and so we use $PV = mRT/MW$. The molecular weight of $CO_2 = 44$:

$$P = \frac{mRT}{V(MW)} = \frac{(5.0 \text{ g})(0.0821 \text{ L} \cdot \text{atm/mole} \cdot \text{K})(350 \text{ K})}{(10 \text{ L})(44 \text{ g/mole})}$$

$$= 0.33 \text{ atm}$$

Problem 5.5 ▬▬▬
A certain quantity of neon gas is under 1.05 atm pressure at 303 K in a 10.0-L vessel. How many moles of neon are present? ∎

EXAMPLE 5.6

If 1.0 L of an unknown gas has 1.15 atm pressure at 40.0°C and the mass of the gas is 3.3 g, what is the molecular weight?

Solution

Convert 40.0°C to absolute temperature: $273 + 40.0 = 313$ K.

$$MW = \frac{mRT}{PV} = \frac{(3.3 \text{ g})(0.0821 \text{ L} \cdot \text{atm/mole} \cdot \text{K})(313 \text{ K})}{(1.15 \text{ atm})(1.0 \text{ L})}$$

$$= 74 \frac{\text{g}}{\text{mole}}$$

Therefore, the molecular weight is 74 amu.

Problem 5.6 ■■■■■

An unknown amount of He gas occupies 30.5 L at 2.00 atm pressure and at 300 K. What is the weight of the gas? ■

5.6 Dalton's Law and Graham's Law

In a mixture of gases, each molecule acts independently of all the others, assuming that the gases do not react with each other. This is why the ideal gas law works for mixtures as well as for pure gases. In particular, the collisions with the walls of the container made by each molecule contribute independently to the total pressure. **Dalton's law of partial pressures** states that the total pressure of a mixture of gases is the sum of the individual partial pressures:

$$P_T = P_1 + P_2 + P_3 + \cdots$$

The **partial pressure** of a single gas in a mixture is the pressure the gas would exert if it were alone in the container. The equation $PV = nRT$ holds separately for each gas in the mixture, as well as for the total mixture.

EXAMPLE 5.7

To a tank already containing N_2 at 2.0 atm and O_2 at 1.0 atm, we add an unknown quantity of CO_2 until the total pressure is 4.6 atm. What is the partial pressure of the CO_2?

Solution

Dalton's law tells us that the addition of CO_2 does not affect the partial pressures of N_2 and O_2 that are already there. Thus, the partial pressures of N_2 and O_2 remain at 2.0 atm and 1.0 atm, respectively, even after the CO_2 is added. The sum of the partial pressures of N_2 and O_2 is

$$2.0 \text{ atm} + 1.0 \text{ atm} = 3.0 \text{ atm}$$

Since the rest of the total pressure of 4.6 atm must be caused by the CO_2, we can determine it by subtracting:

$$\underset{\substack{\text{Total} \\ \text{pressure}}}{4.6 \text{ atm}} - \underset{\substack{\text{Partial pressure} \\ \text{of } N_2 \text{ and } O_2}}{3.0 \text{ atm}} = \underset{\substack{\text{Partial pressure} \\ \text{of } CO_2}}{1.6 \text{ atm}}$$

Breathing and Dalton's Law

A gas naturally flows from an area of higher pressure to one of lower pressure. This fact is responsible for respiration, though we must look not at the total pressure of air but only at the partial pressure of O_2 and CO_2. Our cells use O_2 to oxidize carbon compounds to CO_2 and thus obtain energy. Respiration is the process by which the blood carries O_2 from the lungs to the tissues, collects the CO_2 manufactured by the cells, and brings it to the lungs, where it is breathed out.

The air we breathe contains about 21% O_2 at a partial pressure of about 159 mm Hg. The partial pressure of O_2 in the alveoli (tiny air sacs) in the lungs is about 100 mm Hg. Because the partial pressure is higher in the inhaled air, O_2 flows to the alveoli. At this point, venous blood (blood from the veins) flows past the alveoli. The partial pressure of O_2 in the venous blood is only 40 mm Hg, and so O_2 flows from the alveoli to the venous blood. This new supply of O_2 raises the partial pressure to about 100 mm Hg, and the blood that now has had its O_2 pressure increased becomes arterial blood and

flows to the body tissues. There, the partial pressure of O_2 is about 30 mm Hg or less (because tissue cells have used up most of their oxygen to provide energy). Oxygen now flows from the arterial blood to the body tissues, giving them a fresh supply. This naturally lowers the partial pressure of O_2 in the blood (to about 40 mm Hg) so that the blood now becomes venous blood again and returns to the lungs for a fresh supply of O_2. At each step of the cycle, O_2 flows from a region of higher partial pressure to one of lower partial pressure.

Meanwhile, the CO_2 goes the opposite way for the same reason. The partial pressure of CO_2 in oxidized body tissue is about 60 mm Hg. Thus, CO_2 flows from the tissues to the arterial blood, which is changed by the process to venous blood and now has a CO_2 pressure of 46 mm Hg. When the venous blood reaches the lungs, the CO_2 pressure in the alveoli is 40 mm Hg, and so the CO_2 flows to the alveoli. Because the partial pressure of CO_2 in air is only 0.3 mm Hg, CO_2 flows from the alveoli and is exhaled.

■ Blood circulation showing partial pressures of O_2 and CO_2 in the different parts of the system.

Problem 5.7 ▬▬▬▬▬
A vessel under 2.015 atm pressure contains nitrogen and water vapor. The partial pressure of the nitrogen is 1.908 atm. What is the partial pressure of the water vapor? ■

When a bottle of ammonia, NH_3, is opened in one corner of a room, within minutes its odor spreads to other parts of the room (Figure 5.8). The process whereby two gases mix is called **diffusion.** In this case, the ammonia molecules diffuse into the air. When a bottle of perfume is opened, its fragrance also fills the room, but the perfume molecules take longer to diffuse than the ammonia molecules. Thomas Graham (1805–1869) observed that molecules with small masses diffuse faster than heavy molecules and formulated what is now known as **Graham's law:**

$$\frac{(\text{Diffusion rate of A})^2}{(\text{Diffusion rate of B})^2} = \frac{MW \text{ of B}}{MW \text{ of A}}$$

Graham's law The rate of diffusion of gas molecules is inversely proportional to the square root of their molecular weight

5.7 Intermolecular Forces

As noted in Section 5.1, the strength of the intermolecular forces in any sample of matter determines whether that matter is in the gaseous, liquid, or solid state. In general, **the closer the molecules are to each other, the greater the effect of the intermolecular forces.** In gases the individual molecules are far apart, so much so that we can ignore intermolecular attraction and treat most gases as ideal. This is valid when the temperature is high (room temperature and above) and the pressure is low (1 atm or less), so that the molecules are far apart. When the temperature decreases or the pressure increases, or both, the distances between molecules decrease so that intermolecular forces can no longer be ignored. As a matter of fact, these forces cause condensation and solidification.

Therefore, before discussing the properties and structures of liquids and solids, we must look at the nature of intermolecular forces.

Intermolecular force A force that holds together molecules in liquids and solids

Dipole-Dipole Attractions

As mentioned in Section 3.6, many molecules are dipoles, and the positive end of one molecule attracts the negative end of another. These forces are called **dipole-dipole attractions** and can exist between two identical polar molecules or between two different polar molecules:

$$\delta^+ (\text{H—Cl}) \delta^- \qquad \delta^+ (\text{H—Cl}) \delta^-$$

$$\delta^+ (\text{H—Cl}) \delta^- \qquad \delta^+ (\text{H—Br}) \delta^-$$

Hydrogen Bonds

An especially strong attraction exists between molecules of water (Figure 5.9). The polarity of the O—H bond shifts the electrons toward the oxygen within the molecule so much that each hydrogen acquires a partial positive charge and the oxygen acquires a partial negative charge. Such a hydrogen is very strongly attracted toward an oxygen bearing partial negative charge in a neighboring water molecule. A **hydrogen bond** is a noncovalent interaction between a hydrogen atom and a very electronegative atom (Table 3.2).

FIGURE 5.8 Graham's law. When open containers of aqueous solutions of NH_3 (left) and HCl (right) are placed side by side, molecules of NH_3 and HCl gases escape and react with each other in the air to form NH_4Cl, which appears in this picture as white fumes. (*Charles D. Winters*)

FIGURE 5.9 Two water molecules joined by a hydrogen bond, indicated by the three red dots.

TABLE 5.2 Approximate Strength of Atomic and Molecular Interactions	
Interaction	**Energy (kcal/mole)**
Covalent bond	50–100
Hydrogen bond	5–10
Dipole-dipole attraction	0.1–1
London dispersion force	0.001–0.2

Hydrogen bonds are not restricted to water. They form whenever two molecules, the same or different, come together, provided that

1. **One molecule has a hydrogen atom covalently bonded to an atom of high electronegativity, namely to fluorine, oxygen, and nitrogen.**

2. **The other molecule has an oxygen, nitrogen, or fluorine atom.**

Therefore, we expect hydrogen bonds to form here:

but not here:

Although hydrogen bonds are among the strongest intermolecular forces, we must keep in mind that they are still much weaker than the covalent bonds within the molecules. Covalent bonds are about ten times stronger than hydrogen bonds (Table 5.2). As we shall see in later chapters, hydrogen bonds play an important role in biological molecules.

EXAMPLE 5.8

Can a hydrogen bond form between
(a) Two molecules of methanol, CH_3OH?
(b) Two molecules of formaldehyde, $H_2C=O$?
(c) One molecule of methanol and one of formaldehyde?

Solution
(a) Yes. One molecule has a hydrogen atom covalently bonded to an oxygen atom (actually, both have this configuration, but only one is necessary), and the other molecule has an oxygen atom:

Hydrogen bonding

(b) No. Neither molecule has a hydrogen covalently bonded to an oxygen, nitrogen, or fluorine.

(c) Yes. One molecule (methanol) has a hydrogen atom attached to an oxygen atom; the other (formaldehyde) has an oxygen atom.

Problem 5.8 ━━━
Will the molecules in each set form a hydrogen bond between them?
(a) A molecule of water and a molecule of methanol, CH_3OH.
(b) Two molecules of methane, CH_4. ∎

London Dispersion Forces

Not all gases contain polar molecules. Yet even nonpolar gases, such as helium and hydrogen, liquefy if the temperature is lowered sufficiently; this prompts us to ask, are there intermolecular forces between molecules that are not dipoles? The answer is yes, though such forces, called **London dispersion forces,** are very weak (Table 5.2). Even though London dispersion forces are very weak, they contribute significantly in large molecules to the attractive forces between the molecules because they act over large surface areas.

Even London dispersion forces have their origin in electrostatic interaction. We can explain the attraction between nonpolar atoms by assuming that, when the two atoms get close to each other, the motion of the electron clouds becomes synchronized. For a very small fraction of a second, the negative charge is located on one side of each electron cloud. If the motion of the electrons in two helium clouds becomes synchronized during that tiny fraction of a second, the charges will be at the same relative position, as shown in Figure 5.10. Thus, the electrons of atom B attract the nucleus of atom A during this short interval. A fraction of a second later, when the electrons have moved to the 3 o'clock position, the electrons of A interact with the nucleus of B. Thus, very short-lived and constantly changing interactions take place between the two atoms—the London dispersion forces.

London dispersion forces were first explained by Fritz London (1900–1954) in 1930.

> **London dispersion forces** The short-lived attraction between molecules due to the constant motion of electrons around the atomic nuclei

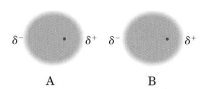

FIGURE 5.10 London dispersion forces. The relative positions of the electron clouds of two helium atoms result in London dispersion forces.

5.8 Liquids

We have seen that the behavior of gases under most circumstances can be described by the ideal gas law, which assumes that there are no intermolecular interactions. As pressure increases, however, the molecules of a gas

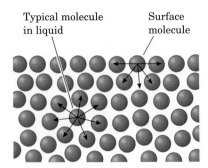

Typical molecule in liquid   Surface molecule

FIGURE 5.11 Surface tension due to intermolecular interactions. Molecules in the interior of a liquid have equal intermolecular attractions in every direction. Molecules in the liquid at the liquid–gas interface experience greater attractions toward the interior of the liquid than toward the gaseous state above them.

A water strider standing on water. *(© Hermann Eisenbeiss / Photo Researchers, Inc.)*

are squeezed into a smaller space, with the result that the distance the average molecule travels between collisions becomes smaller. Over short distances intermolecular forces become significantly stronger, and so molecules have a greater tendency to stick together.

If the distances between molecules decrease so that almost all the molecules touch (or almost touch) each other, the gas becomes a liquid. In contrast to gases, liquids do not fill all the available space. They have a definite volume, irrespective of the container. Because gases have a lot of empty space between molecules, it is easy to compress them into a smaller volume. In liquids there is very little empty space; consequently, liquids are hard to compress. A great increase in pressure is needed to cause a very small decrease in the volume of a liquid. Liquids are, for all practical purposes, incompressible. The density of liquids is much greater than that of gases because the same mass occupies a much smaller volume in liquids than in gases.

The position of the molecules in the liquid state is random, and there is some irregular empty space into which molecules can slide. The molecules thus constantly change their position with respect to neighboring molecules. This property causes liquids to be fluid; this is why they do not have a constant shape, only a constant volume.

In the interior of liquids, each molecule is surrounded by other molecules, and so the intermolecular attractions extend in every direction equally. Not so on the interface between a liquid and a gas (Figure 5.11). A molecule in the liquid state at the surface is attracted more to the molecules beneath it than to the molecules in the gaseous state above it. Therefore, there is a preferential pull toward the center of the liquid. **Surface tension** is the energy of the attraction between the molecules at the surface of a liquid. The pull toward the center crowds the molecules on the surface, thereby establishing a layer, like an elastic skin, that is tough to penetrate.

The surface tension of a liquid is directly related to the intermolecular interaction. Water has a high surface tension because of strong hydrogen bonding among water molecules. A steel needle can easily be made to float on the surface of water. However, the same needle pushed below the elastic skin into the interior of the liquid sinks to the bottom. Similarly, water bugs gliding on the surface of a pond also appear to be walking on an elastic skin of water.

5.9 Evaporation and Condensation. Boiling Point

An important property of liquids is that they evaporate. A few hours after a heavy rain, most of the puddles have dried up. The water has evaporated, gone into the air. The same thing occurs if we leave a glass of water out in the open. Why does this happen?

In any liquid, there is a distribution of velocities among the molecules. Some of the molecules have high kinetic energy and are moving rapidly. Others have low kinetic energy and are moving slowly. Whether fast or slow, a molecule in the center or at the bottom of the liquid cannot go very far before it hits another molecule and has its speed and direction changed by the collision. A molecule at the surface, however, is in a different situation (Figure 5.12). If it is moving slowly, it cannot escape from the liquid because of the attractions of neighboring molecules. If, however, it is moving rapidly (a high-energy molecule) and is moving upward, it can escape from the liquid and enter the gaseous space above it.

Molecules from air (O₂, N₂)   Molecules of vapor

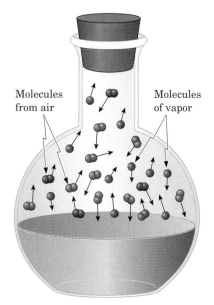

Molecules from air   Molecules of vapor

FIGURE 5.12 Evaporation. Some molecules at the surface of a liquid are moving fast enough to escape into the gaseous space.

FIGURE 5.13 Evaporation and condensation. In a closed container, molecules of liquid escape into the vapor phase, and vapor molecules are recaptured by the liquid.

If the container is open, this process continues until eventually all the molecules have left the liquid and it has dried up. However, if the liquid is in a closed container, as in Figure 5.13, this does not happen because the water **vapor** molecules in the gaseous state cannot diffuse away (as they do if the container is open). They remain in the air space above the liquid, where they behave, of course, like any other gas molecules, meaning that they move rapidly in straight lines until they collide with something. Some of these water vapor molecules move downward and thus can be recaptured by the liquid.

At this point we have an **equilibrium.** As long as the temperature does not change, the amount of water in the vapor no longer increases or decreases. After equilibrium is reached, the space above the liquid shown in Figure 5.13 contains air and water vapor, and it is possible to measure the partial pressure of the water vapor. We call this the **vapor pressure** of the liquid. Note that we measure the partial pressure of a gas but call it the vapor pressure of the liquid.

The vapor pressure of a liquid is a physical property of the liquid and is a function of temperature (Figure 5.14). A liquid has the same vapor pressure no matter what gas is above it (air or any other gas) and no matter what the pressure of that gas is. Vapor pressure also is independent of the volume of the liquid.

The molecules that evaporate from a liquid surface are the more energetic molecules. When they go into the gas phase, the molecules left behind are the less energetic ones. Therefore, the temperature of the liquid drops as a result of evaporation. This is what causes the cooling effect of perspiration. The liquid that comes out of the pores evaporates, so the skin temperature drops. The same cooling effect is felt when you come out of a swimming pool and the layer of water starts to evaporate from your skin.

Figure 5.14 shows that vapor pressure increases as temperature increases. Eventually the liquid boils. The **boiling point** of a liquid is the

Vapor Gas molecules in equilibrium with molecules in liquid or solid state

At equilibrium the rate of vaporization is equal to the rate of liquefaction.

Vapor pressure The partial pressure of gas molecules in equilibrium with a liquid

Boiling is a special form of evaporation during which bubbles form inside the liquid so that the surface of the liquid is greatly increased.

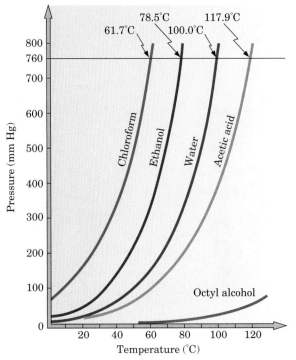

FIGURE 5.14 The change in vapor pressure with temperature for five liquids. The four temperatures at 760 mm Hg (1 atm) are the normal boiling points of each of the four liquids.

temperature at which its vapor pressure is equal to the pressure of the atmosphere in contact with its surface. The boiling point when the atmospheric pressure is 1 atm is called the **normal boiling point.** Thus, 100°C is the normal boiling point of water because that is the temperature at which water boils at 1 atm pressure.

The use of a pressure cooker is an example of boiling water at higher temperatures. In this type of pot, food is cooked at, say, 2 atm, where the boiling point of water is 121°C. Because the food has been raised to such a high temperature, it cooks much faster than it would in an open pot, in which boiling water cannot get hotter than 100°C. Conversely, at low pressures water boils at lower temperatures. For example, at the top of a mountain where the atmospheric pressure is less than 760 mm Hg, the boiling point of water might be 95°C.

As Figure 5.14 shows, different liquids have different normal boiling points. Two factors cause liquids to have high or low boiling points:

1. **Molecular weight and shape**   When intermolecular forces are similar, the boiling point increases with increasing molecular weight. For example, consider methane, CH_4, a gas whose normal boiling point is -164°C, and hexane, C_6H_{14}, a liquid with a normal boiling point of 69°C. Both compounds are subject to fairly similar London dispersion forces and essentially no other intermolecular forces. The difference in boiling points is caused by the difference in molecular weights (16 for CH_4 and 86 for C_6H_{14}). Aside from the effect of molecular weight, the shape of the molecule also is important because the shape helps to determine the surface area. For example, compare pentane, a chain-like molecule, with 2,2-dimethylpropane, a spherically shaped molecule (Figure 5.15). These compounds have the same molecular weight (they are both C_5H_{12}), but the boiling point of pentane is 36.1°C, which is significantly higher than 9.5°C, the boiling point of 2,2-di-

BOX 5G

The Densities of Ice and Water

The hydrogen-bonded superstructure of ice (Figure 5G.1) has empty spaces in the middle of each hexagon, so the molecules in ice are not as closely packed as those in liquid water. Because of this, ice has a lower density (0.917 g/cm³) than liquid water (1.00 g/cm³). More of the hydrogen bonds are broken in water, and the hexagonal superstructure is collapsed into a more densely packed organization (Figure 5G.2). This is why ice floats on top of water instead of sinking to the bottom. Such behavior is highly unusual; most substances are denser in the solid state than in the liquid state. The lower density of ice keeps fish and microorganisms alive in many rivers and lakes that would freeze solid each winter if the ice sank to the bottom. The presence of ice on top insulates the remaining water and keeps it from freezing.

The fact that ice has a lower density than water means that it has a greater molar volume (that is, a given mass of ice takes up more space than the same mass of water). This explains the damage done to biological tissues by freezing. When parts of the body (usually fingers, toes, noses, and ears) are subjected to extreme cold, they develop a condition called frostbite. The water in the cells freezes despite the blood's attempt to keep the temperature at 37 °C. The frozen water expands and in doing so ruptures the cell walls, causing damage. In some cases, frostbitten fingers or toes must be amputated.

Cold weather can also damage plants in a similar way. Many plants are killed when the air temperature drops below the freezing point of water for several hours. Trees can survive cold winters because they have a low water content inside their trunks and branches.

Slow freezing is often more damaging to plant and animal tissues than quick freezing. In slow freezing only a few crystals are formed, and these can grow to large sizes, rupturing the cells. In quick freezing, such as can be achieved in liquid nitrogen (at a temperature of −196 °C), many tiny crystals are formed. Since these do not grow much, tissue damage may be minimal.

■ **FIGURE 5G.1** The structure of ice.

■ **FIGURE 5G.2** The structure of liquid water.

Pentane
(bp 36.2°C)

2,2-Dimethylpropane
(bp 9.5°C)

FIGURE 5.15 Molecular models of pentane and 2,2-dimethylpropane.

BOX 5H

Ethyl Chloride

Ethyl chloride spray is used routinely by athletes to alleviate pain from minor muscular injuries and bruises. Ethyl chloride is sold in canisters, where it is under pressure and it is a liquid. When it is sprayed, this compound evaporates very quickly so that the skin is cooled rapidly. The net result is a freezing effect. This minimizes swelling from bruises and reduces the pain suffered in the injury by deadening nerves.

methylpropane. In the chain-like molecule, the London dispersion forces act over a larger molecular surface than in the spherical molecule of the same molecular weight. Thus, the pentane molecules need a higher temperature (more kinetic energy) than the 2,2-dimethylpropane molecules to move them from the liquid to the gaseous state.

2. **Intermolecular forces**   Water ($MW = 18$) and methane ($MW = 16$) have about the same molecular weight. The normal boiling point of water is $100°C$, and that of methane is $-164°C$. The difference in boiling points is due to the fact that the CH_4 molecules must overcome only London dispersion forces to enter the vapor phase (low boiling point), whereas water molecules, being hydrogen-bonded to each other, need more kinetic energy (and a higher boiling temperature) to escape into the vapor phase.

5.10 Solids

Even in the solid state, molecules and ions do not stop moving completely. They vibrate around fixed points.

> **Crystalline solid** A solid whose atoms, molecules, and ions are rigidly held in an ordered geometric arrangement

When liquids are cooled or compressed (or both), the intermolecular forces stop the random motion of the particles, and a solid is formed. In the solid state, each molecule or ion has stopped moving through space and has become part of an orderly three-dimensional array called a **crystal lattice.** Such a formation of solid from liquid is called crystal formation or **crystallization.**

All crystals have a regular shape that, in many cases, is obvious to the eye (Figure 5.16). This regular shape often reflects the arrangement of the particles within the crystal. For example, table salt is made of small crystalline cubes, and the Na^+ and Cl^- ions in the crystal also are arranged in a cubic system. Metals also consist of particles in a regular crystal lattice (generally not cubic), but here the particles are atoms rather than ions. Because particles in a solid are usually closer together, solids almost always have a higher density than the corresponding liquids.

As can be seen in Figure 5.16, crystals have characteristic shapes and symmetries. We are familiar with the cubic nature of table salt and the hexagonal ice crystals in snow flakes. It is less well known, however, that

(*a*) Garnet

(*b*) Sulfur

(*c*) Quartz

(*d*) Pyrite

FIGURE 5.16   Some crystals. (*Beverly March*)

(a) Graphite

(b) Diamond

(c) "Buckyball"

(d) Soot

(e) Nanotubes

FIGURE 5.17 Solid carbon structures:
(a) graphite, (b) diamond, (c) "buckyball,"
(d) soot, (e) nanotubes. (*Scanning electron
micrograph courtesy of Z. F. Ren et al.
SUNY-Buffalo, NY*)

the same compound can have more than one type of solid state.
The best known example of this is the element carbon (Figure 5.17). Carbon has a crystalline form, diamond, that occurs when solidification takes
place under very high pressure (thousands of atmospheres). Another crystalline form of carbon is the graphite in a pencil. The carbon atoms are
packed differently in high-density, hard diamonds than they are in low-
density, soft graphite.

When two or more forms of a solid
are possible, one is more stable than
the others.

In a third form of carbon, each molecule contains 60 carbon atoms
arranged in a structure having 12 pentagons and 20 hexagons as faces, resembling a soccer ball [Figure 5.17(c)]. Because the famous architect Buck-
minster Fuller (1895–1983) invented domes of a similar structure (he
called them *geodesic domes*), the C-60 substance was named buckminsterfullerene, or "buckyball" for short. This discovery has generated a whole
new area of carbon chemistry. Similar cage-like carbon structures containing 72, 80, and even larger numbers of carbon have been synthesized. As a
group, they are called fullerenes.

Richard E. Smalley (1943–),
Robert F. Curl, Jr. (1933–), and
Harold Kroto (1939–) were
awarded the 1996 Nobel Prize in
Chemistry for the discovery of these
compounds.

New variations on the fullerenes are **nanotubes.** The *nano-* part of the
name comes from the fact that the cross section of each tube is only nanome-

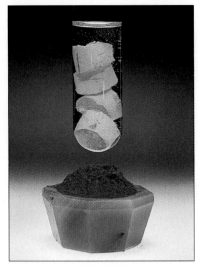

The element phosphorus is one of those that exist in more than one solid state. Like the forms of carbon, white and red phosphorus have different structures. *(Charles D. Winters)*

TABLE 5.3 Types of Solids

Type	Made up of	Characteristics	Examples
Ionic	ions in a crystal lattice	high melting points	NaCl, K_2SO_4
Molecular	molecules in a crystal lattice	low melting points	aspirin, ice
Metallic	metal atoms in a crystal lattice	shiny, low to high melting points; soft to hard	sodium, iron, gold
Polymeric	giant molecules; can be crystalline, semicrystalline, or amorphous	low melting points or cannot be melted; soft to hard	rubber, plastics, proteins
Network	a very large number of atoms connected by covalent bonds	very hard; very high melting points or cannot be melted	diamond, quartz
Amorphous	randomly arranged atoms or molecules	mostly soft, can be made to flow, but no melting point	soot, tar glass

ters (10^{-9} m) across. They come in a variety of forms. Single-walled carbon nanotubes can vary in diameters from 1 to 3 nm and are about 20 μm long. These compounds have generated great industrial interest because of their optical and electronic properties. They may play a role in miniaturization of instruments, giving rise to a new generation of nanoscale devices.

Soot is still a fourth form of solid carbon. This substance solidifies directly out of carbon vapor and is an **amorphous solid;** that is, its atoms have no set pattern and are arranged randomly [Figure 5.17(d)]. Another example of an amorphous solid is glass. In essence, glass is a frozen liquid.

As we have now seen, some crystalline solids consist of orderly arrays of ions (ionic solids; Figure 3.2), and others of molecules (molecular solids). Ions are held in the crystal lattice by ionic bonds (Section 3.3). Molecules are held only by intermolecular forces, which are much weaker than ionic bonds. Therefore, molecular solids generally have much lower melting points than ionic solids.

Besides these, there are other types of solids. Some are extremely large molecules; each molecule may have 10^{22} or 10^{23} atoms, all connected by covalent bonds. In such a case, the entire crystal is one big molecule. We call such molecules **network solids** or **network crystals.** A good example is diamond [Figure 5.17(b)]. When you hold a diamond in your hand, you are holding a single gigantic molecule. Like ionic crystals, network crystals have very high melting points—if they can be liquefied at all. In many cases they cannot be. We can be fairly confident, then, that any solid with a relatively low melting point (less than 300°C) is a molecular solid rather than an ionic crystal or a network crystal.

The types of solids are summarized in Table 5.3.

Amorphous solid A solid whose atoms, molecules, and ions are not in an orderly arrangement

There are other very large molecules, called polymers, that contain about 10^4 to 10^7 atoms and do not form network crystals (Section 12.9).

5.11 Phase Changes

Imagine the following experiment: We heat a piece of ice that is initially at, let's say, −20°C. At first we don't see any change in the physical state of the ice. The temperature of the ice is increasing, but its appearance does not change. At 0°C, the ice melts and liquid water appears. While we continue heating, more and more ice melts, but the temperature stays constant at 0°C until all the ice disappears. After all the ice has changed to liquid, the

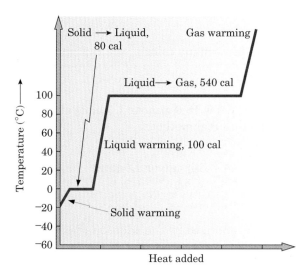

FIGURE 5.18 The heating curve of ice. The graph shows the effect of adding heat to 1 g of ice initially at $-20°C$.

temperature of the water again begins to increase as more heat is added. These changes in the state of matter are called **phase changes**. A **phase** is any part of a system that looks uniform (homogeneous) throughout. Ice (solid) is one phase, and water (liquid) is another.

The diagram describing this experiment (Figure 5.18) is called a heating curve. It shows that from -20 to $0°C$, added heat raises the temperature of the ice. After the ice reaches $0°C$, additional heat does not increase the temperature. Instead, the heat melts the ice by breaking some of the intermolecular hydrogen bonds that provide the rigid hexagonal structure (Figure 5G.1) of solid ice. The heat necessary to melt 1 g of any solid is called its **heat of fusion.** For ice, this is a rather large value: 80 cal/g. During melting, the two phases (solid and liquid) coexist side by side. Only after the ice has completely melted does the temperature of the water rise if we continue to add heat. For every gram of water, it takes 100 cal to raise the temperature from 0 to $100°C$, the normal boiling point. (Contrast this with the 80 cal that only melted the gram of ice but did not raise the temperature.)

The important aspect of this phase change is that it is reversible. If we start with liquid water, let's say at room temperature, and cool it by immersing the container in a Dry Ice bath, the reverse process is observed. The temperature drops until it reaches $0°C$, and then the ice starts to crystallize. During the phase change, also called phase transition, the temperature stays constant but heat is given off. The amount of heat given off is exactly the same as the amount of heat absorbed when the 1 g of ice melted.

Similar considerations apply to the remaining part of the heating curve in Figure 5.18. When the temperature of the liquid reaches the boiling point, $100°C$, another phase change takes place: The liquid is vaporized. While vaporization is taking place, the temperature stays constant at $100°C$. The heat supplied at the boiling point does not increase the temperature; it changes the liquid to a vapor. The amount of heat necessary to vaporize 1 g of any liquid is called its **heat of vaporization.** For water this value is 540 cal/g (as with the heat of fusion, this is a very high value). As long as any liquid water remains, the temperature does not go above $100°C$ (at 1 atm pressure).

Transition from the solid state directly into the vapor state without going through the liquid state is called **sublimation** [Figure 2.6(c)]. Solids usually sublime only at reduced pressures (less than 1 atm). At high altitudes, where the atmospheric pressure is low, snow sublimes. Solid CO_2 (Dry Ice) sublimes at $-78.5°C$ under 1 atm pressure. At 1 atm pressure, CO_2 can exist only as a solid or as a gas, never as a liquid.

> **Change of state** The change of a substance from one state (gas, liquid, or solid) to another

The criterion of uniformity is the way it appears to our eyes and not as it is on the molecular level.

Specific heat was discussed in Section 1.9.

EXAMPLE 5.9

To melt 1.0 g of ice requires 80 cal. How many calories are required to melt 1.0 mole of ice?

Solution
Since the molecular weight of ice (H_2O) is 18.0 g/mole, 1.0 mole of ice has a mass of 18.0 g. Using the factor-label method:

$$\frac{80 \text{ cal}}{1 \text{ g ice}} \times 18.0 \text{ g ice} = 1.4 \times 10^3 \text{ cal, or } 1.4 \text{ kcal}$$

Problem 5.9 ▬
What mass of water at 100°C can be vaporized by the addition of 45.0 kcal of heat? ■

EXAMPLE 5.10

What will the final temperature be if we add 1000 cal of heat to 10.0 g of ice at 0°C?

Solution
The first thing the added heat will do is melt the ice. This will use up 10.0 g × 80 cal/g = 800 cal. The remaining 200 cal will be used to heat the liquid water. The number of calories required for this is calculated by the equation:

$$\text{Heat} = SH \times m \times (T_2 - T_1)$$

$$200 \text{ cal} = (1 \text{ cal/g} \cdot °C) \times 10.0 \text{ g} \times (T_2 - T_1)$$

$$(T_2 - T_1) = \frac{200 \text{ cal}}{(1 \text{ cal/g} \cdot °C) \times 10.0 \text{ g}}$$

$$= 20°C$$

Thus, the temperature of the liquid water will have risen by 20°C from 0°C, and it is now 20°C.

Problem 5.10 ▬
The specific heat of iron is 0.11 cal/g · °C. The heat of fusion of iron is 63.7 cal/g. Iron melts at 1530°C. How much heat must be added to 1.0 g of iron at 25°C to completely melt it? ■

All the phase changes mentioned so far can be illustrated by a simple **phase diagram;** the case of water is shown in Figure 5.19. Temperature is plotted on the *x*-axis and pressure on the *y*-axis. Three areas with different colors are labeled solid, liquid, and gas. Within these areas water exists either as ice or liquid water or water vapor. The line (A-B) separating the solid phase from the liquid phase contains all the freezing (melting) points of water, for example, 0°C at 1 atm and 0.005°C at 400 mm Hg.

At the melting point both solid and liquid phases are present. They coexist. The line separating the liquid phase from the gas phase (A-C) contains all the boiling points of water, for example, 100°C at 1 atm and 84°C at 400 mm Hg. At the boiling points, the liquid phase coexists with the gas phase. Finally, the line separating the solid phase from the gas phase (A-D) contains all the sublimation points. At the sublimation points, solid and gas coexist.

There is a unique point on the phase diagram where all three phases coexist (A); it is called the **triple point.** For water this occurs at 0.01°C and at 4.58 mm Hg pressure.

A phase diagram illustrates how one may go from one phase to another. For example, we have water vapor at 95°C and at 660 mm Hg (E). We want to condense it to liquid water. We can cool the water to 70°C without changing the pressure (moving horizontally from E to F). Or we may increase the pressure to 760 mm Hg without changing the temperature (moving vertically from E to G). Or we can change both temperature and pressure (moving from E to H). Any of these processes will condense the water vapor to liquid water, albeit the liquids will be at different pressures and temperatures. The phase diagram allows us to visualize what will happen to the phase of a substance when we change the experimental conditions from one set of temperature and pressure to another set.

EXAMPLE 5.11

What will happen to ice at 0°C if the pressure drops from 1 atm to 0.03 atm?

Solution
According to Figure 5.19, when the pressure drops while the temperature stays constant, we travel vertically from 1 atm to 0.03 atm. During this process we cross the boundary separating the solid phase from the vapor phase. Thus, when the pressure drops to 0.03 atm, the ice sublimes and becomes vapor.

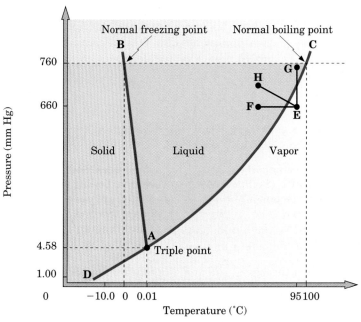

FIGURE 5.19 Phase diagram of water. Temperature and pressure scales are greatly reduced.

SUMMARY

Matter can exist in three different states: gaseous, liquid, and solid. Attractive forces between molecules tend to hold matter together, whereas the kinetic energy of the molecules tends to disorganize matter. At high temperatures (high kinetic energy), matter is in the gaseous state. The **kinetic molecular theory** explains the behavior of gases. Molecules in the gaseous state move randomly. Gases fill all the available space of their container. Gas molecules, in their random motion, collide with the walls of the container and thereby exert pressure.

The pressure of the atmosphere is measured with a barometer. The units of pressure are 1 mm Hg = 1 torr and 760 mm Hg = 1 atm.

Boyle's, Charles's, and **Gay-Lussac's laws** describe the behavior of gases under different conditions. These laws are combined and expressed as the combined gas law:

$$\frac{P_1 V_1}{T_1} = \frac{P_2 V_2}{T_2}$$

Avogadro's law states that, under constant temperature and pressure, equal volumes of different gases contain the same number of molecules. The **ideal gas law,** $PV = nRT$, incorporates Avogadro's law into the combined gas law. Different gases in a mixture exert their own pressures independently. The total pressure is the sum of the partial pressures of the gases **(Dalton's law).** The rate of diffusion of a gas is inversely related to its mass **(Graham's law).**

Intermolecular forces are responsible for the condensation of gases into the liquid and solid states. They are, in order of increasing strength, **London dispersion forces, dipole-dipole attractions,** and **hydrogen bonds. Surface tension** is the energy of intermolecular attractive forces at the surface of a liquid.

Liquids have **vapor pressures.** The most energetic molecules of liquids escape from the surface into the vapor phase and exert their own pressure. The vapor pressure of a liquid increases with increasing temperature. When the temperature of a liquid is raised so that the vapor pressure becomes equal to the atmospheric pressure, the liquid boils. The temperature at which this occurs is the **boiling point.** The boiling point of a liquid is determined by its molecular weight and shape and by intermolecular forces.

Solids crystallize in well-formed geometrical shapes that often reflect the patterns in which the atoms are arranged within the crystals. In amorphous solids atoms or molecules are arranged randomly. When a solid is heated, its temperature is raised until the **melting point** is reached. At the melting point a phase change occurs: The solid becomes a liquid. After all the solid has melted, further heating raises the temperature of the liquid. At the boiling point a second phase change takes place: The liquid becomes a gas. Further heating can only increase the temperature of the gas. The heat necessary to convert 1 g of any liquid to vapor is the **heat of vaporization** of the liquid. The heat necessary to melt 1 g of any solid to liquid is its **heat of fusion.**

A **phase diagram** allows the visualization of what happens to the phase of a substance when experimental conditions in pressure and temperature change. A phase diagram contains all the melting points, boiling points, and sublimation points where two phases coexist. It also contains a unique **triple point** where all three phases coexist.

KEY TERMS

Amorphous solid (Section 5.10)
Avogadro's law (Section 5.5)
Barometer (Section 5.3)
Boiling point (Sections 5.1, 5.9)
Boyle's law (Section 5.4)
Charles's law (Section 5.4)
Combined gas law (Section 5.4)
Crystal lattice (Section 5.10)
Dalton's law (Section 5.6)
Diffusion (Section 5.6)
Dipole-dipole attraction (Section 5.7)
Freezing point (Section 5.11)
Gay-Lussac's law (Section 5.4)
Graham's law (Section 5.6)

Heat of fusion (Section 5.11)
Heat of vaporization (Section 5.11)
Hydrogen bond (Section 5.7)
Ideal gas (Section 5.2)
Ideal gas law (Section 5.5)
Intermolecular forces
    (Sections 5.1, 5.7)
Kinetic molecular theory
    (Section 5.2)
London dispersion forces
    (Section 5.7)
Manometer (Section 5.3)
Melting point (Sections 5.1, 5.11)
Network solid (Section 5.10)

Nanotubes (Section 5.10)
Normal boiling point (Section 5.9)
Partial pressure (Section 5.6)
Phase (Section 5.11)
Phase diagram (Section 5.11)
Pressure (Section 5.3)
STP (Section 5.5)
Sublimation (Section 5.11)
Surface tension (Section 5.8)
Triple point (Section 5.11)
Universal gas constant, R
    (Section 5.5)
Vapor pressure (Section 5.9)
Volume (Section 5.2)

CONCEPTUAL PROBLEMS

5.A Draw two boxes. In box A balls are randomly dispersed throughout the box. In box B the same number of balls occupy only the lower half of the box. The transition from box A to box B represents:
(a) Heating the gas at constant pressure
(b) Reducing the pressure at constant temperature
(c) Cooling the gas and increasing its pressure

5.B Which forces are stronger, intramolecular covalent bonds or intermolecular hydrogen bonds?

5.C Refer to Figure 5.18. How many calories of heat are required to bring 1 g of solid ice at 0°C to a liquid state at 100°C?

5.D Using the phase diagram of water (Figure 5.19), describe the process by which you can sublime 1 g of ice at −10°C and at 1 atm pressure to water vapor at the same temperature.

PROBLEMS

Difficult problems are designated by an asterisk.

5.11 What will happen to water vapor if it is cooled from 100 to −30°C while the pressure stays at 1 atm?

Gases and Pressure

5.12 In terms of the kinetic molecular theory, what causes (a) the pressure and (b) the temperature of gases?

5.13 A sphygmomanometer (Box 5C) reads a pressure of 120 mm Hg. Draw a diagram indicating the level of mercury (Hg) in the evacuated reference arm and in the arm connected to the inflating cuff.

5.14 Express in atmospheres:
(a) 7835 mm Hg   (b) 25 mm Hg
(c) 22.7 in. Hg   (d) standard pressure

***5.15** Under certain weather conditions (just before rain), the air becomes less dense. How does this affect the barometric pressure reading?

Gas Laws

5.16 A gas in a bulb as in Figure 5.3 registers a pressure of 833 mm Hg in the manometer in which the reference arm of the U-shaped tube (A) is sealed and evacuated. What will the reading of the difference in the mercury level be if the reference arm of the U-shaped tube is open to atmospheric (760 mm Hg) pressure?

5.17 Methane gas is compressed from 20 to 2.5 L at a constant temperature. The final pressure is 12.2 atm. What was the original pressure?

5.18 A sample of 23.0 L of NH_3 gas at 10.0°C is heated at constant pressure until it fills a volume of 50.0 L. What is the new temperature in °C?

5.19 If a sample of 4.17 L of C_2H_6 gas at 725°C is cooled to 175°C at constant pressure, what is the new volume?

5.20 A sample of SO_2 gas has a volume of 5.2 L. It is heated at constant pressure from 30 to 90°C. What is the new volume?

5.21 A sample of B_2H_6 gas in a 35-mL container is at a pressure of 450 mm Hg and a temperature of 625°C. If the gas is allowed to cool at constant volume until the pressure is 375 mm Hg, what is the new temperature in °C?

***5.22** A certain quantity of helium gas is at a temperature of 27°C and a pressure of 1.00 atm. What will the new temperature be if its volume is doubled at the same time that its pressure is dropped to one half of its original value?

5.23 A 480-mL balloon is filled with helium at 730 mm Hg barometric pressure. The balloon is released and climbs to an altitude where the barometric pressure is 395 mm Hg. What will the volume of the balloon be if, during the ascent, the temperature drops from 20 to 3°C?

5.24 An ideal gas occupies 56.44 L at 2.00 atm and 310.0 K. If the gas is compressed to 23.52 L and the temperature is lowered to 281.3 K, what is the new pressure?

5.25 A sample of 30.0 mL of Kr gas is at 756 mm Hg at 25.0°C. What is the new volume if the pressure is decreased to 325 mm Hg and the temperature is decreased to −12.5°C?

5.26 An oxygen tank contains 25.0 L of gas at 20 atm pressure at 300 K. The gas is allowed to expand to fill a room of 800 L at 1 atm pressure. What must be the temperature of the gas that these conditions should be attained?

5.27 A 26.4-mL sample of C_2H_4 gas has a pressure of 2.50 atm at 2.5°C. If the volume is increased to 36.2 mL and the temperature raised to 10°C, what is the new pressure?

5.28 The inside pressure of an automobile tire is 1.80 atm at a temperature of 20°C. What will the pressure in the tire be if, after 10 miles of driving, the temperature of the tire increases to 47°C?

5.29 In an autoclave, a constant amount of steam is generated at a constant volume. Under 1.00 atm pressure the steam temperature is 100°C. What pressure setting should be used to obtain a 165°C steam temperature for the sterilization of surgical instruments?

Avogadro's Law and the Ideal Gas Law

5.30 A sample of a gas at 77°C and 1.33 atm occupies a volume of 50.3 L.
   (a) How many moles of the gas are present?
   (b) Does your answer depend on knowing what gas it is?

5.31 A gas weighing 8.00 g occupies 22.4 L at 2.00 atm at 273 K. What is the molecular weight of the gas?

5.32 What volume is occupied by 5.8 g of propane gas (C_3H_8) at 23°C and 1.15 atm pressure?

5.33 If 5.50 moles of BrCl gas occupies 20.2 L at 200.0°C, what is the pressure in mm Hg?

5.34 At a mountain top where the pressure is 650 mm Hg, 0.500 mole of F_2 gas occupies 8.51 L. What is its temperature in °C?

5.35 What volume, in milliliters, does 0.275 g of UF_6 gas occupy at 425°C and 365 torr?

5.36 A hyperbaric chamber has a volume of 200 L.
   (a) How many moles of oxygen are needed to fill the chamber at room temperature (300 K) to 3.00 atm pressure?
   (b) How many grams of oxygen are needed?

5.37 One gulp of air has a volume of 2 L at STP. If air contains 21.5% oxygen, how many molecules of oxygen are in one gulp?

*5.38 At STP, 1 mole of every ideal gas occupies 22.4 L. What are the densities, in g/L, of SO_2, CH_4, and F_2 at STP?

5.39 A sample of cyanogen gas weighing 6.802 g occupies 3.74 L at 25.0°C and 650.0 mm Hg. Calculate the molecular weight of cyanogen.

5.40 How many molecules of CO are there in 100 L of CO at STP?

5.41 The density of C_2H_2 gas in a 4-L container at 0°C and 2 atm pressure is 0.02 g/mL. What would be the density of the gas under identical temperature and pressure if the container is partitioned into two 2-L compartments?

Dalton's and Graham's Laws

5.42 Three gases in open vessels are all at one end of a room. One is a perfume with the molecular formula $C_{10}H_{18}O_2$, the second is NH_3, and the third is HCl. If you sit in the corner farthest from the samples, which odor reaches you first? Which last?

5.43 The partial pressures of a mixture of gases were as follows: oxygen, 210 mm Hg; nitrogen, 560 mm Hg; and carbon dioxide, 15 mm Hg. The total pressure of the gas mixture was 790 mm Hg. Was there another gas present in the mixture?

*5.44 The rate of diffusion of an unknown gas is three times lower than that of helium. If helium has a molecular weight of 4.0 and a diffusion rate of 60 cm/min, what is the molecular weight of the unknown gas?

Intermolecular Forces

5.45 Under which condition does water vapor behave most ideally:
   (a) 0.5 atm, 400 K   (b) 4 atm, 500 K
   (c) 0.01 atm, 500 K?

5.46 Which molecule has the strongest intermolecular interaction:
   (a) CO   (b) HF   (c) Cl_2?

5.47 Can water and dimethyl sulfoxide $(CH_3)_2S{=}O$ form hydrogen bonds between them?

5.48 What kind of intermolecular interactions are there in liquid (a) H—F, (b) CCl_4, and (c) CO? Which will have the highest surface tension?

Evaporation, Condensation, and Boiling Point

5.49 Can dipole-dipole interactions ever be weaker than London dispersion forces? Explain.

5.50 Which compound has a higher boiling point: butane, C_4H_{10}, or hexane, C_6H_{14}?

5.51 Two compounds, octane and 2,2,3,3-tetramethylbutane, have the same molecular weight. Yet octane has a boiling point of 126°C, while 2,2,3,3,-tetramethylbutane boils at 107°C. Explain.

Octane

2,2,3,3-Tetramethylbutane

5.52 Using Figure 5.14, estimate the vapor pressure of ethanol at (a) 30°C (b) 40°C (c) 60°C.

5.53 CH_4 and H_2O have about the same molecular weights. Which will have the higher vapor pressure at room temperature? Why?

5.54 The normal boiling point of a substance depends on both the mass of the molecule and the intermolecular interactions. Considering these, rank in order of increasing boiling point and explain your answer:
   (a) HCl, HBr, HI   (b) O_2, HCl, H_2O_2

*5.55 Kinetic energy increases with increasing temperature, whereas potential energy stays constant. On the basis of this simple principle, explain how a gas condenses to a liquid when the temperature is decreased.

Solids and Phase Changes

5.56 Define: (a) crystal (b) amorphous solid

5.57 How many 1-nm diameter nanotubes laid down side by side measure 1 cm?

5.58 When iodine vapor hits a cold surface, iodine crystals are formed. Name the phase change that is the reverse of this condensation.

5.59 Draw the heating curve for acetic acid if its melting point is 17°C and its normal boiling point 118°C.

5.60 Explain what will happen when you heat a quantity of ice from −10°C at 1 atm to 20°C and at the same time reduce the pressure to 0.1 atm?

Boxes

5.61 (Box 5A) Which has lower entropy, a gas at 100°C or one at 200°C? Explain.

5.62 (Box 5A) Which form of carbon presented in Figure 5.17 has the highest entropy?

5.63 (Box 5B) What is an early symptom of mercury poisoning?

5.64 (Box 5C) In a sphygmomanometer one listens to the first tapping sound as the constrictive pressure of the arm cuff is slowly released. What is the significance of this tapping sound?

5.65 (Box 5D) What happens when we lower the diaphragm in our chest cavity?

5.66 (Box 5E) In carbon monoxide poisoning, the hemoglobin is incapable of transporting oxygen to the tissues. How does the oxygen get delivered to the cells when the patient is put into a hyperbaric chamber?

5.67 (Box 5F) In circulating blood, where is the pressure of CO_2 the highest? The lowest?

5.68 (Box 5G) Why is the damage by severe frost bite irreversible?

5.69 (Box 5G) If you fill a bottle with water, cap it, and freeze to −10°C, the bottle will crack. Explain.

Additional Problems

5.70 In what phase(s) will water at 0.2 atm pressure and at 120°C be?

5.71 An ideal gas occupies 387 mL at 275 mm Hg and 75°C. If the pressure is changed to 1.36 atm and the temperature is increased to 105°C, what is the new volume?

5.72 Why is a form of carbon called "nanotube"?

5.73 Which compound has greater intermolecular interactions, CO or CO_2? Assume that O=C=O is linear.

5.74 On the basis of what you have learned about intermolecular forces, predict which liquid has the highest boiling point:
(a) pentane, C_5H_{12} (b) chloroform, $CHCl_3$
(c) water, H_2O

*5.75 A container is holding 10 L of N_2 under 35 in. Hg pressure. How many moles of N_2 do you have to add to your container to raise the pressure to 60 in. Hg? Assume a constant temperature of 27°C.

5.76 Explain why gases are transparent.

*5.77 The density of a gas is 0.003 00 g/cm^3 at 100°C and 1.00 atm. What is the molecular weight of the gas?

*5.78 Because natural gas (methane) has no odor, gas companies add a warning gas so that leaks can be detected. Two candidates for this are NH_3 and CH_3SH, because of their strong odors. Which could be used to warn the household faster of an impending disaster? How much faster?

5.79 The normal boiling point of hexane is 69°C, and that of pentane is 36°C. Predict which of these has a higher vapor pressure at 20°C.

5.80 Which gas molecules diffuse faster, HCl or C_3H_8?

*5.81 What is the density of CH_4 at 1 atm and at 35°C?

5.82 If 60.0 g of NH_3 occupies 35.1 L under 77.2 in. Hg pressure, what is the temperature of the gas, in °C?

5.83 Water is a liquid at room temperature and 1 atm pressure. Yet H_2S, a heavier molecule, is a gas under the same conditions. Explain.

5.84 Why does the temperature of a liquid drop as a result of evaporation?

Solutions and Colloids

6.1 Introduction

In Chapter 5, we discussed pure substances—systems having only one component. Such systems are the easiest to study, so it was convenient to begin with them. In our daily lives, however, we more frequently encounter systems having more than one component. Such systems—for example, air, smoke, sea water, milk, blood, rocks—are **mixtures** (Section 2.2).

If a mixture is uniform throughout at the molecular level, we call it a **homogeneous mixture,** or, more commonly, a **solution.** Air and sea water are solutions. They are clear and transparent.

In contrast, in most rocks, we can see distinct regions separated from each other by well-defined boundaries. Such rocks are **heterogeneous mixtures.** One other example is a mixture of sand and sugar. We can easily distinguish between the two components; the mixing is not at the molecular level [Figure 2.2(c)].

Thus, mixtures are classified on the basis of how they look to the unaided eye. There are systems, however, that fall between homogeneous and heterogeneous mixtures. Cigarette smoke, milk, and blood plasma may look homogeneous, but they do not have the transparency of air or sea water. These mixtures are classified as **colloidal dispersions.** We will deal with such systems in Section 6.7.

Making a homogeneous solution. A green solid, nickel nitrate, is stirred into water where it dissolves with stirring to form a homogeneous solution. (*Charles D. Winters*)

Above: Human red blood cells. (© *David Scharf/Peter Arnold, Inc.*)

TABLE 6.1 Types of Solutions

Solute		Solvent	Appearance of Solution	Example
Gas	in	Liquid	Liquid	Carbonated water
Liquid	in	Liquid	Liquid	Wine
Solid	in	Liquid	Liquid	Salt water (saline solution)
Gas	in	Gas	Gas	Air
Solid	in	Solid	Solid	14-Carat gold

Beer is a solution in which a liquid (alcohol), a solid (malt), and a gas (CO_2) are dissolved in the solvent water. *(Beverly March)*

Although mixtures can contain many components, we shall restrict our discussion mainly to two-component systems, with the understanding that everything we say can be extended to multicomponent systems.

6.2 Types of Solutions

When we think of a solution, we normally think of a liquid. Liquid solutions are the most common kind, but there are also solutions that are gases or solids. In fact, all mixtures of gases are solutions. Because gas molecules are far apart from each other and there is much empty space between them, two or more gases can mix with each other in any proportions. Because the mixing is at the molecular level, a true solution always forms; that is, there are no heterogeneous mixtures of gases.

With solids, we are at the other extreme. Anytime we mix solids, we get a heterogeneous mixture. Since even microscopic pieces of solid still contain many billions of particles (molecules, ions, or atoms), there is no way to obtain mixing at the molecular level. Homogeneous mixtures of solids do exist, but we make them by first melting the solids, then mixing the molten components, and allowing the mixture to solidify.

The five most common types of solutions are shown in Table 6.1. Examples of other types (gas in solid, liquid in gas, and so on) are also known but are much less important.

When a solution consists of a solid or a gas dissolved in a liquid, the liquid is called the **solvent,** and the solid or gas is called the **solute.** A solvent may have several solutes dissolved in it, even of different types. A common example is spring water, in which gases (carbon dioxide and oxygen) and solids (salts) are dissolved in the solvent, water.

When one liquid is dissolved in another, there is a question about which is the solvent and which the solute. In most cases, the one present in the greater amount is called the solvent, but there is no rigid rule about this.

Many alloys are solid solutions; one example is stainless steel, which is mostly iron but also contains carbon, chromium, and other elements.

We normally do not use the terms "solute" and "solvent" in talking about solutions of gases in gases or solids in solids.

6.3 Characteristics of Solutions

The following are some properties of solutions.

1. The distribution of particles in a solution is uniform. Every part of the solution has exactly the same composition and properties as every other part. That, in fact, is the definition of homogeneous. This means that we cannot usually tell a solution from a pure solvent simply by looking at it. A glass of pure water looks the same as a glass of water containing dissolved salt or sugar. In some cases we can tell by looking, for example, if the solution is colored and we know that the solvent is colorless.

2. The components of a solution do not separate on standing.   A solution of vinegar (acetic acid in water), for example, can remain on the shelf for many years without separating out.

3. A solution cannot be separated into its components by filtration. Both the solvent and the solute pass through the filter paper.

4. For any given solute and solvent, it is possible to make solutions of many different compositions.   For example, we can easily make a solution of 1 g of glucose in 100 g of water, or 2 g, or 6 g, or 8.7 g, or any other amount up to the solubility limit (Section 6.4).

5. Solutions are almost always transparent.   They may be colorless or colored, but you can usually see through them. Solid solutions are exemptions.

6. Solutions can be separated into pure components.   Common separation methods are distillation and chromatography, which you may learn about in the laboratory portion of this course. The separation of a solution into its components is a physical, not a chemical, change.

6.4 Solubility

Suppose we wish to make a solution of table salt (NaCl) in water. We take some water, add a few grams of salt, and stir. At first we see the particles of salt suspended in the water, but soon all the salt dissolves. Now let us add more salt and continue to stir. Again the salt all dissolves. Can we repeat this process indefinitely? The answer is no; there is a limit. The solubility of table salt at 25°C is 36.2 g per 100 g of water. If we add more salt than that, the excess solid does not dissolve but remains suspended as long as we keep stirring; it sinks to the bottom after we stop.

The **solubility** of a solid in a liquid is **the maximum amount of that solid that dissolves in a given amount of liquid at a given temperature.** Solubility is a physical constant, like melting point or boiling point. Each solid has a different solubility in every liquid. Some solids have a very low solubility in a particular solvent. We often call these solids insoluble. Others have a much higher solubility. We call these soluble, but even for soluble solids, there is always a solubility limit. The same is true for gases dissolved in liquids.

Some liquids are essentially insoluble in other liquids (gasoline in water), and others are soluble to a limit. For example, 100 g of water dissolves about 6 g of diethyl ether (another liquid). If we add more ether than that, we see two layers (Figure 6.1). Some liquids, however, are completely soluble in other liquids, no matter how much is present. An example is ethanol and water, which form a solution no matter what quantities of each are mixed. We say that the water and ethanol are *miscible* in all proportions.

When a solvent contains all the solute it can hold at a given temperature, we call the solution **saturated.** Any solution containing a lesser amount of solute is **unsaturated.** A **supersaturated** solution holds more solute in the solvent than it can normally hold at a given temperature. A supersaturated solution is not stable; when disturbed in any way, by stirring or shaking, the excess solute precipitates, and the solution becomes saturated.

If more solute is added to a saturated solution at constant temperature, it looks like none of the additional solid dissolves, since the solution already holds all the solute that it can. But there is actually an equilibrium here, similar to the one discussed in Section 5.9. Some particles of the additional solute dissolve, whereas an equal quantity of dissolved solute comes out of solution. So, even though the concentration of dissolved solute does

We use the word "clear" to mean transparent. A solution of copper sulfate in water is blue, but clear.

Tables of solubility values are found in handbooks.

We use the word "miscible" to refer to a liquid dissolving in a liquid.

Diethyl ether

Water

FIGURE 6.1   Diethyl ether and water form two layers. A separatory funnel permits the bottom layer to be drawn off.

not change, the solute particles themselves are constantly going in and out of solution.

Whether a particular solute dissolves in a particular solvent depends on several factors.

Nature of the Solvent and Solute

Here the rule is **"like dissolves like."** The more similar two compounds are, the more likely it is that one is soluble in the other. This is not an absolute rule, but it does apply in most cases.

When we say "like," we mostly mean alike in *polarity*. That is, polar compounds dissolve in polar compounds, and nonpolar compounds dissolve in nonpolar compounds. For example, the liquids benzene and carbon tetrachloride are nonpolar compounds. They dissolve in each other, and other nonpolar materials, such as gasoline and camphor, dissolve in them. On the other hand, ionic compounds (such as NaCl) and polar compounds (such as table sugar) are insoluble in all these solvents.

The most important polar solvent is water, and we have already seen that most ionic compounds are soluble in water, as are small covalent compounds that can form hydrogen bonds with water. It is worth noting that even polar molecules are usually insoluble in water if they cannot either react with the water or form hydrogen bonds with water molecules. Water as a solvent is discussed in Section 6.6.

Polar compounds dissolve in polar compounds because the positive end of one molecule attracts the negative end of the other.

Temperature

For most solids and liquids that dissolve in liquids, the rule is **solubility increases with increasing temperature.** Sometimes the increase is great. For example, the solubility of glycine, a building block of proteins, is 52.8 g in 100 g of water at 80°C but only 33.2 g at 30°C.

If, for instance, we prepare a saturated solution of glycine in 100 g of water at 80°C, it will hold 52.8 g of glycine. If we now allow it to cool to 30°C, where the solubility is 33.2 g, we might expect the excess glycine, 19.6 g, to come out of solution and form crystals. It often does, but on many occasions it does not. Even though the solution contains more glycine than the water can normally hold at 30°C, the excess may well stay in solution. This is an example of the supersaturated solution referred to earlier. The excess glycine stays in solution because the molecules need a surface on which to begin crystallizing, and no such surface is present.

Supersaturated solutions are not indefinitely stable, however. If we shake or stir the solution, we may find that all the excess solid precipitates at once. Another way to crystallize the excess solute is to add a grain of glycine crystal. This is called **seeding** (Figure 6.2). The seed crystal provides the surface onto which the solute molecules can converge.

For gases, **solubility in liquids almost always decreases with increasing temperature.** This explains why a bottle of beer or a carbonated soft drink foams up when opened at room temperature. When the beverage is cold, the CO_2 it contains is more soluble. At room temperature it is less soluble, and more of it comes out of solution. This additional gaseous CO_2 increases the pressure. When the pressure is released, the gas pushes out some of the liquid.

The effect of temperature on the solubility of gases in water can have important consequences for fish. Oxygen is only slightly soluble in water, but fish need that oxygen to live. When the temperature of the water is increased, perhaps by the output from a power plant, the solubility of the oxygen decreases and may become so low that fish die. This is called *thermal pollution.*

FIGURE 6.2 When a supersaturated solution is disturbed, the dissolved salt (here sodium acetate, $CH_3COO^-Na^+$) rapidly crystallizes. *(Charles D. Winters)*

Acid Rain

The water vapor evaporated by the sun from oceans, lakes, and rivers condenses and forms clouds of water vapor that eventually fall as rain. The raindrops contain small amounts of CO_2, O_2, and N_2. Table 6A shows that, of these gases, CO_2 is the most soluble in water. When CO_2 dissolves in water, it forms carbonic acid, H_2CO_3.

The acidity caused by the CO_2 is not harmful; however, a serious acid rain problem is caused by contaminants that result from industrial pollution. Burning coal or oil that contains sulfur generates sulfur dioxide, which has a high solubility in water. Sulfur dioxide in the air is oxidized to sulfur trioxide. The combination of water and sulfur dioxide results in sulfurous acid, and sulfur trioxide and water give sulfuric acid.

Smelting industries produce other soluble gases as well. The result is that in many parts of the world, especially downwind from heavily industrialized areas, acid rain pours down on forests and lakes. This damages vegetation and kills fish. Such is the situation in the northeastern United States, where in the Adirondacks and parts of New England, as well as in eastern Canada, acid rain has been observed with increasing frequency.

TABLE 6A	Solubility of Some Gases in Water
Gas	**Solubility (g/kg H_2O at 20°C and 1 atm)**
O_2	0.0434
N_2	0.0190
CO_2	1.688
H_2S	3.846
SO_2	112.80
NO_2	0.0617

■ Trees killed by acid rain in eastern Europe. *(Simon Fraser/ SPL/Photo Researchers, Inc.)*

Pressure

Pressure has little effect on the solubility of liquids or solids. For gases, however, **the higher the pressure, the greater the solubility of a gas in a liquid.** This is the basis of the hyperbaric medicine discussed in Box 5E. When the pressure is increased, more oxygen dissolves in the blood plasma and reaches the tissues at higher-than-normal pressures (2 to 3 atm).

Henry's law states that the solubility of a gas in a liquid is directly proportional to the pressure.

6.5 Concentration Units

The amount of a solute dissolved in a given quantity of solvent, called the **concentration** of the solution, can be expressed in a number of ways. Some concentration units are better suited than others for some purposes. Sometimes qualitative terms are good enough. We say a solution is *dilute* or *concentrated*. This tells us little about the concentration, but we

BOX 6B

The Bends

Deep-sea divers encounter high pressures. In order for them to breathe properly under such conditions, oxygen must be supplied under pressure. At one time, this was achieved with compressed air. As pressure increases, the solubility of the gases in the blood increases. This is especially true for nitrogen, which constitutes almost 80 percent of the air.

When divers come up and the pressure on their bodies decreases, the solubility of nitrogen in the blood decreases as well. As a consequence, the previously dissolved nitrogen in the blood and in the tissue starts to form small bubbles, especially in the veins. The formation of gas bubbles, called the bends, can prevent blood circulation. If allowed to develop uncontrolled, the resulting pulmonary embolism can be fatal.

If the ascent is gradual, the dissolved gases are removed by regular exhalation and diffusion through the skin. Divers use decompression chambers, where the high pressure is gradually reduced to normal pressure.

If decompression disease develops after a dive, the patients are put into hyperbaric chambers (see Box 5E) in which they breathe pure oxygen at 2.8 atm pressure. Over a period of 6 hours, the pressure is reduced to 1 atm. This is the standard form of treatment. Nitrogen has also a narcotic effect in dives with compressed air below 40 m. This "rapture of the deep" is similar to alcohol-induced intoxication.

Because of the problem caused by nitrogen, divers often use a helium-oxygen mixture instead of air. The solubility of helium in blood is less affected by pressure than is that of nitrogen. Sudden decompression and ensuing bends are important not only in deep-sea diving but also in high-altitude flight, especially orbital flight.

■ Ascending too rapidly will cause the dissolved nitrogen bubbles to be released and form bubbles in the blood.
(Brian Parker/Tom Stack & Associates)

know that a concentrated solution contains more solute than a dilute solution.

For most purposes, we need quantitative concentrations. For example, a nurse must know how much glucose to give to a patient. Many methods of expressing concentration exist, but in this chapter we deal only with the three most important: percent concentration, molarity, and parts per million (ppm).

Percent Concentration

There are three ways of representing **percent concentration.** The most common is weight of solute per volume of solution. If 10 g of table sugar is dissolved in enough water so that the total volume is 100 mL, the concentration is 10 percent w/v. Note that we need to know the total volume of the solution, *not* the volume of the solvent. For this purpose, we use volumetric flasks, which come in different sizes (Figures 6.3 and 6.4).

> **Percent concentration (% w/v).** The number of g of solute in 100 mL of solution

EXAMPLE 6.1

How would we make 500 mL of a 1.5 percent w/v solution of KCl in water?

Solution

We find out how many g of KCl there are in this solution by taking the percentage:

$$\frac{1.5 \text{ g KCl}}{100 \text{ mL}} \times 500 \text{ mL} = 7.5 \text{ g KCl}$$

We then put 7.5 g of KCl into a 500-mL volumetric flask, add some water, and swirl until the KCl is all dissolved. We then fill with water to the mark. We don't know exactly how much water we have added, but we do know that the total volume is 500 mL.

Problem 6.1 ▬▬▬

How would you make 250 mL of a 4.4 percent w/v KBr solution in water? Assume that a 250-mL volumetric flask is available. ■

EXAMPLE 6.2

If 6.0 g of NaCl is dissolved in enough water to make a 300-mL solution, first estimate and then calculate the w/v percentage of NaCl.

Solution

To estimate, we note that percentage is based on a 100-mL solution. We have 300 mL, which is three times larger than 100 mL. Therefore, in 100 mL, we will have only one third as much NaCl, which is 2 g. Thus, the solution is about 2 percent w/v.

To calculate, we divide the weight of the solute by the volume of the solution and multiply by 100:

$$(6.0 \text{ g}/300 \text{ mL}) \times 100 = 2.0\% \text{ w/v}$$

Problem 6.2 ▬▬▬

If 7.7 g of lithium iodide is dissolved in enough water to make 400.0 mL of solution, what is the w/v percentage if LiI? ■

A second way to represent percent concentration is weight of solute per weight of solution (w/w). Calculations are essentially the same as with w/v except that the weight of solution is used instead of the volume. A volumetric flask is not used for these solutions. (Why not?)

Finally, there is volume of solute per volume of solution (v/v). This unit is used only for solutions of liquids in liquids, most notably alcoholic beverages. For example, 40 percent v/v ethanol in water means that 40 mL of ethanol has been added to enough water to make 100 mL of solution.*

A 40 percent v/v solution of ethanol in water is 80 proof. Proof in the United States is twice the percent concentration (v/v) of ethanol in water.

---

* A 40 percent v/v ethanol solution contains not 60 mL of water for every 40 mL of alcohol, but somewhat less than that. *Volumes of liquids are not always additive.* When one liquid is dissolved in another, the new volume may be more or less than the sum of the volumes of the individual liquids because the molecules move either closer to each other (because of greater attraction) or farther apart.

Molarity

For many purposes, it is easiest to express concentration by the weight or volume percentage methods just discussed, but when we want to focus on the number of molecules present, we need another way. For example, a 5 percent solution of sugar in water does not contain the same number of solute molecules as a 5 percent solution of ethanol in water. That is why chemists often use molarity. The **molarity, *M*,** of a solution is defined as the number of moles of solute dissolved in 1 L of solution. The units of molarity are moles per liter.

$$\text{Molarity } (M) = \frac{\text{moles } (n)}{\text{liter } (V)}$$

Thus, in the same volume of solution, a 0.2-molar solution of sugar in water contains the same number of molecules of solute as a 0.2-molar solution of ethanol in water. In fact, this holds true for equal volumes of any solution, as long as the molarities are the same.

We can prepare a given volume of a given molarity in essentially the same way that we prepare a solution of given w/v concentration, except that we use moles instead of grams in our calculations. We can always find out how many moles of solute there are in any volume of a solution of known molarity by the formula

$$\text{Molarity} \times \text{Volume in liters} = \text{Number of moles}$$
$$M \times V = \text{moles}$$

The solution is then prepared as shown in Figure 6.3.

(*a*) (*b*) (*c*)

FIGURE 6.3 (*a*) A 0.0100-*M* solution of $KMnO_4$ is made by adding enough water to 0.395 g of $KMnO_4$ to make 0.250 L of solution. (*b*) To ensure the correct solution volume, the $KMnO_4$ is placed in a volumetric flask and dissolved in a small amount of water. (*c*) After dissolving is complete, sufficient water is added to fill the flask to the mark. The flask now contains 0.250 L of solution. (*Charles D. Winters*)

EXAMPLE 6.3

How do we prepare 2.0 L of a 0.15 *M* solution of NaOH?

Solution

First, we find out how many moles of NaOH there will be in this solution:

$$M \times V = \text{moles}$$

$$\frac{0.15 \text{ mole NaOH}}{1 \text{ L}} \times 2.0 \text{ L} = 0.30 \text{ mole NaOH}$$

We need 0.30 mole of NaOH to make up this solution. How many g is this? To convert moles to g, we multiply by the appropriate conversion factor (Section 4.3). The formula weight of NaOH = 40.0, so

$$0.30 \text{ mole NaOH} \times \frac{40.0 \text{ g NaOH}}{1 \text{ mole NaOH}} = 12 \text{ g NaOH}$$

We therefore weigh 12 g of NaOH, put it into a 2-L volumetric flask, add some water, swirl until the solid has dissolved, and then fill the flask with water to the 2-L mark.

Problem 6.3 ▬▬

How would you prepare 2.0 L of a 1.06 *M* aqueous solution of KCl? ■

EXAMPLE 6.4

We dissolve 18.0 g Li_2O (molecular weight = 30.0) in sufficient water to make it 500 mL. Estimate and then calculate the molarity of the solution.

Solution

We are looking for molarity that is moles/L. To estimate, we note that 18.0 g is about half of the molecular weight, 30. Therefore, we have about 0.5 mole of Li_2O in 500 mL. But molarity is based on 1 L, which is twice as much as 500 mL. Therefore, there will be 1.0 mole of Li_2O in 1 L; that is, the solution will be 1.0 M. To calculate, we use the conversion factor: 1 mole of Li_2O/30.0 g LiO_2:

$$18.0 \text{ g } Li_2O \times \frac{1 \text{ mole } Li_2O}{30 \text{ g } Li_2O} = 0.60 \text{ mole } Li_2O$$

To find liters of solution, we divide mL by 1000:

$$500 \text{ mL} \times \frac{1 \text{ L}}{1000 \text{ mL}} = 0.500 \text{ L}$$

To find the moles of Li_2O per L of solution, we divide:

$$(0.60 \text{ mole } Li_2O/0.500 \text{ L}) = 1.2 \text{ } M$$

The estimate of 1.0 *M* is reasonably close to the calculated value of 1.2 *M*.

Problem 6.4 ■
If 0.440 g of KSCN is dissolved in enough water to make 340 mL of solution, what is the molarity of the solution? ■

EXAMPLE 6.5

The concentration of sodium chloride in blood serum is approximately 0.14 M. What volume of blood serum contains 2.0 g of NaCl?

Solution
Since we know the concentration in moles per liter, we must first find out how many moles there are in 2.0 g of NaCl (formula weight = 58.5) by multiplying by the proper conversion factor:

$$2.0 \text{ g NaCl} \times \frac{1 \text{ mole NaCl}}{58.5 \text{ g NaCl}} = 0.034 \text{ mole NaCl}$$

We next find liters:

$$0.034 \text{ mole NaCl} \times \frac{1 \text{ L}}{0.14 \text{ mole NaCl}} = 0.24 \text{ L} = 240 \text{ mL}$$

Blood serum is the liquid part of the blood that remains after the cellular particulates and the fibrinogen have been removed.

Blood (left to right): serum, coagulated, and whole. (© *Stan Elems / Visuals Unlimited*)

Problem 6.5 ■
If a 0.300 M glucose ($C_6H_{12}O_6$) solution is available for intravenous infusion, how many mL are needed to deliver 10.0 g of glucose? ■

EXAMPLE 6.6

How many g of HCl are there in 225 mL of concentrated HCl, whose concentration is 6.00 M?

Solution
First, we find out how many moles of HCl there are:

$$M \times V = \text{moles}$$

$$\frac{6.00 \text{ moles}}{1 \text{ L}} \times 0.225 \text{ L} = 1.35 \text{ moles}$$

We get g by multiplying by the conversion factor 1 mole = 36.5 g:

$$1.35 \text{ moles HCl} \times \frac{36.5 \text{ g HCl}}{1 \text{ mole HCl}} = 49.3 \text{ g HCl}$$

Problem 6.6 ■
How many g of H_2SO_4 are there in 50.0 mL of a 3.0 M H_2SO_4 solution? ■

Dilution

We frequently prepare solutions by diluting concentrated solutions rather than by weighing out pure solute (Figure 6.4). Since we are adding only solvent, the number of moles of solute remains unchanged. Before we dilute, the equation that applies is

$$M_1V_1 = \text{moles}$$

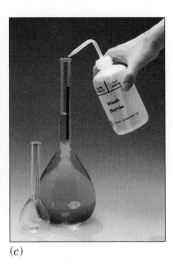

(a)   (b)   (c)

FIGURE 6.4 Making a solution by dilution. (a) A 100-mL volumetric flask is filled to the mark with 0.100 M $K_2Cr_2O_7$. (b) This is transferred to a 1.00-L volumetric flask. (c) The 1.00-L volumetric flask is filled to the mark with distilled water. The concentration of the now diluted $K_2Cr_2O_7$ is 0.0100 M. *(Charles D. Winters)*

After we dilute, the volume and molarity have both changed and we have

$$M_2V_2 = \text{moles}$$

but the number of moles of solute is the same before and after, so we can say that

$$M_1V_1 = M_2V_2$$

This is a handy equation that we can use for dilution problems.

EXAMPLE 6.7

How do we prepare 200 mL of a 3.5 M solution of acetic acid if we have a bottle of concentrated acetic acid (6.0 M)?

Solution
We use the equation

$$M_1V_1 = M_2V_2$$

$$\frac{6.0 \text{ moles}}{1 \text{ L}} \times V_1 = \frac{3.5 \text{ moles}}{1 \text{ L}} \times 0.200 \text{ L}$$

$$V_1 = \frac{(3.5 \text{ moles/L})(0.200 \text{ L})}{6.0 \text{ moles/L}} = 0.12 \text{ L}$$

We put 0.12 L, or 120 mL, of the concentrated acetic acid in a 200-mL volumetric flask, add some water and mix, and then fill to the mark with water.

Problem 6.7 ▬▬▬
A concentrated HCl solution is 12.0 M. You have to make up 300 mL of 0.600 M solution. How would you prepare it? ■

A similar equation can be used for dilution problems involving percent concentrations:

$$\%_1 V_1 = \%_2 V_2$$

EXAMPLE 6.8

How do we prepare 500 mL of a 0.5 percent w/v solution of NaOH if we have a solution of 50 percent w/v NaOH on hand?

Solution

$$(50\%)V_1 = (0.5\%)(500 \text{ mL})$$

$$V_1 = \frac{(0.5\%)(500 \text{ mL})}{50\%} = 5 \text{ mL}$$

We add 5 mL of the concentrated solution to a 500-mL volumetric flask, add some water and mix, and then fill to the mark with water.

Problem 6.8

A concentrated solution of 15 percent w/v KOH solution is available. How would you prepare 20.0 mL of a 0.10 percent w/v KOH solution? ∎

Parts Per Million

Sometimes we need to deal with very dilute solutions, for example, 0.0001 percent. In such cases, it is more convenient to use **parts per million (ppm)**. For example, if drinking water is polluted with lead ions to the extent of 1 ppm, it means that there is 1 mg of lead ions in 1 kg (1 L) of the water. Some solutions are so dilute that we use **parts per billion (ppb).** Modern methods of analysis allow us to detect concentrations this small. Some substances are harmful even at concentrations measured in ppb. One such substance is dioxin, an impurity in the 2,4,5-T herbicide sprayed by the United States as a defoliant in Vietnam.

ppm = [(g solute)/(g solution)] × 10^6

ppb = [(g solute)/(g solution)] × 10^9

6.6 Water As a Solvent

Water covers about 75 percent of the Earth's surface in the form of oceans, ice caps, glaciers, lakes, and rivers. Water vapor is always present in the atmosphere. Life evolved in water, and without it life as we know it could not exist. The human body is about 60 percent water. This water is located both inside the cells of the body (intracellular) and outside (extracellular). Most of the important chemical reactions in living tissue occur in aqueous solution; water serves as a solvent to transport reactants and products from one place in the body to another. Water is also itself a reactant or product in many biochemical reactions. The properties that make water such a good solvent are its polarity and its hydrogen-bonding capacity.

How Water Dissolves Ionic Compounds

We learned in Section 3.3 that ionic compounds in the solid state are composed of a regular array of ions. The crystal is held together by ionic bonds, which are attractions between positive and negative ions. Water, of course, is a polar molecule. When a solid ionic compound is added to water, the ions at the surface of the crystal become surrounded by water molecules.

FIGURE 6.5   Water molecules remove anions and cations from the surface of an ionic solid.

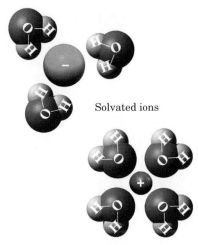

Solvated ions

FIGURE 6.6   Anions and cations solvated by water.

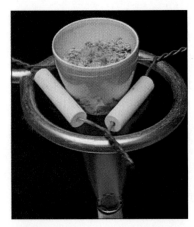

FIGURE 6.7   When blue hydrated copper sulfate is heated, it changes to a white anhydrous salt. *(Charles D. Winters)*

The dot in the formula indicates that the H_2O is present in the crystal, but it is not covalently bonded to the ions of Ca^{2+} or SO_4^{2-}.

Hygroscopic substances Compounds able to absorb water vapor from the air

If we want hygroscopic compounds to remain anhydrous, we must place them in sealed containers that contain no water vapor.

The negative ions (anions) attract the positive ends of water molecules, and the positive ions (cations) attract the negative ends of water molecules (Figure 6.5). Each ion attracts two to four water molecules, and the combined force of attraction is usually enough for the ion to be completely dislodged from its position in the crystal. Once the ion is removed from the crystal, it is completely surrounded by water molecules (Figure 6.6). We call these ions **hydrated.** A more general term, covering all solvents, is **solvated.** The solvation layer—that is, the surrounding shell of solvent molecules—acts as a cushion. It prevents a solvated anion from colliding directly with a solvated cation and thus keeps the solvated ions in solution.

Not all ionic solids are soluble in water. Some rules for predicting solubilities were given in Section 4.7.

Solid Hydrates

The attraction between ions and water molecules is so strong in some cases that water molecules are an integral part of the crystal structure of many solids. Water molecules in a crystal are called **water of hydration.** The substances that contain water in their crystals are themselves called **hydrates.** For example, gypsum and plaster of Paris are hydrates of calcium sulfate: Gypsum is $CaSO_4 \cdot 2H_2O$, calcium sulfate dihydrate, and plaster of Paris is $(CaSO_4)_2 \cdot H_2O$, calcium sulfate monohydrate.

Some of these crystals hold their water tenaciously. To remove it, the crystals must be heated for some time at a high temperature. The crystal without its water is called **anhydrous.** In many cases, the anhydrous crystals are so strongly attracted to water that they take it from the water vapor in the air. That is, some anhydrous crystals become hydrated on standing in air. Crystals that do this are called **hygroscopic.**

Hydrated crystals often look different from the anhydrous forms. For example, copper sulfate pentahydrate, $CuSO_4 \cdot 5H_2O$, is blue, but the anhydrous form is white (Figure 6.7).

Hydrates and Air Pollution: The Decay of Buildings and Monuments

Many buildings and monuments in urban areas throughout the world are decaying, ruined by air pollution. The main culprit in this process is gypsum. The most common stones used for buildings and stone monuments are limestone and marble, both of which are largely calcium carbonate. In the absence of polluted air, these stones can last for thousands of years, and many statues and buildings from ancient times (Babylonian, Egyptian, Greek, and others) survived until recently with little change, and indeed are still intact in rural areas. But in urban areas, the air is polluted with SO_2 and SO_3, which mostly come from the combustion of coal and petroleum products containing small amounts of sulfur compounds as impurities (Box 6A). They react with the calcium carbonate at the surface of the stones to form calcium sulfate. When calcium sulfate interacts with water in the rain, it forms the dihydrate gypsum. The problem is that gypsum has a larger volume than the original marble or limestone, causing the surface of the stone to expand, which in turn results in flaking. This is how statues such as those in the Parthenon (in Athens, Greece) become noseless and later faceless.

■ Acid rain damage to stonework on the walls of York Minster. *(Martin Bond/Science Photo Library/ Photo Researchers, Inc.)*

The difference between hydrated and anhydrous crystals can sometimes have an effect in the body. For example, the compound sodium urate in the anhydrous form exists as spherical crystals, but in the monohydrate the crystals are needle-shaped (Figure 6.8). The deposition of sodium urate monohydrate in the joints (mostly in the big toe) causes gout.

(a)   (b)

FIGURE 6.8 (a) The needle-shaped sodium urate monohydrate crystals that cause gout. *(Courtesy of P. A. Dieppe, P. A. Bacon, A. N. Bamji, and I. Watt and the Gower Medical Publishing Co., Ltd., London)* (b) The pain of gout as depicted by a cartoonist. *(Courtesy of National Library of Medicine)*

Anode
(positive electrode)

Cathode
(negative electrode)

Battery

FIGURE 6.9 Conductance by an electrolyte.

Electrolytes

Ions in water migrate from one place to another, maintaining their charge. Because of this, solutions of ions conduct electricity. They can do this because the ions in the solution migrate independently of one another. As shown in Figure 6.9, cations migrate to the negative electrode, called the **cathode,** and anions migrate to the positive electrode, called the **anode.** The movement of ions constitutes an electric current. Thus, the migration of ions completes the circuit initiated by the battery, and the electric bulb lights up.

Substances that conduct an electric current when dissolved in water or when in the molten state are called **electrolytes.** Sodium chloride is an electrolyte. Hydrated Na^+ ions carry positive charges, and hydrated Cl^- ions carry negative charges; the bulb in Figure 6.9 lights brightly. Substances that do not conduct electricity are called **nonelectrolytes.** Distilled water is a nonelectrolyte. The light bulb shown in Figure 6.9 does not light up if only distilled water is placed in the beaker. However, when tap water is placed in the beaker, the bulb lights dimly. Tap water contains enough ions to carry electricity, but the concentration of these ions is so low that only a small amount of electricity is conducted.

This experiment shows that electric conductance depends on the concentration of ions. The higher the ion concentration, the greater the electric conductance. However, there are differences in electrolytes. If we take a $0.1\,M$ aqueous solution of NaCl and compare it with a $0.1\,M$ HF solution, we find that the NaCl solution lights a bulb brightly, but the HF solution lights it only dimly. We might have expected the two solutions to behave similarly. They have the same concentration, $0.1\,M$, and each "molecule" provides two ions, a cation and an anion (Na^+ and Cl^-; H^+ and F^-). The reason they behave differently is that, whereas Na^+Cl^- dissociates completely to two ions (each hydrated and each moving independently), in the case of HF only a few molecules are dissociated into ions. Most of the HF molecules do not dissociate, and undissociated molecules do not conduct electricity. Compounds that dissociate completely are called **strong electrolytes,** and those that dissociate into ions only partially are called **weak electrolytes.**

Electrolytes are important components of the body because they help to maintain the acid-base balance and the water balance. The most important cations in body tissues are Na^+, K^+, Ca^{2+}, and Mg^{2+}. The most abundant anions in the body are bicarbonate, HCO_3^-, chloride, Cl^-, monohydrogen phosphate, HPO_4^{2-}, and dihydrogen phosphate, $H_2PO_4^-$.

How Water Dissolves Covalent Compounds

Water is a good solvent not only for ionic compounds but for many covalent compounds, too. In a few cases, the covalent compounds dissolve because they react with the water. An important example is the covalent compound HCl. HCl is a gas (with a penetrating, choking odor) that attacks the mucous membranes of eyes, nose, and throat. When dissolved in water, the HCl molecules react with water to give ions:

$$HCl(g) + H_2O(l) \longrightarrow Cl^-(aq) + H_3O^+(aq)$$
$$\text{Hydronium ion}$$

Another example is the gas sulfur trioxide, which reacts as follows:

$$SO_3(g) + 2H_2O(l) \longrightarrow HSO_4^-(aq) + H_3O^+(aq)$$

Because HCl and SO_3 are completely converted to ions in dilute aqueous solution, these solutions are ionic solutions and behave just as other electrolytes do (they conduct a current), even though HCl and SO_3 are themselves covalent compounds, unlike salts such as NaCl.

However, most covalent compounds that dissolve in water do not react with water. They dissolve because the water molecules surround the entire covalent molecule and solvate it. For example, when methanol, CH_3OH, is dissolved in water, the methanol molecules are solvated by the water molecules (Figure 6.10).

There is a simple way to predict which covalent compounds dissolve in water and which do not. Covalent compounds dissolve in water if they can form hydrogen bonds with water, provided that the solute molecules are fairly small. You may recall from Section 5.7 that hydrogen bonding is possible between two molecules if one of them contains an O, N, or F atom and the other contains an O—H, N—H, or F—H bond. Every water molecule contains an O atom and O—H bonds. Therefore, water can form hydrogen bonds with any molecule that contains an O, N, or F atom or an O—H, N—H, or F—H bond. If these molecules are small enough, they will be soluble in water. How small? In general, they should have no more than about three carbon atoms for each O, N, or F.

For example, acetic acid, CH_3COOH, is soluble in water, but benzoic acid, C_6H_5COOH, is not. Similarly, ethanol, C_2H_5OH, is soluble, but dipropyl ether, $C_6H_{14}O$, is not. Table sugar, $C_{12}H_{22}O_{11}$ (Section 19.6), is very soluble in water. Though it contains a large number of carbon atoms, it has so many oxygen atoms that it forms many hydrogen bonds with water molecules; thus, this molecule is very well solvated. Covalent molecules that do not contain O, N, or F atoms are almost always insoluble in water. For example, methanol, CH_3OH, is infinitely soluble in water, but methyl

Sports drinks help to maintain the body's electrolytic balance. *(George Semple)*

In aqueous solutions, H^+ does not exist; it combines with water and forms hydronium ion, H_3O^+.

FIGURE 6.10 Solvation of a polar covalent compound by water. The dotted lines represent hydrogen bonds.

chloride, CH_3Cl, is not. The exception is the rare case where a covalent compound reacts with water, for instance, HCl.

Water in the Body

Water is important in the body not only because it dissolves ionic substances as well as some covalent compounds, but also because it hydrates all polar molecules in the body. It thus serves as a vehicle to carry most of the organic compounds, nutrients, and fuels the body uses, as well as waste material. Blood and urine are but two examples of aqueous body fluids.

Beyond that, the hydration of macromolecules such as proteins, nucleic acids, and polysaccharides allows the proper motions within these molecules, which are necessary for such functions as enzyme activity (see Chapter 22).

6.7 Colloids

Up to now we have been discussing solutions. In a true solution, the maximum diameter of the solute particles is about 1 nm. If the diameter of the solute particles is larger than this, we no longer have a true solution—we have a **colloid.** In a colloid (also called a colloidal dispersion or colloidal system), the diameter of the solute particles ranges from about 1 to 1000 nm. Particles of this size usually have a very large surface area, and that accounts for the two basic characteristics of colloidal systems:

1. They scatter light and therefore appear turbid, cloudy, or milky.

2. Although colloidal particles are large, they form stable dispersions; that is, they do not form separate phases that settle out. As with true solutions, colloids can exist in a variety of phases (Table 6.2).

All colloids exhibit the following characteristic effect. When we shine light through a colloid and look at the system from a 90° angle, we see the pathway of the light without seeing the individual particles (Figure 6.11). This is called the **Tyndall effect** and is due to light scattering. Smoke, serum, and fog, to name a few examples, all exhibit the Tyndall effect. We are all familiar with the sunbeams that can be seen when sunlight passes through dusty air. This, too, is an example of the Tyndall effect. Again, we do not see the individual solid particles in dusty air, we see only the light scattered by them. Ordinary solutions do not show this effect.

Colloidal systems are stable. Mayonnaise stays emulsified and does not separate into oil and water. When the size of colloidal particles is larger

1 nm = 1×10^{-9} m

> **Colloid** A homogeneous mixture containing particles ranging from 1 to 1000 nm in diameter

Colloids are neither true solutions nor heterogeneous mixtures. They constitute a third kind of mixture.

> **Tyndall effect** Light passing through and scattered by a colloid viewed at a right angle

FIGURE 6.11 Light scattering by a colloidal suspension. A narrow beam of light passes through a colloidal mixture (left), then through a salt solution, and finally through another colloidal mixture. This illustrates the light scattering ability of the colloid-sized particles. *(Charles D. Winters)*

TABLE 6.2 Types of Colloidal Systems

Type	Example
Gas in gas	None
Gas in liquid	Whipped cream
Gas in solid	Marshmallows
Liquid in gas	Clouds, fog
Liquid in liquid	Milk, mayonnaise
Liquid in solid	Cheese, butter
Solid in gas	Smoke
Solid in liquid	Jelly, plasma extender
Solid in solid	Dried paint

Emulsions and Emulsifying Agents

Oil and water do not mix. Even when we stir them vigorously and the oil droplets are dispersed in the water, the two phases separate as soon as we stop stirring. There are, however, a number of stable colloidal systems made of oil and water; these are **emulsions.** For example, the oil droplets in milk are dispersed in an aqueous solution. This is possible because, in milk, there is a protective colloid, the milk pro- tein called casein. Casein molecules surround the oil droplets, and, since they are polar and carry a charge, they protect and stabilize the oil droplets. Casein is thus an emulsifying agent. A similar emulsifying agent is egg yolk, which is the ingredient in mayonnaise that coats the oil droplets and prevents them from separating.

than about 1000 nm, the system is unstable and separates into phases. Such systems are called **suspensions.** For example, if we take a lump of soil and disperse it in water, we get a muddy suspension. The soil particles are anywhere from 10^3 to 10^6 nm in diameter. The muddy mixture scatters light and, therefore, appears turbid, but it is not a stable system. If left alone, the soil particles soon settle to the bottom, with clear water above the sediment. Therefore, soil in water is a suspension, not a colloidal system.

The properties of solutions, colloids, and suspensions are summarized in Table 6.3.

What makes a colloidal dispersion stable? To answer this question, we must first realize that the colloidal particles are in constant motion. Just look at the dust particles dancing in a ray of sunlight that enters your room. Actually, you are not seeing the dust particles themselves; they are too small to see. What you are seeing are flashes of scattered light. The motion of the dust particles dispersed in air is a random, chaotic motion. It is the motion of any colloidal particle suspended in a solvent. This is called **Brownian motion** (Figure 6.12).

The cause of Brownian motion is the buffeting and collisions the parti- cles suffer from the solvent molecules (in the case of the dust particles, the solvent is air). This constant Brownian motion creates favorable conditions for collisions between particles. When such large particles collide, they stick together, combine to give larger particles, and finally settle out of the solution. This is what happens in a suspension.

So why do colloidal particles remain in solution despite all the colli- sions due to their Brownian motion? There are two reasons.

1. Most colloidal particles carry a large solvation layer. If the solvent is water, as in the case of protein molecules in the blood, the colloidal par- ticles are surrounded by a large number of water molecules, which move together with the colloidal particles and cushion them. When two

Ordinary household dust particles, magnified 2200 times. *(Courtesy of David Scharf)*

Brownian motion is named for Robert Brown (1773–1858), the Scottish botanist who first observed it in 1827.

TABLE 6.3 Properties of Three Types of Mixtures

Property	Solutions	Colloids	Suspensions
Particle size (nm)	0.1–1.0	1–1000	>1000
Filterable with ordinary paper	No	No	Yes
Homogeneous	Yes	Borderline	No
Settles on standing	No	No	Yes
Behavior to light	Transparent	Tyndall effect	Translucent or opaque

FIGURE 6.12 Brownian motion.

Freshly made wines are often cloudy because of colloidal particles (left), and removing the particles clarifies the wine (right). *(NALCO Chemical Company)*

Particles with the dimension of colloids have many new practical applications. Nanotubes, nanowires, and nanopores in composite coatings have unusual electronic, optical properties based on the enormous surface areas. For example, titanium oxide particles smaller than 20 nm are used to coat surfaces of plastics, glass, and other materials. These thin coatings have self-cleaning, antifogging, antifouling, and sterilizing properties.

Colligative property A property of a solution that depends only on the number of solute particles and not on the chemical identity of the solute

Note that in preparing a solution for this purpose we do not use molarity. That is, we do not need to measure the total volume of the solution.

colloidal particles collide as a result of Brownian motion, they do not actually touch each other; only the solvent layers collide. In this way, the particles do not stick together and thus stay in solution.

2. The large surface area of colloidal particles acquires charges from the solution. All colloids in a particular solution acquire the same kind of charge, for example, negative charge. This leaves a net positive charge in the solvent. When a charged colloidal particle encounters another colloidal particle, the two repel each other because of their like charge.

Thus, the combined effect of solvation layer and surface charge keeps colloidal particles in a stable dispersion. By using these effects, chemists can either increase or decrease the stability of a colloidal system. If we want to get rid of a colloidal dispersion, we can remove either the solvation layer or the surface charge, or both. For example, proteins in the blood form a colloidal dispersion. If we want to isolate a protein from blood, we may want to precipitate it. We can accomplish this in two ways: either remove the hydration layer or remove the surface charges. If we add solvents such as ethanol or acetone, which have great affinity for water, the water will be removed from the solvation layer of the protein, and then the unprotected protein molecules will stick together when they collide and form a sediment. Similarly, by adding electrolytes such as NaCl to the solution, we can remove the charges from the surface of the proteins (by a mechanism too complicated to discuss here). Without their protective charges, two protein molecules no longer repel each other. Instead, when they collide, they stick together and precipitate from the solution.

6.8 Colligative Properties

Any property of a solution that depends only on the *number* of solute particles dissolved in a given amount of solvent and not on the *nature* of these particles is called a **colligative property.**

Freezing-Point Depression

One colligative property is **freezing-point depression.** One mole of any particle dissolved in 1000 g of water lowers the freezing point of the water by 1.86°C. The nature of the solute does not matter, only the number of particles.

This principle is used in a number of practical ways. In the winter, we use salts (sodium chloride and calcium chloride) to melt snow and ice on our streets. The salts dissolve in the melting snow and ice, and this lowers the freezing point of the water. Another application is the use of antifreeze in automobile radiators. Since water expands on freezing (Box 5G), the ice formed in a car's cooling system when the outside temperature falls below 0°C, can crack the engine block. The addition of antifreeze prevents this because it makes the water freeze at a much lower temperature. The most common antifreeze is ethylene glycol, $C_2H_6O_2$.

EXAMPLE 6.9

If we add 275 g of ethylene glycol per 1000 g of water in a car radiator, what will the freezing point of the solution be?

Solution
Since $C_2H_6O_2$ has a molecular weight of 62.0, 275 g contains

$$275 \text{ g } C_2H_6O_2 \times \frac{1 \text{ mole } C_2H_6O_2}{62.0 \text{ g } C_2H_6O_2} = 4.44 \text{ moles } C_2H_6O_2$$

Each mole lowers the freezing point by 1.86°C, so the freezing point is lowered by

$$\frac{1.86°C}{1 \text{ mole}} \times 4.44 \text{ moles} = 8.26°C$$

The freezing point of the water will be lowered from 0°C to −8.26°C, and the radiator will not crack if the outside temperature remains above −8.26°C (17.18°F).

Salting lowers the freezing point of ice. *(Charles D. Winters)*

Problem 6.9
If we add 215 g of CH_3OH to 1000 g of water, what will be the freezing point of the solution? ■

If a solute is ionic, then each mole of solute dissociates to more than 1 mole of particles. For example, if we dissolve 1 mole (58.5 g) of NaCl in 1000 g of water, the solution contains 2 moles of solute particles: 1 mole each of Na^+ and Cl^-. The freezing point of water will be lowered by twice 1.86°C, or 3.72°C.

EXAMPLE 6.10

What will be the freezing point of the solution if we dissolve 1 mole of K_2SO_4 in 1000 g of water?

Solution
One mole of K_2SO_4 dissociates to produce 3 moles of ions: 2 moles of K^+ and 1 mole of SO_4^{2-}. The freezing point will be lowered by $3 \times 1.86°C = 5.58°C$, and the solution will freeze at −5.58°C.

Problem 6.10
Which aqueous solution would give the lowest freezing point?
(a) 6.2 *M* NaCl (b) 2.1 *M* $Al(NO_3)_3$ (c) 4.3 *M* K_2SO_3 ■

Besides freezing-point depression, there are several other colligative properties, including vapor-pressure lowering, boiling-point elevation, and osmotic pressure. We discuss only the last of these because biologically it is the most important.

Osmotic Pressure

In order to understand osmotic pressure, let us consider the beaker shown in Figure 6.13(b). In this beaker, two compartments are separated by an osmotic semipermeable membrane. A **semipermeable membrane** is a thin slice of some material, such as cellophane, that contains very tiny holes (far too small for us to see) that are big enough to let small solvent molecules pass through but not big enough to let large solute molecules pass. In the right compartment of the beaker is pure solvent—in this example, water; in the left compartment is a dispersion of starch (Section 19.8) in water.

Osmotic pressure The amount of external pressure applied to the more concentrated solution to stop the passage of solvent molecules across the semipermeable membrane

An osmotic membrane is a very selective semipermeable membrane that allows only solvent molecules, and nothing else, to pass through.

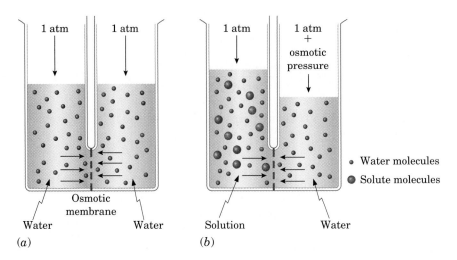

FIGURE 6.13 Osmotic pressure. (*a*) Two compartments separated by an osmotic semipermeable membrane both contain only solvent molecules that can pass through the membrane. The liquid level on each side is the same. (*b*) The compartment on the right contains only solvent; the one on the left contains both solute and solvent. Solute molecules cannot pass through the membrane. In an effort to dilute the solution, more solvent molecules move from the right to the left compartment than vice versa, raising the liquid level on that side.

The starch molecules are too big to go through the membrane and thus are forced to remain in the left compartment. The water molecules, however, easily go back and forth; as far as they are concerned, there is no membrane. Therefore, water molecules go from the right to the left side because the concentration of water in the right compartment is higher (100 percent) than in the left compartment. Molecules will always diffuse from an area of higher concentration to an area of lower concentration, and that is what the water molecules do. If there were no membrane separating the two solutions, the starch molecules would also follow this rule, moving from the higher concentration area (in this case the left side) to the lower one. The membrane, of course, prevents such movement. However, it does not hinder the water molecules from moving into the left compartment. This passage of solvent molecules from the right (solvent or dilute solution) to the left (a more concentrated solution) across a semipermeable membrane is called **osmosis.** When a significant number of water molecules have moved from the right to the left side, the liquid level on the right side goes down and that on the left side goes up, as shown in Figure 6.13(b).

The process cannot continue indefinitely because gravity prevents the difference in levels from becoming too great. Eventually the process stops with the levels unequal. The levels can be made equal again if we apply an external pressure to the higher side. The amount of external pressure necessary to equalize the levels is called the **osmotic pressure.**

Although this discussion assumes that one compartment contains pure solvent, the same principle applies if both compartments contain solutions, as long as the concentrations are different. The solution of higher concentration always has a higher osmotic pressure than the one of lower concentration, which means that the flow of solvent is always from the more dilute to the more concentrated side. Of course, the number of particles is what is important so we must remember that, in ionic solutions, each mole of solvent gives rise to more than 1 mole of particles. For convenience in calculation, we define a new term, **osmolarity (osmol),** as **molarity multiplied by the number of particles produced by each molecule of solute.**

Some ions are small but still do not go through the membrane because they are solvated by a shell of water molecules (Fig. 6.6).

EXAMPLE 6.11

An 0.89 percent w/v NaCl solution is referred to as a physiological saline solution because it has the same concentration of salts as normal human blood, though blood contains several salts and saline solution has only NaCl. What is the osmolarity of this solution?

Solution

$$0.89\% \text{ w/v NaCl} = 8.9 \text{ g NaCl in 1 L of solution}$$

$$\frac{8.9 \text{ g NaCl}}{1 \text{ L solution}} \times \frac{1 \text{ mole NaCl}}{58.5 \text{ g NaCl}} = \frac{0.15 \text{ mole NaCl}}{\text{L solution}} = 0.15 \, M$$

Because each NaCl dissociates into two particles,

$$\text{Osmolarity} = 0.15 \times 2 = 0.30 \text{ osmol}$$

Problem 6.11 ▬▬▬

What is the osmolarity of a 3.3 percent w/v Na_3PO_4 solution? ■

An isotonic saline solution. *(Charles D. Winters)*

Osmotic pressure is a colligative property. The osmotic pressure generated by a solution across the semipermeable membrane—the difference between the heights of the two columns in Figure 6.13(b)—depends on the osmolarity of the solution. If the osmolarity increases by a factor of 2, the osmotic pressure will also increase by a factor of 2.

Osmotic pressure is very important in biological organisms because cell membranes are semipermeable membranes. Therefore, biological fluids must have the proper osmolarity. For example, the red blood cells in the body are suspended in a medium called plasma, which must have the same osmolarity as the red blood cells. Two solutions with the same osmolarity are called **isotonic,** and plasma is said to be isotonic with the red blood cells. This means that no osmotic pressure is generated across the cell membrane.

What would happen if we suspended red blood cells in distilled water instead of in plasma? Inside the red blood cells the osmolarity is about the same as in a physiological saline solution, 0.30 osmol. Distilled water has zero osmolarity. As a consequence, water flows into the red blood cells. The volume of the cells increases, and the cells swell, as shown in Figure 6.14(b). The membrane cannot resist the osmotic pressure, and the red

BOX 6E

Reverse Osmosis and Desalinization

In osmosis, the solvent flows spontaneously from the dilute solution compartment into the concentrated solution compartment. In reverse osmosis, the opposite happens. One must apply pressures greater than the osmotic pressure in order for this to occur. Reverse osmosis is used to make drinkable water from sea water or brackish water. In large plants in the Persian Gulf countries, more than 100 atm pressure is applied to sea water containing 35 000 ppm salt. The water that passes through the semipermeable membrane under this pressure contains only 400 ppm salt. This is well within the limits set by the World Health Organization for drinkable water.

■ An emergency hand-operated water desalinator that works by reverse osmosis. It can produce 4.5 L of pure water per hour from sea water, which can save someone adrift at sea. *(Courtesy of Recovery Engineering, Inc.)*

The effect of osmosis on a carrot. The carrot on the left is in a solution that contains a high concentration of NaCl. Water flowing out of the carrot has caused it to shrink. The carrot on the right, in pure water, is slightly swollen. *(Charles D. Winters)*

Iso-osmotic, hypo-osmotic, and hyper-osmotic are recent synonyms for isotonic, hypotonic, and hypertonic, respectively.

A 5.5 percent glucose solution is also isotonic and is used in intravenous feeding.

(*a*) Isotonic solution

(*b*) Hypertonic solution

(*c*) Hypotonic solution

FIGURE 6.14   Red blood cells in solutions of different tonicity. *(Science Source / Photo Researchers, Inc.)*

blood cells eventually burst, spilling their contents into the water. We call this process **hemolysis.**

Solutions with an osmolarity (and hence osmotic pressure) lower than that of suspended cells are called **hypotonic solutions.** Obviously, it is very important that we always use isotonic solutions and never hypotonic solutions in intravenous feeding and blood transfusion. Hypotonic solutions would simply kill the red blood cells by hemolysis.

Equally important, we should not use **hypertonic solutions** either. A hypertonic solution has a greater osmolarity (and greater osmotic pressure) than the reference cells. If red blood cells are placed in a hypertonic solution (for example, 0.5 osmol glucose solution), water flows from the cells into the glucose solution through the semipermeable cell membrane. This process, called **crenation,** shrivels the cells, as shown in Figure 6.14(c).

As already mentioned, 0.89 percent NaCl (physiological saline) is isotonic with red blood cells and is used in intravenous injections.

EXAMPLE 6.12

Is a 0.50 percent w/v solution of KCl (a) hypertonic, (b) hypotonic, or (c) isotonic compared to red blood cells?

Solution

$$0.50\% \text{ KCl contains } 5.0 \text{ g KCl in } 1.0 \text{ L of solution}$$

$$\frac{5.0 \text{ g KCl}}{1.0 \text{ L soln}} \times \frac{1.0 \text{ mole KCl}}{74.6 \text{ g KCl}} = \frac{0.067 \text{ mole KCl}}{\text{L}} = 0.067 \, M$$

Because each KCl yields two particles, the osmolarity is $0.067 \times 2 = 0.13$ osmol; this is smaller than the osmolarity of the red blood cells, which is 0.30 osmol. Therefore, the KCl solution is hypotonic.

Problem 6.12

Which solution is isotonic compared to red blood cells? (a) $0.1 \, M \, \text{Na}_2\text{SO}_4$, (b) $1.0 \, M \, \text{Na}_2\text{SO}_4$, or (c) $0.2 M \, \text{Na}_2\text{SO}_4$ ∎

Dialysis

In osmosis, there is an osmotic semipermeable membrane that allows only solvent and not solute molecules to pass. If, however, the openings in the membrane are somewhat larger, then small solute molecules can also get through, but large solute molecules, such as macromolecular and colloidal particles, cannot. This process is called **dialysis.** For example,

Dialysis A process in which a solution containing particles of different size is placed in a bag made of a semipermeable membrane. The bag is placed into a solvent or solution containing only small molecules. The solution in the bag reaches equilibrium with the solvent outside, allowing the small molecules to diffuse across the membrane but retaining the large molecules.

ribonucleic acids are important biological molecules that we study in Chapter 24. When biochemists prepare ribonucleic acid solutions, they must remove small molecules, such as NaCl and ethanol, from the solution in order to get a pure nucleic acid preparation. This is done by placing the nucleic acid solution in a dialysis bag (cellophane) of sufficient pore size to allow all the small molecules to diffuse and retain only the large nucleic acid molecules. If the dialysis bag is suspended in flowing distilled water, all the NaCl and small molecules leave the bag. After a certain amount of time, the bag contains only the pure nucleic acids dissolved in water.

This is also the way our kidneys work. The millions of nephrons, or kidney cells, have very large surface areas in which the capillaries of the blood vessels are in contact with the nephrons. The kidneys serve as a gigantic filtering machine. The waste products of the blood dialyse out through semipermeable membranes in the glomeruli (see Sections 31.5 and 31.6) and enter collecting tubes that carry the urine to the ureter. Large protein molecules and cells are retained in the blood.

A portable dialysis unit. *(SIU/Photo Researchers, Inc.)*

The glomeruli of the kidney are fine capillary blood vessels in which the body's waste products are removed from the blood.

BOX 6F

Hemodialysis

The kidneys' main function is to remove toxic waste products from the blood. When the kidneys are not functioning properly, these waste products may threaten life. **Hemodialysis** is a process that performs the same filtration function (see Figure 25.5).

The patient's blood is circulated through a long tube of cellophane membrane suspended in an isotonic solution and then returned to the patient's vein. The cellophane membrane retains all the large molecules (for example, proteins) but allows the small ones, including the toxic wastes, to pass through. Thus, the wastes are removed from the blood by dialysis.

If the cellophane tube were suspended in distilled water, other small molecules and ions, such as Na^+, Cl^-, and glucose, would also be removed from the blood. This is something you don't want to happen. The isotonic solution used in hemodialysis consists of 0.6 percent NaCl, 0.04 percent KCl, 0.2 percent $NaHCO_3$, and 0.72 percent glucose (all w/v). In this manner, no glucose or Na^+ is lost from the blood.

A patient is usually on an artificial kidney machine for 4 to 7 hours. During this time, the isotonic bath is changed every 2 hours. Kidney machines allow people with kidney failure to lead a fairly normal life, though they must take these hemodialysis treatments regularly.

■ A schematic diagram of the hollow fiber (or capillary) dialyzer, the most commonly used artificial kidney. The blood flows through small tubes constructed of semipermeable membrane; these tubes are bathed in the dialyzing solution.

SUMMARY

Systems containing more than one component are **mixtures. Homogeneous mixtures** are uniform throughout. **Heterogeneous mixtures** exhibit well-defined boundaries between phases. When a solution consists of a solid or gas dissolved in a liquid, the liquid is the **solvent,** and the solid or gas is the **solute.** When one liquid is dissolved in another, the one liquid present in greater amount is the solvent. The **solubility** of a substance is the maximum amount of the substance that dissolves in a given amount of solvent at a given temperature.

"Like dissolves like" means that polar molecules are soluble in polar solvents and that nonpolar molecules are soluble in nonpolar solvents. The solubility of solids and liquids in liquids usually increases with increasing temperature; the solubility of gases in liquids usually decreases with an increase in temperature.

Percent concentration is given in either weight per unit volume of solution (w/v), weight per unit weight of solvent (w/w), or volume of solute per unit volume of solution (v/v). **Molarity** is the number of moles of solute in 1 L of solution.

Water is the most important solvent because it dissolves polar compounds and ions through hydrogen bonding and dipole interactions. Hydrated ions are surrounded by water molecules (solvation layer) that move together with the ion and cushion it from colli-

sions with other ions. Aqueous solutions of ions and molten salts are **electrolytes** and conduct electricity.

Colloids exhibit a scattering of light called the **Tyndall effect.** Colloids exhibit a chaotic random motion, called **Brownian motion.** Colloids are stable mixtures in spite of the large size of the colloidal particles (1 to 1000 nm). The stability is due to a solvation layer that cushions the colloids from direct collisions and to an electric charge on the surface of colloidal particles.

Colligative properties are properties that depend only on the number of solute particles present. **Freezing-point depression** and **osmotic pressure** are two examples of colligative properties. Osmotic pressure operates across an osmotic semipermeable membrane that allows only solvent molecules to pass but screens out all other molecules. In osmotic pressure calculations, concentration is measured in **osmolarity,** which is the molarity of the solution multiplied by the number of particles produced by dissociation of the solute.

Red blood cells in **hypotonic solution** swell and burst; this is called **hemolysis.** Red blood cells in **hypertonic solution** shrink; this is called **crenation.** Some semipermeable membranes allow small solute molecules to pass along with solvent molecules. In **dialysis,** such membranes are used to separate large molecules from smaller ones.

KEY TERMS

Acid rain (Box 6A)
Anhydrous (Section 6.6)
Anode (Section 6.6)
Brownian motion (Section 6.7)
Cathode (Section 6.6)
Colligative property (Section 6.8)
Colloid (Section 6.7)
Colloidal dispersions (Section 6.1)
Concentration (Section 6.5)
Crenation (Section 6.8)
Dialysis (Section 6.8)
Dilute (Section 6.5)
Electrolyte (Section 6.6)
Emulsion (Box 6D)
Freezing-point depression
   (Section 6.8)
Hemodialysis (Box 6F)

Hemolysis (Section 6.8)
Henry's law (Section 6.4)
Heterogeneous mixture
   (Section 6.1)
Homogeneous mixture
   (Section 6.1)
Hydrates (Section 6.6)
Hygroscopic (Section 6.6)
Hypertonic (Section 6.8)
Hypotonic (Section 6.8)
Isotonic (Section 6.8)
Miscible (Section 6.4)
Molarity (Section 6.5)
Nonelectrolytes (Section 6.6)
Osmolarity (Section 6.8)
Osmosis (Section 6.8)
Osmotic membrane (Section 6.8)

Osmotic pressure (Section 6.8)
Parts per million (Section 6.5)
Percent concentration (Section 6.5)
Saturated solution (Section 6.4)
Seeding (Section 6.4)
Semipermeable membrane
   (Section 6.8)
Solubility (Section 6.4)
Solute (Section 6.2)
Solution (Section 6.1)
Solvated ions (Section 6.6)
Solvent (Section 6.2)
Supersaturated solution (Section 6.4)
Suspension (Section 6.7)
Tyndall effect (Section 6.7)
Unsaturated solution (Section 6.4)
Water of hydration (Section 6.6)

CONCEPTUAL PROBLEMS

6.A Two bottles of water are carbonated, CO_2 gas added, under 2 atm pressure and capped. One bottle is stored at room temperature; the other is stored in the refrigerator. When the bottle stored at room temperature is opened, large bubbles escaped and with it a third of the water. The bottle stored in the refrigerator was opened without frothing or bubbles escaping. Explain.

6.B A solution of protein is transparent at room temperature. When it is cooled to 10°C, it becomes turbid. What causes this change in appearance?

6.C A 4 *M* acetic acid (CH_3COOH) solution lowered the freezing point by 8°; a 4 *M* KF solution yields 15° freezing-point depression. What can account for this difference?

6.D Using a semipermeable membrane, a 0.005 *M* glucose (small molecule) solution yielded an osmotic pressure of 10 mm Hg. What kind of osmotic pressure change would you expect if instead of a semipermeable membrane you used a dialysis membrane?

PROBLEMS

Difficult problems are designated by an asterisk.

Solutions and Their Characteristics

6.13 Vinegar is a homogeneous aqueous solution containing 6 percent acetic acid. Which is the solvent?

6.14 A solution is made by dissolving glucose in water. Which is the solvent and which is the solute?

6.15 In each of the following, tell whether the solutes and solvents are gases, liquids, or solids.
(a) Bronze (see Box 2D)　(b) Cup of coffee
(c) Car exhaust　(d) Champagne

6.16 Give a familiar example of solutions of each of these types.
(a) Liquid in liquid　(b) Solid in liquid
(c) Gas in liquid　(d) Gas in gas

6.17 Are mixtures of gases true solutions or heterogeneous mixtures? Explain.

Solubility

6.18 You dissolved 0.32 g of aspartic acid in 115.0 mL of water and obtained a clear solution. After two days standing at room temperature, you notice a white powder at the bottom of the beaker. What may have happened?

6.19 The solubility of a compound is 2.5 g in 100 mL of aqueous solution at 25°C. If you put 1.12 g of the compound in a 50-mL volumetric flask at 25°C and add sufficient water to fill it to the 50-mL mark, what kind of solution do you get, saturated or unsaturated? Explain.

6.20 To a separatory funnel with two layers—the nonpolar diethyl ether and the polar water—is added a small amount of solid. After shaking the separatory funnel, in which layer will you find each of the following solids? (a) NaCl (b) camphor (c) KOH

6.21 On the basis of polarity and hydrogen bonding, which solute would be the most soluble in benzene, C_6H_6?
(a) CH_3OH
(b) H_2O

(c) $CH_3CH_2CH_2CH_3$
(d) H_2SO_4

6.22 Suppose that you are trying to dissolve 2.0 g of a certain solid drug in water. You add it to 100 mL of water, and part of it, but not all of it, dissolves. Give two procedures you might try to dissolve the entire 2.0 g.

6.23 How can you convert a saturated solution at room temperature to a supersaturated one?

6.24 The solubility of aspartic acid in water is 0.500 g in 100 mL at 25°C. If you dissolve 0.251 g aspartic acid in 50.0 mL water at 50°C and let the solution cool to 25°C without stirring, shaking, or otherwise disturbing the solution, would you get a saturated, unsaturated, or supersaturated solution? Explain.

6.25 Near power plants, warm water is discharged into a river. Sometimes dead fish are observed in the area. Why do fish die from the warm water?

6.26 If a bottle of beer is allowed to stand for several hours after being opened, it becomes "flat" (it loses CO_2). Explain.

6.27 Would you expect the solubility of ammonia gas in water at 2 atm pressure to be (a) greater, (b) the same, or (c) smaller than at 0.5 atm pressure?

Concentration Units

6.28 Describe how you would make the following solutions.
(a) 500.0 g of 5.32 percent w/w H_2S solution in water
(b) 342.0 g of 0.443 percent w/w benzene in toluene
(c) 12.5 g of 34.2 percent w/w dimethyl sulfoxide in acetone

6.29 Describe exactly how you would make the following solutions.
(a) 280 mL of a 27 percent v/v solution of ethanol, C_2H_5OH, in water
(b) 435 mL of a 1.8 percent v/v solution of ethyl acetate, $C_4H_8O_2$, in water
(c) 1.65 L of an 8.00 percent v/v solution of benzene, C_6H_6, in chloroform, $CHCl_3$

6.30 Describe exactly how you would make the following solutions.
 (a) 250 mL of a 3.6 percent w/v solution of NaCl in water
 (b) 625 mL of a 4.9 percent w/v solution of glycine, $C_2H_5NO_2$, in water
 (c) 43.5 mL of a 13.7 percent w/v solution of Na_2SO_4 in water
 (d) 518 mL of a 2.1 percent w/v solution of acetone, C_3H_6O, in water

6.31 Calculate the w/v percentage of each of these solutes.
 (a) 623 mg of casein in 15.0 mL of milk
 (b) 74 mg vitamin C in 250 mL of orange juice
 (c) 3.25 g of sucrose in 186 mL coffee

6.32 Describe how you would prepare 250 mL of 0.10 *M* NaOH from solid NaOH and water.

6.33 Assuming that the appropriate volumetric flasks are available, describe exactly how you would make these solutions.
 (a) 175 mL of a 1.14 *M* solution of NH_4Br in water
 (b) 1.35 L of a 0.825 *M* solution of NaI in water
 (c) 330 mL of a 0.16 *M* solution of ethanol, C_2H_5OH, in water

6.34 What is the molarity of each solution?
 (a) 47 g of KCl dissolved in enough water to give 375 mL of solution
 (b) 82.6 g of sucrose, $C_{12}H_{22}O_{11}$, dissolved in enough water to give 725 mL of solution
 (c) 9.3 g of $(NH_4)_2SO_4$ dissolved in enough water to give 2.35 L of solution

6.35 A teardrop with a volume of 0.5 mL contains 5.0 mg NaCl. What is the molarity of the NaCl in the teardrop?

6.36 The concentration of stomach acid, HCl, is approximately 0.10 *M*. What volume of stomach acid contains 0.25 mg of HCl?

6.37 If 79.49 g of H_2SO_4 is dissolved in enough solvent to make 2.00 L of solution, what is the molarity?

6.38 If 3.18 g of $BaCl_2$ is dissolved in enough solvent to make 500.0 mL of solution, what is the molarity?

6.39 In the reaction

$$CO_2(g) + H_2O(l) \longrightarrow H_2CO_3(aq)$$

1.25 g of CO_2 was added to enough water to make 50.0 mL of solution. What is the molarity of the H_2CO_3 formed? Assume that the reaction goes to completion.

6.40 A toothpaste contains 0.17 g NaF in 75 mL toothpaste. What is the percent w/v and the molarity of NaF in the toothpaste?

*6.41 A student has a bottle labeled 0.750 percent albumin solution. The bottle contained exactly 5.00 mL. How much water must the student add to make the concentration of albumin 0.125 percent?

6.42 How many g of solute are present in each of the following aqueous solutions?
 (a) 575 mL of a 2.00 *M* solution of HNO_3
 (b) 1.65 L of a 0.286 *M* solution of alanine, $C_3H_7NO_2$
 (c) 320 mL of a 0.0081 *M* solution of $CaSO_4$

6.43 A student has a stock solution of 30.0 percent w/v H_2O_2 (hydrogen peroxide). Describe how the student should prepare 250 mL of a 0.25 percent H_2O_2 solution.

6.44 To make 5.0 L of a fruit punch that contains 10 percent v/v ethanol, how much 95 percent ethanol v/v must be mixed with how much fruit juice?

6.45 A pill weighing 325 mg contains the following. What is the concentration of each in ppm?
 (a) 12.5 mg Captopril   (b) 22 mg Mg^{2+}
 (c) 0.27 mg Ca^{2+}

6.46 One slice of enriched bread weighing 80 g contains 70 μg folic acid. How much is this in ppm and ppb?

*6.47 Dioxin is considered to be poisonous in concentrations above 2 ppb. If a lake containing 1×10^7 L has been contaminated by 0.1 g of dioxin, did the concentration reach a dangerous level?

*6.48 An industrial waste water contained 3.60 ppb cadmium, Cd^{2+}. How many mg of Cd^{2+} could be recovered from a ton (1016 kg) of waste water?

Water As a Solvent

6.49 Define (a) hygroscopic (b) water of hydration (c) electrolyte (d) anhydrous.

6.50 Considering polarities, electronegativities, and similar concepts learned in Chapter 3, classify the following as strong electrolyte, weak electrolyte, or nonelectrolyte.
 (a) KCl   (b) C_2H_5OH (ethanol)   (c) NaOH
 (d) HF   (e) $C_6H_{12}O_6$ (glucose)

6.51 Which would produce the brightest light in the conductance apparatus in Figure 6.9?
 (a) 0.1 *M* KCl   (b) 0.1 *M* $(NH_4)_3PO_4$
 (c) 0.5 *M* sucrose

6.52 Ethanol, C_2H_5OH, is very soluble in water. Describe how water dissolves ethanol.

6.53 Predict which of these covalent compounds is soluble in water.
 (a) C_2H_6 (b) CH_3OH (c) HF
 (d) NH_3 (e) CCl_4

Colloids

6.54 If the dust in the air of a room shows the Tyndall effect, what is the average size of the dust particles?

6.55 On the basis of Tables 6.1 and 6.2, classify the following systems as homogeneous, heterogeneous, or colloidal mixtures.
 (a) Physiological saline solution   (b) Orange juice
 (c) A cloud   (d) Wet sand
 (e) Suds   (f) Milk

6.56 Table 6.2 shows no examples of a gas-in-gas colloidal system. Considering the definition of a colloid, explain why this is so.

6.57 What makes the water molecules in the solvation layer of a colloid different from the water molecules in the bulk?

6.58 What gives the nanotubes their unique optical and electrical properties?

Colligative Properties

6.59 Calculate the freezing points of solutions made by dissolving 1 mole of each of the following ionic solutes in 1000 g of H_2O.
(a) NaCl   (b) $MgCl_2$
(c) $(NH_4)_2CO_3$ (d) $Al(HCO_3)_3$

6.60 If we add 175 g of ethylene glycol, $C_2H_6O_2$, per 1000 g of water to a car radiator, what will be the freezing point of the solution?

6.61 Methanol is used as an antifreeze. How many g of methanol, CH_3OH, do you need per 1000 g of water for an aqueous solution to stay liquid at −20°C?

6.62 In winter, after a snow storm, salt, NaCl, is spread to melt the ice on roads. How many g of salt per 1000 g of ice is needed to make it liquid at −5.0°C?

6.63 How many moles of glycerol, $C_3H_8O_3$, which does not dissociate in aqueous solution, must you add to 1000 g of water to make a solution that freezes at −10°C?

Osmosis

6.64 What is the difference between an osmotic semipermeable membrane and a semipermeable membrane?

6.65 In each case, tell which side (if either) rises and why. The solvent is water.

Osmotic membrane

	Ⓐ	Ⓑ
(a)	1% glucose	5% glucose
(b)	0.1 M glucose	0.5 M glucose
(c)	1 M NaCl	1 M glucose
(d)	1 M NaCl	1 M K_2SO_4
(e)	3% NaCl	3% NaI
(f)	1 M NaBr	1 M NaCl

6.66 An osmotic semipermeable membrane that allows only water to pass separates two compartments, A and B. Compartment A contains 0.9 percent NaCl and B contains 3 percent glycerol, $C_3H_8O_3$.
(a) In which compartment will the level of solution rise?
(b) Which compartment has the higher osmotic pressure (if either)?

6.67 Calculate the osmolarity of each of the following solutions.
(a) 0.39 M Na_2CO_3 (b) 0.62 M $Al(NO_3)_3$
(c) 4.2 M LiBr   (d) 0.009 M K_3PO_4

6.68 Two compartments are separated by a semipermeable osmotic membrane through which only water molecules can pass. Compartment A contains a 0.3 M KCl solution; compartment B contains a 0.2 M Na_3PO_4 solution. Predict from which compartment the water will flow to the other compartment.

*6.69 A 0.9 percent NaCl solution is isotonic with blood plasma. Which solution would crenate red blood cells?
(a) 0.3% NaCl   (b) 0.9 M glucose
(c) 0.9% glucose ($MW = 180$)

Boxes

6.70 (Box 6A) In 1992, a volcano in the Philippines emitted large quantities of CO_2 and SO_2 gases as high as 2 to 3 miles into the atmosphere. Would you expect an increase in acid rain locally or globally? Explain.

6.71 (Box 6A) Oxides of nitrogen (NO, NO_2, N_2O_3) are also responsible for acid rain. Which acids do you suppose can be formed from these oxides?

6.72 (Box 6A) What makes normal rain water slightly acidic?

6.73 (Box 6A) In a carbonated drink of 1 L volume, 3.456 g CO_2 was dissolved under pressure. When you removed the cap, CO_2 gas escaped. How many g of CO_2 were lost when you opened the bottle at 20°C?

6.74 (Box 6B) Why do deep-sea divers use a helium-oxygen mixture in the tank instead of air?

6.75 (Box 6B) What is nitrogen narcosis?

*6.76 (Box 6C) Write balanced equations (two steps) for the conversion of marble to gypsum dihydrate.

6.77 (Box 6C) What is the chemical formula for the main component of limestone and marble?

6.78 (Box 6D) What is the protective colloid in milk?

6.79 (Box 6E) What is the minimum pressure on a sea water that will force water flow from the concentrated solution into the dilute solution?

*6.80 (Box 6F) The artificial kidney machine uses a solution containing 0.6 percent NaCl, 0.04 percent KCl, 0.2 percent $NaHCO_3$, and 0.72 percent glucose (all w/v). Show that this is an isotonic solution.

Additional Problems

6.81 If 10.0 g of ethanol and 30.0 g of water are mixed and the solution has a final volume of 38.0 mL, what is the concentration of ethanol (a) in percent w/v and (b) in percent w/w?

6.82 When a cucumber is put into a saline solution to pickle it, the cucumber shrinks; when a prune is put into the same solution, the prune swells. Explain what happens in each case.

6.83 A solution of As_2O_3 has a molarity of 2×10^{-5} M. What is this concentration in ppm? (Assume that the density of the solution is 1.00 g/mL.)

6.84 What effect does increasing the temperature have on the solubility of most substances?

*6.85 Both methanol, CH_3OH, and ethylene glycol, $C_2H_6O_2$, are used as antifreeze. Which is more efficient; that is, which produces a lower freezing point if equal weights of each are added to the same weight of water?

6.86 You know that a 0.89 percent saline (NaCl) solution is isotonic with blood. In a real-life emergency, you run out of physiological saline solution and have only KCl as a salt, and distilled water. Would it be all right to make a 0.89 percent aqueous KCl solution and use it for intravenous infusion? Explain.

*6.87 Carbon dioxide and sulfur dioxide are soluble in water because they react with the water. Write possible equations for these reactions.

6.88 A reagent label shows that it contains 0.05 ppm lead as a contaminant. How many μg of lead are in 5.0 g of the reagent?

6.89 A concentrated nitric acid solution contains 35 percent HNO_3. How would you prepare 300 mL of 4.5 percent solution?

6.90 Which will have greater osmotic pressure.
(a) A 0.9 percent w/v NaCl solution
(b) A 25 percent w/v solution of a nondissociating dextran with a molecular weight of 15 000

6.91 Government regulations permit a 6-ppb concentration of a certain pollutant. How many g of pollutant are allowed in 1 ton (1016 kg) of water?

*6.92 The average osmolarity of sea water is 1.18. How much pure water would have to be added to 1.0 mL of sea water for it to achieve the osmolarity of blood (0.30 osmol)?

*6.93 A swimming pool containing 20 000 L of water is chlorinated to have a final Cl_2 concentration of 0.00500 M. What is the Cl_2 concentration in ppm? How many kg of Cl_2 was added to the swimming pool to reach this concentration?

Reaction Rates and Chemical Equilibrium

7.1 Introduction

Some chemical reactions take place rapidly; others are very slow. For example, hydrogen gas and oxygen gas react with each other to form water:

$$2H_2(g) + O_2(g) \longrightarrow 2H_2O(l)$$

This reaction is extremely slow. A tank containing a mixture of H_2 and O_2 shows no measurable change even after many years.

In contrast to such a slow reaction, consider what happens when you take one or two aspirin tablets for a slight headache. Very often, the pain disappears in half an hour or so. This shows that the aspirin must have reacted with compounds in the body within that time.

Many reactions are much faster still. For example, if we add a solution of silver ions to a solution of chloride ions (Section 4.7), a precipitate of silver chloride forms almost instantaneously:

$$Aq^+(aq) + Cl^-(aq) \longrightarrow AgCl(s)$$

In the course of several years, a few molecules of H_2 and O_2 will react, but not enough for us to detect within a laboratory period.

Above: Calcium oxide reacting with water. *(Charles D. Winters)*

This reaction is essentially complete in considerably less than 1 s.

The study of reaction rates is called **chemical kinetics.** The **rate of a reaction** is **the change in concentration of a reactant (or product) per unit time.** Every reaction has its own rate, which must be measured in the laboratory.

Consider the reaction carried out in the solvent acetone.

$$CH_3{-}Cl + I^- \xrightarrow[\text{acetone}]{} CH_3{-}I + Cl^-$$

This is a net ionic reaction. Spectator ions are not shown.

The rate could also be determined by following the decrease in concentration of CH_3Cl, if that is more convenient.

To determine the reaction rate, we can measure the concentration of iodomethane, CH_3I, in the acetone at periodic time intervals, say every 10 min. The rate of the reaction is the increase in the concentration of iodomethane divided by the time interval. For example, the concentration might increase from 0 to 0.12 mole/L over a period of 30 min. If so, the rate of the reaction during that time is

$$\frac{0.12 \text{ mole } CH_3I/L - 0 \text{ mole } CH_3I/L}{30 \text{ min}} = \frac{0.12 \text{ mole } CH_3I/L}{30 \text{ min}}$$

$$= \frac{0.0040 \text{ mole } CH_3I/L}{\text{min}}$$

This unit is read "0.0040 mole per liter per minute." During each minute of the reaction, an average of 0.0040 mole of chloromethane, CH_3Cl, is converted to iodomethane for each liter of solution.

EXAMPLE 7.1

Another way of determining the rate of the preceding reaction is to measure the disappearance of the I^- from the solution. Suppose that the concentration of I^- was 0.24 mol I^-/L at the start of the reaction. At the end of 20 min, the concentration dropped to 0.16 mole I^-/L. What is the rate of reaction?

Solution
The rate of the reaction is

$$\frac{0.16 \text{ mol } I^-/L - 0.24 \text{ mol } I^-/L}{20 \text{ min}} = \frac{-0.08 \text{ mol } I^-/L}{20 \text{ min}}$$

$$= -0.0040 \text{ mol } I^-/L/\text{min}$$

Since the stoichiometry of the components is 1:1 in this reaction, we are getting the same numerical answer for the rate whether we monitor a reactant or a product. Note, however, that when we measure the concentration of a reactant that disappears with time, the rate of reaction is always negative.

Problem 7.1 ■━━━
In the reaction

$$2HgO(s) \longrightarrow 2Hg(l) + O_2(g)$$

you measure the evolution of O_2 gas. At the beginning of the reaction, at 0 min, there is 0.020 L of O_2 present. You find that after 15 min the volume of O_2 gas is 0.35 L. What is the rate of reaction? ■

The rates of chemical reactions, both the ones that we carry out in the laboratory and the ones that take place inside our bodies, are very important to us. A reaction that goes more slowly than we need may be useless, whereas a reaction that goes too fast may be dangerous. Thus, we would like to have some idea as to what causes the enormous variety in reaction rates. In the next three sections, we examine this question.

7.2 Molecular Collisions

In order for two molecules or ions to react with each other, they must first collide. As we saw in Chapter 5, molecules in gases and liquids are constantly in motion and frequently collide with each other. If we want a reaction to take place between two compounds, A and B, we mix them together, either by dissolving them in the same solvent or, if they are gases, by simply allowing them to mix. In either case, we can be sure that the constant motion of all the molecules will cause frequent collisions between molecules of A and B. In fact, we can even calculate how many such collisions will take place in a given period of time.

When we make such calculations, we generally get a surprising result. There are so many collisions between A and B molecules that most reactions should be over in considerably less than one second. Because the actual reactions are generally much slower, we must conclude that *most collisions do not result in a reaction.* In most cases, when a molecule of A collides with a molecule of B, they simply bounce apart without reacting. Every once in a while, though, a molecule of A and one of B collide and react to form a new compound. **A collision that results in a reaction** is called an **effective collision.**

Why are some collisions effective and others not? There are two main reasons.

1. In most cases, for a reaction to take place between A and B, one or more covalent bonds must be broken in A or B or both, and energy is required for this to happen. This energy comes from the collision between A and B. If the energy of the collision is large enough, the bonds will break and the reaction will take place. If the collision energy is too low, the molecules will bounce apart without reacting. **The minimum energy necessary for the reaction to happen** is called the **activation energy.**

The energy of any collision depends on the relative speed of the colliding objects and on the angle of approach. Much greater damage is done in a head-on collision of two cars both going 40 mph than in a collision in which a car going 20 mph sideswipes one going 10 mph. It is the same with molecules, as Figure 7.1 shows.

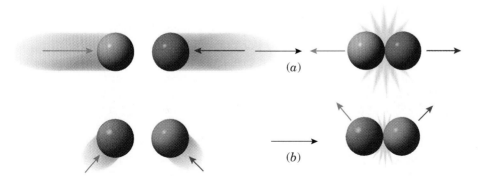

FIGURE 7.1 The energy of molecular collisions varies. Two fast-moving molecules colliding head-on (*a*) have a higher collision energy than two slower-moving molecules colliding at an angle (*b*).

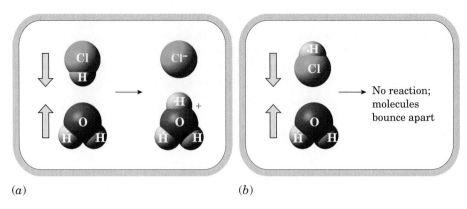

FIGURE 7.2 (*a*) The two molecules are properly oriented for a reaction to take place. (*b*) The orientation is such that a reaction cannot take place, because the Cl, not the H, collides with the O. The colored arrows show the path of the molecules.

2. Even if two molecules collide with an energy greater than the activation energy, a reaction still may not take place if the molecules are not oriented properly when they collide. Consider, for example, the reaction between H_2O and HCl:

$$H_2O + HCl \longrightarrow H_3O^+ + Cl^-$$

For this reaction to take place, the molecules must collide in such a way that the H of the HCl hits the O of the water, as shown in Figure 7.2(a). A collision in which the Cl hits the O, as shown in Figure 7.2(b), cannot lead to reaction, even if there is sufficient energy.

Coming back to the example at the beginning of this chapter, we can now see why the reaction between H_2 and O_2 is so slow. The O_2 and H_2 molecules are constantly colliding, but the percentage of effective collisions is extremely tiny at room temperature.

7.3   Activation Energy and Energy Diagrams

Figure 7.3 shows a typical energy diagram for an exothermic reaction. The products have a lower energy than the reactants; we might, therefore, expect the reaction to take place rapidly. As the curve shows, however, the reactants cannot be converted to products without the necessary activation energy. The activation energy is like a hill. If we are in a mountainous region, we may find that the only way to go from one point to another is to climb over a hill. It is the same in a chemical reaction. Even though the products may have a lower energy than the reactants (H_2O has a lower energy than $H_2 + O_2$), the products cannot be formed unless the reactants "go over the hill," that is, gain the necessary activation energy.

Molecules tend to go spontaneously from higher to lower energy states.

Let us look into this more closely. In a typical reaction, bonds are broken, and new bonds are formed. For example, when H_2 reacts with N_2 to give NH_3, six covalent bonds must break, and six new covalent bonds must be formed (Figure 7.4). Breaking a bond requires an input of energy, but forming a bond releases energy. In a "downhill" reaction of the type shown in Figure 7.3, the amount of energy released in making the new bonds is greater than that required to break the original bonds. In other words, the reaction is exothermic. Yet it may well have a substantial activation energy, or energy barrier, because, in most cases, at least one bond must break before any new bonds can form. This means energy must be put in

"Downhill" reactions are exothermic. "Uphill" reactions are endothermic.

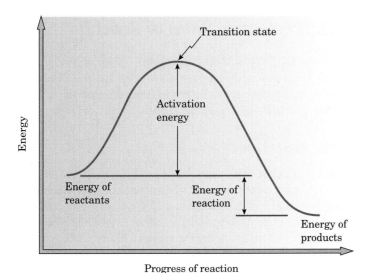

FIGURE 7.3 Energy diagram for an exothermic reaction. The energy of the reactants is greater than the energy of the products.

FIGURE 7.4 The reaction $3H_2 + N_2 \rightarrow 2NH_3$. In this reaction, six covalent bonds (three single bonds and one triple bond) are broken, and six new covalent bonds are formed.

before we get any back. This is like the situation you might find yourself in if somebody offered to let you buy into a business from which, for an investment of \$100 000, you could get an income of \$40 000 per year, beginning in one year. In the long run, you would do very well, but first you need to put up the \$100 000 right now (the activation energy).

Every reaction has a different energy diagram. In some cases, the energy of the products is higher than that of the reactants ("uphill" reactions, Figure 7.5), but in almost all cases there is an energy hill—the activation energy. The activation energy is inversely related to the rate of the reaction. **The lower the activation energy, the faster the reaction; the higher the activation energy, the slower the reaction.**

The top of the energy hill is called the **transition state.** When the reacting molecules reach this point, one or more original bonds are partially broken, and one or more new bonds may be in the process of formation. A typical situation is shown in Figure 7.6.

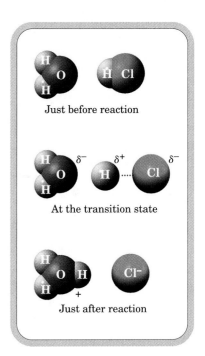

FIGURE 7.6 The position of two reacting molecules before, at, and after the transition state.

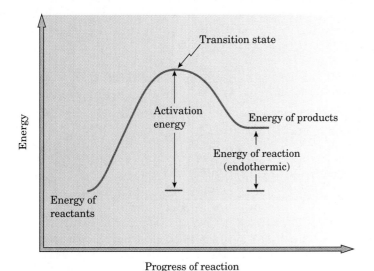

FIGURE 7.5 Energy diagram for an endothermic reaction. The energy of the products is greater than that of the reactants.

7.4 Factors Affecting Rates of Reaction

In the previous section, we saw that reactions occur as a result of collisions between fast-moving molecules possessing a certain minimum energy (the activation energy). In this section, we examine some of the factors that affect activation energies and reaction rates.

Nature of the Reactants

In general, reactions that take place between ions in aqueous solution (Section 4.7) are extremely rapid, almost instantaneous. Activation energies for these reactions are very low because usually no covalent bonds need to be broken. As we might expect, reactions between covalent molecules, whether in aqueous solution or not, are much slower. Many of these require anywhere from 15 min to 24 h or more for most of the starting compounds to be converted to the products. There are, of course, reactions that take a good deal longer, but such reactions are seldom useful.

Concentration

For reactions in the gas phase, an increase in pressure usually increases the rate.

Consider a reaction $A + B \rightarrow C + D$. In most cases, a reaction rate increases when the concentration of either or both reactants is increased (Figure 7.7). For many reactions, though by no means all, there is a direct relationship between concentration and rate; that is, when the concentration of a reactant is doubled, the rate also doubles. All this is easily understandable on the basis of the collision theory. If we double the concentra-

(a)   (b)

FIGURE 7.7 (*a*) When heated in air, steel wool glows but does not burn rapidly because the O_2 concentration of the air is only about 20 percent. (*b*) When the glowing steel wool is put into 100 percent O_2, it burns vigorously. (*a, Charles D. Winters; b, Leon Lewandowski*) tion of A, there are twice as many molecules of A in the same volume, so the molecules of B in that volume now collide with twice as many A molecules per second as before. Since the reaction rate depends on the number of collisions per second, the rate doubles.

The relationship between rate and concentration can be expressed mathematically. For the reaction

$$2H_2O_2 \longrightarrow 2H_2O + O_2$$

the rate was determined to be -0.01 mole $H_2O_2/L \cdot min$ at a constant temperature when the initial concentration of $[H_2O_2]$ was 1 mole/L. In other words, every minute 0.01 mole/L of hydrogen peroxide was used up. Researchers also found that every time the concentration of H_2O_2 was doubled, the rate also doubled. This means that the rate is directly proportional to the concentration of H_2O_2. We can write this relationship as

$$\text{rate} = k[H_2O_2]$$

where the brackets around the H_2O_2 indicate the concentration and k is a constant, called the **rate constant.**

In the case where one of the reactants is a solid, the rate is affected by the surface area of the solid, so a substance in powder form reacts faster than the same substance in the form of large chunks.

The brackets [ ] stand for the concentration of the chemical species whose formula is given between the brackets.

EXAMPLE 7.2

Calculate the rate constant for the reaction

$$2H_2O_2 \longrightarrow 2H_2O + O_2$$

using the rate and the initial H_2O_2 concentration mentioned in the preceding discussion.

Solution
We solve the equation Rate $= k[H_2O_2]$ for k.

$$k = \text{Rate}/[H_2O_2] = \frac{-0.01 \text{ mole } H_2O_2/L/min}{1 \text{ mole } H_2O_2/L} = -0.01/min.$$

Problem 7.2 ▬▬▬
Calculate the rate of reaction for the reaction in Example 7.2 when the initial concentration of H_2O_2 is 0.36 mole/L. ■

Temperature

In virtually all cases, reaction rates increase with increasing temperature. An approximate rule for many reactions is that every time the temperature goes up by 10°C, the rate of reaction doubles. This rule is far from exact, but it is not far from the truth in many cases. As you can see, this is quite a large effect and says, for example, that if we run a reaction at 90°C instead of at room temperature (20°C), the reaction will go about 128 times faster. Putting it another way, if it takes 20 h to convert 100 g of reactant A to product C at 20°C, then it would take only about 10 min at 90°C. Temperature, therefore, is a powerful tool that lets us increase the rates of reactions that are inconveniently slow. It also lets us decrease the rates of reactions that are inconveniently fast. In some cases, we choose to run reac-

There are seven 10° increments between 20° and 90°, and $2^7 = 128$.

Why High Fever Is Dangerous

Box 1C points out that a sustained body temperature of 41.7°C (107°F) is invariably fatal. We can now see why a high fever is dangerous. Normal body temperature is 37°C (98.6°F), and all the many reactions in the body—including respiration, digestion, and the synthesis of various compounds—take place at that temperature. If an increase of 10°C causes the rates of most reactions approximately to double, then an increase of even 1°C makes them go significantly faster than normal.

Fever is a protective mechanism, and a small increase in temperature allows the body to kill germs faster by mobilizing the immune defense mechanism. But this increase must be small: An increase of 1°C brings the temperature to 38°C (100.4°F); an increase of 3°C brings it to 40°C (104°F). A temperature higher than 104°F increases reaction rates to the danger point.

One can easily detect the increase in reaction rates when a patient has a high fever. The pulse rate increases and breathing is faster in an effort to supply increased amounts of oxygen for the fast reactions. A marathon runner may become overheated on a hot and humid day. After a time, perspiration can no longer cool his body effectively, and he may suffer hyperthermia or heat stroke, which, if not treated properly, can cause brain damage.

tions at low temperatures because explosions would result or the reactions would otherwise be out of control at room temperature.

What causes reaction rates to increase with increasing temperature? Once again, we turn to collision theory. Here temperature has two effects.

1. In Section 5.2, we learned that temperature is related to the average kinetic energy of molecules. When the temperature increases, molecules move faster, which means that they collide more frequently. More frequent collisions mean higher reaction rates. However, this factor is much less important than the second factor.

2. Recall from Section 7.2 that a reaction between two molecules takes place only if there is an effective collision—a collision with an energy equal to or greater than the activation energy. However, when the temperature increases, not only is the average speed of the molecules greater, but there is also a different distribution of speeds. The number of very fast molecules increases much more than the average speed (Figure 7.8). This means that the number of effective collisions rises

FIGURE 7.8 Distribution of kinetic energies (molecular velocities) at two temperatures. The kinetic energy on the *x*-axis designated by E_a indicates the energy (molecular velocity) necessary to pass through the activation energy barrier. The shaded areas represent the fraction of molecules that have kinetic energies (molecular velocities) greater than the activation energy.

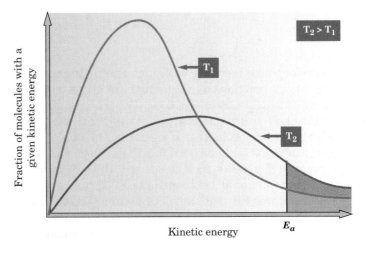

BOX 7B

The Effects of Lowering Body Temperature

As with a significant increase in body temperature, a substantial decrease below 37°C (98.6°F) is also harmful because reaction rates are lower than they should be. However, it is sometimes possible to take advantage of this effect. In some heart operations, it is necessary to stop the flow of oxygen to the brain for a considerable time. At 37°C (98.6°F), the brain cannot survive without oxygen for more than about 5 min without permanent damage. When the patient's body temperature is deliberately lowered to about 28 to 30°C (82.4 to 86°F), however, the oxygen flow can be stopped for a considerable time without causing damage because reaction rates are lowered. At 25.6°C (78°F), the body's oxygen consumption is reduced by 50 percent.

■ An operating table unit monitors a patient packed in ice.

(© 1998 Michael English, M.D./Custom Medical Stock Photo)

even more than the total number of collisions. Not only are there more collisions, but the percentage of collisions that have an energy greater than the activation energy also rises. This factor is mainly responsible for the sharp increase in reaction rates with increasing temperature.

Presence of a Catalyst

Any substance that increases the rate of a reaction without itself being used up is called a **catalyst.** Many catalysts are known: some that increase the rate of only one reaction and others that can affect several reactions. Although we have seen that reactions can be speeded up by increasing the temperature, in some cases they are still too slow even at the highest temperatures we can conveniently reach. In other cases it is not feasible to increase the temperature, for example, because other, unwanted, reactions would also be speeded up. In such cases, a catalyst, if we can find the right one for a given reaction, can prove very valuable. Many important industrial processes rely on catalysts (see Box 7E), and virtually all reactions that take place in living organisms are catalyzed by enzymes (Chapter 22).

Catalysts work by allowing the reaction to take a different pathway, one with a lower activation energy. Without the catalyst, the reactants would have to get over the higher energy hill shown in Figure 7.9. The catalyst provides a lower hill. As we have seen, a lower activation energy means a higher reaction rate.

Each catalyst has its own way of providing an alternative pathway. Many catalysts provide a surface on which the reactants can meet. For example, the reaction between ethylene gas, C_2H_4, and hydrogen gas, H_2, to give ethane gas, C_2H_6, goes so slowly without a catalyst that it is not practical, even if we increase the temperature to a reasonable level. But if the mixture of gases is shaken with finely divided solid platinum, the reaction takes place at a convenient rate (Section 12.6). The C_2H_4 molecules

In this dish, Cl^- ions act as a catalyst in the decomposition of NH_4NO_3. *(Charles D. Winters)*

Homogeneous catalyst A catalyst in the same phase as the reactants—for example, enzymes in body tissues

Heterogeneous catalyst A catalyst in a separate phase from the reactants—for example, the solid platinum in the reaction between the gases C_2H_4 and H_2

BOX 7C

Timed–Release Medication

It is often desirable that a particular medicine act slowly and maintain its action evenly in the body for 24 h. We know that a solid in a powder form reacts faster than the same weight in the form of a pill because the powder has a greater amount of surface area at which the reaction can take place. To slow the reaction and to have the drug delivered evenly to the tissues, pharmaceutical companies coat beads of some of their drugs. The coating prevents the drug from reacting for a time. The thicker the coating the longer it takes the drug to react. A drug with a smaller bead size has more surface area than a drug with a larger bead size; hence, drugs packaged in a smaller bead size will react faster. By combining the proper bead size with the proper amount of coating, the drug can be designed to deliver its effect over a 24-h period. In this way, the patient needs to take only one pill a day. Coating can also prevent the often experienced effect of stomach irritation. For example, aspirin is noted for its ability to cause stomach ulceration or bleeding in some people. Enteric-coated aspirin tablets have a polymeric coat that is acid-resistant. Thus, the drug does not dissolve until it reaches the intestines where it causes no harm.

■ Two packages of timed-release medicines. *(Beverly March)*

and the H_2 molecules meet each other on the surface of the platinum, where the proper bonds can be broken and the reaction can proceed:

We often write the catalyst over or under the arrow.

$$C_2H_4(g) + H_2(g) \xrightarrow{\text{Pt}} C_2H_6(g)$$

FIGURE 7.9 Energy diagram for a catalyzed reaction. The dashed line shows the energy curve for the uncatalyzed process. The catalyst provides an alternate pathway whose activation energy is lower.

Activation energy for catalyzed reaction

Energy

Progress of reaction

7.5 Reversible Reactions and Equilibrium

Many reactions are irreversible. When a piece of paper is completely burned, the products are CO_2 and H_2O. Anybody who takes pure CO_2 and H_2O and tries to make them react to give paper and oxygen will not succeed. A tree, of course, turns CO_2 and H_2O into wood and oxygen, and we, in sophisticated factories, make paper from the wood, but this is not the same as directly combining CO_2, H_2O, and energy in a single process to get paper and oxygen. Therefore, we can certainly consider the burning of paper an irreversible reaction.

Other reactions, however, are reversible. **A reversible reaction is one that can be made to go in either direction.** For example, if carbon monoxide is mixed with water in the gas phase at a high temperature, hydrogen and carbon dioxide are produced:

$$CO(g) + H_2O(g) \longrightarrow CO_2(g) + H_2(g)$$

But, if we desire, we can also make the reaction take place the other way: We can mix hydrogen and carbon dioxide to get carbon monoxide and water vapor:

$$CO_2(g) + H_2(g) \longrightarrow CO(g) + H_2O(g)$$

Let us see what happens when we run a reversible reaction. We will add some carbon monoxide to water vapor in the gas phase. The two compounds begin to react at a certain rate (the forward reaction):

$$CO(g) + H_2O(g) \longrightarrow CO_2(g) + H_2(g)$$

As the reaction proceeds, the concentrations of CO and H_2O gradually decrease because both reactants are being used up. This means that the rate of the reaction also gradually decreases because the rate of a reaction depends on the concentration of the reactants (Section 7.4).

But what is happening in the other direction? Before we added the carbon monoxide, there was no carbon dioxide or hydrogen present. As soon as the forward reaction began, however, it produced small amounts of these substances, and we now have some CO_2 and H_2. These two compounds will now, of course, begin reacting with each other (the reverse reaction):

$$CO_2(g) + H_2(g) \longrightarrow CO(g) + H_2O(g)$$

At first the reverse reaction is very slow, but as the concentrations of H_2 and CO_2 (produced by the forward reaction) gradually increase, the rate of the reverse reaction also gradually increases.

We have a situation, then, in which the rate of the forward reaction gradually decreases, while the rate of the reverse reaction (which began at zero) gradually increases. Eventually the two rates become equal. When this point is reached, we say that the process is in **dynamic equilibrium** or just in **equilibrium.**

What happens in the reaction container once we reach equilibrium? If we measure the concentrations of the substances in the container, we find that **no change in concentration takes place after equilibrium is reached** (Figure 7.10). Whatever the concentrations of all the substances are at equi-

The units of concentration in the gas phase are the same as in solution: moles per liter.

Chemical equilibrium A state in which the rate of the forward reaction equals the rate of the reverse reaction

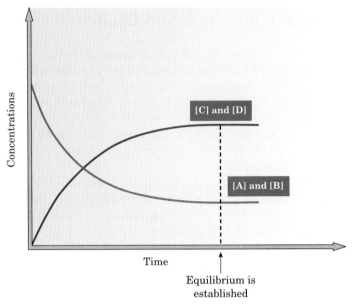

FIGURE 7.10   Changes in the concentrations of A, B, C, and D in the A + B $\rightleftharpoons$ C + D system, as equilibrium is approached. Only A and B are present at the beginning of the reaction.

Another way to look at it is that the concentration of carbon monoxide (and of the other three compounds) does not change because it is being used up as fast as it is being formed.

librium, they remain the same forever unless something is done to disturb the equilibrium (this is discussed in Section 7.8). This does not mean that all the concentrations must be the same—they can, in fact, be all different and usually are—but it does mean that, whatever they are, they no longer change once equilibrium has been reached, no matter how long we wait.

Given the fact that the concentrations of all the substances, both reactants and products, no longer change, can we say that nothing is happening? No, we know that both reactions are going on; all the molecules are constantly reacting—the CO and H_2O are being changed to CO_2 and H_2, and the CO_2 and H_2 are being changed to CO and H_2O. But because the rates of the forward and reverse reactions are the same, none of the concentrations change.

In the example just discussed, we approached equilibrium by adding carbon monoxide to water vapor. We could also have added carbon dioxide to hydrogen. In either case, we eventually get an equilibrium mixture containing the same four compounds (Figure 7.11).

At the start:
A mixture of
CO and H_2O

At equilibrium:
All four substances
present

At the start:
A mixture of
CO_2 and H_2

FIGURE 7.11   An equilibrium can be approached from either direction.

It is not necessary to begin with equal amounts. We could, for example, take 10 moles of carbon monoxide and 0.2 mole of water vapor. We would still arrive at an equilibrium mixture of all four compounds.

7.6 Equilibrium Constants

Chemical equilibria can be treated by a simple mathematical expression. First, let us write

$$aA + bB + \cdots \rightleftharpoons cC + dD + \cdots$$

as the general equation for all reversible reactions. In this equation, the capital letters stand for substances—H_2O and CO_2, for instance—and the small letters are the coefficients of the balanced equation. The double arrow is used to show that the reaction is reversible. The dots mean that more than two substances can be present on either side.

Once equilibrium is reached, the following equation is valid:

$$K = \frac{[C]^c[D]^d \cdots}{[A]^a[B]^b \cdots} \quad \text{The \textbf{equilibrium expression}}$$

Let us examine this equation. The K is a constant called the **equilibrium constant.** What this expression tells us is that, when we multiply the concentrations of the substances on the right-hand side of the chemical equation and divide this product by the concentrations of the substances on the left-hand side (after raising each number to the appropriate power), we get the equilibrium constant, a number that does not change.

Let us look at examples of how to set up equilibrium expressions.

EXAMPLE 7.3

Write the equilibrium expression for the reaction

$$CO(g) + H_2O(g) \rightleftharpoons CO_2(g) + H_2(g)$$

Solution

$$K = \frac{[CO_2][H_2]}{[CO][H_2O]}$$

This expression tells us that, at equilibrium, the concentration of carbon dioxide multiplied by the concentration of hydrogen and divided by the concentrations of water and carbon monoxide is a constant, K. Note that no exponent is written in this equation because all the coefficients of the balanced chemical equation are *one,* and, by convention, an exponent of *1* is not written. Mathematically, it would be just as correct to write the left-hand compounds on top, but the universal custom is to write them as shown here, with the products on top and the reactants on the bottom.

Problem 7.3 ▬▬▬
Write the equilibrium expression for the reaction

$$SO_3 + H_2O \rightleftharpoons H_2SO_4 \quad ■$$

We have already met two examples of dynamic equilibrium in this book: the equilibrium between a liquid and its vapor (Section 5.9) and the equilibrium involving solute particles in saturated solutions (Section 6.4).

This expression was first derived in 1863 by Cato Guldberg (1836–1902) and Peter Waage (1833–1900), two Norwegian chemists

Equilibrium constant A value calculated from the equilibrium expression for a given reaction indicating in which direction the reaction goes

It is understood that the concentration of the species within the brackets is always expressed in moles per liter.

The concentration terms in brackets in the expression for K are for species in solution, either liquid or gaseous.

EXAMPLE 7.4

Write the equilibrium expression for the reaction

$$O_2(g) + 4ClO_2(g) \rightleftharpoons 2Cl_2O_5(g)$$

Solution

$$K = \frac{[Cl_2O_5]^2}{[O_2][ClO_2]^4}$$

In this case, the chemical equation has coefficients other than unity, so the equilibrium expression contains powers.

Problem 7.4 ▬▬▬

Write the equilibrium expression for the reaction

$$2NH_3(g) \rightleftharpoons N_2(g) + 3H_2(g) \quad \blacksquare$$

Now let us see how K is calculated.

EXAMPLE 7.5

Some H_2 is added to I_2 at 427°C, and the following reaction is allowed to come to equilibrium:

$$I_2(g) + H_2(g) \rightleftharpoons 2HI(g)$$

When equilibrium is reached, the concentrations are found to be $[I_2] = 0.42$ mole/L, $[H_2] = 0.025$ mole/L, and $[HI] = 0.76$ mole/L. Calculate K at 427°C.

Solution
The equilibrium expression is

$$K = \frac{[HI]^2}{[I_2][H_2]}$$

Substituting the concentrations, we get

$$K = \frac{(0.76)^2 M^2}{(0.42)M \cdot (0.025)M} = 55$$

The equilibrium constant of this reaction is unitless because the molarities cancel. Equilibrium constants of other reactions may have different units, molarities on different powers (exponent), depending on the stoichiometry of the reaction.

Problem 7.5 ▬▬▬

What is the equilibrium constant for the following reaction if the equilibrium concentrations are $[PCl_3] = 1.66\ M$, $[Cl_2] = 1.66\ M$, and $[PCl_5] = 1.66\ M$?

$$PCl_3 + Cl_2 \rightleftharpoons PCl_5 \quad \blacksquare$$

The last example shows us that the reaction between I_2 and H_2 to give HI has an equilibrium constant of 55. What does this mean? At constant temperature, equilibrium constants remain the same no matter what concentrations we have. That is, at 427°C, if we begin by adding, say, 5 moles of H_2 to 5 moles of I_2, the two reactions take place, and equilibrium is reached. At that point, the concentration of HI, squared, divided by the concentrations of I_2 and H_2 equals 55. If we begin instead, at 427°C, with 7 moles of H_2 and 2 moles of I_2, or 0.4 mole of H_2 and 17 moles of I_2, or any number of moles of HI, once equilibrium is reached, the value of $[HI]^2/[I_2][H_2]$ is again 55. It makes no difference what concentrations of the three substances we begin with. At 427°C, as long as all three substances are present and equilibrium has been reached, the concentrations of the three substances so adjust themselves that the value of the equilibrium expression is 55.

The equilibrium constant is different for every reaction. Some reactions have a large K; others have a small K. A reaction with a very large K proceeds almost to completion (to the right). For example, K for the following reaction is about 10 000 000, or 10^7 at 25°C:

$$HCl(aq) + H_2O(l) \rightleftharpoons H_3O^+(aq) + Cl^-(aq)$$

This symbol $\rightleftharpoons$ means that the equilibrium lies far to the right.

This value of K means that, at equilibrium, $[H_3O^+$ and $Cl^-]$ must be very large and $[HCl]$ and $[H_2O]$ very small so that $[H_3O^+][Cl^-]/[HCl][H_2O]$ can be equal to 10^7. This means that if we add HCl to H_2O we can be certain that, when equilibrium has been reached, an essentially complete reaction has taken place.

On the other hand, a reaction with a very small K, say 10^{-7}, hardly goes at all. Equilibrium effects are most obvious in reactions with K values between 10^3 and 10^{-3}. In such cases, the reaction goes part way, and significant concentrations of all substances are present at equilibrium. An example is the reaction between carbon monoxide and water discussed in Section 7.5, for which K is equal to 10 at 600°C.

The equilibrium constant for a given reaction remains the same no matter what happens to the concentrations, but this is not true for changes in temperature. The value of K does change when the temperature is changed.

7.7 Equilibrium and Reaction Rates

As pointed out in the previous section, the equilibrium expression is valid only after equilibrium has been reached. Before that, there is no equilibrium, and the equilibrium expression is not valid. But how long does it take for a reaction to reach equilibrium? For this there are no easy answers. Some reactions, if the reactants are well mixed, reach equilibrium in less than one second; others will still not get there even after millions of years.

There is no relationship between the rate of a reaction (how long it takes to reach equilibrium) and the value of K. It is possible to have a large K and a low rate, as in the reaction between H_2 and O_2 to give H_2O, which does not reach equilibrium for many years (Section 7.1), or a small K and a high rate, as well as reactions in which both the rate and K are large or both are small.

7.8 Le Chatelier's Principle

When a reaction has reached equilibrium, the forward and reverse reactions are taking place at the same rate, and the concentrations of all the components do not change as long as we don't do anything to the system. But what happens if we do? In 1888 Henri Le Chatelier (1850–1936) put forth the statement known as **Le Chatelier's principle:** If an external stress is applied to a system in equilibrium, the system reacts in such a way as to partially relieve the stress. Let us look at four types of stress that can be put on chemical equilibria.

Addition of Reaction Components

Suppose that the reaction

$$CH_3COOH + C_2H_5OH \xrightleftharpoons{HCl} CH_3COOC_2H_5 + H_2O$$
Acetic acid   Ethanol   Ethyl acetate

The HCl is a catalyst, but this does not affect the equilibrium.

has reached equilibrium. This means that the reaction flask contains all these substances and that the concentrations no longer change. We now disturb the system by adding some acetic acid. The result is that the concentration of acetic acid suddenly increases, which increases the rate of the forward reaction. As a consequence, the concentrations of products (ethyl acetate and water) begin to increase. At the same time, the concentrations of reactants are decreasing. Now, an increase in the concentrations of products causes the rate of the reverse reaction to increase, but the rate of the forward reaction is decreasing, so eventually the two rates will be equal again and a new equilibrium will be established.

When that happens, the concentrations once again remain constant, but they are not the same as they were before. The concentrations of ethyl acetate and water are higher now, and the concentration of ethanol is lower. The concentration of acetic acid is higher because we added some, but it is less than it was immediately after we made the addition.

The equilibrium constant K of course remains the same.

This always happens when we add more of any component to a system in equilibrium. The addition constitutes a stress; the system relieves the stress by increasing the concentrations of the components on the other side of the equation. We say that the equilibrium shifts in the opposite direction:

Adding
CH_3COOH $CH_3COOH + C_2H_5OH \xrightleftharpoons{HCl} CH_3COOC_2H_5 + H_2O$

The addition of acetic acid, on the left-hand side, causes the rate of the forward reaction to increase and the reaction to move toward the right: More ethyl acetate and water are formed and some of the acetic acid and ethanol are used up. The same thing happens if we add ethanol, C_2H_5OH.

On the other hand, if we add water or ethyl acetate, the rate of the reverse reaction increases, and the reaction shifts to the left:

$CH_3COOH + C_2H_5OH \xrightleftharpoons{HCl} CH_3COOC_2H_5 + H_2O$ Adding
$CH_3COOC_2H_5$

We can summarize by saying that the addition of any component causes the equilibrium to shift to the opposite side.

EXAMPLE 7.6

When dinitrogen tetroxide, N_2O_4, a colorless gas, is enclosed in a vessel, a color indicating the formation of brown nitrogen dioxide, NO_2, soon appears (Figure 7.12). The intensity of the brown color indicates the amount of nitrogen dioxide formed. The equilibrium reaction is

$$N_2O_4(g) \rightleftharpoons 2NO_2(g)$$
Dinitrogen tetroxide (colorless)   Nitrogen dioxide (brown)

When more N_2O_4 is added to the equilibrium mixture, the brownish color becomes darker. Explain what happened.

Solution
The darker color indicates that more nitrogen dioxide is formed. This happens because the addition of the reactant shifts the equilibrium to the right, forming more product, NO_2.

Problem 7.6
What happens to the equilibrium reaction

$$2NOBr(g) \rightleftharpoons 2NO(g) + Br_2(g)$$

when Br_2 gas is added to the equilibrium mixture? ∎

Removal of a Reaction Component

It is not always as easy to remove a component from a reaction mixture as it is to add one, but there are often ways to do it. The removal of a component, or even a decrease in its concentration, lowers the corresponding reaction rate and changes the position of the equilibrium. If we remove

FIGURE 7.12 Effect of temperature on the equilibrium.

The tube at the left contains a saturated solution of silver acetate (Ag^+ ions and CH_3COO^- ions) in equilibrium with solid silver acetate. When silver ions in the form of silver nitrate solution are added, the equilibrium

$$Ag^+(aq) + CH_3COO^-(aq) \rightleftharpoons CH_3COOAg(s)$$

shifts to the right, producing more solid silver acetate, as can be seen in the tube on the right. *(Charles D. Winters)* a reactant, the reaction is shifted to the left, toward the side from which the reactant was removed. If we remove a product, the reaction is shifted to the right, toward the side from which the product was removed.

In the case of the acetic acid–ethanol equilibrium, ethyl acetate has the lowest boiling point of the four components and can be removed by distillation. If this is done, the equilibrium shifts to that side so that more ethyl acetate is produced to compensate for the removal:

$$CH_3COOH + C_2H_5OH \underset{HCl}{\rightleftharpoons} CH_3COOC_2H_5 + H_2O$$

 Removing $CH_3COOC_2H_5$

The concentrations of CH_3COOH and C_2H_5OH decrease, and that of H_2O increases. The effect of removing a component is thus the opposite of adding one. The removal of a component causes the equilibrium to shift to the side the component was removed from.

No matter what happens to the individual concentrations, the value of the equilibrium constant remains unchanged.

EXAMPLE 7.7

The beautiful stone we know as marble is mostly calcium carbonate. When acid rain containing sulfuric acid attacks marble, the following equilibrium reaction can be written

$$CaCO_3(s) + H_2SO_4(aq) \rightleftharpoons CaSO_4(s) + CO_2(g) + H_2O(l)$$

How does the fact that CO_2 is a gas influence the equilibrium?

Solution
The gaseous CO_2 diffuses away from the reaction site, meaning that this product is removed from the equilibrium mixture. The equilibrium shifts to the right, so that the statue keeps eroding more and more.

Problem 7.7 ▬▬
In the reaction

$$2H_2O_2(aq) \rightleftharpoons 2H_2O(l) + O_2(g)$$

oxygen has limited solubility in water (see Table 6A). What happens to the equilibrium after the solution becomes saturated with oxygen? ∎

Headstone damaged by acid rain.
(John D. Cunningham/Visuals Unlimited)

Change in Temperature

The effect of a change in temperature on a reaction that has reached equilibrium depends on whether the reaction is exothermic (gives off heat) or endothermic (requires heat). Let us look at an exothermic reaction:

$$2H_2(g) + O_2(g) \rightleftharpoons 2H_2O(l) + 137\ 000\ \text{cal}$$

If we look upon the heat as a product of this reaction, then we can use the same type of reasoning as we did before. An increase in temperature means that we are adding heat. Since heat is a product, its addition pushes the equilibrium to the opposite side. We can therefore say that, if the reaction is at equilibrium and we increase the temperature, the reaction goes to the left—the concentrations of H_2 and O_2 increase and that of H_2O decreases. This is true of all exothermic reactions.

BOX 7D

Sunglasses and Le Chatelier's Principle

Heat is not the only form of energy that affects equilibria. The statements made in the text regarding endothermic or exothermic reactions can be generalized to endergonic and exergonic reactions. A practical illustration of this is the use of sunglasses with adjustable shading. The compound silver chloride, AgCl, is incorporated in the glasses. This compound, on exposure to sunlight, produces metallic silver, Ag.

$$\text{Light} + 2Ag^+ + 2Cl^- \rightleftharpoons 2Ag + Cl_2$$

The more silver produced, the darker the glasses. At night, or when the wearer goes indoors, the reaction is reversed according to Le Chatelier's principle. In this case, addition of sunlight drives the equilibrium to the right; its removal drives it to the left.

An increase in temperature drives an exothermic reaction toward the reactants (to the left).

A decrease in temperature drives an exothermic reaction toward the products (to the right).

For an endothermic reaction, of course, the opposite is true.

An increase in temperature drives an endothermic reaction toward the products.

A decrease in temperature drives an endothermic reaction toward the reactants.

Remember (Section 7.4) that a change in temperature changes not only the position of equilibrium but the value of K, the equilibrium constant, as well.

EXAMPLE 7.8

The conversion of nitrogen dioxide to dinitrogen tetroxide is an exothermic reaction:

$$2NO_2(g) \rightleftharpoons N_2O_4(g) + 13\ 700 \text{ cal}$$

NO_2 is a brown gas; N_2O_4 is a colorless gas. In Figure 7.12 we see that at 50°C the brown color is darker; hence, the NO_2 concentration is higher than at 0°C. Why?

Solution
To go from 0 to 50°C, heat must be added. But heat is a product of this equilibrium reaction, as written. The addition of heat, therefore, will shift the equilibrium to the left. This shift produces more NO_2, hence the darker brown color.

Problem 7.8 ■
In a reaction $A \rightleftharpoons B$, the amount of B increased when the temperature of the reaction was raised. Was this an exothermic or an endothermic reaction? Explain. ■

The Haber Process

Human beings and animals need proteins and other nitrogen-containing compounds in order to live. Ultimately, the nitrogen in these compounds comes from the plants that we eat. There is plenty of N_2 in the atmosphere, but the only way that nature converts this N_2 to compounds usable by biological organisms is by certain bacteria that have the ability to "fix" atmospheric N_2, that is, convert it to ammonia, NH_3. Most of these bacteria live in the roots of certain plants such as clover, alfalfa, peas, and beans. However, the amount of N_2 fixed by such bacteria each year is far less than the amount necessary to feed all the humans and animals in the world.

The world today can support its population only by using fertilizers made by artificial fixing, primarily by the **Haber process,** which converts N_2 to NH_3 by the reaction

$$N_2(g) + 3H_2(g) \rightleftharpoons 2NH_3(g) + 22\,000 \text{ cal}$$

Early workers on the problem of fixing nitrogen were troubled by a conflict between equilibrium and rate. Since this is an exothermic reaction, an increase in temperature drives the equilibrium to the left, so the best results (largest possible yield of NH_3) should be obtained at low temperatures. But at low temperatures, the rate is too low to produce any meaningful amounts of NH_3. In 1908 Fritz Haber (1868–1934) solved this problem when he discovered a catalyst that permits the reaction to take place at a convenient rate at 500°C.

■ Ammonia is produced by bacteria living in these root nodules. *(© C. P. Vance/Visuals Unlimited)*

The NH_3 produced by the Haber process is converted to fertilizers, which are used all over the world. Without these fertilizers, food production would diminish so much that widespread starvation would result.

Addition of a Catalyst

As we saw in Section 7.4, a catalyst increases the rate of a reaction without itself being changed. For a reversible reaction, catalysts always increase the rates of both the forward and reverse reactions to the same extent. Therefore, the addition of a catalyst has no effect on the position of equilibrium. However, adding a catalyst to a system not yet at equilibrium causes it to reach equilibrium faster than it would without the catalyst.

SUMMARY

Some reactions are fast; others are slow. The **rate of a reaction** is the change in concentration of a reactant or a product per unit time. Reactions take place when molecules or ions collide. The rate of a reaction increases with an increasing number of **effective collisions,** that is, ones that lead to a reaction. The energy necessary for a reaction to begin is the **activation energy.** Effective collisions are those that have (a) more than the activation energy and (b) the proper orientation in space.

The lower the activation energy, the faster the reaction. An energy diagram shows the progress of a reaction. The position at the top of the curve is the **transition state.**

Reaction rates generally increase with increasing concentration and temperature and also depend on the nature of the reactants. The rate of some reactions can be increased by the addition of a **catalyst,** which provides a pathway with a lower activation energy.

The rate constant that gives the relationship between the rate of the reaction and the concentration of the reactants is a constant at a constant temperature.

Many reactions are reversible and eventually reach equilibrium—some slowly, some quickly. At **equilibrium,** the forward and reverse reactions take place at equal rates, and concentrations do not change. Every equilibrium has an **equilibrium expression** and an **equilibrium constant,** K, which does not change when concentrations change but does change when the temperature changes. The equilibrium constant is large for some reactions and small for others. There is no relationship between K and the rate of a reaction.

Le Chatelier's principle tells us what happens when we put stress on a system in equilibrium. Addition of a component causes the equilibrium to shift to the opposite side. Removal of a component causes the equilibrium to shift to the side the component is taken from. Increasing the temperature drives an exothermic equilibrium to the side of the reactants. Addition of a catalyst has no effect on the position of equilibrium.

KEY TERMS

Activation energy (Section 7.2)
Catalyst (Section 7.4)
Chemical kinetics (Section 7.1)
Dynamic equilibrium (Section 7.5)
Effective collision (Section 7.2)
Equilibrium (Section 7.5)

Equilibrium constant (Section 7.6)
Equilibrium expression (Section 7.6)
Haber process (Box 7E)
Heterogeneous catalyst (Section 7.4)
Homogeneous catalyst (Section 7.4)
Kinetics (Section 7.1)

Le Chatelier's principle (Section 7.8)
Rate constant (Section 7.4)
Rate of a reaction (Section 7.1)
Reversible reaction (Section 7.5)
Transition state (Section 7.3)

CONCEPTUAL PROBLEMS

7.A Two gases react according to the equation: A + B → AB. Only the effective collisions between A and B will yield the product. The two gases enter the reaction chamber through two jets set 180° to each other. What can you do to have a maximum rate of reaction at a set constant temperature?

7.B A reaction has a high rate constant but a small equilibrium constant. What does this mean in terms of producing an industrial product?

7.C What can you do, using the Le Chatelier principle, to make the reaction in Problem 7.B economically feasible?

7.D Draw a diagram similar to Figure 7.5. Draw a second line of energy profile starting and ending at the same level as the first but having a smaller peak than the first line. Label them 1 and 2. What may have occurred to change the energy profile of a reaction from 1 to 2?

PROBLEMS

Difficult problems are designated by an asterisk.

Reaction Rates

7.9 In the reaction

$$CH_3Br + I^- \longrightarrow CH_3I + Br^-$$

we start the reaction with an initial I^- concentration of 0.450 M. After 10 min the I^- concentration is 0.268 M. What is the rate of reaction?

7.10 If the concentration of a product increases from 0.320 mole/L to 0.750 mole/L over a period of 1 h and 10 min, what is the rate of reaction?

Molecular Collisions and Activation Energy

7.11 Consider the following reaction:

$$H_2(g) + Br_2(g) \longrightarrow 2HBr(g)$$

How many covalent bonds are broken and how many new ones are formed?

7.12 Two kinds of gas molecules are reacted at a set temperature. The gases are blown into the reaction vessel from two tubes. In setup (a) the two tubes are aligned parallel to each other; in (b) they are 90° to each other; and in (c) they are aligned against each other. Which setup would yield the most effective collisions?

7.13 Why are reactions between ions in aqueous solution generally much faster than reactions between covalent molecules?

7.14 A certain reaction is exothermic by 9 kcal/mole and has an activation energy of 14 kcal/mole. Draw an energy diagram for this reaction, and label the transition state.

Factors Affecting Rates of Reaction _____

7.15 A quart of milk quickly spoils if left at room temperature but keeps for several days in a refrigerator. Explain.

7.16 Explain why the rates of reactions increase when the concentration of reactants is increased.

*7.17 If a certain reaction takes 16 h to go to completion at 10°C, at about what temperature should we run it if we want it to go to completion in 1 h?

7.18 In most cases, when we run a reaction by mixing a fixed quantity of A with a fixed quantity of B, the rate of the reaction begins at a maximum and then decreases as time goes by. Explain.

7.19 If you were running a reaction and wished it to go faster, what three things might you try to accomplish this?

7.20 What factors determine whether a reaction run at a given temperature will be fast or slow?

7.21 Explain how a catalyst increases the rate of a reaction.

7.22 If you add a piece of marble, $CaCO_3$, to a 6 M HCl solution at room temperature, you will see some bubbles form around the marble as CO_2 gas slowly rises. If you crush another piece of marble and add it to the same solution at the same temperature, you will see vigorous gas formation, so much so that the solution appears to be boiling. Explain.

Reversible Reactions and Equilibrium _____

7.23 Burning a piece of paper is an irreversible reaction. Give some other examples of irreversible reactions?

7.24 If the reaction $PCl_3 + Cl_2 \rightleftharpoons PCl_5$ is at equilibrium, are the concentrations of PCl_3, Cl_2, and PCl_5 necessarily equal? Is the concentration of PCl_3 necessarily equal to that of Cl_2? Explain.

Equilibrium Constants _____

7.25 Write equilibrium expressions for these reactions.
(a) $2H_2O_2 \rightleftharpoons 2H_2O + O_2$
(b) $2N_2O_5 \rightleftharpoons 2N_2O_4 + O_2$
(c) $6H_2O + 6CO_2 \rightleftharpoons C_6H_{12}O_6 + 6O_2$

7.26 Write the chemical equations corresponding to the following equilibrium expressions.
(a) $K = \dfrac{[H_2CO_3]}{[CO_2][H_2O]}$ (b) $K = \dfrac{[P_4][O_2]^5}{[P_4O_{10}]}$

(c) $K = \dfrac{[F_2]^3[PH_3]}{[HF]^3[PF_3]}$ (d) $K = \dfrac{[Ca]^2[O_2]}{[CaO]^2}$

7.27 When a mixture containing H_2, CO_2, H_2O, and CO at 987°C reached equilibrium, the concentrations were

$[H_2] = 3.37\ M$ $[CO_2] = 0.133\ M$

$[H_2O] = 0.72\ M$ $[CO] = 0.933\ M$

What is the equilibrium constant at 987°C for the reaction

$$H_2(g) + CO_2(g) \rightleftharpoons H_2O(g) + CO(g)$$

*7.28 For the reaction

$$2NO_2 \rightleftharpoons N_2O_4$$

the equilibrium constant at 325 K was found to be 172. When a sample was taken of the equilibrium mixture, it was found to contain 0.0714 M NO_2. What was the equilibrium concentration of N_2O_4?

7.29 When a mixture containing NOCl, NO, and Cl_2 at 25°C reached equilibrium, the concentrations were

$[NO] = 1.4\ M$ $[NOCl] = 2.6\ M$ $[Cl_2] = 0.34\ M$

What is the equilibrium constant at 25°C for the reaction

$$2NOCl(g) \rightleftharpoons 2NO(g) + Cl_2(g)$$

7.30 The following reaction is very fast but has an equilibrium constant of 10^{-34}. Would you be wise to try to synthesize NO from this reaction?

$$N_2(g) + O_2(g) \rightleftharpoons 2NO(g)$$

7.31 Here are equilibrium constants for several reactions. Which of them favor the formation of products and which favor the formation of reactants?
(a) 4.5×10^8 (b) 32 (c) 4.5
(d) 3×10^{-7} (e) 0.0032

7.32 A particular reaction has an equilibrium constant of 1.13 under one set of conditions and 1.72 under a different set of conditions. Which conditions would be more advantageous in an industrial process to obtain the maximum amount of products? Explain.

Equilibrium and Reaction Rates _____

7.33 (a) If a reaction is very exothermic—that is, the products have a much lower energy than the reactants—can we be reasonably certain that it will take place rapidly? (b) If a reaction is very endothermic—that is, the products have a much higher energy than the reactants—can we be reasonably certain that it will take place extremely slowly or not at all? (Think carefully about this one.)

Le Chatelier's Principle _____

7.34 Assume that the following exothermic reaction is at equilibrium:

$$H_2(g) + I_2(g) \rightleftharpoons 2HI(g)$$

Tell whether the position of equilibrium will be shifted to the right or the left if we
(a) Remove some HI.
(b) Add some H_2.
(c) Remove some I_2.

(d) Increase the temperature.

(e) Add a catalyst.

7.35 The following reaction is endothermic:

$$3O_2(g) \rightleftharpoons 2O_3(g)$$

If the reaction is at equilibrium, tell whether the equilibrium will be shifted to the right or the left if we

(a) Remove some O_3.

(b) Remove some O_2.

(c) Add some O_3.

(d) Decrease the temperature.

(e) Add a catalyst.

7.36 The following reaction is exothermic:

$$2NO + Br_2 = 2NOBr$$

After it reaches equilibrium, we add a few drops of Br_2.

(a) What will happen to the equilibrium?

(b) What will happen to the equilibrium constant?

7.37 The following reaction is exothermic:

$$4NO_2(g) + 6H_2O(g) \rightleftharpoons 7O_2(g) + 4NH_3(g)$$

What happens to the position of equilibrium if we raise the reaction temperature from 350 to 420°C?

7.38 Is there any change in conditions that changes the equilibrium constant K of a given reaction?

*7.39 The equilibrium constant at 1127°C for the following endothermic reaction is 571:

$$2H_2S(g) \rightleftharpoons 2H_2(g) + S_2(g)$$

If the mixture is at equilibrium, what happens to K if we

(a) Add some H_2S?

(b) Add some H_2?

(c) Lower the temperature to 1000°C?

Boxes

7.40 (Box 7A) In a bacterial infection, body temperature may rise to 101°F. Does this body defense kill the bacteria directly by heat or by another mechanism? If so, which mechanism?

7.41 (Boxes 7A and 7B) Why is a high fever dangerous? Why is a low body temperature dangerous?

7.42 (Box 7B) Why do surgeons sometimes lower body temperatures during heart operations?

7.43 (Box 7C) A pain killer, for example, Tylenol, can be purchased in two forms both containing the same amount of drug. One form is a solid coated pill, the other is a capsule of the same coat that contains tiny beads. Which medication will act faster? Explain.

7.44 (Box 7D) What reaction takes place when sunlight hits the compound silver chloride?

7.45 (Box 7E) If the equilibrium for the Haber process is unfavorable at high temperatures, why do the factories nevertheless use high temperatures?

Additional Problems

7.46 In the reaction

$$H_2(g) + Cl_2(g) \longrightarrow 2HCl(g)$$

a 10°C increase in temperature doubles the rate of reaction. If the rate of reaction at 15°C is 2.8 moles of HCl per liter per s, what is the rate at −5°C and at 45°C?

*7.47 Draw an energy diagram for an exothermic reaction that yields 75 kcal/mole. The energy of activation is 30 kcal/mole.

7.48 Write equilibrium expressions for these ionic equilibria:

(a) $2OH^-(aq) + H_2PO_4^-(aq) \rightleftharpoons$
$$PO_4^{3-}(aq) + 2H_2O(l)$$

(b) $H_2O(l) + HNO_3(aq) \rightleftharpoons H_3O^+(aq) + NO_3^-(aq)$

(c) $2H_2O(l) \rightleftharpoons OH^-(aq) + H_3O^+(aq)$

7.49 For the reaction of

$$COCl_2(g) \longrightarrow CO(g) + Cl_2(g)$$

the rate of the reaction was −2.3 moles $COCl_2$/L/h when the initial $COCl_2$ concentration was 6.2 moles $COCl_2$/L. What is the rate constant of the reaction?

7.50 The equilibrium constant for

$$COCl_2(g) \rightleftharpoons CO(g) + Cl_2(g)$$

is 25. A measurement made on the equilibrium mixture found that the concentrations of CO and Cl_2 were each 0.80 M. What is the concentration of $COCl_2$ at equilibrium?

7.51 In the reaction

$$N_2O_4(g) \longrightarrow 2NO_2(g)$$

the concentration of N_2O_4 was measured in moles/L at the end of the times shown:

Time(s)	$[N_2O_4]$
0	0.200
10	0.180
20	0.160
30	0.140

What is the rate of reaction?

*7.52 How could you increase the rate of a gaseous reaction without adding more reactants or a catalyst and without changing the temperature?

*7.53 The equilibrium constant for the reaction α-D-glucose $\rightleftharpoons$ β-D-glucose is 1.5 at 30°C. This reaction takes place only in aqueous solution. If you begin with a fresh solution of 1 M α-D-glucose in water, what will be its concentration when equilibrium is reached?

*7.54 In an endothermic reaction, A → B, the activation energy was 100 kcal/mole. Would the activation energy of the reverse reaction, B → A, also be 100 kcal/mole, or would it be more or less? Explain with the aid of a diagram.

7.55 The reaction

$$C_2H_4(g) + Cl_2(g) \rightleftharpoons C_2H_4Cl_2(g)$$

has an equilibrium constant of 10 M^{-1}. What is the equilibrium concentration of $C_2H_4Cl_2$ if the equilibrium concentration of C_2H_4 is 0.2 M and that of Cl_2 is 0.4 M?

7.56 For the reaction

$$CH_3CH\!=\!CHCH_3 + Br_2 \longrightarrow CH_3CH\!-\!CHCH_3$$
$$\underset{Br}{|} \quad \underset{Br}{|}$$

the rate of reaction at 300K was found to be 0.22 M/min. What would be the approximate rate of reaction at 330K?

*7.57 Assume that there are two different reactions taking place at the same temperature. In A, two different spherical molecules collide to yield a product. In B, the shape of the colliding molecules is rod-like. Each reaction has the same number of collisions per second and the same activation energy. Which reaction goes faster?

*7.58 Is it possible for an endothermic reaction to have zero activation energy?

*7.59 In the reaction $2HI(g) \rightarrow H_2(g) + I_2(g)$ the rate of appearance of I_2 is measured. The following concentrations in mole/L were found at the end of the times shown:

Time (min)	[I$_2$]
0	0
10	0.30
20	0.60
30	0.90

What is the rate of reaction?

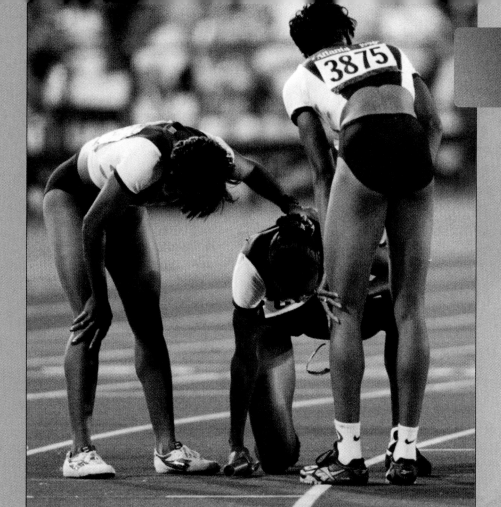

Acids and Bases

8.1   Introduction

We frequently encounter acids and bases in our daily lives. Oranges, lemons, and vinegar are examples of acidic foods, and sulfuric acid is in our automobile batteries. As for bases, we take antacid tablets for heartburn and use household ammonia as a cleaning agent. What do these substances have in common? Why are acids and bases usually discussed together?

In 1884, a young Swedish chemist named Svante Arrhenius (1859–1927) proposed an answer to the first of these questions by giving what was then a new definition of acids and bases. In modern terms, his definitions are

> An **acid** is a substance that produces H_3O^+ ions in aqueous solution.
>
> A **base** is a substance that produces OH^- ions in aqueous solution.

This definition of acid is a slight modification of what Arrhenius said, which was that an acid produces hydrogen ions, H^+. Today we know that H^+ ions cannot exist in water. An H^+ ion is a bare proton, and a charge of $+1$ is too concentrated for such a tiny particle (Section 3.2). Because of

Above: These U.S. runners have just won the gold medal in the 4 × 400 m relay race at the 1996 Olympic games. The buildup of an organic acid called lactic acid has caused severe muscle pain. In a short time the pain will be gone. (© RB-GM-J.O. Atlanta/Liaison Agency)

BOX 8A

Some Important Acids and Bases

Strong Acids Sulfuric acid, H_2SO_4, is used in a great many industrial processes. More tons of sulfuric acid are manufactured in the United States than any other chemical, organic or inorganic. It is a strong acid and a powerful oxidizing agent, as well as a dehydrating agent (capable of removing water from a substance). The dissolving of H_2SO_4 in water gives off a lot of heat. The acid should always be added to water, and not water to the acid, because of the danger of splashing.

Hydrochloric acid, HCl, is an important acid in chemistry laboratories. Pure HCl is a gas, and the HCl in laboratories is an aqueous solution. HCl is the acid in the gastric fluid in your stomach, where it is secreted at a strength of about 5 percent (w/v).

Nitric acid, HNO_3, is a strong oxidizing agent. A drop of it causes the skin to turn yellow. This is due to a reaction with skin protein. A yellow color with nitric acid has long been a test for proteins.

Weak Acids Acetic acid, CH_3COOH, is present in vinegar (about 5 percent w/v). Vinegar is made by allowing wine to oxidize in air, whereupon the ethanol in the wine is oxidized to acetic acid (Section 13.5). Pure acetic acid is called glacial acetic acid because of its melting point of 17 °C, which means that it freezes on a moderately cold day.

Boric acid, H_3BO_3, is a solid. Solutions of boric acid in water were once used as antiseptics, especially for eyes. Boric acid is toxic when swallowed.

Phosphoric acid, H_3PO_4, is one of the strongest of the weak acids. The ions produced from it — $H_2PO_4^-$, HPO_4^{2-}, and PO_4^{3-} — are important in biochemistry (see also Section 31.6).

Strong Bases Sodium hydroxide, NaOH, also called lye, is the most important of the strong bases. It is a solid whose aqueous solutions are used in many industrial processes, including the manufacture of glass and soap.

Potassium hydroxide, KOH, also a solid, is used for many of the same purposes as NaOH.

Weak Bases Ammonia, NH_3, the most important weak base, is a gas with many industrial uses. One of its chief uses is for fertilizers. A 5 percent (w/v) solution is sold in supermarkets as a cleaning agent, and weaker solutions are used as "spirits of ammonia" to revive people who have fainted.

Magnesium hydroxide, $Mg(OH)_2$, is a solid that is insoluble in water. A suspension of about 8 percent (w/v) $Mg(OH)_2$ in water is called milk of magnesia and is used as a laxative. $Mg(OH)_2$ is also used to treat waste water in metal-processing plants and as a flame retardant in plastics.

■ Reaction of nitric acid and copper. *(L. S. Stepanowicz/Visuals Unlimited)*

Although all chemists know that acidic aqueous solutions do not contain H^+ ions, they frequently use the terms "H^+" and "proton" when they really mean "H_3O^+." The three terms are generally interchangeable.

this, any H^+ ion in water immediately combines with an H_2O molecule to give a **hydronium ion, H_3O^+:**

$$H^+(aq) + H_2O(l) \longrightarrow H_3O^+(aq)$$

Apart from this modification, the Arrhenius definitions of acid and base are still valid and useful today, as long as we are talking about aqueous solutions. When we dissolve an acid in water, it reacts with the water to produce H_3O^+. For example, hydrogen chloride, HCl, in its pure state is a poisonous, choking gas. When dissolved in water, it reacts as follows:

$$H—\overset{..}{\underset{..}{Cl}}: + H—\overset{..}{\underset{..}{O}}—H \longrightarrow \left[H—\overset{..}{\underset{\underset{H}{|}}{O}}—H \right]^+ + :\overset{..}{\underset{..}{Cl}}:^-$$

The bottle on the shelf labeled "HCl" is actually not HCl but a solution of H_3O^+ and Cl^- ions in water.

With bases, the situation is slightly different. Many bases are solid metallic hydroxides, such as KOH, NaOH, $Mg(OH)_2$, and $Ca(OH)_2$. These compounds are ionic even in the solid state. When dissolved in water, the ions merely separate, and each ion is solvated by water molecules (Section 6.6). For example,

$$NaOH(s) \xrightarrow{H_2O} Na^+(aq) + OH^-(aq)$$

Other bases, though, are not hydroxides. These bases produce OH^- ions in water by reacting with the water. The most important example of this kind of base is ammonia, NH_3. Like HCl, ammonia is a poisonous, choking gas. When dissolved in water, it reacts as follows:

$$NH_3(aq) + H_2O(l) \rightleftharpoons NH_4^+(aq) + OH^-(aq)$$

Ammonia produces OH^- ions by taking H^+ from water molecules and leaving OH^- behind. This equilibrium lies well to the left. In a 1 M solution of NH_3 in water, only about 4 molecules of NH_3 out of every 1000 have reacted to form NH_4^+. Nevertheless, some OH^- ions are produced, so NH_3 is, therefore, a base.

Bottles of NH_3 in water are sometimes labeled "ammonium hydroxide" or "NH_4OH," but this gives a false impression of what is really in the bottle. Most of the NH_3 molecules have not reacted with the water, so the bottle contains mostly NH_3 and H_2O and only a little NH_4^+ and OH^-.

8.2 Acid and Base Strength

All acids are not equally strong. According to the Arrhenius definition of acids, a **strong acid** is one that **reacts almost completely with water to form H_3O^+ ions.** The six common acids listed in Table 8.1 react completely, or almost completely, when dissolved in water and are referred to as strong acids. All other acids produce a much smaller concentration of H_3O^+ ions and are referred to as **weak acids.** For example, acetic acid in water exists primarily as acetic acid molecules. The following equilibrium lies well to the left, and only a few acetic acid molecules (4 out of every 1000) are converted to acetate ions. Many acids are even weaker than acetic acid.

$$\underset{\text{Acetic acid}}{CH_3COOH(aq)} + H_2O(l) \rightleftharpoons \underset{\text{Acetate ion}}{CH_3COO^-(aq)} + H_3O^+(aq)$$

TABLE 8.1 Six Strong Acids and Four Strong Bases

Acids		Bases	
Formula	*Name*	*Formula*	*Name*
$HClO_4$	Perchloric acid	NaOH	Sodium hydroxide
HI	Hydroiodic acid	KOH	Potassium hydroxide
HBr	Hydrobromic acid	LiOH	Lithium hydroxide
H_2SO_4	Sulfuric acid	$Ba(OH)_2$	Barium hydroxide
HNO_3	Nitric acid		
HCl	Hydrochloric acid		

The word "alkali" is an older term for "base"; it means the same thing.

The boiling point of ammonia is $-33°C$.

We show the equation with equilibrium arrows because it is reversible.

Another reason not to use these labels is that NH_4OH cannot be isolated away from water as a pure substance.

Strong acid An acid that gives up H^+ ion easily and dissociates almost completely in water

Weak acid An acid that gives up H^+ ion with difficulty and dissociates less than 100% in water

Some strong acids and strong bases. *(Beverly March)*

Strong base A base that readily accepts H$^+$

These bases are strong because they are soluble in water and dissociate completely to OH$^-$.

Weak base A base that is less willing to accept H$^+$

A similar situation exists with bases. There are only four common **strong bases** (Table 8.1), all of the metal hydroxide type. As we saw in Section 8.1, ammonia is a **weak base** because the equilibrium lies to the left.

It is important to understand that the strength of an acid or base is not related to its concentration. HCl is a strong acid, whether it is concentrated or dilute, because in its reaction with water the equilibrium lies well to the right:

$$HCl(aq) + H_2O(l) \rightleftharpoons H_3O^+(aq) + Cl^-(aq)$$

Acetic acid is a weak acid, whether it is concentrated or dilute, because, in its reaction with water, the equilibrium lies well to the left.

In Section 6.6, we saw that electrolytes (substances that produce ions in solution) can be strong or weak. The strong acids and bases in Table 8.1 are strong electrolytes. Almost all other acids and bases are weak electrolytes.

8.3 Brønsted–Lowry Acids and Bases

Johannes Brønsted (1879–1947) was a Danish chemist, and Thomas Lowry (1874–1936) was an English chemist.

The Arrhenius definitions of acid and base are very useful in aqueous solutions. But what if water is not involved? In 1923 new definitions of acids and bases were proposed simultaneously but independently by Brønsted and Lowry. These definitions are more general and do not require that the solvent be water:

> An acid is a proton donor.
>
> A base is a proton acceptor.
>
> An acid-base reaction is the transfer of a proton.

As noted in Section 8.1, this is precisely how acids react with water. An acid gives a proton, H$^+$, to water:

$$HCl + H_2O \rightleftharpoons H_3O^+ + Cl^-$$
$$\text{Acid} \quad \text{Base} \qquad \text{Acid} \quad \text{Base}$$

TABLE 8.2 Some Acids and Their Conjugate Bases, in Decreasing Order of Acid Strength

	Acid		Conjugate Base		
Strong Acids	HI	Hydroiodic acid	I$^-$	Iodide ion	Weak Bases
	H$_2$SO$_4$	Sulfuric acid	HSO$_4^-$	Hydrogen sulfate ion	
	HCl	Hydrochloric acid	Cl$^-$	Chloride ion	
	HNO$_3$	Nitric acid	NO$_3^-$	Nitrate ion	
	H$_3$O$^+$	Hydronium ion	H$_2$O	Water	
	HSO$_4^-$	Hydrogen sulfate ion	SO$_4^{2-}$	Sulfate ion	
	H$_3$PO$_4$	Phosphoric acid	H$_2$PO$_4^-$	Dihydrogen phosphate ion	
	CH$_3$COOH	Acetic acid	CH$_3$COO$^-$	Acetate ion	
	H$_2$CO$_3$	Carbonic acid	HCO$_3^-$	Bicarbonate ion	
	H$_2$S	Hydrogen sulfide	HS$^-$	Hydrogen sulfide ion	
	H$_2$PO$_4^-$	Dihydrogen phosphate ion	HPO$_4^{2-}$	Hydrogen phosphate ion	
	NH$_4^+$	Ammonium ion	NH$_3$	Ammonia	
	C$_6$H$_5$OH	Phenol	C$_6$H$_5$O$^-$	Phenoxide ion	
	HCO$_3^-$	Bicarbonate ion	CO$_3^{2-}$	Carbonate ion	
	HPO$_4^{2-}$	Hydrogen phosphate ion	PO$_4^{3-}$	Phosphate ion	
	H$_2$O	Water	OH$^-$	Hydroxide ion	
Weak Acids	C$_2$H$_5$OH	Ethanol	C$_2$H$_5$O$^-$	Ethoxide ion	Strong Bases
	NH$_3$	Ammonia	NH$_2^-$	Amide ion	

We have, in this equation, two new ideas:

1. The water, because it accepts a proton, is a base by the Brønsted-Lowry definition.

2. The hydronium ion is an acid because it can now, at least in theory, give a proton back to the Cl$^-$. Similarly, the Cl$^-$ is a base because it can take the proton back.

To repeat, an acid in the Brønsted-Lowry picture is a proton donor. When it gives up its proton, it always becomes a base. This base is called the **conjugate base** of the acid. Some examples are given in Table 8.2.

Similarly, a base is a proton acceptor. When it accepts a proton, it always becomes an acid called the **conjugate acid** of that base. The examples in Table 8.2 illustrate this also. Every acid and the species that remains when it loses its proton are called a conjugate acid-base pair. Thus, HCl and Cl$^-$ are a conjugate acid-base pair, as are H$_3$O$^+$ and H$_2$O, and CH$_3$COOH and CH$_3$COO$^-$.

Note the following points from the examples in Table 8.2.

1. An acid can be positively or negatively charged, or it can be neutral. A base can be negatively charged, or it can be neutral.

2. Some acids can give up two or three protons (other acids, not shown in Table 8.2, can give up even more). Such acids are called **diprotic** or **triprotic.** For example, H$_2$SO$_4$ can give up two protons to become SO$_4^{2-}$ and H$_3$PO$_4$ can give up three to become PO$_4^{3-}$. In the Brønsted-Lowry picture, however, each acid is considered to give up only one proton **(monoprotic).** Thus, H$_2$SO$_4$ is treated as if it gives up a single proton to become HSO$_4^-$, and HSO$_4^-$ is regarded as a separate acid in its own right (note that HSO$_4^-$ is listed separately in Table 8.2).

3. Some substances appear on both sides of Table 8.2. For example, HCO$_3^-$, the bicarbonate ion, can give up a proton to become CO$_3^{2-}$ (thus, it is an acid) or accept a proton to become H$_2$CO$_3$ (thus, it is a base). A substance that can act as either an acid or a base is called **amphipro-**

A box of Arm & Hammer baking soda (sodium bicarbonate). Sodium bicarbonate is composed of Na$^+$ and HCO$_3^-$, the amphiprotic bicarbonate ion. *(Charles D. Winters)*

Amphoteric is an older term for amphiprotic.

tic. The most important amphiprotic substance in Table 8.2 is water, which accepts a proton to become H_3O^+ or loses one to become OH^-.

4. A substance cannot be a Brønsted-Lowry acid unless it contains a hydrogen atom, but not all hydrogen atoms can in fact be given up. For example, acetic acid, CH_3COOH, has four hydrogens but is monoprotic. It gives up only one of them. Similarly, phenol, C_6H_5OH, gives up only one of its six hydrogens.

We have seen that HCl reacts with water as follows:

$$HCl + H_2O \rightleftharpoons H_3O^+ + Cl^-$$

Acid   Base   Acid   Base

The equilibrium is far to the right. Out of every 10 000 HCl molecules dissolved in water, only one is not converted to Cl^- ions. However, H_3O^+ is an acid, and Cl^- is a base. Why doesn't H_3O^+ give a proton to Cl^- to produce HCl and H_2O? What we are really asking here is, why does the equilibrium lie far to the right in this case?

We can answer this question by using the concept of acid and base strength. In the Brønsted-Lowry definition, **acid strength is the willingness to give up a proton** and **base strength is the willingness to accept a proton.** Because HCl is more willing to give up its proton, it is a stronger acid than H_3O^+. Looking at it the other way, H_2O is a stronger base than Cl^- and, therefore, is more willing to accept the proton. In the battle for the proton, the stronger base always wins. Because H_2O is a stronger base than Cl^-, it gets the proton.

The stronger an acid is, the weaker its conjugate base must be, since the more willing an acid is to donate the proton, the less willing the conjugate base is to take it back.

The acids in Table 8.2 are listed in order of decreasing strength (strongest at the top, weakest at the bottom). It follows that the conjugate bases are listed in order of increasing strength (strongest at the bottom, weakest at the top). We can, therefore, use this table (or any other table that lists acids with the strongest at the top) to predict whether or not any acid will donate its proton to any base. The rule is that **any acid in Table 8.2 donates its proton to the conjugate base of any acid below it but does not donate its proton to the conjugate base of any acid above it.**

These reactions are all equilibria, and what we are really doing is predicting in which direction the equilibria lie.

EXAMPLE 8.1

Which of these reactions proceed from left to right? For each reaction, label each reactant and product as being the stronger acid (SA), stronger base (SB), weaker acid (WA), or weaker base (WB).
(a) $H_3O^+(aq) + CH_3COO^-(aq) \rightleftharpoons H_2O(l) + CH_3COOH(aq)$
(b) $H_3O^+(aq) + I^-(aq) \rightleftharpoons HI(aq) + H_2O(l)$
(c) $H_2CO_3(aq) + OH^-(aq) \rightleftharpoons HCO_3^-(aq) + H_2O(l)$
(d) $HPO_4^{2-}(aq) + NH_3(aq) \rightleftharpoons PO_4^{3-}(aq) + NH_4^+(aq)$

Solution
All we need to do is find the acids and bases in Table 8.2. Reactions (a) and (c) proceed because H_3O^+ is above CH_3COOH and H_2CO_3 is above H_2O. Thus, in reactions (a) and (c), the labels are SA + SB = WA + WB. Reactions (b) and (d) do not proceed because H_3O^+ is below HI and HPO_4^{2-} is below NH_4^+. Thus, for these reactions, the labels are WA + WB = SA + SB.

8.4 Acid Dissociation Constants **201**

Problem 8.1

In which direction will each of these reactions proceed? Label each reactant and product as being SA, SB, WA, or WB.
(a) $CH_3COOH(aq) + NH_3(aq) \rightleftharpoons CH_3COO^-(aq) + NH_4^+(aq)$
(b) $H_2S(aq) + NH_3(aq) \rightleftharpoons HS^-(aq) + NH_4^+(aq)$ ∎

8.4 Acid Dissociation Constants

In Section 8.2, we learned that acids vary in the extent to which they produce H_3O^+ when added to water. Since the reactions of weak acids are all equilibria, we can use equilibrium constants to tell us quantitatively just how strong any weak acid actually is. The reaction that takes place when a weak acid is added to water is

$$HA(aq) + H_2O(l) \rightleftharpoons H_3O^+(aq) + A^-(aq)$$

HA is the general formula for an acid.

The equilibrium expression (Section 7.6) is

$$K_{eq} = \frac{[H_3O^+][A^-]}{[HA][H_2O]} \tag{8.1}$$

Notice that this expression contains the concentration of water. Since water is the solvent, its concentration changes very little when we add HA,

TABLE 8.3 Some K_a and pK_a Values for Weak Acids, in Decreasing Order of Strength

Name	Formula	K_a	pK_a
Phosphoric acid	H_3PO_4	7.5×10^{-3}	2.12
Hydrofluoric acid	HF	3.5×10^{-4}	3.45
Formic acid	HCOOH	1.8×10^{-4}	3.75
Lactic acid	$CH_3CH(OH)COOH$	1.4×10^{-4}	3.85
Acetic acid	CH_3COOH	1.8×10^{-5}	4.75
Carbonic acid	H_2CO_3	4.3×10^{-7}	6.37
Dihydrogen phosphate ion	$H_2PO_4^-$	6.2×10^{-8}	7.21
Boric acid	H_3BO_3	7.3×10^{-10}	9.14
Ammonium ion	NH_4^+	5.6×10^{-10}	9.25
Phenol	C_6H_5OH	1.3×10^{-10}	9.89
Bicarbonate ion	HCO_3^-	5.6×10^{-11}	10.25
Hydrogen phosphate ion	HPO_4^{2-}	2.2×10^{-13}	12.66

so we can treat $[H_2O]$ as constant and rearrange Equation 8.1 to bring the constant terms together on the left-hand side:

$$K_{eq}[H_2O] = \frac{[H_3O^+][A^-]}{[HA]} \tag{8.2}$$

When we multiply one constant, K_{eq}, by another, $[H_2O]$, we get a third constant, which we call K_a:

$$K_a = K_{eq}[H_2O] = \frac{[H_3O^+][A^-]}{[HA]} \tag{8.3}$$

Acid dissociation constant (K_a) An equilibrium constant for the dissociation of an acid into H^+ and an anion in water

K_a is also called the **acidity constant** and the **ionization constant.**

The constant K_a is called the **acid dissociation constant,** and every weak acid has one. Table 8.3 lists some weak acids and their dissociation constants.

The larger the K_a is, the stronger the acid is. We can see this by looking at the equation

$$HA(aq) + H_2O(l) \rightleftharpoons H_3O^+(aq) + A^-(aq)$$

A stronger acid produces more H_3O^+ ions (and A^- ions also) than a weaker acid. This increases the numerator of Equation 8.3 so that K_a is larger for the stronger acid.

The importance of K_a is that it immediately tells us how strong an acid is. For example, Table 8.3 shows us that although acetic acid, formic acid, and phenol are all weak acids, their strengths as acids are not the same. Formic acid, with a K_a of 1.8×10^{-4}, is stronger than acetic acid ($K_a = 1.8 \times 10^{-5}$), whereas phenol, whose K_a is 1.3×10^{-10}, is much weaker than acetic acid.

8.5 Some Properties of Acids and Bases

Chemists stopped tasting chemicals when some of them got sick or died from doing this.

Chemists of today do not taste the substances they work with, but 150 and 200 years ago it was routine to do so. That is how we know that acids taste sour and bases taste bitter. The sour taste of lemons, vinegar, and many other foods is due to the acids they contain.

Both strong acids and strong bases damage living tissue. Externally, they cause chemical burns. Internally, they are even more harmful and can cause severe damage to the digestive tract. One way they cause internal damage is by breaking down the organism's protein molecules.

The most important property of acids and bases is that they react with each other in a process called **neutralization.** This is an appropriate name because, when a strong corrosive acid such as hydrochloric acid reacts with a strong corrosive base such as sodium hydroxide, the product (a solution of ordinary table salt in water) has neither acidic nor basic properties. We call such a solution neutral. Neutralization reactions are further discussed in the next section.

8.6 Reactions of Acids

Acids have several characteristic reactions, and in this section we look at three types.

Reactions with Metals, Metal Hydroxides, and Oxides

Acids react with certain metals (called active metals) to produce hydrogen (H_2) and a salt (Figure 8.1):

$$\underset{\text{Active metal}}{Mg(s)} + \underset{\text{Acid}}{2HCl(aq)} \longrightarrow \underset{\text{A salt}}{MgCl_2(aq)} + \underset{\text{Hydrogen}}{H_2(g)}$$

Acids react with metal hydroxides to give water and a salt:

$$\underset{\text{Acid}}{HNO_3} + \underset{\text{Base}}{KOH} \longrightarrow H_2O + \underset{\text{A salt}}{KNO_3}$$

A **salt** is an ionic compound. We may consider all salts to be composed of a negative ion (anion) from an acid and a positive ion (cation) from a base. The actual equation for the reaction between HNO_3 and KOH is (Section 4.7)

$$H_3O^+ + NO_3^- + K^+ + OH^- \longrightarrow H_2O + H_2O + NO_3^- + K^+$$

This can be simplified by omitting the spectator ions (K^+ and NO_3^-):

$$H_3O^+ + OH^- \longrightarrow H_2O + H_2O$$

This is the net ionic equation when any strong acid reacts in water with any strong base. Strong acids also react with metal oxides; for example,

$$2H_3O^+(aq) + CaO(s) \longrightarrow 3H_2O(l) + Ca^{2+}(aq)$$

Reactions with Carbonates and Bicarbonates

When a strong acid is added to a carbonate or bicarbonate, such as sodium carbonate or potassium bicarbonate, bubbles of gas are rapidly given off. The gas, carbon dioxide, is formed in this manner:

$$2H_3O^+(aq) + CO_3^{2-}(aq) \longrightarrow H_2CO_3(aq) + 2H_2O(l)$$
$$H_2CO_3(aq) \longrightarrow CO_2(g) + H_2O(l)$$

or

$$H_3O^+(aq) + HCO_3^-(aq) \longrightarrow H_2CO_3(aq) + H_2O(l)$$
$$H_2CO_3(aq) \longrightarrow CO_2(g) + H_2O(l)$$

All these fruits and fruit drinks contain organic acids. *(Charles D. Winters)*

In the laboratory, you must be careful not to get any strong acids or bases on your skin or, even worse, in your eyes, where they can cause blindness.

FIGURE 8.1 A ribbon of magnesium metal reacts with aqueous HCl to give H_2 gas and aqueous $MgCl_2$. *(Charles D. Winters)*

BOX 8C

Drugstore Antacids

Stomach fluid is normally quite acidic because of its HCl content. We have all undoubtedly gotten "heartburn" caused by excess stomach acidity. At that time, we may have taken an antacid, which, as the name implies, is a substance that neutralizes acids—in other words, it is a base.

The word "antacid" is a medical term, not one used by chemists. It is, however, found on the labels of many medications available in drugstores and supermarkets. Almost all of them use weak bases (hydroxides and/or carbonates) to decrease the acidity of the stomach. The active ingredients of some of these antacids are as follows (some brands are sold in more than one formulation, but we list only one in each case):

Alka-Seltzer: sodium bicarbonate, $NaHCO_3$; citric acid; and aspirin

Brioschi: $NaHCO_3$

Chooz: calcium carbonate, $CaCO_3$

Di-gel: $CaCO_3$; magnesium hydroxide, $Mg(OH)_2$; and simethicone (an antiflatulent)

Gaviscon: aluminum hydroxide, $Al(OH)_3$; magnesium carbonate, $MgCO_3$

Gelusil: $Al(OH)_3$, $Mg(OH)_2$, and simethicone

Maalox: $Mg(OH)_2$, $Al(OH)_3$, and simethicone

Mylanta: $Al(OH)_3$, $Mg(OH)_2$, and simethicone

Phillips' Milk of Magnesia: $Mg(OH)_2$

■ Commercial remedies for excess stomach acid. *(Charles D. Winters)*

Riopan: magnesium aluminum hydrate (magaldrate)

Rolaids: $CaCO_3$, $Mg(OH)_2$

Tempo: $CaCO_3$, $Al(OH)_3$, and simethicone

Tums: $CaCO_3$

Also in the supermarkets and drugstores are nonprescription drugs, labeled "acid reducers," which lower stomach acidity without being bases themselves. Among these are Zantac, Tagamet, Pepcid, and Axid. Instead of neutralizing acidity, these compounds restrict the secretion of acid in the stomach. In larger doses (sold only with a prescription), some of these drugs are used in the treatment of stomach ulcers.

In either case, the direct product of the proton transfer is carbonic acid, H_2CO_3, but this compound is very unstable and breaks down to CO_2 and H_2O. As shown in Tables 8.2 and 8.3, carbonic acid is a weak acid. By the rule given in Section 8.3, any acid stronger than H_2CO_3 reacts with carbonates or bicarbonates to give CO_2 gas.

Reactions with Ammonia and Amines

Any acid stronger than NH_4^+ (see Table 8.2) is strong enough to react with NH_3 in this way:

$$HCl + NH_3 \longrightarrow NH_4^+ + Cl^-$$

In Chapter 16 we meet a family of compounds called amines, which are similar to ammonia except that one or more of the three hydrogen atoms are replaced by carbon. A typical example is methylamine, CH_3NH_2. The base strength of most amines is similar to that of NH_3, which means that amines react with acids in the same way:

$$HCl + CH_3{-}NH_2 \longrightarrow CH_3{-}NH_3^+ + Cl^-$$

BOX 8D

Rising Cakes and Fire Extinguishers

When an acid is added to the bicarbonate ion, CO_2 gas is given off:

$$HCO_3^-(aq) + H_3O^+(aq) \rightleftharpoons H_2O(l) + H_2CO_3(aq)$$
$$H_2CO_3(aq) \rightleftharpoons CO_2(g) + H_2O(l)$$

This reaction is used in a number of ways, but we discuss two.

Cake Rising Baking powder consists of sodium bicarbonate, $NaHCO_3$, and a solid acidic compound. When the powder is dry, little or no reaction takes place; however, in the presence of water, the acid and bicarbonate ion react to produce CO_2 gas, which is what causes a cake to rise. Potassium acid tartrate, $KC_4H_5O_6$, called cream of tartar, was the earliest acidic compound used for this purpose and is still one of the best; however, because of its high cost, it has been largely replaced in recent years by calcium dihydrogen phosphate, $Ca(H_2PO_4)_2$.

Because of its use in baking powders, $NaHCO_3$ is often called baking soda. Baking powder is seldom used for bread. Bread rising is accomplished by yeast, which produces CO_2 by fermenting the sugars present in the dough.

Fire Extinguishers There are several types of fire extinguishers, but one type uses the reaction between $NaHCO_3$ and an acid to produce CO_2. The pressure of the CO_2 forces water out of the fire extinguisher. In another type of extinguisher, CO_2 itself, in the form of a foam, is used to extinguish the flames.

■ This extinguisher uses CO_2 to put out the fire. *(Charles D. Winters)*

Neither NH_3 nor amines survive as such in the presence of an acid. They are converted to the corresponding positive ammonium ions (conjugate acids). This is very important in the chemistry of the body, as we shall see in later chapters.

8.7 Self-Ionization of Water

We have seen that an acid produces H_3O^+ ions in water and that a base produces OH^- ions. Suppose that we have absolutely pure water, with no added acid or base. Surprisingly enough, even pure water contains a very small number of H_3O^+ and OH^- ions. They come from the reaction

$$\underset{\text{Acid}}{H_2O} + \underset{\text{Base}}{H_2O} \rightleftharpoons \underset{\text{Acid}}{H_3O^+} + \underset{\text{Base}}{OH^-}$$

This reaction is called the self-ionization of water.

in which one water molecule acts as an acid and gives a proton to a second water molecule, which is, therefore, acting as a base.

What is the extent of this reaction? How far does it go? The answer is, not very far. As shown by the arrows, it is an equilibrium lying far to the left. We shall soon see exactly how far, but first let us write the equilibrium expression:

$$K_{eq} = \frac{[H_3O^+][OH^-]}{[H_2O]^2}$$

This expression contains the concentration of water, but since only a very few water molecules react, the concentration of water remains essentially constant. Since this is so, it is convenient to change the expression to

$$K_{eq}[H_2O]^2 = [H_3O^+][OH^-]$$

As we did with the acid dissociation constant (Section 8.4), we can replace $K_{eq}[H_2O]^2$ by a new constant, K_w. Our equilibrium expression is now

$$K_w = [H_3O^+][OH^-]$$

The K_w is called the **ion product of water.** At room temperature, K_w has the value of 1.0×10^{-14}. Since in pure water H_3O^+ and OH^- are formed by the same process (called the **self-ionization of water**), their concentrations must be equal. This means that, in pure water,

$$[H_3O^+] = 1 \times 10^{-7} \text{ mole/L}$$

$$[OH^-] = 1 \times 10^{-7} \text{ mole/L}$$

Thus, in pure water, each liter of H_2O contains only 1×10^{-7} mole of H_3O^+ and 1×10^{-7} mole of OH^-. These are very small concentrations, not enough to make pure water a conductor of electricity. Water is not an electrolyte.

The equation $K_w = [H_3O^+][OH^-]$ is important because it does not apply only to pure water; *it is valid in any water solution.* It says that the product of $[H_3O^+]$ and $[OH^-]$ in any aqueous solution is equal to 1×10^{-14}. If, for example, we add 0.01 mole of HCl to 1 L of pure water, it reacts essentially completely to give H_3O^+ ions. The concentration of H_3O^+ will be $0.01\,M$, or $1 \times 10^{-2}\,M$. This means that the $[OH^-]$ *must* be $1 \times 10^{-12}\,M$ because the product must be 1×10^{-14}.

EXAMPLE 8.2

The $[OH^-]$ of an aqueous solution is $1 \times 10^{-4}\,M$. What is the $[H_3O^+]$?

Solution
We substitute into the equation:

$$K_w = [H_3O^+][OH^-] = 1 \times 10^{-14}$$

$$[H_3O^+](10^{-4}) = 1 \times 10^{-14}$$

$$[H_3O^+] = \frac{1 \times 10^{-14}}{1 \times 10^{-4}} = 1 \times 10^{-10}\,M$$

Problem 8.2 ▬
The [OH$^-$] of an aqueous solution is $1 \times 10^{-12}\,M$. What is the [H$_3$O$^+$]? ▪

Aqueous solutions can have a very high [H$_3$O$^+$], but the [OH$^-$] must then be very low, and vice versa. Any solution with a [H$_3$O$^+$] higher than $1 \times 10^{-7}\,M$ is **acidic.** In such solutions, of course, [OH$^-$] must be lower than $1 \times 10^{-7}\,M$. The higher the [H$_3$O$^+$], the more acidic the solution is. Similarly, any solution with an [OH$^-$] higher than $1 \times 10^{-7}\,M$ is **basic.**

Pure water, in which [H$_3$O$^+$] and [OH$^-$] are equal (they are both $1 \times 10^{-7}\,M$), is **neutral,** that is, neither acidic nor basic.

Pure water is a neutral solution, as is any aqueous solution in which [H$_3$O$^+$] = [OH$^-$] = $1 \times 10^{-7}\,M$.

8.8 pH, pOH, and pK

Because the [H$_3$O$^+$] of any aqueous solution is such an important characteristic, chemists, workers in the health sciences, and other scientists must frequently talk about it. When they do so, it is inconvenient to keep saying, for example, "$6 \times 10^{-11}\,M$", "$2 \times 10^{-2}\,M$," or other numbers of this form. Fortunately, a shorter way exists. Chemists have defined a number, called the pH, to express these concentrations in a more convenient way. The mathematical definition of pH is

$$\text{pH} = -\log[\text{H}_3\text{O}^+]$$

When the [H$_3$O$^+$] is of the form $1 \times 10^{-A}\,M$, where A stands for any number, we don't have to worry about the mathematical definition—the pH is simply A. Instead of saying that a [H$_3$O$^+$] is $1 \times 10^{-4}\,M$, we simply say that the pH is 4. As you can see, all we do is take the exponent of the [H$_3$O$^+$] and drop the minus sign.

EXAMPLE 8.3

The [H$_3$O$^+$] of a certain liquid soap is 1×10^{-9} M. What is the pH?

Solution
The pH is 9.

Problem 8.3 ▬
The [H$_3$O$^+$] of an acidic solution is $1 \times 10^{-3}\,M$. What is the pH? ▪

EXAMPLE 8.4

The pH of black coffee is 5. What is the [H$_3$O$^+$]?

Solution
The [H$_3$O$^+$] is $1 \times 10^{-5}\,M$.

Problem 8.4 ▬
The pH of tomato juice is 4. What is the [H$_3$O$^+$]? ▪

pH is a convenient way to designate the concentration of [H$_3$O$^+$]. In a similar manner, we can refer to the concentration of the [OH$^-$]:

$$\textbf{pOH} = -\log[\text{OH}^-]$$

The notation p, which stands for −log, is also used in connection with the equilibrium constant:

$$pK = -\log K$$

In aqueous solutions, the ion product, K_w, is 1.0×10^{-14} and thus $pK_w = -\log K_w = 14$.

Therefore, we can write

$$14 = pH + pOH$$

Once we know the pH of a solution, we can easily calculate the pOH.

EXAMPLE 8.5

The $[OH^-]$ in a strongly alkaline solution is 1×10^{-2}. What is the pOH and the pH?

Solution
The pOH is 2, and the pH = 14 − 2 = 12.

Problem 8.5 ■■■■■■
The $[OH^-]$ of a solution is 1.0×10^{-4}. What is the pOH and pH? ■

Suppose that the $[H_3O^+]$ is 7.0×10^{-4}. What do we do then? That's when we need to use logarithms. Now that inexpensive scientific calculators are readily available, this is no problem at all. We simply take the logarithm of 7×10^{-4} with the calculator and get the answer −3.15. Just drop the minus sign (because pH = $-\log[H_3O^+]$, and you have the pH = 3.15.

EXAMPLE 8.6

What is the pH of a solution whose $[H_3O^+]$ is $3.0 \times 10^{-2}\ M$?

Solution
Enter 3.0×10^{-2} into your calculator, and press log. The answer is −1.52. Since pH = $-\log[H_3O^+]$, we drop the minus sign, and the pH is 1.52.

Problem 8.6 ■■■■■■
What is the pH of a solution whose $[H_3O^+]$ is $6.0 \times 10^{-9}\ M$? ■

EXAMPLE 8.7

A solution has a pH of 11.70. What is the $[H_3O^+]$?

Solution
Since this is the reverse of the method just described, we first convert 11.70 to −11.70 and then find the antilogarithm on the calculator (inverse log or 10^x on some calculators). When we do this, we get $[H_3O^+] = 2.0 \times 10^{-12}\ M$.

Problem 8.7 ■■■■■■
The pH of blood is 7.4. What is the $[H_3O^+]$? ■

As already indicated, pH is universally used to express the acidity of aqueous solutions (you probably heard of pH even before taking this course). **The lower the pH, the more acidic the solution; the higher the pH, the more basic the solution.** Pure water, with a $[H_3O^+]$ of $1 \times 10^{-7}\ M$, has a pH of 7 and is neutral.

> Acidic solutions have a pH below 7
>
> Basic solutions have a pH above 7.

All fluids in the human body are aqueous; that is, the only solvent is water. Consequently, all body fluids have a pH value. Some of them have a narrow pH range; others have a wide range. The pH of blood must be between 7.35 and 7.45 (slightly basic). If it goes outside these limits, illness and even death will result (Box 8E). In contrast, the pH of urine can vary from 5.5 to 7.5. Table 8.4 gives the pH values of some body fluids, as well as of some other common materials (Figure 8.2).

One thing we must remember when we see pH values is that an increase (or decrease) of one pH unit means a tenfold decrease (or increase) in the $[H_3O^+]$. For example, a pH of 3 does not sound very different from a pH of 4; however, the first means a $[H_3O^+]$ of $10^{-3}\ M$, and the second means a $[H_3O^+]$ of $10^{-4}\ M$. The $[H_3O^+]$ of the pH 3 solution is ten times the $[H_3O^+]$ of the pH 4 solution.

There are two ways to measure the pH of an aqueous solution. In one method, pH paper is used (Figure 8.3). When a drop of solution is placed on this paper, the paper turns a certain color. The pH is determined by comparing the color of the paper with the colors on a chart supplied with the paper.

pH paper is made by soaking plain paper with a mixture of pH indicators. **A pH indicator is a substance that changes color at a certain pH** (Figure 8.2). An example is the compound methyl orange. When a drop of methyl orange solution is added to any aqueous solution with a pH of 3.2 or lower, the

Although the pH of pure water is 7, tap water usually has a lower pH, generally about 6. The reason is that tap water normally contains dissolved CO_2 (from the atmosphere), which reacts with water to give the dissolved acid H_2CO_3.

Litmus paper turns red in acidic solution and blue in basic solution. This indicator is often used as an approximate guide to acidity or basicity.

FIGURE 8.2 The indicators show that vinegar has a pH of about 3, a carbonated beverage has a pH of about 4 to 5, and a household cleaner is strongly basic. *(Charles D. Winters)*

TABLE 8.4	pH Values for Common Materials
Material	**pH**
Battery acid	0.5
Gastric juice	1.0–3.0
Soft drinks	2.0–4.0
Lemon juice	2.2–2.4
Vinegar	2.4–3.4
Tomatoes	4.0–4.4
Black coffee	5.0–5.1
Urine	5.5–7.5
Rain (unpolluted)	6.2
Milk	6.3–6.6
Saliva	6.5–7.5
Pure water	7.0
Blood	7.35–7.45
Bile	6.8–7.0
Pancreatic fluid	7.8–8.0
Seawater	8–9
Soap	8–10
Milk of magnesia	10.5
Household ammonia	11.7
Lye (1 M NaOH)	14

FIGURE 8.3 Rolls of pH paper. *(Beverly March)*

methyl orange turns red (and turns the whole solution red). When added to an aqueous solution with a pH of 4.4 or higher, it turns yellow (and turns the whole solution yellow). These are the particular limits and colors for this indicator. Other indicators have other limits and colors, as Figure 8.4 shows.

The other method for determining pH is more accurate. In this method a pH meter is used (Figure 8.5). The electrode of the pH meter is dipped into the unknown solution, and the pH is read on a dial to the nearest tenth of a pH unit.

In Section 8.4, we saw that K_a values for weak acids are given in exponential notation. pK_a values are often used instead. Table 8.3 lists pK_a as well as K_a values for some weak acids.

The higher the pK_a is, the weaker the acid is.

EXAMPLE 8.8

The K_a for benzoic acid is 6.5×10^{-5}. What is the pK_a of this acid?

Solution
This is done exactly the same way as the calculation of pH from $[H_3O^+]$. Enter the number 6.5×10^{-5} into a scientific calculator and press the log key. The answer is -4.19. Drop the minus sign, and the pK_a is 4.19.

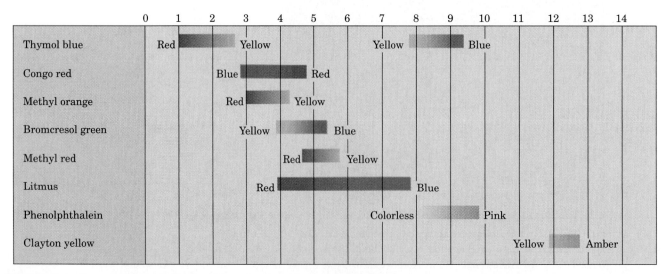

FIGURE 8.4 Some acid-base indicators. Note that some have two color changes.

Problem 8.8

The K_a for hydrocyanic acid, HCN, is 4.9×10^{-10}. What is its pK_a? ∎

8.9 The pH of Aqueous Salt Solutions

When an acid is added to water, the solution becomes acidic (pH below 7); when a base is added to water, the solution becomes basic (pH above 7). Suppose we add a salt to water? What happens to the pH then? At first thought, you might suppose that all salts are neutral and that adding any salt to pure water leaves the pH at 7. This is in fact true for some salts. Many salts are acidic or basic, however, and cause the pH of pure water to change. In order to see why, let us look at sodium acetate, a typical basic salt.

Like almost all other salts, sodium acetate is 100 percent ionized in water so that, when we add CH_3COONa to water, we get a solution containing Na^+ and CH_3COO^- ions. The Na^+ ions do not react with the water, but the CH_3COO^- ions do:

$$CH_3COO^-(aq) + H_2O(l) \rightleftharpoons CH_3COOH(aq) + OH^-(aq)$$

This is an equilibrium reaction lying well to the left. In a 0.1 M solution of sodium acetate, only about one out of every 13 500 acetate ions is con-

FIGURE 8.5 A pH meter. The blue solution is 0.1 M $CuSO_4$. This is the salt of a strong acid and a weak base, so the pH is 4.3 (Section 8.9). *(Charles D. Winters)*

Because this is a reaction with water, it is sometimes called **hydrolysis.**

BOX 8E

Acidosis and Alkalosis

The pH of blood is normally between 7.35 and 7.45. If the pH goes lower than this, the condition is called **acidosis.** A blood pH higher than 7.45 is called **alkalosis.** Both of these are abnormal conditions. Acidosis leads to depression of the nervous system. Mild acidosis can result in fainting; a more severe case can cause coma. Alkalosis leads to overstimulation of the nervous system, muscle cramps, and convulsions. If the acidosis or alkalosis persists for a sufficient period of time, or if the pH gets too far away from 7.35 to 7.45, the patient dies.

Acidosis has several causes. One type, called respiratory acidosis, is caused by difficulty in breathing **(hypoventilation).** An obstruction in the windpipe or diseases such as pneumonia, emphysema, asthma, or congestive heart failure diminish the amount of CO_2 that leaves the body through the lungs. (You can even produce mild acidosis by holding your breath.) The pH of the blood is decreased because the CO_2, unable to escape fast enough, remains in the blood, where it decreases the $[HCO_3^-]/[H_2CO_3]$ ratio.

Respiratory alkalosis is the reverse. It arises from rapid or heavy breathing, called **hyperventilation.** This can come about from fever, infection, the action of certain drugs, or even hysteria. Here the excessive loss of CO_2 raises the $[HCO_3^-]/[H_2CO_3]$ ratio and the pH.

Acidosis caused by other factors is called metabolic acidosis. Two causes of this condition are starvation (or fasting) and heavy exercise. When the body doesn't get enough food, it burns its own fat, and the products of this reaction are acidic compounds that get into the blood. This sometimes happens to people on fad diets. Heavy exercise causes the muscles to produce excessive

■ High blood pH brought on by severe anxiety can cause alkalosis. One remedy is to breathe into a paper bag. *(Charles D. Winters)*

amounts of lactic acid—that is what makes muscles feel tired and sore. Metabolic acidosis is also caused by a number of metabolic irregularities. For example, the disease diabetes mellitus produces acidic compounds called ketone bodies (Section 27.7).

Metabolic alkalosis can also result from various metabolic irregularities and, in addition, can be caused by excessive vomiting. The contents of the stomach is strongly acidic (pH 1.0), and loss of substantial amounts of this acidic material raises the pH of the blood.

verted to acetic acid and OH^- ions at equilibrium. Nevertheless, enough OH^- ions are being produced to cause the pH to rise to about 8.9. Because this is higher than 7, the solution is basic, and sodium acetate is called a basic salt. It is true that acetic acid is also being produced, but the pH of an aqueous solution depends only on the concentration of H_3O^+ (and of OH^-) and not on the concentration of any other acid or base.

We can predict whether a salt will be acidic, neutral, or basic in water solution when we know something about the strength of the acid and base. We have four cases.

1. A salt of a strong base and a weak acid. As we can see from the sodium acetate example, such salts are **basic salts** and raise the pH of pure water.

2. A salt of a strong acid and a weak base. We have the opposite behavior here. An example is ammonium chloride, NH_4Cl. The Cl^- ion does not react with H_2O, but the NH_4^+ ion does:

$$NH_4^+(aq) + H_2O(l) \rightleftharpoons H_3O^+(aq) + NH_3(aq)$$

The newly formed H_3O^+ lowers the pH, as it does for all salts in this category. Such salts are **acidic salts** (see Figure 8.5).

3. A salt of a strong acid and a strong base. In this case, neither ion reacts with water, and the solution remains neutral. These are **neutral salts.** Examples are $NaCl$, KNO_3, and Na_2SO_4.

4. A salt of a weak acid and a weak base. Both ions react with water. Sometimes the effects cancel each other, and sometimes they do not. We will not try to make predictions in such cases.

8.10 Buffers

What Is a Buffer and How Does One Work?

In Box 8E we saw that the body must keep the pH of blood between 7.35 and 7.45. Yet we frequently eat acidic foods such as oranges, lemons, sauerkraut, and tomatoes, and doing so eventually adds considerable quantities of H_3O^+ to the blood. How does the body manage to keep the pH of blood so constant? It does it with buffers. A **buffer solution** is **a solution whose pH does not change very much when H_3O^+ or OH^- ions are added to it.**

How do we make a buffer solution? All we need to do is to add to water approximately equal quantities of a weak acid and its conjugate base. For example, if we dissolve 1 mole of acetic acid (a weak acid) and 1 mole of its conjugate base (in the form of sodium acetate) in 1 L of water, we have a very good buffer solution.

How does it work? It works because we have substantial concentrations of both CH_3COOH and CH_3COO^-. If H_3O^+ is added, the CH_3COO^- ions remove it by reacting with it. If OH^- is added, the CH_3COOH molecules perform the same function

$$H_3O^+(aq) + CH_3COO^-(aq) \rightleftharpoons CH_3COOH(aq) + H_2O(l)$$

$$OH^-(aq) + CH_3COOH(aq) \rightleftharpoons CH_3COO^-(aq) + H_2O(l)$$

(Note from Table 8.2 that both equilibria lie well to the right.) The important thing here is that, when the CH_3COO^- ions remove H_3O^+, they are converted to CH_3COOH molecules. Since a substantial amount of

CH_3COOH is already present, nothing much happens: The $[CH_3COO^-]$ goes down a bit, and the $[CH_3COOH]$ goes up a bit.

We saw in Section 8.4 that the following equilibrium expression holds for acetic acid (a similar one holds for any other weak acid):

$$K_a = \frac{[H_3O^+][CH_3COO^-]}{[CH_3COOH]}$$

If we rearrange this, we get

$$[H_3O^+] = K_a \frac{[CH_3COOH]}{[CH_3COO^-]}$$

This equation tells us that the $[H_3O^+]$ is equal to the constant K_a multiplied by the ratio of $[CH_3COOH]$ to $[CH_3COO^-]$. We've just seen, however, that H_3O^+ ions added to the buffer solution make only a slight change in this ratio: The value of $[CH_3COOH]$ goes up a bit, and the value of $[CH_3COO^-]$ down a bit. Therefore, the $[H_3O^+]$, and the pH, are only slightly changed.

The same thing happens in the opposite direction if we add OH^- to the buffer. The OH^- ions are removed by the CH_3COOH, which in the process is converted to CH_3COO^-. Thus, the $[CH_3COO^-]$ goes up a bit, and the $[CH_3COOH]$ down a bit, so that the $[H_3O^+]$, and the pH, are only slightly changed.

In essence, a buffer solution works because it is able to neutralize, and hence remove from the solution, most of any OH^- or H_3O^+ ions that are added. This effect can be quite powerful. For example, pure water has a pH of 7. If we add 0.1 mole of HCl to 1 L of pure water with no buffer, the pH drops to 1. If we add 0.1 mole of NaOH to 1 L of pure water, the pH rises to 13. A buffer solution containing $1\,M\,H_2PO_4^-$ ion and $1\,M\,HPO_4^{2-}$ ion (its conjugate base) has a pH of 7.21. If we add 0.1 mole of HCl to 1 L of this solution, the pH drops only to 7.12. If we add 0.1 mole of NaOH, the pH rises only to 7.30. In summary,

Pure H_2O + HCl   pH 7 $\longrightarrow$ 1

Buffer + HCl   pH 7.21 $\longrightarrow$ 7.12

Pure H_2O + NaOH   pH 7 $\longrightarrow$ 13

Buffer + NaOH   pH 7.21 $\longrightarrow$ 7.30

It is obvious from this example that a buffer really does the job: It holds the pH relatively constant (Figure 8.6).

Every buffer solution has two characteristics—its pH and its capacity.

Buffer pH

In the example just given, the pH of the $H_2PO_4^-/HPO_4^{2-}$ buffer solution is 7.21 as long as the concentrations of $H_2PO_4^-$ and HPO_4^{2-} are equal. If you look at Table 8.3, you see that 7.21 is the pK_a of the $H_2PO_4^-$ ion. This is not a coincidence. If we make a buffer solution by mixing equimolar concentrations of any weak acid and its conjugate base, the pH of the solution is always equal to the pK_a of the acid.

This allows us to prepare buffer solutions to maintain almost any pH. For example, if we want to maintain a pH of 9.14, we could make a buffer solution from boric acid and its conjugate base (see Table 8.3).

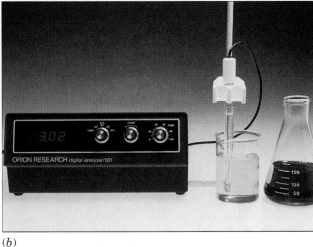

(*a*)   (*b*)

FIGURE 8.6   Buffer solutions. (*a*) The pH meter shows the pH of water that contains a trace of acid (and bromcresol green indicator, see Figure 8.4). The solution at the right is a buffer that has the same pH as human blood, 7.4, and contains the same indicator. (*b*) When 5 mL of 0.10 *M* HCl is added to each solution, the pH of the water drops several units, while the pH of the buffer stays constant. (*Charles D. Winters*)

EXAMPLE 8.9

What is the pH of buffer systems containing equimolar quantities of
(a) H_3PO_4 and $H_2PO_4^-$ (b) H_2CO_3 and HCO_3^-?

Solution
The pH is equal to the pK_a of the acid, which we find in Table 8.3:
(a) pH = 2.1 (b) pH = 6.37.

Problem 8.9
What is the pH of buffer systems containing equimolar quantities of
(a) NH_4^+ and NH_3 (b) CH_3COOH and CH_3COO^-? ■

Suppose that the concentrations of the weak acid and its conjugate base are not equal. In that case, the pH of the buffer will be a bit higher or lower (but not by very much) as long as significant amounts of both acid and conjugate base are present (Box 8F). The most efficient buffers do contain equimolar concentrations, and buffer efficiency decreases as the ratio becomes more unequal. If the acid-to-conjugate-base ratio is more than about 15 to 1 or less than about 1 to 15, the solution is essentially no longer a buffer.

Buffer Capacity

Buffer capacity is the extent to which a buffer solution can prevent a significant change in pH upon addition of an acid or base. We could make a buffer solution by dissolving 1 mole each of CH_3COONa and CH_3COOH in 1 L of H_2O, or we could use only 0.1 mole of each. Both solutions have the same pH: 4.75. However, the former has a buffer capacity ten times that of the latter. If we add 0.2 mole of HCl to the former solution, it performs the way we expect—the pH drops only a little (to 4.57). If we add 0.2 mole of

The Henderson-Hasselbalch Equation

Whenever we work with buffers, a useful equation is

$$pH = pK_a + \log [A^-]/[HA]$$

where $[HA]$ is the concentration of the acid and $[A^-]$ is that of its conjugate base. This equation, called the **Henderson-Hasselbalch equation,** is easily derived from Equation 8.3 but uses pK_a instead of K_a. Suppose that we wish to make a buffer solution whose pH does not correspond exactly to the pK_a of any acid we may have on hand. We can make up such a solution by choosing an acid with a pK_a that is not too far from the desired pH and adjusting the concentrations of the weak acid $[HA]$ and its conjugate base $[A^-]$ to obtain the desired pH. The Henderson-Hasselbalch equation allows us to calculate the proper ratio.

EXAMPLE 8.10

How can we make a buffer solution with a pH of 7.99, using an acid-base pair from Table 8.3?

Solution
The closest pK_a value in Table 8.3 is 7.21, the value for $H_2PO_4^-$. We therefore use $H_2PO_4^-/HPO_4^{2-}$ as our buffer and adjust the $[HPO_4^{2-}]/[H_2PO_4^-]$ ratio to get a pH of 7.99:

$$pH = pK_a + \log [HPO_4^{2-}]/[H_2PO_4^-]$$

$$7.99 = 7.21 + \log [HPO_4^{2-}]/[H_2PO_4^-]$$

$$0.78 = \log [HPO_4^{2-}]/[H_2PO_4^-]$$

On a calculator, we find that the antilogarithm of 0.78 is 6.0. Taking the antilogarithm of both sides, we have

$$6.0 = [HPO_4^{2-}]/[H_2PO_4^-]$$

Thus, to get a buffer solution of pH 7.99, we must use a molar concentration of HPO_4^{2-} that is six times that of $H_2PO_4^-$.

Problem 8.10
How can we make a buffer solution of pH 4.15, using an acid-base pair from Table 8.3? ■

We can also use the Henderson-Hasselbalch equation to calculate the pH if we know the $[A^-]/[HA]$ ratio.

EXAMPLE 8.11

What is the pH of an $H_2PO_4^-/HPO_4^{2-}$ buffer system in which $[HPO_4^{2-}]$ is twice $[H_2PO_4^-]$?

Solution

$$pH = pK_a + \log \frac{[HPO_4^{2-}]}{[H_2PO_4^-]}$$

$$pH = 7.21 + \log \frac{2}{1}$$

The log of 2 (from a calculator) is 0.30. Thus,

$$pH = 7.21 + 0.30 = 7.51$$

Problem 8.11
What is the pH of a buffer solution in which $[CH_3COO^-] = 0.20\ M$ and $[CH_3COOH] = 0.60\ M$? ■

HCl to the latter solution, however, the pH drops all the way to 1.0 because the buffer has been swamped out. The amount of H_3O^+ added has exceeded the buffer capacity. The first 0.1 mole of HCl completely neutralizes essentially all the CH_3COO^- present. After that, the solution contains only CH_3COOH and is no longer a buffer, so that the second 0.1 mole of HCl is able to lower the pH to 1.0.

Blood Buffers

The average pH of human blood is 7.4. Any change larger than 0.1 pH unit in either direction is pathological. If the pH goes below 6.8 or above 7.8, death may result.

To hold the pH of the blood at close to 7.4, the body uses three buffer systems: carbonate, phosphate, and proteins (proteins are discussed in Chapter 21). The most important of these is the carbonate system. The acid is carbonic acid, H_2CO_3; the base is the bicarbonate ion, HCO_3^-. The pK_a of H_2CO_3 is 6.37 (Table 8.3). Since the pH of an equimolar mixture of acid and conjugate base is equal to its pK_a, a buffer made of equal concentrations of H_2CO_3 and HCO_3^- has a pH of 6.37.

Blood, however, has a pH of 7.4. The carbonate buffer can maintain this pH only if $[H_2CO_3]$ and $[HCO_3^-]$ are not equal. In fact, the necessary $[HCO_3^-]/[H_2CO_3]$ ratio is about 10 to 1. The normal concentrations in blood are about 0.025 M HCO_3^- and 0.0025 M H_2CO_3. The buffer works, as all buffers do, because any added H_3O^+ is neutralized by the HCO_3^- and any added OH^- is neutralized by the H_2CO_3.

The fact that the $[HCO_3^-]/[H_2CO_3]$ ratio is 10 to 1 means that this is a better buffer for acids, which lower the ratio and thus improve buffer efficiency, than for bases, which raise the ratio. This is in harmony with the actual functioning of the body because, under normal conditions, larger amounts of acidic than basic substances enter the blood. The 10-to-1 ratio is easily maintained under normal conditions because the body can very quickly increase or decrease the amount of CO_2 entering the blood.

The second buffering system of the blood is made up of hydrogen phosphate ion, HPO_4^{2-}, and dihydrogen phosphate ion, $H_2PO_4^-$. In this case, a 1.6-to-1 $[HPO_4^{2-}]/[H_2PO_4^-]$ ratio is necessary to maintain a pH of 7.4. This is well within the limits of good buffering action.

8.11   Titration, Equivalents, and Normality

In laboratories, the question frequently arises as to how much acid or base is present in a given amount of solution. To measure this, we use a method called titration. If a solution is acidic, **titration** consists of adding base to it until all the acid is neutralized. To do this, we need two things: (1) a means of measuring how much base is added and (2) a means of telling just when the acid is completely neutralized. Titration is shown in Figure 8.7. The buret is the means of measuring the amount of base. An indicator in the solution tells us when the acid is exactly neutralized (we could also use a pH meter) because at that point there is a sudden large change in pH. This is called the **end point** of the titration.

It is important to understand that titration is not a method of determining the acidity (or basicity) of a solution. If we want to do that, we measure the pH, which is the only measurement of solution acidity or basicity. Titration is a method for determining the total acid or base concentration of a solution, which is not the same as the acidity. For example, a 0.1 M solution of HCl in water has a pH of 1, but a 0.1 M solution of acetic acid has a pH of 2.9. These two solutions have very different acidities but the same concentration, and each of these solutions neutralizes the same amount of NaOH solution.

When we have reached the end point in a titration, we can easily read directly on the buret exactly how many milliliters of NaOH solution were required to neutralize the acid. How do we then calculate the amount of acid? Before we can answer that, we must first define equivalents and normality.

We have seen that some acids are diprotic (H_2SO_4, H_2CO_3) and that some are triprotic (H_3PO_4). Similarly, there are some bases where each molecule produces two or three OH^- ions in solution—for example, $Ba(OH)_2$,

(a)   (b)   (c)

FIGURE 8.7   Titration. An acid of unknown concentration is in the Erlenmeyer flask. When base from the buret is added, the acid is neutralized. The end point is reached when the indicator color change becomes permanent. *(Charles D. Winters)*

$Mg(OH)_2$, and $Al(OH)_3$. For acids and bases, we define the **equivalent weight (EW)** as the **formula weight (FW) divided by the number of H^+ or OH^- ions produced.**

$Al(OH)_3$ yields $3OH^-$ ions only under optimal reaction conditions.

EXAMPLE 8.12

What is the equivalent weight of these acids and bases?
(a) HNO_3, FW = 63 amu
(b) NaOH, FW = 40 amu
(c) H_2SO_4, FW = 98 amu
(d) $Ba(OH)_2$, FW = 171.3 amu

Solution
In each case we divide the formula weight by the number of H^+ or OH^- ions produced by the acid or base.
(a) HNO_3, EW = 63/1 = 63 amu
(b) NaOH, EW = 40/1 = 40 amu
(c) H_2SO_4, EW = 98/2 = 49 amu
(d) $Ba(OH)_2$, EW = 171.3/2 = 85.6 amu

Problem 8.12 ■■■■■■
What is the equivalent weight of these acids and bases?
(a) $Mg(OH)_2$, FW = 58.34 amu
(b) HI, FW = 127.93 amu
(c) $Co(OH)_3$ FW = 109.93 amu ■

Because we often deal with very small quantities, we use milliequivalents: 1 meq = 1/1000 eq.

We define an **equivalent (eq)** of any substance as its **equivalent weight expressed in grams.** Thus, 1 eq of HNO_3 is 63 g, and 1 eq of H_2SO_4 is 49 g. This definition is similar to that of the mole, which is the formula weight expressed in g, and we deal with equivalents in a manner similar to the way we deal with moles. That is, if 49 g is 1 eq of H_2SO_4, then 4.9 g = 0.1 eq, 98 g = 2 eq, and so forth.

EXAMPLE 8.13

How many equivalents are there in (a) 18 g of H_2SO_4 (b) 51 g of $Mg(OH)_2$?

Solution

We multiply the number of g by the appropriate conversion factor:

(a) $18 \text{ g } H_2SO_4 \times \dfrac{1 \text{ eq } H_2SO_4}{49 \text{ g } H_2SO_4} = 0.37 \text{ eq } H_2SO_4$

(b) $51 \text{ g } Mg(OH)_2 \times \dfrac{1 \text{ eq } Mg(OH)_2}{29 \text{ g } Mg(OH)_2} = 1.8 \text{ eq } Mg(OH)_2$

Problem 8.13 ▬▬

How many equivalents are there in (a) 63.5 g of H_2SO_3 (b) 3.68 g of $Ba(OH)_2$? ■

The normalities discussed in this section are limited to acid-base reactions. Other systems of normality, not used in this book, apply to other kinds of reactions.

Why have chemists defined equivalent weight and equivalent? The answer to this question is present in the word used: "equivalent." In an acid-base reaction, **one equivalent of any acid exactly neutralizes one equivalent of any base.** Because most acid-base reactions occur in aqueous solution, we can go one step further and define a new concentration unit, **normality (*N*),** as **equivalents of solute per liter of solution.**

EXAMPLE 8.14

What is the normality of a solution made by dissolving 8.5 g of H_2SO_4 in enough water to make 500 mL of solution?

Solution

First, we need the number of equivalents of H_2SO_4/L of solution, and we know that the equivalent weight of H_2SO_4 is 49 amu and 1 eq of sulfuric acid is 49 g. Thus,

$$8.5 \text{ g } H_2SO_4 \times \frac{1 \text{ eq } H_2SO_4}{49 \text{ g } H_2SO_4} = 0.17 \text{ eq } H_2SO_4$$

To find normality, we divide

$$\frac{0.17 \text{ eq } H_2SO_4}{0.500 \text{ L}} = 0.34 \frac{\text{eq } H_2SO_4}{\text{L}} = 0.34 \, N \, H_2SO_4$$

Problem 8.14 ▬▬

What is the normality of a solution made by dissolving 3.4 mg of $Mg(OH)_2$ in enough water to make 1250 mL of solution? ■

There is a simple relationship between the normality and the molarity (Section 6.5) of an acidic or basic solution. Normality is simply the molarity multiplied by the number of H^+ or OH^- ions produced by one molecule of acid or base. For example, in HCl only one H^+ ion is produced. This means that, for all HCl solutions, the normality is equal to the molarity. Sulfuric acid produces two H^+ ions, so for H_2SO_4 solutions the normality is always twice the molarity; thus, a $6\ M$ solution of H_2SO_4 is $12\ N$. *Normality is always equal to or greater than molarity.*

Now that we have defined normality, we can come back to titration and see how to calculate an unknown acid concentration. We know that, at the end point, the number of equivalents of acid equals the number of equivalents of base:

$$Eq_{acid} = Eq_{base}$$

Since normality (eq/L) multiplied by volume (L) gives the number of equivalents, the following equation is valid:

$$V_{acid}N_{acid} = V_{base}N_{base}$$

If we are titrating an acid with a base, we begin by putting the acid in an Erlenmeyer flask. In doing so, we must carefully measure the volume of acid added (because this is V_{acid}). At the end point, we read the buret to find the volume of base added (V_{base}). The normality of the base (N_{base}) must be known before we do the titration. It is then simple to calculate the unknown (N_{acid}).

A pipet is often used for this.

EXAMPLE 8.15

If 50.0 mL of an acid solution is titrated with 0.32 N base, and it takes 24.6 mL of base to reach the end point, what is the normality of the acid?

Solution

$$V_{acid}N_{acid} = V_{base}N_{base}$$

$$(50.0\ \text{mL})(N_{acid}) = (24.6\ \text{mL})(0.32\ N)$$

$$N_{acid} = \frac{(24.6\ \cancel{\text{mL}})(0.32\ N)}{50.0\ \cancel{\text{mL}}} = 0.16\ N$$

Problem 8.15 ▬▬▬▬

If 50.00 mL of an acid solution is titrated with 0.152 N base, and it takes 27.5 mL of base to reach the end point, what is the normality of the acid? ■

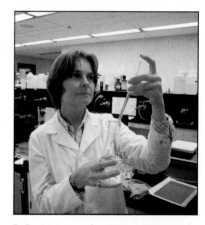

Laboratory worker measuring a volume with a pipet. *(FourByFive)*

Note that this answer is correct no matter which acid is in the flask and no matter which base is in the buret. The advantage of using normalities is that the simple equation $V_{acid}N_{acid} = V_{base}N_{base}$ works for any acid and any base, no matter how many H^+ or OH^- ions they produce.

The same procedure and calculations can be used to determine an unknown base concentration. In this case, a known amount of the base is put in the flask, and the buret is filled with an acid solution of known concentration.

SUMMARY

By the **Arrhenius** definitions, **acids** are substances that produce H_3O^+ ions in water, and **bases** are substances that produce OH^- ions in water. A strong acid or base produces high concentrations of these ions. The **Brønsted-Lowry** definitions expand this concept beyond water: An acid is a proton donor, and a base is a proton acceptor. Every acid has a **conjugate base.** An **amphiprotic** substance, such as water, can act as either an acid or a base.

The strengths of weak acids are expressed by their **dissociation constant, K_a;** the higher the K_a is, the stronger the acid is. Acids react with metals, metal hydroxides, and oxides to give **salts,** which are ionic compounds made up of cations from the base and anions from the acid. Acids also react with carbonates, bicarbonates, ammonia, and amines.

In pure water, a small percentage of molecules undergo the reaction $2H_2O \rightleftharpoons H_3O^+ + OH^-$. As a result, pure water has a concentration of 10^{-7} M for each of H_3O^+ and OH^-. The **ion product** of water, $K_w = [H_3O^+][OH^-]$, is equal to 1.0×10^{-14} in any aqueous solution. Hydronium ion concentrations are generally expressed in **pH units,** with pH = $-\log[H_3O^+]$. Solutions with pH less than 7 are acidic; those with pH higher than 7 are basic. Neutral solutions have a pH of 7. The pH is measured with indicators or with a pH meter.

Salts of strong acids and weak bases are acidic; salts of weak acids and strong bases are basic; salts of strong acids and strong bases are neutral.

A **buffer solution** does not change its pH very much when an acid or base is added. Buffers are made up of approximately equal concentrations of a weak acid and its conjugate base. Every buffer solution has a pH and a capacity. The most important buffers for blood are carbonate and phosphate.

The concentration of aqueous solutions of acids and bases can be measured by **titration,** a technique in which a base of known concentration is added to an acid of unknown concentration (or vice versa) until an end point is reached, at which point the solution is completely neutralized. The **equivalent weight** of an acid or base is its formula weight divided by the number of H^+ or OH^- ions one molecule of acid or base yields. One equivalent of an acid or base is the equivalent weight expressed in g. **Normality** of an acid or base is the number of equivalents in 1 L of solution.

The equation used to calculate the concentration of an acid or base from a titration is

$$V_{acid}N_{acid} = V_{base}N_{base}$$

KEY TERMS

Acid (Section 8.1)
Acid dissociation constant (Section 8.4)
Acidic salt (Section 8.9)
Acidic solution (Section 8.7)
Acidity constant (Section 8.4)
Acidosis (Box 8E)
Alkali (Section 8.1)
Alkalosis (Box 8E)
Amphiprotic (Section 8.3)
Base (Section 8.1)
Basic salt (Section 8.9)
Basic solution (Section 8.7)
Brønsted-Lowry acid and base (Section 8.3)

Buffer solution (Section 8.10)
Buffer capacity (Section 8.10)
Conjugate acid and base (Section 8.3)
Diprotic acid (Section 8.3)
End point (Section 8.11)
Equivalent (Section 8.11)
Equivalent weight (Section 8.11)
Henderson-Hasselbalch equation (Box 8F)
Hydronium ion, H_3O^+ (Section 8.1)
Hyperventilation (Box 8E)
Hypoventilation (Box 8E)
Indicator (Section 8.8)
Ion product (Section 8.7)
Ionization constant (Section 8.4)

K_a (Section 8.4)
K_w (Section 8.7)
Monoprotic acid (Section 8.3)
Neutral salt (Section 8.9)
Neutralization (Section 8.5)
Normality (Section 8.11)
pH (Section 8.8)
pK_a (Section 8.4)
pOH (Section 8.8)
Salt (Section 8.6)
Self-ionization of water (Section 8.7)
Strong acid or base (Sections 8.2, 8.3)
Titration (Section 8.11)
Triprotic acid (Section 8.3)
Weak acid or base (Section 8.2)

CONCEPTUAL PROBLEMS

8.A What single parameter would tell you whether an acid is strong or weak?

8.B Can an aqueous solution have a pH of 0? Explain your answer using HCl (MW = 36) as an example.

8.C How can you change the pH of a buffer? How can you change the buffer capacity of a buffer?

8.D When are the molarity and the normality of an acid the same?

PROBLEMS

Difficult problems are designated by an asterisk.

Arrhenius Acids and Bases

8.16 Write an equation for the reaction that takes place when each acid is added to water. Which reactions proceed nearly to completion?
(a) HF   (b) HBr   (c) H_2SO_3
(d) H_2SO_4 (e) HCO_3^-

8.17 Write an equation for the reaction that takes place when each base is added to water.
(a) LiOH   (b) $(CH_3)_2NH$

Brønsted-Lowry Theory

8.18 Write formulas for the conjugate bases of each acid.
(a) H_2SO_4 (b) H_3BO_3 (c) HI   (d) H_2O
(e) H_3O^+ (f) NH_3 (g) HPO_4^{2-}
(h) $H_2PO_4^-$ (i) H_2S (j) HCO_3^-

8.19 Write formulas for the conjugate acids of these bases.
(a) OH^- (b) NH_2^- (c) NH_3
(d) $C_6H_5O^-$ (e) CO_3^{2-} (f) HCO_3^-
(g) H_2O (h) HPO_4^{2-} (i) $CH_3—NH_2$
(j) PO_4^{3-}

8.20 Tell which of these acids is monoprotic, diprotic, triprotic, and/or amphiprotic.
(a) $H_2PO_4^-$ (b) HBO_3^{2-} (c) $HClO_4$
(d) HSO_3^- (e) HS^- (f) H_2CO_3

8.21 Is there a contradiction between the Brønsted-Lowry and Arrhenius definitions of acid strength? Explain.

8.22 Which of these reactions proceed from left to right? Label in each reaction which is the stronger acid, SA, and stronger base, SB.
(a) $H_3PO_4 + OH^- \rightleftharpoons H_2PO_4^- + H_2O$
(b) $H_2O + Cl^- \rightleftharpoons HCl + OH^-$
(c) $HCO_3^- + OH^- \rightleftharpoons H_2O + CO_3^{2-}$
(d) $C_6H_5OH + C_2H_5O^- \rightleftharpoons C_6H_5O^- + C_2H_5OH$
(e) $HCO_3^- + H_2O \rightleftharpoons CO_3^{2-} + H_3O^+$
(f) $CH_3COOH + H_2PO_4^- \rightleftharpoons CH_3COO^- + H_3PO_4$

Acid Dissociation Constants

8.23 Looking at the values in Table 8.3, indicate which is the stronger acid in each of the following pairs.
(a) $H_2PO_4^-$ and HCOOH
(b) HF and H_3BO_3
(c) H_2CO_3 and CH_3COOH

Acid and Base Properties

8.24 Which of the following acids will react with NH_3?
(a) HNO_3 (b) $H_2PO_4^-$
(c) HPO_4^{2-} (d) HCO_3^-

8.25 Write equations for the reaction of HCl with the following.
(a) Na_2CO_3 (b) Mg   (c) NaOH
(d) NH_3 (e) $CH_3—NH_2$ (f) $NaHCO_3$

Self-Ionization of Water

8.26 Calculate the $[OH^-]$ of each aqueous solution, given the following values of $[H_3O^+]$.
(a) $10^{-11} M$ (b) $10^{-4} M$ (c) $10^{-7} M$ (d) 10 M

8.27 Calculate the $[H_3O^+]$ of each of the following aqueous solutions, given the following values of $[OH^-]$.
(a) $10^{-10} M$ (b) $10^{-2} M$ (c) $10^{-7} M$ (d) $10^0 M$

pH and pK$_a$

8.28 What is the pH of each solution, given the following values of $[H_3O^+]$? Which solutions are acidic, which are basic, and which are neutral?
(a) $10^{-8} M$ (b) $10^{-10} M$ (c) $10^{-2} M$
(d) $10^0 M$ (e) $10^{-7} M$

8.29 Determine the pH and pOH of a solution that has the following $[OH^-]$ values.
(a) $10^{-3} M$ (b) $10^{-1} M$ (c) $10^{-5} M$ (d) $10^{-7} M$

8.30 What is the pH of a solution that has the following $[H_3O^+]$ values? Which solutions are acidic, which are basic, and which are neutral?
(a) $3 \times 10^{-9} M$ (b) $6 \times 10^{-2} M$
(c) $8 \times 10^{-12} M$ (d) $5 \times 10^{-7} M$

8.31 What is the $[H_3O^+]$ of an aqueous solution that has the following pH values? Which of these solutions are acidic, which are basic, and which are neutral?
(a) 12   (b) 7   (c) 2   (d) 13
(e) 8.40   (f) 12.10   (g) 3.52

8.32 What is the $[OH^-]$ and pOH of a solution that has the following pH values?
(a) 2   (b) 12   (c) 5   (d) 11

8.33 Three acids have the following pK_a values: butanoic acid, 4.82; barbituric acid, 5.00; lactic acid, 3.85.
(a) What are the K_a values?
(b) Which is the strongest of the three and which is the weakest?

The pH of Aqueous Salt Solutions

8.34 Give three examples of acidic salts and three of basic salts.

8.35 If we dissolve some solid sodium carbonate, Na_2CO_3, in pure water, the resulting solution is basic, that is, the pH rises above 7. Explain how this happens.

8.36 Tell whether each salt is acidic, basic, or neutral.
(a) Na_3PO_4 (b) $Al_2(SO_4)_3$ (c) MgI_2
(d) K_2SO_4 (e) NH_4Br (f) Li_2CO_3
(g) $BaCl_2$ (h) $Fe(ClO_4)_3$

Buffers

***8.37** What is the connection between buffer action and Le Chatelier's principle?

8.38 Write equations to show what happens when, to a buffer solution containing equimolar amounts of CH_3COOH and CH_3COO^-, we add (a) H_3O^+ (b) OH^-.

8.39 Write equations to show what happens when, to a buffer solution containing equimolar amounts of HPO_4^{2-} and $H_2PO_4^-$, we add (a) H_3O^+ (b) OH^-.

8.40 What is the pH of a buffer solution made by dissolving 0.1 mole of formic acid, HCOOH, and 0.1 mole of sodium formate, HCOONa, in 1 L of water?

8.41 What is meant by buffer capacity?

8.42 The pH of a solution made by dissolving 1 mole of

propanoic acid and 1 mole of sodium propanoate in 1 L of water is 4.85.
- (a) What would the pH be if we used 0.1 mole of each (in 1 L of water) instead of 1 mole?
- (b) With respect to buffer action, how would the two solutions differ?

8.43 What is the lowest pH in human blood below which death will occur?

Titration, Equivalents, and Normality

8.44 What is the purpose of titration?

8.45 What is the weight in g of 0.2 eq of each of the following?
(a) HCl   (b) H_3PO_4 (c) HCOOH   (d) H_2SO_4

8.46 How many equivalents are there in the following?
(a) 72.5 g of KOH
(b) 37 g of oxalic acid $H_2C_2O_4$ (a diprotic acid)
(c) 124 g of H_3PO_4
(d) 2.45 g of $Al(OH)_3$

8.47 What is the normality of a solution made by dissolving 12.7 g of HCl in enough water to make 1 L of solution?

8.48 What is the normality of a solution made by dissolving 3.4 g of $Ba(OH)_2$ in enough water to make 450 mL of solution?

8.49 What is the normality of a solution made by dissolving 5.83 mg of $Mg(OH)_2$ in enough water to make 250 mL of solution?

8.50 Describe exactly how you would prepare each of the following solutions (in each case assume that you have the solid bases).
(a) 400.0 mL of a 0.75 N solution of NaOH
(b) 1.0 L of a 0.071 N solution of $Ba(OH)_2$

8.51 What is the normality of each solution?
(a) 3.0 M HNO_3 (b) 4.5 M H_2SO_4
(c) 0.34 M KOH   (d) 2.7×10^{-3} M $Ca(OH)_2$

8.52 If 25.0 mL of an aqueous solution of H_2SO_4 requires 19.7 mL of 0.72 M NaOH to reach the end point, what is the normality of the H_2SO_4 solution?

8.53 A sample of 27.0 mL of a 0.310 N solution of NaOH is titrated with a 0.740 N H_2SO_4 solution. How many mL of the H_2SO_4 solution are required to reach the end point?

8.54 A 0.300 M solution of H_2SO_4 was used to titrate 10.00 mL of an unknown base; 15.00 mL of acid was required to neutralize the basic solution. What was the normality of the base?

8.55 A solution of an unknown base was titrated with 0.150 N HCl, and 22.0 mL of acid was needed to reach the end point of the titration. How many equivalents of the unknown base were in the solution?

8.56 The normality of the following solutions is listed as (a) 6.0×10^{-5} N $Mg(OH)_2$ (b) 2.84 N H_2CO_3 (c) 4.71×10^{-6} N $Co(OH)_3$. What is their molarity?

*8.57 The usual concentration of HCO_3^- ions in blood plasma is approximately 24 meq/L. How would you make up 1 L of a solution containing this concentration of HCO_3^- ions?

Boxes

8.58 (Box 8A) Which weak base is used as a flame retardant in plastics?

8.59 (Box 8B) What is the most important immediate first aid in any chemical burn of the eyes?

8.60 (Box 8B) With respect to corneal burns, are strong acids or strong bases more dangerous?

8.61 (Box 8C) Which of the antacids listed in Box 8C contain no hydroxides?

8.62 (Box 8D) What compound in baking powder produces the sponginess of a cake?

8.63 (Box 8D) What compounds generate the CO_2 in fire extinguishers?

8.64 (Box 8E) What causes (a) hypoventilation (b) respiratory alkalosis (c) metabolic alkalosis?

8.65 (Box 8F) Calculate the pH of an aqueous solution containing the following.
(a) 0.8 M lactic acid and 0.4 M lactate ion
(b) 0.3 M NH_3 and 1.50 M NH_4^+
(c) 0.42 M HF and 1.38 M F^-

*8.66 (Box 8F) You wish to make a buffer whose pH is 8.21. You have available a stock solution of 1.00 L of 0.100 M NaH_2PO_4, as well as some solid Na_2HPO_4. How many g of the solid Na_2HPO_4 must be added to the stock solution to accomplish this task? (Assume that the volume remains at 1.00 L.)

8.67 (Box 8F) In the past, boric acid has been used to rinse an inflamed eye. How would you make up a $H_3BO_3/H_2BO_3^-$ buffer solution that has a pH of 8.4?

*8.68 (Box 8F) You want to make a CH_3COOH/CH_3COO^- buffer solution with a pH of 5.60. The acetic acid concentration is to be 0.10 M. What should the acetate ion concentration be?

Additional Problems

8.69 Assume that you have a dilute solution of HCl (0.1 M) and a concentrated solution of acetic acid (5 M). Which solution is more acidic? Explain.

*8.70 If the $[OH^-]$ of a solution is 1×10^{-14},
(a) What is the pH of the solution?
(b) What is the $[H_3O^+]$?

8.71 What is the $[OH^-]$ of a solution that has a pH of 5.3?

8.72 Which solution will have the greatest normality?
(a) 9.8% w/v H_3PO_4 (b) 1.7% w/v H_2SO_4
(c) 2.9% w/v KOH

8.73 What is the normality of a solution made by dissolving 0.583 g of the diprotic acid oxalic acid, $H_2C_2O_4$, in enough water to make 1.75 L of solution?

8.74 The pK_a value of barbituric acid is 5.00. If the H_3O^+ and the barbiturate ion concentrations are each 0.00300 M, what is the concentration of the undissociated barbituric acid?

8.75 If pure water self-ionizes to give H_3O^+ and OH^- ions, why doesn't pure water conduct an electric current?

*8.76 A scale of K_b values for bases could be set up in a manner similar to that for the K_a scale for acids. However, this is generally considered unnecessary. Explain.

8.77 Do a 1 M CH$_3$COOH solution and a 1 M HF solution have the same pH? Explain.

8.78 In the reaction

$$CH_3COOH + HCO_3^- \rightleftharpoons CH_3COO^- + H_2CO_3$$

which is the stronger acid and which is the weaker acid? In which direction will the equilibrium lie?

*8.79 We dissolved 4.00 g of an unknown acid in 1.00 L of solution. This was titrated with 0.600 N NaOH, and 38.7 mL of the NaOH solution was needed to neutralize the acid. What was the equivalent weight of the acid?

8.80 Write equations to show what happens when, to a buffer solution containing equal amounts of HCOOH and HCOO$^-$, we add (a) H$_3$O$^+$ (b) OH$^-$.

8.81 If a solution of NH$_3$ in water (aqueous ammonia) is added to a solution of sodium ethoxide (containing ethoxide ions C$_2$H$_5$O$^-$), will an acid-base reaction take place? If so, write an ionic equation for the reaction.

*8.82 If we add 0.1 mole of NH$_3$ to 0.5 mole of HCl dissolved in enough water to make 1 L of solution, what happens to the NH$_3$? Will any of it remain? Explain.

Becquerel and the Curies were awarded the Nobel Prize in Physics in 1903 for this discovery. Rutherford won the Nobel Prize in Chemistry in 1908.

Above: PET (positron emission tomography) scan of the human brain. (CEA-ORSAY/CNRI/Science Photo Library/Photo Researchers, Inc.)

Nuclear Chemistry

9.1 Introduction

Every so often, a scientist makes the kind of discovery that changes the future of the world in some significant way. In 1896, a French scientist Henri Becquerel (1852–1908) made one of the most important of these discoveries. He found, completely by accident, that a certain rock was giving off mysterious radiation. The discovery was totally unexpected; nobody had ever had the faintest idea that anything of the kind was happening, and yet that rock, and others like it, had been giving off radiation since long before there were humans on Earth. It was simply that, before 1896, nobody knew it.

Becquerel's rock contained uranium. The radiation given off by the uranium caused photographic plates to "fog" even though they were wrapped in black paper. This meant that the radiation coming from the rock must be more energetic than visible light because the former could pass through the paper but the latter couldn't.

Shortly after Becquerel's discovery, other scientists, most notably Pierre (1859–1906) and Marie Curie (1867–1934) and Ernest Rutherford (1871–1937), began to study this radiation which Marie Curie called radioactivity. They soon discovered other types of radioactive materials with three types of radiation, which they named after the first three letters of the Greek alphabet, alpha, beta, and gamma.

BOX 9A

Rutherford's Experiment

The names alpha, beta, and gamma were assigned by Ernest Rutherford. Rutherford placed a piece of radium, which is a radioactive material, in a lead container that had a small opening. The radium emitted radiation that passed through the opening and then between the two charged plates (Figure 9A). One ray was deflected to the left, as negatively charged particles would be, and another was deflected to the right, as would be expected from positively charged particles. A third ray passed through without deflection, indicating no charge. The three rays were detected when they darkened a photographic plate. Rutherford named the positively charged rays alpha rays, the negatively charged rays beta rays, and the undeflected rays gamma rays.

■ **FIGURE 9A** Rutherford's experiment.

Alpha particles are helium nuclei. Each contains two protons and two neutrons and has an atomic number of 2.

Beta particles are electrons. Beta "radiation" is just a stream of electrons.

Gamma rays are not particles; they are a high-energy form of electromagnetic radiation.

Gamma rays are only one form of **electromagnetic radiation.** There are many others, including visible light. All consist of waves (Figure 9.1).

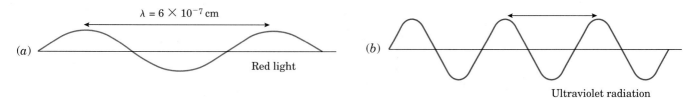

FIGURE 9.1 Two electromagnetic waves with different wavelengths.

FIGURE 9.2   The electromagnetic spectrum.

The only difference between one form of electromagnetic radiation and another is the **wavelength** (λ), which is the distance from one wave crest to the next. All electromagnetic radiation travels at the same speed in vacuum (the speed of light, which is 3.0×10^8 m/s). Because the wavelengths differ but the speed is always constant, the different forms of electromagnetic radiation have different frequencies. The **frequency** (ν) of a radiation is the number of crests that pass a given point in one second. **The longer the wavelength is, the lower the frequency is.** Mathematically, they are related by the following equation, where c is the speed of light:

UV stands for ultraviolet, and IR stands for infrared.

$$\lambda = \frac{c}{\nu}$$

There is also a direct relationship between the frequency of electromagnetic radiation and its energy: **The higher the frequency, the higher the energy.** Electromagnetic radiation comes in packets; the smallest units are called **photons.**

Figure 9.2 shows the position of various types of radiations of the **electromagnetic spectrum.** Gamma rays are electromagnetic radiation of very high frequency. We cannot see them because our eyes are not sensitive to waves of this frequency, but they can be detected by instruments (Section 9.4). Another kind of radiation called **x-rays,** has frequencies (and energies) higher than those of visible light but less than those of gamma rays.

The only radiation we know of with even higher frequency (and energy) than gamma rays are cosmic rays.

Materials that emit radiation (alpha, beta, or gamma) are called **radioactive. Radioactivity comes from the atomic nucleus and not from the electron cloud.** Table 9.1 summarizes the properties of the particles and rays that come out of radioactive nuclei, along with the properties of some other particles and rays. Note that x-rays are not considered to be a form of radioactivity. They do not come out of the nucleus but are generated in other ways.

We have said that humans cannot see gamma rays. We cannot see alpha or beta particles either. Not only that, but we also cannot hear them, smell them, or feel them. They are undetectable by our senses. Radioactivity can be detected only by instruments. We will discuss this subject in Section 9.4.

TABLE 9.1 Particles and Rays Frequently Encountered in Radiation

Particle or Ray	Common Name of Radiation	Symbol	Charge	Atomic Mass Units	Penetrating Power[a]	Energy Range[b]
Proton	Proton beam	$_1^1\text{H}$	+1	1	1–3 cm	60 MeV
Electron	Beta particle	$_{-1}^0\text{e}$ or β	−1	$0.00055 \left(\dfrac{1}{1835}\right)$	0–4 mm	1–3 MeV
Neutron	Neutron beam	$_0^1\text{n}$	0	1	—	—
Positron	—	$_{+1}^0\text{e}$ or β^+	+1	0.00055	—	—
Helium nucleus	Alpha particle	$_2^4\text{He}$ or α	+2	4	0.02–0.04 mm	3–9 MeV
Gamma ray		γ	0	0	1–20 cm	0.01–10 MeV
X-ray			0	0	0.01–1 cm	0.1–10 keV

[a] Distance at which half the radiation has been stopped in water.
[b] 1 MeV = 1.602×10^{-13} joules = 3.829×10^{-14} cal. The electron-volt (eV) is a non-SI energy unit used frequently in nuclear chemistry.

9.2 What Happens When a Nucleus Emits Radioactivity?

As mentioned in Section 2.4, different nuclei are made up of different numbers of protons and neutrons. It is customary to indicate these numbers with subscripts and superscripts placed to the left of the atomic symbol. The number of protons (the atomic number of the element) is shown as a subscript, and the total number of protons and neutrons (the mass number) as a superscript. For example, $_1^1\text{H}$, $_1^2\text{H}$, and $_1^3\text{H}$ are the symbols for the three isotopes of hydrogen. We call them hydrogen-1, hydrogen-2, and hydrogen-3, respectively.

Hydrogen-2 is also called deuterium, and hydrogen-3 is called tritium.

Radioactive and Stable Nuclei

Some isotopes are radioactive, and others are stable. There are more than 300 naturally occurring isotopes. Of these, 264 are stable, meaning that the nuclei of these isotopes never give off any radioactivity. As far as we can tell, they will last forever. The remainder are radioactive—they do give off radioactivity. Furthermore, more than 1000 artificial isotopes have been made in laboratories.

Isotopes in which the number of protons and neutrons are balanced seem to be stable. In the lighter elements, this balance occurs when the numbers of protons and neutrons are approximately equal. For example, $_6^{12}\text{C}$ is a stable isotope (six protons and six neutrons) and so are $_8^{16}\text{O}$ (eight protons and eight neutrons) and $_{10}^{20}\text{Ne}$ (ten protons and ten neutrons). Among the heavier elements, stability requires more neutrons than protons. Lead-206, one of the most stable isotopes of lead, contains 82 protons and 124 neutrons.

Radioactive isotope A radiation-emitting isotope of an element

A radioactive isotope is often called a **radioisotope.**

All artificial isotopes are radioactive.

The role of neutrons seems to be to provide binding energy to overcome the repulsion between protons.

Beta Emission

If there is a serious imbalance in the proton-to-neutron ratio, a nucleus will emit an alpha or beta particle and thereby make the ratio more favorable and the nucleus more stable. For example, if a nucleus has more neutrons than it needs for stability, it can stabilize itself by converting a neutron to a proton and an electron:

Nuclear reaction A reaction that changes atomic nuclei of elements (usually to that of other elements)

Note that for balancing purposes the electron is given an atomic number zero and a charge of −1.

$$_0^1\text{n} \longrightarrow {_1^1\text{H}} + {_{-1}^0\text{e}}$$

The electrons of the beta emission come from the nucleus and not from the surrounding electron cloud.

This process is called **beta emission** because the electrons (beta particles) that are produced are emitted from the nucleus. For example, phosphorus-32 is a beta emitter:

$$^{32}_{15}\text{P} \longrightarrow {}^{32}_{16}\text{S} + {}^{0}_{-1}\text{e}$$

When the unstable phosphorus-32 is converted to sulfur-32, stability is achieved (16 protons and 16 neutrons). Every time **a nucleus emits a beta particle, it is transformed into another nucleus with the same mass number but an atomic number one unit greater.** The changing of one element into another is called **transmutation.** It happens naturally every time an element gives off beta particles.

In balancing nuclear equations, **the sum of the mass numbers (superscripts) must be equal on both sides of the equation. The same is true for the atomic numbers (subscripts).** To find the proper symbol for a nucleus corresponding to a certain atomic number, use the Periodic Table on the inside front cover of this book. In writing nuclear equations we consider the nucleus only and disregard the surrounding electrons.

EXAMPLE 9.1

What nucleus is $^{14}_{6}\text{C}$, a beta emitter, converted to?

$$^{14}_{6}\text{C} \longrightarrow ? + {}^{0}_{-1}\text{e}$$

Solution

The $^{14}_{6}\text{C}$ nucleus has six protons and eight neutrons. One of its neutrons is converted to a proton and an electron. The electron is emitted (this is the beta particle), but the proton remains. The nucleus now has seven protons (therefore it must be nitrogen) and seven neutrons:

$$^{14}_{6}\text{C} \longrightarrow {}^{14}_{7}\text{N} + {}^{0}_{-1}\text{e}$$

Note that this equation is balanced: The sum of the mass numbers is 14 on both sides, and the sum of the atomic numbers is 6 on both sides.

Problem 9.1 ▬▬▬

Iodine-139 is a beta emitter. What is the reaction product? ■

Alpha Emission

Transmutation takes place also in alpha emission. For heavy elements, the loss of alpha particles is an especially important stabilization process. For example,

$$^{238}_{92}\text{U} \longrightarrow {}^{234}_{90}\text{Th} + {}^{4}_{2}\text{He}$$

$$^{210}_{84}\text{Po} \longrightarrow {}^{206}_{82}\text{Pb} + {}^{4}_{2}\text{He} + \gamma$$

The symbol γ, the Greek letter gamma, represents a gamma ray.

In the second example, stable lead-206 is obtained after alpha emission. Note that the polonium-210 emits both alpha particles and gamma rays.

Can you deduce from these examples the general rule for alpha emission? **The new nucleus always has a mass number four units lower and an atomic number two units lower than the original.**

Both alpha and beta emission can be either "pure" or mixed with gamma rays.

EXAMPLE 9.2

Polonium-218 is an alpha emitter. Write an equation for this alpha emission and identify the product.

Solution

The atomic number of polonium is 84, and so the equation is

$$^{218}_{84}\text{Po} \longrightarrow {}^{4}_{2}\text{He} + ?$$

Subtract 4 from the mass number 218 to get the new mass number, 214. Subtract 2 from atomic number 84 to get the new atomic number, 82. In the Periodic Table, we find that the element with an atomic number of 82 is lead, Pb. Therefore, the product is $^{214}_{82}\text{Pb}$.

Problem 9.2 ━━

Thorium-223 is an alpha emitter. Identify the isotope that is formed. ■

Positron Emission

A **positron** is a particle that has the same mass as an electron but a charge of $+1$ rather than -1. Its symbol is β^+ or $_{+1}^{0}e$. Positron emission is much less frequent than beta emission, but some nuclei do emit positrons. Since a positron has no appreciable mass when it is emitted, the nucleus is transmuted into another nucleus with the same mass number, but with an atomic number one unit less. An example is

$$^{11}_{6}\text{C} \longrightarrow {}^{0}_{+1}e + {}^{11}_{5}\text{B}$$

EXAMPLE 9.3

Nitrogen-13 is a positron emitter. Write an equation for this nuclear reaction, and identify the product.

Solution

The mass number of the new isotope will still be 13 because a positron has no appreciable mass. To get the atomic number, we subtract 1 from 7, the atomic number of nitrogen. The new atomic number is therefore 6. The element with an atomic number of 6 is carbon. Therefore, the equation is

$$^{13}_{7}\text{N} \longrightarrow {}^{0}_{+1}e + {}^{13}_{6}\text{C}$$

Problem 9.3 ━━

Arsenic-74 is a positron emitter used in locating brain tumors. What is the product nucleus formed? ■

Gamma Emission

Besides alpha, beta, and positron emitters, one occasionally encounters pure gamma emitters:

$$^{11m}_{5}\text{B} \longrightarrow {}^{11}_{5}\text{B} + \gamma$$

In this equation, the $^{11m}_{5}\text{B}$ symbolizes the boron nucleus in a high-energy (excited) metastable state. In this case, there is no transmutation. The boron is still boron, but it is in a lower energy form (more stable) after the emission of excess energy in the form of gamma rays. When all excess energy has been omitted (sometimes in a discrete, stepwise manner), the nucleus returns to its most stable, lowest energy state — the **ground state.**

9.3 Half-Life

When a nucleus gives off radiation, it is said to decay.

Suppose that we have 40 g of a radioactive isotope, say $^{90}_{38}Sr$. It contains many atoms, of course (in this case, about 2.7×10^{23}). These atoms are unstable and decay. In this case, they give off beta particles and thereby become $^{90}_{39}Y$ atoms. But do they all decay at once? No, they do not. They decay one at a time, at a fixed rate. For strontium-90, the decay rate is such that one half of our sample (about 1.35×10^{23} atoms) will have decayed by the end of 29 years. **The time it takes for one half of any sample of radioactive material to decay** is called the **half-life,** $t_{1/2}$.

It does not matter how big or small a sample is. For example, in the case of our 40 g of strontium-90, 20 g will be left at the end of 29 years (the rest will have been converted to yttrium-90). It will then take another 29 years for half of the remainder to decay, so that at the end of that time we will have 10 g. If we wait still a third period of 29 years, then 5 g will be left. If we had begun with 100 g, then 50 g would be left after the first 29-year period.

Figure 9.3 shows the radioactive decay curve of $^{131}_{53}I$. Inspection of this graph shows that, at the end of eight days, half of the original $^{131}_{53}I$ has disappeared. Thus, the half-life of iodine-131 is eight days. It would take a total of 16 days, or two half-lives, for three fourths of the original iodine-131 to decay.

EXAMPLE 9.4

If 10 mg of $^{131}_{53}I$ is administered to a patient, how much is left in the body after 32 days?

Solution
We know that 32 days corresponds to four half-lives because $t_{1/2}$ for this isotope is eight days. After one half-life, 5 mg is left; after two half-lives, 2.5 mg; after three half-lives, 1.25 mg; and after four half-lives, 0.63 mg.

Problem 9.4 ■
Barium-122 has a half-life of 2 min. You just obtained a sample weighing 10 g. It takes 10 min to set up the experiment in which barium-122 will be used. How many g of barium-122 will be left when you begin the experiment? ■

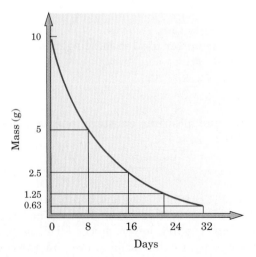

FIGURE 9.3 The decay curve of iodine-131.

BOX 9B

Radioactive Dating

Carbon-14, with a half-life of 5730 years, can be used to date archeological objects as old as 60 000 years. The dating is based on the principle that an organism, plant or animal, as long as it lives, maintains a constant carbon-12/carbon-14 ratio. When the organism dies, the carbon-12 level remains constant but the carbon-14 decays through the reaction

$$^{14}_{6}\text{C} \longrightarrow {}^{14}_{7}\text{N} + {}^{0}_{-1}\text{e}$$

Thus, the changing carbon-12/carbon-14 ratio can be used to determine the date of an artifact.

For example, in charcoal made from a tree that has recently died, the carbon-14 gives a radioactive count of 13.70 disintegrations/min/g of carbon. In a piece of charcoal found in a cave in France near some ancient cave paintings, the carbon-14 count was 1.71 disintegrations/min/g of carbon. From this information the cave paintings can be dated. After one half-life, the number of disintegrations/min/g is 6.85; after two half-lives, it is 3.42; and after three half-lives, it is 1.71. Therefore, three half-lives have passed since the paintings were done. Since carbon-14 has a half-life of 5730 years, the paintings are about $3 \times 5730 = 17\,190$ years old.

The famous Shroud of Turin, a piece of linen cloth with the image of a man's head on it, was believed by many to be the original cloth that was wrapped around the body of Jesus Christ. However, radioactive dating showed, with 95 percent certainty, that the plants from which the linen was obtained were alive sometime between 1260 and 1380 CE, proving that

■ Cro-Magnon cave painting. (Sci-Vu/Visuals Unlimited)

it could not have been the shroud of Christ. It was not necessary to destroy the shroud in order to make the tests. Scientists in different laboratories used just a total of a few square centimeters of cloth from its edge.

Rock samples can be dated on the basis of their lead-206 and uranium-238 content. The assumption is that all the lead-206 has come from the decay of uranium-238,* which has a half-life of 4.5 billion years. One of the oldest rocks found on Earth is a granite outcrop in Greenland, dated to be 3.7×10^9 years old.

On the basis of dating meteorites, the estimated age of the solar system is 4.6×10^9 years.

---

*Uranium-238 does not decay directly to lead-206. It first goes to thorium-234, which then also decays to radium-230, and so on. Lead-206 is the final result of a series of such transmutations.

(a)　　　　　　　　　　　(b)

■ The Shroud of Turin (a) as it appears to the eye, and (b) as a photographic negative, which brings out the image more clearly. (Santa Visalli/The IMAGE BANK)

TABLE 9.2 Half-Lives of Some Radioactive Nuclei

Name	Symbol	Half-Life	Radiation
Hydrogen-3 (tritium)	$^{3}_{1}H$	12.26 years	Beta
Carbon-14	$^{14}_{6}C$	5730 years	Beta
Phosphorus-28	$^{28}_{15}P$	0.27 s	Positrons
Phosphorus-32	$^{32}_{15}P$	14.3 days	Beta
Potassium-40	$^{40}_{19}K$	1.28×10^{9} years	Beta + gamma
Scandium-42	$^{42}_{21}Sc$	0.68 s	Positrons
Cobalt-60	$^{60}_{27}Co$	5.3 years	Gamma
Strontium-90	$^{90}_{38}Sr$	29 years	Beta
Technetium-99m	$^{99m}_{43}Tc$	6.0 h	Gamma
Indium-116	$^{116}_{49}In$	14 s	Beta
Iodine-131	$^{131}_{53}I$	8 days	Beta + gamma
Mercury-197	$^{197}_{80}Hg$	64 h	Gamma
Polonium-210	$^{210}_{84}Po$	138 days	Alpha
Radon-205	$^{205}_{86}Rn$	1.8 min.	Alpha
Radon-222	$^{222}_{86}Rn$	3.8 days	Alpha
Uranium-238	$^{238}_{92}U$	4×10^{9} years	Alpha

It must be noted that, in theory, it would take an infinite amount of time for all of a radioactive sample to disappear. However, most of the radioactivity disappears after five half-lives; by that time, only 3 percent of the original is left. After ten half-lives, less than 0.1 percent of the activity remains.

It is important to note that the half-life of an isotope is independent of temperature and pressure, and indeed of all other conditions, and is a property of the particular isotope only. It does not depend on what kind of other atoms surround the particular nucleus, that is, what kind of molecule the nucleus is part of. **We do not know any way to speed up radioactive decay or slow it down.**

Some half-lives are given in Table 9.2. Even this brief sampling indicates that there are tremendous differences in half-lives. Some isotopes, such as technetium-99m, decay and disappear in a day; others, such as uranium-238, will be radioactive for billions of years. Very short-lived isotopes, especially the artificial heavy elements (Section 9.7) with atomic numbers greater than 100, have half-lives of the order of seconds.

The usefulness or inherent danger in radioactive isotopes is related to their half-lives, as well as to their mode of decay. In assessing long-range health effects of atomic-bomb damage or of nuclear power plant accidents such as what happened at Three Mile Island, Pennsylvania, in 1979, or Chernobyl (in the former Soviet Union) in 1986 (Box 9G), we can see that radioactive isotopes with long half-lives, such as $^{85}_{36}Kr$ ($t_{1/2} = 10$ years) or $^{60}_{27}Co$ ($t_{1/2} = 5.3$ years), are more important than the short-lived ones. On the other hand, when a radioactive isotope is used in medical diagnosis or therapy, short-lived isotopes are more useful because they disappear faster from the body, for example, $^{99m}_{43}Tc$, $^{32}_{15}P$, $^{131}_{53}I$, and $^{197}_{80}Hg$.

9.4 Characteristics of Radiation

As already noted, our senses cannot detect radioactivity. We cannot see it, hear it, feel it, or smell it. Then how do we know it is there? Alpha, beta, gamma, and x-rays, as well as proton and positron beams, all have a prop-

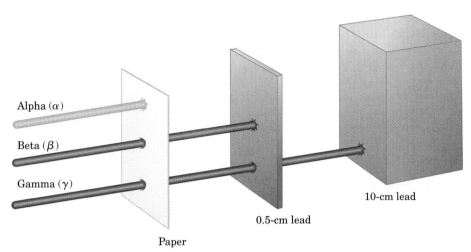

Alpha (α)

Beta (β)

Gamma (γ)

Paper

0.5-cm lead

10-cm lead

Alpha particles penetrate the least. They are stopped by a piece of paper. Beta particles easily penetrate paper but cannot go through a 0.5-cm sheet of lead. Gamma rays have the greatest penetrating power. It takes 10 cm of lead to stop them.

erty we can use to detect them: When they interact with matter, they usually knock electrons out of the electron cloud surrounding the atomic nucleus and thereby create positively charged ions from neutral atoms. Because of this, we call all of them **ionizing radiation.**

Radiation is characterized by two physical measurements: (a) its **intensity,** which is the number of particles or photons emerging per unit time, and (b) the **energy** of each particle or photon.

Intensity

To measure intensity, we take advantage of the ionizing property of radiation. Instruments such as the **Geiger-Müller counter** (Figure 9.4) or the **proportional counter** contain a gas such as helium or argon. When radioactive nuclei emit beta particles (electrons), these particles ionize the gas, and the instrument registers this by indicating that an electric current is passing between two electrodes. In this way, the instrument counts particle after particle.

A Geiger-Müller counter. *(Charles D. Winters)*

Thin window

Glass

Coating of conductive material (negative electrode)

Collector electrode (positive electrode)

FIGURE 9.4 The Geiger-Müller counter. The counter is made of a glass cylinder, the inner surface of which is coated with a material that conducts electricity. In the center is the collector electrode, made of tungsten wire. This is positively charged to about 1200 volts. The space between is filled with helium or argon gas at a low pressure. A thin Mylar plate (window) at the end keeps the gas in and at the same time allows the ionizing radiation to pass through. When the radiation enters the tube, it converts an argon or helium atom to a positive ion, which goes to the negative electrode and is counted.

A television picture tube works on a
similar principle.

Other measuring devices, such as **scintillation counters,** have a
material called a phosphor that emits a unit of light when an alpha or beta
particle or gamma ray strikes it. Again the particles are counted one by
one. The quantitative measure of radiation intensity can be reported in
counts/min or counts/s.

The intensity of any radiation decreases with the square of the distance:

$$\frac{I_1}{I_2} = \frac{d_2{}^2}{d_1{}^2}$$

EXAMPLE 9.5

mCi and other units of radiation are
defined in the next section.

If the intensity of radiation is 28 mCi at a distance of 1.0 m, what is
the intensity at a distance of 2.0 m?

Solution
From the preceding equation, we have

$$\frac{28 \text{ mCi}}{I_2} = \frac{2.0^2}{1.0^2}$$

$$I_2 = \frac{28 \text{ mCi}}{4.0} = 7.0 \text{ mCi}$$

Problem 9.5 ▬▬▬
If the intensity of radiation 1.0 cm from a source is 300 mCi, what is the
intensity of 3.0 m? ■

Energy

The energy of different particles or photons varies. As we have seen in
Table 9.1, each particle has a certain range of energy. For example, beta
particles have an energy range of 1–3 MeV. This range may overlap with
the energy range of another type of radiation, for example, gamma rays.
The penetrating power of a radiation depends on its energy as well as on
the mass of its particles. Since the photons of gamma radiation have no
mass, but the beta particles do have a mass, the penetrating power of
gamma rays is greater than that of beta particles.

One easy way to protect against ionizing radiation is to wear lead
aprons to cover sensitive organs. This is done routinely when diagnostic
x-rays are taken. Another way to lessen the damage from ionizing radiation is to move farther away from the source.

9.5 Radiation Dosimetry and Human Health

The term "activity of a radiation"
is the same as "intensity of a
radiation."

A common unit of radiation intensity is the **curie** (Ci), named in honor
of Marie Curie, whose lifelong work with radioactive materials greatly
helped our understanding of nuclear phenomena. One curie is defined as
3.7×10^{10} disintegrations per second (or counts per second). This is radiation of very high intensity, the amount a person would get from 1 g of pure
$^{286}_{88}$Ra. This intensity is too high for regular medical use; the most common
units used in the health sciences are small fractions of this:

1 millicurie (mCi) = 3.7×10^7 counts/s

1 microcurie (μCi) = 3.7×10^4 counts/s

EXAMPLE 9.6

A radioactive isotope with an intensity of 100 mCi per vial is sold to a hospital. The vial contains 10 mL of liquid. The instruction is to administer 2.5 mCi intravenously. How much of the liquid should one use?

Solution

The intensity of a sample is directly proportional to the amount present, so

$$2.5 \text{ mCi} \times \frac{10 \text{ mL}}{100 \text{ mCi}} = 0.25 \text{ mL}$$

Problem 9.6

A radioactive isotope in a 9.0-mL vial has an intensity of 300 mCi. A patient is required to take 50 mCi intravenously. How much liquid should be used for the injection? ∎

Another, albeit a much smaller, unit of radiation activity (intensity) is the **becquerel** (Bq). One becquerel is one disintegration per second, and so 1 Bq = 2.8×10^{-11} Ci.

In studying the effect of radiation on the body, neither the energy of the radiation (in kcal/mole) nor its intensity (in Ci) alone or in combination is of particular importance. The critical question is what kind of effects such radiation produces in the body. Three different units are used to describe the effects of radiation: roentgens, rads, and rems.

Roentgens   Roentgens (R) measure the ability of the radiation to affect matter, and in particular human beings and animals. One **roentgen** is the amount of radiation that produces ions having 2.58×10^{-4} coulombs per kilogram (a coulomb is a unit of electrical charge). The number of roentgens is the **energy delivered** by the radiation and is therefore a measure of exposure to a particular form of radiation.

Wilhelm Roentgen (1845–1923) discovered x-rays in 1895.

Rads   Exposure, or delivered energy, does not take into account the effect on tissue. Radiation damages body tissue by causing ionization. In order for ionization to occur, the delivered energy must be absorbed by the tissue. The common measure of the amount of **radiation absorbed** is the **rad,** which stands for **r**adiation **a**bsorbed **d**ose. The SI unit is the **gray** (Gy), where 1 Gy = 100 rad.

The relationship between delivered dose in roentgens and absorbed dose in rads is that exposure to 1 roentgen yields 0.97 rad of absorbed radiation in water, 0.96 rad in muscle, and 0.93 rad in bone. This relationship holds for high-energy photons. For lower energy photons, such as "soft" x-rays, each roentgen yields 3 rads of absorbed dose in bone. This is the principle of diagnostic x-rays, wherein soft tissue lets the radiation through to strike a photographic plate but the bones absorb the radiation and cast a shadow on the plate.

Rem   The most important biological measure of radiation is the rem, which stands for **r**oentgen **e**quivalent **m**an. One rem is a measure of the effect of the radiation when 1 roentgen is absorbed by a person. Other units are the **millirem** (mrem, 1 mrem = 1×10^{-3} rem) and the **sievert** (Sv, 1 Sv = 100 rem). The sievert is the SI unit. The need for such units arises because 1 rad of absorbed energy from different sources can cause different damage to the tissues. For example, 1 rad from alpha rays or neutrons causes ten times more damage than 1 rad from x-rays or gamma rays. The various radiation units are summarized in Table 9.3.

TABLE 9.3 Radiation Dosimetry

Measurement	International Unit	Conventional Unit
Intensity	1 becquerel (Bq) = 1 disintegration/s	1 curie (Ci) = 3.7×10^{10} disintegrations/s
Production of ions	1 roentgen (R) = 2.58×10^{-4} coulomb/kg	1 roentgen (R) = 2.58×10^{-4} coulomb/kg
Absorbed dose	1 gray (Gy) = 1 joule/kg	1 rad = 0.01 Gy
Absorbed dose by humans	1 sievert (Sv) = 1 joule/kg	1 rem = 0.01 Sv

Although they cause more damage than x-rays or gamma rays, alpha particles have a very low penetrating power and cannot pass through the skin (Table 9.1). Consequently, they are not harmful to humans or animals as long as they do not get into the body some other way. If they do get in, however, they are quite harmful. They can get inside, for example, if a person swallows or inhales a small particle of a substance that emits alpha particles. Beta particles are less damaging to tissue than alpha particles, but they penetrate farther and so are generally more harmful. Gamma rays, which can completely penetrate the skin, are by far the most dangerous and harmful form of radiation.

Therefore, for comparison purposes and for determining exposure from all kinds of sources, the **equivalent dose** is an important measure. If an organ receives radiation from different sources, the total effect can be summed up in rems (or mrem or Sv). For example, 10 mrem of alpha particles and 15 mrem of gamma radiation give a total of 25 mrem absorbed equivalent dose. Table 9.4 shows the amount of radiation exposure an aver-

Cosmic rays High-energy particles, mainly protons, from outer space bombarding the Earth

TABLE 9.4 Average Exposure to Radiation from Common Sources

Naturally Occurring Radiation	Dose (mrem/year)
Cosmic rays	27
Terrestrial radiation (rocks, buildings)	28
Inside human body (K-40 and Ra-226 in the bones)	39
Radon in the air	200
Total	294
Artificial Radiation	
Medical x-rays[b]	39
Nuclear medicine	14
Consumer products	10
Nuclear power plants	0.5
All others	1.5
Total	65.0
Grand total	359[c]

[a] Source: National Council on Radiation Protection and Measurements, NCRP Report No. 93 (1993).

[b] Individual medical procedures may expose certain parts of the body to much higher doses. For instance, one chest x-ray gives 27 mrem, and a diagnostic GI series gives 1970 mrem.

[c] The federal safety standard for allowable occupational exposure is about 5000 mrem/year. However, it has been suggested that this be lowered to 4000 mrem/year or even lower, to reduce the risk of cancer stemming from low levels of radiation.

BOX 9C

The Indoor Radon Problem

Most of our exposure to ionizing radiation comes from natural sources (Table 9.4). Radon gas is the main cause of this exposure. Radon has more than twenty isotopes, all radioactive, but the most important is radon-222, an alpha particle emitter. Radon-222 is a natural decay product of uranium-238, which is widely distributed in Earth's crust. Radon poses a particular health hazard among radioactive elements because it is a gas at normal temperatures and pressures. This means that radon enters our lungs with the air we breathe and gets trapped in the mucous lining of the lung. Radon-222 has a half-life of 3.8 days. It decays naturally and produces, among other isotopes, two harmful alpha emitters: polonium-218 and polonium-214. The Po isotopes are solid and do not leave the lung with exhalation. In the long run, they can cause lung cancer. The Environmental Protection Agency sets a standard of 4 pCi/L (one picocurie, pCi, is 10^{-12} Ci) as a safe exposure level. A survey of single-family homes in the United States showed that 7 percent exceeded this level. Most radon seeps into dwellings through cracks in cement foundations and around pipes. It accumulates mostly in basements. The remedy is to ventilate basements as well as houses enough to reduce the radiation levels. In a notorious case, in Grand Junction, Colorado, houses were built from bricks made from uranium tailings. Obviously, the radiation levels in these buildings were unacceptably high. They could not be controlled, and the buildings had to be destroyed. In our modern radiation-conscious age, more and more home buyers request a certification of radon levels before buying a house.

■ Testing devices are available to determine whether radon is building up in a home. *(Charles D. Winters)*

BOX 9D

How Radiation Damages Tissues. Free Radicals

As mentioned in the text, high-energy radiation damages tissue by causing ionization. This means that the radiation knocks electrons out of molecules that make up the tissue (generally one electron per molecule), thereby forming unstable ions. For example, interaction of high-energy radiation with water forms an unstable cation:

$$\text{Energy} + \text{H—O—H} \longrightarrow \text{H—}\overset{\cdot\cdot}{\underset{\cdot\cdot}{\text{O}}}\text{—H}^+ + {}_{-1}^{0}e$$

The positive charge on the H_2O^+ means that one of the electrons normally present in the water molecule, either in a covalent bond or an unshared pair, is missing in this cation; it has been knocked out. Once formed, the cation is unstable and decomposes:

$$\text{H—}\overset{\cdot\cdot}{\underset{\cdot\cdot}{\text{O}}}\text{—H}^+ \longrightarrow \text{H}^+ + \overset{\cdot\cdot}{:}\overset{\cdot\cdot}{\text{O}}\text{—H}$$

The ·OH has an unpaired electron (indicated by the dot), which makes it extremely reactive. Compounds that have unpaired electrons are called **free radicals.** Because they are so reactive, they rapidly interact with other molecules, causing chemical reactions that damage the tissues. These reactions are all the more serious if they occur inside the cell nucleus and damage genetic material. They affect rapidly dividing cells more than they do stationary cells. Thus, the damage is greater to embryonic cells, cells of the bone marrow and intestines, and cells in the lymph. Symptoms of radiation sickness are nausea, vomiting, a decrease in the white blood cell count, and loss of hair.

BOX 9E

CT, mri, Ultrasound, and PET Scans

One particular x-ray diagnostic technique, **computer-assisted tomography (CT scan)** has seen a phenomenal growth in the more than 25 years since its invention. Allan M. Cormack of Tufts University and Godfrey N. Hounsfield, an English scientist, shared the 1979 Nobel Prize in Medicine for this invention.

CT scans can reveal small differences in the density of tissues because different tissues slow down x-rays to different extents. A modern CT scan needs one rotation of the x-ray source around the body. A large number of detectors measure simultaneously the x-rays passing through the body. A computer then clearly reconstructs the internal structure of the body. CT scans are used mainly to detect brain tumors because no other technique is so accurate in locating tissues within the skull. Not only are internal structures displayed (Figure 9E.1), but one can also distinguish among the various fluid-filled cavities in the brain and between gray and white matter. CT scans are also extensively used to detect the damage caused by strokes. Their use is not restricted to the brain—any part of the body can be examined with this tool.

Another diagnostic tool, **magnetic resonance imaging (mri),** is especially good for the soft tissues that contain many hydrogen atoms, mostly in the form of water or fat. It probes the nuclear spin of atoms, mainly protons, which at a certain frequency invert their spin when subjected to radio-

■ *FIGURE 9E.2* An mri scan.
(© Debbie Breen/Phototake)

frequency waves. An advantage of mri is that bones contain few hydrogens, so that they are almost transparent in this imaging. Thus the soft tissues with their differing water contents stand out in the image (Figure 9E.2). Furthermore, mri has the advantage that it does not use ionizing radiation (Figure 9E.3), as CT (and PET) scans do.

Another diagnostic scan employing nonionizing radiation is the **ultrasound scan** (sonograph). Ultrasound can image soft tissues with high water content such as kidney cysts, or tumors or stones (Figure 9E.4).

The fourth important type of imaging is called **positron emission tomography (PET) scan.** It is based on the prop-

■ *FIGURE 9E.1* A CT scan of human brain. The upper-right corner indicates the section of the skull that was scanned. *(Courtesy of Lakeville-Nassau CT Center, NY)*

■ *FIGURE 9E.3* A medical mri imager.
(Beverly March; courtesy of Long Island Jewish Hospital)

erty that certain isotopes (such as carbon-11) have of emitting positrons (Section 9.2). Positrons have very short lives because when they collide with an electron, they annihilate each other, resulting in gamma rays. Because electrons are present in every atom, there are always lots of them around, so positrons generated in the body cannot live long. For example, if one is fed glucose labeled with carbon-11, the glucose gets into the blood and from there into the brain. Detectors for the gamma rays can pick up the signals that come from the areas where the glucose is used up. In this way one can see which areas of the brain are involved when we process visual information, for example (Figure 9E.5). PET scans can diagnose early stages of epilepsy and other diseases that involve abnormal glucose metabolism, such as schizophrenia.

The latest technique employs a combination of CT and PET scans. The CT scan provides a high-resolution three-dimensional image of tissues with different densities. The PET scan highlights the tissues with the highest metabolic rates indicating the position of possible malignancy.

■ *FIGURE 9E.4* Ultrasound scan of a human kidney showing a tumor. The size of the tumor is indicated on the scan. *(Courtesy of Radiology Department, Sub-urban Hospital, Bethesda, MD)*

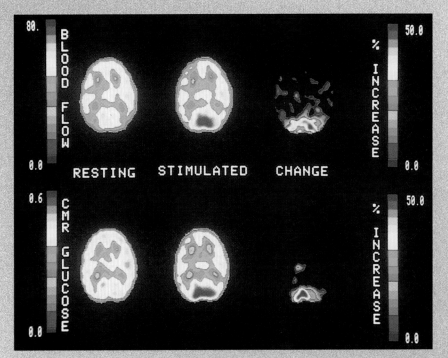

■ *FIGURE 9E.5* Positon emission tomography brain scans showing that glucose can cross the blood-brain barrier. The second scan in each case shows the result of visual stimulation. Both the local cerebral blood flow (48 percent increase) and the glucose metabolism (56 percent increase) go up during visual stimulation. *(Courtesy of Dr. Peter T. Fox, Washington University School of Medicine)*

FIGURE 9.5 Worker with a radiation badge. *(Yoav Levy/Phototake)*

age person obtains yearly from both natural and artificial sources. People who work in nuclear medicine are of course exposed to greater amounts. In order to ensure that exposures do not get too high, they wear radiation badges (Figure 9.5). A single whole-body irradiation of 25 rem is noticeable in the blood count, and 100 rem causes the typical symptoms of radiation sickness. A dose of 400 rem causes death within one month in 50 percent of exposed persons, and 600 rem is almost invariably lethal within a short time. It should be noted that up to 50 000 rem is needed to kill bacteria and up to 10^6 rem to inactivate viruses. It is not so much the energy content of a radiation that is harmful, but rather the ionization created when radiation interacts with matter. It is the extreme reactivity of free radicals (Box 9D) created in this interaction that causes damage in tissues.

Most of us, fortunately, never get a single dose of more than a few rem and so never suffer from any form of radiation sickness. This does not mean, however, that small doses are totally harmless. The harm comes in two ways:

1. Small doses of radioactivity or ionizing radiation from all sources over a period of years can cause cancer, especially blood cancers such as leukemia. Repeated exposure to sunlight also carries risk of damage to the tissues. Most of the high-energy UV radiation of the sun is absorbed by the protective ozone layer in the stratosphere. But in tanning the repeated overexposure to UV radiation can cause skin cancer. Nobody knows how many cancers have been caused in this way because the doses are so small and continue for so many years that they cannot generally be measured. Also, because there seem to be so many other causes of cancer, it is difficult or impossible to decide if any particular case is caused by radiation.

2. If any form of radiation strikes an egg or sperm cell, it can cause a change in the genes (Box 24E). Such changes are called mutations. If this particular egg or sperm cell mates, grows, and becomes a new individual, that individual may have mutated characteristics, which are usually harmful and frequently lethal.

Because radiation carries so much potential for harm, it would be nice if we could totally escape it. But can we? Table 9.4 shows that this is impossible. Naturally occurring radiation, which we call **background radiation,** is present everywhere on Earth. As Table 9.4 shows, this background radiation is much more than the average radiation level from artificial sources (which are mostly diagnostic x-rays). Therefore, if we stopped all forms of artificial radiation, including medical uses, we would still be exposed to the background radiation.

9.6 Medical Uses of Radioactive Materials

Radioactive isotopes have two main uses in medicine: diagnosis and therapy (Table 9.5).

Diagnosis

Diagnosis is the more widely used aspect of nuclear medicine. Chemically and metabolically, a radioactive isotope in the body behaves exactly the same way as the nonradioactive isotopes of the same element. In the simplest form of diagnosis, a radioactive isotope is injected intravenously, and a technician uses the various types of detectors to monitor how the radiation is distributed in the body of the patient. Most of the radioactive isotopes used in diagnostic nuclear medicine are gamma emitters. We already said

TABLE 9.5 Some Medically Useful Radioactive Isotopes

	Isotope	Use
$^{3}_{1}H$	Tritium	Measure water content of body
$^{11}_{6}C$	Carbon-11	Brain scan with positron emission tomography (PET) to trace glucose pathways
$^{14}_{6}C$	Carbon-14	Radioimmunoassay
$^{32}_{15}P$	Phosphorus-32	Detection of eye tumors
$^{51}_{24}Cr$	Chromium-51	Albumism diagnosis, size and shape of spleen, gastrointestinal disorders
$^{59}_{26}Fe$	Iron-59	Bone marrow function, diagnosis of anemias
$^{60}_{27}Co$	Cobalt-60	Treatment of cancer
$^{67}_{31}Ga$	Gallium-67	Whole-body scan for tumors
$^{75}_{34}Se$	Selenium-75	Pancreas scan
$^{81m}_{36}Kr$	Krypton-81m	Lung ventilation scan
$^{85}_{38}Sr$	Strontium-85	Bone scan for bone diseases, including cancer
$^{99m}_{43}Tc$	Technetium-99m	Brain, liver, kidney, bone marrow scans; diagnosis of damaged heart muscles
$^{125}_{53}I$	Iodine-125	Prostate cancer treatment
$^{131}_{53}I$	Iodine-131	Diagnosis of thyroid malfunction; treatment of hyperthyroidism and thyroid cancer
$^{197}_{80}Hg$	Mercury-197	Kidney scan
$^{201}_{81}Tl$	Thallium-201	Heart scan and exercise stress test

A scan of the radiation released by radioactive iodine concentrated in thyroid tissue gives an image of the thyroid gland. *(CNRI/SPL/Photo Researchers, Inc.)*

Normal

Meningioma (brain tumor)

"Brain death"

Scalp tumor

FIGURE 9.6 A comparison of dynamic scan patterns for normal and pathological brains. The studies were performed by injecting technetium-99m into blood vessels. *(From* CRC Handbook in Clinical Laboratory Science, Vol. 1, Nuclear Medicine. *CRC Press, Inc.)*

This is called tagging a drug.

that this is the most dangerous form of radiation because of its penetrating power. It is also the form of radiation that can, once administered, get out of the body and, therefore, be measured with the greatest accuracy. Therefore, only very small doses of a gamma-emitting radioisotope are needed for its presence to be detected.

The use of iodine-131 to diagnose malfunctioning thyroid glands is a good example. The thyroid glands in the neck produce a hormone, thyroxine, that controls the overall rate of metabolism (use of food) in the body. Thyroxine contains iodine atoms. When radioactive iodine-131 is administered into the blood stream, the thyroid glands take it up and incorporate it into the thyroxine (Box 14E). A normally functioning thyroid absorbs about 12 percent of the administered iodine within a few hours. An overactive thyroid (hyperthyroidism) absorbs and localizes iodine-131 in the glands faster, and an underactive thyroid (hypothyroidism) does this much more slowly than normal. Thus, by counting the gamma rays emitted from the neck, we can determine the rate of uptake of iodine-131 into the glands and diagnose hyper- or hypothyroidism.

Most organ scans are similarly based on the preferential uptake of some radioactive isotopes by a particular organ (Figure 9.6).

Another important use of radioactive isotopes is to learn what happens to ingested material. The foods and drugs swallowed or otherwise taken in by the body are transformed, decomposed, and excreted. To understand the pharmacology of a drug, it is important to know how these processes occur. For example, a certain drug may be effective in treating certain bacterial infections. Before this drug can be used in a clinical trial, the manufacturer must prove that the drug is not harmful to humans. In a typical case this is done first in animal studies. The drug in question is synthesized, and some radioactive isotope, such as hydrogen-3, carbon-14, or phosphorus-32, is incorporated into its structure. The drug is administered to the test animals, and after a certain period the animals are sacrificed. The fate of the drug is then followed by isolating from the body any radioactive compounds formed.

A typical pharmacological experiment was the study of the effects of tetracycline. This powerful antibiotic tends to accumulate in bones and is not given to pregnant women because it is transferred to the bones of the fetus. A particular tetracycline was tagged with radioactive hydrogen-3, and its uptake in rat bones was monitored in the presence and absence of a sulfa drug. With the aid of a scintillation counter, the radiation intensity of the maternal and fetal bones was determined, and it was found that the sulfa drug helped to minimize this undesirable side effect of tetracycline, that is, its accumulation in the fetal bones.

The metabolic fate of essential chemicals in the body can also be followed with radioactive tracers. A number of normal and pathological body functions have been elucidated by the use of radioactive isotopes.

Therapy

The main use of radioactive isotopes in therapy is the selective destruction of pathological cells and tissues. Remember that radiation, whether from gamma rays or other sources, is detrimental to cells. Ionizing radiation damages cells, especially those that divide rapidly. The damage may be great enough to destroy the cells or to sufficiently alter the genes so that multiplication of the cells slows down.

Cancerous cells are the main target for such ionizing radiation. Radiation is usually used when the cancer is not well localized; therefore, surgical removal is not the complete answer. It is also used when the cancerous cells spread and are in a metastatic state as well as for preventive purposes when one wants to eliminate any possible remaining cancerous cells after surgery has been performed. The idea, of course, is to kill cancerous cells but not normal ones. Therefore, radiation such as x-rays or gamma rays from cobalt-60 is highly collimated, that is, aimed at a small part of the body where the cancerous cells are suspected. Besides the gamma rays from cobalt-60, other ionizing radiation is also used to treat inoperable tumors. Proton beams from cyclotrons have been used to treat ocular melanoma and tumors of the skull base and spine.

Despite this pinpointing, the radiation kills normal cells along with the cancerous cells. Because the radiation is most effective against rapidly dividing cancer cells rather than normal cells, and because the radiation is aimed at a specific location, the damage to healthy tissues is minimized. Figure 9.7 shows a patient receiving radiation treatment.

Another way to localize radiation damage in therapy is to use specific radioactive isotopes. In the case of thyroid cancer, large doses of iodine-131 are given and taken up by the glands. The isotope, which has high radioactivity, kills all the cells of the gland, the cancerous as well as the healthy ones, but it does not appreciably damage other organs. Another radioisotope, iodine-125, is used in the treatment of prostate cancer. Seeds of iodine-125, a gamma emitter, are implanted in the cancerous area of the prostate gland while being imaged with ultrasound. The seeds deliver 160 Gy (16 000 rad) over their lifetime. In a similar manner, localization can be achieved by implanting a tiny grain of yttrium-90 in the pituitary gland. This beta emitter destroys the gland but nothing much beyond that. The hope is that this treatment will slow the growth of tumors elsewhere in the body because the pituitary gland stimulates cell reproduction, and without the growth hormone secreted by the pituitary, the cell proliferation of tumor cells will cease.

A metastatic state exists when the cancerous cells have broken off from their primary site(s) and are moving to other parts of the body.

Because of this, radiation sickness is a side effect of radiation therapy.

Iodine-131 therapy may be given as a single dose of 27 mCi or a cumulative dose of up to 1 Ci.

FIGURE 9.7 Patient receiving radiation treatment. *(Beverly March; courtesy of Long Island Jewish Hospital)*

BOX 9F

Energy from Mass

When a deuterium nucleus and a tritium nucleus are transmuted into a helium nucleus and a neutron, a large amount of energy is given off. Where does this energy come from? If we weighed the nuclei of the two hydrogen isotopes and compared this with the weight of the helium nucleus plus that of a neutron, we would find that the combined mass of the former is greater than the combined mass of the latter by 60 mg/mole ^{3}H. When the deuterium and tritium nuclei are converted to helium and a neutron, the extra mass has to go somewhere. Where does it go? It is converted to energy. We even know, from the equation of Albert Einstein (1879–1955), how much energy we can get from conversion of any amount of mass.

$$E = mc^2$$

The equation says that the mass (m) that was lost, in kg, multiplied by the square of the velocity of light (c^2)*, in m^2/s^2 is equal to the amount of energy created (E), in joules. For example, 1 g of matter completely converted to energy would produce 8.8×10^{13} joules, which is enough energy to boil 34 000 000 L of water initially at 20°C. This is equivalent to the amount of water you may find in an Olympic-size swimming pool. As you can see, we get a lot of energy from a little bit of mass.

*$c = 3.0 \times 10^8$ m/s.

9.7 Nuclear Fusion

It is estimated that 98 percent of all matter in the universe is made up of hydrogen and helium. This is explained by the "big bang" theory of the formation of the universe. This theory postulates that our universe started with an explosion (big bang) in which matter was formed out of energy, and, at the beginning, only the lightest element, hydrogen, was in existence. Later, as the universe expanded, stars were born when the hydrogen clouds collapsed under gravitational forces. In the cores of these stars, hydrogen nuclei fused together and formed helium.

The transmutation of hydrogen nuclei into a helium nucleus liberates a very large amount of energy in the form of photons (Box 9F), largely by the reaction:

$$^2_1\text{H} + ^3_1\text{H} \longrightarrow ^4_2\text{He} + ^1_0\text{n} + \text{energy}$$

This process, called **nuclear fusion,** is how the sun makes its energy. Uncontrolled fusion is employed in the "hydrogen bomb." If we can ever achieve a controlled version of this fusion reaction, we shall be able to solve our energy problems. However, this may not occur for a long time.

All the **transuranium elements** (elements with atomic numbers greater than 92) from 93 to 118 are artificial and have been prepared by a fusion process in which heavy nuclei are bombarded with light ones. Many, as their names indicate, were first prepared at the Lawrence Laboratory of the University of California, Berkeley, by Glenn Seaborg (1912–1999), Nobel laureate in chemistry, 1951) and his colleagues:

The reactions occurring in the sun are essentially the same as those occurring in the hydrogen bomb.

Nuclear fusion Joining together atomic nuclei in forming nuclides heavier than the starting material

$$^{244}_{96}\text{Cm} + ^4_2\text{He} \longrightarrow ^{245}_{97}\text{Bk} + ^1_1\text{H} + 2\,^1_0\text{n}$$

$$^{238}_{92}\text{U} + ^{12}_6\text{C} \longrightarrow ^{246}_{98}\text{Cf} + 4\,^1_0\text{n}$$

$$^{252}_{98}\text{Cf} + ^{10}_5\text{B} \longrightarrow ^{257}_{103}\text{Lr} + 5\,^1_0\text{n}$$

These transuranium elements are unstable, and most have very short half-lives. For example, that of $^{257}_{103}$Lr is 0.65 s.

9.8 Nuclear Fission. Atomic Energy

In the 1930s, Enrico Fermi (1901–1954) and his colleagues in Rome and Otto Hahn (1879–1968), Lisa Meitner (1878–1968), and Fritz Strassman (1902–   ) in Germany tried to produce new transuranium elements by bombarding uranium-235 with neutrons. To their surprise, Hahn and his coworkers found that, rather than fusion, they obtained **nuclear fission** (fragmentation of large nuclei into smaller pieces):

$$^{235}_{92}\text{U} + ^{1}_{0}\text{n} \longrightarrow ^{139}_{56}\text{Ba} + ^{94}_{36}\text{Kr} + 3\ ^{1}_{0}\text{n} + \gamma$$

In this reaction, the uranium-235 nucleus is broken into smaller particles, but the most important product is energy, which is produced because the products have less mass than the starting materials (Box 9F). This form of energy is called **atomic energy** and has been used for both war (in the atomic bomb) and peace.

In the fission of uranium-235, each fission produces three neutrons, which in turn can generate more fission by colliding with other uranium-235 nuclei. If even one of these neutrons produces new fission, the process becomes a self-propagating **chain reaction** (Figure 9.8) that continues at

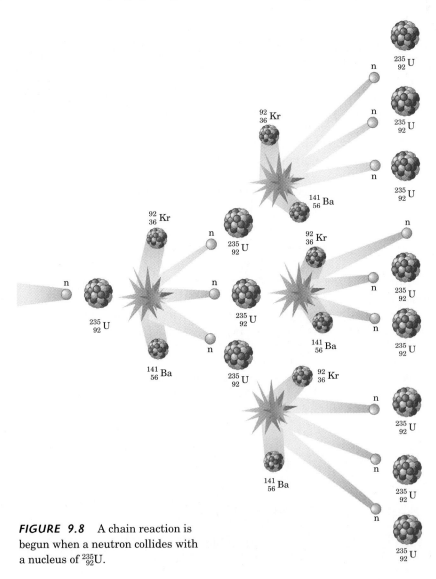

FIGURE 9.8 A chain reaction is begun when a neutron collides with a nucleus of $^{235}_{92}\text{U}$.

a constant rate. If all three neutrons are allowed to produce new fission, the rate of the reaction increases constantly and eventually culminates in a nuclear explosion. The rate of reaction can be controlled by inserting boron control rods to absorb neutrons.

In nuclear plants, the energy produced by fission is sent to heat exchangers and used to generate steam, which drives a turbine to produce electricity (Figure 9.9). Today more than 15 percent of the electrical energy in the United States is supplied by nuclear power plants. The opposition to nuclear plants is based on safety considerations and on the unsolved problems of waste disposal. Nuclear plants in general have had good safety records, but accidents such as those at Chernobyl (Box 9G) and Three Mile Island, Pennsylvania, have caused concern. The location of reactors in highly populated areas is definitely objectionable. An inadequate emergency evacuation plan led to the denial of an operating license to the Shoreham nuclear power plant on Long Island in the 1980s.

FIGURE 9.9 The nuclear power plant at Salem, New Jersey. *(Courtesy Public Service Electric & Gas Co.)*

BOX 9G

Radioactive Fallout from Nuclear Accidents

On April 26, 1986, an accident occurred at the nuclear reactor in the town of Chernobyl in the former Soviet Union. It was a clear reminder of the dangers involved in this industry and of the far-reaching contamination that such accidents create. In Sweden, more than 500 miles away from the accident, the radioactive cloud increased the background radiation from 4 to 15 times normal. The radioactive cloud reached England, about 1300 miles away, a week later. As a consequence, the natural background radiation was increased by 15 percent.

Radioactivity from iodine-131 was 400 Bq/L in milk and 200 Bq/kg in leafy vegetables. Even some 4000 miles away in Spokane, Washington, iodine-131 activity of 242 Bq/L was found in rain water, and smaller activities—1.03 Bq/L of ruthenium-103 and 0.66 Bq/L of cesium-137—were also recorded. These levels are not harmful.

Closer to the source of the nuclear accident, however, in neighboring Poland, potassium iodide pills were given to children. This was done to prevent radioactive iodine-131 (which might come from contaminated food) from concentrating in their thyroid glands, which could lead to cancer. Soon after the accident, about 115 000 people were evacuated from the immediate area and nearby fallout zones, but even that was not sufficient, and more than three years later many additional villages in the fallout path were evacuated.

■ Map showing areas most greatly affected by the Chernobyl accident.

■ The nuclear power plant at Chernobyl after the accident. *(Novosti/Gamma Liaison)*

FIGURE 9.10 Storing nuclear wastes in a storage room carved out of an underground salt bed. *(Courtesy U.S. Department of Energy)*

Waste disposal is a long-range problem. The fission products of nuclear reactors are highly radioactive themselves, with long half-lives. Thus, storing them under water or burying them underground (Figure 9.10) just postpones the disposal for generations to come. The fear is that corrosion and leakage from such storage tanks may contaminate water supplies. Remember that up to now, no one has ever discovered a way to speed up or slow down radioactive decay.

The only proposed permanent disposal site of both civil and military nuclear waste in the United States is located in Yucca Mountain of Nevada. The feasibility of this site is under study, and the Department of Energy is mandated to seek a license by the year 2001 to build a repository there. Environmental concerns abound. The site, which would not open until 2010, cannot be guaranteed to stay dry for centuries. Moisture may corrode the steel cylinders and even the inner glass/ceramic cylinders surrounding the nuclear waste. The fear is that leakage from such storage tanks may escape when carbon-14 is oxidized to radioactive carbon dioxide or, in a less likely case, when other radioactive nuclides may contaminate the ground water, which lies far below the desert rock of Yucca Mountain.

To keep all these nuclear problems in perspective, it is well to remember that most other ways of generating large amounts of electrical power also have their environmental problems. For example, burning coal or oil contributes to the accumulation of CO_2 in the atmosphere and to acid rain (see Box 6A).

SUMMARY

Certain isotopes are radioactive; others are stable. The three major types of radioactivity are **alpha particles** (helium nuclei), **beta particles** (electrons), and **gamma rays** (high-energy photons). **Radioactivity** comes from the nucleus, not from the electron cloud. When a nucleus emits a beta particle, it is changed to a nucleus of another element that has the same mass number but an atomic number one unit higher. When an alpha particle is emitted, the new element has an atomic number two units lower and a mass number four units lower. **Positron** (positive electron) emission leads to a new element having the same mass number but an atomic number one unit lower. In gamma emission, there is no transmutation, only the energy of the isotope is lowered.

Each radioactive isotope decays at a fixed rate described by its **half-life,** which is the time required for half the sample to decay.

Radiation is detected and counted by instruments such as Geiger-Müller counters. The main unit of intensity of radiation is the **curie,** which is 3.7×10^{10} disintegrations/s. For medical purposes and to measure potential radiation damage, the absorbed dose, measured in **rads,** is an important factor. Different particles damage body tissues differently. The **rem** is a measure of the damage.

Radioactive nuclei are used in medicine for both diagnosis and therapy.

Helium is synthesized in the interiors of stars by **fusion** of hydrogen nuclei. The energy released in this process is the energy of our sun. Artificial elements are created by fusion, bombarding atomic nuclei with high-energy nuclei. With certain nuclei, such bombardment causes the nucleus to split (**fission).** Nuclear fission releases large amounts of energy, which can be either controlled (nuclear reactors) or uncontrolled (nuclear weapons).

KEY TERMS

Alpha emission (Section 9.2)
Alpha particle (Section 9.1)
Atomic energy (Section 9.8)
Background radiation (Section 9.5)
Becquerel (Section 9.5)
Beta emission (Section 9.2)
Beta particle (Section 9.1)
Chain reaction (Section 9.8)
CT scan (Box 9E)
Curie (Section 9.5)
Equivalent dose (Section 9.5)
Electromagnetic radiation
    (Section 9.1)
Electromagnetic spectrum
    (Section 9.1)
Free radicals (Box 9D)

Frequency (Section 9.1)
Gamma emission (Section 9.2)
Gamma rays (Section 9.1)
Geiger-Müller counter (Section 9.4)
Gray (Section 9.5)
Half-life (Section 9.3)
Ionizing radiation (Section 9.4)
Magnetic resonance imaging (mri)
    (Box 9E)
Nuclear fission (Section 9.8)
Nuclear fusion (Section 9.7)
Photon (Section 9.1)
Positron (Section 9.2)
Positron emission tomography (PET)
    scans (Box 9E)

Proportional counter (Section 9.4)
Rad (Section 9.5)
Radioactive dating (Box 9B)
Radioactive isotope (Section 9.2)
Radioactivity (Section 9.1)
Radioisotope (Section 9.2)
Rem (Section 9.5)
Roentgen (Section 9.5)
Scintillation counter (Section 9.4)
Sievert (Section 9.5)
Transmutation (Section 9.2)
Transuranium element
    (Section 9.7)
Wavelength (Section 9.1)
x-ray (Section 9.1)

CONCEPTUAL PROBLEMS

9.A The half-life of a radioactive nucleus is two months. An experiment was started, and 2 g of the radioactive material was stored in a container. Somebody forgot about the container, which was discovered after a while; 0.125 g of radioactive material was still there. How much time passed since the start of the experiment?

9.B Federal regulations allow 5000 mrem/year exposure. (5000 mrem is equivalent to 5 rem = 0.05 Sv = 0.05 joules/kg body weight.) Thus, the allowed exposure of a 50 kg (110 lb) body is 2.5 joules. This is about 0.5 cal of energy, which is only sufficient to raise the temperature of 1 g of water by 0.5 degree. What is so special about radiation that we worry about such small amounts of energy deposited in our body?

9.C Draw Box 1 with two large and two medium-sized spheres. Draw Box 2 with two very large spheres and

four very small spheres in it. What nuclear reaction can be associated with going from Box 1 to Box 2?

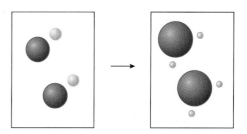

9.D In the first three steps in the decay of uranium-238, the following isotopic species appear: uranium-238 → thorium-234 → protactinium-234 → uranium-234. What kind of emission occurs in each of these steps?

PROBLEMS

Difficult problems are designated by an asterisk.

Radioactivity and Electromagnetic Radiation

9.7 Do gamma rays travel with the speed of light in a vacuum?

9.8 How did Becquerel discover radioactivity?

*9.9 Microwaves are a form of electromagnetic radiation that is used for the rapid heating of foods. What is the frequency of a microwave with a wavelength of 5.8 cm?

*9.10 In each case, given the frequency, determine the wavelength in cm or nm (10^{-9} m) and tell what kind of radiation it is.

(a) 7.5×10^{14}/s (b) 1.0×10^{10}/s
(c) 1.1×10^{15}/s (d) 1.5×10^{18}/s

9.11 Which color in the visible light spectrum has the lowest frequency?

9.12 Which has the longest wavelength: (a) infrared (b) ultraviolet (c) x-rays? Which has the highest energy?

Radioactive and Stable Isotopes

9.13 Write the symbol for a nucleus with
(a) 9 protons and 10 neutrons.
(b) 15 protons and 17 neutrons.
(c) 37 protons and 50 neutrons.

9.14 In each pair, tell which isotope is more likely to be radioactive.
(a) Nitrogen-14 and nitrogen-13
(b) Phosphorus-31 and phosphorus-33
(c) Lithium-7 and lithium-9
(d) Calcium-39 and calcium-40

9.15 How many artificial stable (nonradioactive) nuclei are known?

Natural Transmutation and Nuclear Reactions

9.16 Samarium-151 is a beta emitter. What is the transmutation product of the nuclear reaction? Write an equation for the reaction.

9.17 The following nuclei turn into new nuclei by emitting beta particles. In each case, write the formula for the new nucleus.
(a) $^{159}_{63}Eu$ (b) $^{141}_{56}Ba$ (c) $^{242}_{95}Am$

*9.18 Chromium-51 is used in diagnosing pathology of the spleen. The nucleus of this isotope captures an electron according to the following equation. What is the transmutation product?

$$^{51}_{24}Cr + {^0_{-1}}e \longrightarrow ?$$

9.19 The following nuclei turn into new nuclei by emitting alpha particles. In each case, write the formula for the new nucleus.
(a) $^{210}_{83}Bi$ (b) $^{238}_{94}Pu$ (c) $^{174}_{72}Hf$

9.20 Californium-246 is formed by bombarding uranium-238 with a light nucleus. During this transmutation four neutrons are emitted. What was the bombarding nucleus?

9.21 Phosphorus-29 is a positron emitter. Write the balanced nuclear equation.

9.22 For each of the following, write a balanced nuclear equation and identify the radiation emitted.
(a) Beryllium-10 changes to boron-10
(b) Europium-151m changes to europium-151
(c) Thallium-195 changes to mercury-195
(d) Plutonium-239 changes to uranium-235

9.23 In the radioactive decay of polonium-214 the first two steps are (1) alpha emission and (2) beta emission. Write the balanced nuclear equations for these two steps.

9.24 What kind of emission does not result in transmutation?

9.25 Complete the following nuclear reactions.
(a) $^{16}_8O + {^{16}_8O} \longrightarrow ? + {^4_2He}$
(b) $^{235}_{92}U + {^1_0n} \longrightarrow {^{90}_{38}Sr} + ? + 3\,{^1_0n}$
(c) $^{13}_6C + {^4_2He} \longrightarrow {^{16}_8O} + ?$
(d) $^{210}_{83}Bi \longrightarrow {_{21}^0e} + ?$
(e) $^{12}_6C + {^1_1H} \longrightarrow ? + \gamma$

Half-Life

9.26 What is a decay curve?

9.27 Polonium-218, a decay product of radon-222 (Box 9C), has a half-life of 3 min. What percent of the polonium-218 formed will remain in the lung 9 min after inhalation?

*9.28 A rock containing 1 mg plutonium-239/kg rock is found in a glacier. The half-life of plutonium-239 is 25 000 years. If this rock was deposited 100 000 years ago during an Ice Age, how much plutonium-239, per kg of rock, was in the rock at that time?

9.29 The element radium is extremely radioactive. If you converted a piece of radium metal to radium chloride (the weight of radium remaining the same), would it become less radioactive?

9.30 In what ways can we increase the rate of radioactive decay? Decrease it?

9.31 If we isolate 50.0 mg of potassium-45, a beta emitter, in pure form, after 1 h only 3.1 mg of the radioactive material would be left. What is the half-life of potassium-45?

*9.32 A patient receives 200 mCi of iodine-131, which has a half-life of eight days.
(a) If 12 percent of this is taken up by the thyroid gland after 2 h, what will be the intensity of the thyroid after 2 h, in mCi and in counts/min?
(b) After 24 days how much intensity will remain in the thyroid gland?

Characteristics of Radiation

9.33 Why is it necessary to use instruments to detect radioactivity?

9.34 If you work in a lab containing radioisotopes emitting all kinds of radiation, from which emission should you seek most protection?

9.35 Do Geiger-Müller counters measure (a) intensity or (b) energy of radiation?

9.36 It is known that radioactivity is being emitted with an intensity of 175 mCi at a distance of 1.0 m from the source. How far, in m, from the source should you stand if you wish to be subjected to no more than 0.20 mCi?

Radiation Dosimetry

9.37 State clearly the properties that are measured in the following units.
(a) Rads (b) Rems (c) Roentgens
(d) Curies (e) Grays (f) Becquerels
(g) Sieverts

9.38 A radioactive isotope with an intensity of 80.0 mCi per vial is delivered to a hospital. The vial contains 7.0 cc of liquid. The instruction is to administer 7.2 mCi intravenously. How many cubic centimeters of liquid should be used for one injection?

*9.39 Why is it that when a hand is exposed to alpha rays the damage to the person is not serious, but if the alpha emitter gets into the lung as an aerosol, the damage to the person's health is very serious?

9.40 A certain radioisotope has an intensity of 1 million Bq at a 1-cm distance from the source. What would be the intensity at 20 cm? Give your answer in Bq as well as μCi.

9.41 Assuming the same amount of effective radiation in rads from alpha particles, beta particles, and gamma rays, which source would be the most damaging to the tissues?

Medical Uses of Radioactive Materials

9.42 Determine which radioactive isotopes are used for each of the following medical sources.
(a) Cancer treatment   (b) Pancreas scans
(c) Bone scans   (d) Brain scans
(e) Kidney scans

9.43 In 1986 the nuclear reactor in Chernobyl had melt down and spewed radioactive nuclei that was carried by the winds for thousands of miles. Today one finds that, among the child survivors, the most common damage is thyroid cancer. What radioactive nucleus do you expect to be responsible for these cancers?

9.44 Two different iodine isotopes are used in cancer therapy. Name the isotopes and the organ cancers for which they are used.

9.45 Cobalt-60, with a half-life of 5.3 years, is used in cancer therapy. The energy of the radiation from cobalt-62 (with a half-life of 14 min) is even higher. Why isn't this isotope also used for cancer therapy?

9.46 Match the radioactive isotope in the left-hand column with its proper use in the right-hand column:

_____ (a) Cobalt-60   1. Heart scan in exercise
_____ (b) Thallium-201   2. Measure water
_____ (c) Hydrogen-3   content of body
_____ (d) Mercury-197   3. Kidney scan
                            4. Cancer therapy

Fusion and Fission

9.47 What is the most abundant nucleus in the universe?

9.48 What is the product of the fusion of hydrogen nuclei?

9.49 Assuming that one proton and two neutrons will be produced in an alpha bombardment fusion reaction, what target nucleus would you use to obtain berkelium-249?

9.50 Element 109 was first prepared in 1982. A single atom of this element, with a mass number of 266 ($^{266}_{109}$Une), was made by bombarding a bismuth-209 nucleus with an iron-58 nucleus. What other products, if any, must have been formed besides $^{266}_{109}$Une?

9.51 What nuclear particles cause chain reactions? How do we control chain reactions?

9.52 Boron-10 is used as control rods in nuclear reactors. It absorbs a neutron while it emits an alpha particle. What is the resulting product in the control rods?

Boxes

9.53 (Box 9A) Explain why, in Rutherford's experiment, the three rays were deflected in different directions.

9.54 (Box 9B) In a recent archeological dig in the Amazon region of Brazil, charcoal paintings were found in a cave. The carbon-14 content of the charcoal was one fourth of what is found in charcoal prepared from this year's tree harvest. How long ago was the cave settled?

9.55 (Box 9B) Carbon-14 dating of the Shroud of Turin indicated that the plant from which the shroud was made was alive around 1350 CE. How many half-lives does this correspond to?

9.56 (Box 9B) The half-life of carbon-14 is 5730 years. The wrapping of an Egyptian mummy gave off 6.85 counts/min/g of carbon. A piece of linen purchased today would give an activity of 13.70 counts/min/g of carbon. How old is the mummy?

9.57 (Box 9C) How does radon-222 produce polonium-218?

9.58 (Box 9D) How does ultraviolet radiation damage the skin?

9.59 (Box 9E) What are the advantages of an mri scan over a CT scan?

9.60 (Box 9E) Can one use CT scans to detect cerebral hemorrhages (bleeding in the brain)? Explain.

9.61 (Box 9E) Why do positrons have a very short lifetime?

9.62 (Box 9F) Explain how the fusion of hydrogen nuclei to produce a helium nucleus gives off energy.

9.63 (Box 9G) In the nuclear accident, one of the radioactive nuclei that people are concerned about is iodine-131. This is so because iodine is easily vaporized, can be carried by the winds, and can cause radioactive fallout hundreds, even thousands of miles away. Why is iodine-131 especially harmful?

Additional Problems

9.64 During the bombardment of argon-40 with protons, one neutron is emitted for each proton absorbed. What is the new element formed?

9.65 Neon-19 and sodium-20 are positron emitters. What are the products in each case?

9.66 The half-life of nitrogen-16 is 7 s. How long does it take for 100 mg of nitrogen-16 to be reduced to 6.25 mg?

9.67 Do curies and becquerels measure the same or different properties of radiation?

*__9.68__ Selenium-75 has a half-life of 120.4 days. This means that it would take 602 days (five half-lives) to diminish to 3 percent of the original quantity. Yet this isotope is used for pancreatic scans without any fear that the radioactivity will cause undue harm to the patient. Suggest a possible explanation.

9.69 Use Table 9.4 to determine the percentage of annual radiation we receive from
(a) Naturally occurring sources.
(b) Diagnostic medical x-rays.
(c) Nuclear power plants.

9.70 A beta emitter sends an electron out of the nucleus, yet there are no electrons in the nucleus. Explain from what particle the electron originates?

9.71 Will x-rays or radar cause more ionization?

9.72 You have an old wristwatch that still has radium paint on its dial. Measurement of the radioactivity of the watch shows a beta-particle count of 0.50 counts/s, and 1.0-μCi radiation of this sort produces 1000 mrem/year. How much radiation, in mrem, do you expect from the wristwatch if you wear it for a year?

9.73 Americium-241 is used in some smoke detectors. It has a half-life of 475 years. It is an alpha emitter. What is the decay product of americium-241? Approximately what percent of the original americium-241 will be still around after 1000 years?

9.74 On rare occasions, a nucleus, instead of emitting a beta particle, captures one. Berkelium-246 is such a nucleus. What is the product of this nuclear transmutation?

9.75 A patient is reported to have been irradiated by a dose of 1 Sv in a nuclear accident. Is he in mortal danger? Explain.

9.76 What is the ground state of a nucleus?

9.77 Explain the following statements.

(a) It is impossible to have a completely pure sample of any radioactive isotope.

(b) Beta emission of a radioactive isotope creates a new isotope with an atomic number one unit higher than that of the radioactive isotope.

9.78 Yttrium-90, which emits beta particles, is used in radiotherapy. What is the decay product of yttrium-90?

*9.79 The half-lives of some oxygen isotopes are: oxygen-14 = 71 s; oxygen-15 = 124 s; oxygen-19 = 29 s; and oxygen-20 = 14 s. Oxygen-16 is the stable nonradioactive isotope. Do the half-lives indicate anything about the stability of the other oxygen isotopes? Explain.

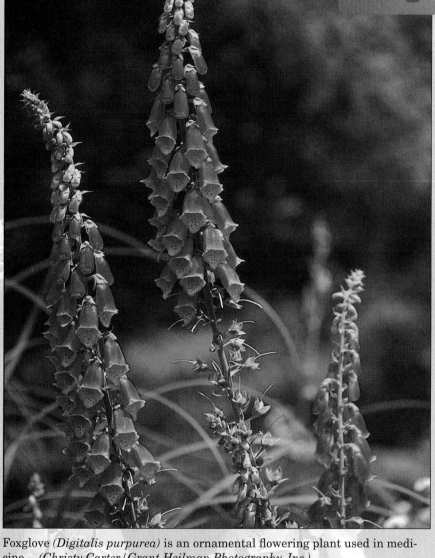

PART 2

Organic Chemistry

Foxglove *(Digitalis purpurea)* is an ornamental flowering plant used in medicine. *(Christy Carter / Grant Heilman Photography, Inc.)*

Organic Chemistry

10.1 Introduction

Organic chemistry The study of the compounds of carbon

Above: The bark of the Pacific yew contains paclitaxel, a substance that has proven effective in treating certain types of ovarian and breast cancer. *(Tom & Pat Leeson/Photo Researchers, Inc.)*

Organic chemistry is the study of the compounds of carbon. Perhaps the most remarkable feature of organic chemistry is that it is the chemistry of carbon and only a few other elements—chiefly, hydrogen, oxygen, and nitrogen. Chemists have discovered or made well over ten million compounds composed of carbon and these three other elements, and an estimated 100 000 new ones are discovered or prepared in the laboratory each year. Organic compounds are everywhere around us—in our foods, flavors, and fragrances; in our medicines, toiletries, and cosmetics; in our plastics, films, fibers, and resins; in our paints and varnishes; in our glues and adhesives; and, of course, in our bodies and those of all other living things.

In the early days of chemistry, scientists thought that there were two classes of compounds: organic and inorganic. Organic compounds were those produced by living organisms, and inorganic compounds were those found in rocks and other nonliving matter. It was the belief at the time that chemists could not synthesize any organic compound starting only from inorganic compounds. They thought that a "vital force," possessed only by living organisms, was necessary to produce organic compounds. This theory was very easy to disprove if indeed it was wrong. All it required was one experiment in which an organic compound was made from inorganic compounds. Friedrich Wöhler (1800–1882) carried out such an experiment in 1828. He heated an aqueous solution of ammonium chloride and silver cyanate, both inorganic compounds and, to his surprise, obtained urea, clearly an "organic" compound found in urine.

$$NH_4Cl + AgNCO \xrightarrow{heat} H_2N-\overset{\overset{\displaystyle O}{\|}}{C}-NH_2 + AgCl$$

Ammonium chloride   Silver cyanate   Urea   Silver chloride

Although this single experiment of Wöhler's was enough to disprove the "doctrine of vital force," it took several years and a number of additional experiments for the entire scientific community to accept the fact that organic compounds could be produced in the laboratory. This meant that the terms "organic" and "inorganic" no longer had real meaning because organic compounds could be obtained from inorganic as well as organic starting materials. A few years later, Friedrich August Kekulé (1829–1896) assigned the modern definition—organic compounds are those containing carbon—and his definition has been accepted ever since.

Enzymes, DNA, RNA, hormones, vitamins, and almost all other important chemicals in living systems are organic compounds, and their reactions are often strikingly similar to those observed occurring in test tubes. For this reason, knowledge of organic chemistry will deepen your understanding of biochemistry.

> Organic and inorganic compounds differ in properties because they differ in structure, not because they obey different natural laws. There is only one set of natural laws for all compounds.

10.2   Sources of Organic Compounds

Chemists obtain organic compounds in two principal ways: isolation from nature and synthesis in the laboratory.

Isolation from Nature

Living organisms are chemical factories. Each plant and animal, even microorganisms such as bacteria, makes thousands of organic compounds by a process called biosynthesis. One way, then, to get organic compounds is to extract them from biological sources. In this text, we will meet many compounds that are or have been obtained in this way. Some important examples are insulin, the penicillins, vitamin E, cholesterol, table sugar, nicotine, quinine, and the anticancer drug Taxol.

Besides plant and animal sources, nature also supplies us with three other important sources of organic compounds: natural gas, petroleum, and coal, which we discuss in Section 11.11.

Sugar cane, Hawaii. *(D. E. Cox/Tony Stone Images)*

Synthesis in the Laboratory

Ever since Wöhler synthesized urea, organic chemists have developed more ways to make the same compounds that nature makes. In recent years, the methods for doing this have become so sophisticated that there are few natural organic compounds, no matter how complicated, that chemists cannot synthesize in the laboratory.

Taxol. Search and Discovery

In the early 1960s, the National Cancer Institute undertook a program to analyze samples of native plant materials in the hope of discovering substances effective in the fight against cancer. Among the materials tested was an extract of the bark of the Pacific yew, *Taxus brevifolia*, a slow-growing tree found in the old-growth forests of the Pacific Northwest. This extract proved to be remarkably effective in treating certain types of ovarian and breast cancer, even in cases where other forms of chemotherapy failed. The structure of the cancer-fighting component of yew bark was determined in 1962, and the compound was named Paclitaxel (Taxol, The Merck Index, 12th ed., #7117).

■ Pacific yew bark being stripped for Taxol extraction. *(© Peter K. Ziminiski/Visuals Unlimited)*

Paclitaxel
(Taxol)

Unfortunately, the bark of a single 100-year-old tree yields only about 1 g of Taxol, not enough for effective treatment of even one cancer patient. Furthermore, getting Taxol means stripping the bark from trees, thus killing them. Fortunately, an alternative source of the drug was found. Researchers in France discovered that the needles of a related plant, *Taxus baccata*, contain a compound that can be converted to Taxol in the laboratory. Because the needles can be gathered without harming the plant, it is not necessary to kill trees to obtain the drug.

Taxol inhibits cell division by acting on microtubules. It does this in two ways; it stimulates microtubule polymerization and stabilizes the resulting structural units. Before cell division can take place, the cell must disassemble these units, and Taxol prevents this disassembly. Because cancer cells are the fastest dividing cells, Taxol effectively controls their spreading.

The remarkable success of Taxol in the treatment of breast and ovarian cancer has stimulated research efforts to discover and synthesize other substances that work the same way in the body and that may be even more effective anticancer agents than Taxol.

Vitamin C in an orange is identical to its synthetic tablet form. *(George Semple)*

Compounds made in the laboratory are identical to those found in nature. The ethanol made by chemists is exactly the same as the ethanol prepared by distilling wine. The molecules of ethanol are the same, and so are all of their physical and chemical properties; there is no way that anyone can tell whether a given sample of ethanol was made by chemists or obtained from nature. Therefore, there is no advantage in paying more money for, say, vitamin C obtained from a natural source than for synthetic vitamin C, because the two are identical in every way.

Organic chemists, however, have not rested with duplicating nature's compounds. They also synthesize compounds not found in nature. In fact, the majority of the more than ten million known organic compounds are purely synthetic and do not exist in living organisms. For example, many modern drugs—Valium, Vasotec, Prozac, Zantac, Cardizem, Lasix, Viagra, and Enovid—are all synthetic organic compounds not found in nature. Even the over-the-counter drugs aspirin and ibuprofen are synthetic organic compounds not found in nature.

10.3 Structure of Organic Compounds

The **Lewis model of bonding** (Section 3.2) enables us to account for the fact that carbon forms four covalent bonds that may be various combinations of single, double, and triple bonds. Furthermore, the valence-shell electron-pair repulsion (VSEPR) model (Section 3.5) enables us to account for the fact that the most common bond angles about carbon atoms in covalent compounds are approximately 109.5°, 120°, and 180°.

Table 10.1 shows several covalent compounds containing carbon bonded to hydrogen, oxygen, nitrogen, and chlorine. From these examples, we see that

- Carbon forms four covalent bonds and has no unshared pairs of electrons.

- Nitrogen forms three covalent bonds and has one unshared pair of electrons.

- Oxygen forms two covalent bonds and has two unshared pairs of electrons.

- Hydrogen forms one covalent bond and has no unshared pairs of electrons.

- Chlorine (and fluorine, bromine, and iodine as well) forms one covalent bond and has three unshared pairs of electrons.

TABLE 10.1   Single, Double, and Triple Bonds in Compounds of Carbon. Bond Angles are Predicted Using the VSEPR Model.

Ethane
(bond angles
109.5°)

Ethylene
(bond angles
120°)

Acetylene
(bond angles
180°)

Ethyl chloride
(bond angles
109.5°)

Methanol
(bond angles
109.5°)

Formaldehyde
(bond angles
120°)

Methylamine
(bond angles
109.5°)

Methyleneimine
(bond angles
120°)

Hydrogen cyanide
(bond angle 180°)

EXAMPLE 10.1

The structural formulas for acetic acid, CH_3COOH, and ethylamine, $CH_3CH_2NH_2$, are

$$
\begin{array}{cc}
\underset{\text{Acetic acid}}{\text{H}-\overset{\overset{\textstyle H}{|}}{\underset{\underset{\textstyle H}{|}}{C}}-\overset{\overset{\textstyle O}{\|}}{C}-\text{O}-\text{H}} &
\underset{\text{Ethylamine}}{\text{H}-\overset{\overset{\textstyle H}{|}}{\underset{\underset{\textstyle H}{|}}{C}}-\overset{\overset{\textstyle H}{|}}{\underset{\underset{\textstyle H}{|}}{C}}-\overset{}{\underset{\underset{\textstyle H}{|}}{N}}-\text{H}}
\end{array}
$$

    Acetic acid   Ethylamine

(a) Complete the Lewis structure for each molecule by adding unshared pairs of electrons so that each atom of carbon, oxygen, and nitrogen has a complete octet.
(b) Using the VSEPR model, predict all bond angles in each molecule.

Solution

(a) Each carbon atom is already surrounded by eight valence electrons and therefore has a complete octet. To complete the octet of each oxygen, add two unshared pairs of electrons. To complete the octet of nitrogen, add one unshared pair of electrons.
(b) To predict bond angles about a carbon, nitrogen, or oxygen atom, count the number of regions of electron density about it. If it is surrounded by four regions of electron density, predict bond angles of 109.5°. If it is surrounded by three regions, predict bond angles of 120°, and if it is surrounded by two regions, predict bond angles of 180°.

Acetic acid

Ethylamine

Problem 10.1 ▬▬

The structural formulas for ethanol, CH_3CH_2OH, and propene, $CH_3CH\!=\!CH_2$, are

$$
\begin{array}{cc}
\underset{\text{Ethanol}}{\text{H}-\overset{\overset{\textstyle H}{|}}{\underset{\underset{\textstyle H}{|}}{C}}-\overset{\overset{\textstyle H}{|}}{\underset{\underset{\textstyle H}{|}}{C}}-\text{O}-\text{H}} &
\underset{\text{Propene}}{\text{H}-\overset{\overset{\textstyle H}{|}}{\underset{\underset{\textstyle H}{|}}{C}}-\overset{}{\underset{\underset{\textstyle H}{|}}{C}}\!=\!\overset{}{\underset{\underset{\textstyle H}{|}}{C}}-\text{H}}
\end{array}
$$

    Ethanol   Propene

(a) Complete the Lewis structure for each molecule showing all valence electrons.

(b) Using the VSEPR model, predict all bond angles in each molecule. ∎

10.4 Functional Groups

Carbon combines with other atoms (chiefly H, N, O, S, halogens) to form structural units called **functional groups.** Functional groups are important for three reasons.

- They are sites of chemical reaction; a particular functional group, in whatever compound it is found, undergoes the same types of chemical reactions.
- They are the units by which we divide organic compounds into classes.
- They serve as a basis for naming organic compounds.

Introduced here are several functional groups that we encounter early in our study of organic chemistry. At this point, our concern is only pattern recognition. We have more to say about the physical and chemical properties of these functional groups in Chapters 11 to 19. A complete list of the major organic functional groups that we study is presented on the inside back cover of this text.

> **Functional group** An atom or group of atoms within a molecule that shows a characteristic set of physical and chemical properties

> **Hydroxyl group** An —OH group bonded to a tetrahedral carbon atom

Alcohols

The functional group of an **alcohol** is an **—OH (hydroxyl) group** bonded to a tetrahedral carbon atom (a carbon having single bonds to four other atoms).

Functional group An alcohol (Ethanol)

Table wine contains about 10 to 13 percent ethanol. *(Charles D. Winters)*

We can also represent this alcohol in a more abbreviated form called a **condensed structural formula.** In a condensed structural formula, CH_3 indicates a carbon bonded to three hydrogens, CH_2 indicates a carbon bonded to two hydrogens, and CH indicates a carbon bonded to one hydrogen. Unshared pairs of electrons are generally not shown in a condensed structural formula. Thus, the condensed structural formula for the alcohol of molecular formula C_2H_6O is $CH_3—CH_2—OH$. It is also common to write these formulas in an even more condensed manner, by omitting all single bonds: CH_3CH_2OH.

Alcohols are classified as **primary (1°), secondary (2°),** or **tertiary (3°)** depending on the number of carbon atoms bonded to the carbon bearing the —OH group.

> **Primary (1°) alcohol** An alcohol in which the carbon atom bearing the —OH group is bonded to only one other carbon group

> **Secondary (2°) alcohol** An alcohol in which the carbon atom bearing the —OH group is bonded to two other carbon groups

> **Tertiary (3°) alcohol** An alcohol in which the carbon atom bearing the —OH group is bonded to three other carbon groups

A 1° alcohol A 2° alcohol A 3° alcohol

EXAMPLE 10.2

Draw Lewis structures and condensed structural formulas for the two alcohols of molecular formula C_3H_8O. Classify each as primary, secondary, or tertiary.

Solution

Begin by drawing the three carbon atoms in a chain. The oxygen atom of the hydroxyl group may be bonded to the carbon chain in two ways: either to an end carbon or to the middle carbon.

C—C—C   C—C—C—OH   C—C—C (with OH on middle carbon)

The chain carbon   The two locations for the —OH group

Finally, add seven more hydrogens for a total of eight shown in the molecular formula. Show unshared electron pairs on the Lewis structures but not on the condensed structural formulas.

Lewis structure   **Condensed structural formula**

$CH_3CH_2CH_2OH$

A primary alcohol

CH_3CHCH_3 (with OH)

A secondary alcohol

The secondary alcohol, whose common name is isopropyl alcohol, is the cooling, soothing component in rubbing alcohol.

Isopropyl alcohol can be used to disinfect cuts and scrapes. *(Charles D. Winters)*

Amino group An —NH_2 group

Primary (1°) amine An amine in which nitrogen is bonded to one carbon and two hydrogens

Secondary (2°) amine An amine in which nitrogen is bonded to two carbons and one hydrogen

Tertiary (3°) amine An amine in which nitrogen is bonded to three carbons

Problem 10.2

Draw Lewis structures and condensed structural formulas for the four alcohols of molecular formula $C_4H_{10}O$. Classify each alcohol as primary, secondary, or tertiary. Hint: First consider the order of attachment of the four carbon atoms; they can be bonded either four in a chain or three in a chain with the fourth carbon as a branch on the middle carbon. ∎

Amines

The functional group of an **amine** is an **amino group**—a nitrogen atom bonded to one, two, or three carbon atoms. In a **primary (1°) amine**, nitrogen is bonded to one carbon group. In a **secondary (2°) amine**, it is bonded to two carbon groups, and in a **tertiary (3°) amine**, it is bonded to three carbon groups.

H—N̈—H	CH₃—N̈—H	CH₃—N̈—H	CH₃—N̈—CH₃
H	H	CH₃	CH₃
Ammonia	Methylamine (a 1° amine)	Dimethylamine (a 2° amine)	Trimethylamine (a 3° amine)

EXAMPLE 10.3

Draw condensed structural formulas for the two primary amines of molecular formula C_3H_9N.

Solution

For a primary amine, draw a nitrogen atom bonded to two hydrogens and one carbon.

$$C-C-C-\underset{\underset{H}{|}}{N}-H \qquad C-\underset{\underset{H}{|}}{\overset{\overset{C}{|}}{C}}-N-H \qquad \longrightarrow \qquad CH_3CH_2CH_2-\underset{\underset{H}{|}}{N}-H \qquad CH_3\underset{\underset{H}{|}}{\overset{\overset{CH_3}{|}}{CH}}-N-H$$

The three carbons may be bonded to nitrogen in two ways.   Add seven hydrogens to give each carbon four bonds and give the correct molecular formula.

Problem 10.3 ▬▬

Draw structural formulas for the three secondary amines of molecular formula $C_4H_{11}N$. ■

Aldehydes and Ketones

Both aldehydes and ketones contain a **C=O (carbonyl) group.** The functional group of an **aldehyde** is a carbonyl group bonded through its carbon to two hydrogens in the case of formaldehyde, CH_2O, the simplest aldehyde, and to another carbon and a hydrogen in all other aldehydes. In a condensed structural formula, the aldehyde group may be written showing the carbon-oxygen double bond as —CH=O, or, alternatively, it may be written —CHO. The functional group of a **ketone** is a carbonyl group bonded to two carbon atoms.

> **Carbonyl group** A C=O group

> **Aldehyde** A compound containing a carbonyl group bonded to a hydrogen (a —CHO group)

> **Ketone** A compound containing a carbonyl group bonded to two carbon groups

Functional group   Acetaldehyde (an aldehyde)   Functional group   Acetone (a ketone)

Acetone is a ketone.
(Charles D. Winters)

EXAMPLE 10.4

Draw condensed structural formulas for the two aldehydes of molecular formula C_4H_8O.

Solution

First draw the functional group of an aldehyde, and then add the remaining carbons. These may be bonded in two ways. Then, add seven hydrogens to complete the four bonds of each carbon.

$$CH_3CH_2CH_2\overset{\overset{\displaystyle O}{\|}}{C}H$$

or

$$CH_3CH_2CH_2CHO$$

$$CH_3\overset{\overset{\displaystyle O}{\|}}{C}H\overset{\displaystyle }{\underset{\displaystyle CH_3}{|}}CH$$

or

$$CH_3\underset{\underset{\displaystyle CH_3}{|}}{C}HCHO$$

Problem 10.4
Draw condensed structural formulas for the three ketones of molecular formula $C_5H_{10}O$. ∎

Carboxylic Acids

Carboxyl group A —COOH group

Carboxylic acid A compound containing a —COOH group

The functional group of a **carboxylic acid** is a —**COOH** (**carboxyl:** *carb*onyl + hydr*oxyl*) **group.** In a condensed structural formula, a carboxyl group may also be written —CO_2H.

Functional group   Acetic acid (a carboxylic acid)

EXAMPLE 10.5

Draw a condensed structural formula for the single carboxylic acid of molecular formula $C_3H_6O_2$.

Solution

The only way the carbon atoms can be written is three in a chain, and the —COOH group must be on an end carbon of the chain.

$$CH_3-CH_2-\overset{\overset{O}{\|}}{C}-O-H \quad \text{or} \quad CH_3CH_2COOH$$

Problem 10.5 ━━━

Draw condensed structural formulas for the two carboxylic acids of molecular formula $C_4H_8O_2$. ■

SUMMARY

Organic chemistry is the study of the compounds of carbon. Chemists obtain organic compounds by isolation from plant and animal sources and by synthesis in the laboratory.

Functional groups are sites of chemical reactivity; a particular functional group, in whatever compound it is found, always undergoes the same types of reactions. In addition, functional groups are the characteristic structural units by which both classify and name organic compounds. Important functional groups for us at this stage in the course are the **hydroxyl group** of 1°, 2°, and 3° alcohols; the **amino group** of 1°, 2°, and 3° amines; the **carbonyl group** of aldehydes and ketones; and the **carboxyl group** of carboxylic acids.

KEY TERMS

Alcohol (Section 10.4)
Aldehyde (Section 10.4)
Amine (Section 10.4)
Amino group (Section 10.4)
Carbonyl group (Section 10.4)
Carboxyl group (Section 10.4)
Carboxylic acid (Section 10.4)

Condensed structural formula (Section 10.4)
Functional group (Section 10.4)
Hydroxyl group (Section 10.4)
Ketone (Section 10.4)
Lewis structure (Section 10.3)
Organic chemistry (Section 10.1)

Primary (1°) alcohol (Section 10.4)
Primary (1°) amine (Section 10.4)
Secondary (2°) alcohol (Section 10.4)
Secondary (2°) amine (Section 10.4)
Tertiary (3°) alcohol (Section 10.4)
Tertiary (3°) amine (Section 10.4)

CONCEPTUAL PROBLEMS

Difficult problems are designated by an asterisk.

*10.A Suppose that you are told that only organic substances are produced by living organisms. How would you rebut this statement?

10.B There are millions of known organic compounds. Although only a few of the possible reactions for the majority of them have been studied, how they will react under new experimental conditions can be predicted quite accurately. Organic chemists can even predict the reactions of compounds yet to be dicovered. What characteristic of organic compounds allows their reactions to be predicted?

*10.C Think about the types of substances in your immediate environment, and make a list of those that are organic, for example, textile fibers. We will ask you to return to this list later in the course and to refine, correct, and possibly expand it.

10.D List the four principal elements that make up organic compounds and the number of bonds each typically forms.

PROBLEMS

Difficult problems are designated by an asterisk.

Review of Lewis Structures

10.6 Write Lewis structures for these compounds. Show all valence electrons. None of them contains a ring of atoms. (Hint: Remember that carbon has four bonds, nitrogen has three bonds and one unshared pair of electrons, oxygen has two bonds and two unshared pairs of electrons, and each halogen has one bond and three unshared pairs of electrons.

(a) H_2O_2
Hydrogen peroxide

(b) N_2H_4
Hydrazine

(c) CH_3OH
Methanol

(d) CH_3SH
Methanethiol

(e) CH_3NH_2
Methylamine

(f) CH_3Cl
Chloromethane

(g) CH_3OCH_3
Dimethyl ether

(h) C_2H_6
Ethane

(i) C_2H_4
Ethylene

(j) C_2H_2
Acetylene

(k) CO_2
Carbon dioxide

(l) CH_2O
Formaldehyde

(m) H_2CO_3
Carbonic acid

(n) CH_3COOH
Acetic acid

10.7 Write Lewis structures for these ions.

(a) HCO_3^-
Bicarbonate ion

(b) CO_3^{2-}
Carbonate ion

(c) CH_3COO^-
Acetate ion

(d) Cl^-
Chloride ion

*10.8 Why are the following molecular formulas impossible?

(a) CH_5 (b) C_2H_7

Review of the VSEPR Model

10.9 Use the VSEPR model to predict bond angles about each highlighted atom.

* 10.10 Following is a structural formula and a ball-and-stick model of benzene, C_6H_6.

(a) Predict each H—C—C and C—C—C bond angle on benzene.

(b) Predict the shape of a benzene molecule.

Sources of Organic Compounds

10.11 Is there any difference between vanillin made synthetically and vanillin extracted from vanilla beans, assuming that both are chemically pure?

10.12 What important experiment was carried out by Wöhler in 1828?

Functional Groups

10.13 Draw Lewis structures for each functional group. Be certain to show all valence electrons. (a) Carbonyl group (b) Carboxyl group (c) Hydroxyl group (d) Primary amino group

10.14 Complete these structural formulas by adding enough hydrogens to complete the tetravalence of each carbon. Then write the molecular formula of each compound.

(a)
$$C-C=C-\overset{\overset{\displaystyle C}{|}}{C}-C$$

(b)
$$C-C-C-\overset{\overset{\displaystyle O}{\|}}{C}-OH$$

(c)
$$C-C-\overset{\overset{\displaystyle O}{\|}}{C}-C$$

(d)
$$C-\overset{\overset{\displaystyle O}{\|}}{C}-\underset{\underset{\displaystyle C}{|}}{C}-H$$

(e)
$$C-\overset{\overset{\displaystyle C}{|}}{\underset{\underset{\displaystyle C}{|}}{C}}-C-C-NH_2$$

(f)
$$C-C-\overset{\overset{\displaystyle O}{\|}}{\underset{\underset{\displaystyle NH_2}{|}}{C}}-OH$$

(g)
$$C-\overset{\overset{\displaystyle OH}{|}}{C}-C-C-C$$

(h)
$$C-\overset{\overset{\displaystyle OH}{|}}{C}-C-\overset{\overset{\displaystyle O}{\|}}{C}-OH$$

(i)
$$C=C-C-OH$$

10.15 Some of these structural formulas are incorrect (that is, they do not represent a real compound) because they have atoms with an incorrect number of bonds. Which structural formulas are incorrect, and which atoms in them have an incorrect number of bonds?

(a)
```
    H   H
    |   |
H — C — N — H
    |   |
    H   H
```

(b)
```
        Cl
        |
H — C = C — H
    |   |
    H   H
```

(c)
```
    H   H
    |   |
H — N — C — C — O — H
    |   |   |
    H   H   H
```

(d)
```
    H   H
    |   |
H — C — C — O
    |   |
    H   H
```

(e)
```
    H   H   O
    |   |   ||
H — O — C — C — C — O — H
        |   |
        H   H
```

(f)
```
    H   H   O
    |   |   ||
H — C — C — C — H
    |   |   |
    H   H   H
```

(g)
```
    H           H
    |           |
H — C — C = C = C — C — H
    |   |       |   |
    H   H       H   H
```

(h)
```
        H
        |
H — C ≡ C — C — H
        |
        H
```

10.16 What is the meaning of the term tertiary (3°) when it is used to classify alcohols?

10.17 Draw a structural formula for the one tertiary (3°) alcohol of molecular formula $C_4H_{10}O$.

10.18 What is the meaning of the term tertiary (3°) when it is used to classify amines?

10.19 Draw a structural formula for the one tertiary (3°) amine of molecular formula $C_4H_{11}N$.

10.20 Identify the functional groups in each compound.

(a)
```
        OH   O
        |    ||
CH₃ — CH — C — OH
```
Lactic acid

(b) HO — CH₂ — CH₂ — OH
Ethylene glycol

(c)
```
            O
            ||
CH₃ — CH — C — OH
        |
        NH₂
```
Alanine

(d)
```
            OH   O
            |    ||
HO — CH₂ — CH — C — H
```
Glyceraldehyde

(e)
```
        O          O
        ||         ||
CH₃ — C — CH₂ — C — OH
```
Acetoacetic acid

(f) H₂NCH₂CH₂CH₂CH₂CH₂CH₂NH₂
1,6-Hexanediamine

10.21 Draw condensed structural formulas for all compounds of molecular formula C_4H_8O that contain a carbonyl group (there are two aldehydes and one ketone).

*10.22 Draw structural formulas for:
(a) The four primary (1°) alcohols of molecular formula $C_5H_{12}O$.
(b) The three secondary (2°) alcohols of molecular formula $C_5H_{12}O$.
(c) The one tertiary (3°) alcohol of molecular formula $C_5H_{12}O$.

10.23 Draw structural formulas for the six ketones of molecular formula $C_6H_{12}O$.

*10.24 Draw structural formulas for the eight carboxylic acids of molecular formula $C_6H_{12}O_2$.

*10.25 Draw structural formulas for
(a) The four primary (1°) amines of molecular formula $C_4H_{11}N$.
(b) The three secondary (2°) amines of molecular formula $C_4H_{11}N$.
(c) The one tertiary (3°) amine of molecular formula $C_4H_{11}N$.

Boxes

10.26 (Box 10A) How was Taxol discovered?

10.27 (Box 10A) In what way does Taxol interfere with cell division?

Additional Problems

10.28 Use the VSEPR model to predict bond angles about each atom of carbon, nitrogen, and oxygen in these molecules. Hint: First add unshared pairs of electrons as necessary to complete the valence shell of each atom, and then make your predictions of bond angles.

(a) CH₃ — CH₂ — CH₂ — OH

(b)
```
            O
            ||
CH₃ — CH₂ — C — H
```
(c) CH₃ — CH = CH₂
(d) CH₃ — C ≡ C — CH₃

(e)
$$CH_3-\overset{\overset{\textstyle O}{\|}}{C}-O-CH_3$$

(f)
$$CH_3-\overset{\overset{\textstyle CH_3}{|}}{N}-CH_3$$

10.29 Silicon is immediately below carbon in the Periodic Table. Predict the C—Si—C bond angle in tetramethylsilane, $(CH_3)_4Si$.

10.30 Draw the structure for a compound of molecular formula
(a) C_2H_6O that is an alcohol.
(b) C_3H_6O that is an aldehyde.
(c) C_3H_6O that is a ketone.
(d) $C_3H_6O_2$ that is a carboxylic acid.

*10.31 Draw structural formulas for the eight aldehydes of molecular formula $C_6H_{12}O$.

*10.32 Draw structural formulas for the three tertiary (3°) amines of molecular formula $C_5H_{13}N$.

10.33 Which of these bonds are polar and which are nonpolar? (Hint: Review Section 3.2.)
(a) C—C　(b) C═C　(c) C—H　(d) C—O
(e) O—H　(f) C—N　(g) N—H　(h) N—O

10.34 Of the bonds in Problem 10.33, which is the most polar? Which is the least polar?

10.35 Using the symbol δ^+ to indicate partial positive charge and δ^- to indicate partial negative charge, indicate the polarity of the most polar bond (or bonds if there are two or more of the same polarity) in each molecule.
(a) CH_3OH　　　　　(b) CH_3NH_2
(c) CH_2O　　　　　(d) CH_3COCH_3
(e) $HSCH_2CH_2NH_2$　(f) CH_3COOH

Alkanes and Cycloalkanes

11.1 Introduction

In this chapter we begin our study of organic compounds with the physical
and chemical properties of alkanes and cycloalkanes, both among the simplest types of organic compounds.

A **hydrocarbon** is a compound composed of only carbon and hydrogen.
A **saturated hydrocarbon** contains only single bonds. Saturated in this
context means that each carbon in the hydrocarbon has the maximum
number of hydrogens bonded to it.

An **alkane** is a saturated hydrocarbon whose carbon atoms are
arranged in an open chain. A **cycloalkane** is a saturated hydrocarbon in
which two carbon atoms of the chain are joined to form a ring.

Alkanes and cycloalkanes are commonly referred to as **aliphatic
hydrocarbons** because the physical properties of the higher members of
this class resemble those of the long carbon-chain molecules we find in animal fats and plant oils (Greek: *aleiphar*, fat or oil).

Above: A petroleum refinery.
Petroleum, along with natural gas,
provides close to 90 percent of the
organic raw materials for the synthesis
and manufacture of synthetic fibers,
plastics, detergents, drugs, dyes, and a
multitude of other products.
(J.L. Bohin/Photo Researchers, Inc.)

FIGURE 11.1   Methane and ethane.

11.2   Structure of Alkanes

Methane, CH_4, and ethane, C_2H_6, are the first two members of the alkane family. Shown in Figure 11.1 are Lewis structures and ball-and-stick models for these molecules. The shape of methane is tetrahedral, and all H—C—H bond angles are 109.5°. Each carbon atom in ethane is also tetrahedral, and all bond angles are approximately 109.5°.

Although the three-dimensional shapes of larger alkanes are more complex than those of methane and ethane, the four bonds about each carbon atom are still arranged in a tetrahedral manner, and all bond angles are approximately 109.5°. The next members of the alkane family are propane, butane, and pentane.

Propane is the chief component of LP or bottled gas.

$CH_3CH_2CH_3$
Propane
(C_3H_8)

$CH_3CH_2CH_2CH_3$
Butane
(C_4H_{10})

$CH_3CH_2CH_2CH_2CH_3$
Pentane
(C_5H_{12})

Structural formulas for alkanes can also be written in a more condensed form. For example, the structural formula of pentane contains three CH_2 (**methylene**) groups in the middle of the chain. They can be grouped together and the structural formula written $CH_3(CH_2)_3CH_3$. The first ten alkanes with unbranched chains are given in Table 11.1. Note that the names of all these alkanes end in "-ane." We will have more to say about naming alkanes in Section 11.4.

TABLE 11.1   The First Ten Alkanes with Unbranched Chains

Name	Molecular Formula	Condensed Structural Formula	Name	Molecular Formula	Condensed Structural Formula
methane	CH_4	CH_4	hexane	C_6H_{14}	$CH_3(CH_2)_4CH_3$
ethane	C_2H_6	CH_3CH_3	heptane	C_7H_{16}	$CH_3(CH_2)_5CH_3$
propane	C_3H_8	$CH_3CH_2CH_3$	octane	C_8H_{18}	$CH_3(CH_2)_6CH_3$
butane	C_4H_{10}	$CH_3(CH_2)_2CH_3$	nonane	C_9H_{20}	$CH_3(CH_2)_7CH_3$
pentane	C_5H_{12}	$CH_3(CH_2)_3CH_3$	decane	$C_{10}H_{22}$	$CH_3(CH_2)_8CH_3$

11.3 Constitutional Isomerism in Alkanes

Constitutional isomers are compounds that have the same molecular formula but different structural formulas. By "different structural formulas," we mean that they differ in the kinds of bonds (single, double, or triple) and/or in the order of connection among their bonds. For the molecular formulas CH_4, C_2H_6, and C_3H_8, only one order of attachment of atoms is possible; therefore, there are no isomers for these molecular formulas. For the molecular formula C_4H_{10}, two structural formulas are possible. In one of these, named butane, the four carbons are bonded in a chain; in the other, named 2-methylpropane, three carbons are bonded in a chain with the fourth carbon as a branch on the chain. We refer to butane and 2-methylpropane as constitutional isomers; they are different compounds and have different physical and chemical properties. In the past, constitutional isomers have also been referred to as **structural isomers.**

> **Isomer** Compounds that have the same molecular formula but different structural formulas

> **Constitutional isomers** Compounds with the same molecular formula but a different order of attachment of their atoms

$CH_3CH_2CH_2CH_3$
Butane
(bp $-0.5°C$)

$$\underset{\substack{| \\ CH_3CHCH_3}}{CH_3}$$

2–Methylpropane
(bp $-11.6°C$)

In Section 10.4, we encountered several examples of constitutional isomers, although we did not call them that at the time. We saw that there are two alcohols of molecular formula C_3H_8O, two primary amines of molecular formula C_3H_9N, two aldehydes of molecular formula C_4H_8O, and two carboxylic acids of molecular formula $C_4H_8O_2$.

To determine whether two or more structural formulas represent constitutional isomers, write the molecular formula of each and then compare

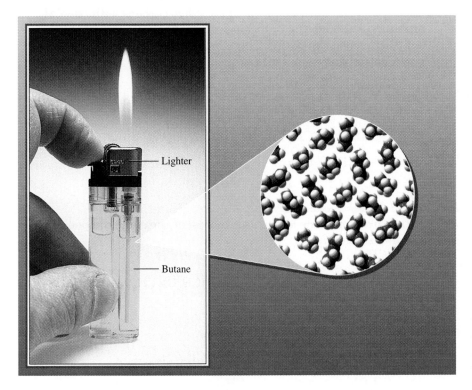

Lighter

Butane

Butane ($CH_3CH_2CH_2CH_3$) is the fuel in this lighter. Butane molecules are present in the liquid and gaseous states in the lighter. (*Charles D. Winters*)

them. All compounds that have the same molecular formula but different structural formulas are constitutional isomers.

EXAMPLE 11.1

Do the structural formulas in each set represent the same compound or constitutional isomers?
(a) $CH_3CH_2CH_2CH_2CH_2CH_3$ and $CH_3CH_2CH_2$ (each is C_6H_{14})
$$CH_2CH_2CH_3$$

(b) CH_3CHCH_2CH with CH_3 branch and $CH_3CH_2CHCHCH_3$ with CH_3 branch (each is C_7H_{16})

$$\underset{CH_3 \quad CH_3}{CH_3CHCH_2CH} \quad \text{and} \quad \underset{CH_3}{CH_3CH_2CHCHCH_3}$$

Solution

First, find the longest chain of carbon atoms. Note that it makes no difference if the chain is drawn straight or bent. As structural formulas are drawn in this problem, there is no attempt to show three-dimensional shapes. Second, number the longest chain from the end nearest the first branch. Third, compare the lengths of the two chains and the size and locations of any branches. Structural formulas that have the same order of attachment of atoms represent the same compound; those that have different orders of attachment of atoms represent constitutional isomers.

(a) Each structural formula has an unbranched chain of six carbons; they are identical and represent the same compound.

$$\overset{1}{C}H_3\overset{2}{C}H_2\overset{3}{C}H_2\overset{4}{C}H_2\overset{5}{C}H_2\overset{6}{C}H_3 \quad \text{and} \quad \overset{1}{C}H_3\overset{2}{C}H_2\overset{3}{C}H_2$$
$$\overset{4}{C}H_2\overset{5}{C}H_2\overset{6}{C}H_3$$

(b) Each structural formula has a chain of five carbons with two CH_3 branches. Although the branches are identical, they are at different locations on the chains. Therefore, these structural formulas represent constitutional isomers.

$$\overset{1}{C}H_3\overset{2}{C}H\overset{3}{C}H_2\overset{4}{C}H \quad \text{and} \quad \overset{5}{C}H_3\overset{4}{C}H_2\overset{3}{C}H\overset{2}{C}H\overset{1}{C}H_3$$

Problem 11.1 ■
Do the structural formulas in each set represent the same compound or constitutional isomers?

(a) $CH_3CHCHCH_3$ with CH_2CH_3 branch and $CH_3CH_2CHCH_2CHCH_3$ with two CH_3 branches

with CH_2CH_3 branch below

(b) $CH_3CHCHCH_3$ with CH_3 branch and CH_2CH_3 branch and $CH_3CHCHCH_2CH_3$ with CH_3 branch and CH_3 branch

EXAMPLE 11.2

Draw structural formulas for the five constitutional isomers of molecular formula C_6H_{14}.

Solution

In solving problems of this type, you should devise a strategy and then follow it. Here is one strategy. First, draw a structural formula for the constitutional isomer with all six carbons in an unbranched chain. Then, draw structural formulas for all constitutional isomers with five carbons in a chain and one carbon as a branch on the chain. Finally, draw structural formulas for all constitutional isomers with four carbons in a chain and two carbons as branches.

Six carbons in an unbranched chain:

$$\overset{1}{C}H_3\overset{2}{C}H_2\overset{3}{C}H_2\overset{4}{C}H_2\overset{5}{C}H_2\overset{6}{C}H_3$$

Five carbons in a chain; one carbon as a branch:

$$\overset{1}{C}H_3\overset{2}{\underset{|}{C}}H\overset{3}{C}H_2\overset{4}{C}H_2\overset{5}{C}H_3 \qquad \overset{1}{C}H_3\overset{2}{C}H_2\overset{3}{\underset{|}{C}}H\overset{4}{C}H_2\overset{5}{C}H_3$$

with CH_3 branches above carbon 2 and carbon 3 respectively

Four carbons in a chain; two carbons as branches:

$$\overset{1}{C}H_3\overset{2}{\underset{|}{C}}\overset{3}{C}H_2\overset{4}{C}H_3 \qquad \overset{1}{C}H_3\overset{2}{\underset{|}{C}}H\overset{3}{C}H\overset{4}{C}H_3$$
$$CH_3 \qquad\qquad CH_3$$

with additional CH_3 branches below carbon 2 and carbon 3 respectively

No constitutional isomers with only three carbons in the longest chain are possible for C_6H_{14}.

Problem 11.2 ▬▬▬▬

Draw structural formulas for the three constitutional isomers of molecular formula C_5H_{12}. ■

The ability of carbon atoms to form strong, stable bonds with other carbon atoms results in a staggering number of constitutional isomers, as the following table shows.

Molecular Formula	Constitutional Isomers
CH_4	1
C_5H_{12}	3
$C_{10}H_{22}$	75
$C_{15}H_{32}$	4 347
$C_{25}H_{52}$	36 797 588
$C_{30}H_{62}$	4 111 846 763

Thus, for even a small number of carbon and hydrogen atoms, a very large number of constitutional isomers is possible. In fact, the potential for structural and functional group individuality among organic molecules made from just the basic building blocks of carbon, hydrogen, nitrogen, and oxygen is practically limitless.

11.4 Nomenclature of Alkanes

The IUPAC System

Ideally every organic compound should have a name from which its structural formula can be drawn. For this purpose, chemists have adopted a set of rules established by an organization called the **International Union of Pure and Applied Chemistry (IUPAC).**

The IUPAC system for naming organic compounds was made official in 1892.

TABLE 11.2	Prefixes Used in the IUPAC System to Show the Presence of One to Ten Carbons in an Unbranched Chain		
Prefix	**Number of Carbon Atoms**	**Prefix**	**Number of Carbon Atoms**
meth-	1	hex-	6
eth-	2	hept-	7
prop-	3	oct-	8
but-	4	non-	9
pent-	5	dec-	10

The IUPAC name of an alkane with an unbranched chain of carbon atoms consists of two parts: (1) a prefix to show the number of carbon atoms in the chain and (2) the suffix "-ane" to show that the compound is a saturated hydrocarbon. Prefixes used to show the presence of one to ten carbon atoms are given in Table 11.2.

The first four prefixes listed in Table 11.2 were chosen by the IUPAC because they were well established in the language of organic chemistry long before the nomenclature of organic compounds was systematized. For example, the prefix "but-" appears in the name butyric acid, a compound of four carbon atoms formed by the air oxidation of butter fat (Latin: *butyrum*, butter). Prefixes to show five or more carbons are derived from Greek or Latin numbers. Refer to Table 11.1 for the names, molecular formulas, and condensed structural formulas for the first ten alkanes with unbranched chains.

IUPAC names of alkanes with branched chains consist of a parent name that shows the longest chain of carbon atoms and substituent names that indicate the groups attached to the parent chain:

$$
\begin{array}{c}
\hspace{2.2cm}CH_3 \longleftarrow \text{substituent}\\[2pt]
\overset{1\ \ \ 2\ \ \ \ 3\ \ \ \ 4|\ \ \ 5\ \ \ \ 6\ \ \ \ 7\ \ \ \ 8}{CH_3CH_2CH_2CHCH_2CH_2CH_2CH_3} \longleftarrow \text{parent chain}\\[2pt]
\text{4-Methyloctane}
\end{array}
$$

A substituent group derived from an alkane by removal of a hydrogen atom is called an **alkyl group.** The symbol **R—** is commonly used to represent an alkyl group. Alkyl groups are named by dropping the "-ane" from the name of the parent alkane and adding the suffix "-yl." Names and condensed structural formulas for eight of the most common alkyl groups are given in Table 11.3. The prefix "*sec-*" is an abbreviation for secondary; the prefix "*tert-*" is an abbreviation for tertiary. The rules of the IUPAC system for naming alkanes are as follows:

1. The general name for a saturated hydrocarbon with an unbranched chain of carbon atoms consists of a prefix showing the number of carbon atoms in the chain and the ending "-ane."

2. For branched-chain alkanes, the longest chain of carbon atoms is taken as the parent chain and its name becomes the root name.

3. Each substituent is given a name and a number. The number shows the carbon atom of the parent chain to which the substituent is bonded.

$$
\begin{array}{c}
\hspace{0.9cm}CH_3\\[2pt]
\overset{1\ \ \ \ 2|\ \ \ 3}{CH_3CHCH_3}\\[2pt]
\text{2-Methylpropane}
\end{array}
$$

Alkyl group A group derived by removing a hydrogen from an alkane; given the symbol R—

R— A symbol used to represent an alkyl group

TABLE 11.3 Names of the Most Common Alkyl Groups

Name	Condensed Structural Formula	Name	Condensed Structural Formula
methyl	$—CH_3$	isobutyl	$—CH_2CHCH_3$ | CH_3
ethyl	$—CH_2CH_3$		
propyl	$—CH_2CH_2CH_3$	*sec*-butyl	$—CHCH_2CH_3$ | CH_3
isopropyl	$—CHCH_3$ | CH_3		
			CH_3 |
butyl	$—CH_2CH_2CH_2CH_3$	*tert*-butyl	$—CCH_3$ | CH_3

4. If there is one substituent, number the parent chain from the end that gives the substituent the lower number. The following alkane must be numbered as shown and named 2-methylpentane. Numbering from the other end of the chain gives the incorrect name 4-methylpentane.

$$
\begin{array}{c}
CH_3 \\
\overset{5}{CH_3}\overset{4}{CH_2}\overset{3}{CH_2}\overset{2}{CH}\overset{1}{CH_3}
\end{array}
$$

2-Methylpentane
(not 4-methylpentane)

5. If the same substituent occurs more than once, number the parent chain from the end that gives the lower number to the substituent encountered first. The number of times the substituent occurs is indicated by a prefix "di-," "tri-," "tetra-," "penta-," "hexa-," and so on.

$$
\begin{array}{c}
CH_3 \quad\quad CH_3 \\
\overset{6}{CH_3}\overset{5}{CH_2}\overset{4}{CH}\overset{3}{CH_2}\overset{2}{CH}\overset{1}{CH_3}
\end{array}
$$

2,4-Dimethylhexane
(not 3,5-dimethylhexane)

A comma is used to separate position numbers; a hyphen is used to connect the number to the name.

6. If there are two or more different substituents, list them in alphabetical order and number the chain from the end that gives the lower number to the substituent encountered first. If there are different substituents in equivalent positions on opposite ends of the parent chain, the substituent of lower alphabetical order is given the lower number.

$$
\begin{array}{c}
CH_3 \\
\overset{1}{CH_3}\overset{2}{CH_2}\overset{3}{CH}\overset{4}{CH_2}\overset{5}{CH}\overset{6}{CH_2}\overset{7}{CH_3} \\
CH_2CH_3
\end{array}
$$

3-Ethyl-5-methylheptane

7. The prefixes "di-," "tri-," "tetra-," and so on are not included in alphabetizing. Neither are the hyphenated prefixes "*sec*-" and "*tert*-." The names of substituents are alphabetized first, and then these prefixes are inserted. In this example, the alphabetizing parts are ethyl and methyl, not ethyl and dimethyl.

$$
\begin{array}{c}
\qquad\quad \overset{\displaystyle CH_3}{|}\ \ \overset{\displaystyle CH_2CH_3}{|} \\
\overset{1}{}\ \overset{2}{}|\ \overset{3}{}\ \overset{4}{}|\ \overset{5}{}\ \ \overset{6}{} \\
CH_3CCH_2CHCH_2CH_3 \\
\overset{\displaystyle |}{CH_3}
\end{array}
$$

4-Ethyl-2,2-dimethylhexane
(not 2,2-dimethyl-4-ethylhexane)

EXAMPLE 11.3

Write IUPAC names for these alkanes.

(a) $CH_3CHCH_2CH_3$ (b) $CH_3CHCH_2CHCH_2CH_2CH_3$
         | | |
        CH_3 CH_3 $CHCH_3$
                                            |
                                          CH_3

Solution

The longest chain in the alkane is numbered from the end nearer the substituent encountered first (rule 4). The substituents in (b) are listed in alphabetical order (rule 6).

(a) $\overset{1}{C}H_3\overset{2}{C}H\overset{3}{C}H_2\overset{4}{C}H_3$ (b) $\overset{1}{C}H_3\overset{2}{C}H\overset{3}{C}H_2\overset{4}{C}H\overset{5}{C}H_2\overset{6}{C}H_2\overset{7}{C}H_3$
         | | |
        CH_3 CH_3 $CHCH_3$
                                            |
                                          CH_3

2-Methylbutane   4-Isopropyl-2-methylheptane

Problem 11.3 ■

Write IUPAC names for these alkanes.

$$
\begin{array}{c}
\quad\ \ \overset{\displaystyle CH_3}{|}\qquad\qquad \overset{\displaystyle CH_3}{|} \\
(a)\ CH_3CHCH_2CH_2CHCHCH_3 \\
\qquad\qquad\qquad\overset{\displaystyle |}{CH_2CH_2CH_3}
\end{array}
$$

$$
\begin{array}{c}
\qquad\quad \overset{\displaystyle CH_2CH_2CH_3}{|} \\
(b)\ CH_3CH_2CH_2CCH_2CH_2CH_3 \\
\qquad\quad\overset{\displaystyle |}{CH_3CHCH_3}
\end{array}
$$

■

Common Names

In the older system of **common nomenclature,** the total number of carbon atoms in an alkane, regardless of their arrangement, determines the name. The first three alkanes are methane, ethane, and propane. All alkanes of formula C_4H_{10} are called butanes, all those of formula C_5H_{12} are called pentanes, and all those of formula C_6H_{14} are called hexanes. For alkanes beyond propane, "iso-" indicates that one end of an otherwise unbranched chain terminates in a $(CH_3)_2CH—$ group. Following are examples of common names.

$$
\begin{array}{cccc}
 & \overset{\displaystyle CH_3}{|} & & \overset{\displaystyle CH_3}{|} \\
CH_3CH_2CH_2CH_3 & CH_3CHCH_3 & CH_3CH_2CH_2CH_2CH_3 & CH_3CH_2CHCH_3 \\
\text{Butane} & \text{Isobutane} & \text{Pentane} & \text{Isopentane}
\end{array}
$$

This system of common names has no good way of handling other branching patterns; for more complex alkanes, it is necessary to use the more flexible IUPAC system.

In this text, we concentrate on IUPAC names. We also use common names, however, especially when the common name is used almost exclusively in the everyday discussions of chemists and biochemists. When both IUPAC and common names are given in the text, we always give the IUPAC name first followed by the common name in parentheses. In this way, you should have no doubt about which name is which.

11.5 Cycloalkanes

A hydrocarbon that contains carbon atoms joined to form a ring is called a **cyclic hydrocarbon.** As noted at the beginning of this chapter, when all carbons of the ring are saturated, the hydrocarbon is called a **cycloalkane.** Cycloalkanes of ring sizes ranging from 3 to over 30 carbon atoms are found in nature, and in principle there is no limit to ring size. Five-membered (cyclopentane) and six-membered (cyclohexane) rings are especially abundant in nature, and for this reason we concentrate on them in this text.

As a matter of convenience, organic chemists often do not show all carbons and hydrogens when writing structural formulas for cycloalkanes. Rather, a ring is represented by a regular polygon having the same number of sides as there are carbon atoms in the ring. For example, cyclobutane is represented by a square, cyclopentane by a pentagon, and cyclohexane by a hexagon (Figure 11.2).

The abbreviated structural formulas shown in Figure 11.2 are called **line-angle drawings.** In a line-angle drawing, each angle and line terminus represents a carbon, each single line represents a C—C bond, each double line represents a C=C bond, and each triple line represents a C≡C bond. Thus, only the carbon framework of the molecule is shown, and you are left to fill in hydrogen atoms as necessary to complete the tetravalence of each carbon.

To name a cycloalkane, prefix the name of the corresponding open-chain alkane with "cyclo-," and name each substituent on the ring. If there is only one substituent on the ring, there is no need to give it a location number. If there are two substituents, number the ring beginning with the substituent of lower alphabetical order.

> **Cycloalkane** A saturated hydrocarbon that contains carbon atoms joined to form a ring

> **Line-angle drawing** An abbreviated way to draw structural formulas in which each angle and line terminus represents a carbon atom and each line represents a bond

FIGURE 11.2 Examples of cycloalkanes.

Remember that a line ending in space (a line terminus) indicates a CH_3 group.

EXAMPLE 11.4

Write the molecular formula and IUPAC name for each cycloalkane.

(a) (b)

Solution

(a) First replace each angle and each line terminus by a carbon and then add hydrogens as necessary so that each carbon has four bonds. The molecular formula of this compound is thus C_8H_{16}. Because there is only one substituent on the ring, there is no need to number the atoms of the ring. The IUPAC name of this cycloalkane is isopropylcyclopentane.

(b) Replacing each angle and line terminus by a carbon and adding hydrogens so that each carbon has four bonds gives the molecular formula $C_{11}H_{22}$. To name this compound, first number the atoms of the cyclohexane ring beginning with *tert*-butyl, the substituent of lower alphabetical order (remember, alphabetical order here is determined by the *b* of butyl, and not by the *t* of *tert*-). The name of this cycloalkane is 1-*tert*-butyl-4-methylcyclohexane.

Problem 11.4

Write the molecular formula and IUPAC name for each cycloalkane.

(a) (b) (c)

11.6 The IUPAC System. A General System of Nomenclature

Sections 11.4 and 11.5 illustrated the application of the IUPAC system of nomenclature to two specific classes of organic compounds, alkanes and cycloalkanes. Now let us describe the general approach of the IUPAC system. In this introduction, we will consider compounds with carbon-carbon single bonds (alkanes), but also those with carbon-carbon double bonds (alkenes) and triple bonds (alkynes) as well as those containing the functional groups presented in Section 10.4.

The IUPAC name assigned to any compound with a chain of carbon atoms consists of three parts: a **prefix,** an **infix** (a modifying element inserted into a word), and a **suffix.** Each part provides specific information about the structural formula of the compound.

1. The prefix shows the number of carbon atoms in the parent chain. Prefixes for one to ten carbon atoms in a chain are given in Table 11.2.

2. The infix shows the nature of the carbon-carbon bonds in the parent chain.

Infix	Nature of the Carbon-Carbon Bonds in the Parent Chain
-an-	all single bonds
-en-	one or more double bonds
-yn-	one or more triple bonds

3. The suffix shows the class to which the compound belongs.

Suffix	Class of Compound
-e	hydrocarbon
-ol	alcohol
-amine	amine
-al	aldehyde
-one	ketone
-oic acid	carboxylic acid

EXAMPLE 11.5

Following are IUPAC names and condensed structural formulas for four compounds. Divide each name into a prefix, infix, and suffix, and specify the information about the structural formula contained in each part of the name.

(a) CH_2=$CHCH_3$ (b) CH_3CH_2OH
     Propene   Ethanol

(c) $CH_3CH_2NH_2$ (d) $CH_3CH_2CH_2CH_2\overset{\overset{\displaystyle O}{\|}}{C}OH$
     Ethanamine   Pentanoic acid

Solution

(a) prop-en-e
  ⌐ a carbon-carbon double bond
  ∟ a hydrocarbon
  ∟ three carbon atoms

(b) eth-an-ol
  ⌐ only carbon-carbon single bonds
  ∟ an —OH (hydroxyl) group
  ∟ two carbon atoms

(continued on page 276)

(c) eth-an-amine

only carbon-carbon single bonds

an NH₂ (amino) group

two carbon atoms

(d) pent-an-oic acid

only carbon-carbon single bonds

a —CO₂H (carboxyl) group

five carbon atoms

Problem 11.5

Combine the proper prefix, infix, and suffix and write the IUPAC name for each compound.

(a) $CH_3\overset{\displaystyle O}{\overset{\|}{C}}CH_3$

(b) $CH_3CH_2CH_2CH_2\overset{\displaystyle O}{\overset{\|}{C}}H$

(c) ⬠=O

(d) ⬡

11.7 Shapes of Alkanes and Cycloalkanes

In this section, we concentrate on ways to visualize molecules as three-dimensional objects and to visualize bond angles and relative distances between atoms and functional groups within a molecule. We urge you to physically build models and to study and manipulate the molecules prepared for you on the CD-ROM supplied with this text. Organic molecules are three-dimensional objects, and it is essential that you become comfortable in dealing with them as such.

Alkanes

Although the VSEPR model tells us the geometry about each carbon atom, it gives us no information about the three-dimensional shape of an entire molecule. The fact is that there is free rotation about each carbon-carbon bond in an alkane and, as a result, even a molecule as simple as ethane has an infinite number of possible three-dimensional shapes, or **conformations.**

Figure 11.3 shows three conformations for a butane molecule. Conformation (a) is the most stable because the methyl groups at the ends of the four carbon chain are farthest apart. Conformation (b) is formed by a rotation of 120° about the single bond joining carbons 2 and 3. In this conformation, there is some crowding of groups as the two methyl groups are brought closer together. Rotation about the C_2—C_3 single bond by another 60° gives conformation (c), which is the most crowded because the two methyl groups are facing each other.

Figure 11.3 shows three possible conformations for a butane molecule. In fact, there are an infinite number of possible conformations that differ only in the angles of rotation about the various C—C bonds within the molecule. Within an actual sample of butane, each molecule is constantly changing conformation as a result of collisions with other butane molecules and with the walls of the container. But at any given time, a majority of

Conformation Any three-dimensional arrangement of atoms in a molecule that results by rotation about a single bond

(a) Least crowded conformation; methyl groups are farthest apart.

(b) Intermediate crowding; methyl groups are closer to each other.

(c) Most crowded conformation; methyl groups are closest to each other.

FIGURE 11.3 Three conformations of a butane molecule.

butane molecules are in the most stable, fully extended conformation. There are the fewest butane molecules in the most crowded conformation.

For any alkane, just as for butane, there are an infinite number of conformations. The majority of molecules in any sample will be in the least crowded conformation; the fewest will be in the most crowded conformation.

Cycloalkanes

We limit our discussion to the conformations of cyclopentanes and cyclohexanes because these are the most common carbon rings in nature.

Cyclopentane

The most stable conformation of cyclopentane is the **envelope conformation** shown in Figure 11.4. In this conformation, four carbon atoms are in a plane, and the fifth is bent out of the plane, like an envelope with its flap bent upward. The C—C—C bond angles in cyclopentane are approximately 109.5°.

Cyclohexane

The most stable conformation of cyclohexane is the **chair conformation** (Figure 11.5), in which all C—C—C bond angles are approximately 109.5°.

In a chair conformation, the twelve C—H bonds are arranged in two different orientations. Six of them are called **axial bonds,** and the other six are called **equatorial bonds.** One way to visualize the difference between these two types of bonds is to imagine an axis through the center of the chair, perpendicular to the seat (Figure 11.6). Axial bonds are parallel to this axis. Three axial bonds point up; the other three point down. Notice also that axial bonds alternate, first up and then down as you move from one carbon to the next.

Equatorial bonds are approximately perpendicular to our imaginary axis and also alternate first slightly up and then slightly down as you move

FIGURE 11.4 Cyclopentane. The most stable conformation is an envelope conformation.

Chair conformation The most stable conformation of a cyclohexane ring; all bond angles are approximately 109.5°

Axial position A position on a chair conformation of a cyclohexane ring that extends from the ring parallel to the imaginary axis of the ring

Equatorial position A position on a chair conformation of a cyclohexane ring that extends from the ring roughly perpendicular to the imaginary axis of the ring

FIGURE 11.5   Cyclohexane. The most stable conformation is a chair conformation.

from one carbon to the next. Notice further that if the axial bond on a carbon points upward, the equatorial bond on that carbon points slightly downward. Conversely, if the axial bond on a particular bond points downward, the equatorial bond on that carbon points slightly upward.

Finally, notice that each equatorial bond is parallel to two ring bonds on opposite sides of the ring. A different pair of equatorial C—H bonds is shown in each of the following structural formulas along with the two ring bonds to which each pair is parallel.

FIGURE 11.6   Chair conformation of cyclohexane showing axial and equatorial C—H bonds.

EXAMPLE 11.6

Following is a chair conformation of methylcyclohexane showing a methyl group and one hydrogen. Indicate by a label whether each group is equatorial or axial.

Solution

The methyl group is axial, and the hydrogen is equatorial.

Problem 11.6 ▬

Following is a chair conformation of cyclohexane with carbon atoms numbered 1 through 6. Draw methyl groups that are equatorial on carbons 1, 2, and 4.

■

Suppose that a —CH₃ or other group on a cyclohexane ring may occupy either an equatorial or axial position. The compound is more stable when the substituent group is equatorial. Perhaps the simplest way to see this is to look at molecular models. Figure 11.7(a) shows a molecular model of methylcyclohexane with the methyl group equatorial. In this position, the methyl group is as far away as possible from other atoms of the ring. When methyl is axial, it quite literally bangs into two hydrogen atoms on the top side of the ring. Thus, the more stable conformation of a substituted cyclohexane ring has the substituent group equatorial.

(*a*)   Methyl is equatorial.   (*b*)   Methyl is axial.

FIGURE 11.7   Methylcyclohexane.

The Poisonous Puffer Fish

Nature is by no means limited to carbon in six-membered rings. Tetrodotoxin (The Merck Index, 12th ed., #9382), one of the most potent toxins known, is composed of a set of interconnected six-membered rings, each in a chair conformation. All but one of these rings have atoms other than carbon in them. Tetrodotoxin is produced in the liver and ovaries of many species of *Tetraodontidae*, especially the puffer fish, so called because it inflates itself to an almost spherical spiny ball when alarmed. It is evidently a species highly preoccupied with defense, but the Japanese are not put off. They regard the puffer, called *fugu* in Japanese, as a delicacy. To serve it in a public restaurant, a chef must be registered as sufficiently skilled in removing the toxic organs so as to make the flesh safe to eat. Tetrodotoxin blocks the sodium ion channels, which are essential for neurotransmission (Section 23.3). This prevents communication between neurons and muscle cells and results in weakness, paralysis, and eventual death.

■ A puffer fish with its body inflated. *(Tim Rock/Animals Animals)*

Tetrodotoxin

11.8 Cis-Trans Isomerism in Cycloalkanes

Cycloalkanes with substituents on two or more carbons of the ring show a type of isomerism called **cis-trans isomerism.** Cycloalkane cis-trans isomers have (1) the same molecular formula, (2) the same order of attachment of their atoms, but (3) a different arrangement of their atoms in space because of restricted rotation around the carbon-carbon bonds of the ring.

Because cis-trans isomers differ in the orientation of their atoms in space, they are **stereoisomers.** Cis-trans isomerism is one type of stereoisomerism. We will study another type, called enantiomerism, in Chapter 15.

Cis-trans isomerism in cycloalkanes can be illustrated by 1,2-dimethylcyclopentane. The cyclopentane ring is drawn as a planar pentagon viewed edge-on. (In determining the number of cis-trans isomers in a substituted cycloalkane, it is adequate to draw the cycloalkane ring as a planar polygon.) Carbon-carbon bonds of the ring that project forward are shown as heavy lines. When viewed from this perspective, substituents attached to the cyclopentane ring project above and below the plane of the ring. In one isomer of 1,2-dimethylcyclopentane, the methyl groups are on the same side of the ring (either both above or both below the plane of the ring); in the other, they are on opposite sides of the ring (one above and one below the plane of the ring).

> **Cis-trans isomers** Isomers that have the same order of attachment of their atoms but a different arrangement of their atoms in space due to the presence of either a ring or a carbon-carbon double bond

Cis-trans isomers have been called geometric isomers, an older term that is still sometimes used.

The prefix **cis-** (Latin: on the same side) is used to indicate that the substituents are on the same side of the ring; the prefix **trans-** (Latin: across) is used to indicate that they are on opposite sides of the ring.

cis-1,2-Dimethylcyclopentane   *trans*-1,2-Dimethylcyclopentane

Cis A prefix meaning on the same side

Trans A prefix meaning across from

Alternatively, the cyclopentane ring can be viewed from above, with the ring in the plane of the paper. Substituents on the ring then either project toward you (that is, they stick up above the page) and are shown by solid wedges, or they project away from you (stick down below the page) and are shown by broken wedges. In the following structural formulas, only the two methyl groups are shown; hydrogen atoms of the ring are not shown.

cis-1,2-Dimethylcyclopentane   *trans*-1,2-Dimethylcyclopentane

Occasionally hydrogen atoms are written before the carbon, H_3C—, to avoid crowding or to emphasize the C—C bond, as in H_3C—CH_3.

Similarly, a cyclohexane ring can be drawn as a planar hexagon and viewed through the plane of the ring. Alternatively, it can be viewed from above with substituent groups pointing toward you shown by solid wedges, and those pointing away from you shown as broken wedges.

trans-1,4-Dimethylcyclohexane   *cis*-1,4-Dimethylcyclohexane

EXAMPLE 11.7

Which cycloalkanes show cis-trans isomerism? For each that does, draw both isomers.

Solution
(a) Methylcyclopentane does not show cis-trans isomerism because it has only one substituent on the ring.
(b) 1,1-Dimethylcyclobutane does not show cis-trans isomerism because only one arrangement is possible for the two methyl groups on the ring; they must be trans.
(c) 1,3-Dimethylcyclobutane shows cis-trans isomerism.

cis-1,3-Dimethylcyclobutane   *trans*-1,3-Dimethylcyclobutane

Problem 11.7 ■■■■■

Which cycloalkanes show cis-trans isomerism? For each that does, draw both isomers.

(a) H₃C⟨⟩CH₃   (b) ⟨⟩—CH₂CH₃   (c) ⟨⟩ with CH₃ and CH₃   ■

11.9   Physical Properties

The most important physical properties of organic compounds are melting point, boiling point, solubility, and density.

Melting and Boiling Points

Because the difference in electronegativity between carbon and hydrogen is so small, alkanes are nonpolar compounds, and interactions between their molecules consist only of very weak London dispersion forces (Section 5.7). Therefore, boiling points of alkanes are lower than those of almost any other type of compound of the same molecular weight. In general, both boiling and melting points of alkanes increase with increasing molecular weight (Table 11.4).

The low-molecular-weight alkanes—those containing up to 5 carbons—are gases at room temperature. Alkanes containing 5 to 17 carbons are colorless liquids. High-molecular-weight alkanes (18 or more carbons) are white, waxy solids. Solid paraffin wax, for example, is a mixture of high-molecular-weight alkanes.

Alkanes that are constitutional isomers are different compounds and have different physical and chemical properties. Table 11.5 lists boiling points of the five constitutional isomers of molecular formula C_6H_{14}. The boiling point of each branched-chain isomer is lower than that of hexane itself, and the more branching there is, the lower the boiling point is. These differences in boiling points are related to molecular shape in the following way. As branching increases, the alkane molecule becomes more compact, and so its surface area decreases. As we learned in Section 5.7, as surface area decreases, the strength of London dispersion forces between molecules

Paraffin wax and mineral oil are mixtures of alkanes. *(Charles D. Winters)*

TABLE 11.4   Physical Properties of Some Unbranched Alkanes

Name	Condensed Structural Formula	Mol wt (g/mol)	mp (°C)	bp (°C)	Density of Liquid (g/mL at 0° C)*
methane	CH_4	16.0	−182	−164	(a gas)
ethane	CH_3CH_3	30.1	−183	−88	(a gas)
propane	$CH_3CH_2CH_3$	44.1	−190	−42	(a gas)
butane	$CH_3(CH_2)_2CH_3$	58.1	−138	0	(a gas)
pentane	$CH_3(CH_2)_3CH_3$	72.2	−130	36	0.626
hexane	$CH_3(CH_2)_4CH_3$	86.2	−95	69	0.659
heptane	$CH_3(CH_2)_5CH_3$	100.2	−90	98	0.684
octane	$CH_3(CH_2)_6CH_3$	114.2	−57	126	0.703
nonane	$CH_3(CH_2)_7CH_3$	128.3	−51	151	0.718
decane	$CH_3(CH_2)_8CH_3$	142.3	−30	174	0.730

* For comparison, the density of H_2O is 1 g/mL at 4°C.

TABLE 11.5 Boiling Points of the Five Isomeric Alkanes of Molecular Formula C₆H₁₄

Name	bp (°C)
hexane	68.7
2-methylpentane	60.3
3-methylpentane	63.3
2,3-dimethylbutane	58.0
2,2-dimethylbutane	49.7

Hexane

2,2-Dimethylbutane

decreases, and so boiling points also decrease. Thus, for any group of alkane constitutional isomers, it is usually observed that the least branched isomer has the highest boiling point and that the most branched isomer has the lowest boiling point.

Solubility

Alkanes are not soluble in water, which dissolves only ionic and polar compounds. Alkanes are soluble in each other, however—an example of "like dissolves like" (Section 6.4). Alkanes are also soluble in other nonpolar organic compounds, such as benzene and diethyl ether.

Density

The average density of the liquid alkanes listed in Table 11.4 is about 0.7 g/mL; that of higher-molecular-weight alkanes is about 0.8 g/mL. All liquid and solid alkanes are less dense than water (1.0 g/mL) and, because they are insoluble in water, float on water.

EXAMPLE 11.8

Arrange the alkanes in each set in order of increasing boiling point.
(a) Butane, decane, and hexane
(b) 2-Methylheptane, octane, and 2,2,4-trimethylpentane

Solution

(a) All compounds are unbranched alkanes. As the number of carbon atoms in the chain increases, London dispersion forces between molecules increase, and boiling points increase (Section 5.9). Decane has the highest boiling point, and butane has the lowest boiling point.

$CH_3CH_2CH_2CH_3$ $CH_3CH_2CH_2CH_2CH_2CH_3$ $CH_3(CH_2)_8CH_3$
Butane   Hexane   Decane
(bp −0.5°C)   (bp 69°C)   (bp 174°C)

(b) These three alkanes are constitutional isomers of molecular formula C_8H_{18}. Their relative boiling points depend on the degree of branching. 2,2,4-Trimethylpentane is the most highly branched isomer and, therefore, has the smallest surface area and the lowest boiling point. Octane, the unbranched isomer, has the largest surface area and the highest boiling point.

(continued on page 284)

$$\begin{array}{ccc}
\underset{\substack{|\\CH_3}}{\overset{\substack{CH_3\quad CH_3\\|\qquad\;\;|}}{CH_3CCH_2CHCH_3}} & \underset{\substack{|\\CH_3}}{CH_3CHCH_2CH_2CH_2CH_2CH_3} & CH_3(CH_2)_6CH_3 \\
\text{2,2,4–Trimethylpentane} & \text{2–Methylheptane} & \text{Octane} \\
\text{(bp 99°C)} & \text{(bp 118°C)} & \text{(bp 126°C)}
\end{array}$$

Problem 11.8

Arrange the alkanes in each set in order of increasing boiling point.
(a) 2-Methylbutane, pentane, and 2,2-dimethylpropane
(b) 3,3-Dimethylheptane, nonane, and 2,2,4-trimethylhexane ∎

11.10 Reactions of Alkanes

Methane burn-off at a sewage treatment plant. *(© R. F. Ashley / Visuals Unlimited)*

Combustion reactions are always exothermic.

Alkanes and cycloalkanes are quite unreactive toward most reagents, a behavior consistent with the fact that they are nonpolar compounds. Under certain conditions, however, they do react with oxygen, O_2. By far their most important reaction with oxygen is oxidation (combustion) to form carbon dioxide and water.

Oxidation of saturated hydrocarbons is the basis for their use as energy sources for heat [natural gas, liquefied petroleum gas (LPG), and fuel oil] and power (gasoline, diesel fuel, and aviation fuel). Following are balanced equations for the complete combustion of methane, the major component of natural gas, and propane, the major component of LPG. The heat liberated when an alkane is oxidized to carbon dioxide and water is called its heat of combustion (Section 4.9).

$$CH_4 + 2O_2 \longrightarrow CO_2 + 2H_2O + 212 \text{ kcal/mol}$$
$$\text{Methane}$$

$$CH_3CH_2CH_3 + 5O_2 \longrightarrow 3CO_2 + 4H_2O + 530 \text{ kcal/mol}$$
$$\text{Propane}$$

11.11 Sources of Alkanes

Petroleum refinery towers.

The two major sources of alkanes are natural gas and petroleum. **Natural gas** consists of approximately 90 to 95 percent methane, 5 to 10 percent ethane, and a mixture of other relatively low-boiling alkanes, chiefly propane, butane, and 2-methylpropane.

Petroleum is a thick, viscous liquid mixture of thousands of compounds, most of them hydrocarbons, formed from the decomposition of marine plants and animals. Petroleum and petroleum-derived products fuel automobiles, aircraft, and trains. They provide most of the greases and lubricants required for the machinery of our highly industrialized society. Furthermore, petroleum, along with natural gas, provides close to 90 percent of the organic raw materials for the synthesis and manufacture of synthetic fibers, plastics, detergents, drugs, dyes, and a multitude of other products.

Gases

Boiling point range
below 20°C

Gasoline (naphthas)

20–200°C

Kerosene

175–275°C

Fuel oil

200–400°C

Lubricating oil

Above 350°C

Crude oil
and vapor are
preheated

Residue (asphalt)

FIGURE 11.8 Fractional distillation of petroleum. The lighter, more volatile fractions are removed from higher up the column, and the heavier, less volatile fractions are removed from lower down.

The fundamental separation process in refining petroleum is fractional distillation (Figure 11.8). Practically all crude oil that enters a refinery goes to distillation units where it is heated to temperatures as high as 370 to 425°C and separated into fractions. Each fraction contains a mixture of hydrocarbons that boils within a particular range.

1. Gases boiling below 20°C are taken off at the top of the distillation column. This fraction is a mixture of low-molecular-weight hydrocarbons, predominantly propane, butane, and 2-methylpropane, substances that can be liquefied under pressure at room temperature. The liquefied mixture, known as liquefied petroleum gas, can be stored and shipped in metal tanks and is a convenient source of gaseous fuel for home heating and cooking.

2. Naphthas, bp 20 to 200°C, are a mixture of C_5 to C_{12} alkanes and cycloalkanes, plus small amounts of benzene, toluene, and other aromatic hydrocarbons (Chapter 14). The light naphtha fraction, bp 20 to 150°C, is a source of gasoline and averages approximately 25 percent of crude petroleum.

3. Kerosene, bp 175 to 275°C, is a mixture of C_9 to C_{15} hydrocarbons.

4. Fuel oil, bp 250 to 400°C, is a mixture of C_{15} to C_{18} hydrocarbons. Diesel fuel is obtained from this fraction.

5. Lubricating oil and heavy fuel oil distill from the column at temperatures above 350°C.

6. Asphalt is the black, tarry residue remaining after removal of the other volatile fractions. It is used for covering roads.

BOX 11B

Octane Rating. What Those Numbers at the Pump Mean

Gasoline is a complex mixture of C_6 to C_{12} hydrocarbons. Its quality as a fuel for internal combustion engines depends on how much it makes an engine "knock." Engine knocking occurs when some of the air-fuel mixture explodes prematurely and independently of ignition by the spark plug. Two compounds were selected for rating gasoline quality. One of them, 2,2,4-trimethylpentane (isooctane), has very good antiknock properties (the fuel-air mixture burns smoothly in the combustion chamber) and was assigned an octane rating of 100. The other, heptane, has poor antiknock properties and was assigned an octane rating of 0.

$$CH_3CCH_2CHCH_3 \qquad CH_3(CH_2)_5CH_3$$

2,2,4-Trimethylpentane
(octane rating 100)

Heptane
(octane rating 0)

The **octane rating** of a particular gasoline is a number equal to the percent isooctane in a mixture of isooctane and heptane that has the same antiknock properties as the gasoline

being rated. For example, the antiknock properties of 2-methylhexane are the same as those of a mixture of 42 percent isooctane and 58 percent heptane; therefore, the octane rating of 2-methylhexane is 42.

Octane itself has an octane rating of -20, which means that it produces even more engine knocking than heptane. Ethanol, which is added to gasoline to produce gasohol, has an octane rating of 105.

■ Typical octane ratings of commonly available gasolines. *(Charles D. Winters)*

SUMMARY

A **hydrocarbon** is a compound that contains only carbon and hydrogen. A **saturated hydrocarbon** contains only single bonds. An **alkane** is a saturated hydrocarbon whose carbon atoms are arranged in an open chain. A **cycloalkane** is a saturated hydrocarbon in which two carbon atoms of the chain are joined to form a ring.

Constitutional isomers have the same molecular formula but a different order of attachment of their atoms.

Alkanes are named according to a set of rules developed by the **International Union of Pure and Applied Chemistry.** The IUPAC name of a compound consists of three parts: (1) a **prefix** that tells the number of carbon atoms in the parent chain, (2) an **infix** that tells the nature of the carbon-carbon bonds in the parent chain, and (3) a **suffix** that tells the class to which the compound belongs. Substituents derived from alkanes by removal of a hydrogen atom are called **alkyl groups** and are given the symbol R—.

An alkane that contains carbon atoms bonded to form a ring is called a **cycloalkane.** To name a

cycloalkane, prefix the name of the open-chain alkane by "cyclo-."

A **conformation** is any three-dimensional arrangement of the atoms of a molecule that results by rotation about a single bond. The lowest-energy conformation of cyclopentane is an **envelope conformation.** The lowest-energy conformation of cyclohexane is a **chair conformation.** In a chair conformation, six bonds are **axial,** and six bonds are **equatorial.**

Cis-trans isomers have (1) the same molecular formula, (2) the same order of attachment of their atoms, (3) but a different orientation of their atoms in space because of the restricted rotation around the C—C bonds of the ring. Among cis-trans isomers of cycloalkanes, **cis** means that substituents are on the same side of the ring; **trans** means that they are on opposite sides of the ring.

Alkanes are nonpolar compounds, and the only forces of attraction between their molecules are **London dispersion forces.** At room temperature, low-molecular-weight alkanes are gases. Higher-molecular-weight alkanes are liquids, and very high-molecular-weight alkanes are solids. For any group of alkane constitutional isomers, the least branched isomer generally has the highest boiling point, and the most branched isomer generally has the lowest boiling point.

All liquid and solid alkanes are less dense than water. Alkanes are insoluble in water but soluble in each other and in other nonpolar organic solvents such as benzene.

Natural gas consists of 90 to 95 percent methane with lesser amounts of ethane and other lower-molecular-weight hydrocarbons. **Petroleum** is a liquid mixture of thousands of different hydrocarbons.

KEY TERMS

Aliphatic hydrocarbon (Section 11.1)
Alkane (Section 11.1)
Alkyl group, R— (Section 11.4)
Axial position (Section 11.7)
Chair conformation (Section 11.7)
Cis (Section 11.8)
Cis-trans isomers (Section 11.8)
Common nomenclature (Section 11.4)
Conformation (Section 11.7)

Constitutional isomer (Section 11.3)
Cycloalkane (Sections 11.1, 11.5)
Envelope conformation (Section 11.7)
Equatorial position (Section 11.7)
Hydrocarbon (Section 11.1)
Infix (Section 11.6)
Isomer (Section 11.3)
IUPAC nomenclature (Section 11.4)
Line-angle drawing (Section 11.5)

Methylene group (Section 11.2)
Natural gas (Section 11.11)
Petroleum (Section 11.11)
Prefix (Section 11.6)
Saturated hydrocarbon (Section 11.1)
Stereoisomers (Section 11.8)
Structural isomers (Section 11.3)
Suffix (Section 11.6)
Trans (Section 11.8)

KEY REACTIONS

1. Oxidation of Alkanes (Section 11.10)
Oxidation of alkanes to carbon dioxide and water is the basis for their use as energy sources of heat and power.

$$CH_3CH_2CH_3 + 5O_2 \longrightarrow 3CO_2 + 4H_2O + energy$$

CONCEPTUAL PROBLEMS

11.A What structural feature of cycloalkanes makes cis-trans isomerism in them possible?

11.B Is cis-trans isomerism possible in alkanes?

11.C Why is it not accurate to describe unbranched alkanes as "straight chain" alkanes?

11.D How are the boiling points of hydrocarbons during petroleum refining related to their size?

PROBLEMS

Difficult problems are designated by an asterisk.

Alkanes and Cycloalkanes

11.9 Define
(a) Hydrocarbon   (b) Alkane
(c) Cycloalkane   (d) Aliphatic hydrocarbon

Constitutional Isomerism

11.10 Which statements are true about constitutional isomers?
(a) They have the same molecular formula.
(b) They have the same molecular weight.
(c) They have the same order of attachment of atoms.
(d) They have the same physical properties.

11.11 For each pair of compounds, tell whether the structures shown represent (1) the same compound, (2) different compounds that are constitutional isomers, or (3) different compounds that are not constitutional isomers.

(a) CH_2-CH_2 / CH_2-CH_2 and $CH_3-CH-CH_3$ with CH_3

(b) $CH_2=C-CH=CH_2$ with CH_3 and $CH_3-CH-C\equiv CH$ with CH_3

(c) H_2C-CH_2 / H_2C CH_2 / O ring   and $CH_3-C-CH_2-CH_3$ with O

(d) H₂C—CH₂ / H₂C   CH₂ \ O (cyclic) and $CH_3—CH_2—O—CH_2—CH_3$

(e) H₂C—CH / H₂C   CH₂ \ N / H (cyclic) and ring with NH₂

(f) H₂C—CH₂ / H₂C   CH₂ \ O (cyclic) and $CH_2=CH—CH_2—CH_2—OH$

11.12 Which structural formulas represent identical compounds and which represent constitutional isomers?

(a) $CH_3CH_2\overset{\underset{|}{OH}}{C}HCH_3$

(b) ◇—OH

(c) $HOCH_2—▷$

(d) $CH_3\overset{\underset{|}{CH_2OH}}{C}HCH_3$

(e) $HOCH_2\overset{\underset{|}{CH_3}}{C}HCH_3$

(f) $CH_3CH_2CH_2CH_2OH$

(g) $CH_3\overset{\underset{|}{CH_2CH_3}}{C}HOH$

(h) $CH_3\overset{\underset{|}{CH_3}}{\underset{\overset{|}{OH}}{C}}CH_3$

*11.13 Name and draw structural formulas for the nine constitutional isomers of molecular formula C_7H_{16}.

*11.14 Draw structural formulas for the following:
  (a) The four alcohols of molecular formula $C_4H_{10}O$.
  (b) The three tertiary (3°) amines of molecular formula $C_5H_{13}N$.
  (c) The two aldehydes of molecular formula C_4H_8O.
  (d) The three ketones of molecular formula $C_5H_{10}O$.
  (e) The four carboxylic acids of molecular formula $C_5H_{10}O_2$.

Nomenclature of Alkanes and Cycloalkanes

11.15 Name these alkyl groups.
  (a) $CH_3CH_2—$
  (b) $CH_3\overset{\underset{|}{CH_3}}{C}H—$
  (c) $CH_3\overset{\underset{|}{CH_3}}{C}HCH_2—$
  (d) $CH_3\overset{\underset{|}{CH_3}}{\underset{\overset{|}{CH_3}}{C}}—$

11.16 Write IUPAC names for isobutane and isopentane.

11.17 Write IUPAC names for these alkanes and cycloalkanes.
  (a) $CH_3\overset{\underset{|}{CH_3}}{C}HCH_2CH_2CH_3$
  (b) $CH_3\overset{\underset{|}{CH_3}}{C}HCH_2CH_2\overset{\underset{|}{CH_3}}{C}HCH_3$
  (c) $CH_3(CH_2)_4\overset{\underset{|}{CH_2CH_3}}{C}HCH_2CH_3$
  (d) $(CH_3)_2CH\overset{\underset{\overset{|}{CH_3}}{C}}{\overset{|}{CH_3}}CH_3$
  (e) ⬠—$CH_2\overset{\underset{|}{CH_3}}{C}HCH_3$
  (f) cyclohexane with CH₂CH₃, H₃C, CH₃ substituents

11.18 Write structural formulas for these alkanes and cycloalkanes.
  (a) 2,2,4-Trimethylhexane
  (b) 2,2-Dimethylpropane
  (c) 3-Ethyl-2,4,5-trimethyloctane
  (d) 5-Butyl-2,2-dimethylnonane
  (e) 4-Isopropyloctane
  (f) 3,3-Dimethylpentane
  (g) *trans*-1,3-Dimethylcyclopentane
  (h) *cis*-1,2-Diethylcyclobutane

The IUPAC System of Nomenclature

11.19 Draw a structural formula for each compound.
  (a) Ethanol
  (b) Ethanal
  (c) Ethanoic acid
  (d) Butanone
  (e) Butanal
  (f) Butanoic acid
  (g) Propanal
  (h) Cyclopropanol
  (i) Cyclopentanol
  (j) Cyclopentene
  (k) Cyclopentanone

*11.20 Write the IUPAC name for each compound.
  (a) $CH_3\overset{\overset{O}{\|}}{C}CH_3$
  (b) $CH_3(CH_2)_3\overset{\overset{O}{\|}}{C}H$
  (c) $CH_3(CH_2)_8\overset{\overset{O}{\|}}{C}OH$
  (d) cyclohexene

(e)

(f) ⬡—OH

Conformations of Alkanes and Cycloalkanes

11.21 The condensed structural formula of butane is $CH_3CH_2CH_2CH_3$. Explain why this formula does not show the geometry of the real molecule.

11.22 Define conformation.

11.23 Draw a conformation of ethane in which hydrogen atoms on adjacent carbons are as far apart as possible. Also draw a conformation in which they are as close together as possible. In a sample of ethane molecules at room temperature, which conformation is the more likely?

Cis-Trans Isomerism in Cycloalkanes

11.24 Name and draw structural formulas for the cis and trans isomers of 1,2-dimethylcyclopropane.

11.25 Name and draw structural formulas for the six cycloalkanes of molecular formula C_5H_{10}. Be certain to include cis-trans isomers as well as constitutional isomers.

11.26 Why is equatorial-methylcyclohexane more stable than axial-methylcyclohexane?

*11.27 Following is a structural formula for 2-isopropyl-5-methylcyclohexanol.

2-Isopropyl-5-methylcyclohexanol

Using a planar hexagon representation for the cyclohexane ring, draw a structural formula for the cis-trans isomer with isopropyl trans to —OH and methyl cis to —OH. If you answered this part correctly, you have drawn the isomer found in nature and given the name menthol.

*11.28 Following is a representation of the glucose molecule. Convert the representation to an alternative representation using the ring on the right.

Physical Properties of Alkanes and Cycloalkanes

11.29 In Problem 11.13, you drew structural formulas for all constitutional isomers of molecular formula C_7H_{16}. Predict which isomer has the lowest boiling point and which has the highest boiling point.

11.30 What unbranched alkane has about the same boiling point as water (see Table 11.4)? Calculate the molecular weight of this alkane, and compare it with that of water.

11.31 What generalizations can you make about the densities of alkanes relative to that of water?

11.32 What generalization can you make about the solubility of alkanes in water?

11.33 Suppose that you had samples of hexane and octane. Could you tell the difference by looking at them? What color would they be? How could you tell which is which?

*11.34 As you can see from Table 11.4, each CH_2 group added to the carbon chain of an alkane increases its boiling point. This increase is greater going from CH_4 to C_2H_6 and from C_2H_6 to C_3H_8 than it is from C_8H_{18} to C_9H_{20} or from C_9H_{20} to $C_{10}H_{22}$. What do you think is the reason for this?

Reactions of Alkanes

11.35 Write balanced equations for combustion of each hydrocarbon. Assume that each is converted completely to carbon dioxide and water. (a) Hexane (b) Cyclohexane (c) 2-Methylpentane

11.36 Following are heats of combustion of methane and propane. On a gram-for-gram basis, which hydrocarbon is the best source of heat energy?

Hydrocarbon	Component of	Heat of Combustion (kcal/mol)
CH_4	natural gas	212
$CH_3CH_2CH_3$	LPG	530

Boxes

*11.37 (Box 11A) How many rings in tetrodotoxin contain only carbon atoms? How many contain nitrogen atoms? How many contain two oxygen atoms?

11.38 (Box 11B) What is "octane rating"?

11.39 (Box 11B) What two reference hydrocarbons are used for setting the scale of octane ratings?

11.40 (Box 11B) Octane has an octane rating of -20. Will it produce more or less engine knocking than heptane?

Additional Problems

11.41 Tell whether the compounds in each set are constitutional isomers.

(a) CH_3CH_2OH and CH_3OCH_3

(b) $CH_3\overset{O}{\overset{\|}{C}}CH_3$ and $CH_3CH_2\overset{O}{\overset{\|}{C}}H$

(c) $CH_3\overset{O}{\overset{\|}{C}}OCH_3$ and $CH_3CH_2\overset{O}{\overset{\|}{C}}OH$

(d) $CH_3\overset{OH}{\overset{|}{C}}HCH_2CH_3$ and $CH_3\overset{O}{\overset{\|}{C}}CH_2CH_3$

(e) ⬠ and $CH_3CH_2CH_2CH_2CH_3$

(f) ⬠ and $CH_2{=}CHCH_2CH_2CH_3$

11.42 Explain why each is an incorrect IUPAC name. Write the correct IUPAC name for the intended compound.
(a) 1,3-Dimethylbutane
(b) 4-Methylpentane
(c) 2,2-Diethylbutane
(d) 2-Ethyl-3-methylpentane
(e) 2-Propylpentane
(f) 2,2-Diethylheptane
(g) 2,2-Dimethylcyclopropane
(h) 1-Ethyl-5-methylcyclohexane

11.43 Which compounds show cis-trans isomerism? For each that does, draw both isomers using solid and dashed wedges to show the orientation in space of the —OH and —CH₃ groups.

(a)

(b)

(c)

*11.44 Following is a structural formula and ball-and-stick model of cholestanol. Rings A, B, and C are all in chair conformations. The only difference in structure between cholestanol and cholesterol (Section 20.9) is that cholesterol has a carbon-carbon double bond in ring B.

Cholestanol

(a) Is the hydroxyl group on ring A axial or equatorial?
(b) Is the methyl group at the junction of rings A/B axial or equatorial to ring A? Is it axial or equatorial to ring B?
(c) Is the methyl group at the junction of rings C/D axial or equatorial to ring C?

*11.45 Following is the structural formula of cholic acid, a component of human bile (Section 20.11), whose function is to aid in the absorption and digestion of dietary fats. Rings A, B, C of cholic acid are all chair conformations.

Cholic acid

(a) There are hydroxyl groups on rings A, B, and C. Tell whether each is axial or equatorial.
(b) Is the methyl group at the junction of rings A/B axial or equatorial to ring A? Is it axial or equatorial to ring B?
(c) Is the methyl group at the junction of rings C/D axial or equatorial to ring C?

11.46 Tetradecane, $C_{14}H_{30}$, is an unbranched alkane with a melting point of 5.9°C and a boiling point of 254°C. Is tetradecane a solid, liquid, or gas at room temperature?

11.47 Dodecane, $C_{12}H_{24}$, is an unbranched alkane. Predict the following:
(a) Will it dissolve in water?
(b) Will it dissolve in hexane?
(c) Will it burn when ignited?
(d) Is it a liquid, solid, or gas at room temperature and atmospheric pressure?
(e) Is it more or less dense than water?

Alkenes and Alkynes

12.1   Introduction

An **unsaturated hydrocarbon** contains one or more carbon-carbon double or triple bonds. The term "unsaturation" shows that there are fewer hydrogens bonded to carbon than in an alkane. There are three classes of unsaturated hydrocarbons: alkenes, alkynes, and arenes. **Alkenes** contain one or more carbon-carbon double bonds, and **alkynes** contain one or more carbon-carbon triple bonds. The simplest alkene is ethylene, and the simplest alkyne is acetylene.

In this chapter we study the structure, nomenclature, and physical properties of alkenes and alkynes. In addition, we study the chemical properties of alkenes.

Above: Carotene is a naturally occurring polyene in carrots and tomatoes (Problem 12.52). *(Charles D. Winters)*

Arene A compound containing
one or more benzene rings

The third class of unsaturated hydrocarbons are the **arenes.** The Lewis structure of benzene, the simplest arene, is

Benzene
(an arene)

The chemistry of benzene and its derivatives is quite different from that of alkenes and alkynes, but, even though we do not study the chemistry of arenes until Chapter 14, we will show structural formulas of compounds containing benzene rings before that time. What you need to remember at this point is that a benzene ring is not chemically reactive under any of the conditions that we describe in this chapter and in Chapter 13.

12.2 Structure

Shapes of Alkenes

Using the VSEPR model (Section 3.5), we predict a value of 120° for the bond angles about each carbon in a double bond. The observed H—C—C bond angle in ethylene is 121.7°, close to that predicted. In other alkenes, deviations from the predicted angle of 120° may be somewhat larger because of interaction between alkyl groups attached to the carbons of the double bond.

Ethylene   Propene

If we look at a molecular model of ethylene, we see that the two carbons of the double bond and the four hydrogens bonded to them all lie in the same plane; this means that ethylene is a flat or planar molecule. Furthermore, chemists have discovered that there is no rotation possible about the carbon-carbon double bond of ethylene or that of any other alkene. Whereas there is free rotation about each carbon-carbon single bond in an alkane (Section 11.7), rotation about the carbon-carbon double bond in an alkene is not possible.

Cis-Trans Isomerism in Alkenes

Cis-trans isomerism Isomers that
have the same order of attachment of their atoms but a different arrangement of their atoms
in space due to the presence of
either a ring (Section 11.8) or a
carbon-carbon double bond

Because of restricted rotation about a carbon-carbon double bond, an alkene in which each carbon of the double bond has two different groups bonded to it shows **cis-trans isomerism.** For example, 2-butene has two cis-trans isomers. In *cis*-2-butene, the two methyl groups are on one side of the double bond, and the two hydrogens are on the other side. In *trans*-2-

butene, the two methyl groups are on opposite sides of the double bond. *cis*-2-Butene and *trans*-2-butene are different compounds and have different physical and chemical properties.

cis-2-Butene
(mp −139°C, bp 4°C)

trans-2-Butene
(mp −106°C, bp 1°C)

Shapes of Alkynes

The simplest alkyne is acetylene, C_2H_2. Acetylene is a linear molecule; all bond angles are 180°.

12.3 Nomenclature

Alkenes are named using the IUPAC system, but as we shall see, some are still referred to by their common names.

Alkenes are often called olefins (oil forming), an older name that is still sometimes used.

IUPAC Names

IUPAC names of alkenes are formed by changing the "-an-" infix of the parent alkane to "-en-" (Section 11.6). Hence, $CH_2{=}CH_2$ is ethene, and $CH_3CH{=}CH_2$ is propene. In higher alkenes, where isomers that differ in location of the double bond exist, a numbering system must be used. The longest carbon chain that contains the double bond is numbered from the end that gives the carbon atoms of the double bond the lower set of numbers. The location of the double bond is indicated by the number of its first carbon. Branched or substituted alkenes are named in a manner similar to alkanes: Carbon atoms are numbered, substituent groups are located and named, the double bond is located, and the main chain is named.

In 1-hexene, the double bond is between carbons 1 and 2, but we show only the 1 in the name. It is understood that the double bond goes from the digit shown to the next higher digit.

$$\overset{6}{C}H_3\overset{5}{C}H_2\overset{4}{C}H_2\overset{3}{C}H_2\overset{2}{C}H{=}\overset{1}{C}H_2$$

1-Hexene

$$\overset{6}{C}H_3\overset{5}{C}H_2\overset{4}{C}H\overset{3}{C}H_2\overset{2}{C}H{=}\overset{1}{C}H_2$$
$$\overset{|}{C}H_3$$

4-Methyl-1-hexene

$$\overset{5}{C}H_3\overset{4}{C}H_2\overset{3}{C}H\overset{2}{C}{=}\overset{1}{C}H_2$$
$$\overset{|}{C}H_2CH_3 \qquad CH_3$$

2-Ethyl-3-methyl-1-pentene

Note that, although there is a chain of six carbon atoms in 2-ethyl-3-methyl-1-pentene, the longest chain that contains the double bond has only five carbons; the parent alkane is pentane, and the molecule is named as a disubstituted 1-pentene.

IUPAC names of alkynes are formed by changing the "-an-" infix of the parent alkane to "-yn-" (Section 11.6). Thus, HC≡CH is ethyne, and CH$_3$C≡CH is propyne. The IUPAC system retains the name acetylene; therefore, there are two acceptable names for HC≡CH: ethyne and acetylene. Of these two names, acetylene is used much more frequently. In higher alkynes, the longest carbon chain that contains the triple bond is numbered from the end that gives the triply bonded carbons the lower set of numbers. The location of the triple bond is indicated by the number of its first carbon atom.

$$\overset{4}{C}H_3\overset{3}{C}H\overset{2}{C}\equiv\overset{1}{C}H$$
$$|$$
$$CH_3$$

3-Methyl-1-butyne

$$\overset{CH_3}{\underset{|}{}}$$
$$\overset{1}{C}H_3\overset{2}{C}H_2\overset{3}{C}\equiv\overset{4}{C}\overset{5}{C}H_2\overset{6}{C}\overset{7}{C}H_3$$
$$|$$
$$CH_3$$

6,6-Dimethyl-3-heptyne

EXAMPLE 12.1

Write the IUPAC name of each unsaturated hydrocarbon.

(a) CH$_2$=CH(CH$_2$)$_5$CH$_3$ (b)

$$\underset{H_3C}{\overset{H_3C}{}}C=C\underset{H}{\overset{CH_3}{}}$$

(c) CH$_3$(CH$_2$)$_2$C≡CCH$_3$

Solution

(a) There are eight carbons in the parent chain; thus, the parent alkane is octane. To show the presence of the carbon-carbon double bond, change "-ane" to "-ene." The chain is numbered beginning with the first carbon of the double bond. This alkene is 1-octene.

(b) Because there are four carbon atoms in the chain containing the carbon-carbon double bond, the parent alkane is butane. The double bond is between carbons 2 and 3 of the chain, and there is a methyl group on carbon 2. This alkene is 2-methyl-2-butene.

(c) There are six carbons in the parent chain, with the carbon-carbon triple bond between carbons 2 and 3 of the chain. This alkyne is 2-hexyne.

Problem 12.1 ▬▬▬

Write the IUPAC name of each unsaturated hydrocarbon.

$$\overset{CH_3}{\underset{|}{}}$$
(a) CH$_3$CH$_2$CCH=CH$_2$ (b) (CH$_3$)$_2$C=C(CH$_3$)$_2$
$$|$$
$$CH_3$$

$$\overset{CH_3}{\underset{|}{}}$$
(c) CH$_3$CC≡CH
$$|$$
$$CH_3$$ ■

Common Names

Despite the precision and universal acceptance of IUPAC nomenclature, some alkenes and alkynes, particularly those of low molecular weight, are known almost exclusively by their common names. Two examples are:

$$\overset{CH_3}{\underset{|}{}}$$

CH$_3$CH=CH$_2$ CH$_3$C=CH$_2$

IUPAC name: Propene   2-Methylpropene
Common name: Propylene   Isobutylene

Common names for alkynes are derived by prefixing the names of the substituents on the carbon-carbon triple bond to the name acetylene.

$$CH_3C{\equiv}CH \qquad CH_3C{\equiv}CCH_3$$

IUPAC name:	Propyne	2-Butyne
Common name:	Methylacetylene	Dimethylacetylene

Cis-Trans Configurations of Alkenes

The orientation of the carbon atoms of the main carbon chain determines whether an alkene is cis or trans. If the carbons of the parent chain are on the same side of the double bond, the alkene is cis; if they are on opposite sides, the alkene is trans. In the first example, they are on opposite sides; thus, this alkene is a trans alkene. In the second example, they are on the same side, so this is a cis alkene.

trans-3-Hexene   cis-3,4-Dimethyl-2-pentene

EXAMPLE 12.2

Name each alkene and specify its configuration.

(a)   (b)

Solution

(a) The chain contains seven carbon atoms and is numbered from the right to give the lower number to the first carbon of the double bond. The carbon atoms of the parent chain are on opposite sides of the double bond. The name of this alkene is *trans*-3-heptene.

(b) The longest chain contains seven carbon atoms and is numbered from the right so that the first carbon of the double bond is carbon 3 of the chain. The carbon atoms of the parent chain are on the same side of the double bond. This alkene is *cis*-4-methyl-3-heptene.

Problem 12.2 ■

Name each alkene and specify its configuration.

(a)   (b)

■

The Case of the Iowa and New York Strains of the European Corn Borer

Although humans communicate largely by sight and sound, chemical signals are the primary means of communication for the vast majority of other species in the animal world. Often, communication within a species is specific for one of two configurational isomers. For example, a member of a given species may respond to a cis isomer of a chemical but not to the trans isomer. Or, alternatively, it might respond to a quite precise blend of cis and trans isomers but not to other blends of these same isomers.

Several groups of scientists have studied the components of the sex **pheromones** of both the Iowa and New York strains of the European corn borer. (A pheromone is a chemical secreted by an organism to influence the behavior of another member of the same species.) Females of these closely related species secrete the sex attractant 11-tetradecenyl acetate. Males of the Iowa strain show maximum response to a mixture containing 96 percent of the cis isomer and 4 percent of the trans isomer. When the pure cis isomer is used alone, males are only weakly attracted. Males of the New York strain show an entirely different response pattern. They respond maximally to a mixture containing 3 percent of the cis isomer and 97 percent of the trans isomer.

trans-11-Tetradecenyl acetate

There is evidence that an optimal response to a narrow range of stereoisomers as we see here is widespread in nature and that most insects maintain species isolation for mating and reproduction by the stereochemistry of their pheromones.

cis-11-Tetradecenyl acetate

■ The European corn borer. *Pyrausta nubilalis*. *(Runk/Schoenberger/Grant Heilman Photography, Inc.)*

Cycloalkenes

In naming **cycloalkenes,** the carbon atoms of the ring double bond are numbered 1 and 2 in the direction that gives the substituent encountered first the lower number. Therefore, it is not necessary to use a location number for the carbons of the double bond. Substituents are listed in alphabetical order.

3-Methylcyclopentene   4-Ethyl-1-methylcyclohexene

EXAMPLE 12.3

Write the IUPAC name for each cycloalkene.

(a) (b) (c) $(CH_3)_2CH$—$\langle \rangle$—CH_3

Solution

(a) 3,3-Dimethylcyclohexene   (b) 1,2-Dimethylcyclopentene
(c) 4-Isopropyl-1-methylcyclohexene

Problem 12.3 ■

Write the IUPAC name for each cycloalkene.

(a) (b) (c) —$C(CH_3)_3$

■

Dienes, Trienes, and Polyenes

Alkenes that contain more than one double bond are named as alkadienes, alkatrienes, and so on. Those that contain several double bonds are also referred to more generally as **polyenes** (Greek: *poly,* many). Following are three dienes:

$CH_2{=}CHCH_2CH{=}CH_2$ $CH_2{=}CCH{=}CH_2$ with CH_3

1,4-Pentadiene   2-Methyl-1,3-butadiene   1,3-Cyclopentadiene
(Isoprene)

We have already seen that two cis-trans isomers are possible for an alkene with one carbon-carbon double bond that can show cis-trans isomerism. For an alkene with n carbon-carbon double bonds, each of which can show cis-trans isomerism, 2^n cis-trans isomers are possible.

EXAMPLE 12.4

How many cis-trans isomers are possible for 2,4-heptadiene?

Solution

This molecule has two carbon-carbon double bonds, each of which shows cis-trans isomerism. As shown in the table, there are $2^2 = 4$ cis-trans isomers. Two of these are drawn here.

trans-2-trans-4-Heptadiene   *trans-2-cis-4*-Heptadiene

Double bond	
$C_2{-}C_3$	$C_4{-}C_5$
trans	trans
trans	cis
cis	trans
cis	cis

BOX 12B

Cis-Trans Isomerism in Vision

The retina, the light-detecting layer in the back of our eyes, contains reddish compounds called visual pigments. Their name, rhodopsin, is derived from the Greek word meaning rose-colored. Each rhodopsin molecule is a combination of one molecule of a protein called opsin and one molecule of 11-*cis*-retinal, a derivative of vitamin A in which the CH_2OH group of carbon 15 is converted to an aldehyde group.

■ Rod cells in a human eye. *(© Omikron/Photo Researchers, Inc.)*

When light hits rhodopsin, 11-*cis*-retinal, the less stable 11-cis double bond is converted to the more stable 11-trans double bond. This change in configuration causes a change in the shape of the opsin part of the rhodopsin molecule, which in turn causes firing of neurons in the optic nerve and produces a visual image.

The retinas of vertebrates have two kinds of cells that contain rhodopsin, rods and cones. Cones, which function in bright light and are used for color vision, are concentrated in the central portion of the retina, called the macula, and are responsible for the greatest visual acuity. The remaining area of the retina consists mostly of rods, which are used for peripheral and night vision. 11-*cis*-Retinal is present in both cones and rods. The opsin in each, however, is somewhat different. Rods have one kind of opsin, whereas cones have three kinds—one for blue, one for green, and one for red color vision.

Problem 12.4 ▬▬▬

Draw structural formulas for the other two cis-trans isomers of 2,4-heptadiene. ■

EXAMPLE 12.5

How many cis-trans isomers are possible for 3,7-dimethyl-2,6-octadien-1-ol?

$$CH_3C=CHCH_2CH_2C=CHCH_2OH$$

with CH_3 groups on carbons 3 and 7

Solution

Cis-trans isomerism is possible only about the double bond between carbons 2 and 3 of the chain (numbered from the —OH group). Thus, $2^1 = 2$ cis-trans isomers are possible.

Problem 12.5 ▰▰▰▰
The trans isomer of 3,7-dimethyl-2,6-octadien-1-ol, named geraniol, is a major component of the oils of rose, citronella, and lemon grass. Draw a structural formula for this compound. ∎

An example of a biologically important compound for which a number of cis-trans isomers is possible is vitamin A. There are four carbon-carbon double bonds in the chain of carbon atoms attached to the substituted cyclohexene ring, and each has the potential for cis-trans isomerism. There are $2^4 = 16$ cis-trans isomers possible for this structural formula. Vitamin A is the all-trans isomer.

Lemon grass from Shelton Herb Farm, North Carolina. *(© David Sieren / Visuals Unlimited)*

Vitamin A (retinol)

12.4 Physical Properties

Alkenes and alkynes are nonpolar compounds, and the only attractive forces between their molecules are the very weak London dispersion forces (Section 5.7). Therefore, their physical properties are similar to those of alkanes (Section 11.9) with the same carbon skeletons. Those that are liquid at room temperature have densities less than 1.0 g/mL (they float on water). They are insoluble in water but soluble in one another and in other nonpolar organic liquids.

12.5 Naturally Occurring Alkenes. The Terpenes

A **terpene** is a compound whose carbon skeleton can be divided into two or more units that are identical with the carbon skeleton of isoprene. Carbon 1 of an **isoprene unit** is called the head, and carbon 4 is called the tail. Terpenes are formed by bonding together the tail of one isoprene unit to the head of another.

> **Terpene** A compound whose carbon skeleton can be divided into two or more units identical to the carbon skeleton of isoprene

$$CH_2=\overset{\overset{\displaystyle CH_3}{|}}{C}-CH=CH_2$$

2-Methyl-1,3-butadiene
(Isoprene)

head $\overset{\overset{\displaystyle C}{|}}{\underset{1}{C}}-\overset{2}{C}-\overset{3}{C}-\overset{4}{C}$ tail

Isoprene unit

Terpenes are among the most widely distributed compounds in the biological world, and a study of their structure provides a glimpse of the wondrous diversity that nature can generate from a simple carbon skeleton.

FIGURE 12.1   Myrcene. (*a*) The structural formula of myrcene and (*b*) a ball-and-stick model divided to show two isoprene units joined by a carbon-carbon bond between the head of one unit and the tail of the other.

California laurel, *Umbellularia californica,* one source of myrcene.
(© *Don Suzio*)

Terpenes also illustrate an important principle of the molecular logic of living systems, namely that in building large molecules, small subunits are bonded together by a series of enzyme-catalyzed reactions and then chemically modified by additional enzyme-catalyzed reactions. Chemists use the same principles in the laboratory, but our methods do not have the precision and selectivity of the enzyme-catalyzed reactions of cellular systems.

Probably the terpenes most familiar to you, at least by odor, are components of the so-called essential oils obtained by steam distillation or ether extraction of various parts of plants. Essential oils contain the relatively low-molecular-weight substances that are in large part responsible for characteristic plant fragrances. Many essential oils, particularly those from flowers, are used in perfumes.

One example of a terpene obtained from an essential oil is myrcene, $C_{10}H_{16}$, a component of bayberry wax and oils of bay and verbena. Myrcene is a triene with a parent chain of eight carbon atoms and two one-carbon branches (Figure 12.1).

Figure 12.2 shows structural formulas of four more terpenes, each consisting of two isoprene units. Geraniol has the same isoprene skeleton as myrcene. In addition, it has an —OH (hydroxyl) group. In the other three terpenes, the carbon atoms present in myrcene and geraniol are cross-linked to give cyclic structures. To help you identify the points of cross linkage and ring formation, the carbon atoms of the geraniol skeleton are numbered 1 through 8. This numbering pattern is used in the remaining terpenes to show points of cross-linking.

Shown in Figure 12.3 are structural formulas of two terpenes divisible into three isoprene units. Derivatives of farnesol and geraniol are intermediates in the biosynthesis of cholesterol (Section 28.4).

Vitamin A (Section 12.3), a terpene of molecular formula $C_{20}H_{30}O$, consists of four isoprene units bonded head-to-tail and cross-linked at one point to form a six-membered ring.

Geraniol
(Rose and other flowers)

Limonene
(Oil of lemon and orange)

Menthol
(Peppermint)

Camphor
(Camphor tree)

carbon skeleton
of geraniol
cross-linked here

carbon skeleton of
geraniol cross-linked
in two places

FIGURE 12.2 Four terpenes, each divisible into two isoprene units.

Farnesol
(Lily–of–the–valley)

Caryophyllene
(Oil of cloves)

FIGURE 12.3 Two terpenes containing three isoprene units.

12.6 Reactions of Alkenes

The most characteristic reaction of alkenes is addition to the carbon-carbon double bond: The double bond is broken, and in its place single bonds are formed to two new atoms or groups of atoms. Several examples of alkene addition reactions are shown in Table 12.1 along with the descriptive name(s) associated with each.

Addition of Hydrogen Halides

The hydrogen halides HCl, HBr, and HI add to alkenes to give haloalkanes (alkyl halides). Addition of HCl to ethylene, for example, gives chloroethane (ethyl chloride).

$$CH_2{=}CH_2 + \boxed{HCl} \longrightarrow \underset{\substack{\text{Chloroethane}\\\text{(Ethyl chloride)}}}{\overset{\boxed{H}\quad\boxed{Cl}}{CH_2{-}CH_2}}$$

Ethylene

In the IUPAC system, halogen substituents are listed in alphabetical order along with other substituents. Common names of haloalkanes consist of the common name of the alkyl group followed by the name of the halide as a separate word.

Addition of HCl to propene gives 2-chloropropane (isopropyl chloride); hydrogen adds to carbon 1 of propene and chlorine adds to carbon 2. If the orientation of addition were reversed, 1-chloropropane (propyl chloride) would be formed. The observed result is that 2-chloropropane is formed to

TABLE 12.1	Characteristic Addition Reactions of Alkenes
Reaction	**Descriptive Name(s)**
$\underset{/}{\overset{\backslash}{C}}{=}\underset{\backslash}{\overset{/}{C}} + \boxed{HCl} \longrightarrow -\underset{\boxed{H}}{\overset{\mid}{C}}-\underset{\boxed{Cl}}{\overset{\mid}{C}}-$	hydrochlorination
$\underset{/}{\overset{\backslash}{C}}{=}\underset{\backslash}{\overset{/}{C}} + \boxed{H_2O} \longrightarrow -\underset{\boxed{H}}{\overset{\mid}{C}}-\underset{\boxed{OH}}{\overset{\mid}{C}}-$	hydration
$\underset{/}{\overset{\backslash}{C}}{=}\underset{\backslash}{\overset{/}{C}} + \boxed{Br_2} \longrightarrow -\underset{\boxed{Br}}{\overset{\mid}{C}}-\underset{\boxed{Br}}{\overset{\mid}{C}}-$	bromination
$\underset{/}{\overset{\backslash}{C}}{=}\underset{\backslash}{\overset{/}{C}} + \boxed{H_2} \longrightarrow -\underset{\boxed{H}}{\overset{\mid}{C}}-\underset{\boxed{H}}{\overset{\mid}{C}}-$	hydrogenation (reduction)

the virtual exclusion of 1-chloropropane. We say that addition of HCl to propene is **regioselective.**

$$CH_3CH{=}CH_2 + \boxed{HCl} \longrightarrow CH_3\overset{\overset{\displaystyle Cl}{|}}{C}H{-}\overset{\overset{\displaystyle H}{|}}{C}H_2 + CH_3\overset{\overset{\displaystyle H}{|}}{C}H{-}\overset{\overset{\displaystyle Cl}{|}}{C}H_2$$

Propene   2-Chloropropane   1-Chloropropane
(not observed)

Regioselective reaction A reaction in which one direction of bond forming or bond breaking occurs in preference to all other directions

This regioselectivity was noted by Vladimir Markovnikov (1838–1904) who made the generalization known as **Markovnikov's rule;** in the addition of HX to an alkene, hydrogen adds to the doubly bonded carbon that has the greater number of hydrogens already bonded to it; halogen adds to the other carbon. Although Markovnikov's rule provides a way to predict the product of many alkene addition reactions, it does not explain why one product predominates over other possible products.

Markovnikov's rule In the addition of HX or H_2O to an alkene, hydrogen adds to the carbon of the double bond having the greater number of hydrogens

Markovnikov's rule is often paraphrased "them that has gets."

EXAMPLE 12.6

Draw a structural formula for the product of each alkene addition reaction.

$$\text{(a)} \quad CH_3\overset{\overset{\displaystyle CH_3}{|}}{C}{=}CH_2 + HI \longrightarrow$$

(b) [cyclopentene with CH₃] + HCl ⟶

Solution
Markovnikov's rule predicts that in (a) hydrogen of HI adds to carbon 1 and iodide adds to carbon 2 to give 2-iodo-2-methylpropane. In (b) H adds to carbon 2 of the ring and Cl adds to carbon 1 to give 1-chloro-1-methylcyclopentane.

$$\text{(a)} \quad CH_3\overset{\overset{\displaystyle CH_3}{|}}{\underset{\underset{\displaystyle I}{|}}{C}}CH_3$$

(b) [cyclopentane ring with positions 2 and 1, Cl and CH₃]

2-Iodo-2-methylpropane   1-Chloro-1-methylcyclopentane

Problem 12.6 ▬▬
Draw a structural formula for the product of each alkene addition reaction.

(a) $CH_3CH{=}CH_2 + HBr \longrightarrow$ (b) [cyclohexane ring]$={=}CH_2 + HBr \longrightarrow$ ■

Chemists account for the addition of HX to an alkene by a two-step **reaction mechanism,** which we illustrate by the reaction of 2-butene with hydrogen chloride to give 2-chlorobutane. Let us first look at this mechanism in overview and then go back and study each step in detail. In overview, Step 1 is the addition of H^+ to 2-butene. To show this addition, chemists use a **curved arrow,** which shows the repositioning of an electron pair from its origin (the tail of the arrow) to its destination (the head of the arrow).

The curved arrow in Step 1 shows that the double bond of the alkene is broken and that its electron pair is used to form a new covalent bond with the hydrogen. This step results in the formation of an organic cation. Step 2 is the reaction of the organic cation with chloride ion to form 2-chlorobutane.

Reaction mechanism A step-by-step description of how a chemical reaction occurs

MECHANISM   Addition of HCl to 2-Butene

Step 1: Reaction of the carbon-carbon double bond of the alkene with H^+ forms the *sec*-butyl cation.

$$CH_3CH=CHCH_3 + H^+ \longrightarrow CH_3\overset{+}{C}H-\overset{\overset{\displaystyle H}{|}}{C}HCH_3$$

sec-Butyl cation
(a 2° carbocation)

Step 2: Reaction of the *sec*-butyl cation with chloride ion completes the valence shell of carbon and gives 2-chlorobutane.

$$:\!\ddot{C}l:^- + CH_3\overset{+}{C}HCH_2CH_3 \longrightarrow CH_3\overset{\overset{\displaystyle :\ddot{C}l:}{|}}{C}HCH_2CH_3$$

Chloride ion   *sec*-Butyl cation

Now let us go back and look at the individual steps in more detail. There is a great deal of important organic chemistry embedded in these two steps, and it is important that you understand it.

Step 1 results in the formation of an organic cation. One carbon atom in this cation has only six electrons in its valence shell and carries a charge of +1. A species containing a positively charged carbon atom is called a **carbocation** (*carbon + cation*). Carbocations are classified as primary (1°), secondary (2°), or tertiary (3°) depending on the number of carbon groups bonded to the carbon bearing the positive charge.

> **Carbocation** A species containing a carbon atom with only three bonds to it and bearing a positive charge

Chemists have determined that a 3° carbocation is more stable and requires a lower activation energy (Section 7.3) for its formation than a 2° carbocation, which, in turn, is more stable and requires a lower activation energy for its formation than a 1° carbocation. Following is the order of stability of four types of alkyl carbocations:

$$H-\overset{\overset{\displaystyle H}{|}}{\underset{\underset{\displaystyle H}{|}}{C}}+ \qquad H_3C-\overset{\overset{\displaystyle H}{|}}{\underset{\underset{\displaystyle H}{|}}{C}}+ \qquad H_3C-\overset{\overset{\displaystyle CH_3}{|}}{\underset{\underset{\displaystyle H}{|}}{C}}+ \qquad H_3C-\overset{\overset{\displaystyle CH_3}{|}}{\underset{\underset{\displaystyle CH_3}{|}}{C}}+$$

| Methyl cation (methyl) | Ethyl cation (1°) | Isopropyl cation (2°) | *tert*-Butyl cation (3°) |

→ Order of increasing carbocation stability

EXAMPLE 12.7

Arrange these carbocations in order of increasing stability.

$$(a)\ CH_3\overset{\overset{\displaystyle CH_3}{|}}{\underset{\underset{\displaystyle CH_3}{|}}{C}}HCCH_3 \qquad (b)\ CH_3\overset{\overset{\displaystyle CH_3}{|}}{\underset{\underset{\displaystyle CH_3}{|}}{\overset{+}{C}}}CHCH_3 \qquad (c)\ CH_3\overset{\overset{\displaystyle CH_3}{|}}{\underset{\underset{\displaystyle CH_3}{|}}{C}}CH_2CH_2^+$$

Solution

The positively charged carbon in (a) has two carbon groups bonded to it; therefore, it is a secondary (2°) carbocation. The positively charged carbon in (b) has three carbon groups bonded to it and, therefore, is a tertiary (3°) carbocation. The positively charged carbon in (c) has only one carbon group bonded to it; it is a primary (1°) carbocation. In order of increasing stability, the three carbocations are c < a < b.

Problem 12.7 ▬▬▬

Arrange these carbocations in order of increasing stability.

(a) ⬡⁺—CH₃   (b) ⬡—CH₃   (c) ⬡—CH₂⁺ ■

EXAMPLE 12.8

Propose a mechanism for the addition of HI to methylenecyclohexane to give 1-iodo-1-methylcyclohexane.

⬡=CH₂ + HI ⟶ ⬡ (I, CH₃)

Methylenecyclohexane   1-Iodo-1-methylcyclohexane

Solution

Propose a two-step mechanism similar to that proposed for the addition of HCl to propene.

Step 1: Reaction of H⁺ with the carbon-carbon double bond gives a 3° carbocation intermediate.

⬡=CH₂ + H⁺ ⟶ ⬡⁺—CH₃

A 3° carbocation
intermediate

Step 2: Reaction of the 3° carbocation intermediate with iodide ion completes the valence shell of carbon and gives the product.

⬡⁺—CH₃ + :Ï:⁻ ⟶ ⬡ (:Ï:, CH₃)

Problem 12.8 ▬▬▬

Propose a mechanism for the addition of HBr to 1-methylcyclohexene to give 1-bromo-1-methylcyclohexane. ■

Addition of Water. Acid-Catalyzed Hydration

In the presence of an acid catalyst (most commonly concentrated sulfuric acid), water adds to the carbon-carbon double bond of an alkene to give an alcohol. Addition of water is called **hydration.** In the case of simple alkenes, hydration follows Markovnikov's rule; —H adds to the carbon of

Hydration Addition of water

the double bond with the greater number of hydrogens and —OH adds to the carbon with the fewer hydrogens.

$$\underset{\text{Propene}}{CH_3CH=CH_2} + \boxed{H_2O} \xrightarrow{H_2SO_4} \underset{\text{2-Propanol}}{CH_3\overset{\overset{\displaystyle OH}{|}}{CH}-\overset{\overset{\displaystyle H}{|}}{CH_2}}$$

$$\underset{\text{2-Methylpropene}}{\overset{\overset{\displaystyle CH_3}{|}}{CH_3C}=CH_2} + \boxed{H_2O} \xrightarrow{H_2SO_4} \underset{\text{2-Methyl-2-propanol}}{\overset{\overset{\displaystyle CH_3}{|}}{CH_3C}-\underset{\underset{\displaystyle HO\ \ H}{}}{CH_2}}$$

Most industrial ethanol is made by the acid-catalyzed hydration of ethylene.

EXAMPLE 12.9

Draw a structural formula for the product of acid-catalyzed hydration of 1-methylcyclohexene.

Solution

Markovnikov's rule predicts that H adds to the carbon with the greater number of hydrogens, which, in this case, is carbon 2 of the cyclohexene ring. OH then adds to carbon 1 of the ring.

1-Methylcyclohexene $+ H_2O \xrightarrow{H_2SO_4}$ 1-Methylcyclohexanol

Problem 12.9

Draw a structural formula for the product of each alkene hydration reaction.

(a) $\underset{}{\overset{\overset{\displaystyle CH_3}{|}}{CH_3C}=CHCH_3} + H_2O \xrightarrow{H_2SO_4}$

(b) $\underset{}{\overset{\overset{\displaystyle CH_3}{|}}{CH_2=CCH_2CH_3}} + H_2O \xrightarrow{H_2SO_4}$ ∎

The mechanism for acid-catalyzed hydration of alkenes is similar to what we already proposed for addition of HCl, HBr, and HI to alkenes and is illustrated by the hydration of propene. This mechanism is consistent with the fact that acid is a catalyst. An H^+ is consumed in Step 1, but another is generated in Step 3.

MECHANISM   Acid-Catalyzed Hydration of Propene

Step 1:   Addition of H^+ to the more substituted carbon of the double bond gives a 2° carbocation intermediate.

$$CH_3CH=CH_2 + H^+ \longrightarrow CH_3\overset{+}{\underset{\underset{\displaystyle \text{a 2° carbocation intermediate}}{}}{CH}}\overset{\overset{\displaystyle H}{|}}{CH_2}$$

Step 2: The carbocation intermediate completes its valence shell by forming a new covalent bond with an unshared pair of electrons of the oxygen atom of H_2O to give an oxonium ion.

$$CH_3\overset{+}{C}HCH_3 + :\overset{\cdot\cdot}{O}-H \longrightarrow \underset{\text{An oxonium ion}}{\underset{|}{\overset{\overset{\displaystyle H\diagdown\overset{\cdot\cdot}{O}_+\diagup H}{|}}{CH_3CHCH_3}}}$$

Step 3: Loss of H^+ from the oxonium ion gives the alcohol and generates a new H^+ catalyst.

$$\underset{CH_3CHCH_3}{\overset{H\diagdown\overset{\cdot\cdot}{O}_+\diagup H}{|}} \longrightarrow \underset{CH_3CHCH_3}{\overset{:\overset{\cdot\cdot}{O}H}{|}} + H^+$$

> **Oxonium ion** An ion in which oxygen is bonded to three other atoms and bears a positive charge

---

EXAMPLE 12.10

Propose a three-step mechanism for the acid-catalyzed hydration of methylenecyclohexane to give 1-methylcyclohexanol.

Solution

The reaction mechanism is similar to that for the acid-catalyzed hydration of propene.

Step 1: Reaction of the carbon-carbon double bond with H^+ gives a 3° carbocation intermediate.

A 3° carbocation intermediate

Step 2: Reaction of the carbocation intermediate with water completes the valence shell of carbon and gives an oxonium ion.

An oxonium ion

Step 3: Loss of H^+ from the oxonium ion completes the reaction and generates a new H^+ catalyst.

Problem 12.10 ▬▬▬
Propose a mechanism for the acid-catalyzed hydration of
1-methylcyclohexene to give 1-methylcyclohexanol. ■

Addition of Bromine and Chlorine

Chlorine, Cl_2, and bromine, Br_2, react with alkenes at room temperature
by addition of halogen atoms to the carbon atoms of the double bond. Reaction is generally carried out either with the pure reagents or by mixing
them in an inert solvent, such as carbon tetrachloride, CCl_4, or dichloromethane, CH_2Cl_2.

$$CH_3CH=CHCH_3 + Br_2 \xrightarrow{CCl_4} CH_3\overset{\overset{\displaystyle Br}{|}}{C}H-\overset{\overset{\displaystyle Br}{|}}{C}HCH_3$$

2-Butene   2,3-Dibromobutane

Addition of bromine and chlorine to a cycloalkene gives a trans 1,2-
dihalocycloalkane; addition of bromine to cyclohexene, for example, gives
trans-1,2-dibromocyclohexane.

Cyclohexene   *trans*-1,2-Dibromocyclohexane

Addition of bromine to an alkene is a particularly useful qualitative test
for the presence of a carbon-carbon double bond. If we dissolve bromine in
carbon tetrachloride, the solution is red. Both alkenes and dibromoalkanes
are colorless. If we mix a few drops of the red bromine solution with an
unknown sample suspected of being an alkene, disappearance of the red
color as bromine adds to the double bond tells us that an alkene is indeed
present.

(a)   *(b)*   *(c)*

(a) A drop of Br_2 dissolved in CCl_4 is added to an unknown liquid. If the color remains
after stirring *(b)*, it indicates the absence of unsaturation. If the color disappears *(c)*, the
unknown is unsaturated. *(Beverly March)*

EXAMPLE 12.11

Complete these reactions.

(a) ⬠ + Br$_2$ $\xrightarrow{\text{CH}_2\text{Cl}_2}$ (b) [cyclohexene with CH$_3$] + Cl$_2$ $\xrightarrow{\text{CH}_2\text{Cl}_2}$

Solution

In addition of Br$_2$ and Cl$_2$ to a cycloalkene, the halogen atoms in the product are trans to each other.

(a) ⬠ + Br$_2$ $\xrightarrow{\text{CH}_2\text{Cl}_2}$ [cyclopentane ring with Br (wedge) and Br (dash)]

(b) [cyclohexene with CH$_3$] + Cl$_2$ $\xrightarrow{\text{CH}_2\text{Cl}_2}$ [cyclohexane ring with CH$_3$ and Cl on same carbon, Cl on adjacent carbon]

Problem 12.11 ▬▬▬▬
Complete these reactions.

(a) $\overset{\text{CH}_3}{\underset{\text{CH}_3}{\text{CH}_3\text{C}}}\text{CH}{=}\text{CH}_2$ + Br$_2$ $\xrightarrow{\text{CH}_2\text{Cl}_2}$ (b) [cyclohexane ring with =CH$_2$] + Cl$_2$ $\xrightarrow{\text{CH}_2\text{Cl}_2}$ ■

Addition of Hydrogen—Reduction

Virtually all alkenes react quantitatively with molecular hydrogen, H$_2$, in the presence of a transition metal catalyst to give alkanes. Commonly used transition metal catalysts include platinum, palladium, ruthenium, and nickel. Because conversion of an alkene to an alkane involves reduction by hydrogen in the presence of a catalyst, the process is called **catalytic reduction** or, alternatively, **catalytic hydrogenation.**

In Section 20.3, we shall see how catalytic hydrogenation is used to solidify liquid vegetable oils to margarines and semisolid cooking fats.

$$\underset{\text{H}}{\overset{\text{H}_3\text{C}}{}}\text{C}{=}\text{C}\underset{\text{CH}_3}{\overset{\text{H}}{}} + \text{H}_2 \xrightarrow[\text{25°C, 3 atm}]{\text{Pd}} \text{CH}_3\text{CH}_2\text{CH}_2\text{CH}_3$$

trans-2-Butene   Butane

[cyclohexene] + H$_2$ $\xrightarrow[\text{25°C, 3 atm}]{\text{Pd}}$ [cyclohexane]

Cyclohexene   Cyclohexane

The metal catalyst is used as a finely powdered solid. Reaction is carried out by dissolving the alkene in ethanol or another nonreacting organic solvent, adding the solid catalyst, and exposing the mixture to hydrogen gas at pressures of from 1 to 100 atm.

12.7 Polymerization of Ethylene and Substituted Ethylenes

Polymer From the Greek *poly,* many and *meros,* parts; any long-chain molecule synthesized by bonding together many single parts called monomers

Monomer From the Greek *mono,* single and *meros,* part; the simplest nonredundant unit from which a polymer is synthesized

From the perspective of the chemical industry, the single most important reaction of alkenes is the formation of **chain-growth polymers** (Greek words: *poly,* many, and *meros,* part). In the presence of certain catalysts called initiators, many alkenes form polymers made by the stepwise addition of **monomers** (Greek words: *mono,* one, and *meros,* part) to a growing polymer chain, as illustrated by the formation of polyethylene from ethylene. In alkene polymers of industrial and commercial importance, n is a large number, typically several thousand.

$$n\mathrm{CH_2}{=}\mathrm{CH_2} \xrightarrow{\text{initiator}} \mathrm{+CH_2CH_2\,}{\overline{)}_n}$$

Ethylene   Polyethylene
(a monomer)   (a polymer)

The structure of a polymer is shown by placing parentheses around the **repeating unit,** which is the smallest molecular fragment that contains all the nonredundant structural features of the chain. Thus, the structure of an entire polymer chain can be reproduced by repeating the enclosed structure in both directions. A subscript n is placed outside the parentheses to indicate that this unit is repeated n times.

Part of an extended polymer chain   The repeating unit

An exception to this notation are the polymers formed from symmetric monomers, such as polyethylene, $\mathrm{+CH_2CH_2\,}{\overline{)}_n}$, and polytetrafluoroethylene, $\mathrm{+CF_2CF_2\,}{\overline{)}_n}$. Although the simplest repeating units are $\mathrm{-CH_2-}$ and $\mathrm{-CF_2-}$, repectively, we show two methylene groups and two difluoromethylene groups because they originate from ethylene ($\mathrm{CH_2}{=}\mathrm{CH_2}$) and tetrafluoroethylene ($\mathrm{CF_2}{=}\mathrm{CF_2}$), the monomer units from which these polymers are synthesized.

The most common method of naming a polymer is to attach the prefix "poly-" to the name of the monomer from which the polymer is synthesized, as for example polyethylene and polystyrene. Where the name of the monomer is more than one word, as for example the monomer vinyl chloride, parentheses are used to enclose the name of the monomer.

Polystyrene   Styrene   Poly(vinyl chloride)   Vinyl chloride
(PS)   (PVC)

Table 12.2 lists several important polymers derived from ethylene and substituted ethylenes along with their common names and most important uses.

Low-Density Polyethylene

The first commercial process for ethylene polymerization used peroxide catalysts at temperatures of 500°C and pressures of 1000 atm and yields a tough, transparent polymer known as **low-density polyethylene**

TABLE 12.2 Polymers Derived from Substituted Ethylenes

Monomer Formula	Common Name	Polymer Name(s) and Common Uses
$CH_2{=}CH_2$	ethylene	polyethylene, Polythene; break-resistant containers and packaging materials
$CH_2{=}CHCH_3$	propylene	polypropylene, Herculon; textile and carpet fibers
$CH_2{=}CHCl$	vinyl chloride	poly(vinyl chloride), PVC; construction tubing
$CH_2{=}CCl_2$	1,1-dichloroethylene	poly(1,1-dichloroethylene); Saran Wrap is a copolymer with vinyl chloride
$CH_2{=}CHCN$	acrylonitrile	polyacrylonitrile, Orlon; acrylics and acrylates
$CF_2{=}CF_2$	tetrafluoroethylene	polytetrafluoroethylene, PTFE; Teflon, nonstick coatings
$CH_2{=}CHC_6H_5$	styrene	polystyrene, Styrofoam; insulating materials
$CH_2{=}CHCOOCH_2CH_3$	ethyl acrylate	poly(ethyl acrylate); latex paints
$CH_2{=}CCOOCH_3$ $\quad\vert$ $\quad CH_3$	methyl methacrylate	poly(methyl methacrylate), Lucite Plexiglas; glass substitutes

(a)

(b)

(c)

(d)

Some articles made from chain-growth polymers. (*a*) Saran Wrap, a copolymer of vinyl chloride and 1,1-dichloroethylene. (*b*) Plastic containers for various supermarket products, mostly made of polyethylene and polypropylene. (*c*) Teflon-coated kitchenware. (*d*) Articles made from polystyrene. (*a, c Beverly March; b, d Charles D. Winters*)

FIGURE 12.4 Fabrication of LDPE film. A tube of melted LDPE with a jet of compressed air is forced through an opening and blown into a gigantic thin-walled bubble. The film is then cooled and taken up onto a roller. This double-walled film can be slit down the side to give LDPE film or it can be sealed at points along its length to make LDPE bags.

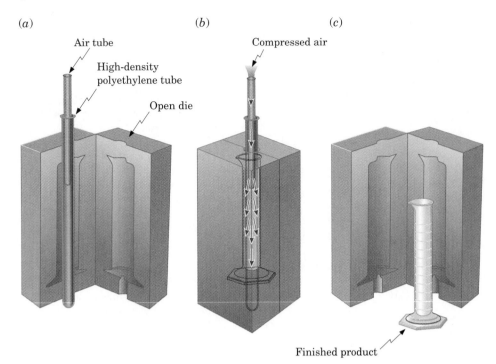

FIGURE 12.5 Blow molding an HDPE container. (*a*) A short length of HDPE tubing is placed in an open die and the die is closed, sealing the bottom of the tube. (*b*) Compressed air is forced into the hot polyethylene/die assembly, and the tubing is literally blown up to take the shape of the mold. (*c*) After the assembly cools, the die is opened, and there is the container!

Polyethylene films are produced by extruding the molten plastic through a ring-like gap and inflating the film into a balloon. *(The Stock Market)*

(LDPE). At the molecular level, LDPE chains are highly branched with the result that they do not pack well together. LDPE softens and melts at about 115°C, which means that it cannot be used in products that are exposed to boiling water.

Today approximately 65 percent of all low-density polyethylene is used for the manufacture of films by a blow-molding technique illustrated in Figure 12.4. LDPE film is inexpensive, which makes it ideal for packaging such consumer items as baked goods and vegetables, and for the manufacture of trash bags.

High-Density Polyethylene

An alternative method for polymerization of alkenes, one that does not involve peroxide catalysts, was developed by Karl Ziegler of Germany and Giulio Natta of Italy in the 1950s. Polyethylene from Ziegler-Natta systems, termed **high-density polyethylene (HDPE),** has little chain branching; consequently, its chains pack together more closely and form microcrystals with stronger intermolecular forces of attraction between them than is the case with low-density polyethylene. HDPE has a higher melting point and is three to ten times stronger than low-density polyethylene; it is also opaque rather than transparent.

Approximately 45 percent of all HDPE is blow molded (Figure 12.5). HDPE is used for consumer items such as milk and water jugs, grocery bags, and squeezable bottles.

BOX 12C

Recycling of Plastics

Polymers are plastics that can be molded when hot and retain their shape when cooled. Because they are durable and lightweight, plastics are probably the most versatile synthetic materials in existence. In fact, their current production in the United States exceeds that of steel. Plastics have come under criticism, however, for their role in the trash crisis. They comprise 21 percent of the volume and 8 percent of the weight of solid wastes, most of which is derived from disposable packaging and wrapping.

Six types of plastics are commonly used for packaging applications. In 1988, manufacturers adopted recycling code letters developed by the Society of the Plastics Industry. Currently only polyethylene terephthalate (PET) and high-density polyethylene are being recycled in large quantities. The synthesis and structure of PET, a polyester, is described in Section 18.8.

Code	Polymer	Common Uses
1 PET	poly(ethylene terephthalate)	soft drink bottles, household chemical bottles, films, textile fibers
2 HDPE	high-density polyethylene	milk and water jugs, grocery bags, squeezable bottles
3 V	poly(vinyl chloride), PVC	shampoo bottles, pipes, shower curtains, vinyl siding, wire insulation, floor tiles
4 LDPE	low-density polyethylene	shrink wrap, trash and grocery bags, sandwich bags, squeeze bottles
5 PP	polypropylene	plastic lids, clothing fibers, bottle caps, toys, diaper linings
6 PS	polystyrene	styrofoam cups, egg cartons, disposable utensils, packaging materials, appliances
7	all other plastics	various

The process for the recycling of most plastics is simple, with separation of the plastic from other contaminants being the most labor-intensive step. For example, PET soft drink bottles usually have a paper label and adhesive that must be removed before the PET can be reused. Recycling begins with hand or machine sorting, after which the bottles are shredded into small chips. An air cyclone then removes paper and other lightweight materials. After any remaining labels and adhesives are eliminated with a detergent wash, the PET chips are dried. The PET produced by this method is 99.9 percent free of contaminants and sells for about half the price of the virgin material.

■ These students are wearing jackets made from recycled PET soda bottles. *(Charles D. Winters)*

SUMMARY

An **alkene** is an unsaturated hydrocarbon that contains a carbon-carbon double bond. An **alkyne** is an unsaturated hydrocarbon that contains a carbon-carbon triple bond.

In the IUPAC system, the presence of a **carbon-carbon double bond** is shown by changing the infix of the parent hydrocarbon from "-an-" to "-en-." The presence of a **carbon-carbon triple bond** is shown by changing the infix of the parent alkane from "-an-" to "-yn-."

The structural feature that makes **cis-trans isomerism** possible in alkenes is restricted rotation about the two carbons of the double bond. The cis or trans configuration of an alkene is determined by the orientation of the atoms of the parent chain about the double bond. If atoms of the parent chain are on the same side of the double bond, the configuration of the alkene is cis; if they are on opposite sides, the configuration is trans.

For compounds containing two or more double bonds, the infix is changed to "-adien-," "-atrien-," and so on. Compounds containing several double bonds are called polyenes.

Because alkenes and alkynes are nonpolar compounds and the only interactions between their molecules are London dispersion forces, their physical properties are similar to those of alkanes of similar carbon skeleton.

The characteristic structural feature of a **terpene** is a carbon skeleton that can be divided into two or more **isoprene units.**

A characteristic reaction of alkenes is **addition,** during which the double bond is broken and bonds to two new atoms or groups of atoms are formed in its place. A **reaction mechanism** is a description of how a chemical reaction occurs, including the role of the catalyst, if one is present. A **carbocation** contains a carbon with only six electrons in its valence shell and bears a positive charge. Carbocations are planar with bond angles of 120° about the positive carbon. The order of stability of carbocations is 3° > 2° > 1° > methyl.

Polymerization is the process of bonding together many small **monomers** into large, high-molecular-weight **polymers. Chain-growth polymerization** proceeds by the sequential addition of monomer units to a growing polymer chain.

KEY TERMS

Alkene (Section 12.1)
Alkyne (Section 12.1)
Carbocation (Section 12.6)
Carbon-carbon double bond
  (Section 12.1)
Carbon-carbon triple bond
  (Section 12.1)
Catalytic reduction (Section 12.6)
Chain-growth polymers
  (Section 12.7)

Cis-trans isomerism in alkenes
  (Section 12.2)
Curved arrow (Section 12.6)
Cycloalkene (Section 12.3)
High-density polyethylene, HDPE
  (Section 12.7)
Hydration (Section 12.6)
Isoprene unit (Section 12.5)
Low-density polyethylene, LDPE
  (Section 12.7)
Markovnikov's rule (Section 12.6)

Monomer (Section 12.7)
Oxonium ion (Section 12.6)
Pheromone (Box 12A)
Polymer (Section 12.7)
Reaction mechanism (Section 12.6)
Regioselective reaction
  (Section 12.6)
Repeating unit (Section 12.7)
Terpene (Section 12.5)
Unsaturated hydrocarbon
  (Section 12.1)

KEY REACTIONS

1. Addition of HX (Section 12.6)

Addition of HX to the carbon-carbon double bond of an alkene follows Markovnikov's rule. Reaction occurs in two steps and involves formation of a carbocation intermediate.

2. Acid-Catalyzed Hydration (Section 12.6)

Addition of H_2O to the carbon-carbon double bond of an alkene follows Markovnikov's rule. Reaction occurs in three steps and involves formation of a carbocation intermediate.

$$CH_3\overset{\underset{\displaystyle CH_3}{|}}{C}{=}CH_2 + H_2O \xrightarrow{H_2SO_4} CH_3\underset{\underset{\displaystyle OH}{|}}{\overset{\overset{\displaystyle CH_3}{|}}{C}}CH_3$$

3. Addition of Bromine and Chlorine (Section 12.6)
Addition to a cycloalkene gives a *trans*-1,2-dihalocycloalkane.

$$\text{(cyclohexene)} + Br_2 \xrightarrow{CCl_4} \text{(trans-1,2-dibromocyclohexane)}$$

4. Reduction. Formation of Alkanes (Section 12.6)
Catalytic reduction involves addition of hydrogen to form two new C—H bonds.

$$\text{(cyclohexene)} + H_2 \xrightarrow[\text{metal catalyst}]{\text{transition}} \text{(cyclohexane)}$$

5. Polymerization of Ethylene and Substituted Ethylenes (Section 12.7)
In polymerization of alkenes, monomer units are bonded together without loss of atoms.

$$n\,CH_2{=}CH_2 \xrightarrow{\text{catalyst}} {+}CH_2CH_2{\rightarrow}_n$$

CONCEPTUAL PROBLEMS

12.A What structural feature in alkenes makes cis-trans isomerism in them possible? What structural feature in cycloalkanes makes cis-trans isomerism possible in them? What do these two structural features have in common?

12.B Knowing what you do about the meaning of the terms "saturated" and "unsaturated" as applied to alkanes and alkenes, what do you think these same terms mean when used to describe animal fats such as those present in butter and animal meats? What might the term "polyunsaturated" mean in this same context?

12.C Writing reaction mechanisms is often described as "electron pushing." Why is this so?

12.D A carbon atom with a positive charge has one fewer than its normal number of bonds, yet a nitrogen atom with a positive charge has one more than its normal number of bonds. Why is this so?

PROBLEMS

Difficult problems are designated by an asterisk.

Structure of Alkenes and Alkynes

12.12 Each carbon atom in ethane and in ethylene is surrounded by eight valence electrons and has four bonds to it. Explain how the VSEPR model predicts a bond angle of 109.5° for a carbon in ethane but one of 120° for a carbon in ethylene.

12.13 Predict all bond angles about each highlighted carbon atom. To make these predictions, use the VSEPR model (Section 3.5).

(a)

(b)

(c) $HC{\equiv}C{-}CH{=}CH_2$

(d)

12.14 What is the difference in structure between a saturated hydrocarbon and an unsaturated hydrocarbon?

Nomenclature of Alkenes and Alkynes

12.15 Draw a structural formula for each compound.
(a) *trans*-2-Methyl-3-hexene
(b) 2-Methyl-3-hexyne
(c) 2-Methyl-1-butene
(d) 3-Ethyl-3-methyl-1-pentyne
(e) 2,3-Dimethyl-2-pentene
(f) *cis*-2-Hexene
(g) 3-Chloropropene
(h) 3-Methylcyclohexene
(i) 1,2-Dimethylcyclohexene
(j) *trans*-3,4-Dimethyl-3-heptene
(k) Cyclopropene
(l) Diethylacetylene

12.16 Name each compound.

(a) $CH_2=CH(CH_2)_4CH_3$ (b)

(c)

(d) $(CH_3)_2CHCH=C(CH_3)_2$

(e) $CH_3(CH_2)_5C\equiv CH$ (f) $CH_3CH_2C\equiv CC(CH_3)_3$

*12.17 Name and draw structural formulas for all alkenes of molecular formula C_5H_{10}. As you draw these alkenes, remember that cis and trans isomers are different compounds and must be counted separately.

12.18 Assign a cis or trans configuration to the carbon-carbon double bond of these compounds, each of which is an intermediate in the citric acid cycle (Section 26.4). Under each is given its common name.

(a)

Fumeric acid

(b)

Aconitic acid

12.19 For each molecule that shows cis-trans isomerism, draw the cis isomer.

(a)

(b)

(c)

(d)

*12.20 Draw structural formulas for all compounds of molecular formula C_5H_{10} that are
(a) alkenes that do not show cis-trans isomerism.
(b) alkenes that do show cis-trans isomerism.
(c) cycloalkanes that do not show cis-trans isomerism.
(d) cycloalkanes that do show cis-trans isomerism.

12.21 β-Ocimene, a triene found in the fragrance of cotton blossoms and several essential oils, has the IUPAC name *cis*-3,7-dimethyl-1,3,6-octatriene. (Cis refers to the configuration of the double bond between carbons 3 and 4, the only double bond in this molecule about which cis-trans isomerism is possible). Draw a structural formula for β-ocimene.

Terpenes

*12.22 Show how the carbon skeleton of farnesol can be coiled and then cross-linked to give the carbon skeleton of caryophyllene (Figure 12.3).

12.23 Show that the structural formula of vitamin A (Section 12.3) can be divided into four isoprene units joined by head-to-tail linkages and cross-linked at one point to form the six-membered ring.

*12.24 α-Santonin, isolated from the flower heads of certain species of Artemisia, is an anthelmintic; that is, it is a drug used to rid the body of worms (helminths). It has been estimated that over one third of the world's population is infested with these parasites.

Santonin

Santonin can be isolated from the flower heads of wormwood, *Artemisia absinthium*. (© *Kenneth J. Stein / Phototake, NYC*)

Locate the three isoprene units in santonin and show how the carbon skeleton of farnesol might be coiled and then cross-linked to give santonin. Two different coiling patterns of the carbon skeleton of farnesol can lead to santonin. Try to find them both.

*12.25 In many parts of South America, extracts of the leaves and twigs of *Montanoa tomentosa* are used as a contraceptive and an abortifacient as well as to stimulate menstruation and to facilitate labor. The plant material responsible for these effects is zoapatanol.

Zoapatanol

(a) Show that the carbon skeleton of zoapatanol can be divided into four isoprene units bonded head-to-tail and then cross-linked in one point along the chain.
(b) How many cis-trans isomers are possible for zoapatanol? Consider the possibilities for cis-trans isomerism in cyclic compounds and about carbon-carbon double bonds.

Addition Reactions of Alkenes

12.26 Define addition reaction. Write an equation for an addition reaction of propene.

12.27 What reagent and/or catalysts are necessary to bring about each conversion?

(a) $CH_3CH=CHCH_3 \longrightarrow CH_3CH_2\overset{\displaystyle Br}{\underset{\displaystyle |}{C}}HCH_3$

(b) $CH_3\overset{\displaystyle CH_3}{\underset{\displaystyle |}{C}}=CH_2 \longrightarrow CH_3\overset{\displaystyle CH_3}{\underset{\displaystyle |}{\underset{\displaystyle |}{\underset{\displaystyle OH}{C}}}}CH_3$

(c) [cyclopentene] $\longrightarrow$ [cyclopentane with I]

(d) $CH_3\overset{\displaystyle CH_3}{\underset{\displaystyle |}{C}}=CH_2 \longrightarrow CH_3\overset{\displaystyle CH_3}{\underset{\displaystyle |}{\underset{\displaystyle |}{\underset{\displaystyle Br}{C}}}}\overset{\displaystyle}{\underset{\displaystyle |}{\underset{\displaystyle Br}{C}}}H_2$

12.28 Complete these equations.

(a) [cyclopentene]$-CH_2CH_3 + HCl \longrightarrow$

(b) [cyclopentene]$-CH_2CH_3 + H_2O \xrightarrow{H_2SO_4}$

(c) $CH_3(CH_2)_5CH=CH_2 + HI \longrightarrow$

(d) [cyclohexane]$-\overset{\displaystyle CH_2}{\underset{\displaystyle CH_3}{C}} + HCl \longrightarrow$

(e) $CH_3CH=CHCH_2CH_3 + H_2O \xrightarrow{H_2SO_4}$

(f) $CH_2=CHCH_2CH_2CH_3 + H_2O \xrightarrow{H_2SO_4}$

12.29 From each pair, select the more stable carbocation.

(a) $CH_3CH_2CH_2{}^+$ or $CH_3\overset{+}{C}HCH_3$

(b) $CH_3\overset{\displaystyle CH_3}{\underset{\displaystyle |}{\underset{\displaystyle +}{C}}}HCHCH_3$ or $CH_3\overset{\displaystyle CH_3}{\underset{\displaystyle |}{\underset{\displaystyle +}{C}}}CH_2CH_3$

12.30 Draw structural formulas for all possible carbocations formed by reaction of each alkene with HCl. Label each carbocation primary, secondary, or tertiary, and state which, if either, of the possible carbocations is more stable.

(a) $CH_3CH_2\overset{\displaystyle CH_3}{\underset{\displaystyle |}{C}}=CHCH_3$

(b) $CH_3CH_2CH=CHCH_3$

(c) [cyclopentene]$-CH_3$

(d) [cyclohexane]$=CH_2$

12.31 Draw a structural formula for the product formed by treatment of 2-methyl-2-pentene with each reagent. (a) HCl (b) H_2O in the presence of H_2SO_4

12.32 Draw a structural formula for the product of each reaction.
(a) 1-Methylcyclohexene + Br_2
(b) 1,2-Dimethylcyclopentene + Cl_2

*12.33 Draw a structural formula for an alkene with the indicated molecular formula that gives the compound shown as the major project. Note that more than one alkene may give the same compound as the major product.

(a) $C_5H_{10} + H_2O \xrightarrow{H_2SO_4} CH_3\overset{\displaystyle CH_3}{\underset{\displaystyle |}{\underset{\displaystyle |}{\underset{\displaystyle OH}{C}}}}CH_2CH_3$

(b) $C_5H_{10} + Br_2 \longrightarrow CH_3\overset{\displaystyle CH_3}{\underset{\displaystyle |}{C}}HCHCH_2 \atop {\underset{\displaystyle Br\ Br}{}}$

(c) $C_7H_{12} + HCl \longrightarrow$ [cyclohexane with CH_3 and Cl]

12.34 Draw a structural formula for an alkene of molecular formula C_5H_{10} that reacts with Br_2 to give each indicated product.

(a) $CH_3\overset{\displaystyle CH_3}{\underset{\displaystyle |}{\underset{\displaystyle |}{\underset{\displaystyle BrBr}{C}}}}-CHCH_3$

(b) $CH_2\overset{\displaystyle CH_3}{\underset{\displaystyle |}{\underset{\displaystyle |}{\underset{\displaystyle BrBr}{C}}}}CH_2CH_3$

(c) $CH_2\overset{}{\underset{\displaystyle |}{\underset{\displaystyle BrBr}{C}}}HCH_2CH_2CH_3$

*12.35 Draw the structural formula for an alkene of molecular formula C_5H_{10} that reacts with HCl to give the indicated chloroalkane as the major product.

(a) $CH_3\overset{\displaystyle CH_3}{\underset{\displaystyle |}{\underset{\displaystyle |}{\underset{\displaystyle Cl}{C}}}}CH_2CH_3$

(b) $CH_3\overset{\displaystyle CH_3}{\underset{\displaystyle |}{\underset{\displaystyle |}{\underset{\displaystyle Cl}{C}}}}HCHCH_3$

(c) $CH_3\overset{}{\underset{\displaystyle |}{\underset{\displaystyle Cl}{C}}}HCH_2CH_2CH_3$

*12.36 Draw the structural formula of an alkene that undergoes acid-catalyzed hydration to give the indicated alcohol as the major product. More than one alkene may give each alcohol as the major product.
(a) 3-Hexanol
(b) 1-Methylcyclobutanol
(c) 2-Methyl-2-butanol
(d) 2-Propanol

*12.37 Terpin is prepared commercially by the acid-catalyzed hydration of limonene (Figure 12.2).

Limonene + 2H$_2$O $\xrightarrow{\text{H}_2\text{SO}_4}$ C$_{10}$H$_{20}$O$_2$ Terpin

(a) Propose a structural formula for terpin.
(b) How many cis-trans isomers are possible for the structural formula you propose?
(c) Terpin hydrate, the isomer of terpin in which the one-carbon and three-carbon groups on the ring are trans to each other, is used as an expectorant in cough medicines. Draw a structural formula for terpin hydrate showing the trans orientation of these groups.

12.38 Draw the product formed by treatment of each alkene with H$_2$/Ni.

(a)

(b)

(c)

(d)

12.39 Hydrocarbon A, C$_5$H$_8$, reacts with 2 moles of Br$_2$ to give 1,2,3,4-tetrabromo-2-methylbutane. What is the structure of hydrocarbon A?

Lycopene

12.40 Show how to convert ethylene to these compounds.
(a) Ethane (b) Ethanol (c) Bromoethane
(d) 1,2-Dibromoethane (e) Chloroethane

12.41 Show how to convert 1-butene to each of these compounds. (a) Butane (b) 2-Butanol (c) 2-Bromobutane
(d) 1,2-Dibromobutane

Boxes

12.42 (Box 12A) What is the meaning of the term pheromone?

12.43 (Box 12A) What is the molecular formula of 11-tetradecenyl acetate? What is its molecular weight?

12.44 (Box 12A) Assume that 1×10^{-12} g of 11-tetradecenyl acetate is secreted by a single corn borer. How many molecules is this?

12.45 (Box 12B) What different functions are performed by the rods and cones in the eye?

12.46 (Box 12B) In which isomer of retinal is the end-to-end distance longer, the all-trans isomer or the 11-cis isomer?

12.47 (Box 12C) What types of consumer products are made of high-density polyethylene? What types of products are made of low-density polyethylene? One type of polyethylene is currently recyclable, and the other is not. Which is which?

12.48 (Box 12C) In recycling codes, what do these abbreviations stand for? (a) V (b) PP (c) PS

Additional Problems

12.49 Which is more unsaturated, an alkene or an alkyne?

12.50 Name and draw structural formulas for all alkenes of molecular formula C$_6$H$_{12}$ that have these carbon skeletons. Remember cis and trans isomers.

(a)

(b)

(c)

12.51 Following is the structural formula of lycopene, C$_{40}$H$_{56}$, a deep-red compound that is partially responsible for the red color of ripe fruits, especially tomatoes. Approximately 20 mg of lycopene can be isolated from 1 kg of fresh ripe tomatoes.

(a) Show that lycopene is a terpene; that is, its carbon skeleton can be divided into two sets of four isoprene units with the units in each set joined head-to-tail.
(b) How many of the carbon-carbon double bonds in lycopene have the possibility for cis-trans isomerism? Lycopene is the all-trans isomer.

12.52 As you might suspect, β-carotene, C$_{40}$H$_{56}$, a precursor to vitamin A, was first isolated from carrots. Dilute solutions of β-carotene are yellow, hence its use as a food coloring. In plants, it is almost always present in combination with chlorophyll to assist in the harvesting of the energy of sunlight. As tree leaves die in the fall, the green of their chlorophyll molecules is replaced by the yellow and reds of carotene and carotene-related molecules. Compare the carbon skeletons of β-carotene and lycopene. What are the similarities? What are the differences?

β-Carotene

12.53 From each pair, select the more stable carbocation.

(a)   or

(b)   or

12.54 Draw the structural formula for a cycloalkene of molecular formula C_6H_{10} that reacts with Cl_2 to give each indicated compound.

(a)

(b)

(c)

(d)

*__12.55__ Draw the structural formula of an alkene that undergoes acid-catalyzed hydration to give each alcohol as the major product. More than one alkene may give each compound as the major product.
(a) Cyclohexanol
(b) 1,2-Dimethylcyclopentanol
(c) 1-Methylcyclohexanol
(d) 2-Methyl-2-butanol

12.56 Show how to convert cyclopentene into these compounds. (a) 1,2-Dibromocyclopentane (b) Cyclopentanol (c) Iodocyclopentane

Alcohols, Ethers, and Thiols

13.1 Introduction

In this chapter, we study the physical and chemical properties of alcohols and ethers, two classes of oxygen-containing organic compounds. We also study thiols, a class of sulfur-containing organic compounds. A thiol is like an alcohol in structure, except that it contains an —SH group rather than an —OH group.

$$CH_3CH_2OH \qquad CH_3CH_2OCH_2CH_3 \qquad CH_3CH_2SH$$
Ethanol   Diethyl ether   Ethanethiol
(an alcohol)   (an ether)   (a thiol)

These three compounds are certainly familiar to you. Ethanol is the fuel additive in gasohol, the alcohol in alcoholic beverages, and an important industrial and laboratory solvent. Diethyl ether, or ether as it is also named, was the first inhalation anesthetic used in general surgery. It is also an important industrial and laboratory solvent. Ethanethiol, like other low-molecular-weight thiols, has a stench. Traces of ethanethiol are added to natural gas so that gas leaks can be detected by the smell of the thiol.

Above: Fermentation vats of wine grapes at the Beaulieu Vineyards, California. *(© Earl Roberge/ Photo Researchers, Inc.)*

FIGURE 13.1 Methanol, CH_3OH. (*a*) Lewis structure and (*b*) ball-and-stick model. The H—C—O bond angle in methanol is 108.6°, very close to the tetrahedral angle of 109.5°.

Methanol, CH_3OH, is the fuel used in cars of the type that race in Indianapolis. (*D. Young / Tom Stack & Associates*)

13.2 Structure

Alcohols

The functional group of an **alcohol** is an **—OH (hydroxyl) group** bonded to a tetrahedral carbon atom (Section 10.5). Figure 13.1 shows a Lewis structure and ball-and-stick model of methanol, CH_3OH, the simplest alcohol.

Ethers

The functional group of an **ether** is an atom of oxygen bonded to two carbon atoms. Figure 13.2 shows a Lewis structure and ball-and-stick model of dimethyl ether, CH_3OCH_3, the simplest ether.

> **Ether** A compound containing an oxygen atom bonded to two carbon atoms

Thiols

The functional group of a **thiol** is an **—SH (sulfhydryl) group** bonded to a tetrahedral carbon atom. Figure 13.3 shows a Lewis structure and a ball-and-stick model of methanethiol, CH_3SH, the simplest thiol.

> **Thiol** A compound containing an —SH (sulfhydryl) group bonded to a tetrahedral carbon atom

13.3 Nomenclature

Alcohols

In the IUPAC system, alcohols are named by selecting the longest carbon chain containing the —OH group as the parent alkane and numbering it from the end closer to the —OH group. To show that the compound is an

FIGURE 13.3 Methanethiol, CH_3SH. (*a*) Lewis structure and (*b*) ball-and-stick model. The H—S—C bond angle is 100.3°, somewhat smaller than the tetrahedral angle of 109.5°.

FIGURE 13.2 Dimethyl ether, CH_3OCH_3. (*a*) Lewis structure and (*b*) ball-and-stick model. The C—O—C bond angle is 110.3°, close to the predicted tetrahedral angle of 109.5°

alcohol, we change the suffix "**-e**" of the parent alkane to "**-ol**" (Section 11.6) and use a number to show the location of the —OH group. When the parent chain is numbered, the location of the —OH group takes precedence over alkyl groups and halogens. For cyclic alcohols, numbering begins at the carbon bearing the —OH group.

To derive common names for alcohols, we name the alkyl group bonded to —OH and then add the word "alcohol." Here are IUPAC names and, in parentheses, common names for eight low-molecular-weight alcohols.

CH₃CH₂OH   CH₃CH₂CH₂OH   CH₃CHCH₃   CH₃CH₂CH₂CH₂OH

Ethanol   1-Propanol   2-Propanol   1-Butanol
(Ethyl alcohol)   (Propyl alcohol)   (Isopropyl alcohol)   (Butyl alcohol)

CH₃CH₂CHCH₃   CH₃CHCH₂OH   CH₃COH

Because —OH is always at the 1 position in a ring, we do not need to show the 1 in the name.

2-Butanol   2-Methyl-1-propanol   2-Methyl-2-propanol   Cyclohexanol
(*sec*-Butyl alcohol)   (Isobutyl alcohol)   (*tert*-Butyl alcohol)   (Cyclohexyl alcohol)

EXAMPLE 13.1

Write the IUPAC name for each alcohol.

(a) CH₃CHCH₂CHCH₃   (b)

Solution
(a) The longest carbon chain contains five carbons; therefore, the parent alkane is pentane. Number the parent chain from the direction that gives the lower number to the carbon bearing the —OH group. The name of this alcohol is 4-methyl-2-pentanol.
(b) The parent cycloalkane is cyclohexane. Number the atoms of the ring beginning with the carbon bearing the —OH group as carbon 1. The name of this alcohol is *trans*-2-methylcyclohexanol.

Problem 13.1

Write the IUPAC name for each alcohol.

(a)   (b)   (c) CH₃CCH₂OH

We classify alcohols as **primary (1°), secondary (2°),** or **tertiary (3°)** depending on the number of carbon groups bonded to the carbon bearing the —OH (Section 10.4).

R—C—OH
|
H

Primary (1°)

R—C—OH
|
H

Secondary (2°)

R—C—OH
|
R″

Tertiary (3°)

(Top structures: Primary has H above and below; Secondary has R′ above, H below; Tertiary has R′ above, R″ below.)

EXAMPLE 13.2

Classify each alcohol as primary, secondary, or tertiary.

(a) [cyclohexyl–C(OH)(CH₃)(H) structure]   (b) CH₃$\overset{\underset{\textstyle CH_3}{|}}{\underset{}{\overset{\textstyle CH_3}{|}}}$COH   (c) [cyclopentyl–CH₂OH structure]

Solution

(a) Secondary (2°); the carbon bearing the —OH group is bonded to two carbon groups.
(b) Tertiary (3°); the carbon bearing the —OH group is bonded to three carbon groups.
(c) Primary (1°); the carbon bearing the —OH group is bonded to only one carbon group.

Problem 13.2

Classify each alcohol as primary, secondary, or tertiary.

(a) CH₃$\overset{\underset{\textstyle CH_3}{|}}{\underset{\textstyle CH_3}{|}}$CCH₂OH   (b) [cyclopropyl]—OH   (c) CH₂=CHCH₂OH   (d) [cyclopentyl with CH₃ and OH]

∎

In the IUPAC system, a compound containing two hydroxyl groups is named as a **diol,** one containing three hydroxyl groups as a **triol,** and so on. In IUPAC names for diols, triols, and so on, the final "-e" (the suffix) in the name of the parent alkane is retained, as for example in the name 1,2-ethanediol.

As with many other organic compounds, common names for certain diols and triols have persisted. Compounds containing two hydroxyl groups on adjacent carbons are often referred to as **glycols.** Ethylene glycol and propylene glycol are synthesized from ethylene and propylene, respectively, hence their common names.

Ethylene glycol is a polar molecule and dissolves readily in the polar solvent water. *(Charles D. Winters)*

> **Diol** A compound containing two —OH (hydroxyl) groups

> **Glycol** A compound with two hydroxyl (—OH) groups on adjacent carbons

CH₂—CH₂
| |
OH   OH

1,2-Ethanediol
(Ethylene glycol)

CH₃—CH—CH₂
| |
OH   OH

1,2-Propanediol
(Propylene glycol)

CH₂—CH—CH₂
| | |
OH   OH   OH

1,2,3-Propanetriol
(Glycerol, Glycerin)

Glycerol is relatively inexpensive because it is a byproduct of the production of soaps from animal fats and vegetable oils (Section 20.3).

BOX 13A

Nitroglycerin, an Explosive and a Drug

In 1847, Ascanio Sobrero (1812–1888) discovered that 1,2,3-propanetriol, more commonly named glycerin, reacts with nitric acid in the presence of sulfuric acid to give a pale yellow, oily liquid called nitroglycerin (The Merck Index, 12th ed., #6704). Sobrero also discovered the explosive properties of this compound; when he heated a small quantity of it, it exploded!

$$
\begin{array}{l}
CH_2\!-\!OH \\
CH\!-\!OH \qquad + \; 3HNO_3 \; \xrightarrow{\; H_2SO_4 \;} \\
CH_2\!-\!OH
\end{array}
$$

1,2,3-Propanetriol
(Glycerol, Glycerin)

$$
\begin{array}{l}
CH_2\!-\!ONO_2 \\
CH\!-\!ONO_2 \qquad + \; 3H_2O \\
CH_2\!-\!ONO_2
\end{array}
$$

1,2,3-Propanetriol trinitrate
(Nitroglycerin)

Nitroglycerin very soon became widely used for blasting in the construction of canals, tunnels, roads, and mines and, of course, for warfare.

A problem with the use of nitroglycerin was soon recognized: It was difficult to handle safely, and accidental explosions were not uncommon. This problem was solved by the Swedish chemist Alfred Nobel (1833–1896), whose brother was killed in 1864 when a nitroglycerin factory in which he was working exploded. In 1866, Nobel discovered that a clay-like substance called diatomaceous earth absorbs nitroglycerin so that it will not explode without a fuse. He named this mixture of nitroglycerine, diatomaceous earth, and sodium carbonate dynamite.

Surprising as it may seem, nitroglycerin is used in medicine to treat angina pectoris, the symptoms of which are sharp chest pains caused by reduced flow of blood in the coronary artery. Nitroglycerin, available in liquid form (diluted with alcohol to render it nonexplosive), tablet form, or paste form, relaxes the smooth muscles of blood vessels, causing dilation of the coronary artery, which, in turn, allows more blood to reach the heart.

When Nobel became ill with heart disease, his physicians advised him to take nitroglycerin to relieve his chest pains. He refused, saying he could not understand how the explosive could relieve chest pains. It took science more than 100 years to find the answer. It is nitric oxide, NO, derived from the nitro groups of nitroglycerin, that actually relieves the pain. See Box 23F—Nitric Oxide As a Secondary Messenger.

(a)

(b)

■ (a) Nitroglycerin is more stable if absorbed onto an inert solid, a combination called *dynamite*. (b) The fortune of Alfred Nobel, 1833–1896, built on the manufacture of dynamite, now funds the Nobel Prizes. (a, Charles D. Winters; b, The Bettmann Archive)

Compounds containing —OH and C=C groups are often referred to as unsaturated alcohols because of the presence of the carbon-carbon double bond. In the IUPAC system, the parent alkane is numbered to give the —OH group the lowest possible number. The double bond is shown by changing the infix of the parent alkane from "-an-" to "-en-" (Section 11.6), and the hydroxyl group is shown by changing the suffix of the parent alkane from "-e" to "-ol." Numbers must be used to show the location of both the double bond and the hydroxyl group.

EXAMPLE 13.3

Write the IUPAC name for each unsaturated alcohol.

(a) CH_2=$CHCH_2OH$ (b)

$$HOCH_2CH_2 \quad CH_2CH_3$$
$$\diagdown \qquad \diagup$$
$$C = C$$
$$\diagup \qquad \diagdown$$
$$H \qquad H$$

Solution

(a) The parent alkane is propane. In numbering this chain, the location of the OH group takes precedence over that of the double bond. The name of this unsaturated alcohol is 2-propen-1-ol.

(b) The parent alkane is hexane. Number the carbon chain to give the —OH group the lower number, which in this case is numbering from the left. The name of this unsaturated alcohol is *cis*-3-hexen-1-ol. It is sometimes called leaf alcohol because of its occurrence in leaves of fragrant plants, including trees and shrubs (The Merck Index, 12th ed., #4737).

Problem 13.3 ■━━━━

Write the IUPAC name for each unsaturated alcohol.

(a) CH_2=$CHCH_2CH_2OH$ (b) ⬠—OH ■

Ethers

Although ethers can be named according to the IUPAC system, chemists almost invariably use common names for low-molecular-weight ethers. Common names are derived by listing the alkyl groups attached to oxygen in alphabetical order and adding the word "ether."

$$CH_3CH_2OCH_2CH_3$$

Diethyl ether

$$CH_3$$
$$|$$
$$CH_3OCCH_3$$
$$|$$
$$CH_3$$

Methyl *tert*-butyl ether (MTBE)

Diethyl ether is also named ethyl ether. Methyl *tert*-butyl ether (note that the two alkyl groups are not given in alphabetical order) is currently used as a gasoline additive but is being phased out because of its environmental and health hazards.

EXAMPLE 13.4

Write the common name for each ether.

(a)
$$CH_3$$
$$|$$
$$CH_3COCH_2CH_3$$
$$|$$
$$CH_3$$
   (b) ⬡—O—⬡

Solution

(a) The groups bonded to the ether oxygen are *tert*-butyl and ethyl. Its common name is *tert*-butyl ethyl ether.

(b) Two cyclohexyl groups are bonded to the ether oxygen. Its common name is dicyclohexyl ether.

Problem 13.4 ▰▰▰
Write the common name for each ether.

$$\underset{\text{(a)}}{} CH_3\overset{\overset{\displaystyle CH_3}{|}}{C}HCH_2OCH_2CH_3 \quad \text{(b)}$$ $-OCH_3$ ∎

Cyclic ether An ether in which the ether oxygen is one of the atoms of a ring

Cyclic ethers are ethers in which the ether oxygen is one of the atoms in a ring. These ethers are also known by their common names.

Ethylene oxide   Tetrahydrofuran   1,4-Dioxane

Thiols

Mercaptan A common name for any molecule containing an —SH group

The sulfur analog of an alcohol is called a thiol ("thi-" from the Greek: *theion,* sulfur) or, in the older literature, a **mercaptan,** which literally means mercury capturing. Thiols react with Hg^{2+} in aqueous solution to give sulfide salts as insoluble precipitates. Thiophenol, C_6H_5SH, for example, gives $(C_6H_5S)_2Hg$.

In the IUPAC system, thiols are named by selecting the longest carbon chain that contains the —SH group as the parent alkane. To show that the compound is a thiol, we retain the final "-e" in the name of the parent alkane and add the suffix "-thiol." A number must be used to locate the —SH group on the parent chain.

Common names for simple thiols are derived by naming the alkyl group attached to —SH and adding the word "mercaptan."

$$CH_3CH_2SH \qquad\qquad CH_3\overset{\overset{\displaystyle CH_3}{|}}{C}HCH_2SH$$

Ethanethiol   2-Methyl-1-propanethiol
(Ethyl mercaptan)   (Isobutyl mercaptan)

Mushrooms, onions, garlic, and coffee all contain sulfur compounds. One of these present in the aroma of coffee is

 CH_2SH

(Charles D. Winters)

EXAMPLE 13.5

Write the IUPAC name for each thiol.

$$\text{(a)}\ CH_3CH_2CH_2CH_2CH_2SH \quad \text{(b)}\ CH_3\overset{\overset{\displaystyle SH}{|}}{C}HCH_2CH_3$$

Solution
(a) The parent alkane is pentane. Show the presence of the —SH group by adding "-thiol" to the name of the parent alkane. The IUPAC name of this thiol is 1-pentanethiol. Its common name is pentyl mercaptan.
(b) The parent alkane is butane. The IUPAC name of this thiol is 2-butanethiol. Its common name is *sec*-butyl mercaptan.

Ethylene Oxide—A Chemical Sterilant

Ethylene oxide is a colorless, flammable gas, bp 11°C. Because it is such a highly strained molecule (the normal tetrahedral bond angles of both C and O are compressed to approximately 60°), ethylene oxide reacts with the amino (—NH₂) and sulfhydryl (—SH) groups present in biological materials.

At sufficiently high concentrations, it reacts with enough molecules in cells to cause the death of microorganisms. This toxic property is the basis for its use as a fumigant in foodstuffs and textiles and its use in hospitals to sterilize surgical instruments.

Problem 13.5
Write the IUPAC name for each thiol.

$$\underset{\text{(a)}}{\text{CH}_3\overset{\displaystyle\overset{\text{CH}_3}{|}}{\text{CH}}\text{CH}_2\text{CH}_2\text{SH}} \qquad \underset{\text{(b)}}{\text{CH}_3\overset{\displaystyle\overset{\text{CH}_3}{|}}{\text{CH}}\underset{\displaystyle\underset{\text{SH}}{|}}{\text{CH}}\text{CH}_3}$$

13.4 Physical Properties

Alcohols

Table 13.1 compares the boiling points and water solubilities of some alcohols and alkanes of similar molecular weight. Notice that, in each group, the alcohol has the higher boiling point and is the more soluble in water.

Because of the large difference in electronegativity (Table 3.2) between oxygen and carbon (3.5–2.5) and between oxygen and hydrogen (3.5–2.1) both the C—O and O—H bonds of an alcohol are polar covalent, and alcohols are polar molecules, as illustrated in Figure 13.4 for methanol. The higher boiling points of alcohols compared with those of alkanes of similar molecular weight are due to the fact that alcohols are polar molecules and are associated with one another in the liquid state by **hydrogen bonding**

Alcohols, like water, are neutral compounds.

Hydrogen bonding A noncovalent interaction between a partial positive charge on a hydrogen bonded to an atom of high electronegativity and a partial negative charge on a nearby oxygen, nitrogen, or fluorine

FIGURE 13.4 Polarity of the C—O—H bonds in methanol. There are partial positive charges on carbon and hydrogen and a partial negative charge on oxygen.

TABLE 13.1 Boiling Points and Solubilities in Water of Groups of Alcohols and Alkanes of Similar Molecular Weight

Structural Formula	Name	Molecular Weight	bp (°C)	Solubility in Water
CH₃OH	methanol	32	65	infinite
CH₃CH₃	ethane	30	−89	insoluble
CH₃CH₂OH	ethanol	46	78	infinite
CH₃CH₂CH₃	propane	44	−42	insoluble
CH₃CH₂CH₂OH	1-propanol	60	97	infinite
CH₃CH₂CH₂CH₃	butane	58	0	insoluble
CH₃CH₂CH₂CH₂OH	1-butanol	74	117	8 g/100 g
CH₃CH₂CH₂CH₂CH₃	pentane	72	36	insoluble
CH₃CH₂CH₂CH₂CH₂OH	1-pentanol	88	138	2.3 g/100 g
CH₃CH₂CH₂CH₂CH₂CH₃	hexane	86	69	insoluble

FIGURE 13.5   The association of ethanol molecules in the liquid state. Each O—H can participate in up to three hydrogen bonds (one through hydrogen and two through oxygen). Only two of these three possible hydrogen bonds per molecule are shown here.

Solutions in which ethanol is the solvent are called tinctures.

(Figure 13.5). The strength of hydrogen bonding between alcohol molecules is approximately 2 to 5 kcal/mol, which means that extra energy is required to separate hydrogen-bonded alcohols from their neighbors.

Because of increased London dispersion forces (Section 5.7) between larger molecules, boiling points of all types of compounds, including alcohols, increase with increasing molecular weight. Compare, for example, the boiling points of ethanol, 1-propanol, 1-butanol, and 1-pentanol.

Ethylene glycol is colorless; the color of most antifreezes comes from additives. Review Section 6.8 for a discussion of freezing point depression.

Alcohols are much more soluble in water than are alkanes, alkenes, and alkynes of similar molecular weight because alcohol molecules interact by hydrogen bonding with water molecules. Methanol, ethanol, and 1-propanol are soluble in water in all proportions. As molecular weight increases, the physical properties of alcohols become more like those of hydrocarbons of similar molecular weight. Alcohols of higher molecular weight are much less soluble in water because of the increase in size of the hydrocarbon portion of their molecules.

Ethers

Ethers are polar compounds in which oxygen bears a partial negative charge and each attached carbon bears a partial positive charge (Figure 13.6). Because of the size of the groups bonded to an ether oxygen, however, only weak forces of attraction exist between ether molecules in the pure liquid; consequently, boiling points of ethers are lower than those of alcohols of similar molecular weight (Table 13.2). Boiling points of ethers are close to those of hydrocarbons of similar molecular weight (compare Tables 11.4 and 13.2).

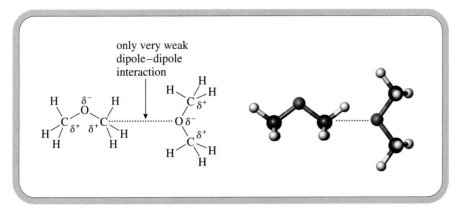

FIGURE 13.6 Ethers are polar molecules, but, because of the size of the groups bonded to the ether oxygen, only weak attractive interactions exist between their molecules in the liquid state.

Ethers are more soluble in water than hydrocarbons of similar molecular weight and shape (compare Tables 11.4 and 13.2). Their greater solubility is due to the fact that the oxygen atom of an ether carries a partial negative charge and forms hydrogen bonds with water.

The effect of hydrogen bonding on physical properties is illustrated dramatically by comparing the boiling points of ethanol (78°C) and its constitutional isomer dimethyl ether (−24°C). The difference in boiling point between these two compounds is due to the presence in ethanol of a polar O—H group, which is capable of forming hydrogen bonds. This hydrogen bonding increases intermolecular associations and thus gives ethanol a higher boiling point than dimethyl ether.

CH_3CH_2OH
Ethanol
(bp 78°C)

CH_3OCH_3
Dimethyl ether
(bp −24°C)

TABLE 13.2 Boiling Points and Solubilities in Water of Some Alcohols and Ethers of Similar Molecular Weight

Structural Formula	Name	Molecular Weight	bp (°C)	Solubility in Water
CH_3CH_2OH	ethanol	46	78	infinite
CH_3OCH_3	dimethyl ether	46	−24	7 g/100 g
$CH_3CH_2CH_2CH_2OH$	1-butanol	74	117	8 g/100 g
$CH_3CH_2OCH_2CH_3$	diethyl ether	74	35	8 g/100 g
$CH_3CH_2CH_2CH_2CH_2OH$	1-pentanol	88	138	2.3 g/100 g
$CH_3CH_2CH_2CH_2OCH_3$	butyl methyl ether	88	71	slight

TABLE 13.3	Boiling Points of Three Thiols and Alcohols of the Same Number of Carbon Atoms		
Thiol	**bp (°C)**	**Alcohol**	**bp (°C)**
methanethiol	6	methanol	65
ethanethiol	35	ethanol	78
1-butanethiol	98	1-butanol	117

The scent of the spotted skunk, native to the Sonoran Desert, is a mixture of two thiols, 3-methyl-1-butanethiol and 2-butene-1-thiol. *(Steven J. Krasemann/Photo Researchers, Inc.)*

Thiols

The most outstanding physical property of low-molecular-weight thiols is their stench. They are responsible for smells such as those from skunks, rotten eggs, and sewage. The scent of skunks is due primarily to these two thiols:

$CH_3CH=CHCH_2SH$
2-Butene-1-thiol

$CH_3CHCH_2CH_2SH$
3-Methyl-1-butanethiol

Because of the small difference in electronegativity between sulfur and hydrogen (2.5 − 2.1 = 0.4), we classify the S—H bond as nonpolar covalent. Because of this lack of polarity, thiols show little association by hydrogen bonding. Consequently, they have lower boiling points and are less soluble in water and other polar solvents than alcohols of similar molecular weights. Table 13.3 gives boiling points for three low-molecular-weight thiols. Shown for comparison are boiling points of alcohols of the same number of carbon atoms.

Earlier we illustrated the importance of hydrogen bonding in alcohols by comparing the boiling points of ethanol (78°C) and its constitutional isomer dimethyl ether (−24°C). By comparison, the boiling point of ethanethiol is 35°C, and that of its constitutional isomer dimethyl sulfide is 37°C. Because the boiling points of these constitutional isomers are almost identical, we know that little or no association by hydrogen bonding occurs between thiol molecules.

CH_3CH_2SH
Ethanethiol
(bp 35°C)

CH_3SCH_3
Dimethyl sulfide
(bp 37°C)

13.5 Reactions of Alcohols

As we prepare for the section on biochemistry beginning in Chapter 20, the two most important reactions of alcohols for us to know are dehydration to alkenes and oxidation to aldehydes, ketones, or carboxylic acids.

Acid-Catalyzed Dehydration of Alcohols to Alkenes

We can convert an alcohol to an alkene by eliminating a molecule of water from adjacent carbon atoms. Elimination of water is called **dehydration**. In the laboratory, dehydration of an alcohol is most often brought about by heating it with either 85 percent phosphoric acid or concentrated sulfuric acid. Primary alcohols are the most difficult to dehydrate and generally require heating in concentrated sulfuric acid at temperatures as high as 180°C. Secondary alcohols undergo acid-catalyzed dehydration at somewhat lower temperatures. Acid-catalyzed dehydration of tertiary alcohols often requires temperatures only slightly above room temperature.

> **Dehydration** Elimination of a molecule of water from an alcohol. An OH is removed from one carbon, and an H is removed from an adjacent carbon

$$CH_3CH_2OH \xrightarrow[180°C]{H_2SO_4} CH_2{=}CH_2 + H_2O$$

Cyclohexanol   Cyclohexene

$$CH_3\underset{\underset{CH_3}{|}}{\overset{\overset{CH_3}{|}}{C}}OH \xrightarrow[50°C]{H_2SO_4} CH_3\overset{\overset{CH_3}{|}}{C}{=}CH_2 + H_2O$$

2-Methyl-2-propanol   2-Methylpropene
(*tert*-Butyl alcohol)   (Isobutylene)

Thus, the ease of acid-catalyzed dehydration of alcohols is in this order:

1° alcohols   2° alcohols   3° alcohols

Ease of dehydration of alcohols →

When isomeric alkenes are obtained in the acid-catalyzed dehydration of an alcohol, the alkene having the greater number of carbon groups on the double bond generally predominates, a generalization known as **Zaitsev's rule**. In the acid-catalyzed dehydration of 2-butanol, for example, the major product is 2-butene, which has two carbon groups (two methyl groups) on its double bond. The minor product is 1-butene, which has only one carbon group (an ethyl group) on its double bond.

> **Zaitsev's rule** Dehydration of an alcohol normally forms the alkene with the more highly substituted double bond

Organic reactions rarely give only one product. Side reactions often yield a variety of products, most in only small quantities. In contrast, most biochemical reactions are very specific and yield only one product.

$$CH_3CH_2\underset{\underset{}{\overset{\overset{OH}{|}}{C}}}HCH_3 \xrightarrow[\text{heat}]{85\% \ H_3PO_4} CH_3CH{=}CHCH_3 + CH_3CH_2CH{=}CH_2$$

2-Butanol   2-Butene   1-Butene
    (80%)   (20%)

EXAMPLE 13.6

Draw structural formulas for the alkenes formed by the acid-catalyzed dehydration of each alcohol. For each part, predict which alkene is the major product.

(a) $\underset{\underset{OH}{|}}{\overset{\overset{CH_3}{|}}{CH_3CHCHCH_3}}$ $\xrightarrow{H_2SO_4}$ (b)   [structure: cyclopentane ring with OH at top and CH₃ to the right] $\xrightarrow{H_2SO_4}$

Solution

(a) Elimination of H_2O from carbons 2-3 gives 2-methyl-2-butene; elimination of H_2O from carbons 1-2 gives 3-methyl-1-butene. 2-Methyl-2-butene has three carbon groups (three methyl groups) on its double bond and is the major product. 3-Methyl-1-butene has only one carbon group (an isopropyl group) on its double bond and is the minor product.

$$\underset{\underset{OH}{|}}{\overset{\overset{CH_3}{|}}{\overset{4\ \ 3|\ \ 2\ \ 1}{CH_3CHCHCH_3}}} \xrightarrow[\substack{\text{acid-catalyzed} \\ \text{dehydration}}]{H_2SO_4} \overset{\overset{CH_3}{|}}{CH_3C}\!=\!CHCH_3 + \overset{\overset{CH_3}{|}}{CH_3CHCH}\!=\!CH_2 + H_2O$$

3-Methyl-2-butanol   2-Methyl-2-butene   3-Methyl-1-butene
                       (major product)

(b) The major product, 1-methylcyclopentene, has three carbon groups on its double bond. The minor product, 3-methylcyclopentene, has only two carbon groups on its double bond.

[reaction: 2-methylcyclopentanol → 1-methylcyclopentene + 3-methylcyclopentene + H_2O]

2-Methylcyclopentanol   1-Methylcyclopentene   3-Methylcyclopentene
                         (major product)

Problem 13.6

Draw structural formulas for the alkenes formed by the acid-catalyzed dehydration of each alcohol. For each, predict which is the major product.

(a) $\underset{\underset{OH}{|}}{\overset{\overset{CH_3}{|}}{CH_3CCH_2CH_3}}$ $\xrightarrow[\substack{\text{acid-catalyzed} \\ \text{dehydration}}]{H_2SO_4}$ (b)   [structure: cyclopentane ring with OH and CH₃] $\xrightarrow[\substack{\text{acid-catalyzed} \\ \text{dehydration}}]{H_2SO_4}$

In Section 12.7, we discussed the acid-catalyzed hydration of alkenes to give alcohols. In this section, we discuss the acid-catalyzed dehydration of alcohols to give alkenes. In fact, hydration-dehydration reactions are reversible. Alkene hydration and alcohol dehydration are competing reactions, and the following equilibrium exists.

$$\overset{}{\underset{}{C}}\!=\!\overset{}{\underset{}{C} + H_2O} \underset{}{\overset{\substack{\text{acid} \\ \text{catalyst}}}{\rightleftharpoons}} \underset{\underset{H\ \ \ OH}{|\ \ \ \ |}}{-\overset{|}{C}-\overset{|}{C}-}$$

An alkene   An alcohol

Large amounts of water (in other words, using dilute aqueous acid) favor alcohol formation, whereas scarcity of water (using concentrated acid) or experimental conditions where water is removed (heating the reaction mixture above 100°C) favor alkene formation. Thus, depending on experimental conditions, it is possible to use the hydration-dehydration equilibrium to prepare either alcohols or alkenes, each in high yields.

Oxidation of Primary and Secondary Alcohols

A primary alcohol can be oxidized to an aldehyde or to a carboxylic acid, depending on the experimental conditions. Following is a series of transformations in which a primary alcohol is oxidized first to an aldehyde and then to a carboxylic acid. The fact that each transformation involves oxidation is indicated by the symbol O in brackets over the reaction arrow.

$$CH_3{-}\underset{\underset{H}{|}}{\overset{\overset{OH}{|}}{C}}{-}H \xrightarrow{[O]} CH_3{-}\overset{\overset{O}{\|}}{C}{-}H \xrightarrow{[O]} CH_3{-}\overset{\overset{O}{\|}}{C}{-}OH$$

A primary alcohol   An aldehyde   A carboxylic acid

The reagent most commonly used in the laboratory for the oxidation of a primary alcohol to a carboxylic acid is potassium dichromate, $K_2Cr_2O_7$, dissolved in aqueous sulfuric acid. Oxidation of 1-octanol, for example, gives octanoic acid. This experimental condition is more than sufficient to oxidize the intermediate aldehyde to a carboxylic acid.

$$CH_3(CH_2)_6CH_2OH \xrightarrow[H_2SO_4]{K_2Cr_2O_7} CH_3(CH_2)_6\overset{\overset{O}{\|}}{C}H \xrightarrow[H_2SO_4]{K_2Cr_2O_7} CH_3(CH_2)_6\overset{\overset{O}{\|}}{C}OH$$

1-Octanol   Octanal   Octanoic acid

Although the usual product of oxidation of a primary alcohol is a carboxylic acid, it is often possible to stop the oxidation at the aldehyde stage by distilling the mixture so that the aldehyde (which usually has a lower boiling point than either the primary alcohol or the carboxylic acid) is removed from the reaction mixture before it can be oxidized further.

Secondary alcohols are oxidized to ketones by potassium dichromate. Menthol, a secondary alcohol (The Merck Index, 12th ed., #5882), is present in peppermint and other mint oils and is used in liqueurs, cigarettes, cough drops, perfumery, and nasal inhalers. Its oxidation product, menthone, is also used in perfumes and artificial flavors.

Peppermint plant *Mentha piperita* is a perennial herb with aromatic qualities used in candies, gums, hot and cold beverages, and garnish for punch and fruit. (© *Kathy Merrifield / Photo Researchers, Inc.*)

2-Isopropyl-5-methyl-
cyclohexanol
(Menthol)

2-Isopropyl-5-methyl-
cyclohexanone
(Menthone)

Menthol in throat sprays and lozenges soothes the respiratory tract.

Ethanol As a Drug and a Poison

Alcoholism is a major social problem in many areas of the world. Contrary to popular belief, ethanol, a physically addictive drug, is a depressant, not a stimulant. An individual may feel stimulated, but sensory perception is impaired and reflexes are slowed. Among its other effects, ethanol can cause cirrhosis of the liver, memory loss, and, in pregnant women, brain damage to the developing fetus.

One treatment for alcohol addiction uses the drug disulfiram (Antabuse, The Merck Index, 12th ed., #3428), which interferes with the body's metabolism of ethanol.

CH₃CH₂ and CH₃CH₂ — N — C(=S) — S — S — C(=S) — N — CH₂CH₃ and CH₂CH₃

Disulfiram
(Antabuse)

The first step in this metabolism is oxidation of ethanol to acetaldehyde, which is then oxidized further.

$$CH_3CH_2OH \xrightarrow{[O]} CH_3CH(=O) \xrightarrow{[O]} \text{other products}$$

Ethanol   Acetaldehyde

In the presence of disulfiram, acetaldehyde is formed, but instead of being oxidized further, it builds up in the blood, causing nausea, sweating, and vomiting. Knowing this will happen is almost always enough to keep a person on disulfiram from drinking. Although Antabuse is effective in alcohol addiction therapy, it is obviously not a cure. A cure is much more complicated and difficult to produce.

Tertiary alcohols are resistant to oxidation because the carbon bearing the —OH is bonded to three carbon atoms and, therefore, cannot form a carbon-oxygen double bond.

$$\xrightarrow[H_2SO_4]{K_2Cr_2O_7} \text{(no oxidation)}$$

1-Methylcyclopentanol

EXAMPLE 13.7

Draw the product formed by oxidation of each alcohol with potassium dichromate.
(a) 1-Hexanol   (b) 2-Hexanol

Solution
1-Hexanol, a primary alcohol, is oxidized to either hexanal or hexanoic acid, depending on experimental conditions. 2-Hexanol, a secondary alcohol, is oxidized to the ketone 2-hexanone.

(a) $CH_3(CH_2)_4CH(=O)$ or $CH_3(CH_2)_4COH(=O)$ (b) $CH_3(CH_2)_3CCH_3(=O)$

Hexanal   Hexanoic acid   2-Hexanone

Problem 13.7
Draw the product formed by oxidation of each alcohol with potassium dichromate.
(a) Cyclohexanol   (b) 2-Pentanol ∎

13.6 Reactions of Ethers

Ethers, R—O—R, resemble hydrocarbons in their resistance to chemical reaction. They do not react with oxidizing agents, such as potassium dichromate. They are not affected by most acids or bases at moderate tem-

Breath-Alcohol Screening Test

Potassium dichromate oxidation of ethanol to acetic acid is the basis for the original breath alcohol screening test used by law enforcement agencies to determine a person's blood alcohol content (BAC). The test is based on the difference in color between the dichromate ion (reddish orange) in the reagent and the chromium(III) ion (green) in the product.

$$CH_3CH_2OH + Cr_2O_7^{-2} \xrightarrow[H_2O]{H_2SO_4}$$

Ethanol   Dichromate ion
(reddish orange)

$$CH_3\overset{\overset{\displaystyle O}{\|}}{C}OH + Cr^{3+}$$

Acetic acid   Chromium(III) ion
(green)

In its simplest form, a breath alcohol screening test is a sealed glass tube that contains a potassium dichromate–sulfuric acid reagent impregnated on silica gel. To administer the test, the ends of the tube are broken off, a mouthpiece is fitted to one end, and the other end is inserted into the neck of a plastic bag. The person being tested then blows into the mouthpiece until the plastic bag is inflated.

As breath containing ethanol vapor passes through the tube, reddish-orange dichromate ion is reduced to green chromium(III) ion. The concentration of ethanol in the breath is then estimated by measuring how far the green color extends along the length of the tube. When it extends beyond the halfway point, the person is judged as having a sufficiently high blood alcohol content to warrant further, more precise testing.

This test measures the alcohol content of the breath. The legal definition of being under the influence of alcohol, however, is based on blood alcohol content, not breath alcohol content. The chemical correlation between these two measurements is based on the fact that air deep within the lungs is in equilibrium with blood passing through the pulmonary arteries, and an equilibrium is established between blood alcohol and breath alcohol. Based on tests in persons drinking alcohol, researchers have determined that 2100 mL of breath contains the same amount of ethanol as 1.00 mL of blood.

Person forces breath through mouthpiece into the tube.

Glass tube containing potassium dichromate–sulfuric acid coated on silica gel particles

As person blows into the tube, the plastic bag becomes inflated.

peratures. Because of their good solvent properties and general inertness to chemical reaction, ethers are excellent solvents in which to carry out many organic reactions.

13.7   Reactions of Thiols

The most common reaction of thiols in biological systems is their oxidation to disulfides. The functional group of a **disulfide** is an —S—S— group.

> **Disulfide** A compound containing an —S—S— group

$$2\,R-S-H \xrightarrow{\text{oxidation}} R-S-S-R$$

A thiol   A disulfide

We derive common names of disulfides by listing the names of the groups bonded to sulfur and adding the word "disulfide."

Thiols are readily oxidized to disulfides by molecular oxygen. In fact, thiols are so susceptible to oxidation that they must be protected from con-

BOX 13E

Ethers and Anesthesia

Before the mid-1800s, surgery was performed only when absolutely necessary because no truly effective general anesthetic was available. More often than not, patients were drugged, hypnotized, or simply tied down. In 1772, Joseph Priestly isolated nitrous oxide, N_2O, a colorless gas, and in 1799, Sir Humphry Davy demonstrated its anesthetic effect, naming it laughing gas. In 1844, an American dentist, Horace Wells, introduced nitrous oxide into general dental practice. One patient awakened prematurely, however, screaming with pain, and another died. Wells was forced to withdraw from practice, became embittered and depressed, and committed suicide at age 33. In the same period, a Boston chemist, Charles Jackson, anesthetized himself with diethyl ether; he also persuaded a dentist, William Morton, to use it. Subsequently, they persuaded a surgeon, John Warren, to give a public demonstration of surgery under anesthesia. The operation was completely successful, and soon general anesthesia by diethyl ether was routine for general surgery.

■ This painting by Robert Hinckley shows the first use of ether as an anesthetic in 1846. Dr. Robert John Collins was removing a tumor from the patient's neck, and dentist W. T. G. Morton—who discovered its anesthetic properties—administered the ether. *(Boston Medical Library in the Francis A. Countway Library of Medicine)*

Diethyl ether is easy to use and causes excellent muscle relaxation. Blood pressure, pulse rate, and respiration are usually only slightly affected. Its chief drawbacks are its irritating effect on the respiratory passages and its aftereffect of nausea. Further, when mixed with air in the right proportions, it is explosive. Modern operating theaters are often a maze of sophisticated electrical equipment that is used throughout most surgical procedures. During pronged ether anesthesia, concentrations of the gas in a patient's fatty tissue, abdominal cavity, and bladder can reach explosive levels. An electrical spark could produce an explosion! Precautionary measures to guard against such accidents became so cumbersome that the incentive was strong to develop alternative nonflammable and nonexplosive anesthetics.

Among the inhalation anesthetics used today are several halogenated ethers, the most important being enflurane (The Merck Index, 12th ed., #3621) and isoflurane (The Merck Index, 12th ed., #5191).

Enflurane
(Ethrane)

Isoflurane
(Forane)

Halothane
(Fluothane)

One of today's most widely used inhalation anesthetics, halothane (The Merck Index, 12th ed., #4634), is not an ether but rather a polyhalogenated derivative of ethane. Halothane does not cause discomfort and is considered safe, although it does depress respiratory and cardiovascular action.

tact with air during storage. Disulfides are, in turn, easily reduced to thiols by several different reducing agents. This easy conversion between thiols and disulfides is very important in protein chemistry, as we shall see in Chapters 21 and 22.

$$2HOCH_2CH_2SH \underset{reduction}{\overset{oxidation}{\rightleftharpoons}} HOCH_2CH_2S-SCH_2CH_2OH$$

A thiol   A disulfide

SUMMARY

The functional group of an **alcohol** is an —**OH (hydroxyl) group** bonded to a tetrahedral carbon atom. The functional group of an **ether** is an atom of oxygen bonded to two carbon atoms. A **thiol** is the sulfur analog of an alcohol; it contains an —**SH (sulfhydryl) group** in place of an —OH group.

IUPAC names of alcohols are derived by changing the suffix of the parent alkane from "-e" to "-ol." The chain is numbered to give the carbon bearing —OH the lower number. Common names for alcohols are derived by naming the alkyl group bonded to —OH and adding the word "alcohol." Alcohols are classified as 1°, 2°, or 3° depending on the number of carbon groups bonded to the carbon bearing the —OH group. Compounds containing two hydroxyl groups on adjacent carbons are called **glycols.**

Common names for ethers are derived by naming the two groups bonded to oxygen followed by the word "ether." A **cyclic ether** is an ether in which the ether oxygen is one of the atoms in a ring.

Thiols are named in the same manner as alcohols, but the suffix "-e" is retained, and "**-thiol**" is added. Common names for thiols are derived by naming the alkyl group bonded to —SH and adding the word **"mercaptan."**

Alcohols are polar compounds with oxygen bearing a partial negative charge and both the carbon and hydrogen bonded to it bearing partial positive charges. Because of intermolecular association by **hydrogen bonding,** the boiling points of alcohols are higher than those of hydrocarbons of similar molecular weight. Because of increased London dispersion forces, the boiling points of alcohols increase with increasing molecular weight. Alcohols interact with water by hydrogen bonding and, therefore, are more soluble in water than are hydrocarbons of similar molecular weight.

Ethers are weakly polar compounds. Their boiling points are close to those of hydrocarbons of similar molecular weight. Because they form hydrogen bonds with water, ethers are more soluble in water than are hydrocarbons of similar molecular weight.

The S—H is nonpolar, and the physical properties of thiols are like those of hydrocarbons of similar molecular weight.

KEY TERMS

Alcohol (Section 13.2)
Cyclic ether (Section 13.3)
Dehydration (Section 13.5)
Diol (Section 13.3)
Disulfide (Section 13.7)
Ether (Section 13.2)

Glycol (Section 13.3)
Hydrogen bonding (Section 13.4)
Hydroxyl (—OH) group (Section 13.2)
Mercaptan (Section 13.3)
Primary (1°) alcohol (Section 13.3)
Secondary (2°) alcohol (Section 13.3)

Sulfhydryl (—SH) group (Section 13.2)
Tertiary (3°) alcohol (Section 13.3)
Thiol (Section 13.2)
Triol (Section 13.3)
Zaitsev's rule (Section 13.5)

KEY REACTIONS

1. **Acid-Catalyzed Dehydration of an Alcohol (Section 13.5)**

   When isomeric alkenes are possible, the major product is generally the more substituted alkene (Zaitsev's rule).

   $$CH_3CH_2\overset{\overset{\displaystyle OH}{|}}{C}HCH_3 \xrightarrow[\text{heat}]{H_3PO_4}$$

   $$CH_3CH{=}CHCH_3 + CH_3CH_2CH{=}CH_2 + H_2O$$
   Major product

2. **Oxidation of a Primary Alcohol (Section 13.5)**

   Oxidation of a primary alcohol by potassium dichromate gives either an aldehyde or a carboxylic acid depending on the experimental conditions.

   $$CH_3(CH_2)_6CH_2OH \xrightarrow[H_2SO_4]{K_2Cr_2O_7}$$

   $$CH_3(CH_2)_6\overset{\overset{\displaystyle O}{\|}}{C}H \xrightarrow[H_2SO_4]{K_2Cr_2O_7} CH_3(CH_2)_6\overset{\overset{\displaystyle O}{\|}}{C}OH$$

3. **Oxidation of a Secondary Alcohol (Section 13.5)**

   Oxidation of a secondary alcohol by potassium dichromate gives a ketone.

   $$CH_3(CH_2)_4\overset{\overset{\displaystyle OH}{|}}{C}HCH_3 \xrightarrow[H_2SO_4]{K_2Cr_2O_7} CH_3(CH_2)_4\overset{\overset{\displaystyle O}{\|}}{C}CH_3$$

4. **Oxidation of a Thiol to a Disulfide (Section 13.7)**

   Oxidation of a thiol by O_2 gives a disulfide. Reduction of a disulfide gives two thiols.

   $$2RS{-}H \underset{\text{reduction}}{\overset{\text{oxidation}}{\rightleftharpoons}} RS{-}SR$$

CONCEPTUAL PROBLEMS

Difficult problems are designated by an asterisk.

13.A Explain in terms of noncovalent interactions why the low-molecular-weight alcohols are soluble in water but the low-molecular-weight alkanes and alkenes are not.

13.B Knowing what you do about electronegativity, the polarity of covalent bonds, and hydrogen bonding, would you expect an N—H···N hydrogen bond to be stronger than, the same strength as, or weaker than an O—H···O hydrogen bond?

*****13.C** Low-molecular-weight alcohols and ethers can both form hydrogen bonds with water. Yet the low-molecular-weight ethers are practically insoluble in water, whereas the low-molecular-weight alcohols are completely soluble in water. How do you account for these differences in their water solubility?

13.D State Le Chatelier's principle (review Section 7.8). Show how our ability to convert an alcohol to an alkene, or the alkene to an alcohol is an application of this principle.

PROBLEMS

Difficult problems are designated by an asterisk.

Structure and Nomenclature

13.8 What is the difference between a primary, secondary, and tertiary alcohol?

13.9 Which are secondary alcohols?

(a) [cyclohexane with CH$_3$ and OH groups] (b) $(CH_3)_3COH$

(c) [structure with OH] (d) [cyclopentane]—OH

13.10 Write the IUPAC name of each compound.
(a) $CH_3CH_2CH_2CH_2CH_2OH$
(b) $HOCH_2CH_2CH_2OH$ (c) $CH_3CH=CHCH_2OH$
(d) $HOCH_2CH_2\overset{\underset{|}{CH_3}}{C}HCH_3$ (e) [cyclohexane with two OH groups]
(f) $CH_3CH_2CH_2CH_2SH$

13.11 Draw a structural formula for each alcohol.
(a) Isopropyl alcohol
(b) Propylene glycol
(c) 5-Methyl-2-hexanol
(d) 2-Methyl-2-propyl-1,3-propanediol
(e) 1-Octanol
(f) Isobutyl alcohol
(g) 1,4-Butanediol
(h) *cis*-5-Methyl-2-hexen-1-ol
(i) *cis*-3-Pentene-1-ol
(j) *trans*-1,4-Cyclohexanediol

13.12 Write the common name for each ether.

(a) [cyclopentane]—O—[cyclopentane] (b) $[CH_3(CH_2)_4]_2O$

(c) $CH_3\overset{\underset{|}{CH_3}}{C}H O \overset{\underset{|}{CH_3}}{C}HCH_3$

13.13 The chemical name for bombykol, the sex pheromone secreted by the female silkworm moth to attract male silkworm moths, is *trans*-10-*cis*-12-hexadecadien-1-ol. (It has one hydroxyl group and two carbon-carbon double bonds in a 16-carbon chain.)
(a) Draw a structural formula for bombykol, showing the correct configuration about each carbon-carbon double bond.
(b) How many cis-trans isomers are possible for this structural formula? All possible cis-trans isomers have been synthesized in the laboratory, but only the one named bombykol is produced by the female silkworm moth, and only it attracts male silkworm moths.

Female silkworm moth *Bombyx mori* lays 300 to 500 eggs and dies shortly afterward. (© *S. Nagendra/Photo Researchers, Inc.*)

Physical Properties

13.14 Arrange these compounds in order of increasing boiling point. Values in °C are −42, 78, 117, and 198.
(a) $CH_3CH_2CH_2CH_2OH$ (b) CH_3CH_2OH
(c) $HOCH_2CH_2OH$ (d) $CH_3CH_2CH_3$

13.15 Arrange these compounds in order of increasing boiling point. Values in °C are 0, 35, and 97.
(a) $CH_3CH_2CH_2OH$　　(b) $CH_3CH_2OCH_2CH_3$
(c) $CH_3CH_2CH_2CH_3$

13.16 Following are structural formulas for 1-butanol and 1-butanethiol. One of these compounds has a boiling point of 98.5°C, the other has a boiling point of 117°C. Which compound has which boiling point?

$$CH_3CH_2CH_2CH_2OH \qquad CH_3CH_2CH_2CH_2SH$$
1-Butanol　　　　　　　　1-Butanethiol

13.17 Explain why methanethiol, CH_3SH, has a lower boiling point (6°C) than methanol, CH_3OH (65°C), even though it has a higher molecular weight.

13.18 2-Propanol (isopropyl alcohol) is commonly used as rubbing alcohol to cool the skin. 2-Hexanol, also a liquid, is not suitable for this purpose. Why not?

13.19 Explain why glycerol is much thicker (more viscous) than ethylene glycol, which in turn is much thicker than ethanol.

13.20 From each pair, select the compound that is more soluble in water.
(a) CH_3OH　or　CH_3OCH_3
(b) $CH_3\overset{OH}{\underset{|}{C}}HCH_3$　or　$CH_3\overset{CH_2}{\underset{||}{C}}CH_3$
(c) $CH_3CH_2CH_2SH$　or　$CH_3CH_2CH_2OH$

13.21 Arrange the compounds in each set in order of decreasing solubility in water.
(a) Ethanol; butane; diethyl ether
(b) 1-Hexanol; 1,2-hexanediol; hexane

Synthesis of Alcohols

13.22 Give the structural formula of an alkene or alkenes from which each alcohol can be prepared.
(a) 2-Butanol　　　　(b) 1-Methylcyclohexanol
(c) 3-Hexanol　　　　(d) 2-Methyl-2-pentanol
(e) Cyclopentanol

Reactions of Alcohols

13.23 Show how to distinguish between cyclohexanol and cyclohexene by a simple chemical test. Hint: Treat each with Br_2 in CCl_4 and watch what happens.

13.24 Write equations for the reaction of 1-butanol, a primary alcohol, with these reagents.
(a) H_2SO_4, heat　　(b) $K_2Cr_2O_7$, H_2SO_4

13.25 Write equations for the reaction of 2-butanol, a secondary alcohol, with these reagents.
(a) H_2SO_4, heat　　(b) $K_2Cr_2O_7$, H_2SO_4

13.26 Complete the equations for these reactions.

(a) $CH_3(CH_2)_6CH_2OH \xrightarrow[H_2SO_4]{K_2Cr_2O_7}$

(b) $HOCH_2CH_2CH_2OH \xrightarrow[H_2SO_4]{K_2Cr_2O_7}$

(c) $\xrightarrow[\text{heat}]{H_2SO_4}$　(d) $+ H_2O \xrightarrow{H_2SO_4}$

*13.27 The citric acid cycle is a metabolic pathway by which organisms use carbon sources to generate energy. In this problem, we focus on just four of the intermediates in this cycle. Name the type of reaction that takes place in each step shown here. By reaction type, we mean oxidation, reduction, dehydration, hydration, and the like.

$$\underset{\text{Citric acid}}{HO-\overset{CH_2-COOH}{\underset{CH_2-COOH}{\overset{|}{\underset{|}{C}}}}-COOH} \xrightarrow{(1)} \underset{\text{Aconitic acid}}{\overset{CH_2-COOH}{\underset{CH-COOH}{\overset{|}{\underset{||}{C}}}}-COOH} \xrightarrow{(2)}$$

$$\underset{\text{Isocitric acid}}{H-\overset{CH_2-COOH}{\underset{HO}{\overset{|}{C}}}-COOH} \xrightarrow{(3)} \underset{\text{Oxalosuccinic acid}}{H-\overset{CH_2-COOH}{\underset{O}{\overset{|}{C}}}-COOH}$$

Reaction of Thiols

13.28 Following is a structural formula for the amino acid cysteine.

$$HS-CH_2-\overset{}{\underset{NH_2}{CH}}-\overset{O}{\overset{||}{C}}-OH$$

(a) Name the three functional groups in cysteine.
(b) In the body, cysteine is oxidized to a disulfide. Draw a structural formula for this disulfide.

13.29 Lipoic acid is a growth factor for many bacteria and protozoa. Although it is not classified as a vitamin for humans, it is present in body tissues in small amounts and is an essential component of several of the enzymes involved in human metabolism.

Lipoic acid

(a) Name the two functional groups in lipoic acid.
(b) At one stage in its function in human metabolism, the disulfide bond of lipoic acid is reduced to two thiol groups. Draw a structural formula for this reduced form of lipoic acid.

Syntheses

*13.30 Show how to convert:
(a) 1-Propanol to 2-propanol in two steps.
(b) Propene to propanone (acetone) in two steps.

*13.31 Show how to convert cyclohexanol to these compounds.
 (a) Cyclohexene   (b) Cyclohexane
 (c) Cyclohexanone

*13.32 Show reagents and experimental conditions to synthesize each compound from 1-propanol. Any derivative of 1-propanol prepared in an earlier part of this problem may then be used for a later synthesis.
 (a) Propanal   (b) Propanoic acid
 (c) Propene   (d) 2-Propanol
 (e) 2-Bromopropane (f) 2-Chloropropane
 (g) Propanone

13.33 Show how to convert the alcohol on the left to compounds (a) and (b).

$$\text{\Large$\bigcirc$}-CH_2OH \quad (a) \quad \text{\Large$\bigcirc$}-\overset{\overset{\textstyle O}{\|}}{C}H \quad (b) \quad \text{\Large$\bigcirc$}-\overset{\overset{\textstyle O}{\|}}{C}OH$$

Boxes

13.34 (Box 13A) When was nitroglycerin discovered? Is this substance a solid, liquid, or gas?

13.35 (Box 13A) What was Alfred Nobel's discovery that made nitroglycerin safer to handle?

13.36 (Box 13A) What is the relationship between the medical use of nitroglycerin to relieve the sharp chest pains (angina) associated with heart disease and the gas nitric oxide, NO?

13.37 (Box 13B) What does it mean to say that ethylene oxide is a highly strained molecule?

13.38 (Box 13C) What property of disulfiram (Antabuse) makes it effective in treating alcohol addiction?

13.39 (Box 13C) Why must pregnant women be particularly careful about alcohol consumption?

13.40 (Box 13D) What is the color of dichromate ion? Of chromium(III) ion? Explain how the conversion of one to the other is used in breath alcohol screening.

13.41 (Box 13D) The legal definition of being under the influence of alcohol is based on blood alcohol content. What is the relationship between breath alcohol content and blood alcohol content?

13.42 (Box 13E) What are the advantages and disadvantages of using diethyl ether as an anesthetic?

13.43 (Box 13E) Show that enflurane and isoflurane are constitutional isomers.

13.44 (Box 13E) Write the IUPAC name of halothane (Fluothane).

Additional Problems

*13.45 Draw structural formulas and write IUPAC names for the eight isomeric alcohols of molecular formula $C_5H_{12}O$.

*13.46 Draw structural formulas and write common names for the six isomeric ethers of molecular formula $C_5H_{12}O$.

13.47 Explain why the boiling point of ethylene glycol, bp 198°C, is so much higher than that of 1-propanol, bp 97°C, even though their molecular weights are about the same.

13.48 Following are structural formulas for three compounds of similar molecular weight. Their boiling points, arranged from lowest to highest, are 69, 138, and 230°C. Which compound has which boiling point?

$HOCH_2CH_2CH_2CH_2OH$ $CH_3CH_2CH_2CH_2CH_2CH_3$
 1,4-Butanediol   Hexane
 (*MW* 90 g/mol)   (*MW* 86 g/mol)

$CH_3CH_2CH_2CH_2CH_2OH$
 1-Pentanol
 (*MW* 88 g/mol)

13.49 Of the three compounds given in Problem 13.48, one is insoluble in water, another has a solubility of 2.3 g/100 g water, and one is infinitely soluble in water. Which compound has which solubility?

13.50 Each compound given in this problem is a common organic solvent. From each pair of compounds, select the solvent with the greater solubility in water.
 (a) CH_2Cl_2 or CH_3CH_2OH
 (b) $CH_3CH_2OCH_2CH_3$ or CH_3CH_2OH
 (c) $CH_3CH_2OCH_2CH_3$ or $CH_3(CH_2)_3CH_3$

*13.51 Show how to prepare each compound from 2-methyl-1-propanol (isobutyl alcohol). For any preparation involving more than one step, show each intermediate compound formed.

(a) $CH_3\overset{\overset{\textstyle CH_3}{|}}{C}{=}CH_2$ (b) $CH_3\overset{\overset{\textstyle CH_3}{|}}{\underset{\underset{\textstyle OH}{|}}{C}}CH_3$ (c) $CH_3\overset{\overset{\textstyle CH_3}{|}}{C}HCOOH$

13.52 Show how to prepare each compound from 2-methylcyclohexanol. For any preparation involving more than one step, show each intermediate compound formed.

(a)   (b)   (c)

Benzene and Its Derivatives

14.1 Introduction

All the hydrocarbons discussed so far—alkanes, alkenes, and alkynes—are called aliphatic hydrocarbons. More than 150 years ago, chemists realized that there is another class of hydrocarbons, one whose properties are quite different from those of aliphatic hydrocarbons. Because some of these new hydrocarbons have pleasant odors, they were called **aromatic compounds.** Today we know that not all aromatic compounds have pleasant odors. Some do, but some have no odor at all, and others have downright unpleasant odors.

The term **arene** is used to describe aromatic hydrocarbons. Just as a group derived by removal of an H from an alkane is called an alkyl group and given the symbol R—, a group derived by removal of an H from an arene is called an **aryl group** and given the symbol **Ar**—.

Aromatic compound A term used to classify benzene and its derivatives

Arene A compound containing one or more benzene rings

14.2 The Structure of Benzene

Benzene, the simplest aromatic hydrocarbon, was discovered by Michael Faraday (1791–1867) in 1825. Its structure presented an immediate problem to chemists of that day. The problem was that benzene has the molecular formula C_6H_6, and a compound with so few hydrogens for its six car-

Above: Peppers of the capsicum family. (See Box 14D.) *(Douglas Brown)*

Benzene is an important compound in both the chemical industry and the laboratory, but it must be handled carefully. Not only is it poisonous if ingested in liquid form, but the vapor is also toxic and can be absorbed either by breathing or through the skin. Long-term inhalation can cause liver damage and cancer.

bons (compare hexane C_6H_{14}) should be unsaturated. But benzene does not behave like an alkene, the only class of unsaturated hydrocarbons known at that time. Whereas 1-hexene, for example, reacts instantly with Br_2 (Section 12.6), benzene does not react at all with this reagent. Nor does benzene react with HBr, H_2O/H_2SO_4, or H_2/Pd, reagents that normally add to carbon-carbon double bonds.

Benzene does undergo chemical reactions, but its characteristic reaction is substitution rather than addition. When benzene is treated with chlorine in the presence of ferric chloride as a catalyst, for example, a compound of molecular formula C_6H_5Cl is formed. In this compound, a chlorine atom is substituted for a hydrogen atom. By comparison, 1-hexene reacts with chlorine by addition to give 1,2-dichlorohexane.

$$C_6H_6 \ + \ Cl_2 \xrightarrow[\text{(substitution)}]{FeCl_3} \ C_6H_5Cl \ + \ HCl$$

Benzene   Chlorobenzene

$$CH_3CH_2CH_2CH_2CH{=}CH_2 + Cl_2 \xrightarrow[\text{(addition)}]{} CH_3CH_2CH_2CH_2\overset{\overset{\displaystyle Cl}{|}}{C}H{-}\overset{\overset{\displaystyle Cl}{|}}{C}H_2$$

1-Hexene   1,2-Dichlorohexane

Kekulé's Structure of Benzene

The first structure for benzene was proposed by Friedrich August Kekulé in 1872 and consisted of a six-membered ring with alternating single and double bonds and with one hydrogen attached to each carbon.

When you see formulas like the line-angle structure for benzene, don't forget that each carbon must be connected to enough hydrogens to give it four bonds. We usually don't show these hydrogens, but they are there just the same.

A Kekulé structure showing all atoms

A Kekulé structure as a line-angle drawing

Although Kekulé's proposal was consistent with many of the chemical properties of benzene, it was contested for years. The major objection was that it did not account for the unusual chemical behavior of benzene. If benzene contains three double bonds, Kekulé's critics asked, why does it not undergo reactions typical of alkenes?

Resonance Structure of Benzene

The concept of resonance, developed by Linus Pauling in the 1930s, provided the first adequate description of the structure of benzene. According to the theory of resonance, certain molecules and ions are best described by writing two or more Lewis structures and considering the real molecule or ion to be a **resonance hybrid** of these structures. Each individual Lewis structure is called a **contributing structure.** We show that the real molecule is a resonance hybrid of the two Lewis structures by positioning a double-headed arrow between them.

Alternative Lewis contributing structures for benzene

We say that benzene is a resonance hybrid of two Lewis contributing structures, often referred to as Kekulé structures. The resonance hybrid has some of the characteristics of each Kekulé structure, one of which is that the carbon-carbon bonds are neither single nor double but something intermediate. The closed loop of six electrons characteristic of a benzene ring is called the **aromatic sextet.**

(The six electrons of the aromatic sextet are shown in color.)

Note that if you flip the left contributing structure for benzene as you would when you turn a page of a book back to a previous page, you can place the flipped structure on top of the right contributing structure, and all atoms and bonds coincide. Then you might ask, "Aren't these two contributing structures really the same?" The answer is no, they are not. A rule of resonance theory is that atoms do not move. The only thing contributing structures represent is alternative ways to pair six electrons to create three double bonds.

Wherever we find resonance, we find stability. The real structure is always more stable than any of the fictitious contributing structures. The benzene ring is greatly stabilized by resonance, which is why it does not undergo addition reactions. To do so would pull two electrons out of the aromatic sextet, resulting in a loss of resonance stability.

(does not happen)

14.3 Nomenclature

One Substituent

Monosubstituted alkylbenzenes are named as derivatives of benzene, as for example ethylbenzene. The IUPAC system retains certain common names for several of the simpler monosubstituted alkylbenzenes. Examples are **toluene** and **styrene.**

Ethylbenzene   Toluene   Styrene

The common names for the following compounds are also retained by the IUPAC system.

Household products containing benzene derivatives. (*Charles D. Winters*)

OH   OCH$_3$ NO$_2$ NH$_2$ $\overset{\displaystyle O}{\overset{\|}{C}}$—H $\overset{\displaystyle O}{\overset{\|}{C}}$—OH

Phenol   Anisole   Nitrobenzene   Aniline   Benzaldehyde   Benzoic acid

> **Phenyl group** C$_6$H$_5$—, the aryl group derived by removing a hydrogen from benzene

The substituent group derived by loss of an H from benzene is a **phenyl group,** the common symbol for which is **Ph—.**

Benzene   Phenyl group (Ph—)

In molecules containing other functional groups, a phenyl group is often named as a substituent.

C$_6$H$_5$CH$_2$CH$_2$OH   PhCH$_2$CH=CH$_2$

2-Phenylethanol   3-Phenyl-1-propene

Two Substituents

When two substituents occur on a benzene ring, three isomers are possible. The substituents may be located by either numbering the atoms of the ring or by using the locators **ortho, meta,** and **para.** The numbers 1,2- are equivalent to *ortho* (Greek: straight); 1,3- is equivalent to *meta* (Greek: after); and 1,4- is equivalent to *para* (Greek: beyond).

> **Ortho (o)** Refers to groups occupying the 1 and 2 positions on a benzene ring

> **Meta (m)** Refers to groups occupying the 1 and 3 positions on a benzene ring

> **Para (p)** Refers to groups occupying the 1 and 4 positions on a benzene ring

When one of the two substituents on the ring imparts a special name to the compound, as for example —CH$_3$, —OH, —NH$_2$, or —COOH, then the compound is named as a derivative of that parent molecule, and that substituent is assumed to occupy ring position number 1. The IUPAC system retains the common name **xylene** for the three isomeric dimethylbenzenes. Where neither substituent imparts a special name, the two substituents are located and listed in alphabetical order before the ending "-benzene." The carbon of the benzene ring with the substituent of lower alphabetical ranking is numbered C-1.

> *p*-Xylene is a starting material for the synthesis of poly(ethylene terephthalate). Consumer products derived from this polymer include Dacron polyester and Mylar films (Section 18.8).

COOH   NH$_2$ CH$_3$ CH$_2$CH$_3$

Br
4-Bromobenzoic acid
(*p*-Bromobenzoic acid)

Cl
3-Chloroaniline
(*m*-Chloroaniline)

CH$_3$
1,3-Dimethylbenzene
(*m*-Xylene)

Cl
1-Chloro-4-ethylbenzene
(*p*-Chloroethylbenzene)

Three or More Substituents

When three or more substituents are present on a benzene ring, their locations are specified by numbers. If one of the substituents imparts a special name, then the molecule is named as a derivative of that parent molecule. If none of the substituents imparts a special name, then the substituents are located, numbered to give the smallest set of numbers, and listed in alphabetical order before the ending "-benzene." In the following examples, the first compound is a derivative of toluene, and the second is a derivative of phenol. Because no substituent in the third compound imparts a special name, its three substituents are listed in alphabetical order followed by the word "benzene."

4-Chloro-2-nitrotoluene   2,4,6-Tribromophenol   2-Bromo-1-ethyl-4-nitrobenzene

EXAMPLE 14.1

Write names for these compounds.

(a)   (b)   (c)

Solution

(a) The parent is toluene, and the compound is 3-iodotoluene or *m*-iodotoluene.
(b) The parent is benzoic acid, and the compound is 3,5-dibromobenzoic acid.
(c) The parent is aniline, and the compound is 4-chloroaniline or *p*-chloroaniline.

Problem 14.1 ■

Write names for these compounds.

(a)   (b)   (c)

Polynuclear Aromatic Hydrocarbons

Polynuclear aromatic hydrocarbons (PAHs) contain two or more aromatic rings, with each pair of rings sharing two carbon atoms. Naphthalene, anthracene, and phenanthrene, the most common PAHs, and substances derived from them are found in coal tar and high-boiling petroleum residues. At one time, naphthalene was used as a moth repellent and insecticide in preserving woolens and furs, but its use has decreased due to the introduction of chlorinated hydrocarbons such as *p*-dichlorobenzene.

> **Polynuclear aromatic hydrocarbon**
> A hydrocarbon containing two or more benzene rings, each pair of which shares two carbon atoms

Naphthalene   Anthracene   Phenanthrene

DDT. A Boon and a Curse

Probably the best known insecticide worldwide is dichloro-diphenyltrichloroethane (not an IUPAC name), commonly abbreviated DDT (The Merck Index, 12th ed., #2898).

Dichlorodiphenyltrichloroethane
(DDT)

This compound was first prepared in 1874, but it was not until the late 1930s that its potential as an insecticide was recognized. It was first used for this purpose in 1939 and was extremely effective in ridding large areas of the world of the insect hosts that transmit malaria and typhus. In addition, it has been so effective in killing crop-destroying insect pests that crop yields in many areas have been increased dramatically.

Widespread use of DDT, however, has been a double-edged sword. Despite DDT's enormous benefits, it has an enormous disadvantage. Because it resists biodegradation, it remains in the soil for years, and this persistence in the environment is the problem. It is estimated that the tissues of adult humans contain on the average 5 to 10 parts per million of DDT.

The dangers associated with the persistence of DDT in the environment were dramatically portrayed by Rachel Carson in her 1962 book *Silent Spring*, which documented the serious decline in the population of eagles and other raptors as well as many other kinds of birds. It was discovered soon thereafter that DDT inhibits the mechanism by which these birds incorporate calcium into their egg shells, with the result that shells become so thin and weak that they break during incubation.

Because of these problems, almost all nations have now banned the use of DDT for agricultural purposes. It is still used, however, in some areas to control the population of disease-spreading insects.

■ DDT was sprayed on crops in the United States until its use was banned in the 1970s. *(Courtesy of U.S. Department of Agriculture)*

Carcinogenic Polynuclear Aromatics and Smoking

A compound that causes cancer is called a **carcinogen.** Although many compounds have now been discovered to be carcinogens, the first to be identified were a group of polynuclear aromatic hydrocarbons, all of which have at least four aromatic rings. Among them is

■ A smoker's lung showing carcinoma. *(Courtesy of Anatomical Collections, National Museum of Health and Medicine, Armed Forces Institute of Pathology, Washington, DC)*

Benzo[a]pyrene

Benzo[a]pyrene is present in automobile exhausts and in cigarette smoke. If you smoke, you greatly increase your chances of getting cancer, especially mouth and/or lung cancers.

14.4 Reactions of Benzene and Its Derivatives

By far the most characteristic reaction of aromatic compounds is substitution at a ring carbon, which is given the name **aromatic substitution.** Some groups that can be introduced directly on the ring are the halogens, the nitro ($-NO_2$) group, and the sulfonic acid ($-SO_3H$) group.

Halogenation

As noted in Section 14.2, chlorine and bromine do not react with benzene, in contrast to their instantaneous reaction with cyclohexene and other alkenes (Section 12.6). In the presence of an iron catalyst, however, chlorine reacts rapidly with benzene to give chlorobenzene and HCl.

Chlorobenzene

Treatment of benzene with bromine in the presence of $FeCl_3$ results in formation of bromobenzene and HBr.

Nitration

When benzene or one of its derivatives is heated with a mixture of concentrated nitric and sulfuric acids, one of the hydrogen atoms bonded to the ring is replaced by a nitro group.

Nitrobenzene

A particular value of nitration is that the resulting $-NO_2$ group can be reduced to a primary amino group, $-NH_2$, by catalytic reduction using hydrogen in the presence of transition-metal catalyst. Notice that, in this example, neither the benzene ring nor the carboxyl group is affected by these experimental conditions.

4-Nitrobenzoic acid
(*p*-Nitrobenzoic acid)

4-Aminobenzoic acid
(*p*-Aminobenzoic acid, PABA)

p-Aminobenzoic acid is required by bacteria for the synthesis of folic acid (Section 29.5), which is in turn required for the synthesis of the heterocyclic aromatic amine bases of nucleic acids (Section 24.2). Whereas bacteria can synthesize folic acid from *p*-aminobenzoic acid, folic acid is a vitamin for humans and must be obtained in the diet.

Iodide Ion and Goiter

One hundred years ago, goiter, an enlargement of the thyroid gland due to iodine deficiency, was common in the central United States and central Canada. The disease results from a deficiency of thyroxine, a hormone synthesized in the thyroid gland. Young mammals require this hormone for normal growth and development. Its lack during fetal development results in mental retardation. Low levels of thyroxine in adults result in hypothyroidism, commonly called goiter, the symptoms of which are lethargy, obesity, and dry skin.

Synthesis of thyroxine begins by reaction of iodide ion with the protein thyroglobulin. In particular, iodide reacts with the phenolic ring of the amino acid tyrosine, one of the 20 amino-acid building blocks of proteins (Section 21.2). In subsequent steps, two iodinated molecules interact by forming an ether bond between their aromatic rings. In the final step, the protein part is cleaved, releasing thyroxine.

■ A patient suffering from goiter.
(Phototake)

Iodine is an element that comes primarily from the sea. Rich sources of it, therefore, are fish and other seafoods. The iodine in our diets that doesn't come from the sea most commonly comes from food additives. Most of the iodide ion in the North American diet comes from table salt fortified with sodium iodide, commonly referred to as iodized salt. Another source is dairy products, which accumulate iodide because of the iodine-containing additives used in cattle feeds and the iodine-containing disinfectants used on milking machines and milk storage tanks.

Sulfonation

Heating an aromatic compound with concentrated sulfuric acid gives a sulfonic acid, all of which are strong acids comparable in strength to sulfuric acid.

$$\text{C}_6\text{H}_5\text{-H} + \text{H}_2\text{SO}_4 \longrightarrow \text{C}_6\text{H}_5\text{-SO}_3\text{H} + \text{H}_2\text{O}$$

Benzenesulfonic acid

A major use of sulfonation is in the preparation of synthetic detergents, an important example of which is sodium 4-dodecybenzenesulfonate. To prepare this type of detergent, a linear alkylbenzene is treated with concentrated sulfuric acid to give an alkylbenzenesulfonic acid. The sulfonic acid is then neutralized with sodium hydroxide.

Dodecylbenzene → Sodium 4-dodecylbenzenesulfonate (an anionic detergent)

Alkylbenzenesulfonate detergents were introduced in the late 1950s, and today they command close to 90 percent of the market once held by natural soaps. We discuss the chemistry and cleansing action of soaps and detergents in Box 20B.

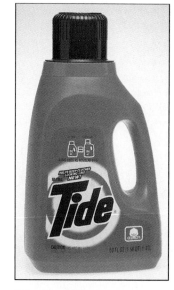

(Charles D. Winters)

14.5 Phenols

Structure and Nomenclature

The functional group of a **phenol** is a hydroxyl group bonded to a benzene ring. Substituted phenols are named either as derivatives of phenol or by common names.

Phenol in crystal form. *(Charles D. Winters)*

Phenol 3-Methylphenol (*m*-Cresol) 1,2-Benzenediol (Catechol) 1,3-Benzenediol (Resorcinol) 1,4-Benzenediol (Hydroquinone)

Phenols are widely distributed in nature. Phenol itself and the isomeric cresols (*o*-, *m*-, and *p*-cresol) are found in coal tar. Thymol and vanillin are important constituents of thyme and vanilla beans, respectively.

Phenol A compound that contains an —OH bonded to a benzene ring

2-Isopropyl-5-methylphenol (Thymol)

4-Hydroxy-3-methoxy-benzaldehyde (Vanillin)

Thymol is a constituent of garden thyme, *Thymus vulgaris*. *(© Connie Toops)*

Plant that yields vanilla beans. *(© John D. Cunningham/Visuals Unlimited)*

Capsaicin. For Those Who Like It Hot

Capsaicin, the pungent principle from the fruit of various species of peppers (*Capsicum* and *Solanum*), was isolated in 1876, and its structure was determined in 1919. Capsaicin contains both a phenol and a phenol ether.

Red chillies being dried.
(Chuck Pefley/Tony Stone Images)

Capsaicin
(from various types of peppers)

The inflammatory properties of capsaicin are well known; as little as one drop in 5 L of water can be detected by the human tongue. We all know of the burning sensation in the mouth and sudden tearing in the eyes caused by a good dose of hot chili peppers. Capsaicin-containing extracts from these flaming foods are also used in sprays to ward off dogs or other animals that might nip at your heels while you are running or cycling.

Ironically, capsaicin is able to cause pain and relieve it as well. Currently, two capsaicin-containing creams, Mioton and Zostrix, are prescribed to treat the burning pain associated with post-herpetic neuralgia, a complication of the disease known as shingles. They are also prescribed for diabetics to relieve persistent foot and leg pain.

Poison ivy. *(Charles D. Winters)*

Cloves. *(Charles D. Winters)*

Phenol, or carbolic acid as it was once called, is a low-melting solid only slightly soluble in water. In sufficiently high concentrations, it is corrosive to all kinds of cells. In dilute solutions, it has some antiseptic properties and was introduced into the practice of surgery by Joseph Lister in 1865. Phenol has been replaced by antiseptics that are both more powerful and have fewer undesirable side effects. Among these are hexylresorcinol, which is widely used in nonprescription preparations as a mild antiseptic and disinfectant. Eugenol, which can be isolated from the flower buds (cloves) of *Eugenia aromatica,* is used as a dental antiseptic and analgesic. Urushiol (The Merck Index, 12th ed., #10028) is the main component of the irritating oil of poison ivy.

Hexylresorcinol

Eugenol

Urushiol

Acidity of Phenols

Phenols and alcohols both contain a hydroxyl group, —OH. Phenols, however, are grouped as a separate class of compounds because their chemical properties are quite different from those of alcohols. One of the most

TABLE 14.1 Relative Acidities of 0.1 M Solutions of Ethanol, Phenol, and HCl

Acid Ionization Equation	$[H_3O^+]$	pH
$CH_3CH_2OH + H_2O \rightleftharpoons CH_3CH_2O^- + H_3O^+$	1×10^{-7}	7.0
$C_6H_5OH + H_2O \rightleftharpoons C_6H_5O^- + H_3O^+$	3.3×10^{-6}	5.4
$HCl + H_2O \rightleftharpoons Cl^- + H_3O^+$	0.1	1.0

important of these differences is that phenols are significantly more acidic than are alcohols, as Table 14.1 shows.

In aqueous solution, alcohols are neutral substances, and the hydrogen ion concentration of 0.1 M ethanol is the same as that of pure water. A 0.1 M solution of phenol is slightly acidic and has a pH of 5.4. By contrast, 0.1 M HCl, a strong acid (completely ionized in aqueous solution), has a pH of 1.0.

Acid-Base Reactions of Phenols

Because they are weak acids, phenols react with strong bases such as NaOH to form water-soluble salts.

Phenol	Sodium	Sodium phenoxide	Water
pK_a 9.95	hydroxide	(weaker base)	pK_a 15.7
(stronger acid)	(stronger base)		(weaker acid)

Most phenols do not react with weak bases such as sodium bicarbonate; they do not dissolve in aqueous sodium bicarbonate.

BOX 14E

Deodorants

Body odor is caused by the action of certain skin bacteria on natural secretions that are part of perspiration. The secretions have no odor, but skin bacteria react with them to produce compounds with unpleasant odors. Throughout most of history, people dealt with these odors by using perfumes and/or soap and water; however, they also simply chose to smell bad. Today in the United States and other developed countries, the use of personal deodorants is widespread. These deodorants act by killing the bacteria that react with skin secretions. The most common antibacterial agent in deodorants today is triclosan (The Merck Index, 12th ed., #9790), which is both an antiseptic and a disinfectant. A typical solid deodorant contains less than 1 percent triclosan, the rest being inactive ingredients (that is, they do not kill bacteria), ethanol or 1,2-propanediol, and water. Often an antiperspirant and a fragrance are added as well.

Triclosan

SUMMARY

Benzene and its alkyl derivatives are classified as **aromatic hydrocarbons** or **arenes.** The first structure for benzene was proposed by Friedrich August Kekulé in 1872. The theory of **resonance,** developed by Linus Pauling in the 1930s, provided the first adequate structure for benzene.

Aromatic compounds are named by the IUPAC system. The C_6H_5— group is named **phenyl.** Two substituents on a benzene ring may be located by either numbering the atoms of the ring or by using the locators **ortho (o), meta (m),** and **para(p).**

Polynuclear aromatic hydrocarbons contain two or more benzene rings, each sharing two carbon atoms with another ring.

A characteristic reaction of aromatic compounds is **aromatic substitution** in which another atom or group of atoms is substituted for a hydrogen atom of the aromatic ring.

The functional group of a **phenol** is an —OH group bonded to a benzene ring. Phenol and its derivatives are weak acids, pK_a approximately 10.0, but are considerably stronger acids than water and alcohols, $pK_a = 16-18$.

KEY TERMS

Ar— (Section 14.1)
Arene (Section 14.1)
Aromatic compound (Section 14.1)
Aromatic sextet (Section 14.2)
Aromatic substitution (Section 14.4)
Aryl group (Section 14.1)
Carcinogen (Box 14B)

Contributing structures
  (Section 14.2)
Kekulé structure (Section 14.2)
Meta (m-) (Section 14.3)
Ortho (o-) (Section 14.3)
Para (p-) (Sections 14.3)

Phenol (Section 14.5)
Phenyl group (Section 14.3)
Polynuclear aromatic hydrocarbons
  (PAHs) (Section 14.3)
Resonance hybrid (Section 14.2)

KEY REACTIONS

1. Halogenation (Section 14.4)
   Treatment of an aromatic compound with Cl_2 or Br_2 in the presence of an $FeCl_3$ catalyst substitutes a halogen for an H.

2. Nitration (Section 14.4)
   Treatment of an aromatic compound with a mixture of concentrated nitric and sulfuric acids substitutes a nitro group for an H.

3. Reduction of a Nitro Group to a Primary Amino Group (Section 14.4)
   Treatment of a nitro group with hydrogen in the presence of a transition-metal catalyst reduces it to a primary amino group.

4. Sulfonation (Section 14.4)
   Treatment of an aromatic compound with concentrated sulfuric acid substitutes a sulfonic acid group for an H.

5. Reaction of Phenols with Strong Bases (Section 14.5)
   Phenols are weak acids and react with strong bases to form water-soluble salts.

CONCEPTUAL PROBLEMS

14.A Can an aromatic compound be a saturated compound?

14.B Account for the fact that the six-membered ring in benzene is planar, but the six-membered ring in cyclohexane is not.

14.C One analogy often used to explain the concept of a resonance hybrid is to relate a rhinoceros to a unicorn and a dragon. Explain the reasoning in this analogy and how it might relate to a resonance hybrid.

14.D We say that naphthalene, anthracene, phenanthrene, and benzo[a]pyrene are polynuclear aromatic hydrocarbons. In this context, what does polynuclear mean? What does aromatic mean? What does hydrocarbon mean?

PROBLEMS

Difficult problems are designated by an asterisk.

Aromatic Compounds

14.2 Define aromatic compound.

14.3 What is the difference in structure between a saturated and an unsaturated compound?

14.4 Write a structural formula and name for the simplest (a) alkane, (b) alkene, (c) alkyne, and (d) aromatic hydrocarbon.

14.5 Do aromatic rings have double bonds? Are they unsaturated? Explain.

14.6 The compound 1,4-dichlorobenzene (*p*-dichlorobenzene) has a rigid geometry that does not allow free rotation. Yet there are no cis-trans isomers for this structure. Explain why not.

Nomenclature and Structural Formulas

14.7 Name these compounds.

(c) $C_6H_5CH_2CH_2CH_2OH$

14.8 Draw structural formulas for these compounds.
(a) 1-Bromo-2-chloro-4-ethylbenzene
(b) 4-Bromo-1,2-dimethylbenzene
(c) 2,4,6-Trinitrotoluene
(d) 4-Phenyl-2-pentanol
(e) *p*-Cresol
(f) 2,4-Dichlorophenol

Acidity of Phenols

14.9 Both phenol and cyclohexanol are only slightly soluble in water. Account for the fact that phenol dissolves in aqueous sodium hydroxide but cyclohexanol does not.

Reactions of Aromatic Compounds

14.10 Three possible products can be formed in the chlorination of bromobenzene. Draw a structural formula for each.

14.11 What reagents and/or catalysts are necessary to carry out each conversion? Each conversion requires only one step.

14.12 What reagents and/or catalysts are necessary to carry out each conversion? Each conversion requires two steps.

(b)

14.13 Aromatic substitution can be done on naphthalene. When naphthalene is sulfonated with concentrated H_2SO_4, two (and only two) different sulfonic acids are formed. Draw structural formulas for each.

Boxes

14.14 (Box 14A) From what parts of its common name are the letters DDT derived?

14.15 (Box 14A) What are the advantages and disadvantages of DDT as an insecticide?

14.16 (Box 14A) Would you expect DDT to be soluble or insoluble in water? Explain.

14.17 (Box 14A) One of the degradation products of DDT is dichlorodiphenyldichloroethylene, abbreviated DDE, which is formed by loss of HCl for DDT. It is DDE that inhibits the enzyme responsible for the incorporation of calcium ion into bird egg shells. Draw a structural formula for DDE.

14.18 (Box 14A) What is meant by the term biodegradable?

14.19 (Box 14B) What is a carcinogen? What kind of carcinogens are found in cigarette smoke?

14.20 (Box 14C) In the absence of iodine in the diet, goiter develops. Explain why goiter is a regional disease.

14.21 (Box 14D) From what types of plants is capsaicin isolated?

14.22 (Box 14D) Find the ether group and the phenol group in capsaicin.

14.23 (Box 14D) In what ways is capsaicin used in medicine?

14.24 (Box 14E) What causes body odor?

14.25 (Box 14E) Name the two oxygen-containing functional groups in triclosan.

14.26 (Box 14E) Would you expect triclosan to be soluble or insoluble in water? Soluble or insoluble in 0.10 *M* sodium hydroxide? Explain.

Additional Problems

*14.27 The structure for naphthalene given in Section 14.3 is only one of three possible resonance structures. Draw the other two.

14.28 Draw structural formulas for these compounds.
(a) 1-Phenylcyclopropanol
(b) Styrene
(c) *m*-Bromophenol

(d) 4-Nitrobenzoic acid
(e) Isobutylbenzene
(f) *m*-Xylene

*14.29 2,6-Di-*tert*-butyl-4-methylphenol, more commonly known as butylated hydroxytoluene or BHT, is used as an antioxidant in foods to "retard spoilage." Explain the rationale behind the name butylated hydroxytoluene.

$(CH_3)_3C$ ⟶ OH ⟶ $C(CH_3)_3$ / CH_3

2,6-Di-*tert*-butyl-4-methylphenol

Food containing the preservative BHT. (*Adrienne Hart-Davis/Science Photo Library/Photo Researchers, Inc.*)

14.30 Write the structural formula for the product of each reaction.

(a) + HNO_3 $\xrightarrow{H_2SO_4}$

(b) CH_3 ... CH_3 + Br_2 $\xrightarrow{FeCl_3}$

(c) Br ... Br + H_2SO_4 ⟶

Chirality

15.1 Introduction

When you look in a mirror, you see a reflection, or **mirror image,** of yourself. Now suppose that your mirror image becomes a three-dimensional object. We could then ask, "What is the relationship between you and your mirror image?" By relationship we mean, "Can your mirror image be superposed on the original 'you' in such a way that every detail corresponds exactly to the original?" The answer is no, you and your mirror image are not superposable. If you have a ring on the little finger of your right hand, for example, your mirror image has the ring on the little finger of its left hand. If you part your hair on the right side, it will be parted on the left side in your mirror image. Simply stated, you and your mirror image are different objects. You cannot superpose one on the other.

In this chapter, we study the relationship between objects and their mirror images; that is, we study isomers called enantiomers and diastereomers. The relationship among these isomers and those we studied in Chapters 11 through 14 are summarized in Figure 15.1.

Mirror image The reflection of an object in a mirror

Above: Median cross section through a shell of a chambered nautilus found in deep waters of the southwest Pacific Ocean. The shell shows handedness; this cross section is a left-handed spiral. *(Lester Lefkowitz/Tony Stone Worldwide)*

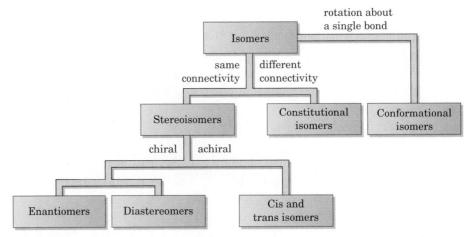

FIGURE 15.1 Relationships among isomers. In this chapter, we study enantiomers and diastereomers.

15.2 Stereocenters and Enantiomers

Objects that are not superposable on their mirror images are said to be **chiral** (pronounced ki-ral, rhymes with spiral; from the Greek: *cheir,* hand); that is, they show handedness. Chirality is encountered in three-dimensional objects of all sorts. Your left hand is chiral, and so is your right hand. A spiral binding on a notebook is chiral. A machine screw with a right-handed twist is chiral. A ship's propeller is chiral. As you examine the objects in the world around you, you will undoubtedly conclude that the vast majority of them are chiral as well.

An object and its mirror image are superposable if one of them can be oriented in space so that all its features (corners, edges, points, designs, etc.) correspond exactly to those on the other member of the pair. If this can be done, the object and its mirror image are identical; the object is said

The threads of a drill or screw twist along the axis of a helix, and some plants climb by sending out tendrils that twist helically. *(Charles D. Winters)*

(a)

(b)

Mirror images. (a) Mirror images of two wood carvings. The mirror image cannot be superposed on the actual statue. The man's right arm is resting on the camera in the mirror image, but in the actual statue the man's left arm is resting on the camera. (b) Left- and right-handed sea shells. If you cup a right-handed shell in your right hand, with your thumb pointing from the narrow end to the wide end, the opening will be on the right. *(Charles D. Winters)*

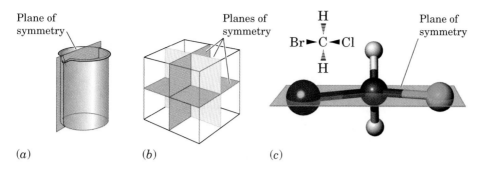

FIGURE 15.2 Planes of symmetry in a beaker, a cube, and CH_2BrCl. The beaker and CH_2BrCl have one plane of symmetry; a cube has several planes of symmetry.

to be **achiral.** Examples of achiral objects are an undecorated cup, an unmarked baseball bat, a regular tetrahedron, a cube, and a sphere.

Every achiral object possesses a **plane of symmetry.** The beaker shown in Figure 15.2 has a single plane of symmetry, as does a bromochloromethane molecule. A cube has several planes of symmetry.

The most common cause of chirality in organic molecules is a tetrahedral carbon atom with four different groups bonded to it. Such a carbon atom is called a **stereocenter.** We can illustrate a stereocenter by considering 2-hydroxypropanoic acid, more commonly named lactic acid. Figure 15.3 shows three-dimensional representations for this molecule and its mirror image. In these representations, all bond angles about the central carbon atom are approximately 109.5°, and the four bonds from it are directed toward the corners of a regular tetrahedron.

A model of lactic acid can be turned and rotated in any direction in space, but as long as bonds on one model are not broken and rearranged, only two of the four groups attached to its central carbon can be made to coincide with those on its mirror image. Because lactic acid and its mirror image are nonsuperposable, they are classified as **enantiomers.** Enantiomers, like gloves and socks, always occur in pairs.

An equimolar mixture of two enantiomers is called a **racemic mixture,** a term derived from the name *racemic acid* (Latin: *racemus,* a cluster of grapes). Racemic acid is the name originally given to an equimolar mixture of the enantiomers of tartaric acid that form as a by-product during the fermentation of grape juice.

Achiral An object that lacks chirality; an object that is superposable on its mirror image

Plane of symmetry An imaginary plane passing through an object, dividing it such that one half is the mirror image of the other half

Stereocenter A tetrahedral carbon atom that has four different groups bonded to it

Enantiomers Stereoisomers that are nonsuperposable mirror images; refers to a relationship between pairs of objects

Racemic mixture A mixture of equal amounts of two enantiomers

FIGURE 15.3 Three-dimensional representations of lactic acid and its mirror image.

EXAMPLE 15.1

Each molecule has one stereocenter. Draw three-dimensional representations for its enantiomers.

(a) $CH_3CHCH_2CH_3$ with OH above the second carbon

(b) a phenyl ring attached to $CHCH_3$ with NH_2 above

Solution

First identify the stereocenter, and then draw that carbon showing the tetrahedral orientation of its four bonds. One way to do this is with two bonds in the plane of the paper, a third bond toward you in front of the plane, and the fourth bond away from you behind the plane. Now place the four groups bonded to the stereocenter on these positions. How you do this is purely arbitrary; there are 12 equivalent ways to do it. To draw the enantiomers in (a), for example, place the —OH group in the vertical position in the plane and the —CH₃ to the left also in the plane. Now place the —H away from you and the —CH₂CH₃ group toward you; this gives the enantiomer of (a) on the left. Its enantiomer is on the right.

(a)

(b)

Problem 15.1

Each molecule has one stereocenter. Draw three-dimensional representations for its enantiomers.

(a) cyclopentane ring attached to $CHCH_3$ with $COOH$ above

(b) $CH_3CHCHCH_3$ with OH above and CH_3 below

15.3 Naming Stereocenters. The R,S System

Because a pair of enantiomers are two different compounds, each must have a different name. For example, 2-butanol exists as a pair of enantiomers shown in Example 15.1(a), and we must have some way of indicating which enantiomer has which configuration. To show configuration at a stereocenter, we use the **R,S system.**

The first step in assigning an R or S configuration to a stereocenter is to arrange the groups bonded to it in order of priority. Priority is based on the atomic numbers of the groups bonded to the stereocenter—the higher the atomic number, the higher the priority. If priority cannot be assigned on the basis of the atoms bonded directly to the stereocenter, look at the next set of atoms and continue until a priority can be assigned. Table 15.1 shows the priorities of the most common groups we deal with in organic

R,S system A set of rules for specifying configuration about a stereocenter

TABLE 15.1 R,S Priorities of Some Common Groups

Group	Reason for Priority; First Point of Difference (Atomic Numbers)
—OH	**oxygen** (8)
—NH$_2$	**nitrogen** (7)
$\overset{\text{O}}{\overset{\|}{—C—OH}}$	carbon to oxygen, oxygen, then **oxygen** (6 → 8,8,8)
$\overset{\text{O}}{\overset{\|}{—C—NH_2}}$	carbon to oxygen, oxygen, then **nitrogen** (6 → 8,8,7)
$\overset{\text{O}}{\overset{\|}{—C—H}}$	carbon to oxygen, oxygen, then **hydrogen** (6 → 8,8,1)
—CH$_2$—OH	carbon to **oxygen** (6 → 8)
—CH$_2$CH$_3$	carbon to **carbon** (6 → 6)
—CH$_2$—H	carbon to **hydrogen** (6 → 1)
—H	**hydrogen** (1)

Increasing priority ↑

chemistry and biochemistry. For the purposes of assigning priorities, the carbon of a C=O group is considered to be bonded to two oxygens by single bonds.

EXAMPLE 15.2

Assign priorities to the groups in each set.
(a) —CH$_2$OH   and   —CH$_2$CH$_2$OH
(b) —CH$_2$CH$_2$OH   and   —CH$_2$NH$_2$

Solution
(a) The first point of difference is O (atomic number 8) of the —OH group compared to C (atomic number 6) of the —CH$_2$OH group.

first point of
difference

—CH$_2$OH   —CH$_2$CH$_2$OH

Higher priority   Lower priority

(b) The first point of difference is C (atomic number 6) of the —CH$_2$OH group compared to N (atomic number 7) of the —NH$_2$ group.

first point of
difference

—CH$_2$CH$_2$OH   —CH$_2$NH$_2$

Lower priority   **Higher priority**

Problem 15.2 ■

Assign priorities to the groups in each set.
(a) —CH$_2$OH   and   —CH$_2$CH$_2$COOH
(b) —CH$_2$NH$_2$ and   —CH$_2$CH$_2$COOH ■

To assign an R or S configuration to a stereocenter:

1. Assign a priority from 1 (highest) to 4 (lowest) to each group bonded to the stereocenter.

2. Orient the molecule in space so that the group of lowest priority (4) is directed away from you as would, for instance, the steering column of a car. The three groups of higher priority (1–3) then project toward you as would the spokes of a steering wheel.

3. Read the three groups projecting toward you in order from highest (1) to lowest (3) priority.

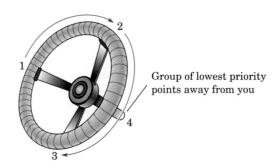

Group of lowest priority points away from you

4. If reading the groups 1-2-3 proceeds in a clockwise direction, the configuration is designated as **R** (Latin: *rectus*, straight); if reading the groups 1-2-3 proceeds in a counterclockwise direction, the configuration is **S** (Latin: *sinister*, left). You can also visualize this as follows: turning the steering wheel to the right equals R and turning it to the left equals S.

R From the Latin *rectus*, meaning straight, correct; used in the R,S system to show that when the group of lowest priority is away from you, the order of priority of groups on a stereocenter is clockwise

S From the Latin *sinister*, meaning left; used in the R,S system to show that when the group of lowest priority is away from you, the order of priority of groups on a stereocenter is counterclockwise

EXAMPLE 15.3

Assign an R or S configuration to each stereocenter.

(a) OH
H₃C—C—H
CH₂CH₃
2-Butanol

(b) H₂N H
H₃C—C—COOH
Alanine

Solution
View each molecule through the stereocenter and along the bond from the stereocenter to the group of lowest priority.
(a) The order of priority is —OH > —CH₂CH₃ > —CH₃ > —H. Therefore, view the molecule along the C—H bond with the H pointing away from you. Reading the other three groups in the order 1-2-3 occurs in the clockwise direction. Therefore, the configuration is R, and this enantiomer is (R)-2-butanol.

(R)-2-Butanol

With —H, the group of lowest priority pointing away from you, this is what you see.

(b) The order of priority is —NH$_2$ > —COOH > —CH$_3$ > —H. View the molecule along the C—H bond with H pointing away from you. Reading the groups in the order 1-2-3 occurs in the clockwise direction; therefore, the configuration is R, and this enantiomer is (R)-alanine.

With —H, the group of lowest priority pointing away from you, this is what you see.

(R)-Alanine

Problem 15.3
Assign an R or S configuration to each stereocenter.

(a)

$$\underset{\underset{H_3C}{\overset{}{\underset{}{|}}}{\overset{CH_3CH_2}{\overset{}{\underset{H}{\diagdown}}}}\overset{\overset{O}{\|}}{C-COH}$$

2-Methylbutanoic acid

(b)

$$\underset{\underset{}{\overset{HO}{\overset{H_{\prime\prime\prime}}{\diagup}}}}{\overset{\overset{\overset{O}{\|}}{C-H}}{C}}\diagdown CH_2OH$$

Glyceraldehyde ∎

The R,S system can be used to specify the configuration of any stereocenter on any molecule. It is not, however, the only system used for this purpose. As we shall see presently, the D,L system is also used, primarily to specify the configuration of carbohydrates (Chapter 19) and amino acids (Chapter 21). In this system, the enantiomers of glyceraldehyde, for example, are named D-glyceraldehyde and L-glyceraldehyde.

15.4 Molecules with Two or More Stereocenters

For a molecule with n stereocenters, the maximum number of stereoisomers possible is 2^n. We have already verified that, for a molecule with one stereocenter, $2^1 = 2$ stereoisomers (one pair of enantiomers) are possible. For a molecule with two stereocenters, $2^2 = 4$ stereoisomers are possible; for a molecule with three stereocenters, $2^3 = 8$ stereoisomers are possible, and so forth.

Acyclic Molecules

We begin our study of acyclic molecules by considering 2,3,4-trihydroxybutanal, a molecule with two stereocenters.

$$\underset{\underset{OH}{|}}{HOCH_2}-\underset{\underset{OH}{|}}{CH}-CH-\overset{\overset{O}{\|}}{C}-H$$

2,3,4-Trihydroxybutanal

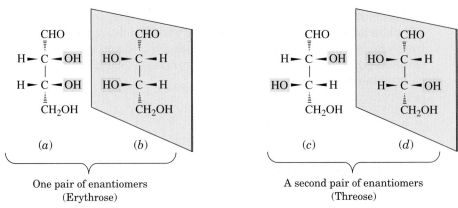

FIGURE 15.4 The four stereoisomers of 2,3,4-trihydroxybutanal, a compound with two stereocenters.

The maximum number of stereoisomers possible for this molecule is $2^2 = 4$, each of which is drawn in Figure 15.4.

Stereoisomers (a) and (b) are nonsuperposable mirror images and are, therefore, a pair of enantiomers. Stereoisomers (c) and (d) are also nonsuperposable mirror images and are a second pair of enantiomers. We describe the four stereoisomers of 2,3,4-trihydroxybutanal by saying that they consist of two pairs of enantiomers. Enantiomers (a) and (b) are named erythrose. Enantiomers (c) and (d) are named threose. Erythrose and threose belong to the class of compounds called carbohydrates, which we discuss in Chapter 19. Erythrose is synthesized in erythrocytes (red blood cells), hence the derivation of its name.

We have specified the relationship between (a) and (b) and that between (c) and (d). What is the relationship between (a) and (c), between (a) and (d), between (b) and (c), and between (b) and (d)? The answer is that they are **diastereomers;** they are stereoisomers that are not mirror images.

> **Diastereomers** Stereoisomers that are not mirror images of each other

EXAMPLE 15.4

1,2,3-Butanetriol has two stereocenters (carbons 2 and 3); therefore, $2^2 = 4$ stereoisomers are possible for it. Following are three-dimensional representations for each.
(a) Which compounds are pairs of enantiomers?
(b) Which compounds are diastereomers?

$$
\begin{array}{cccc}
\text{CH}_2\text{OH} & \text{CH}_2\text{OH} & \text{CH}_2\text{OH} & \text{CH}_2\text{OH} \\
\text{H}{-}\text{C}{-}\text{OH} & \text{H}{-}\text{C}{-}\text{OH} & \text{HO}{-}\text{C}{-}\text{H} & \text{HO}{-}\text{C}{-}\text{H} \\
\text{HO}{-}\text{C}{-}\text{H} & \text{H}{-}\text{C}{-}\text{OH} & \text{HO}{-}\text{C}{-}\text{H} & \text{H}{-}\text{C}{-}\text{OH} \\
\text{CH}_3 & \text{CH}_3 & \text{CH}_3 & \text{CH}_3 \\
(1) & (2) & (3) & (4)
\end{array}
$$

Solution
(a) Enantiomers are stereoisomers that are nonsuperposable mirror images. Compounds (1) and (4) are one pair of enantiomers, and compounds (2) and (3) are a second pair of enantiomers.
(b) Diastereomers are stereoisomers that are not mirror images. Compounds (1) and (2), (1) and (3), (2) and (4), and (3) and (4) are diastereomers.

Problem 15.4 ▬▬▬▬▬▬

3-Amino-2-butanol has two stereocenters (carbons 2 and 3); thus, $2^2 = 4$
stereoisomers are possible for it.
(a) Which compounds are pairs of enantiomers?
(b) Which compounds are diastereomers?

CH₃	CH₃	CH₃	CH₃

$$\begin{array}{cccc}
\text{CH}_3 & \text{CH}_3 & \text{CH}_3 & \text{CH}_3 \\
\text{H}-\text{C}-\text{OH} & \text{H}-\text{C}-\text{OH} & \text{HO}-\text{C}-\text{H} & \text{HO}-\text{C}-\text{H} \\
\text{H}_2\text{N}-\text{C}-\text{H} & \text{H}-\text{C}-\text{NH}_2 & \text{H}-\text{C}-\text{NH}_2 & \text{H}_2\text{N}-\text{C}-\text{H} \\
\text{CH}_3 & \text{CH}_3 & \text{CH}_3 & \text{CH}_3 \\
(1) & (2) & (3) & (4)
\end{array}$$
■

Cyclic Molecules

In this section, we concentrate on derivatives of cyclopentane and cyclohexane containing two stereocenters. We can analyze chirality in these cyclic
compounds in the same way we analyzed it in acyclic compounds. For a molecule with n stereocenters, the maximum number of stereoisomers is 2^n.

Let us start with 2-methylcyclopentanol, a compound with two stereocenters, here marked by asterisks.

OH

CH₃

2-Methylcyclopentanol

Using the 2^n rule, we predict a maximum of $2^2 = 4$ stereoisomers for this
compound. We begin with the cis and trans isomers, each of which is chiral.
The cis isomer exists as one pair of enantiomers, and the trans isomer
exists as a second pair of enantiomers. These are drawn here using a planar pentagon to represent the five-membered ring.

cis-2-Methylcyclopentanol   *trans*-2-Methylcyclopentanol
(a pair of enantiomers)   (a pair of enantiomers)

As an example of a disubstituted cyclohexane, let us consider 2-methyl-
cyclohexanol. This compound has two stereocenters and can exist as $2^2 = 4$ stereoisomers.

OH

CH₃

2-Methylcyclohexanol

The cis isomer exists as one pair of enantiomers, and the trans isomer
exists as a second pair of enantiomers. These are drawn here, using a planar hexagon to represent the six-membered ring.

cis-2-Methylcyclohexanol
(one pair of enantiomers)

trans-2-Methylcyclohexanol
(a second pair of enantiomers)

EXAMPLE 15.5

How many stereoisomers exist for 3-methylcyclopentanol?

Solution
Carbons 1 and 3 of this compound are stereocenters, and $2^2 = 4$ stereoisomers are possible. The cis isomer exists as one pair of enantiomers; the trans isomer as a second pair of enantiomers.

cis-3-Methylcyclopentanol
(a pair of enantiomers)

trans-3-Methylcyclopentanol
(a pair of enantiomers)

Problem 15.5 ■
How many stereoisomers exist for 3-methylcyclohexanol? ■

EXAMPLE 15.6

Mark all stereocenters in each compound with an asterisk. How many stereoisomers are possible for each?

(a) [structure: cyclopentanone with CH₃]

(b) [structure: cyclohexane with CH₃, OH, CH₃]

(c) $CH_3-CH-CH-COH$ with OH, NH₂, O

Solution
Each stereocenter is marked with an asterisk, and under each compound is given the number of stereoisomers possible for it.

(a) [structure with * marked]

(b) [structure with * marked]

(c) $CH_3-\overset{*}{C}H-\overset{*}{C}H-COH$ with OH, NH₂, O

$2^1 = 2$ $2^3 = 8$ $2^2 = 4$

BOX 15A

Chiral Drugs

Some common drugs used in human medicine, for example, aspirin (Section 18.3), are achiral. Others, such as the penicillin and erythromycin classes of antibiotics and the drug captopril, are chiral and are sold as single enantiomers. Captopril (The Merck Index, 12th ed., #1817) is very effective for the treatment of high blood pressure and congestive heart failure. It is manufactured and sold as the (S,S)-stereoisomer.

A large number of chiral drugs, however, are sold as racemic mixtures. The popular analgesic ibuprofen (the active ingredient in Motrin, Advil, and many other non-aspirin analgesics) is an example. Only the S enantiomer of ibuprofen is biologically active. (S)-Ibuprofen reaches therapeutic concentrations in the human body in 12 minutes, whereas the racemic mixture takes 30 minutes. However, the inactive R enantiomer is not wasted. The body converts it to the active S enantiomer, but that takes time.

■ Ibuprofen (for example, Advil) is sold as a racemic mixture and naproxen (for example, Aleve) is sold as the S enantiomer. *(Charles D. Winters)*

Recently, the U.S. Food and Drug Administration established new guidelines for the testing and marketing of chiral drugs. After reviewing these guidelines, many drug companies have decided to develop only single enantiomers of new chiral drugs. In addition to regulatory pressure, there are patent considerations. If a company has a patent on a racemic mixture of a drug, a new patent can often be taken out on one of its enantiomers.

Captopril

(S)-Ibuprofen

Problem 15.6

Mark all stereocenters in each compound with an asterisk. How many stereoisomers are possible for each?

(a) HO, HO — CH$_2$CHCOOH with NH$_2$

(b) CH$_2$=CHCHCH$_2$CH$_3$ with OH

(c) cyclohexane with OH and NH$_2$

15.5 Optical Activity. How Chirality Is Detected in the Laboratory

As we have already established, the two members of a pair of enantiomers are different compounds, and we must expect, therefore, that they differ in some properties. One such property is their effect on the plane of polarized light. Each member of a pair of enantiomers rotates the plane of polarized light, and for this reason, each enantiomer is said to be **optically active.**

Optically active Showing that a compound rotates the plane of polarized light

FIGURE 15.5 Schematic diagram of a polarimeter with its sample tube containing a solution of an optically active compound. The analyzing filter has been turned clockwise by α degrees to restore the dark field.

To understand how optical activity is detected in the laboratory, we must first understand plane-polarized light and a polarimeter, the instrument used to detect optical activity.

Ordinary light consists of waves vibrating in all planes perpendicular to its direction of propagation. Certain materials, such as a Polaroid sheet (a plastic film like the ones used in polarized sunglasses), selectively transmit light waves vibrating only in parallel planes. Electromagnetic radiation vibrating in only parallel planes is said to be **plane polarized.**

A **polarimeter** consists of a light source emitting unpolarized light, a polarizing filter, an analyzing filter, and a sample tube (Figure 15.5). If the sample tube is empty, the intensity of light reaching the detector (in this case, your eye) is at its maximum when the polarizing axes of the two filters are parallel to each other. If the analyzing filter is turned either clockwise or counterclockwise, less light is transmitted. When the axis of the analyzing filter is at right angles to the axis of the polarizing filter, the field of view is dark.

When a solution of an optically active compound is placed in the sample tube, it rotates the plane of the polarized light. If it rotates the plane clockwise, we say it is **dextrorotatory;** if it rotates the plane counterclockwise, we say it is **levorotatory.** Each member of a pair of enantiomers rotates the plane of polarized light the same number of degrees, but the directions are opposite. If one enantiomer is dextrorotatory, the other is levorotatory.

The number of degrees by which an optically active compound rotates the plane of polarized light is called its **specific rotation** and is given the symbol $[\alpha]$. A dextrorotatory compound is indicated by a plus sign in parentheses, $(+)$, and a levorotatory compound is indicated by a minus sign in parentheses, $(-)$. It is common to report the temperature (in °C) at which the measurement is made and the wavelength of light used. The most common wavelength of light used in polarimetry is the sodium D line, the same wavelength responsible for the yellow color of sodium-vapor lamps. Following are specific rotations for the enantiomers of lactic acid measured at 21°C and using the D line of a sodium-vapor lamp as the light source.

Plane-polarized light Light vibrating in only parallel planes

Polarimeter An instrument for measuring the ability of a compound to rotate the plane of polarized light

Dextrorotatory Clockwise (to the right) rotation of the plane of polarized light in a polarimeter

Levorotatory Counterclockwise (to the left) rotation of the plane of polarized light in a polarimeter

The $(+)$ enantiomer of lactic acid is produced by muscle tissue in humans. The $(-)$ enantiomer is found in sour milk.

COOH
|
C•••••H
H₃C   OH
(S)-(+)-Lactic acid
$[\alpha]_D^{21}$ +2.6

COOH
|
H•••••C
HO   CH₃
(R)-(−)-Lactic acid
$[\alpha]_D^{21}$ −2.6

You should note that there is no relationship between the direction of rotation of plane polarized light for an enantiomer, that is, whether it is dextrorotatory or levorotatory, and its R or S configuration. The S enantiomer of lactic acid, for example, is dextrorotatory and is designated (S)-(+)-lactic acid. Conversion of the carboxyl group to its sodium salt in no way affects the configuration of the stereocenter, and yet the specific rotation changes from dextrorotatory to levorotatory. Each compound has the same configuration at its stereocenter; yet, one is dextrorotatory, and the other is levorotatory.

$$\underset{\substack{\text{(S)-(+)-Lactic acid}\\ [\alpha]_D^{21}\ +2.6}}{\overset{\text{COOH}}{\underset{\text{H}_3\text{C}}{\text{C}}\text{H} \atop \text{OH}}} + \text{NaOH} \longrightarrow \underset{\substack{\text{(S)-(-)-Sodium lactate}\\ [\alpha]_D^{21}\ -13.5}}{\overset{\text{COO}^-\ \text{Na}^+}{\underset{\text{H}_3\text{C}}{\text{C}}\text{H} \atop \text{OH}}} + \text{H}_2\text{O}$$

15.6 The Significance of Chirality in the Biological World

Except for inorganic salts and a relatively few low-molecular-weight organic substances, the molecules in living systems, both plant and animal, are chiral. Although these molecules can exist as a number of stereoisomers, almost invariably only one stereoisomer is found in nature. Of course, instances do occur in which more than one stereoisomer is found, but these rarely exist together in the same biological system.

Chirality in Biomolecules

Perhaps the most conspicuous examples of chirality among biological molecules are the enzymes, all of which have many stereocenters. An illustration is chymotrypsin, an enzyme in the intestines of animals that catalyzes the digestion of proteins (Chapter 22). Chymotrypsin has 251 stereocenters. The maximum number of stereoisomers possible is 2^{251}, a staggeringly large number, almost beyond comprehension. Fortunately, nature does not squander its precious energy and resources unnecessarily; only one of these stereoisomers is produced and used by any given organism. Because enzymes are chiral substances, most either produce or react only with substances that match their stereochemical requirements.

The horns on this African gazelle show chirality. *(Courtesy of William Brown)*

How an Enzyme Distinguishes Between a Molecule and Its Enantiomer

An enzyme catalyzes a biological reaction of molecules by first positioning them at a **binding site** on its surface. An enzyme with specific binding sites for three of the four groups on a stereocenter can distinguish between a molecule and its enantiomer or one of its diastereomers. Assume, for example, that an enzyme involved in catalyzing a reaction of glyceraldehyde has three binding sites—one specific for —H, a second specific for —OH, and a third specific for —CHO. Assume further that the three sites are arranged on the enzyme surface as shown in Figure 15.6. The enzyme can distinguish (R)-glyceraldehyde (the natural or biologically active form) from its enantiomer because the natural enantiomer can be absorbed with three groups interacting with their appropriate binding sites; for the S enantiomer, at best only two groups can interact with these binding sites.

(R)-Glyceraldehyde
fits the three binding
sites on surface.

(S)-Glyceraldehyde
fits only two of the
three binding sites.

FIGURE 15.6 A schematic diagram of an enzyme surface capable of interacting with (R)-glyceraldehyde at three binding sites, but with (S)-glyceraldehyde at only two of these sites.

Because interactions between molecules in living systems take place in a chiral environment, it should be no surprise that a molecule and its enantiomer or one of its diastereomers elicit different physiological responses.

SUMMARY

A **mirror image** is the reflection of an object in a mirror. **Enantiomers** are a pair of stereoisomers that are mirror images of each other. A **racemic mixture** is a mixture of equal amounts of two enantiomers.

An object that is not superposable on its mirror image is said to be **chiral**. An **achiral** object is one that lacks chirality; that is, it has a superposable mirror image. An object is achiral if it possesses a **plane of symmetry,** which is an imaginary plane passing through the object dividing it such that one half is the reflection of the other half.

A most common cause of chirality in organic molecules is a tetrahedral carbon atom with four different groups bonded to it. Such a carbon is called a **stereocenter.** The **configuration** at a stereocenter can be specified using the **R,S system.** For a molecule with n stereocenters, the maximum number of stereoisomers possible is 2^n.

Light that vibrates in only parallel planes is said to be **plane polarized. A polarimeter** is an instrument used to measure optical activity. A compound is said to be **optically active** if it rotates the plane of polarized light. If it rotates the plane clockwise, it is **dextrorotatory;** if it rotates the plane counterclockwise, it is **levorotatory.** Each member of a pair of enantiomers rotates the plane of polarized light an equal number of degrees, but the directions are opposite.

An enzyme catalyzes biological reactions of molecules by first positioning them at **binding sites** on its surface. An enzyme with binding sites specific for three of the four groups on a stereocenter can distinguish between a molecule and its enantiomer or one of its diastereomers.

KEY TERMS

CONCEPTUAL PROBLEMS

Difficult problems are designated by an asterisk.

*15.A Next time you have the opportunity to view a collection of sea shells that have a helical twist, study the chirality (handedness) of their twists. For each kind of shell, do you find an equal number of left-handed and right-handed twists or, for example, do they all have the same handedness? What about the handedness among various kinds of shells with helical twists?

*15.B Think about the helical coil of a telephone cord or the spiral binding on a notebook and suppose that you view the spiral from one end and find that it has a left-handed twist. If you view the same spiral from the other end, does it have a left-handed twist from that end as well, or does it have a right-handed twist?

*15.C The presence of a stereocenter is a sufficient condition for chirality, but it is not a necessary condition. Explain how it is possible to have chirality without a stereocenter.

PROBLEMS

Difficult problems are designated by an asterisk.

Chirality

15.7 Which of these objects are chiral (assume that there is no label or other identifying mark)? (a) Pair of scissors (b) Tennis ball (c) Paper clip (d) Beaker (e) The swirl created in water as it drains out of a sink or bathtub.

Enantiomers

15.8 Which compounds contain stereocenters? (a) 2-Chloropentane (b) 3-Chloropentane (c) 3-Chloro-1-butene (d) 1,2-Dichloropropane

15.9 Using only C, H, and O, write structural formulas for the lowest-molecular-weight chiral: (a) Alkane (b) Alcohol (c) Aldehyde (d) Ketone (e) Carboxylic acid

15.10 Draw mirror images for these molecules:

(a)
$$H_3C\overset{OH}{\underset{H}{\overset{|}{\underset{}{C}}}}COOH$$

(b)
$$H\overset{CHO}{-}\overset{|}{C}-OH$$
$$\overset{|}{CH_2OH}$$

(c)
$$H_2N\overset{COOH}{-}\overset{|}{C}-H$$
$$\overset{|}{CH_3}$$

(d)

(e)

(f)

15.11 Mark each stereocenter in these molecules with an asterisk. Note that not all contain stereocenters.

(a)
$$CH_3CCH{=}CH_2$$
with CH_3 above and OH below

(b)
$$HCOH$$
with $COOH$ above and CH_3 below

(c)
$$CH_3CHCHCOOH$$
with CH_3 above and NH_2 below

(d)
$$CH_3CCH_2CH_3$$
with O double bond above

(e)
$$HCOH$$
with CH_2OH above and CH_2OH below

(f)
$$CH_3CH_2CHCH{=}CH_2$$
with OH above

(g)
$$HOCCOOH$$
with CH_2COOH above and CH_2COOH below

Designation of Configuration. The R,S System

15.12 Assign priorities to the groups in each set.
(a) —H —CH$_3$ —OH —CH$_2$OH
(b) —CH$_3$ —H —COOH —NH$_2$
(c) —CH$_3$ —CH$_2$SH —NH$_2$ —COOH

15.13 Which molecules have R configurations?

(a) with CH$_3$ top, Br and CH$_2$OH bottom, H wedge

(b) with CH$_3$ top, HOCH$_2$ and Br bottom, H wedge

(c) with CH$_2$OH top, H$_3$C and Br bottom, H wedge

(d) with CH$_2$OH top, Br and CH$_3$ bottom, H wedge

Molecules with Two or More Stereocenters _____

15.14 For centuries, Chinese herbal medicine has used extracts of *Ephedra sinica* to treat asthma. The asthma-relieving component of this plant is ephedrine, a very potent dilator of the air passages of the lungs. The naturally occurring stereoisomer is levorotatory and has the following structure. Assign an R or S configuration to each stereocenter.

Ephedrine $[\alpha]_D^{21}$ −41

15.15 The specific rotation of naturally occurring ephedrine, shown in Problem 15.14 is −41. What is the specific rotation of its enantiomer?

*15.16 Label each stereocenter in these molecules with an asterisk. How many stereoisomers are possible for each molecule?

(a) $CH_3CHCHCOOH$
$\qquad$ | $\quad$ |
$\qquad$ HO $\;$ OH

(b) CH_2-COOH
$\qquad$ |
$\qquad$ CH$-$COOH
$\qquad$ |
$\quad$ HO$-$CH$-$COOH

(c) OH ... CH$_3$

(d) CH$_3$ ... H$_3$C $\quad$ CH$_3$

(e) CH$_3$ ... OH ... H$_3$C $\quad$ CH$_3$

(f) CH$_3$ ... O ... COOH

(g) H$_3$C $\quad$ CH$_3$ ... O ... OH ... OH

(h) CH$_3$ ... O

Boxes _____

15.17 (Box 15A) What does it mean to say that a drug is chiral? If a drug is chiral, will it be optically active; that is, will it rotate the plane of polarized light?

15.18 (Box 15A) What is a racemic mixture? Is a racemic mixture optically active; that is, will it rotate the plane of polarized light?

15.19 (Box 15A) How many stereocenters are there in ibuprofen? How many stereoisomers are possible for this molecule?

15.20 (Box 15A) Why does (s)-ibuprofen reach a therapeutic concentration in the human body more quickly than the racemic mixture of ibuprofen?

Additional Problems _____

15.21 Which of the eight alcohols of molecular formula $C_5H_{12}O$ are chiral? For the structural formulas of these eight alcohols, review your answer to Problem 13.45.

15.22 Which carboxylic acids of molecular formula $C_6H_{12}O_2$ are chiral? For the structural formulas of these eight carboxylic acids, review your answer to Problem 10.24.

15.23 Write the structural formula of an alcohol of molecular formula $C_6H_{14}O$ that contains two stereocenters.

*15.24 Label the eight stereocenters in cholesterol. How many stereoisomers are possible for a molecule of this structural formula?

Cholesterol

*15.25 Label the four stereocenters in amoxicillin. Which belongs to the family of semisynthetic penicillins?

Amoxicillin

*15.26 In answer to Problem 11.28, you drew a chair conformation of glucose in which all groups on the six-membered ring are equatorial.

(a) Identify the five stereocenters in this molecule?

(b) How many stereoisomers are possible?

(c) How many pairs of enantiomers are possible?

15.27 Consider a cyclohexane ring substituted with one hydroxyl group and one methyl group. Draw a structural formula for a compound of this composition that

(a) Does not show cis-trans isomerism and has no stereocenters

(b) Shows cis-trans isomerism but has no stereocenters

(c) Shows cis-trans isomerism and has two stereocenters

15.28 Which of these molecules are chiral? For each that is, tell how many stereocenters it has and how many stereoisomers are possible.

(a) 2,2,4-Trimethylpentane (Box 11B)

(b) 11-Tetradecenyl acetate (Box 12A)

(c) 11-*cis*-Retinal (Box 12B)

(d) Myrcene (Figure 12.1)

(e) Menthol (Figure 12.2)

(f) Lycopene (Problem 12.26)

(g) Nitroglycerin (Box 13A)

(h) Halothane (Box 13E)

(i) Benzene (Section 12.2)

(j) DDT (Box 14A)

(k) Thyroxine (Box 14C)

(l) Capsaicin (Box 14D)

15.29 Which of the molecules in Problem 15.28 show cis-trans isomerism?

15.30 Triamcinolone acetonide, the active ingredient in Azmacort Inhalation Aerosol, is a steroid used to treat bronchial asthma. Notice that this molecule is unusual in that it contains one fluorine atom.

Triamcinolone acetonide

(a) Label the eight stereocenters in this molecule.

(b) How many stereoisomers are possible for it? (Of this number, only one is the active ingredient in Azmacort.)

Chapter 16

The R— groups in these primary, secondary, and tertiary amines may be the same or different; they also may be aliphatic or aromatic.

Above: Opium poppies. *(Frank Orel/ Tony Stone Images)*

Amines

16.1 Introduction

Carbon, hydrogen, and oxygen are the three most common elements in organic compounds. Because of the wide distribution of amines in the biological world, nitrogen is the fourth most common element of organic compounds. Amino groups are found in thousands of biomolecules from neurotransmitters to vitamins, and from proteins to nucleic acids. The most important chemical property of amines is their basicity.

16.2 Structure and Classification

Amines are classified as **primary (1°)**, **secondary (2°)**, or **tertiary (3°)**, depending on the number of carbon groups bonded to nitrogen (Section 10.4).

$$R-NH_2 \qquad R-NH \qquad R-N-R$$
$$\quad\;\; \overset{\textstyle R}{|} \qquad\quad \overset{\textstyle R}{|}$$

A 1° amine   A 2° amine   A 3° amine

Amines are further classified as aliphatic or aromatic. In an **aliphatic amine,** all the carbons bonded to nitrogen are derived from alkyl groups; in an **aromatic amine,** one or more of the groups bonded to nitrogen are aryl groups.

> **Aliphatic amine** An amine in which nitrogen is bonded only to alkyl groups

> **Aromatic amine** An amine in which nitrogen is bonded to one or more aryl groups

Aniline
(a 1° aromatic amine)

N-Methylaniline
(a 2° aromatic amine)

Benzyldimethylamine
(a 3° aliphatic amine)

BOX 16A

Amphetamines (Pep Pills)

Amphetamine (*The Merck Index*, 12th ed., #623), methamphetamine (*The Merck Index*, 12th ed., #6015), and phentermine (*The Merck Index*, 12th ed., #7415), all synthetic amines, are powerful stimulants of the central nervous system. Like most other amines, these three are stored and administered as their salts. The sulfate salt of amphetamine is prescribed as Benzedrine; the hydrochloride salt of the s enantiomer of methamphetamine, as Methedrine; and the hydrochloride salt of phentermine, as Fastin.

Amphetamine   Methamphetamine

Phentermine

These three substances are members of a class of compounds all of which have similar physiological effects and are referred to by the general name amphetamines. Structurally they have in common a benzene ring with a three-carbon side chain and an amine nitrogen on the second carbon of the chain. Physiologically, they have in common that they reduce fatigue and diminish hunger by raising the glucose level of the blood. Because of these properties, amphetamines are widely prescribed to counter mild depression, reduce hyperactivity in children, and suppress appetites for people trying to lose weight. They are also used illegally to reduce fatigue and elevate mood. Abuse of amphetamines can have severe effects on both body and mind. They are addictive, concentrate in the brain and nervous system, and

■ Epinephrine (adrenalin) anaphylactic shock kit. *(© 1999 SIU Biomed Comm/Custom Medical Stock Photo)*

can lead to long periods of sleeplessness, loss of weight, and paranoia.

The action of amphetamines is similar to that of epinephrine, commonly called adrenalin (*The Merck Index*, 12th ed., #3656).

Epinephrine
(Adrenalin)

Epinephrine, secreted by the adrenal gland, is both a neurotransmitter in the brain (Section 23.5) and a hormone in the bloodstream (Section 23.5). When an individual feels excitement or fear, epinephrine helps to make glucose available to tissues that need it for immediate action, for example, leg muscles that must be used to run from danger.

> **Heterocyclic aliphatic amine**
> A heterocyclic amine in which nitrogen is bonded only to alkyl groups

An amine in which the nitrogen atom is part of a ring is classified as a **heterocyclic amine.** When the ring is saturated, the amine is classified as a **heterocyclic aliphatic amine.** When the nitrogen is part of an aromatic ring (Section 14.2), the amine is classified as a **heterocyclic aromatic amine.**

Pyrrolidine   Piperidine
(heterocyclic aliphatic amines)

Pyrrole   Pyridine
(heterocyclic aromatic amines;
the six electrons of the aromatic
sextet are shown in color)

> **Heterocyclic amine** An amine in which nitrogen is one of the atoms of a ring

> **Heterocyclic aromatic amine** An amine in which nitrogen is one of the atoms of an aromatic ring

EXAMPLE 16.1

How many hydrogens does piperidine have? How many does pyridine have? Write the molecular formula of each amine.

Solution
Remember that in line-angle drawings, hydrogen atoms are not shown. Piperidine has 11 hydrogens, and its molecular formula is $C_5H_{11}N$. Pyridine has 5 hydrogens; its molecular formula is C_5H_5N.

Problem 16.1 ■■■■
How many hydrogens does pyrrolidine have? How many does pyrrole have? Write the molecular formula of each amine. ■

16.3 Nomenclature

Systematic names for aliphatic amines are derived just as they are for alcohols. The suffix "-e" of the parent alkane is dropped and replaced by "-amine." The location of the amino group on the parent chain is indicated by a number.

$$CH_3CHCH_3$$
$$\overset{\displaystyle NH_2}{|}$$

2-Propanamine

(S)-1-Phenylethanamine

$$H_2N(CH_2)_6NH_2$$

1,6-Hexanediamine

BOX 16B

Alkaloids

Alkaloids are basic nitrogen-containing compounds found in the roots, bark, leaves, berries, or fruits of plants. In almost all alkaloids, the nitrogen atom is part of a ring. The name "alkaloid" was chosen because these compounds are alkali-like (alkali is an older term for a basic substance) and react with strong acids to give water-soluble salts. Thousands of alkaloids have been isolated from plant sources, and several of these are used in modern medicine.

When administered to animals, including humans, alkaloids have pronounced physiological effects. Whatever their individual effects, most alkaloids are toxic in large enough doses. For some, the toxic dose is very small!

Ingestion of coniine (*The Merck Index*, 12th ed., #2569), isolated from water hemlock (a member of the carrot family), can cause weakness, labored respiration, paralysis, and eventually death. It is the toxic substance in "poison hemlock" used in the death of Socrates. Water hemlock is easily confused with Queen Anne's lace, a type of wild carrot (Figure 16B.1), and this mistake has killed numerous people.

Nicotine (*The Merck Index*, 12th ed., #6611) occurs in the tobacco plant (Figure 16B.2). In small doses, it is an addictive stimulant. In larger doses, it causes depression, nausea, and vomiting. In still larger doses, it is a deadly poison. Solutions of nicotine in water are used as insecticides.

Cocaine (*The Merck Index*, 12th ed., #2516) is a central nervous system stimulant obtained from the leaves of the coca plant. In small doses, it decreases fatigue and gives a sense of well-being. Prolonged use leads to physical addiction and depression.

(S)-Coniine

(S)-Nicotine

Cocaine

■ **FIGURE 16B.1** Queen Anne's lace plant. *(© Michael P. Gadomski/Photo Researchers, Inc.)*

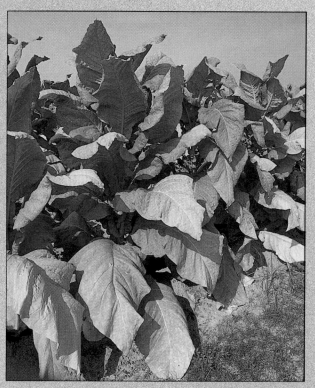

■ **FIGURE 16B.2** Tobacco plants. *(© Inga Spence/Visuals Unlimited)*

The Komodo dragon, which eats both live and dead prey, is not harmed by the toxic cadaverine found in this dead pig. *(Science VU/Visuals Unlimited)*

EXAMPLE 16.2

Write the IUPAC name for each amine. Be certain to specify the configuration of the stereocenter in (c).

(a) $CH_3(CH_2)_5NH_2$ (b) $H_2N(CH_2)_5NH_2$ (c)

$$C_6H_5CH_2$$
$$\underset{\underset{H_3C}{H}}{\overset{}{C}}-NH_2$$

Solution

(a) The parent alkane has six carbon atoms and is therefore hexane. The amino group is on carbon 1, giving the IUPAC name 1-hexanamine.

(b) The parent chain has five carbon atoms and is therefore pentane. There are amino groups on carbons 1 and 5, giving the IUPAC name 1,5-pentanediamine. The common name of this diamine is cadaverine (*The Merck Index,* 12th ed., #1645), which should give you a hint of where it occurs in nature and of its odor. Cadaverine is one of the end products of decaying flesh and is quite poisonous.

(c) The parent chain has three carbon atoms and is therefore propane. To have the lowest numbers possible, number from the end that places the phenyl group on carbon 1 and the amino group on carbon 2. The priority for determining R or S configuration are $NH_2 > CH_2C_6H_5 > CH_3 > H$. Its systematic name is (S)-1-phenyl-2-propanamine. Its common name is amphetamine.

Problem 16.2 ▬▬
Write a structural formula for each amine. (a) 2-Methyl-1-propanamine (b) Cyclohexanamine (c) 1,4-Butanediamine ■

IUPAC nomenclature retains the common name **aniline** for $C_6H_5NH_2$, the simplest aromatic amine. Its simple derivatives are named using the prefixes *o-*, *m-*, and *p-* or, alternatively, numbers to locate substituents. Several derivatives of aniline have common names that are still widely used. Among these is **toluidine** for a methyl-substituted aniline.

NH_2	NH_2	NH_2
	NO_2	CH_3
Aniline	4-Nitroaniline (*p*-Nitroaniline)	4-Methylaniline (*p*-Toluidine)

Unsymmetrical secondary and tertiary amines are commonly named as *N*-substituted primary amines. The largest group bonded to nitrogen is taken as the parent amine; the smaller group(s) bonded to nitrogen are named, and their location is indicated by the prefix *N* (indicating that they are bonded to nitrogen).

—NHCH₃

$$-N\overset{CH_3}{\underset{CH_3}{}}$$

N-Methylaniline

N,N-Dimethyl-cyclopentanamine

Common names for most aliphatic amines list the groups bonded to nitrogen in alphabetical order in one word ending in the suffix "-amine."

CH_3NH_2 $CH_3\underset{\underset{CH_3}{|}}{\overset{\overset{CH_3}{|}}{C}}NH_2$ $CH_3CH_2\overset{\overset{CH_3}{|}}{N}CH_2CH_3$

Methylamine   *tert*-Butylamine   Diethylmethylamine   Dicyclopentylamine

EXAMPLE 16.3

Write a structural formula for each amine. (a) Isopropylamine (b) Cyclohexylmethylamine (c) Triethylamine

Solution

(a) $(CH_3)_2CHNH_2$ (b) —NHCH$_3$ (c) $(CH_3CH_2)_3N$

Problem 16.3 ■

Write a structural formula for each amine. (a) Isobutylamine (b) Diphenylamine (c) Diisopropylamine ■

When four atoms or groups of atoms are bonded to a nitrogen atom, the compound is named as a salt of the corresponding amine. The ending "-amine" (or aniline or pyridine or the like) is replaced by "-ammonium" (or anilinium or pyridinium or the like) and the name of the anion (chloride, acetate, and so on) is added. Cetylpyridinium chloride is used as a topical antiseptic and disinfectant in some mouthwashes (*The Merck Index,* 12th ed., #20174).

$$H-\overset{\overset{H}{|}}{\underset{\underset{H}{|}}{N^+}}-H \quad Cl^-$$

$$H_3C-\overset{\overset{CH_3}{|}}{\underset{\underset{CH_3}{|}}{N^+}}-CH_3 \quad Cl^-$$

$$\overset{+}{N}-CH_2(CH_2)_{14}CH_3 \quad Cl^-$$

Ammonium chloride   Tetramethylammonium chloride   Hexadecylpyridinium chloride (Cetylpyridinium chloride)

Several over-the-counter mouthwashes contain an *N*-alkylpyridinium chloride as an antibacterial agent. *(Charles D. Winters)*

16.4 Physical Properties

Amines are polar compounds, and both primary and secondary amines form intermolecular hydrogen bonds with one another (Figure 16.1). Tertiary amines do not have a hydrogen bonded to N and, therefore, do not form hydrogen bonds with one another.

An $N-H\cdots N$ hydrogen bond is weaker than an $O-H\cdots O$ hydrogen bond because the difference in electronegativity between nitrogen and hydrogen $(3.0 - 2.1 = 0.9)$ is less than that between oxygen and hydrogen $(3.5 - 2.1 = 1.4)$. The effect of intermolecular hydrogen bonding between amine molecules versus that between alcohol molecules can be illustrated by comparing the boiling points of methylamine and methanol. Both compounds have polar molecules and interact in the pure liquid by hydrogen bonding.

FIGURE 16.1 Intermolecular association by hydrogen bonding between two molecules of a secondary amine.

Tranquilizers

In our modern, high-tech society, the pressures and conflicts of daily life are much greater than they were in earlier times, when everything moved at a slower pace. One way of coping is to use tranquilizers, drugs that provide relief of the symptoms of anxiety or tension.

The first modern tranquilizers were derivatives of a compound called benzodiazepine. The first of these, chlorodiazepoxide, better known as Librium (*The Merck Index*, 12th ed., #2132), was introduced in 1960 and was soon followed by more than two dozen related compounds. Of these, diazepam, better known as Valium (*The Merck Index*, 12th ed., #3042), became one of the most widely used.

Diazepam
(Valium)

Benzodiazepine

Chlorodiazepoxide
(Librium)

Librium, Valium, and other benzodiazepines are central nervous system sedative/hypnotics. As sedatives, they diminish activity and excitement and thereby have a calming effect. As hypnotics, they produce drowsiness and sleep. They affect pathways in the central nervous system (CNS) that are mediated by the inhibitory neurotransmitter 4-aminobutanoic acid (GABA, Section 23.4). By increasing the affinity of GABA for its neuroreceptors, the benzodiazepines decrease CNS activity mediated by GABA, with the accompanying reduction in anxiety and alertness.

Methanol has the higher boiling point because hydrogen bonding between its molecules is stronger than that between molecules of methylamine.

	CH_3NH_2	CH_3OH
MW (g/mol)	31.1	32.0
bp (°C)	−6.3	65.0

All classes of amines form hydrogen bonds with water and are more soluble in water than are hydrocarbons of comparable molecular weight. Most low-molecular-weight amines are completely soluble in water (Table 16.1), but higher molecular-weight amines are only moderately soluble in water or are insoluble.

16.5 Basicity

Like ammonia, amines are weak bases, and aqueous solutions of amines are basic. The following acid-base reaction between an amine and water is written using curved arrows to emphasize that, in this proton-transfer reaction, the unshared pair of electrons on nitrogen forms a new covalent bond with hydrogen and displaces hydroxide ion.

TABLE 16.1 Physical Properties of Selected Amines

Name	Structural Formula	mp (°C)	bp (°C)	Solubility in Water
ammonia	NH_3	−78	−33	very soluble
Primary Amines				
methylamine	CH_3NH_2	−95	−6	very soluble
ethylamine	$CH_3CH_2NH_2$	−81	17	very soluble
propylamine	$CH_3CH_2CH_2NH_2$	−83	48	very soluble
cyclohexylamine	$C_6H_{11}NH_2$	−17	135	slightly soluble
Secondary Amines				
dimethylamine	$(CH_3)_2NH$	−93	7	very soluble
diethylamine	$(CH_3CH_2)_2NH$	−48	56	very soluble
Tertiary Amines				
trimethylamine	$(CH_3)_3N$	−117	3	very soluble
triethylamine	$(CH_3CH_2)_3N$	−114	89	slightly soluble
Aromatic Amines				
aniline	$C_6H_5NH_2$	−6	184	slightly soluble
Heterocyclic Aromatic Amines				
pyridine	C_5H_5N	−42	116	very soluble

Methylamine
(a base)

Methylammonium hydroxide

The equilibrium constant, K_{eq}, for the reaction of an amine with water has the following form, illustrated for the reaction of methylamine with water to give methylammonium hydroxide:

$$K_{eq} = \frac{[CH_3NH_3^+][OH^-]}{[CH_3NH_2][H_2O]}$$

Because the concentration of water in dilute aqueous solutions of amines is essentially constant and so has little effect on the equilibrium, we combine $[H_2O]$ with K_{eq} to give a new constant called a base dissociation constant, K_b, where $K_b = K_{eq}[H_2O]$. pK_b is defined as the negative logarithm of K_b.

$$K_b = \frac{[CH_3NH_3^+][OH^-]}{[CH_3NH_2]} = 4.37 \times 10^{-4}$$

$$pK_b = -\log 4.37 \times 10^{-4} = 3.360$$

All aliphatic amines have about the same base strength, pK_b 3.0–4.0, and are slightly stronger bases than ammonia (Table 16.2). Aromatic amines and heterocyclic aromatic amines have values of pK_b in the range 8.5–9.5 and are considerably weaker bases than ammonia and aliphatic amines.

TABLE 16.2 Approximate Base Strengths of Amines

Amine	pK_b
Aliphatic	3.0–4.0
Ammonia	4.74
Aromatic	8.5–9.5

BOX 16D

Morphine As a Clue in Drug Design and Discovery

The analgesic, soporific, euphoriant, and cough-suppressant properties of the dried juice obtained from unripe seed pods of the opium poppy *Papaver somniferum* have been known for centuries. The active principal is morphine, a powerful drug that is still widely used in medicine today, mainly to ease pain following surgery. Also occurring in the opium poppy is codeine, a monomethyl ether of morphine.

Dextromethorphan

Morphine

Codeine

Meperidine
(structure drawn to show
its relationship to morphine)

Meperidine
(more common way to
draw its structure)

It was hoped that meperidine and related synthetic drugs would be free of many of the undesirable side effects that occur with morphine. It is now clear, however, that they are not. Meperidine, for example, is definitely addictive. In spite of much determined research, there are as yet no agents that are as effective as morphine for the relief of severe pain but that are absolutely free of the risk of addiction.

Even though morphine is one of modern medicine's most effective pain killers, it has two serious side effects: it is addictive, and it depresses the respiratory control center of the central nervous system. Large doses of morphine can lead to death by respiratory failure. For these reasons, chemists for years have sought to synthesize compounds structurally related to morphine in the hope that they would be equally effective medicines but with reduced side effects.

One of these synthetic compounds, dextromethorphan (the "dextro-" comes from the fact that the compound is optically active and dextrorotatory), has no analgesic activity. However, it shows approximately the same cough-suppressing activity as morphine and, in the form of its hydrobromide salt, is used extensively in over-the-counter cough remedies (*The Merck Index*, 12th ed., #8274). Another synthetic relative of morphine is meperidine, the hydrochloride salt of which is the widely used analgesic Demerol (*The Merck Index*, 12th ed., #5894).

■ Opium poppy showing an unripe seed capsule. *(Dr. Jeremy Burgess/Science Photo Library/Photo Researchers, Inc.)*

TABLE 16.3 Basicities of 0.1 *M* Solutions of Three Bases

Base Dissociation Equation	[OH⁻]	pH
$NaOH \rightleftharpoons Na^+ + OH^-$	0.1	13.0
$CH_3NH_2 + H_2O \rightleftharpoons CH_3NH_3^+ + OH^-$	6.6×10^{-3}	11.8

$$\bigcirc\!\!-\!NH_2 + H_2O \rightleftharpoons \bigcirc\!\!-\!NH_3^+ + OH^- \quad 6.5 \times 10^{-6} \quad 8.8$$

Increasing base strength ↑

Another way to compare the relative base strengths of aliphatic and aromatic amines is to look at the hydroxide ion concentration and pH of 0.1 *M* aqueous solutions of each (Table 16.3). For comparison, the hydroxide ion concentration and pH of 0.1 *M* NaOH are also included. A 0.1 *M* solution of an aromatic amine, such as aniline, is slightly basic and has a pH of 8.8. A 0.1 *M* solution of an aliphatic amine, such as methylamine, is more basic and has a pH of 11.8. By contrast, a 0.1 *M* solution of NaOH, a strong base (completely ionized in aqueous solution), has a pH of 13.

Even more importantly, we need to know which form of an amine exists in body fluids, such as blood plasma, cerebrospinal fluid, cellular fluids, and urine. For example, in a normal, healthy person, the pH of blood plasma is approximately 7.40, which is very slightly basic. If an aliphatic amine is dissolved in blood plasma, is it present as the amine or as the ammonium ion? The answer is that aliphatic amines are present in blood and cerebrospinal fluids predominantly as their protonated (ammonium ion) forms. Thus, even though we may write the structural formula of the neurotransmitter dopamine as shown on the left, it is present in neurons as the protonated (ammonium ion) form shown on the right. It is important to realize, however, that because the amine and ammonium ion forms are always in equilibrium, there is still some of the unprotonated form present in solution.

HO—⬡—NH₂   HO—⬡—NH₃⁺

Dopamine   Protonated or ammonium ion form

Aromatic amines, on the other hand, are considerably weaker bases than aliphatic amines and are present in blood and cerebrospinal fluid largely in the unprotonated form.

EXAMPLE 16.4

Select the stronger base in each pair of amines.

(a) (A) or (B)

(b) (C) or (D)

(continued on page 382)

Solution

(a) Morpholine (B) is the stronger base because it is an aliphatic amine. Pyridine (A), a heterocyclic aromatic amine, is the weaker base.

(b) Benzylamine (D) is the stronger base because it is an aliphatic amine. Even though this amine contains an aromatic ring, it is not an aromatic amine because the benzene ring is not bonded to nitrogen. *o*-Toluidine (C), an aromatic amine, is the weaker base.

Problem 16.4 ▪■■

Select the stronger base from each pair of amines.

(a) [structure] N   or   [structure] —NH₂

      (A)   (B)

(b) NH₃   or   [structure] —NH₂

   (C)   (D) ■

16.6 Reaction with Acids

Amines, whether soluble or insoluble in water, react quantitatively with strong acids to form water-soluble salts as illustrated by the reaction of (R)-norepinephrine (noradrenaline, *The Merck Index,* 12th ed., #6788) with aqueous HCl to form a hydrochloride salt.

       (R)-Norepinephrine   (R)-Norepinephrine hydrochloride
  (only slightly soluble in water)   (a water-soluble salt)

Norepinephrine, secreted by the medulla of the adrenal gland, is a neurotransmitter (Section 23.5). It has been suggested that it is a neurotransmitter in those areas of the brain that mediate emotional behavior.

Two drugs that are amine salts labeled as hydrochlorides. *(Beverly March)*

> **EXAMPLE 16.5**
>
> Complete each acid-base reaction, and name the salt formed.
>
> (a) (CH₃CH₂)₂NH + HCl ⟶   (b) [pyridine structure] N + CH₃COOH ⟶
>
> **Solution**
>
> (a) (CH₃CH₂)₂NH₂⁺ Cl⁻   (b) [pyridinium structure]   CH₃COO⁻
>   Diethylammonium chloride   Pyridinium acetate

BOX 16E

The Solubility of Drugs in Body Fluids

Many drugs have "·HCl" or some other acid as part of their chemical formulas and occasionally as part of their generic names. Invariably these drugs are high-molecular-weight amines that are insoluble in aqueous body fluids such as blood plasma and cerebrospinal fluid. In order for the administered drug to be effective in its target organs or cells, it must be absorbed and carried by body fluids. Therefore, the drug is treated with an acid to form a water-soluble ammonium salt. Each of the drugs shown here is listed by its chemical name and, under it in parentheses, is given a name under which it is commonly marketed. For example, Neo-Synephrine, Novocain, and Methadone are marketed as water-soluble hydrochloride salts. Proventil, a prescription bronchodilator used to treat bronchial asthma, is marketed as a water-soluble hydrogen sulfate salt.

There is another reason besides increased water-solubility for preparing these and other amine drugs as salts. Amines themselves are very susceptible to oxidation and decomposition by atmospheric oxygen, with a corresponding loss of biological activity. Their amine salts, however, are far less susceptible to oxidation; consequently, they retain their effectiveness for a far longer time.

Phenylephrine · HCl
(Neo-Synephrine,
a nasal decongestant)

Procaine · HCl
(Novocain, a local anesthetic)

Methadone · HCl
(a narcotic analgesic)

Albuterol · H_2SO_4
(Proventil, a bronchodilator)

■ This inhaler delivers puffs of albuterol (Proventil), a potent bronchodilator whose structure is patterned after that of epinephrine (adrenaline). *(Key Pharmaceuticals, Inc. All rights reserved.)*

Problem 16.5

Complete each acid-base reaction and name the salt formed.

(a) $(CH_3CH_2)_3N + HCl \longrightarrow$

(b) ⬡NH + $CH_3COOH \longrightarrow$

SUMMARY

Amines are classified as **primary, secondary,** or **tertiary,** depending on the number of carbon atoms bonded to nitrogen. In an **aliphatic amine,** all carbon atoms bonded to nitrogen are derived from alkyl groups. In an **aromatic amine,** one or more of the groups bonded to nitrogen are aryl groups. A **heterocyclic amine** is one in which the nitrogen atom is part of a ring.

In systematic nomenclature, aliphatic amines are named by changing the final "-e" of the parent alkane to "-amine" and using a number to indicate the location of the amino group on the parent chain.

In the common system of nomenclature, aliphatic amines are named by listing each carbon group in alphabetical order in one word ending in the suffix "-amine."

Amines are polar compounds, and primary and secondary amines associate by intermolecular hydrogen bonding. All classes of amines form hydrogen bonds with water and are more soluble in water than hydrocarbons of comparable molecular weight.

Amines are weak bases, and aqueous solutions of amines are basic. The base ionization constant for an amine in water is given the symbol K_b.

KEY TERMS

Aliphatic amine (Section 16.2)
Alkaloid (Box 16B)
Amine (Section 16.2)
Aromatic amine (Section 16.2)
Heterocyclic aliphatic amine
 (Section 16.2)

Heterocyclic amine (Section 16.2)
Heterocyclic aromatic amine
 (Section 16.2)

Primary (1°) amine (Section 16.2)
Secondary (2°) amine (Section 16.2)
Tertiary (3°) amine (Section 16.2)

KEY REACTIONS

1. **Basicity of Aliphatic Amines (Section 16.5)**
 Most aliphatic amines have about the same basicity (pK_b 3.0–4.0) and are slightly stronger bases than ammonia.

$$CH_3NH_2 + H_2O \rightleftharpoons CH_3NH_3^+ + OH^- \qquad pK_b = 3.36$$

2. **Basicity of Aromatic Amines (Section 16.5)**
 Most aromatic amines (pK_b 8.5–9.5) are considerably weaker bases than ammonia and aliphatic amines.

$$pK_b = 9.37$$

3. **Reaction with Acids (Section 16.6)**
 All amines react quantitatively with strong acids to form water-soluble salts.

Insoluble in water   A water-soluble salt

CONCEPTUAL PROBLEMS

16.A We use the terms primary, secondary, and tertiary to classify both alcohols and amines. Cyclohexanol, for example, is classified as a secondary alcohol, and cyclohexanamine is classified as a primary amine. In each compound, the functional group is attached to a carbon of the cyclohexane ring. Explain why

one compound is classified as secondary, whereas the other is primary.

16.B Why are amines more basic than alcohols?

16.C Arrange these compounds in order of increasing ability to form intermolecular hydrogen bonds: CH_3OH; CH_3SH; $(CH_3)_2NH$.

PROBLEMS

Difficult problems are designated by an asterisk.

Structure and Nomenclature

16.6 What is the difference in structure between primary, secondary, and tertiary amines? Between primary, secondary, and tertiary alcohols?

16.7 What is the difference in structure between an aliphatic amine and an aromatic amine?

16.8 Unlike benzene, pyrrole has only five atoms in its ring. Why is it considered to be an aromatic compound?

16.9 Draw a structural formula for each amine.
(a) 2-Butanamine
(b) 1-Octanamine
(c) 2,2-Dimethyl-1-propanamine
(d) 1,5-Pentanediamine
(e) 2-Bromoaniline
(f) Tributylamine
(g) *N,N*-Dimethylaniline
(h) Dicyclohexylamine
(i) *sec*-Butylamine
(j) 2,4-Dimethylaniline

16.10 Classify each amino group as primary, secondary, or tertiary; as aliphatic or aromatic.

(a)

Serotonin
(a neurotransmitter)

(b)

Benzocaine
(a topical anesthetic)

(c)

Chloroquine
(a drug for the
treatment of malaria)

16.11 Epinephrine is a hormone secreted by the adrenal medulla. Among its actions, it is a bronchodilator. Albuterol, sold under several trade names, including Proventil and Salbumol, is one of the most effective and widely prescribed antiasthma drugs. The R enantiomer of albuterol is approximately 68 times more effective in the treatment of asthma than the S enantiomer.

(R)-Epinephrine
(Adrenaline)

(R)-Albuterol

(a) Classify each amino group as primary, secondary, or tertiary.
(b) List the similarities and differences between the structural formulas of these two compounds.

*16.12 There are eight constitutional isomers of molecular formula $C_4H_{11}N$. Name and draw structural formulas for each. Classify each amine as primary, secondary, or tertiary.

Physical Properties

16.13 Propylamine, ethylmethylamine, and trimethylamine are constitutional isomers of molecular formula C_3H_9N. Account for the fact that trimethylamine has the lowest boiling point of the three.

$$CH_3CH_2CH_2NH_2 \qquad CH_3CH_2NHCH_3 \qquad (CH_3)_3N$$
bp 48°C   bp 37°C   bp 3°C

16.14 Account for the fact that 1-butanamine has a lower boiling point than 1-butanol.

$$CH_3CH_2CH_2CH_2OH \qquad CH_3CH_2CH_2CH_2NH_2$$
bp 117°C   bp 78°C

Basicity of Amines

16.15 Write structural formulas for these amine salts.
(a) Dimethylammonium iodide
(b) Ethyltrimethylammonium hydroxide
(c) Tetramethylammonium chloride
(d) Anilinium bromide

16.16 Name these salts. (a) $CH_3CH_2NH_3^+ Cl^-$
(b) $(CH_3CH_2)_2NH_2^+ Cl^-$

(c) $\langle\!\!\!\bigcirc\!\!\!\rangle\!-NH_3^+HSO_4^-$

16.17 Explain why the drug morphine is often administered in the form of its salt morphine sulfate.

16.18 From each pair of compounds, select the stronger base.

(a)   or

(b)   —N(CH₃)₂   or   —N(CH₃)₂

(c)   [NH₂ / CH₃]   or   [CH₂NH₂]

16.19 Following are two structural formulas for alanine (2-aminopropanoic acid), one of the building blocks of proteins (Chapter 21). Is alanine better represented by structural formula (A) or structural formula (B)? Explain.

$$CH_3CHCOH \rightleftharpoons CH_3CHCO^-$$
$$\quad NH_2 \qquad\qquad NH_3^+$$
$$\quad (A) \qquad\qquad\quad (B)$$

16.20 Complete the following acid-base reactions.

(a)   CH₃COH + [pyridine] ⟶

        Acetic acid   Pyridine

(b)   [benzene]—CH₂CHNH₂ + HCl ⟶
                           |
                          CH₃

        1-Phenyl-2-propanamine
        (Amphetamine)

(c)   [benzene]—CH₂CHNHCH₃ + H₂SO₄ ⟶
                           |
                          CH₃

        (Methamphetamine)

*16.21 The pK_b of amphetamine [Problem 16.20(c)] is approximately 3.2.
   (a) Which form of amphetamine would you expect to be present at pH 1.0, such as might be present in stomach acid?
   (b) Which form of amphetamine would you expect to be present at pH 7.40, the pH of blood plasma?

*16.22 Pyridoxamine is one form of vitamin B₆.

        Pyridoxamine
        (Vitamin B₆)

        HO   [ring]   CH₂NH₂
                      CH₂OH
        H₃C   N

(a) Which nitrogen atom of pyridoxamine is the stronger base?
(b) Draw the structural formula of the hydrochloride salt formed when pyridoxamine is treated with 1 mole of HCl.

*16.23 Epibatidine, a colorless oil isolated from the skin of the Ecuadorian poison frog *Epipedobates tricolor,* has several times the analgesic potency of morphine. It is the first chlorine-containing, nonopioid (nonmorphine-like in structure) analgesic ever isolated from a natural source.

Poison arrow frog. *(Tom McHugh / Photo Researchers, Inc.)*

(a) Which of the two nitrogen atoms of epibatidine is the more basic?
(b) Mark all stereocenters in this molecule (there are three of them).

        Epibatidine

16.24 Many tumors of the breast are correlated with estrogen levels. Drugs that interfere with estrogen binding have antitumor activity and may even help prevent tumor occurrence. A widely used antiestrogen drug is tamoxifen.

        H₃C   —OCH₂CH₂N(CH₃)₂

        Tamoxifen

(a) Name the functional groups in tamoxifen.

(b) Classify the amino group in tamoxifen as primary, secondary, or tertiary.

(c) How many stereoisomers are possible for tamoxifen?

Boxes

16.25 (Box 16A) What are the differences in structure between the natural hormone epinephrine and the synthetic pep pill amphetamine? Between amphetamine and methamphetamine?

16.26 (Box 16A) What are the possible negative effects of illegal use of amphetamines such as methamphetamine?

16.27 (Box 16B) What is an alkaloid? Are all alkaloids basic to litmus?

16.28 (Box 16B) Identify all stereocenters in coniine and nicotine. How many stereoisomers are possible for each?

16.29 (Box 16B) Of the two nitrogen atoms in nicotine, which is converted to its salt by reaction with 1 mole of HCl? Draw the structural formula of this salt.

*16.30 (Box 16B) Cocaine has four stereocenters. Identify each. Draw the structural formula for the salt formed by treatment of cocaine with 1 mole of HCl.

16.31 (Box 16C) What structural feature is common to all benzodiazepines?

16.32 (Box 16C) Is Librium chiral? Is Valium chiral?

16.33 (Box 16C) Benzodiazepines affect neural pathways in the central nervous system mediated by GABA, whose IUPAC name is 4-aminobutanoic acid. Draw a structural formula of GABA.

16.34 (Box 16D) Name the three oxygen-containing functional groups and the one nitrogen-containing functional group in morphine. Classify the amine as primary, secondary, or tertiary.

16.35 (Box 16D) Morphine, dextromethorphan, and meperidine have in common a quaternary carbon (a carbon bonded to four other carbons) separated by two carbon atoms from a tertiary amine. Identify this structural feature in these three drugs.

16.36 (Box 16D) Besides the structural feature named in Problem 16.35, what other structural features do morphine, dextromethorphan, and meperidine have in common?

16.37 (Box 16E) If you saw this label on a decongestant: phenylephrine·HCl, would you worry about being exposed to a strong acid such as HCl? Explain.

16.38 (Box 16E) Give two reasons why amine-containing drugs are most commonly administered as their salts.

Additional Problems

*16.39 Draw a structural formula for each compound of the given molecular formula.

(a) A 2° aromatic amine, C_7H_9N

(b) A 3° aromatic amine, $C_8H_{11}N$

(c) A 1° aliphatic amine, C_7H_9N

(d) A chiral 1° amine, $C_4H_{11}N$

(e) A 3° heterocyclic amine, $C_5H_{11}N$

(f) A trisubstituted 1° aromatic amine, $C_9H_{13}N$

(g) A chiral quaternary ammonium salt, $C_9H_{22}NCl$

16.40 Arrange these compounds in order of increasing boiling point:

$$CH_3CH_2CH_2CH_3 \qquad CH_3CH_2CH_2OH \qquad CH_3CH_2CH_2NH_2$$

16.41 Account for the fact that amines have about the same solubility in water as alcohols of the same molecular weight.

16.42 If you dissolved $CH_3CH_2CH_2OH$ and $CH_3CH_2CH_2NH_2$ in the same container of water and lowered the pH of the solution to 2 by adding HCl, would anything happen to the structures of these compounds? If so, write the formula of the species present in solution at pH 2.

16.43 The compound phenylpropanolamine hydrochloride is used both as a decongestant and an anorexic. The chemical name of this compound is 1-phenyl-2-amino-1-propanol.

(a) Draw the structural formula of 1-phenyl-2-amino-1-propanol.

(b) How many stereocenters are present in this molecule? How many stereoisomers are possible for it?

(c) Draw the structural formula for the salt formed by reaction of this compound with HCl.

16.44 Procaine was one of the first local anesthetics for infiltration and regional anesthesia. Its hydrochloride salt is marketed as Novocain.

Procaine

(a) Is procaine chiral? Does it contain a stereocenter?

(b) Which nitrogen atom of procaine is the stronger base?

(c) Draw the formula of the salt formed by treating procaine with 1 mole of HCl, showing which nitrogen is protonated and bears the positive charge.

*16.45 Several poisonous plants, including *Atropa belladonna*, contain the alkaloid atropine. The name "belladonna" (which means beautiful lady) probably comes from the fact that Roman women used extracts from this plant to make themselves more attractive. Atropine is widely used by ophthalmologists and optometrists to dilate the pupils for eye examination.

(a) Classify the amino group in atropine as primary, secondary, or tertiary.

(b) Locate the one stereocenter in atropine.

(c) Draw a structural formula for atropine sulfate (atropine plus 1 mole of H_2SO_4).

(d) Account for the fact that atropine is almost insoluble in water (1 g in 455 mL cold water), but atropine sulfate is very soluble in water (1 g in 5 mL of cold water).

(e) Account for the fact that a dilute solution of atropine is basic (pH approximately 10.0).

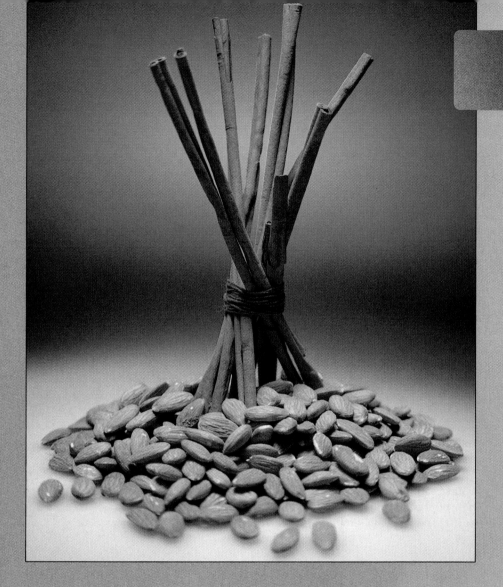

Aldehydes and Ketones

17.1 Introduction

In this and the following chapter, we study the physical and chemical properties of compounds containing the **carbonyl group, C=O.** Because this group is the functional group of aldehydes, ketones, and carboxylic acids and their derivatives, it is one of the most important functional groups in organic chemistry. Its chemical properties are straightforward, and an understanding of its characteristic reaction themes leads very quickly to an understanding of a wide variety of organic and biochemical reactions.

17.2 Structure and Bonding

The functional group of an **aldehyde** is a carbonyl group bonded to a hydrogen atom (Section 10.4). In methanal, the simplest aldehyde, the carbonyl group is bonded to two hydrogen atoms. In other aldehydes, it is

Aldehyde A compound containing a carbonyl group bonded to hydrogen (a CHO group)

Above: Benzaldehyde is found in the kernels of bitter almonds, and cinnamaldehyde is found in Ceylonese and Chinese cinnamon oils. *(Charles D. Winters)*

Ketone A compound containing a carbonyl group bonded to two carbon groups

bonded to one hydrogen atom and one carbon atom. The functional group of a **ketone** is a carbonyl group bonded to two carbon atoms (Section 10.4).

HCH
Methanal
(Formaldehyde)

CH₃CH
Ethanal
(Acetaldehyde)

CH₃CCH₃
Propanone
(Acetone)

Don't confuse RCHO with alcohols, which are ROH.

Because they always contain at least one hydrogen bonded to the C=O group, aldehydes are often written RCHO to save space. Similarly, ketones are often written RCOR′.

17.3 Nomenclature

IUPAC Names

The IUPAC names for aldehydes and ketones follow the familiar pattern of selecting as the parent alkane the longest chain of carbon atoms that contains the functional group (Section 11.6). That the compound is an aldehyde is shown by changing the suffix "-e" of the parent alkane to "-al." Because the carbonyl group of an aldehyde can appear only at the end of a parent chain and numbering must start with it as carbon-1, there is no need to use a number to locate the aldehyde group.

For **unsaturated aldehydes,** the presence of a carbon-carbon double bond is indicated by the infix "-en-" and a number to indicate the first carbon of the double bond. As with other molecules with both an infix and a suffix, the location of the suffix determines the numbering pattern.

$$CH_3(CH_2)_4CH$$
Hexanal

$$CH_3CHCH_2CH$$
3-Methylbutanal

$$CH_2{=}CHCH$$
2-Propenal
(Acrolein)

The IUPAC system retains common names for some aldehydes, including benzaldehyde and cinnamaldehyde.

—CHO
Benzaldehyde

C=C
trans-3-Phenyl-2-propenal
(Cinnamaldehyde)

In the IUPAC system, ketones are named by selecting as the parent alkane the longest chain that contains the carbonyl group and then indicating the presence of this group by changing the suffix "-e" to "-one" (Section 11.6). The parent chain is numbered from the direction that gives the carbonyl carbon the smaller number.

$$CH_3CH_2CCH_2CHCH_3$$
5-Methyl-3-hexanone

CH₃
2-Methylcyclohexanone

The IUPAC system retains the common names acetone, acetophenone, and benzophenone.

| Acetone | Acetophenone | Benzophenone |

EXAMPLE 17.1

Write the IUPAC name for each compound:

(a) $CH_3CH_2\overset{\displaystyle CH_3}{\overset{|}{CH}}\overset{\displaystyle CH_2CH_3}{\underset{|}{CH}}CHO$ (b) (c)

Solution

(a) The longest chain has six carbons, but the longest chain that contains the carbonyl group has five carbons. The name is 2-ethyl-3-methylpentanal.
(b) Number the six-membered ring beginning with the carbonyl carbon. The IUPAC name is 3-methyl-2-cyclohexenone.
(c) This molecule is derived from benzaldehyde. Its IUPAC name is 2-ethylbenzaldehyde.

Problem 17.1 ▬▬

Write the IUPAC name for each compound. Specify the R or S configuration of (c).

(a) $CH_3\overset{\displaystyle CH_3}{\overset{|}{\underset{|}{C}}}CH_2CHO$ (b) (c) $C_6H_5-\overset{\displaystyle CHO}{\underset{\displaystyle CH_3}{C}}{\overset{}{\cdots H}}$ ■
      $\quad\ CH_3$

EXAMPLE 17.2

Write structural formulas for all ketones of molecular formula $C_6H_{12}O$ and give each its IUPAC name. Which of these ketones is/are chiral?

Solution

There are six ketones of this molecular formula. Only 3-methyl-2-pentanone has a stereocenter and is chiral.

| 2-Hexanone | 3-Hexanone | 4-Methyl-2-pentanone |

| 3-Methyl-2-pentanone | 2-Methyl-3-pentanone | 3,3-Dimethyl-2-butanone |

Problem 17.2

Write structural formulas for all aldehydes of molecular formula $C_6H_{12}O$ and give each its IUPAC name. Which of these aldehydes is/are chiral? ∎

IUPAC Names for More Complex Aldehydes and Ketones

Order of precedence of functional groups A system for ranking functional groups in order of priority for the purposes of IUPAC nomenclature

In naming compounds that contain more than one functional group that might be indicated by a suffix, the IUPAC system has established an **order of precedence of functional groups.** The order of precedence for the functional groups we concentrate on is given in Table 17.1.

EXAMPLE 17.3

Write the IUPAC name for each compound. Specify the R or S configuration for (b).

(a) CH_3CCH_2CH (b) [structure with OH, H, H₃C, C, CH₂CH₂CCH₃]

Solution

(a) The aldehyde has a higher precedence than a ketone, so the suffix is "-al," indicating the aldehyde. The presence of the ketone is indicated by the prefix "oxo-." The IUPAC name of this compound is 3-oxobutanal.

(b) The C=O group has higher precedence than the —OH group. The —OH group is indicated by the prefix "hydroxy-." The IUPAC name of this compound is (R)-5-hydroxy-2-hexanone.

Problem 17.3

Write the IUPAC name for each compound.

(a) CH_3CHCH (b) [benzene ring with CHO and NH₂] (c) $H_2NCH_2CH_2CH_2CCH_3$
     $\quad$ OH ∎

Common Names

The common name for an aldehyde is derived from the common name of the corresponding carboxylic acid. The word "acid" is dropped, and the suffix "-ic" or "-oic" is changed to "-aldehyde." Because we have not yet studied

TABLE 17.1 Order of Precedence of Five Functional Groups

	Class of Compound	Functional Group	Suffix if Higher in Precedence	Prefix if Lower in Precedence
Increasing precedence ↑	carboxylic acid	—COOH	-oic acid	—
	aldehyde	—CHO	-al	oxo-
	ketone	>C=O	-one	oxo-
	alcohol	—OH	-ol	hydroxy-
	amine	—NH₂	-amine	amino-

BOX 17A

Some Important Naturally Occurring Aldehydes and Ketones

Benzaldehyde
(oil of almonds)

Cinnamaldehyde
(oil of cinnamon)

Citronellal
(citronella oils; also in
lemon and lemon grass oils)

Muscone
(from the musk deer;
used in perfumes)

Vanillin
(vanilla bean)

β-Ionone
(from violets)

■ West Indian vanilla, *Vanilla pompona.*
(Jane Grushow from Grant Heilman)

■ Violets. *(© Tom & Pat Leeson/Photo Researchers, Inc.)*

common names for carboxylic acids, we are not in a position to discuss common names for aldehydes. We can, however, illustrate how they are derived by reference to two common names with which you are familiar. The name formaldehyde is derived from formic acid, and the name acetaldehyde is derived from acetic acid.

Formaldehyde   Formic acid   Acetaldehyde   Acetic acid

Common names for ketones are derived by naming each alkyl or aryl group bonded to the carbonyl group as a separate word, followed by the word "ketone."

$$
\underset{\substack{\displaystyle | \\ CH_3}}{CH_3CHCCH_2CH_3} \quad \overset{O}{\overset{\|}{}}
$$

$$
CH_3CHCCH_2CH_3 \quad CH_3CH_2CCH_2CH_3
$$

Ethyl isopropyl ketone   Diethyl ketone   Dicyclohexyl ketone

17.4   Physical Properties

Electronegativity is discussed in Section 3.2, and the polarity of covalent bonds is discussed in Section 3.6.

Oxygen is more electronegative than carbon (3.5 compared with 2.5; Table 3.2); therefore, a carbon-oxygen double bond is polar, with oxygen bearing a partial negative charge and carbon bearing a partial positive charge.

$$
\underset{/}{\overset{\backslash}{C}} \overset{\delta^+ \quad \delta^-}{=} \ddot{\underset{..}{O}}
$$

Polarity of a
carbonyl group

Aldehydes and ketones are polar compounds and, as a result, have higher boiling points than ethers, hydrocarbons, and nonpolar compounds of comparable molecular weight (Table 17.2). In a liquid aldehyde or ketone, the intermolecular attractions are between the partially positively charged carbonyl carbon of one molecule and the partially negatively charged carbonyl oxygen of another molecule. There is no hydrogen bonding between aldehyde or ketone molecules, and this lack of hydrogen bonding is why these compounds have lower boiling points than alcohols (Section 13.4) and carboxylic acids (Section 18.2), compounds in which there is hydrogen bonding between molecules.

Because the oxygen atom of their carbonyl groups are hydrogen bond acceptors, the low-molecular-weight aldehydes and ketones are more soluble in water than are nonpolar compounds of comparable molecular weight (Table 17.3). As the hydrocarbon portion of the molecule increases in size, aldehydes and ketones become less soluble in water.

2-Butanone, or as it is more commonly known, methyl ethyl ketone (MEK), is used as a solvent for paints and varnishes.

TABLE 17.2   Boiling Points of Six Compounds of Comparable Molecular Weight

Name	Structural Formula	Molecular Weight	bp (°C)
diethyl ether	$CH_3CH_2OCH_2CH_3$	74	34
pentane	$CH_3CH_2CH_2CH_2CH_3$	72	36
butanal	$CH_3CH_2CH_2CHO$	72	76
2-butanone	$CH_3CH_2COCH_3$	72	80
1-butanol	$CH_3CH_2CH_2CH_2OH$	74	117
propanoic acid	CH_3CH_2COOH	72	141

TABLE 17.3 Physical Properties of Selected Aldehydes and Ketones

IUPAC Name	Common Name	Structural Formula	bp (°C)	Solubility (g/100 g water)
methanal	formaldehyde	HCHO	−21	infinite
ethanal	acetaldehyde	CH_3CHO	20	infinite
propanal	propionaldehyde	CH_3CH_2CHO	49	16
butanal	butyraldehyde	$CH_3CH_2CH_2CHO$	76	7
hexanal	caproaldehyde	$CH_3(CH_2)_4CHO$	129	slight
propanone	acetone	CH_3COCH_3	56	infinite
2-butanone	ethyl methyl ketone	$CH_3COCH_2CH_3$	80	26
3-pentanone	diethyl ketone	$CH_3CH_2COCH_2CH_3$	101	5

Most aldehydes and ketones have strong odors. The odors of ketones are generally pleasant, and many are used in perfumes and as flavoring agents (Box 17A). The odors of aldehydes vary. You may be familiar with the smell of formaldehyde from biology laboratories. If so, you know that it is not pleasant. Many higher aldehydes, however, have pleasant odors and are also used in perfumes.

BOX 17B

Formaldehyde—A Case Study of Toxicity and LD_{50}

Formaldehyde, a colorless, pungent-smelling gas, is an important chemical used widely by industry in the manufacture of building materials and numerous household products. It is present in tobacco smoke and may also be present in the exhausts of unvented fuel-burning appliances such as gas stoves and kerosene space heaters.

The most significant sources of formaldehyde in the home are likely to be pressed wood products (hardwood plywood wall paneling, particleboard, and fiberboard) made with adhesives that contain urea-formaldehyde resins.

During the 1970s, many homeowners had urea-formaldehyde foam insulation (UFFI) installed in the wall cavities of their homes as an energy conservation measure. However, many of these homes were found to have relatively high indoor concentrations of formaldehyde soon after the UFFI installation. Few homes are now being insulated with this product. Studies show that formaldehyde emissions from UFFI decline with time; therefore, homes in which UFFI was installed many years ago are unlikely to have high levels of formaldehyde now.

Exposure to formaldehyde can cause watering of the eyes, burning sensations in the eyes and throat, nausea, and difficulty in breathing in some persons exposed to elevated levels (above 0.1 ppm). High concentrations may trigger attacks in people with asthma. There is evidence that some people can develop a sensitivity to formaldehyde.

A crude measure of the relative toxicity of formaldehyde or any other chemical is its LD_{50} value, where "LD" stands for lethal dose. LD_{50} values are normally given as mg of substance per kg of body weight. An LD_{50} value is the single dose that, when administered to a group of test animals, kills 50 percent of them within 14 days. Test animals usually used for this purpose are mice and rats.

The LD_{50} value depends on the mode of administration, the most common of which are oral, intraperitoneal (i.p.) injection, and intravenous (i.v.) injection. The LD_{50} values of nicotine for mice, for example, under these three modes of administration are 0.3 mg/kg i.v., 9.5 mg/kg i.p., and 230 mg/kg oral. This means that if a group of mice whose average body weight is 0.070 kg (about 2.5 oz) were fed a single oral dose of 16 mg of nicotine, half of them would be dead in 14 days. For botulin toxin, one of the most potent bacterial neurotoxins known and the one implicated in many cases of food poisoning, the minimum lethal dose (MLD) in mice is 3×10^{-4} $\mu g/kg$ (3×10^{-7} mg/kg).

As we have said, LD_{50} is only a crude indicator of toxicity. For one thing, the lethal dose determined in rats or mice may be completely different in other species, including humans. For obvious reasons, an LD_{50} test cannot be carried out on humans. Moreover, death is not the only type of toxicity with which we must be concerned.

The oral LD_{50} for formaldehyde for rats is 800 mg/kg. But formaldehyde has other toxicities as well. It has been found that rats inhaling air containing 14.3 ppm of formaldehyde over an 8-h exposure period develop nasal cancers. There is, however, no evidence that such exposure causes nasal cancer in humans. Nonetheless, U. S. government regulations now set the limit for formaldehyde at 1.5 ppm per 8-h exposure period. All industrial countries have such limits, although they vary somewhat from country to country.

17.5 Addition of Alcohols

Hemiacetal A molecule containing a carbon bonded to an —OH and an —OR or —OAr group; the product of adding one molecule of alcohol to the carbonyl group of an aldehyde or ketone

Addition of a molecule of alcohol to the carbonyl group of an aldehyde or ketone forms a **hemiacetal** (a half-acetal). The positive part of the alcohol molecule (—H) adds to the carbonyl oxygen and the negative part (RO—) adds to the carbonyl carbon.

$$\underset{\textstyle}{CH_3\overset{O}{\overset{\|}{C}}CH_3} + \underset{\textstyle}{\overset{H}{\underset{|}{O}}CH_2CH_3} \rightleftharpoons \underset{\textstyle\underset{CH_3}{|}}{CH_3\overset{OH}{\overset{|}{C}}OCH_2CH_3}$$

A hemiacetal

The functional group of a hemiacetal is a carbon bonded to an —OH group and an —OR or —OAr group.

$$\underset{\text{from an aldehyde}}{} \quad R\overset{OH}{\underset{H}{\overset{|}{\underset{|}{C}}}}OR' \qquad R\overset{OH}{\underset{R''}{\overset{|}{\underset{|}{C}}}}OR' \quad \underset{\text{from a ketone}}{}$$

Hemiacetals

Hemiacetals are generally unstable and are only minor components of an equilibrium mixture except in one very important type of molecule. When a hydroxyl group is part of the same molecule that contains the carbonyl group and a five- or six-membered ring can form, the compound exists almost entirely in a cyclic hemiacetal form. In this case, the —OH group adds to the C=O group of the same molecule. We shall have much more to say about cyclic hemiacetals when we consider the chemistry of carbohydrates in Chapter 19.

$$\underset{\underset{OH}{|}}{\overset{5\quad4\quad3\quad2\quad1}{CH_3CHCH_2CH_2}\overset{O}{\overset{\|}{CH}}} \xrightarrow[\text{close to each other}]{\text{redraw to show —OH and CHO}} \quad \rightleftharpoons$$

4-Hydroxypentanal A cyclic hemiacetal (major form present at equilibrium)

Acetal A molecule containing two —OR or —OAr groups bonded to the same carbon

Hemiacetals can react further with alcohols to form **acetals** plus water. This reaction is acid-catalyzed.

$$\underset{\underset{CH_3}{|}}{CH_3\overset{OH}{\overset{|}{C}}OCH_2CH_3} + CH_3CH_2OH \underset{}{\overset{H^+}{\rightleftharpoons}} \underset{\underset{CH_3}{|}}{CH_3\overset{OCH_2CH_3}{\overset{|}{C}}OCH_2CH_3} + H_2O$$

A hemiacetal A diethyl acetal

The functional group of an acetal is a carbon bonded to two —OR or —OAr groups.

from an aldehyde

$$\underset{\substack{\text{from an}\\\text{aldehyde}}}{} \quad \overset{\displaystyle OR'}{\underset{\displaystyle H}{R-C-OR'}} \qquad \overset{\displaystyle OR'}{\underset{\displaystyle R''}{R-C-OR'}} \quad \underset{\substack{\text{from a}\\\text{ketone}}}{}$$

Acetals

All steps in hemiacetal and acetal formation are reversible. As with any other equilibrium, we can make this one go in either direction by using Le Chatelier's principle. If we want to drive this equilibrium reaction to the right (formation of the acetal), we either use a large excess of alcohol or remove water from the equilibrium mixture. If we want to bring about **hydrolysis** of the acetal, that is, convert the acetal to the original aldehyde and alcohol, or ketone and alcohol, we use a large excess of water.

> **Hydrolysis** A reaction with water (hydro-) in which one or more bonds are broken (-lysis) and the —H and —OH of water add to the ends of the bond or bonds broken

EXAMPLE 17.4

Show the reaction of each ketone with one molecule of alcohol to form a hemiacetal and then with a second molecule of alcohol to form an acetal. Note in (b) that ethylene glycol is a diol and that one molecule of it provides both —OH groups.

(a) $\underset{\displaystyle}{CH_3CH_2\overset{\displaystyle O}{\overset{\|}{C}}CH_3} + 2CH_3CH_2OH \overset{H^+}{\rightleftharpoons}$

(b) ⬠=O + HOCH₂CH₂OH $\overset{H^+}{\rightleftharpoons}$

Solution
Given are structural formulas of the hemiacetal and then the acetal:

(a) $\underset{\displaystyle CH_3}{\overset{\displaystyle OH}{CH_3CH_2\overset{|}{\underset{|}{C}}OCH_2CH_3}} \longrightarrow \underset{\displaystyle CH_3}{\overset{\displaystyle OCH_2CH_3}{CH_3CH_2\overset{|}{\underset{|}{C}}OCH_2CH_3}}$

 A hemiacetal　　　　　　An acetal

(b) A hemiacetal　　　　　A cyclic acetal

Problem 17.4

Show the reaction of each aldehyde with one molecule of alcohol to form a hemiacetal and then with a second molecule of alcohol to form an acetal.

(a) ⬡—$\overset{\displaystyle O}{\overset{\|}{CH}}$ + 2CH₃OH $\overset{H^+}{\rightleftharpoons}$

(b) ⬡—$\overset{\displaystyle O}{\overset{\|}{CH}}$ + HOCH₂CH₂OH $\overset{H^+}{\rightleftharpoons}$

EXAMPLE 17.5

Identify all hemiacetals and acetals in the following structures, and tell whether they are formed from aldehydes or ketones.

(a) CH₃CH₂C(OCH₂CH₃)(OCH₂CH₃)CH₃

$$\text{(a) } CH_3CH_2\overset{\overset{\displaystyle OCH_2CH_3}{|}}{\underset{\underset{\displaystyle OCH_2CH_3}{|}}{C}}CH_3$$

(b) $CH_3CH_2OCH_2CH_2OH$

(c) [pyranyl ring with OH]

Solution

Both hemiacetals and acetals have two oxygen atoms bonded to the same carbon. If one of the oxygens is bonded to an —H, the compound is a hemiacetal; if neither oxygen is bonded to an —H, then the compound is an acetal. If the carbon bonded to the two oxygens is bonded to an —H, the original carbonyl group was an aldehyde. If this carbon is bonded to two carbons, the original carbonyl group was a ketone. Compound (a) is an acetal derived from a ketone. Compound (b) is neither a hemiacetal nor an acetal because it does not have a carbon bonded to two oxygens; its functional groups are an ether and a primary alcohol. Compound (c) is a hemiacetal derived from an aldehyde.

$$\text{(a) } CH_3CH_2\overset{\overset{\displaystyle OCH_2CH_3}{|}}{\underset{\underset{\displaystyle OCH_2CH_3}{|}}{C}}CH_3 + H_2O \xrightarrow{H^+} CH_3CH_2\overset{\overset{\displaystyle O}{\|}}{C}CH_3 + 2CH_3CH_2OH$$

An acetal   2-Butanone

(c) [pyranyl ring with OH] ⟶ [open chain with CHO and OH] $\xrightarrow[\text{the carbon chain}]{\text{redraw to show}}$ $HOCH_2CH_2CH_2CH_2\overset{\overset{\displaystyle O}{\|}}{C}H$

A hemiacetal   5-Hydroxypentanal

Problem 17.5 ▬▬▬

Identify all hemiacetals and acetals in the following structures, and tell whether they are formed from aldehydes or ketones.

$$\text{(a) } CH_3CH_2\overset{\overset{\displaystyle OH}{|}}{\underset{\underset{\displaystyle OCH_2CH_3}{|}}{C}}CH_2CH_3$$

(b) $CH_3OCH_2CH_2OCH_3$

(c) [pyranyl ring with OCH₃] ■

17.6   Keto-Enol Tautomerism

A carbon atom adjacent to a carbonyl group is called an α-carbon, and a hydrogen atom bonded to it is called an α-hydrogen.

$$\underset{\alpha\text{-carbons}}{\overset{\alpha\text{-hydrogens}}{CH_3-\overset{\overset{\displaystyle O}{\|}}{C}-CH_2-CH_3}}$$

Enol A molecule containing an —OH group bonded to a carbon of a carbon-carbon double bond

A carbonyl compound that has a hydrogen on an α-carbon is in equilibrium with a constitutional isomer called an **enol.** The name enol is derived from the IUPAC designation of it as both an alkene ("-en-") and an alcohol ("-ol").

$$CH_3-\overset{\overset{\displaystyle O}{\|}}{C}-CH_3 \rightleftharpoons CH_3-\overset{\overset{\displaystyle OH}{|}}{C}=CH_2$$

Acetone   Acetone
(keto form)   (enol form)

Keto and enol forms are examples of **tautomers,** constitutional isomers in equilibrium with each other that differ in location of a hydrogen atom relative to an O or N. This type of isomerism is called **tautomerism.**

For most simple aldehydes and ketones, the keto form is more stable than the enol form, and considerably less than 0.01 percent of the molecules are present as the enol form. But there are some molecules for which the enol form is more stable than the keto form. One such molecule is 2,4-pentanedione. The enol form of this molecule is stabilized by hydrogen bonding between the —OH and C=O groups.

> **Tautomers** Constitutional isomers that differ in the location of hydrogen and a double bond relative to an O or N

$$CH_3-\overset{\overset{\displaystyle O}{\|}}{C}-CH_2-\overset{\overset{\displaystyle O}{\|}}{C}-CH_3 \rightleftharpoons CH_3-\overset{\overset{\displaystyle O \cdots H \overset{\delta^+}{} \cdots O}{}}{C}=CH-\overset{}{C}-CH_3$$

keto form   enol form
(less stable)   (more stable)

Hydrogen bonding

EXAMPLE 17.6

Write two enol forms for each ketone.

(a)

(b) $CH_3\overset{\overset{\displaystyle O}{\|}}{C}CH_2(CH_2)_2CH_3$

Solution

(a)

(b) $CH_2=\overset{\overset{\displaystyle OH}{|}}{C}CH_2(CH_2)_2CH_3 \rightleftharpoons CH_3\overset{\overset{\displaystyle OH}{|}}{C}=CH(CH_2)_2CH_3$

Problem 17.6

Draw the structural formula for the keto form of each enol.

(a) (b) (c)

17.7 Oxidation

The body uses nicotinamide adenine dinucleotide, NAD⁺, for this type of oxidation (Section 26.3).

Aldehydes are oxidized to carboxylic acids by a variety of oxidizing agents, including potassium dichromate.

$$CH_3(CH_2)_4CH \overset{O}{\underset{}{\parallel}} \xrightarrow[H_2SO_4]{K_2Cr_2O_7} CH_3(CH_2)_4COH \overset{O}{\underset{}{\parallel}}$$

Hexanal   Hexanoic acid

Many other oxidizing agents will also do this, including the oxygen in the air. In fact, aldehydes that are liquid at room temperature are so sensitive to oxidation that they must be protected from contact with air during storage. Often this is done by sealing the aldehyde in a container under an atmosphere of nitrogen.

$$2 \, C_6H_5-\overset{O}{\overset{\parallel}{C}}H + O_2 \longrightarrow 2 \, C_6H_5-\overset{O}{\overset{\parallel}{C}}OH$$

Benzaldehyde   Benzoic acid

Ketones, on the other hand, resist oxidation by most oxidizing agents, including potassium dichromate and molecular oxygen.

The fact that aldehydes are so easy to oxidize and ketones are not allows us to use simple tests to distinguish between these compounds in the laboratory. Suppose that we have a compound we know is either an aldehyde or ketone. All we need to do is treat it with a mild oxidizing agent. If it can be oxidized, it is an aldehyde; otherwise, it is a ketone. One reagent that has been used for this purpose is **Tollens' reagent.**

Tollens' reagent contains silver nitrate and ammonia in water. When these two compounds are mixed, the silver ion combines with NH_3 to form the ion $Ag(NH_3)_2^+$. When this solution is added to an aldehyde, the aldehyde acts as a reducing agent and reduces the complexed silver ion to silver metal. If this reaction is carried out properly, the silver metal precipitates as a smooth, mirror-like deposit, hence the name **silver-mirror test.** If the remaining solution is then acidified with HCl, the carboxylic anion formed when the aldehyde is oxidized is converted to the carboxylic acid, RCOOH.

A silver mirror has been deposited on the inside of this flask by the reaction of an aldehyde with Tollens' reagent. *(Charles D. Winters)*

$$R-\overset{O}{\overset{\parallel}{C}}-H + 2Ag(NH_3)_2^+ + 3OH^- \longrightarrow R-\overset{O}{\overset{\parallel}{C}}-O^- + 2Ag + 4NH_3 + 2H_2O$$

Aldehyde   Tollens' reagent   Carboxylic anion   Silver mirror

Today, silver(I) is rarely used for the oxidation of aldehydes because of the cost of silver and because other, more convenient methods exist for this oxidation. This reaction, however, is still used for silvering mirrors.

EXAMPLE 17.7

Draw a structural formula for the product formed by treating each compound with Tollens' reagent followed by acidification with aqueous HCl. (a) Pentanal (b) 4-Hydroxybenzaldehyde

Solution

The aldehyde group in each compound is oxidized to a carboxylic anion, —COO⁻. Acidification with HCl converts the anion to a carboxylic acid, —COOH.

(a) $CH_3(CH_2)_3\overset{O}{\overset{\|}{C}}OH$ (b) $HO—\bigcirc—\overset{O}{\overset{\|}{C}}OH$

  Pentanoic acid   4-Hydroxybenzoic acid

Problem 17.7 ▬▬

Complete these oxidations.

(a) Hexanedial + $O_2 \rightarrow$ (b) 3-Phenylpropanal + Tollens' reagent → ■

17.8 Reduction

In Section 12.6, we saw that the C=C double bond of an alkene can be reduced by hydrogen in the presence of a transition-metal catalyst to a C—C single bond. The same is true of the C=O double bond of an aldehyde or ketone. Aldehydes are reduced to primary alcohols, and ketones are reduced to secondary alcohols.

The body does not use H_2 to reduce the carbonyl groups of aldehydes and ketones; it uses NADH (Section 26.3).

$$\overset{O}{\overset{\|}{R}CH} \xrightarrow[\text{catalyst}]{\overset{H_2}{\text{metal}}} RCH_2OH \qquad \overset{O}{\overset{\|}{R}CR'} \xrightarrow[\text{catalyst}]{\overset{H_2}{\text{metal}}} \overset{OH}{\overset{|}{R}CHR'}$$

An aldehyde   A primary alcohol   A ketone   A secondary alcohol

The reduction of a C=O double bond under these conditions is slower than reduction of a C=C double bond. Thus, if there is a C=C in the same molecule with the C=O, the C=C is reduced first.

EXAMPLE 17.8

Complete these reductions.

(a) $CH_3CH_2CH_2\overset{O}{\overset{\|}{C}}H \xrightarrow[\text{catalyst}]{\overset{H_2}{\text{metal}}}$ (b) $CH_3O—\bigcirc—\overset{O}{\overset{\|}{C}}CH_3 \xrightarrow[\text{catalyst}]{\overset{H_2}{\text{metal}}}$

Solution

The carbonyl group of the aldehyde in (a) is reduced to a primary alcohol and that of the ketone in (b) is reduced to a secondary alcohol.

(a) $CH_3CH_2CH_2CH_2OH$ (b) $CH_3O—\bigcirc—\overset{OH}{\overset{|}{C}}HCH_3$

Problem 17.8 ▬▬

What aldehyde or ketone gives these alcohols on reduction with H_2/metal catalyst?

(a) $\bigcirc—OH$ (b) $CH_3O—\bigcirc—CH_2CH_2OH$

(c) $CH_3\overset{OH}{\overset{|}{C}}H(CH_2)_3\overset{OH}{\overset{|}{C}}HCH_3$ ■

Reduction of aldehydes and ketones also takes place in biological organisms, in which case organic reducing agents are used. One of the most important is the reduced form of nicotinamide adenine dinucleotide (NADH) (Section 26.3).

SUMMARY

An **aldehyde** contains a carbonyl group bonded to a hydrogen atom. A **ketone** contains a carbonyl group bonded to two carbon atoms. An aldehyde is named by changing "-e" of the parent alkane to "-al." A ketone is named by changing "-e" of the parent alkane to "-one" and using a number to locate the carbonyl group. In naming compounds that contain more than one functional group, the IUPAC system has established an **order of precedence of functional groups.** If the carbonyl group of an aldehyde or ketone is lower in precedence than other functional groups in the molecule, it is indicated by the prefix "-oxo-."

Aldehydes and ketones are polar compounds, have higher boiling points, and are more soluble in water than nonpolar compounds of comparable molecular weight.

KEY TERMS

Acetal (Section 17.5)
Aldehyde (Section 17.2)
Carbonyl group, C=O (Section 17.1)
Enol (Section 17.6)
Hemiacetal (Section 17.5)

Hydrolysis (Section 17.5)
Ketone (Section 17.2)
Order of precedence of functional groups (Section 17.3)

Silver mirror test (Section 17.7)
Tautomerism (Section 17.6)
Tollens' reagent (Section 17.7)

KEY REACTIONS

1. Addition of Alcohols to Form Hemiacetals (Section 17.5)
Hemiacetals are only minor components of an equilibrium mixture of aldehyde or ketone and alcohol, except where the —OH and C=O groups are parts of the same molecule and a five- or six-membered ring can form.

$$CH_3CHCH_2CH_2CH \rightleftharpoons$$

OH

4-Hydroxypentanal   A cyclic hemiacetal

2. Addition of Alcohols to Form Acetals (Section 17.5)
Formation of acetals is catalyzed by acid. Acetals are hydrolyzed in aqueous acid.

$$=O + HOCH_2CH_2OH \xrightarrow{H^+} \quad + H_2O$$

3. Keto-Enol Tautomerism (Section 17.6)
The keto form generally predominates at equilibrium.

$$CH_3CCH_3 \rightleftharpoons CH_3C=CH_2$$

Keto form   Enol form
(approx. 99.9%)

4. Oxidation of an Aldehyde to a Carboxylic Acid (Section 17.7)
The aldehyde group is among the most easily oxidized functional groups. Oxidizing agents include $K_2Cr_2O_7$, Tollens' reagent, and O_2.

$$CH + Ag(NH_3)_2^+ \longrightarrow CO^- + Ag$$

Tollens' reagent

5. Reduction (Section 17.8)
Aldehydes are reduced to primary alcohols, and ketones are reduced to secondary alcohols using H_2 in the presence of a transition metal catalyst such as Pt or Ni.

$$=O \xrightarrow[\text{metal catalyst}]{H_2} \quad -OH$$

CONCEPTUAL PROBLEMS

17.A In predicting the geometry of molecules containing carbonyl groups, the VSEPR model does not differentiate between single and double bonds. What experimental evidence supports this?

17.B What is the difference in structure between an aromatic aldehyde and an aliphatic aldehyde?

17.C Is it possible for the carbon atom of a carbonyl group to be a stereocenter? Explain.

17.D What is the difference in meaning between the terms hydration and hydrolysis? Between hydration and dehydration? Give an example of each.

PROBLEMS

Difficult problems are designated by an asterisk.

Preparation of Aldehydes and Ketones

Aldehydes can be prepared by the oxidation of primary alcohols, whereas ketones can be prepared by the oxidation of secondary alcohols (Section 13.5).

17.9 Complete these reactions.

(a)

(b)

17.10 Write an equation for each conversion.
(a) 1-Pentanol to pentanal
(b) 1-Pentanol to pentanoic acid
(c) 2-Pentanol to 2-pentanone
(d) 2-Propanol to acetone
(e) Cyclohexanol to cyclohexanone

Structure and Nomenclature

17.11 What is the difference in structure between an aldehyde and a ketone?

17.12 Which compounds contain carbonyl groups?

(a) CH_3CHCH_3 with OH

(b) CH_3CH_2CH with =O

(c) benzene-$COCH_2CH_3$ with =O

(d) cyclopentane=O

(e)

(f) $CH_3CH_2CH_2COH$ with =O

17.13 Draw structural formulas for the four aldehydes of molecular formula $C_5H_{10}O$. Which of these aldehydes is/are chiral?

17.14 Draw structural formulas for these aldehydes:
(a) Formaldehyde
(b) Propanal
(c) 3,7-Dimethyloctanal
(d) Decanal
(e) 4-Hydroxybenzaldehyde
(f) 2,3-Dihydroxypropanal

17.15 Draw structural formulas for these ketones.
(a) Ethyl isopropyl ketone
(b) 2-Chlorocyclohexanone
(c) Acetophenone
(d) Diisopropyl ketone
(e) Acetone
(f) 2,5-Dimethylcyclohexanone

17.16 Name these compounds.

(a) $(CH_3CH_2CH_2)_2C=O$

(b)

(c)

(d)

(e)

(f) $HC(CH_2)_4CH$ with =O on both ends

Physical Properties

17.17 In each pair of compounds, select the one with the higher boiling point:
(a) CH_3CHO or CH_3CH_2OH

(b) CH_3CCH_3 with =O or $CH_3CH_2CCH_2CH_3$ with =O

(c) $CH_3(CH_2)_3CHO$ or $CH_3(CH_2)_3CH_3$

(d) $CH_3CH_2\overset{\overset{\displaystyle O}{\|}}{C}CH_3$ or $CH_3CH_2\overset{\overset{\displaystyle OH}{|}}{C}HCH_3$

17.18 Acetone is completely soluble in water, but 4-heptanone is completely insoluble. Explain.

*__17.19__ Why does acetone have a higher boiling point (56°C) than ethyl methyl ether (11°C), even though their molecular weights are almost the same?

17.20 Show how acetaldehyde can form hydrogen bonds with water.

17.21 Why can't two molecules of acetone form a hydrogen bond with each other?

Addition of Alcohols

17.22 What is the characteristic structural formula of a hemiacetal? Of an acetal?

*__17.23__ Which compounds are hemiacetals, which are acetals, and which are neither?

(a) phenyl–$CHOCH_2CH_3$ with OCH_2CH_3

(b) $CH_3CH_2\overset{\overset{\displaystyle OH}{|}}{C}HOCH_3$

(c) $CH_3OCH_2OCH_3$

(d)

(e)

(f)

*__17.24__ Which compounds are hemiacetals, which are acetals, and which are neither?

(a) $CH_3\overset{\overset{\displaystyle OCH_3}{|}}{C}CH_2CH_2CH_3$

(b) phenyl–CCH_2CH_3 with OCH_3 and OCH_3

(c)

(d)

(e) –OCH₃

(f)

17.25 Draw the hemiacetal and then the acetal formed in each reaction.

(a) $CH_3CH_2CHO + 2CH_3CH_2OH \xrightarrow{H^+}$

(b) –$CHO + HOCH_2CH_2OH \xrightarrow{H^+}$

(c) $=O + 2CH_3CH_2OH \xrightarrow{H^+}$

(d) $=O + HOCH_2CH_2OH \xrightarrow{H^+}$

*__17.26__ Draw the structures of the aldehydes or ketones and alcohols formed when these acetals are hydrolyzed:

(a) $CH_3CH_2\overset{\overset{\displaystyle OCH_2CH_3}{|}}{\underset{\underset{\displaystyle OCH_2CH_3}{|}}{C}}CH_2CH_3$

(b) CH_3O––$\overset{\overset{\displaystyle OCH_3}{|}}{C}HOCH_3$

(c)

(d) –OCH_2CH_3

Keto-Enol Tautomerism

17.27 Which of these compounds undergo keto-enol tautomerism?

(a) $CH_3\overset{\overset{\displaystyle O}{\|}}{C}H$

(b) $CH_3\overset{\overset{\displaystyle O}{\|}}{C}CH_3$

(c) –$\overset{\overset{\displaystyle O}{\|}}{C}H$

(d) –$\overset{\overset{\displaystyle O}{\|}}{C}CH_3$

(e) –$\overset{\overset{\displaystyle O}{\|}}{C}$–

(f)

17.28 Draw all enol forms of each aldehyde and ketone.

(a) CH$_3$CH$_2$CH (with =O above)
$$\overset{O}{\overset{\|}{\text{CH}_3\text{CH}_2\text{CH}}}$$

(b) $$\overset{O}{\overset{\|}{\text{CH}_3\text{CCH}_2\text{CH}_3}}$$

(c) [benzene ring]—$$\overset{O}{\overset{\|}{\text{CCH}_2\text{CH}_3}}$$

(d) [cyclopentanone ring with CH$_3$ substituent]

17.29 Draw the structural formula of the keto form of each enol.

(a) CH$_2$=CH—OH

(b) $$\overset{OH}{\overset{|}{\text{CH}_3\text{C}}}\text{=CHCH}_2\text{CH}_2\text{CH}_3$$

(c) [benzene ring]—$$\text{CH=}\overset{OH}{\overset{|}{\text{CCH}_3}}$$

(d) [cyclopentene ring with OH]

Oxidation/Reduction of Aldehydes and Ketones

17.30 Draw the structural formula for the principal organic product formed when each compound is treated with K$_2$Cr$_2$O$_7$/H$_2$SO$_4$. If there is no reaction, say so.

(a) $$\overset{O}{\overset{\|}{\text{CH}_3\text{CH}_2\text{CH}_2\text{CH}}}$$

(b) [benzene ring]—$$\overset{O}{\overset{\|}{\text{CH}}}$$

(c) [cyclohexane ring]=O

(d) [cyclohexane ring]—OH

17.31 Draw the structural formula for the principal organic product formed when each compound in Problem 17.30 is treated with Tollens' reagent. If there is no reaction, say so.

17.32 If you were given a compound that could be either pentanal or 3-pentanone, what simple laboratory test could you use to tell which it is?

17.33 Suppose that you take a bottle of benzaldehyde (a liquid, bp 179°C) from a shelf and find a white solid in the bottom of the bottle. The solid turns litmus red; that is, it is acidic. Yet aldehydes are neutral compounds. How can you explain these observations?

17.34 Explain why reduction of an aldehyde always gives a primary alcohol and reduction of a ketone always gives a secondary alcohol.

17.35 Write the structural formula for the principal organic product formed by treating each compound with H$_2$/metal catalyst.

(a) $$\overset{O}{\overset{\|}{\text{CH}_3\text{CCH}_2\text{CH}_3}}$$

(b) $$\text{CH}_3\text{(CH}_2\text{)}_4\overset{O}{\overset{\|}{\text{CH}}}$$

(c) [cyclopentanone ring with CH$_3$ substituent]

(d) [benzene ring with CH=O and OH substituents]

*17.36 1,3-Dihydroxy-2-propanone, more commonly known as dihydroxyacetone, is the active ingredient in artificial tanning agents such as Man-Tan and Magic Tan.
(a) Write a structural formula for this compound.
(b) Would you expect it to be soluble or insoluble in water?
(c) Write the structural formula of the product formed by its reduction with H$_2$/metal catalyst.

(© George Semple)

17.37 Draw a structural formula for the product formed by treatment of butanal with each set of reagents.
(a) H$_2$/metal catalyst
(b) Ag(NH$_3$)$_2$$^+$ (Tollens' reagent)
(c) K$_2$Cr$_2$O$_7$/H$_2$SO$_4$

17.38 Draw a structural formula for the product formed by treatment of acetophenone, C$_6$H$_5$COCH$_3$, with each set of reagents in Problem 17.37.

Synthesis

*17.39 Starting with cyclohexanone, show how to prepare these compounds. In addition to the given starting

material, use any other organic or inorganic reagents as necessary.
(a) Cyclohexanol
(b) Cyclohexene
(c) *trans*-1,2-Dibromocyclohexane
(d) Cyclohexane
(e) Chlorocyclohexane

17.40 Draw the structural formula of an aldehyde or ketone that can be reduced to produce each alcohol. If none exists, say so.

(a) CH$_3$CHCH$_3$ with OH

(b) [benzene ring]—CH$_2$OH

(c) CH$_3$OH

(d) [cyclohexane ring with OH and CH$_3$]

(e) [cyclopentane ring with OH]

*17.41 1-Propanol can be prepared by reduction of an aldehyde, but it cannot be prepared by acid-catalyzed hydration of an alkene. Explain why it can't be prepared from an alkene.

17.42 Show how to bring about these conversions. In addition to the given starting material, use any other organic or inorganic reagents necessary.

(a) C$_6$H$_5$CCH$_2$CH$_3$ $\longrightarrow$ C$_6$H$_5$CHCH$_2$CH$_3$ $\longrightarrow$ C$_6$H$_5$CH=CHCH$_3$

(b) [cyclopentanone]=O $\longrightarrow$ [cyclopentane]—OH $\longrightarrow$ [cyclopentene] $\longrightarrow$ [cyclopentane]—Cl

*17.43 Show how to bring about these conversions. In addition to the given starting material, use any other organic or inorganic reagents as necessary.
(a) 1-Pentene to 2-pentanone
(b) Cyclohexene to cyclohexanone

Boxes

17.44 (Box 17B) What is the meaning of LD$_{50}$?

17.45 (Box 17B) What are the three most common modes of administering a chemical for the purposes of determining its LD$_{50}$?

17.46 (Box 17B) For mice, the oral LD$_{50}$ of arsenic, administered as arsenic trioxide, As$_2$O$_3$, is 15.1 mg/kg. What dose per mouse will kill, within 14 days, half of a group of mice whose average body weight is 0.35 kg?

17.47 (Box 17B) Why is LD$_{50}$ not a complete indication of the toxicity of a substance?

Additional Problems

17.48 Indicate the aldehyde or ketone group in these compounds.

(a) HCCH$_2$CH$_2$CH$_2$CCH$_3$

(b) [cyclohexanone ring with CH group]

(c) HOCH$_2$CHCH with HO and O

(d) [bicyclic naphthalenone structure]

(e) [benzene ring]—CCH$_2$CH$_3$

(f) HO—[benzene ring with CH$_3$O]—CH

17.49 Draw a structural formula for the (a) one ketone and (b) two aldehydes of molecular formula C$_4$H$_8$O.

17.50 Draw structural formulas for these compounds.
(a) 1-Chloro-2-propanone
(b) 3-Hydroxybutanal
(c) 4-Hydroxy-4-methyl-2-pentanone
(d) 3-Methyl-3-phenylbutanal
(e) 1,3-Cyclohexanedione
(f) 3-Methyl-3-buten-2-one
(g) 5-Oxohexanal
(h) 3-Oxobutanoic acid

*17.51 Why does acetone have a lower boiling point (56°C) than 2-propanol (82°C), even though their molecular weights are almost the same?

17.52 Propanal (bp 49°C) and 1-propanol (bp 97°C) have about the same molecular weight, yet their boiling points differ by almost 50°C. Explain this fact.

*17.53 5-Hydroxyhexanal forms a six-membered cyclic hemiacetal, which predominates at equilibrium in aqueous solution.

CH$_3$CHCH$_2$CH$_2$CH$_2$CH $\overset{H^+}{\rightleftharpoons}$ a cyclic hemiacetal, with OH and O

5-Hydroxyhexanol

(a) Draw a structural formula for this cyclic hemiacetal.

(b) How many stereoisomers are possible for 5-hydroxyhexanal?

(c) How many stereoisomers are possible for this cyclic hemiacetal?

*17.54 The following molecule belongs to a class of compounds called enediols; each carbon of the double bond carries an —OH group. Draw structural formulas for the α-hydroxyketone and the α-hydroxyaldehyde with which this enediol is in equilibrium.

$$\text{α-hydroxyaldehyde} \rightleftharpoons \underset{\underset{\displaystyle CH_3}{|}}{\overset{\overset{\displaystyle HC—OH}{\|}}{C}}—OH \rightleftharpoons \text{α-hydroxyketone}$$

An enediol

17.55 Explain why the compound $CH_2\!\!=\!\!CHOH$ has never been isolated.

17.56 If you were given a compound that could be either 1-pentanol or pentanal, what simple laboratory test could you use to tell which it is?

17.57 Alcohols can be prepared by acid-catalyzed hydration of alkenes (Section 12.6) and by reduction of aldehydes and ketones. Show how you might prepare each alcohol by (1) acid-catalyzed hydration of an alkene and (2) by reduction of an aldehyde or ketone.

(a) CH_3CH_2OH

(b)

(c) $CH_3\overset{\overset{\displaystyle OH}{|}}{C}HCH_3$

(d)

Carboxylic Acids, Anhydrides, Esters, and Amides

18.1 Introduction

In this chapter, we study carboxylic acids, another class of organic compounds containing the carbonyl group. We also study three classes of organic compounds derived from carboxylic acids: anhydrides, esters, and amides. Under the general formula of each carboxylic acid derivative is a drawing to help you see how the functional group of the derivative is formally related to a carboxyl group. Loss of —OH from a carboxyl group and H— from ammonia, for example, gives an amide.

Above: Citrus fruits are sources of citric acid, a tricarboxylic acid. *(Charles D. Winters)*

$$\underset{\text{A carboxylic acid}}{\overset{\displaystyle O}{\overset{\|}{RCOH}}} \qquad \underset{\text{An anhydride}}{\overset{\displaystyle O \quad O}{\overset{\| \quad \|}{RCOCR'}}} \qquad \underset{\text{An ester}}{\overset{\displaystyle O}{\overset{\|}{RCOR'}}} \qquad \underset{\text{An amide}}{\overset{\displaystyle O}{\overset{\|}{RCNH_2}}}$$

$$\overset{O}{\overset{\|}{RC}}-OH \quad H-\overset{O}{\overset{\|}{OCR'}} \qquad \overset{O}{\overset{\|}{RC}}-OH \quad H-\overset{O}{\overset{\|}{OR'}} \qquad \overset{O}{\overset{\|}{RC}}-OH \quad H-NH_2$$

Of these three carboxylic derivatives, anhydrides are by far the most reactive; they are so reactive that they are rarely found in nature. Esters are less reactive than anhydrides, and amides are even less reactive. Unlike anhydrides, esters and amides are universally present in the biological world.

18.2   Carboxylic Acids

Structure and Nomenclature

The functional group of a **carboxylic acid** is a **carboxyl group** (Section 10.4), which can be represented in any one of three ways.

Carboxyl group A —COOH group

$$
\begin{matrix} & O \\ & \| \\ -C & -OH \end{matrix} \qquad -COOH \qquad -CO_2H
$$

The IUPAC name of a carboxylic acid is derived from the name of the longest carbon chain that contains the carboxyl group. The final "-e" is dropped from the name of the parent alkane and the suffix "-oic" is added followed by the word "acid" (Section 11.6). The chain is numbered beginning with the carbon of the carboxyl group. Because the carboxyl carbon is understood to be carbon 1, there is no need to give it a number. In the following examples, the common name is given in parentheses.

$$
\begin{matrix} & & & & & & O \\ & & & & & & \| \\ CH_3CH_2CH_2CH_2CH_2COH \end{matrix}
$$

Hexanoic acid
(Caproic acid)

$$
\begin{matrix} CH_3 & & O \\ | & & \| \\ CH_3CHCH_2COH \end{matrix}
$$

3-Methylbutanoic acid
(Isovaleric acid)

In the IUPAC system, a carboxyl group takes precedence over most other functional groups (Table 17.1). When a carboxyl group is present in a molecule, an —OH group of an alcohol is indicated by the prefix "hydroxy-," and an —NH$_2$ group of a primary amine is indicated by "amino-."

$$
\begin{matrix} OH & & O \\ | & & \| \\ CH_3CHCH_2CH_2CH_2COH \end{matrix}
$$

5-Hydroxyhexanoic acid

$$
\begin{matrix} & & & O \\ & & & \| \\ H_2NCH_2CH_2CH_2COH \end{matrix}
$$

4-Aminobutanoic acid

Dicarboxylic acids are named by adding the suffix "-dioic acid" to the name of the carbon chain that contains both carboxyl groups. The numbers of the carboxyl carbons are not indicated because they can be only at the ends of the parent chain. Following are IUPAC names and common names for four important aliphatic dicarboxylic acids. The name oxalic acid (*The Merck Index*, 12th ed. #7043) is derived from one of its sources in the biological world—plants of the genus *Oxalis*, one of which is rhubarb. It also occurs in human and animal urine, and calcium oxalate is a component of one kind of kidney stone. Succinic acid is an intermediate in the citric acid cycle (Section 26.4). Adipic acid is one of the two monomers required for the synthesis of the polymer nylon 66 (Section 18.8).

Structure	Name
$\underset{\text{Ethanedioic acid}}{\underset{\text{(Oxalic acid)}}{HOC-COH}}$	$\underset{\text{Propanedioic acid}}{\underset{\text{(Malonic acid)}}{HOCCH_2COH}}$

Ethanedioic acid (Oxalic acid)   Propanedioic acid (Malonic acid)   Butanedioic acid (Succinic acid)   Hexanedioic acid (Adipic acid)

HOC—COH   HOCCH₂COH   HOCCH₂CH₂COH   HOCCH₂CH₂CH₂CH₂COH

Leaves of the rhubarb plant contain the poison oxalic acid as its potassium or sodium salt. *(© Hans Reinhard/OKAPIA/Photo Researchers, Inc.)*

The unbranched carboxylic acids having between 4 and 20 carbon atoms are known as fatty acids. We study them in Chapter 20.

Certain aromatic carboxylic acids have common names by which they are more usually known. For example, 2-hydroxybenzoic acid is more often called salicylic acid, a name derived from the fact that this compound was first obtained from the bark of the willow, a tree of the genus *Salix*. It is from salicylic acid that aspirin (Box 18A) is derived. Terephthalic acid is one of the two organic components required for the synthesis of the textile fiber known as Dacron polyester (Section 18.8).

Benzoic acid   2-Hydroxybenzoic acid (Salicylic acid)   1,4-Benzenedicarboxylic acid (Terephthalic acid)

Aliphatic carboxylic acids, many of which were known long before the development of IUPAC nomenclature, are named according to their source or for some characteristic property. Table 18.1 lists several of the unbranched aliphatic carboxylic acids found in the biological world along with the common name of each. Those of 16, 18, and 20 carbon atoms are particularly abundant in animal fats and vegetable oils (Section 20.2) and the phospholipid components of biological membranes (Section 20.6).

When common names are used, the Greek letters alpha (α), beta (β), gamma (γ), delta (δ), and so forth, are often added as a prefix to locate substituents.

Formic acid was first obtained in 1670 from the destructive distillation of ants, whose Latin genus is *Formica*. It is one of the components of the venom injected by stinging ants. *(Ted Nelson/Dembinsky Photo Associates)*

TABLE 18.1 **Several Aliphatic Carboxylic Acids and Their Common Names**

Structure	IUPAC Name	Common Name	Derivation
HCOOH	methanoic acid	formic acid	Latin: *formica*, ant
CH₃COOH	ethanoic acid	acetic acid	Latin: *acetum*, vinegar
CH₃CH₂COOH	propanoic acid	propionic acid	Greek: *propion*, first fat
CH₃(CH₂)₂COOH	butanoic acid	butyric acid	Latin: *butyrum*, butter
CH₃(CH₂)₃COOH	pentanoic acid	valeric acid	Latin: *valere*, to be strong
CH₃(CH₂)₄COOH	hexanoic acid	caproic acid	Latin: *caper*, goat
CH₃(CH₂)₆COOH	octanoic acid	caprylic acid	Latin: *caper*, goat
CH₃(CH₂)₈COOH	decanoic acid	capric acid	Latin: *caper*, goat
CH₃(CH₂)₁₀COOH	dodecanoic acid	lauric acid	Latin: *laurus*, laurel
CH₃(CH₂)₁₂COOH	tetradecanoic acid	myristic acid	Greek: *myristikos*, fragrant
CH₃(CH₂)₁₄COOH	hexadecanoic acid	palmitic acid	Latin: *palma*, palm tree
CH₃(CH₂)₁₆COOH	octadecanoic acid	stearic acid	Greek: *stear*, solid fat
CH₃(CH₂)₁₈COOH	eicosanoic acid	arachidic acid	Greek: *arachis*, peanut

$$\underset{5}{\overset{\delta}{C}}-\underset{4}{\overset{\gamma}{C}}-\underset{3}{\overset{\beta}{C}}-\underset{2}{\overset{\alpha}{C}}-\underset{1}{\overset{O}{\overset{\|}{C}}}-OH$$

$$\underset{\overset{|}{OH}}{CH_2CH_2CH_2}\overset{O}{\overset{\|}{C}}OH$$

4-Hydroxybutanoic acid
(γ-Hydroxybutyric acid)

$$H_3C-\underset{\underset{H}{\overset{\cdots}{}}\ NH_2}{\overset{O}{\overset{\|}{C}}}-OH$$

2-Aminopropanoic acid
(α-Aminopropanoic
acid; Alanine)

EXAMPLE 18.1

Write the IUPAC name for each carboxylic acid:

(a)
$$CH_3(CH_2)_7 \qquad (CH_2)_7COOH$$
$$C=C$$
$$H \qquad\qquad H$$

(b)

(c)
$$\underset{CH_3}{\overset{OH}{\underset{|}{\overset{|}{H\overset{\cdots}{\diagup}\ \underset{\diagdown}{C}}}}} COOH$$

(d) $ClCH_2COOH$

Solution

Given first is the IUPAC name and then, in parentheses, the common name.

(a) The carbon chain contains 18 atoms; therefore, the parent alkane is octadecane. There is a cis double bond between carbons 9 and 10. Its name is *cis*-9-octadecenoic acid (its common name is oleic acid).

(b) 4-Hydroxybenzoic acid (*p*-hydroxybenzoic acid)

(c) (R)-2-Hydroxypropanoic acid [(R)-lactic acid]

(d) Chloroethanoic acid (chloroacetic acid)

Problem 18.1

Each of these compounds has a well-recognized common name. A derivative of glyceric acid is an intermediate in glycolysis (Section 27.2). Maleic acid is an intermediate in the citric acid cycle (Section 26.4). Mevalonic acid is an intermediate in the biosynthesis of steroids (Section 28.4). Write the IUPAC name for each compound.

(a)
$$\underset{\overset{|}{CH_2OH}}{\overset{\overset{\displaystyle COOH}{|}}{CHOH}}$$
Glyceric acid

(b)
$$HOOC \qquad COOH$$
$$C=C$$
$$H \qquad\quad H$$
Maleic acid

(c) $HOCH_2CH_2\underset{\underset{CH_3}{\overset{|}{}}}{\overset{\overset{\displaystyle OH}{|}}{C}}CH_2COOH$
Mevalonic acid ■

Physical Properties

As Table 18.2 shows, carboxylic acids have significantly higher boiling points than other types of organic compounds of comparable molecular weight. Their higher boiling points are a result of their polarity and of the fact that they form very strong intermolecular hydrogen bonds.

TABLE 18.2	Boiling Points and Solubilities in Water of Two Groups of Carboxylic Acids, Alcohols, and Aldehydes of Comparable Molecular Weight			
Structure	**Name**	**Molecular Weight**	**Boiling Point (°C)**	**Solubility (g/100 mL H₂O)**
CH_3COOH	acetic acid	60.5	118	infinite
$CH_3CH_2CH_2OH$	1-propanol	60.1	97	infinite
CH_3CH_2CHO	propanal	58.1	48	16
$CH_3(CH_2)_2COOH$	butanoic acid	88.1	163	infinite
$CH_3(CH_2)_3CH_2OH$	1-pentanol	88.1	137	2.3
$CH_3(CH_2)_3CHO$	pentanal	86.1	103	slight

Acetic acid molecules can interact through hydrogen bonds. Shown here is partly solid glacial acetic acid. Like most substances (water being a notable exception), the solid is denser than the liquid. *(Photo by Charles D. Winters; model by S. M. Young)*

Hydrophilic From the Greek meaning water-loving

Hydrophobic From the Greek meaning water-hating

hydrogen bonding between two molecules

$$H_3C-C \quad \overset{\delta^-}{O}\cdots\overset{\delta^+}{H}-O \quad C-CH_3$$
$$\underset{\delta^+ \quad \delta^-}{O-H\cdots O}$$

Carboxylic acids also interact with water molecules by hydrogen bonding through both the carbonyl and hydroxyl groups. Because of these hydrogen-bonding interactions, carboxylic acids are more soluble in water than are alcohols, ethers, aldehydes, and ketones of comparable molecular weight.

The solubility of a carboxylic acid in water decreases as its molecular weight increases. We account for this trend in the following way. A carboxylic acid consists of two regions of distinctly different polarity: a polar **hydrophilic** carboxyl group that increases water solubility and, except for formic acid, a nonpolar **hydrophobic** hydrocarbon chain that decreases water solubility. Other polar hydrophilic groups that increase water solubility are the hydroxyl (—OH), amino (—NH₂), and carbonyl (C=O) groups.

The first four aliphatic carboxylic acids (methanoic, ethanoic, propanoic, and butanoic) are infinitely soluble in water because the hydrophilic character of the carboxyl group more than counterbalances the hydrophobic character of the hydrocarbon chain. As the size of the hydrocarbon chain increases relative to that of the carboxyl group, water solubility decreases. The solubility of hexanoic acid (six carbons) in water is 1.0 g/100 g water.

One other physical property of carboxylic acids must be mentioned. The liquid carboxylic acids from propanoic acid to decanoic acid have extremely foul odors, about as bad, though different, as those of thiols. Butanoic acid is found in stale perspiration and is a major component of "locker room odor." Valeric acid smells even worse, and goats, which secrete C_6, C_8, and C_{10} acids, are not famous for their pleasant odors.

Acidity

Carboxylic acids are weak acids. Values of K_a for most unsubstituted aliphatic and aromatic carboxylic acids fall within the range 10^{-4} to 10^{-5} ($pK_a = 4.0$ to 5.0). The value of K_a for acetic acid, for example, is 1.74×10^{-5}. Its pK_a is 4.76.

$$CH_3COOH \rightleftharpoons CH_3COO^- + H^+ \qquad K_a = \frac{[CH_3COO^-][H^+]}{[CH_3COOH]} = 1.74 \times 10^{-5}$$

$$pK_a = 4.76$$

Substituents of high electronegativity, especially F and Cl, near the carboxyl group increase the acidity of carboxylic acids, often by several orders of magnitude. Compare, for example, the acidities of acetic acid and the chlorine-substituted acetic acids. Both dichloroacetic acid and trichloroacetic acid are stronger acids than H_3PO_4 ($pK_a = 2.1$). Dichloroacetic acid (*The Merck Index,* 12th ed., #3100) is used as a topical astringent and to treat genital warts in males. A 50 percent aqueous solution of trichloroacetic acid is used by dentists to cauterize gums. This strong acid kills the bleeding and diseased tissue and allows the growth of healthy gum tissue. Trichloroacetic acid solution is also used to cauterize canker sores.

Formula:	CH_3COOH	$ClCH_2COOH$	$Cl_2CHCOOH$	Cl_3CCOOH
Name:	Acetic acid	Chloroacetic acid	Dichloroacetic acid	Trichloroacetic acid
pK_a:	4.76	2.86	1.48	0.70

Increasing acid strength →

Reaction with Bases

All carboxylic acids, whether soluble or insoluble in water, react with NaOH, KOH, and other strong bases to form water-soluble salts.

Sodium benzoate and calcium propanoate are used as preservatives in baked goods. *(Charles D. Winters)*

$$\text{C}_6\text{H}_5\text{—COOH} + NaOH \xrightarrow{H_2O} \text{C}_6\text{H}_5\text{—COO}^-Na^+ + H_2O$$

Benzoic acid
(slightly soluble in water)

Sodium benzoate
(60 g/100 mL water)

Sodium benzoate (*The Merck Index,* 12th ed., #8725), a fungal growth inhibitor, is often added to baked goods "to retard spoilage." Calcium propanoate (*The Merck Index,* 12th ed., #1745) is also used for the same purpose. Carboxylic acids also form water-soluble salts with ammonia and amines.

$$\text{C}_6\text{H}_5\text{—COOH} + NH_3 \xrightarrow{H_2O} \text{C}_6\text{H}_5\text{—COO}^-NH_4^+$$

Benzoic acid
(slightly soluble in water)

Ammonium benzoate
(20 g/100 mL water)

Carboxylic acids react with sodium bicarbonate and sodium carbonate to form water-soluble sodium salts and carbonic acid. Carbonic acid then decomposes to give water and carbon dioxide, which evolves as a gas.

$$CH_3COOH + Na^+HCO_3^- \longrightarrow CH_3COO^-Na^+ + CO_2 + H_2O$$

Salts of carboxylic acids are named in the same manner as the salts of inorganic acids: the cation is named first and then the anion. The name of the anion is derived from the name of the carboxylic acid by dropping the suffix "-ic acid" and adding the suffix "-ate." For example, $CH_3CH_2COO^-Na^+$ is named sodium propanoate, and $CH_3(CH_2)_{14}COO^-Na^+$ is named sodium hexadecanoate (sodium palmitate).

EXAMPLE 18.2

Complete each acid-base reaction and name the salt formed.
(a) $CH_3(CH_2)_2COOH + NaOH \longrightarrow$

(b) $CH_3\overset{\overset{\displaystyle OH}{|}}{C}HCOOH + NaHCO_3 \longrightarrow$

Solution
Each carboxylic acid is converted to its sodium salt. In (b), carbonic acid is formed; it then decomposes to carbon dioxide and water.

(a) $CH_3(CH_2)_2COOH + NaOH \longrightarrow CH_3(CH_2)_2COO^-Na^+ + H_2O$
  Butanoic acid   Sodium butanoate

(b) $CH_3\overset{\overset{\displaystyle OH}{|}}{C}HCOOH + NaHCO_3 \longrightarrow$
  2-Hydroxypropanoic acid
  (Lactic acid)

$CH_3\overset{\overset{\displaystyle OH}{|}}{C}HCOO^-Na^+ + H_2O + CO_2$
  Sodium 2-hydroxypropanoate
  (Sodium lactate)

Problem 18.2 ■
Write an equation for the reaction of each acid in Example 18.2 with ammonia and name the salt formed. ■

Because carboxylic acids are acidic, the form in which they exist in an aqueous solution depends on the pH of the solution. If we dissolve a carboxylic acid in water, the solution will be slightly acidic due to ionization of the carboxyl group. If we now add enough HCl to the solution to bring its pH to 2.0 or lower (more acidic), most of the carboxylic acid molecules are present in their un-ionized (—COOH) form. When the pH of the solution is equal to the pK_a of the acid, the un-ionized and ionized forms are present in approximately equal amounts. When the pH of the solution is brought to about 8.0 or higher (more basic), most carboxylic molecules are present in their ionized (—COO⁻) form.

at pH 2.0 or lower (more acidic) / at pH = pK_a = 4.0 − 5.0 both forms are present in approximately equal amounts / at pH 8.0 or higher (more basic)

Ester A compound in which the H of a carboxyl group, RCOOH, is replaced by an —OR′ group, RCOOR′

Fischer Esterification

Treatment of a carboxylic acid with an alcohol in the presence of an acid catalyst, most commonly concentrated sulfuric acid, gives an **ester.**

$$CH_3\overset{\displaystyle O}{\overset{\|}{C}}OH \ + \ CH_3CH_2OH \ \underset{}{\overset{H_2SO_4}{\rightleftharpoons}} \ CH_3\overset{\displaystyle O}{\overset{\|}{C}}OCH_2CH_3 \ + \ H_2O$$

Ethanoic acid   Ethanol   Ethyl ethanoate
(Acetic acid)   (Ethyl alcohol)   (Ethyl acetate)

Preparing an ester in this manner is given the special name **Fischer esterification** after the German chemist, Emil Fischer (1852–1919).

  Fischer esterification is reversible, and generally, at equilibrium, the quantities of carboxylic acid and alcohol present are appreciable. By controlling reaction conditions, however, it is possible to use Fischer esterification to prepare esters in high yields. If the alcohol is inexpensive, a large excess of it can be used to drive the equilibrium to the right and increase the amount of carboxylic acid converted to its ester (remember Le Chatelier's principle, Section 7.8). Alternatively, water may be removed from the reaction mixture as it is formed.

> **Fischer esterification** The process of forming an ester by refluxing a carboxylic acid and an alcohol in the presence of an acid catalyst, commonly sulfuric acid

Some products that contain ethyl acetate, a common carboxylic ester. *(Charles D. Winters)*

EXAMPLE 18.3

Complete these Fischer esterification reactions:

(a) $\bigcirc\!\!-\!\!\overset{\displaystyle O}{\overset{\|}{C}}OH + CH_3OH \overset{H^+}{\rightleftharpoons}$

(b) $HO\overset{\displaystyle O}{\overset{\|}{C}}CH_2CH_2\overset{\displaystyle O}{\overset{\|}{C}}OH + 2CH_3CH_2OH \overset{H^+}{\rightleftharpoons}$

Solution
Here is a structural formula for the ester produced in each reaction.

(a) $\bigcirc\!\!-\!\!\overset{\displaystyle O}{\overset{\|}{C}}OCH_3$ (b) $CH_3CH_2O\overset{\displaystyle O}{\overset{\|}{C}}CH_2CH_2\overset{\displaystyle O}{\overset{\|}{C}}OCH_2CH_3$

Problem 18.3

Complete these Fischer esterification reactions:

(a) $HOCH_2CH_2CH_2\overset{\displaystyle O}{\overset{\|}{C}}OH \overset{H^+}{\rightleftharpoons}$ (a cyclic ester)

(b) $CH_3\overset{\displaystyle O}{\underset{\underset{\displaystyle CH_3}{|}}{\overset{\|}{C}H}}\!COH + HO\!\!-\!\!\bigcirc \overset{H^+}{\rightleftharpoons}$

18.3   Carboxylic Anhydrides

Structure and Nomenclature

The functional group of a **carboxylic anhydride** is two carbonyl groups bonded to the same oxygen atom. The anhydride may be symmetrical (from two identical acyl groups), or it may be mixed (from two different acyl groups).

> **Anhydride** A compound in which two carbonyl groups are bonded to a single oxygen, RCO—O—COR′

BOX 18A

From Willow Bark to Aspirin and Beyond

The first drug developed for widespread use was aspirin, today's most common pain reliever. Americans alone consume approximately 80 billion tablets of aspirin a year! The story of the development of this modern pain reliever goes back more than 2000 years. In 400 B.C.E., the Greek physician Hippocrates recommended chewing bark of the willow tree to alleviate the pain of childbirth and to treat eye infections.

The active component of willow bark was found to be salicin (*The Merck Index*, 12th ed., #8476), a compound composed of salicyl alcohol bonded to a unit of β-D-glucose (Section 19.2). Hydrolysis of salicin in aqueous acid gives salicyl alcohol, which can be oxidized to salicylic acid. Salicylic acid proved to be an even more effective reliever of pain, fever, and inflammation than salicin, and without its extremely bitter taste. Unfortunately, patients quickly recognized salicylic acid's major side effect: it causes severe irritation of the mucous membrane lining of the stomach.

Salicin   Salicyl alcohol

Salicylic acid

In the search for less irritating but still effective derivatives of salicylic acid, chemists at the Bayer division of I. G. Farben in Germany in 1883 prepared acetylsalicylic acid and gave it the name aspirin.

Salicylic acid   Acetic anhydride

Acetylsalicylic acid (Aspirin)

■ Willow trees are a source of salicylic acid. *(Charles D. Winters)*

$$CH_3C-O-CCH_3$$
Acetic anhydride

Reaction with Water. Hydrolysis

Anhydrides react readily with water to form two molecules of carboxylic acid. In fact, they react so readily with water that they must be protected from moisture during storage.

Aspirin proved to be less irritating to the stomach than salicylic acid and also more effective in relieving the pain and inflammation of rheumatoid arthritis. Aspirin, however, is still irritating to the stomach, and frequent use can cause duodenal ulcers in susceptible persons.

In the 1960s, in a search for even more effective and less irritating analgesics and anti-inflammatory drugs, chemists at the Boots Pure Drug Company in England, who were studying compounds structurally related to salicylic acid, discovered an even more potent compound, which they named ibuprofen (*The Merck Index*, 12th ed., #4925). Soon thereafter, Syntex Corporation in the United States developed naproxen (*The Merck Index*, 12th ed., #6504), the active ingredient in Aleve. Each compound has one stereocenter and can exist as a pair of enantiomers. For each drug, the physiologically active form is the s enantiomer.

In the 1960s, it was discovered that aspirin acts by inhibiting cyclooxygenase (COX), a key enzyme in the conversion of arachidonic acid to prostaglandins (Box 20J). With this discovery, it became clear why only one enantiomer of ibuprofen and naproxen is active: Only the s enantiomer of each has the correct handedness to bind to COX and inhibit its activity.

(S)-Ibuprofen

(S)-Naproxen

■ Two nonprescription pain relievers. *(Charles D. Winters)*

$$CH_3\overset{\displaystyle O}{\overset{\|}{C}}O\overset{\displaystyle O}{\overset{\|}{C}}CH_3 + H_2O \longrightarrow CH_3\overset{\displaystyle O}{\overset{\|}{C}}OH + HO\overset{\displaystyle O}{\overset{\|}{C}}CH_3$$

Reaction with Alcohols

Anhydrides react with alcohols to give an ester and a carboxylic acid. Thus, this reaction is a useful method for the synthesis of esters.

$$\underset{\text{Acetic anhydride}}{CH_3\overset{\displaystyle O}{\overset{\|}{C}}O\overset{\displaystyle O}{\overset{\|}{C}}CH_3} + \underset{\text{Ethanol}}{HOCH_2CH_3} \longrightarrow \underset{\text{Ethyl acetate}}{CH_3\overset{\displaystyle O}{\overset{\|}{C}}OCH_2CH_3} + \underset{\text{Acetic acid}}{CH_3\overset{\displaystyle O}{\overset{\|}{C}}OH}$$

Aspirin is synthesized on an industrial scale by reaction of acetic anhydride and salicylic acid (Box 18A).

Reaction with Ammonia and Amides

Anhydrides react with ammonia and with 1° and 2° amines to form amides. Two moles of ammonia or amine are required: one to form the amide and one to neutralize the carboxylic acid byproduct.

$$
\underset{\text{Acetic anhydride}}{CH_3\overset{O}{\overset{\|}{C}}O\overset{O}{\overset{\|}{C}}CH_3} + \underset{\text{Ammonia}}{2NH_3} \longrightarrow \underset{\substack{\text{Ethanamide}\\\text{(Acetamide)}}}{CH_3\overset{O}{\overset{\|}{C}}NH_2} + \underset{\text{Ammonium acetate}}{CH_3\overset{O}{\overset{\|}{C}}O^-NH_4^+}
$$

18.4   Carboxylic Esters

Structure and Nomenclature

Carboxylic esters are neutral compounds because the —COOH hydrogen has been replaced by an R group. Their boiling points are lower than both the carboxylic acids and alcohols from which they are made.

The functional group of a **carboxylic ester** is a carbonyl group bonded to —OR or —OAr. Both IUPAC and common names of esters are derived from the names of the parent carboxylic acids. The alkyl or aryl group bonded to oxygen is named first, followed by the name of the acid in which the suffix "-ic acid" is replaced by the suffix "-ate."

$$
\underset{\substack{\text{Ethyl ethanoate}\\\text{(Ethyl acetate)}}}{CH_3\overset{O}{\overset{\|}{C}}OCH_2CH_3} \qquad \underset{\substack{\text{Diethyl butanedioate}\\\text{(Diethyl succinate)}}}{CH_3CH_2O\overset{O}{\overset{\|}{C}}CH_2CH_2\overset{O}{\overset{\|}{C}}OCH_2CH_3}
$$

Lactone A cyclic ester

Cyclic esters are called **lactones.** The IUPAC system has developed a set of rules for naming these compounds. Nonetheless, the simplest lactones are still named by dropping the suffix "-ic" or "-oic acid" from the name of the parent carboxylic acid and adding the suffix "-olactone."

4-Butanolactone   6-Hexanolactone

EXAMPLE 18.4

Write the IUPAC name of each ester.

(a) $CH_3\overset{\overset{\displaystyle CH_3}{|}}{C}HCH_2\overset{O}{\overset{\|}{C}}OCH_3$ (b) $CH_3CH_2CH_2\overset{O}{\overset{\|}{C}}OCH_2CH_3$

Solution

Given first are IUPAC names and then, in parentheses, common names.

(a) Methyl 3-methylbutanoate (methyl isovalerate, from isovaleric acid)

(b) Ethyl butanoate (ethyl butyrate, from butyric acid)

From Moldy Clover to a Blood Thinner

In 1933, a disgruntled farmer delivered a bale of moldy clover, a pail of unclotted blood, and a dead cow to the laboratory of Dr. Carl Link at the University of Wisconsin. Six years and many bales of moldy clover later, Link and his collaborators isolated the anticoagulant dicoumarol (*The Merck Index*, 12th ed., #3140), a substance that delays or prevents blood clotting. When cows are fed moldy clover, they ingest dicoumarol, their blood clotting is inhibited, and they bleed to death from minor cuts and scratches. Within a few years after its discovery, dicoumarol became widely used to treat victims of heart attack and others at risk for developing blood clots.

Dicoumarol is a derivative of coumarin, the substance that gives sweet clover its pleasant smell. Coumarin is converted to dicoumarol as sweet clover becomes moldy. Notice that both coumarin and dicoumarol contain lactones. In addition, dicoumarol contains two enols (Section 17.6).

■ The powerful anticoagulant dicoumarol was first isolated from moldy clover. *(Grant Heilman/Grant Heilman Photography, Inc.)*

Coumarin
(from sweet clover)

Dicoumarol
(an anticoagulant formed
in moldy sweet clover)

Warfarin
(a synthetic anticoagulant)

In a search for anticoagulants even more potent, Link developed warfarin (named after the Wisconsin Alumni Research Foundation), now used primarily as a rat poison, with an LD_{50} value 100 mg/kg for male rats. When rats consume it, their blood fails to clot, and they bleed to death. Warfarin is also used as a blood thinner in humans. The s enantiomer shown here is more active than the R enantiomer. The commercial product is sold as a racemic mixture.

Problem 18.4 ■

Write the IUPAC name for each ester.

(a) $\bigcirc$—C(=O)OCH$_2$CH$_3$ (b) CH$_3$COCHCH$_2$CH$_3$
                                          |
                                          CH$_3$

■

Reaction with Water. Hydrolysis

Esters are hydrolyzed only very slowly, even in boiling water. Hydrolysis becomes considerably more rapid, however, when the ester is heated in aqueous acid or base. We discussed acid-catalyzed Fischer esterification in Section 18.2 and pointed out that it is an equilibrium reaction. Hydrolysis of esters in aqueous acid, also an equilibrium reaction, is the reverse of Fischer esterification. If there is a large excess of water present, equilibrium is driven to the right to form the carboxylic acid and alcohol (Le Chatelier's Principle, Section 7.8).

$$R-\overset{\overset{\displaystyle O}{\|}}{C}-OCH_3 + H_2O \overset{H^+}{\rightleftharpoons} R-\overset{\overset{\displaystyle O}{\|}}{C}-OH + CH_3OH$$

Saponification Hydrolysis of an ester in aqueous NaOH or KOH to an alcohol and the sodium or potassium salt of a carboxylic acid

Hydrolysis of an ester may also be carried out using a hot aqueous base, such as aqueous NaOH. This reaction is often called **saponification,** a reference to its use in the manufacture of soaps (Box 20B). The carboxylic acid formed in the hydrolysis reacts with hydroxide ion to form a carboxylic acid anion. Thus, each mole of ester hydrolyzed requires 1 mole of base as shown in the following balanced equation:

$$R\overset{\overset{\displaystyle O}{\|}}{C}OCH_3 + NaOH \overset{H_2O}{\longrightarrow} R\overset{\overset{\displaystyle O}{\|}}{C}O^-Na^+ + CH_3OH$$

EXAMPLE 18.5

Complete the equation for each hydrolysis. Show the products as they are ionized under these experimental conditions.

(a) $-\overset{\overset{\displaystyle O}{\|}}{C}OCH(CH_3)_2 + NaOH \overset{H_2O}{\longrightarrow}$

(b) $CH_3\overset{\overset{\displaystyle O}{\|}}{C}OCH_2CH_2O\overset{\overset{\displaystyle O}{\|}}{C}CH_3 + 2NaOH \overset{H_2O}{\longrightarrow}$

Solution
The products of hydrolysis of (a) are benzoic acid and 2-propanol. In aqueous NaOH, benzoic acid is converted to its sodium salt. Therefore, 1 mole of NaOH is required for hydrolysis of each mole of this ester. Compound (b) is a diester of ethylene glycol. Two moles of NaOH are required for its hydrolysis.

(a) $-\overset{\overset{\displaystyle O}{\|}}{C}O^-Na^+ +$ $HOCH(CH_3)_2$

       Sodium benzoate   2-Propanol
                       (Isopropyl alcohol)

(b) $2CH_3\overset{\overset{\displaystyle O}{\|}}{C}O^-Na^+ + HOCH_2CH_2OH$

    Sodium acetate   1,2-Ethanediol
                  (Ethylene glycol)

Problem 18.5 ▬▬▬▬

Complete the equation for each hydrolysis. Show all products as they are ionized under these experimental conditions.

(a) $+ 2NaOH \overset{H_2O}{\longrightarrow}$

(b) $CH_3\overset{\overset{\displaystyle O}{\|}}{C}CH_2CH_2CH_2\overset{\overset{\displaystyle O}{\|}}{C}OCH_2CH_3 + H_2O \overset{HCl}{\longrightarrow}$ ■

BOX 18C

Barbiturates

In 1864, Adolph von Baeyer (1835–1917) discovered that heating the diethyl ester of malonic acid with urea in the presence of sodium ethoxide (like sodium hydroxide, a strong base) gives a cyclic compound which he named barbituric acid (*The Merck Index*, 12th ed., #990).

Diethyl propanedioate (Diethyl malonate) + Urea $\xrightarrow[\text{CH}_3\text{CH}_2\text{OH}]{\text{CH}_3\text{CH}_2\text{O}^-\text{Na}^+}$ Barbituric acid + $2\text{CH}_3\text{CH}_2\text{OH}$

It is said that Baeyer named this compound after a friend named Barbara. It is also said that he named it after St. Barbara, the patron saint of artillerymen. Whatever the case, a number of derivatives of barbituric acid have powerful sedative and hypnotic effects. One such derivative is pentobarbital (*The Merck Index*, 12th ed., #7272). As with other derivatives of barbituric acid, pentobarbital is quite insoluble in water and body fluids. To increase its solubility in these fluids, pentobarbital is converted to its sodium salt, which is given the name Nembutal.

Pentobarbital

Sodium pentobarbital (Nembutal)

Other examples of barbiturates are secobarbital (*The Merck Index*, 12th ed., #8563) and thiopental (*The Merck Index*, 12th ed., #9487), each of which is most commonly administered as its sodium salt. Notice that thiopental is similar in structure to pentobarbital, except that an atom of sulfur is substituted for an oxygen in one of the C=O groups of the six-membered ring.

Secobarbital (Seconal)

Thiopental (Pentothal)

Technically speaking, only the sodium salts of these compounds should be called barbiturates, but in practice, all derivatives of barbituric acid are called barbiturates, whether in the un-ionized form or the ionized, water-soluble salt form.

Barbiturates have two principal effects. In small doses, they are sedatives (tranquilizers); in larger doses they induce sleep. Pentothal is used as a general anesthetic. Pentobarbital and secobarbital are often used as a preanesthetic to prepare patients for surgery. Barbituric acid has none of these effects.

Barbiturates are dangerous because they are addictive, which means that a regular user will suffer withdrawal symptoms when their use is stopped. Barbiturates are especially dangerous when taken with alcohol because the combined effect (called a synergistic effect) is usually greater than the sum of the effects of either drug taken separately.

Reaction with Ammonia and Amines

Esters react with ammonia and with 1° and 2° amines to form amides. The nitrogen atom of the amine substitutes at the carbonyl carbon for the —OR group of the ester.

$$C_6H_5CH_2\overset{\displaystyle O}{\overset{\|}{C}}OCH_2CH_3 + NH_3 \longrightarrow C_6H_5CH_2\overset{\displaystyle O}{\overset{\|}{C}}NH_2 + CH_3CH_2OH$$

Ethyl phenylacetate   Phenylacetamide

EXAMPLE 18.6

Complete these equations. The stoichiometry of each is given in the equation.

(a) $CH_3(CH_2)_4\overset{\displaystyle O}{\overset{\|}{C}}OCH_3 + NH_3 \longrightarrow$
Methyl hexanoate

(b) $CH_3CH_2O\overset{\displaystyle O}{\overset{\|}{C}}OCH_2CH_3 + 2NH_3 \longrightarrow$
Diethyl carbonate

Solution

(a) $CH_3(CH_2)_4\overset{\displaystyle O}{\overset{\|}{C}}NH_2 + CH_3OH$ (b) $H_2N\overset{\displaystyle O}{\overset{\|}{C}}NH_2 + 2CH_3CH_2OH$
    Hexanamide   Urea

Problem 18.6

Complete these equations. The stoichiometry of each is given in the equation.

(a) $CH_3\overset{\displaystyle O}{\overset{\|}{C}}O\!-\!\!\langle\;\;\rangle\!\!-\!O\overset{\displaystyle O}{\overset{\|}{C}}CH_3 + 2NH_3 \longrightarrow$ (b) $\quad + NH_3 \longrightarrow$ ∎

18.5 Carboxylic Amides

Structure and Nomenclature

> **Amide** A compound in which a carbonyl group is bonded to a nitrogen atom, RCONR′₂; one or both of the R′ groups may be H, alkyl, or aryl groups

We discuss proteins, which are polyamides, in Chapter 21.

The functional group of a **carboxylic amide** is a carbonyl group bonded to a nitrogen atom. Amides are named by dropping the suffix "-oic acid" from the IUPAC name of the parent acid, or "-ic acid" from its common name, and adding "-amide." If the nitrogen atom of the amide is bonded to an alkyl or aryl group, the group is named and its location on nitrogen is indicated by "*N*-." Two alkyl or aryl groups on nitrogen are indicated by "*N,N*-di."

$$CH_3\overset{\displaystyle O}{\overset{\|}{C}}NH_2 \qquad CH_3\overset{\displaystyle O}{\overset{\|}{C}}NHCH_3 \qquad H\overset{\displaystyle O}{\overset{\|}{C}}N(CH_3)_2$$

Acetamide   *N*-Methylacetamide   *N,N*-Dimethylformamide
(a 1° amide)   (a 2° amide)   (a 3° amide)

Cyclic amides are called **lactams.** Their common names are derived in a manner similar to those of lactones, with the difference being that the suffix "-olactone" is replaced by "-olactam."

3-Butanolactam   6-Hexanolactam

EXAMPLE 18.7

Write the IUPAC name for each amide.

$$\text{(a) } CH_3CH_2CH_2\overset{\displaystyle O}{\overset{\|}{C}}NH_2 \quad \text{(b) } H_2N\overset{\displaystyle O}{\overset{\|}{C}}CH_2CH_2CH_2CH_2\overset{\displaystyle O}{\overset{\|}{C}}NH_2$$

Solution

Given first is the IUPAC name and then, in parentheses, the common name.
(a) Butanamide (butyramide, from butyric acid)
(b) Hexanediamide (adipamide, from adipic acid)

Problem 18.7 ▬▬▬▬

Draw a structural formula for each amide.
(a) *N*-Cyclohexylacetamide   (b) Benzamide ■

Reaction with Water. Hydrolysis

Amides require more vigorous conditions for hydrolysis in both acid and base than do anhydrides or esters. Hydrolysis in hot aqueous acid gives a carboxylic acid and an ammonium ion. Hydrolysis is driven to completion by the acid-base reaction between ammonia or the amine and acid to form an ammonium ion. One mole of acid is required per mole of amide.

$$CH_3CH_2CH_2\overset{\displaystyle O}{\overset{\|}{C}}NH_2 + H_2O + HCl \xrightarrow[\text{heat}]{H_2O} CH_3CH_2CH_2\overset{\displaystyle O}{\overset{\|}{C}}OH + NH_4^+ Cl^-$$

Butanamide   Butanoic acid

In aqueous base, the products of amide hydrolysis are a carboxylic acid salt and ammonia or an amine. This hydrolysis is driven to completion by the acid-base reaction between the carboxylic acid and base to form a salt. One mole of base is required per mole of amide.

$$CH_3\overset{\displaystyle O}{\overset{\|}{C}}NH\text{—}\bigcirc + NaOH \xrightarrow[\text{heat}]{H_2O} CH_3\overset{\displaystyle O}{\overset{\|}{C}}O^-Na^+ + H_2N\text{—}\bigcirc$$

N-Phenylethanamide   Sodium acetate   Aniline
(*N*-Phenylacetamide,
Acetanilide)

The Penicillins and Cephalosporins.
β-Lactam Antibiotics

The **penicillins** were discovered in 1928 by the Scottish bacteriologist Sir Alexander Fleming. As a result of the brilliant experimental work of Sir Howard Florey, an Australian pathologist, and Ernst Chain, a German chemist who fled Nazi Germany, penicillin G was introduced into the practice of medicine in 1943. For their pioneering work in developing one of the most effective antibiotics of all time, Fleming, Florey, and Chain were awarded the Nobel Prize in Medicine and Physiology in 1945.

The mold from which Fleming discovered penicillin was *Penicillium notatum,* a strain that gives a relatively low yield of penicillin. It was replaced in commercial production of the antibiotic by *P. chrysogenum,* a strain cultured from a mold found growing on a grapefruit in a market in Peoria, Illinois. The structural feature common to all penicillins is a **β-lactam** ring bonded to a five-membered sulfur-containing ring. The penicillins owe their antibacterial activity to a common mechanism that inhibits the biosynthesis of a vital part of bacterial cell walls.

The penicillins differ in the group bonded to the acyl carbon

Penicillin G
(a β-lactam antibiotic)

Soon after the penicillins were introduced into medical practice, penicillin-resistant strains of bacteria began to appear and have since proliferated. One approach to combating resistant strains is to synthesize newer, more effective penicillins. Among those developed are ampicillin, methicillin, and amoxicillin. Another approach is to search for newer, more effective β-lactam antibiotics. The most effective of these discovered so far are the **cephalosporins,** the first of which was isolated from the fungus *Cephalosporium acremonium.* This class of β-lactam antibiotics has an even

The cephalosporins differ in the group bonded to the acyl carbon and the side chain attached to the thiazine ring

A cephalosporin, a newer
generation β-lactam antibiotic

■ Macrophotograph of the fungus *Penicillium notatum* growing in a petri dish culture on Whickerham's agar. This fungus is a species that was used as an early source of the antibiotic penicillin. *(Andrew McClenaghan/Science Photo Library/Photo Researchers, Inc.)*

broader spectrum of antibacterial activity than the penicillins and is effective against many penicillin-resistant bacterial strains.

EXAMPLE 18.8

Write a balanced equation for the hydrolysis of each amide in concentrated aqueous HCl. Show all products as they exist in aqueous HCl.

(a) $CH_3\overset{O}{\overset{||}{C}}N(CH_3)_2$ (b) [structure: six-membered lactam ring with C=O and NH]

Solution

(a) Hydrolysis of *N,N*-dimethylacetamide gives acetic acid and dimethylamine. Dimethylamine, a base, reacts with HCl to form dimethylammonium ion, shown here as dimethylammonium chloride.

$$CH_3\overset{O}{\overset{||}{C}}N(CH_3)_2 + H_2O + HCl \xrightarrow[heat]{} CH_3\overset{O}{\overset{||}{C}}OH + (CH_3)_2NH_2^+Cl^-$$

(b) Hydrolysis of this lactam gives the protonated form of 5-aminopentanoic acid.

[structure: six-membered lactam ring with C=O and NH] $+ H_2O + HCl \xrightarrow[heat]{} HO\overset{O}{\overset{||}{C}}CH_2CH_2CH_2CH_2NH_3^+Cl^-$

Problem 18.8

Write a balanced equation for the hydrolysis of the amides in Example 18.8 in concentrated aqueous NaOH. Show all products as they exist in aqueous NaOH. ∎

18.6 Interconversion of Functional Groups

A useful way to think about the relative reactivities of carboxylic acids and their three functional derivatives is shown in Figure 18.1. Any compound lower in this figure can be prepared from any compound above it by treatment with an appropriate oxygen or nitrogen compound. An anhydride, for

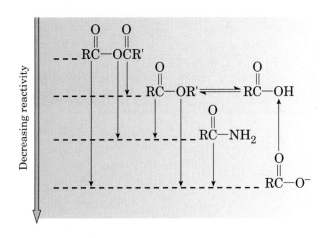

FIGURE 18.1 Relative reactivities of carboxylic acid derivatives toward substitution at the carbonyl carbon. A more reactive derivative may be converted to a less reactive derivative by treatment with an appropriate reagent.

From Cocaine to Procaine and Beyond

Cocaine (*The Merck Index*, 12th ed., #2517) is an alkaloid present in the leaves of the South American coca plant *Erythroxylon coca*. It was first isolated in 1880, and soon thereafter its property as a local anesthetic was discovered. Cocaine was introduced into medicine and dentistry in 1884 by two young Viennese physicians, Sigmund Freud and Karl Koller. Unfortunately the use of cocaine can create a dependence, as Freud himself observed when he used it to wean a colleague from morphine and thereby produced one of the first documented cases of cocaine addiction.

Procaine
(Novocain)

Lidocaine
(Xylocaine)

Etidocaine
(Duranest)

Cocaine

Cocaine reduces fatigue, permits greater physical endurance, and gives a feeling of tremendous confidence and power. In some of the Sherlock Holmes stories, the great detective injects himself with a 7 percent solution of cocaine to overcome boredom.

After determining cocaine's structure, chemists could ask: How is the structure of cocaine related to its anesthetic effects? Is it possible to separate the anesthetic effects from the habituation? If these questions could be answered, it might be possible to prepare synthetic drugs with the structural features essential for the anesthetic activity but without those giving rise to the undesirable effects. Chemists focused on three structural features of cocaine: its benzoic ester, its basic nitrogen atom, and something of its carbon skeleton. This search resulted in 1905 in the synthesis of procaine (*The Merck Index*, 12th ed., #7937), which almost immediately replaced cocaine in dentistry and surgery. Lidocaine (*The Merck Index*, 12th ed., #5505) was introduced in 1948 and today is one of the most widely used local anesthetics. More recently, other members of the "caine" family of local anesthetics have been introduced, for example, etidocaine (*The Merck Index*, 12th ed., #3907). All these local anesthetics are administered as their water-soluble hydrochloride salts.

Thus, seizing on clues provided by nature, chemists have been able to synthesize drugs far more suitable for a specific function than anything known to be produced by nature itself.

■ Nineteenth-century advertisement for cocaine.
(National Library of Medicine)

example, can be converted to an ester, an amide, or a carboxylic acid. An ester or amide, however, does not react with a carboxylic acid to give an anhydride.

18.7 Phosphoric Esters and Anhydrides

Phosphoric Anhydrides

Because of the special importance of phosphoric anhydrides in biochemical systems, we include them here to show the similarity between them and the anhydrides of carboxylic acids. The functional group of a **phosphoric anhydride** is two phosphoryl (P=O) groups bonded to the same oxygen atom. Shown here are structural formulas for two anhydrides of phosphoric acid and the ions derived by ionization of the acidic hydrogens of each.

$$\underset{\substack{\text{Diphosphoric acid}\\ \text{(Pyrophosphoric acid)}}}{HO-\overset{\overset{\displaystyle O}{\|}}{\underset{\underset{\displaystyle OH}{|}}{P}}-O-\overset{\overset{\displaystyle O}{\|}}{\underset{\underset{\displaystyle OH}{|}}{P}}-OH} \qquad \underset{\substack{\text{Diphosphate ion}\\ \text{(Pyrophosphate ion)}}}{^-O-\overset{\overset{\displaystyle O}{\|}}{\underset{\underset{\displaystyle O^-}{|}}{P}}-O-\overset{\overset{\displaystyle O}{\|}}{\underset{\underset{\displaystyle O^-}{|}}{P}}-O^-}$$

At physiological pH (approximately 7.3 to 7.4), pyrophosphoric acid exists as the pyrophosphate ion and triphosphoric acid exists as the triphosphate ion.

$$\underset{\text{Triphosphoric acid}}{HO-\overset{\overset{\displaystyle O}{\|}}{\underset{\underset{\displaystyle OH}{|}}{P}}-O-\overset{\overset{\displaystyle O}{\|}}{\underset{\underset{\displaystyle OH}{|}}{P}}-O-\overset{\overset{\displaystyle O}{\|}}{\underset{\underset{\displaystyle OH}{|}}{P}}-OH} \qquad \underset{\text{Triphosphate ion}}{^-O-\overset{\overset{\displaystyle O}{\|}}{\underset{\underset{\displaystyle O^-}{|}}{P}}-O-\overset{\overset{\displaystyle O}{\|}}{\underset{\underset{\displaystyle O^-}{|}}{P}}-O-\overset{\overset{\displaystyle O}{\|}}{\underset{\underset{\displaystyle O^-}{|}}{P}}-O^-}$$

Phosphoric Esters

Phosphoric acid has three —OH groups and forms mono-, di-, and triphosphoric esters, which are named by giving the name(s) of the alkyl or aryl group(s) bonded to oxygen followed by the word "phosphate," as for example dimethyl phosphate. In more complex **phosphoric esters,** it is common to name the organic molecule and then indicate the presence of the phosphoric ester by either the word "phosphate" or the prefix "phospho-." Dihydroxyacetone phosphate, for example, is an intermediate in glycolysis (Section 27.2). Pyridoxal phosphate is one of the metabolically active forms of vitamin B_6. The last two phosphoric esters are shown as they are ionized at pH 7.4, the pH of blood plasma.

Vitamin B_6, pyridoxal. *(Charles D. Winters)*

$$\underset{\text{Dimethyl phosphate}}{CH_3O-\overset{\overset{\displaystyle O}{\|}}{\underset{\underset{\displaystyle OCH_3}{|}}{P}}-OH} \qquad \underset{\text{Dihydroxyacetone phosphate}}{\overset{\overset{\displaystyle CH_2OH}{|}}{\underset{\underset{\displaystyle CH_2}{\|}}{\underset{\underset{\displaystyle O^-}{|}}{\overset{C=O}{}}}-O-\overset{\overset{\displaystyle O}{\|}}{\underset{\underset{\displaystyle O^-}{|}}{P}}-O^-}} \qquad \underset{\text{Pyridoxal phosphate}}{}$$

Step-growth polymerization
A polymerization in which chain growth occurs in a stepwise manner between difunctional monomers, as for example between adipic acid and hexamethylenediamine to form nylon 66

18.8 Step-Growth Polymerizations

Polymerizations in which chain growth occurs in a stepwise manner are called **step-growth** or **condensation polymerizations.** Step-growth polymers are formed by reaction between molecules containing two functional groups, with each new bond created in a separate step. In this section, we discuss three types of step-growth polymers: polyamides, polyesters, and polycarbonates.

Polyamides

In the early 1930s, chemists at E. I. DuPont de Nemours & Company began fundamental research into the reactions between dicarboxylic acids and diamines to form **polyamides** and in 1934 synthesized nylon 66, the first purely synthetic fiber. Nylon 66 is so named because it is synthesized from two different monomers, each containing six carbon atoms.

In the synthesis of nylon 66, hexanedioic acid and 1,6-hexanediamine are dissolved in aqueous ethanol and then heated in an autoclave to 250°C and an internal pressure of 15 atm. Under these conditions, —COOH and —NH₂ groups react by loss of H₂O to form a polyamide. Nylon 66 formed under these conditions melts at 250 to 260°C and has a molecular weight range of 10 000 to 20 000 g/mol.

> **Polyamide** A polymer in which each monomer unit is joined to the next by an amide bond, as for example nylon 66

$$n\,HOC(CH_2)_4COH \;+\; n\,H_2N(CH_2)_6NH_2 \xrightarrow[-H_2O]{heat} \left(C(CH_2)_4CNH(CH_2)_6NH \right)_n$$

Hexanedioic acid (Adipic acid)   1,6-Hexanediamine (Hexamethylenediamine)   Nylon 66 (a polyamide)

The nylons are a family of polymers, the members of which have subtly different properties that suit them to one use or another. The two most widely used members of this family are nylon 66 and nylon 6, the latter so named because it is synthesized from hexanolactam, a six-carbon monomer. In this synthesis, the lactam is partially hydrolyzed to 6-aminohexanoic acid and then heated to 250°C to bring about polymerization. Nylon 6 is fabricated into fibers, brush bristles, rope, high-impact moldings, and tire cords.

$$n \quad \xrightarrow[\text{2. heat}]{\text{1. partial hydrolysis}} \left(NH(CH_2)_5C \right)_n$$

Hexanolactam (Caprolactam)   Nylon 6 (a polyamide)

Based on extensive research into the relationships between molecular structure and bulk physical properties, scientists at DuPont reasoned that a polyamide containing benzene rings would be stiffer and stronger than either nylon 66 or nylon 6. The result was a polyamide, which DuPont named Kevlar.

1,4-Benzenedicarboxylic acid (Terephthalic acid)   1,4-Benzenediamine (*p*-Phenylenediamine)   Kevlar (a polyaromatic amide)

One of the remarkable features of Kevlar is that it weighs less than other materials of similar strength. For example, a cable woven of Kevlar has a strength equal to that of a similarly woven steel cable. Yet the Kevlar cable has only 20 percent of the weight of the steel cable! Kevlar now finds use in such articles as anchor cables for offshore drilling rigs and reinforcement fibers for automobile tires. It is also woven into a fabric that is so tough that it can be used for bulletproof vests, jackets, and raincoats.

Polyesters

The first **polyester,** developed in the 1940s, involved polymerization of 1,4-benzenedicarboxylic acid with 1,2-ethanediol to give poly(ethylene terephthalate), abbreviated PET. Virtually all PET is now made from the diester by the following reaction.

Bulletproof vests have a thick layer of Kevlar. *(Charles D. Winters)*

$$n\text{CH}_3\text{OC}\!-\!\!\langle \rangle\!\!-\!\text{COCH}_3 + n\text{HOCH}_2\text{CH}_2\text{OH} \xrightarrow{\text{heat}}$$

Dimethyl terephthalate   1,2-Ethanediol
(Ethylene glycol)

$$\left(\!\!\text{C}\!-\!\!\langle \rangle\!\!-\!\text{COCH}_2\text{CH}_2\text{O}\!\!\right)_{\!n} + 2n\text{CH}_3\text{OH}$$

Poly(ethylene terephthalate)
(Dacron, Mylar)

> **Polyester** A polymer in which each monomer unit is joined to the next by an ester bond, as for example poly(ethylene terephthalate)

The crude polyester can be melted, extruded, and then drawn to form the textile fiber Dacron polyester, outstanding features of which are its stiffness (about four times that of nylon 66), very high strength, and remarkable resistance to creasing and wrinkling. Because the early Dacron polyester fibers were harsh to the touch due to their stiffness, they were usually blended with cotton or wool to make acceptable textile fibers. Newly developed fabrication techniques now produce less harsh Dacron polyester textile fibers. PET is also fabricated into Mylar films and recyclable plastic beverage containers.

Polycarbonates

A **polycarbonate,** the most familiar of which is Lexan, is formed by reaction between the disodium salt of bisphenol A and phosgene. Phosgene is a derivative of carbonic acid, H_2CO_3, in which both —OH groups have been replaced by chlorine atoms.

> **Polycarbonate** A polyester in which the carboxyl groups are derived from carbonic acid

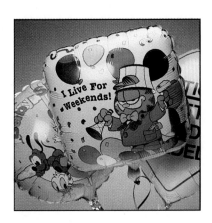

Mylar can be made into extremely strong films. Because the film has very tiny pores, it is used for balloons that can be inflated with helium; the helium atoms diffuse only slowly through the pores of the film. *(Charles D. Winters)*

A polycarbonate hockey mask. *(Charles D. Winters)*

BOX 18F

Stitches That Dissolve

As the technological capabilities of medicine have grown, the demand for synthetic materials that can be used inside the body has increased as well. Polymers have many of the characteristics of an ideal biomaterial: they are lightweight and strong, are inert or biodegradable depending on their chemical structure, and have physical properties (softness, rigidity, elasticity) that are easily tailored to match those of natural tissues.

Even though most medical uses of polymeric materials require biostability, applications have been developed that require them to be biodegradable. An example is the use of

two organic polymers—poly(glycolic acid) and poly(lactic acid)—as absorbable sutures, which go under the trade name of Lactomer.

Traditional suture materials such as catgut must be removed by a health care specialist after they have served their purpose. Stitches of Lactomer, however, are hydrolyzed slowly over a period of approximately 2 weeks. By the time the torn tissues have healed, the stitches have hydrolyzed, and no suture removal is necessary. Glycolic and lactic acids formed during this hydrolysis are metabolized and excreted by the body.

$$
\underset{\text{Glycolic acid}}{\text{HOCH}_2\text{COH}} + \underset{\text{Lactic acid}}{\underset{\overset{|}{\text{CH}_3}}{\text{HOCHCOH}}} \xrightarrow[-n\text{H}_2\text{O}]{\text{polymerization}} \underset{\text{A polymer of}}{\underset{\underset{\text{poly(glycolic acid)-poly(lactic acid)}}{}}{\left(\!\!\text{OCH}_2\text{COCHC}\underset{\overset{|}{\text{CH}_3}}{}\right)_{\!\!n}}}
$$

Lexan is a tough, transparent polymer that has high impact and tensile strengths and retains its properties over a wide temperature range. It is used in sporting equipment (helmets and face masks); to make light, impact-resistant housings for household appliances; and in the manufacture of safety glass and unbreakable windows.

SUMMARY

The functional group of a **carboxylic acid** is the **carboxyl group, —COOH (—CO$_2$H).** IUPAC names of carboxylic acids are derived from the parent alkane by dropping the suffix "-e" and adding "-oic acid." Dicarboxylic acids are named as "-dioic acids."

Carboxylic acids are polar compounds and have higher boiling points and are more soluble in water than alcohols, aldehydes, ketones, and ethers of comparable molecular weight. A carboxylic acid consists of two regions of distinctly different polarity: a polar **hydrophilic** carboxyl group, which increases solubility in water, and a nonpolar **hydrophobic** hydrocarbon chain, which decreases solubility in water.

A **carboxylic anhydride** contains two carbonyl groups bonded to the same oxygen, $RCO-O-COR'$. A **carboxylic ester** contains a carbonyl group bonded to an $-OR'$ group, $RCOOR'$; R' may be an alkyl or aryl group. A **carboxylic amide** contains a carbonyl group bonded to a nitrogen atom, $RCONR'_2$; one or both of the R' groups may be H, an alkyl group, or an aryl group.

Step-growth polymerizations involve the step-wise reaction of difunctional monomers. Important commercial polymers synthesized through step-growth processes include polyamides, polyesters, and polycarbonates.

KEY TERMS

Carboxyl group (Section 18.1)
Carboxylic acid (Section 18.2)
Carboxylic amide (Section 18.5)
Carboxylic anhydride (Section 18.3)
Carboxylic ester (Section 18.4)
Cephalosporins (Box 18D)
Ester (Section 18.2)
Fischer esterification (Section 18.2)

Hydrophilic (Section 18.2)
Hydrophobic (Section 18.2)
Lactam (Section 18.5)
β-Lactam antibiotics (Box 18D)
Lactone (Section 18.4)
Penicillins (Box 18D)
Phosphoric anhydride (Section 18.7)

Phosphoric ester (Section 18.7)
Polyamide (Section 18.8)
Polycarbonate (Section 18.8)
Polyester (Section 18.8)
Saponification (Section 18.4)
Step-growth polymerizations
(Section 18.8)

KEY REACTIONS

1. Acidity of Carboxylic Acids (Section 18.2)
Values of pK_a for most unsubstituted aliphatic and aromatic carboxylic acids are within the range 4 to 5.

$$CH_3COH \rightleftharpoons CH_3CO^- + H^+ \qquad pK_a = 4.76$$

2. Reaction of Carboxylic Acids with Bases (Section 18.2)
Carboxylic acids form water-soluble salts with alkali metal hydroxides, carbonates, bicarbonates, as well as with ammonia and amines.

$$C_6H_5COH + NaOH \xrightarrow{H_2O} C_6H_5CO^-Na^+ + H_2O$$

3. Fischer Esterification (Section 18.2)
Fischer esterification is reversible, and to achieve high yields of ester, it is necessary to force the equilibrium to the right. One way to accomplish this is to use an excess of the alcohol. Another way is to remove water as it is formed.

$$CH_3COH + CH_3CH_2CH_2OH \underset{}{\overset{H_2SO_4}{\rightleftharpoons}} CH_3COCH_2CH_2CH_3 + H_2O$$

4. Hydrolysis of a Carboxylic Anhydride (Section 18.3)
Low-molecular-weight anhydrides react readily with water. Higher-molecular-weight anhydrides react less rapidly.

$$CH_3COCCH_3 + H_2O \longrightarrow CH_3COH + HOCCH_3$$

5. Reaction of a Carboxylic Anhydride with an Alcohol (Section 18.3)
Treatment of an anhydride with an alcohol gives an ester and a carboxylic acid.

$$CH_3COCCH_3 + HOCH_2CH_3 \longrightarrow CH_3COCH_2CH_3 + CH_3COH$$

6. Reaction of a Carboxylic Anhydride with Ammonia or an Amine (Section 18.3)

Reaction requires 2 moles of ammonia or amine: 1 mole to form the amide and 1 mole to neutralize the carboxylic acid byproduct.

$$CH_3\overset{O}{\overset{||}{C}}O\overset{O}{\overset{||}{C}}CH_3 + 2NH_3 \longrightarrow CH_3\overset{O}{\overset{||}{C}}NH_2 + CH_3\overset{O}{\overset{||}{C}}O^- NH_4^+$$

7. Hydrolysis of a Carboxylic Ester (Section 18.4)

Esters are hydrolyzed rapidly only in the presence of acid or base. Acid is a catalyst. Base is required in an equimolar amount.

CH₃CO— (cyclohexyl) + NaOH $\xrightarrow{H_2O}$

CH₃CO⁻Na⁺ + HO— (cyclohexyl)

8. Reaction of a Carboxylic Ester with Ammonia or an Amine (Section 18.4)

Treatment of an ester with ammonia, a primary amine, or a secondary amine gives an amide.

(phenyl)—CH₂COCH₃ + NH₃ ⟶

(phenyl)—CH₂CNH₂ + CH₃OH

9. Hydrolysis of a Carboxylic Amide (Section 18.5)

Either acid or base is required in an amount equivalent to that of the amide.

$$CH_3CH_2CH_2\overset{O}{\overset{||}{C}}NH_2 + H_2O + HCl \xrightarrow[\text{heat}]{H_2O}$$

$$CH_3CH_2CH_2\overset{O}{\overset{||}{C}}OH + NH_4^+\ Cl^-$$

$$CH_3CH_2CH_2\overset{O}{\overset{||}{C}}NH_2 + NaOH \xrightarrow[\text{heat}]{H_2O}$$

$$CH_3CH_2CH_2\overset{O}{\overset{||}{C}}O^-\ Na^+ + NH_3$$

CONCEPTUAL PROBLEMS

Difficult problems are designated by an asterisk.

18.A Alcohols, phenols, and carboxylic acids all contain an —OH group. Which are the strongest acids? Which are the weakest acids?

18.B What type of structural feature do the anhydrides of phosphoric acid and carboxylic acids have in common?

*18.C Write short sections of two parallel chains (each chain running in the same direction) of nylon 66 and show how it is possible to align them such that there is hydrogen bonding between the N—H groups of one chain and C=O groups of the parallel chain.

18.D Propose an explanation for the fact that substitution of an atom of high electronegativity on the carbon adjacent to a carboxyl group increases the acidity of the carboxyl group. Given your explanation, which would you predict to be the stronger acid, trichloroacetic acid or trifluoroacetic acid?

PROBLEMS

Difficult problems are designated by an asterisk.

Structure and Nomenclature

18.9 Name and draw structural formulas for the four carboxylic acids of molecular formula $C_5H_{10}O_2$. Which of these carboxylic acids is/are chiral?

*18.10 Write the IUPAC name for each compound.

(a) H₃C—CH(CH₃)—CH₂—CH₂—CH₂—CH(CH₃)—C(=O)—OH

(b) CH₃CH₂CH₂CH(NH₂)COOH

(c) $CH_3(CH_2)_4COOH$

        OH
        |
(d) $HOOCCHCH_2COOH$

18.11 Draw a structural formula for each carboxylic acid.
(a) 4-Nitrophenylacetic acid
(b) 4-Aminobutanoic acid
(c) 4-Phenylbutanoic acid
(d) *cis*-3-Hexenedioic acid
(e) 2,3-Dihydroxypropanoic acid
(f) 3-Oxohexanoic acid

18.12 Draw structural formulas for these salts.
(a) Sodium benzoate   (b) Lithium acetate
(c) Ammonium acetate   (d) Disodium adipate
(e) Sodium salicylate   (f) Calcium butanoate

18.13 Calcium oxalate is a major component of kidney stones. Draw a structural formula of this compound.

18.14 The monopotassium salt of oxalic acid is present in certain leafy vegetables, including rhubarb. Both oxalic acid and its salts are poisonous in high concentrations. Draw the structural formula of monopotassium oxalate.

Physical Properties

18.15 Draw the structural formula for the dimer formed when two molecules of formic acid form hydrogen bonds with each other.

*__18.16__ Propanedioic (malonic) acid forms an internal hydrogen bond (the H of one COOH group forms a hydrogen bond with an O of the other COOH group). Draw a structural formula to show this internal hydrogen bond.

18.17 Hexanoic (caproic) acid has a solubility in water of about 1 g/100 g of water. Which part of the molecule contributes to water solubility, and which part prevents solubility?

18.18 Arrange the compounds in each set in order of increasing boiling point.

          O   O
          ‖   ‖
(a) $CH_3(CH_2)_6COH$ $CH_3(CH_2)_6CH$

                               $CH_3(CH_2)_6CH_2OH$

          O
          ‖
(b) CH_3CH_2COH $CH_3CH_2CH_2CH_2OH$

                             $CH_3CH_2OCH_2CH_3$

Preparation of Carboxylic Acids

18.19 Complete these oxidations.

(a) $CH_3(CH_2)_4CH_2OH \xrightarrow[H_2SO_4]{K_2Cr_2O_7}$

        CHO
(b) [benzene ring with CHO, OCH_3, OH substituents] $+ Ag(NH_3)_2^+ \longrightarrow$

(c) $HO-$[cyclohexane ring]$-CH_2OH \xrightarrow[H_2SO_4]{K_2Cr_2O_7}$

18.20 Draw the structural formula of a compound of the given molecular formula that, on oxidation by potassium dichromate, gives the carboxylic acid or dicarboxylic acid shown.

                               O
                               ‖
(a) $C_6H_{14}O \xrightarrow{oxidation} CH_3(CH_2)_4COH$

                               O
                               ‖
(b) $C_6H_{12}O \xrightarrow{oxidation} CH_3(CH_2)_4COH$

                         O   O
                         ‖   ‖
(c) $C_6H_{14}O_2 \xrightarrow{oxidation} HOC(CH_2)_4COH$

Acidity of Carboxylic Acids

18.21 Complete these acid-base reactions:

(a) [benzene ring]$-CH_2COOH + NaOH \longrightarrow$

(b) $CH_3CH=CHCH_2COOH + NaHCO_3 \longrightarrow$

(c) [benzene ring with COOH and OCH_3 substituents] $+ NaHCO_3 \longrightarrow$

        OH
        |
(d) $CH_3CHCOOH + H_2NCH_2CH_2OH \longrightarrow$

(e) $CH_3CH=CHCH_2COO^-Na^+ + HCl \longrightarrow$

18.22 What form(s) of acetic acid are present at pH 2.0 or lower? At pH 4.0 to 5.0? At pH 8.0 or higher?

*__18.23__ The normal pH range for blood plasma is 7.35 to 7.45. Under these conditions, would you expect the carboxyl group of lactic acid ($pK_a = 4.07$) to exist primarily as a carboxyl group or as a carboxylic anion? Explain.

*__18.24__ The pK_a of ascorbic acid (Box 19B) is 4.10. Would you expect ascorbic acid dissolved in blood plasma, pH 7.35 to 7.45, to exist primarily as ascorbic acid or as ascorbate anion? Explain.

Reactions of Carboxylic Acids

*__18.25__ You are given a mixture of octanoic acid and 1-octanol, neither of which is soluble in water. Describe a simple method for separating them.

18.26 Give the expected organic products formed when phenylacetic acid, $C_6H_5CH_2COOH$, is treated with each reagent.
(a) $NaHCO_3, H_2O$ (b) $NaOH, H_2O$
(c) NH_3, H_2O (d) $CH_3OH + H_2SO_4$ (catalyst)

18.27 Complete these examples of Fischer esterification. Assume an excess of the alcohol.

(a) $CH_3COOH + HOCH_2CH_2CH(CH_3)_2 \overset{H^+}{\rightleftharpoons}$

(b) [benzene ring with two COOH substituents] $+ CH_3OH \overset{H^+}{\rightleftharpoons}$

(c) $HOOC(CH_2)_2COOH + CH_3CH_2OH \overset{H^+}{\rightleftharpoons}$

18.28 Methyl 2-hydroxybenzoate (methyl salicylate) has the odor of oil of wintergreen. This ester is prepared by Fischer esterification of 2-hydroxybenzoic acid (salicylic acid) with methanol. Draw the structural formula of methyl 2-hydroxybenzoate.

18.29 From what carboxylic acid and alcohol is each ester derived?

(a) CH_3CO—⬡—$OCCH_3$ (each C=O)

(b) $CH_3OCCH_2CH_2COCH_3$ (each C=O)

(c) ⬡—$COCH_3$ (C=O)

(d) $CH_3CH_2CH{=}CHCOCH(CH_3)_2$ (C=O)

18.30 When 5-hydroxypentanoic acid is treated with an acid catalyst, it forms a lactone (a cyclic ester). Draw the structural formula of this lactone.

Structure and Nomenclature of Anhydrides, Esters, and Amides

18.31 Draw a structural formula for each compound.
(a) Dimethyl carbonate
(b) *p*-Nitrobenzamide
(c) Ethyl 3-hydroxybutanoate
(d) Diethyl oxalate
(e) Ethyl *cis*-2-pentenoate
(f) Butanoic anhydride

18.32 Write the IUPAC name for each compound.

(a) ⬡—COC—⬡ (O O)

(b) $CH_3(CH_2)_8COCH_3$ (C=O)

(c) $CH_3(CH_2)_4CNHCH_3$ (C=O)

(d) H_2N—⬡—CNH_2 (C=O)

(e) CH_3CO—⬠ (C=O)

(f) $CH_3CCHCOCH_2CH_3$ (O O)

18.33 When oil from the head of the sperm whale is cooled, spermaceti, a translucent wax with a white, pearly luster crystallizes from the mixture. Spermaceti, which makes up 11 percent of whale oil, is composed mainly of hexadecyl hexadecanoate. (Hexadecane has 16 carbons in an unbranched chain.) At one time, spermaceti was widely used to make cos-

metics, fragrant soaps, and candles. Draw a structural formula of spermaceti.

Physical Properties of Esters

18.34 Acetic acid and methyl formate are constitutional isomers. Both are liquids at room temperature: one with a boiling point of 32°C and the other with a boiling point of 118°C. Which of the two has the higher boiling point?

18.35 Flavoring agents are the largest class of food additives. Each ester listed here is a synthetic flavor additive, with a flavor very close to the natural flavor. While these esters are the major components of the natural flavors, each natural flavor contains many more components, some in only trace amounts. Draw the structural formula of each ester. (Isopentane is the common name for 2-methylbutane.)
(a) Ethyl formate (rum)
(b) Isopentyl acetate (banana)
(c) Octyl acetate (orange)
(d) Methyl butanoate (apple)
(e) Ethyl butanoate (pineapple)
(f) Methyl 2-aminobenzoate (grape)

Reactions of Anhydrides, Esters, and Amides

18.36 Which of these types of compounds will produce bubbles of CO_2 when added to an aqueous solution of sodium bicarbonate? (a) A carboxylic acid (b) A carboxylic ester (c) The sodium salt of a carboxylic acid

18.37 Propanoic acid and methyl acetate are constitutional isomers, and both are liquids at room temperature. One of these compounds has a boiling point of 141°C; the other has a boiling point of 57°C. Which compound has which boiling point?

CH_3CH_2COH (C=O) CH_3COCH_3 (C=O)
Propanoic acid   Methyl acetate

18.38 Show how to synthesize each ester from a carboxylic acid and an alcohol by Fischer esterification

(a) ⬡—$OC(CH_2)_4CH_3$ (C=O)

(b) $(CH_3)_2CHCOCH_2CH_3$ (C=O)

18.39 Show how to prepare these amides by reaction of an ester with ammonia or an amine.

(a) ⬡—$NHC(CH_2)_4CH_3$ (C=O)

(b) $(CH_3)_2CHCN(CH_3)_2$ (C=O)

(c) $H_2NC(CH_2)_4CNH_2$ (O O)

18.40 Write the product(s) of treatment of propanoic anhydride with each reagent.
(a) Ethanol (1 equivalent)
(b) Ammonia (2 equivalents)

18.41 Complete these reactions.

(a) $CH_3O-\!\!\!\bigcirc\!\!\!-NH_2$ + $CH_3\overset{O}{\overset{\|}{C}}O\overset{O}{\overset{\|}{C}}CH_3$ ⟶

(b) $CH_3\overset{O}{\overset{\|}{C}}OCH_3$ + $HN\!\!\bigcirc$ ⟶

18.42 The analgesic phenacetin is synthesized by treating 4-ethoxyaniline with acetic anhydride. Write an equation for the formation of phenacetin.

$CH_3CH_2O-\!\!\!\bigcirc\!\!\!-NH_2$ 4-Ethoxyaniline

18.43 Nicotinic acid, more commonly named niacin, is one of the B vitamins. Show how nicotinic acid can be converted to ethyl nicotinate and then to nicotinamide.

Nicotinic acid $\xrightarrow{?}$ Ethyl nicotinate $\xrightarrow{?}$ Nicotinamide

Nicotinic acid (Niacin) — Ethyl nicotinate — Nicotinamide

18.44 Define and give an example of saponification.

18.45 What product is formed when ethyl benzoate is treated with these reagents?
(a) H_2O, NaOH, heat
(b) $CH_3CH_2CH_2CH_2NH_2$

18.46 What product is formed when benzamide is treated with these reagents.
(a) H_2O, HCl, heat
(b) NaOH, H_2O, heat

*18.47 Meprobamate is a tranquilizer prescribed under one or more of 58 different trade names. Phenobarbital is a long-acting sedative, hypnotic, and anticonvulsant.
(a) Name the functional groups in each compound.
(b) Draw structural formulas of the products from complete hydrolysis of all functional groups in aqueous NaOH.

Meprobamate

Phenobarbital

*18.48 *N,N*-Diethyl-*m*-toluamide (Deet) the active ingredient in several common insect repellents, is synthesized from 3-methylbenzoic acid (*m*-toluic acid) and diethylamine. Show how this synthesis can be accomplished by first converting *m*-toluic acid to an ester and then to Deet.

N,N-Diethyl-*m*-toluamide (Deet)

18.49 Following are structural formulas for two local anesthetics. Lidocaine was introduced in 1948 and is now the most widely used local anesthetic for infiltration and regional anesthesia. Its hydrochloride is marketed under the name Xylocaine. Mepivacaine (hydrochloride marketed as Carbocaine) is faster and somewhat longer in duration than lidocaine.

Lidocaine (Xylocaine)

Mepivacaine (Carbocaine)

(a) Name the functional groups in each compound.
(b) What similarities in structure do you find between these compounds?

Anhydrides and Esters of Phosphoric Acid

18.50 Draw structural formulas for the mono-, di-, and triethyl esters of phosphoric acid.

18.51 Dihydroxyacetone and phosphoric acid form a monoester called dihydroxyacetone phosphate, which is an intermediate in glycolysis (Section 27.2). Draw a structural formula for this monophosphate ester.

$$HO-CH_2-\overset{\overset{\displaystyle O}{\|}}{C}-CH_2-OH$$

1,3-Dihydroxy-2-propanone
(Dihydroxyacetone)

18.52 Show how triphosphoric acid can be formed from three molecules of phosphoric acid. How many H_2O molecules are split out?

18.53 Write an equation for the hydrolysis of trimethyl phosphate to dimethyl phosphate and methanol.

Boxes

18.54 (Box 18A) What is the compound in willow bark that is responsible for its ability to relieve pain? How is this compound related to salicylic acid?

18.55 (Box 18A) Name the two functional groups in aspirin.

18.56 (Box 18A) What is the structural relationship between aspirin and ibuprofen? Between aspirin and naproxen?

18.57 (Box 18B) Name the oxygen-containing functional group in coumarin. Name the two types of oxygen-containing functional groups in dicoumarol.

18.58 (Box 18B) What is the structural relationship between warfarin and dicoumarol?

18.59 (Box 18B) What is a medical use of warfarin?

18.60 (Box 18C) Urea is very soluble in water. Show how water can form hydrogen bonds with this molecule.

18.61 (Box 18C) Barbiturates are derived from urea. Identify the portion of the structure of pentobarbital, secobarbital, and thiopental that is derived from urea.

18.62 (Box 18D) Identify the β-lactam portion of penicillin G. Identify the three stereocenters in penicillin G.

18.63 (Box 18E) Cocaine has two esters. Identify each. Draw the structural formulas of the products formed by hydrolysis of each ester group in cocaine in aqueous acid.

18.64 (Box 18E) Name the three functional groups in procaine.

18.65 (Box 18E) Of the two nitrogen atoms in procaine, which is the stronger base? Draw the structural formula of the product formed by reaction of procaine with 1 mole of HCl.

18.66 (Box 18E) Name the two functional groups in lidocaine; in etidocaine.

18.67 (Box 18F) Why do Lactomer stitches dissolve within 2 to 3 weeks following surgery?

Additional Problems

18.68 Megatomoic acid, the sex attractant of the female black carpet beetle, has the following structure. The name of the parent alkane with 14 carbon atoms is tetradecane.

$$CH_3(CH_2)_7CH=CHCH=CHCH_2COOH$$
Megatomoic acid

(a) What is its IUPAC name?
(b) State the number of stereoisomers possible for this compound.

18.69 The IUPAC name of ibuprofen is 2-(4-isobutyl-phenyl)propanoic acid. Draw the structural formula of ibuprofen.

*__*18.70__ Potassium sorbate is added as a preservative to certain foods to prevent bacteria and molds from causing spoilage and to extend the products' shelf lives. The IUPAC name of potassium sorbate is potassium *(trans,trans)*-2,4-hexadienoate. Draw a structural formula of potassium sorbate.

18.71 Zinc 10-undecenoate is used to treat certain fungal infections, particularly *tinea pedis* (athlete's foot). Draw a structural formula of this zinc salt. Undecanoic acid has 11 carbons in its parent chain.

18.72 1-Heptanol, heptanal, and hexanoic acid all have approximately the same molecular weight.
(a) Draw a structural formula for each molecule.
(b) Their boiling points, from lowest to highest, are 153, 176, and 205°C. Which compound has which boiling point? Explain.

*__*18.73__ Excess ascorbic acid (pK_a 4.10) is excreted in the urine, the pH of which is normally in the range 4.8 to 8.4. What form of ascorbic acid would you expect to be present in urine of pH 8.4, ascorbic acid or ascorbate anion? Explain.

*__*18.74__ The pH of human gastric juice is normally in the range 1.0 to 3.0. What form of lactic acid (pK_a = 4.07), would you expect to be present in the stomach, lactic acid or its anion? Explain.

18.75 Benzocaine, a topical anesthetic, is prepared by treatment of 4-aminobenzoic acid with ethanol in the presence of an acid catalyst followed by neutralization. Draw the structural formula of benzocaine.

18.76 You are given a mixture of octanoic acid and ethyl octanoate, neither of which is soluble in water. Describe a simple test to tell which compound is which.

18.77 The analgesic acetaminophen is synthesized by treatment of 4-aminophenol with one equivalent of acetic anhydride. Write an equation for the formation of acetaminophen. (Hint: The —NH_2 group is more reactive with acetic anhydride than the —OH group.)

*__*18.78__ Procaine (its hydrochloride is marketed as Novocain) was one of the first local anesthetics for infiltration and regional anesthesia. Show how to synthesize procaine using the given reagents as sources of carbon atoms.

p-Aminobenzoic
acid

2-Diethylaminoethanol

Procaine
(Novocaine)

18.79 Write structural formulas for the acid form and the sodium salt form of dimethyl phosphate.

*__18.80__ Following is a structural formula for 1,3-diphospho-glycerate, an intermediate in glycolysis (Section 27.2). This molecule contains a mixed anhydride (an anhydride of a carboxylic acid and phosphoric acid) and one phosphoric ester. Draw structural formulas for products formed by hydrolysis of the anhydride and ester bonds. Show each product as it would exist in solution at pH 7.40.

1,3-Diphosphoglycerate

Ribosome in action. The lower yellow half represents the 30S ribosome; the upper blue half represents the 50S ribosome. The yellow and green twisted cones are tRNA's, and the chain of beads stand for mRNA. *(Courtesy of Dr. J. Frank, Wadsworth Center, Albany, New York)*

PART 3

Biochemistry

Chapter 19

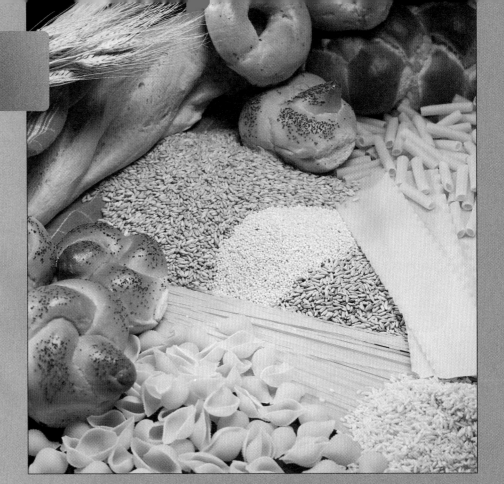

Carbohydrates

19.1 Introduction

Carbohydrates are the most abundant organic compounds in the plant world. They act as storehouses of chemical energy (glucose, starch, glycogen); are components of supportive structures in plants (cellulose), crustacean shells (chitin), and connective tissues in animals (acidic polysaccharides); and are essential components of nucleic acids (D-ribose and 2-deoxy-D-ribose). Carbohydrates make up about three fourths of the dry weight of plants. Animals (including humans) get their carbohydrates by eating plants, but they do not store much of what they consume. Less than 1 percent of the body weight of animals is made up of carbohydrates.

The name "carbohydrate," means hydrate of carbon and derives from the formula $C_n(H_2O)_m$. Two examples of carbohydrates with molecular formulas that can be written alternatively as hydrates of carbon are

glucose (blood sugar): $C_6H_{12}O_6$, which can be written as $C_6(H_2O)_6$

sucrose (table sugar): $C_{12}H_{22}O_{11}$, which can be written as $C_{12}(H_2O)_{11}$

Not all carbohydrates, however, have this general formula. Some contain too few oxygen atoms to fit this formula, and some others contain too many oxygens. Some also contain nitrogen. The term "carbohydrate" has become

Above: Breads, grains, and pasta are sources of carbohydrates. *(Charles D. Winters)*

firmly rooted in chemical nomenclature, and although not completely accurate, it persists as the name for this class of compounds.

At the molecular level, most **carbohydrates** are polyhydroxyaldehydes, polyhydroxyketones, or compounds that yield them after hydrolysis. The simpler members of the carbohydrate family are often referred to as **saccharides** because of the sweet taste of sugars (Latin: *saccharum,* sugar). Carbohydrates are classified as monosaccharides, oligosaccharides, or polysaccharides depending on their size.

> **Carbohydrate** A polyhydroxy-aldehyde or polyhydroxyketone or a substance that gives these compounds on hydrolysis

> **Saccharide** A simpler member of the carbohydrate family, such as glucose

19.2 Monosaccharides

Structure and Nomenclature

Monosaccharides have the general formula $C_nH_{2n}O_n$, with one of the carbons being the carbonyl group of either an aldehyde or a ketone. The most common monosaccharides have three to eight carbon atoms. The suffix "-ose" indicates that a molecule is a carbohydrate, and the prefixes "tri-," "tetr-," "pent-," and so forth indicate the number of carbon atoms in the chain. Monosaccharides containing an aldehyde group are classified as **aldoses;** those containing a ketone group are classified as **ketoses.**

> **Monosaccharide** A carbohydrate that cannot be hydrolyzed to a simpler compound

> **Aldose** A monosaccharide containing an aldehyde group

> **Ketose** A monosaccharide containing a ketone group

Monosaccharides Classified by Number of Carbon Atoms

Name	*Formula*
triose	$C_3H_6O_3$
tetrose	$C_4H_8O_4$
pentose	$C_5H_{10}O_5$
hexose	$C_6H_{12}O_6$
heptose	$C_7H_{14}O_7$
octose	$C_8H_{16}O_8$

There are only two trioses: the aldotriose glyceraldehyde and the ketotriose dihydroxyacetone.

$$
\begin{array}{ll}
\text{CHO} & \text{CH}_2\text{OH} \\
| & | \\
\text{CHOH} & \text{C}=\text{O} \\
| & | \\
\text{CH}_2\text{OH} & \text{CH}_2\text{OH} \\
\text{Glyceraldehyde} & \text{Dihydroxyacetone} \\
\text{(an aldotriose)} & \text{(a ketotriose)}
\end{array}
$$

Often the designations "aldo-" and "keto-" are omitted, and these molecules are referred to simply as trioses, tetroses, and the like.

Glyceraldehyde is a common name; the IUPAC name for this monosaccharide is 2,3-dihydroxypropanal. Similarly, dihydroxyacetone is a common name; the IUPAC name for this compound is 1,3-dihydroxypropanone. The common names for these and other monosaccharides, however, are so firmly rooted in the literature of organic chemistry and biochemistry that they are used almost exclusively whenever these compounds are referred to. Therefore, throughout our discussions of the chemistry and biochemistry of carbohydrates, we use the names most common in the literature of chemistry and biochemistry.

Fischer Projection Formulas

Glyceraldehyde contains a stereocenter and therefore exists as a pair of enantiomers.

$$\begin{array}{c} CHO \\ H-C-OH \\ CH_2OH \end{array}$$
(R)-Glyceraldehyde

$$\begin{array}{c} CHO \\ HO-C-H \\ CH_2OH \end{array}$$
(S)-Glyceraldehyde

Fischer projection A two-dimensional representation for showing the configuration of a stereocenter; horizontal lines represent bonds projecting forward from the stereocenter, and vertical lines represent bonds projecting to the rear

Chemists commonly use two-dimensional representations called **Fischer projections** to show the configuration of carbohydrates. To write a Fischer projection, draw a three-dimensional representation of the molecule oriented so that the vertical bonds from the stereocenter are directed away from you and the horizontal bonds from it are directed toward you. Then write the molecule as a two-dimensional figure with the stereocenter indicated by the point at which the bonds cross.

$$\begin{array}{c} CHO \\ H-C-OH \\ CH_2OH \end{array} \xrightarrow[\text{Fischer projection}]{\text{convert to a}} \begin{array}{c} CHO \\ H-\!\!\!\!-\!\!\!\!-OH \\ CH_2OH \end{array}$$

(R)-Glyceraldehyde
(three-dimensional
representation)

(R)-Glyceraldehyde
(Fischer projection)

The horizontal segments of this Fischer projection represent bonds directed toward you, and the vertical segments represent bonds directed away from you. The only atom in the plane of the paper is the stereocenter.

D- and L-Monosaccharides

Emil Fischer, who in 1902 became the second Nobel Prize winner in chemistry, made many fundamental discoveries in the chemistry of carbohydrates, proteins, and other areas of organic and biochemistry.

Even though the R,S system is widely accepted today as a standard for designating configuration, the configuration of carbohydrates as well as those of amino acids and many other compounds in biochemistry is commonly designated by the D,L system proposed by Emil Fischer in 1891. At that time, it was known that one enantiomer of glyceraldehyde has a specific rotation of +13.5°; the other has a specific rotation of −13.5°. Fischer proposed that these enantiomers be designated D and L, but at the time there was no experimental way to determine which enantiomer had which specific rotation. Therefore, he did the only possible thing—he made an arbitrary assignment. He assigned the dextrorotatory enantiomer the following configuration and named it D-glyceraldehyde. He named its enantiomer L-glyceraldehyde. Fischer could have been wrong, but by a stroke of good fortune, he wasn't. In 1952, his assignment of configuration to the enantiomers of glyceraldehyde was proved correct by a special application of x-ray crystallography.

$$\begin{array}{c} CHO \\ H-\!\!\!\!-\!\!\!\!-OH \\ CH_2OH \end{array}$$
D-Glyceraldehyde
$[\alpha]_D^{25} = +13.5°$

$$\begin{array}{c} CHO \\ HO-\!\!\!\!-\!\!\!\!-H \\ CH_2OH \end{array}$$
L-Glyceraldehyde
$[\alpha]_D^{25} = -13.5°$

D-Glyceraldehyde and L-glyceraldehyde serve as reference points for the assignment of relative configuration to all other aldoses and ketoses. The reference point is the stereocenter farthest from the carbonyl group. Because this stereocenter is always the next to the last carbon on the chain, it is called the **penultimate carbon**. A **D-monosaccharide** has the same configuration at its penultimate carbon as D-glyceraldehyde (its —OH is on the right); an **L-monosaccharide** has the same configuration at its penultimate carbon as L-glyceraldehyde (its —OH is on the left).

Tables 19.1 and 19.2 show names and Fischer projections for all D-aldo- and D-2-ketotetroses, pentoses, and hexoses. Each name consists of three parts. The D specifies the configuration at the stereocenter farthest from

Penultimate carbon The stereo-center of a monosaccharide farthest from the carbonyl group, as for example carbon-5 of glucose

D-Monosaccharide A monosac-charide that, when written as a Fischer projection, has the —OH on its penultimate carbon to the right

TABLE 19.1 Configurational Relationships Among the Isomeric D-Aldotetroses, D-Aldopentoses, and D-Aldohexoses

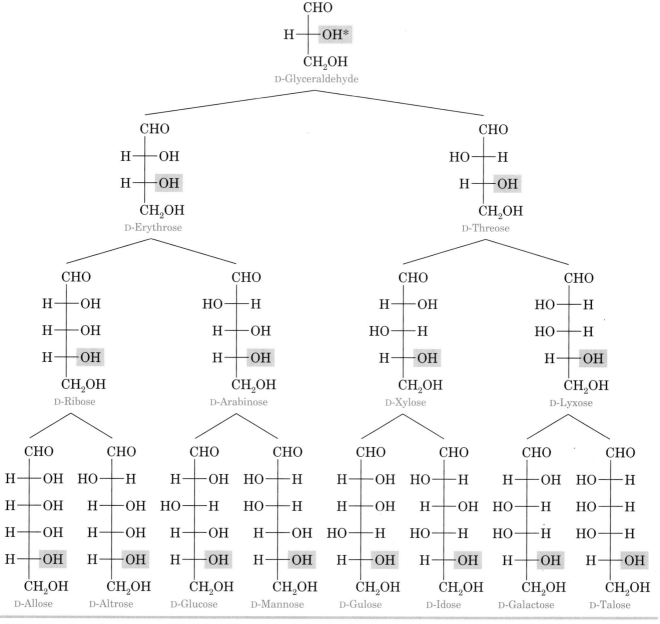

* The configuration of the reference —OH on the penultimate carbon is shown in color.

L-Monosaccharide A monosaccharide that, when written as a Fischer projection, has the —OH on its penultimate carbon to the left

TABLE 19.2 **Configurational Relationships Among the D-2-Ketopentoses and D-2-Ketohexoses**

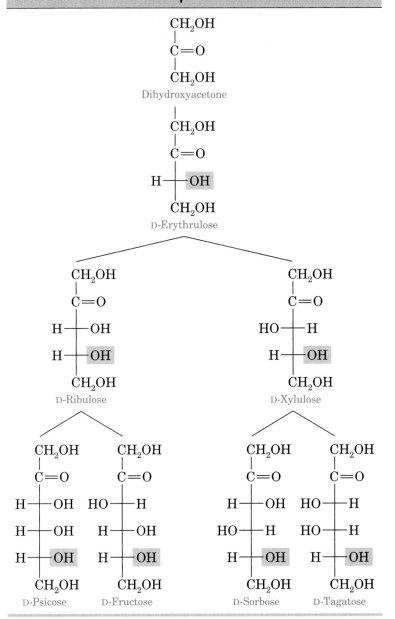

the carbonyl group. Prefixes such as "rib-," "arabin-," and "gluc-" specify the relative configuration of all other stereocenters. The suffix "-ose" shows that the compound is a carbohydrate.

The three most abundant hexoses in the biological world are D-glucose, D-galactose, and D-fructose. The first two are D-aldohexoses; the third is a D-2-ketohexose. Glucose, by far the most common hexose, is also known as dextrose because it is dextrorotatory. Other names for this monosaccharide are grape sugar and blood sugar. Human blood normally contains 65 to 110 mg of glucose/100 mL of blood.

Glucose is synthesized by chlorophyll-containing plants using sunlight as a source of energy. In the process called **photosynthesis,** plants convert carbon dioxide from the air and water from the soil to glucose and oxygen (Box 28A).

$$6CO_2 + 6H_2O + energy \xrightarrow[\text{chlorophyll}]{\text{sunlight}} C_6H_{12}O_6 + 6O_2$$

Carbon   Water   Glucose   Oxygen
dioxide

Photosynthesis. *(Charles D. Winters)*

It is the chlorophyll in leaves and grass that makes them green. It is chlorophyll in the bark of palo verde trees that makes their bark green.

EXAMPLE 19.1

Draw Fischer projections for the four aldotetroses. Which are D-monosaccharides, which are L-monosaccharides, and which are enantiomers? Refer to Table 19.1 and write the name of each aldotetrose.

Solution

Following are Fischer projections for the four aldotetroses. The D- and L- refer to the configuration of the penultimate carbon, which, in the case of aldotetroses, is carbon 3. In the Fischer projection of a D-aldotetrose, the —OH on carbon 3 is on the right and, in an L-aldotetrose, it is on the left.

| one pair of enantiomers | | a second pair of enantiomers | |

D-Erythrose   L-Erythrose   D-Threose   L-Threose

Problem 19.1

Draw Fischer projections for all 2-ketopentoses. Which are D-ketopentoses, which are L-ketopentoses, and which are enantiomers? Refer to Table 19.2 and write the name of these ketopentoses. ■

Amino Sugars

Amino sugars contain an —NH$_2$ group in place of an —OH group. Only three amino sugars are common in nature: D-glucosamine, D-mannosamine, and D-galactosamine.

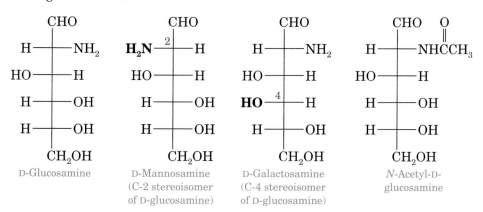

D-Glucosamine   D-Mannosamine   D-Galactosamine   N-Acetyl-D-
               (C-2 stereoisomer   (C-4 stereoisomer   glucosamine
               of D-glucosamine)   of D-glucosamine)

Galactosemia

One infant of every 18 000 is born with a genetic defect that makes it unable to utilize the monosaccharide galactose. Galactose is part of milk sugar, lactose, and when the body cannot utilize it, it accumulates in the blood and urine. Accumulation of galactose in the blood is harmful because it can lead to mental retardation, failure to grow, formation of cataracts in the eye, and, in severe cases, death due to liver damage. When galactose accumulation is due to a deficiency of the enzyme galactokinase, the disorder known as galactosuria has only mild symptoms. When the enzyme galactose-1-phosphate uridinyltransferase is deficient, the disorder is called galactosemia, and its symptoms are severe.

The deleterious effects of galactosemia can be avoided by giving the infant a milk formula in which sucrose is substituted for lactose. Because sucrose contains no galactose, the infant is thus being fed a galactose-free diet. A galactose-free diet is critical only in infancy. With maturation, children develop another enzyme capable of metabolizing galactose. They are thus able to tolerate galactose as they mature.

Northern lobster. *(© Andrew J. Martinez 1993/Photo Researchers, Inc.)*

N-Acetyl-D-glucosamine, a derivative of D-glucosamine, is a component of many polysaccharides, including chitin, the hard shell-like exoskeleton of lobsters, crabs, shrimp, and other shellfish. Several other amino sugars are components of naturally occurring antibiotics.

19.3 The Cyclic Structure of Monosaccharides

We saw in Section 17.5 that aldehydes and ketones react with alcohols to form **hemiacetals.** We also saw that cyclic hemiacetals form very readily when hydroxyl and carbonyl groups are part of the same molecule and their interaction can form a five- or six-membered ring. For example, 4-hydroxypentanal forms a five-membered cyclic hemiacetal. Note that 4-hydroxypentanal contains one stereocenter and that a second stereocenter is generated at carbon 1 as a result of hemiacetal formation.

$$\underset{\text{4-Hydroxypentanal}}{\overset{4}{CH_3}\underset{|}{\overset{}{CH}}CH_2CH_2\overset{1}{\overset{O}{\|}}{CH}} \;\equiv\; \underset{\substack{\text{(redrawn to show how} \\ \text{the cyclic hemiacetal forms)}}}{} \;\rightleftharpoons\; \underset{\text{A cyclic hemiacetal}}{}$$

Monosaccharides have hydroxyl and carbonyl groups in the same molecule. As a result, they too exist almost exclusively as five- and six-membered cyclic hemiacetals.

Haworth Projections

A common way of representing the cyclic structure of monosaccharides is the **Haworth projection,** named after the English chemist Sir Walter N. Haworth (Nobel Prize in Chemistry, 1937). In a Haworth projection, a five- or six-membered cyclic hemiacetal is represented as a planar pentagon or hexagon, as the case may be, lying perpendicular to the plane of the paper. Groups attached to the carbons of the ring then lie either above or below

Haworth projection A way to view furanose and pyranose forms of monosaccharides; the ring is drawn flat and viewed through its edge with the anomeric carbon on the right and the oxygen atom of the ring in the rear

FIGURE 19.1 Haworth projections for α-D-glucopyranose and β-D-glucopyranose.

the plane of the ring. The new carbon stereocenter created in forming the cyclic structure is called an **anomeric carbon.** Stereoisomers that differ in configuration only at the anomeric carbon are called **anomers.** The anomeric carbon of an aldose is C-1; that of the most common ketoses is C-2.

Haworth projections are most commonly written with the anomeric carbon to the right and the hemiacetal oxygen to the back (Figure 19.1).

In the terminology of carbohydrate chemistry, the designation β means that the —OH on the anomeric carbon of the cyclic hemiacetal is on the same side of the ring as the terminal —CH_2OH. Conversely, the designation α means that the —OH on the anomeric carbon of the cyclic hemiacetal is on the side of the ring opposite the terminal —CH_2OH.

A six-membered hemiacetal ring is indicated by the infix "-pyran-," and a five-membered hemiacetal ring is indicated by the infix "-furan-." The terms **furanose** and **pyranose** are used because monosaccharide five- and six-membered rings correspond to the heterocyclic compounds furan and pyran.

Furan   Pyran

Because the α and β forms of glucose are six-membered cyclic hemiacetals, they are named α-D-glucopyranose and β-D-glucopyranose. These infixes are not always used in monosaccharide names, however. Thus, the glucopyranoses, for example, are often named simply α-D-glucose and β-D-glucose.

You would do well to remember the configuration of groups on the Haworth projections of α-D-glucopyranose and β-D-glucopyranose as reference structures. Knowing how the open-chain configuration of any other aldohexoses differs from that of D-glucose, you can then construct its Haworth projection by reference to the Haworth projection of D-glucose.

Anomeric carbon The hemiacetal carbon of the cyclic form of a monosaccharide

Anomers Monosaccharides that differ in configuration only at their anomeric carbons

Furanose A five-membered cyclic hemiacetal form of a monosaccharide

Pyranose A six-membered cyclic hemiacetal form of a monosaccharide

EXAMPLE 19.2

Draw Haworth projections for the α and β anomers of D-galactopyranose.

Solution

One way to arrive at these projections is to use the α and β forms of D-glucopyranose as reference and to remember (or discover by looking at Table 19.1) that D-galactose differs from D-glucose only in the

(continued on page 448)

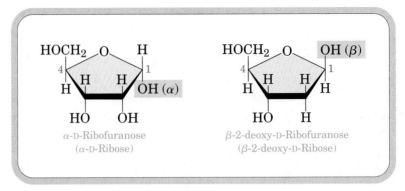

FIGURE 19.2   Haworth projections for two D-ribofuranoses.

configuration at carbon 4. Thus, begin with the Haworth projections shown in Figure 19.1 and then invert the configuration at carbon 4.

Configuration differs from that of D-glucose at C-4.

α-D-Galactopyranose
(α-D-Galactose)

β-D-Galactopyranose
(β-D-Galactose)

Problem 19.2

Mannose exists in aqueous solution as a mixture of α-D-mannopyranose and β-D-mannopyranose. Draw Haworth projections for these molecules. ■

Thus far we have looked only at aldohexoses. Pentoses also form cyclic hemiacetals. The most prevalent forms of D-ribose and other pentoses in the biological world are furanoses. Shown in Figure 19.2 are Haworth projections for α-D-ribofuranose (α-D-ribose) and β-2-deoxy-D-ribofuranose (β-2-deoxy-D-ribose). The prefix "2-deoxy-" indicates the absence of oxygen at carbon 2. Units of D-ribose and 2-deoxy-D-ribose in nucleic acids and most other biological molecules are found almost exclusively in the β-configuration.

Other monosaccharides also form five-membered cyclic hemiacetals. Shown in Figure 19.3 are the five-membered cyclic hemiacetals of fructose. The β-D-fructofuranose form is found in the disaccharide sucrose (Section 19.6).

α-D-Fructofuranose
(α-D-Fructose)

D-Fructose

β-D-Fructofuranose
(β-D-Fructose)

anomeric carbon

FIGURE 19.3   Furanose forms of D-fructose at equilibrium in aqueous solution.

L-Ascorbic Acid (Vitamin C)

The structure of L-ascorbic acid (vitamin C) resembles that of a monosaccharide. In fact, this vitamin is synthesized both biochemically by plants and some animals and commercially from D-glucose. Humans do not have the enzymes required for this synthesis; therefore, we must obtain it in the food we eat or as a vitamin supplement (Section 29.5). Approximately 66 million kg of vitamin C are synthesized every year in the United States.

Biochemical and industrial syntheses

D-Glucose

L-Ascorbic acid
(Vitamin C)

L-Ascorbic acid is very easily oxidized to L-dehydroascorbic acid, a diketone. Both L-ascorbic acid and L-dehydroascorbic acid are physiologically active and are found together in most body fluids.

oxidation ⇌ reduction

L-Ascorbic acid
(Vitamin C)

L-Dehydroascorbic acid

■ Oranges are a major source of vitamin C. *(Charles D. Winters)*

Conformation Representations

A five-membered ring is so close to being planar that Haworth projections are adequate representations of furanoses. For pyranoses, however, the six-membered ring is more accurately represented as a **chair conformation** (Section 11.7). Structural formulas for α-D-glucopyranose and β-D-glucopyranose are drawn as chair conformations in Figure 19.4. Also shown is the open-chain or free aldehyde form with which the cyclic hemiacetal forms are in equilibrium in aqueous solution. Notice that each group, including the anomeric —OH, on the chair conformation of β-D-glucopyranose is equatorial. Notice also that the —OH group on the

FIGURE 19.4 Chair conformations of β-D-glucopyranose and α-D-glucopyranose.

anomeric carbon is axial in α-D-glucopyranose. Because of the equatorial orientation of the —OH on its anomeric carbon, β-D-glucopyranose is more stable and predominates in aqueous solution.

At this point, you should compare the relative orientations of groups on the D-glucopyranose ring in the Haworth projection and chair conformation. The orientations of groups on carbons 1 through 5 of β-D-glucopyranose, for example, are up, down, up, down, and up in both representations.

β-D-Glucopyranose
(Haworth projection)

β-D-Glucopyranose
(chair conformation)

EXAMPLE 19.3

Draw chair conformations for α-D-galactopyranose and β-D-galactopyranose. Label the anomeric carbon in each.

Solution

D-Galactose differs in configuration from D-glucose only at carbon 4. Therefore, draw the α and β forms of D-glucopyranose and then interchange the positions of the —OH and —H groups on carbon 4.

β-D-Galactopyranose
(β-D-Galactose)

D-Galactose

α-D-Galactopyranose
(α-D-Galactose)

TABLE 19.3 Specific Rotations for α and β Anomers of
D-Glucopyranose and D-Galactopyranose Before
and After Mutarotation

Monosaccharide	Specific Rotation (degrees)	Specific Rotation After Mutarotation (degrees)	Percent Present at Equilibrium
α-D-glucose	+112.0	+52.7	36
β-D-glucose	+18.7	+52.7	64
α-D-galactose	+150.7	+80.2	28
β-D-galactose	+52.8	+80.2	72

Problem 19.3 ■
Draw chair conformations for α-D-mannopyranose and
β-D-mannopyranose. Label the anomeric carbon in each. ■

Mutarotation

Mutarotation is the change in specific rotation that accompanies the
equilibration of α- and β-anomers in aqueous solution. As an example, a
solution prepared by dissolving crystalline α-D-glucopyranose in water has
an initial specific rotation of +112° (Table 19.3), which gradually decreases
to an equilibrium value of +52.7° as α-D-glucopyranose reaches an equilibrium with β-D-glucopyranose. A solution of β-D-glucopyranose also undergoes mutarotation, during which the specific rotation changes from +18.7°
to the same equilibrium value of +52.7°. The equilibrium mixture consists
of 64 percent β-D-glucopyranose and 36 percent α-D-glucopyranose with
only a trace (0.003 percent) of the open-chain form. Mutarotation is common to all carbohydrates that exist in hemiacetal forms.

> **Mutarotation** The change in
> specific rotation that occurs
> when an α or β form of a carbo-
> hydrate is converted to an equi-
> librium mixture of the two
> forms

β-D-Glucopyranose
$[\alpha]_D^{25} + 18.7°$ | Open-chain form | α-D-Glucopyranose
$[\alpha]_D^{25} + 112°$

19.4 Physical Properties of Monosaccharides

Monosaccharides are colorless, crystalline solids. Because hydrogen bonding is possible between their polar —OH groups and water, all monosaccharides are very soluble in water. They are only slightly soluble in ethanol
and are insoluble in nonpolar solvents such as diethyl ether, chloroform,
and benzene.

Although all monosaccharides are sweet to the taste, some are sweeter
than others (Table 19.4). D-Fructose tastes the sweetest, even sweeter than
sucrose (table sugar, Section 19.6). The sweet taste of honey is due largely
to D-fructose and D-glucose. Lactose (Section 19.6) has almost no sweetness
and is sometimes added to foods as a filler. Some people cannot tolerate
lactose well and should avoid these foods (Box 19A).

We have no mechanical way to measure sweetness. It is done by having
a group of people taste solutions of
varying sweetnesses.

In the production of sucrose, sugar cane or sugar beet is boiled with water, and the resulting solution is cooled. Sucrose crystals separate and are collected. Subsequent boiling to concentrate the solution followed by cooling yields a dark, thick syrup known as molasses.

TABLE 19.4 Relative Sweetness of Some Carbohydrate and Artificial Sweetening Agents

Carbohydrate	Sweetness Relative to Sucrose	Artificial Sweetener	Sweetness Relative to Sucrose
fructose	1.74	saccharin	450
invert sugar	1.25	acesulfame-K	200
sucrose (table sugar)	1.00	aspartame	160
honey	0.97		
glucose	0.74		
maltose	0.33		
galactose	0.32		
lactose (milk sugar)	0.16		

19.5 Reactions of Monosaccharides

Formation of Glycosides (Acetals)

We saw in Section 17.5 that treatment of an aldehyde or ketone with one molecule of alcohol gives a hemiacetal, and treatment of the hemiacetal with a molecule of alcohol gives an acetal. Treatment of monosaccharides, all of which exist almost exclusively in a cyclic hemiacetal form, also gives acetals, as illustrated by the reaction of β-D-glucopyranose with methanol.

β-D-Glucopyranose (β-D-Glucose)   Methyl β-D-glucopyranoside (Methyl β-D-glucoside)   Methyl α-D-glucopyranoside (Methyl α-D-glucoside)

Glycoside A carbohydrate in which the —OH on its anomeric carbon is replaced by —OR

Glycosidic bond The bond from the anomeric carbon of a glycoside to an —OR group

A cyclic acetal derived from a monosaccharide is called a **glycoside**, and the bond from the anomeric carbon to the —OR group is called a **glycosidic bond.** Mutarotation is not possible in a glycoside because an acetal is no longer in equilibrium with the open-chain carbonyl-containing compound. Glycosides are stable in water and aqueous base, but like other acetals (Section 17.5), they are hydrolyzed in aqueous acid to an alcohol and a monosaccharide.

Glycosides are named by listing the alkyl or aryl group bonded to oxygen followed by the name of the carbohydrate in which the ending "-e" is replaced by "-ide." For example, the methyl glycoside derived from β-D-glucopyranose is named methyl β-D-glucopyranoside; that derived from β-D-ribofuranose is named methyl β-D-ribofuranoside.

EXAMPLE 19.4

Draw a structural formula for each glycoside. In each, label the anomeric carbon and the glycosidic bond.
(a) Methyl β-D-ribofuranoside (methyl β-D-riboside); draw a Haworth projection.

(b) Methyl α-D-galactopyranoside (methyl α-D-galactoside); draw both a Haworth projection and a chair conformation.

Solution

(a)

Methyl β-D-ribofuranoside
(Methyl β-D-riboside)

(b)

Methyl α-D-galactopyranoside
(Methyl α-D-galactoside)

Problem 19.4

Draw a structural formula for each glycoside. In each, label the anomeric carbon and the glycosidic bond.

(a) Methyl β-D-fructofuranoside (methyl β-D-fructoside); draw a Haworth projection.

(b) Methyl α-D-mannopyranoside (methyl α-D-mannoside); draw both a Haworth projection and a chair conformation. ■

Reduction to Alditols

The carbonyl group of a monosaccharide can be reduced to an hydroxyl group by a variety of reducing agents, including hydrogen in the presence of a transition metal catalyst. The reduction products are known as **alditols.** Reduction of D-glucose gives D-glucitol, more commonly known as D-sorbitol. Note that D-glucose is shown here in the open-chain form. Only a small amount of this form is present in solution, but, as it is reduced, the equilibrium between cyclic hemiacetal forms and the open-chain form shifts to replace it.

Alditol The product formed when the C=O group of a monosaccharide is reduced to a CHOH group

Sorbitol is found in the plant world in many berries and in cherries, plums, pears, apples, seaweed, and algae. It is about 60 percent as sweet as

Many "sugar free" products contain sugar alcohols, such as D-sorbital and xylitol. *(Gregory Smolin)*

sucrose (table sugar) and is used in the manufacture of candies and as a sugar substitute for diabetics.

Other alditols common in the biological world are erythritol, D-mannitol, and xylitol.

$$CH_2OH$$

Erythritol   D-Mannitol   Xylitol

Xylitol is used as a sweetening agent in "sugarless" gum, candy, and sweet cereals.

Oxidation to Aldonic Acids. Reducing Sugars

As we saw in Section 17.7, aldehydes (RCHO) are oxidized to carboxylic acids (RCOOH) by several oxidizing agents, including oxygen, O_2. Similarly, the aldehyde group of an aldose can be oxidized, under basic conditions, to a carboxylate group. Under these conditions, the cyclic form of an aldose is in equilibrium with the open-chain form, which is then oxidized by the mild oxidizing agent. D-Glucose, for example, is oxidized to D-gluconate (the anion of D-gluconic acid).

β-D-Glucopyranose (β-D-Glucose)   D-Glucose   D-Gluconate

Reducing sugar A carbohydrate that reacts with a mild oxidizing agent under basic conditions to give an aldonic acid; the carbohydrate reduces the oxidizing agent

Any carbohydrate that reacts with an oxidizing agent to form an aldonic acid is classified as a **reducing sugar** (it reduces the oxidizing agent).

Surprisingly, 2-ketoses are also reducing sugars. Carbon 1 (a CH_2OH group) of a ketose is not oxidized directly. Rather, under the basic conditions of this oxidation, a 2-ketose is in equilibrium with an aldose by way of an enediol intermediate. The aldose is then oxidized by the mild oxidizing agent.

A 2-ketose   An enediol   An aldose   An aldonate

Oxidation to Uronic Acids

Enzyme-catalyzed oxidation of the primary hydroxyl group at C-6 of a hexose yields a uronic acid. Enzyme-catalyzed oxidation of D-glucose, for example, yields D-glucuronic acid (*The Merck Index,* 12th ed., #4474), shown here in both its open-chain and cyclic hemiacetal forms.

D-Glucose

Fischer projection   Chair conformation

D-Glucuronic acid
(a uronic acid)

D-Glucuronic acid is widely distributed in both the plant and animal world. In humans, it is an important component of the acidic polysaccharides of connective tissues (Section 19.8). It is also used by the body to detoxify foreign hydroxyl-containing compounds, that is, phenols and alcohols. In the liver, these compounds are converted to glycosides of glucuronic acid (glucuronides) and excreted in the urine. The intravenous anesthetic propofol (*The Merck Index,* 12th ed., #8020), for example, is converted to the following glucuronide and excreted in the urine.

Propofol   A urine-soluble glucuronide

Phosphoric Esters

Mono- and diphosphoric esters are important intermediates in the metabolism of monosaccharides. For example, the first step in glycolysis (Section 27.2) is conversion of glucose to glucose 6-phosphate. Note that phosphoric

BOX 19C

Testing for Glucose

The analytical procedure most often performed in a clinical chemistry laboratory is the determination of glucose in blood, urine, or other biological fluids. This is true because of the high incidence of diabetes mellitus. Approximately 2 million known diabetics live in the United States, and it is estimated that another 1 million are undiagnosed.

Diabetes mellitus is characterized by insufficient blood levels of the hormone insulin or insufficient insulin receptors on cells (Box 23G). If the blood concentration of insulin is too low, muscle and liver cells do not absorb glucose from the blood, which, in turn, leads to increased levels of blood glucose (hyperglycemia), impaired metabolism of fats and proteins, ketosis, and possible diabetic coma. A rapid test for blood glucose levels is critical for early diagnosis and effective management of this disease. In addition to being rapid, a test must also be specific for D-glucose; it must give a positive test for glucose but not react with any other substance normally present in biological fluids.

Blood glucose levels are now measured by an enzyme-based procedure using the enzyme glucose oxidase. This enzyme catalyzes the oxidation of β-D-glucose to D-gluconic acid.

form. Fortunately, this interconversion is rapid and complete in the short time required for the test.

Molecular oxygen, O_2, is the oxidizing agent in this reaction and is reduced to hydrogen peroxide, H_2O_2. In one procedure, hydrogen peroxide formed in the glucose oxidase-catalyzed reaction is used to oxidize colorless o-toluidine to a colored product in a reaction catalyzed by the enzyme peroxidase. The concentration of the colored oxidation product is determined spectrophotometrically and is proportional to the concentration of glucose in the test solution.

$$\text{2-Methylaniline} \quad + H_2O_2 \xrightarrow{\text{peroxidase}} \text{colored product}$$

2-Methylaniline
(o-Toluidine)

Several commercially available test kits use the glucose oxidase reaction for qualitative determination of glucose in urine.

$$\beta\text{-D-Glucopyranose} \quad + O_2 + H_2O \xrightarrow{\substack{\text{glucose} \\ \text{oxidase}}}$$

β-D-Glucopyranose
(β-D-Glucose)

$$\begin{array}{c}
\text{COOH} \\
\text{H}\!-\!\!-\!\text{OH} \\
\text{HO}\!-\!\!-\!\text{H} \\
\text{H}\!-\!\!-\!\text{OH} \\
\text{H}\!-\!\!-\!\text{OH} \\
\text{CH}_2\text{OH}
\end{array} \quad + \quad \begin{array}{c} H_2O_2 \\ \text{Hydrogen} \\ \text{peroxide} \end{array}$$

D-Gluconic acid

Glucose oxidase is specific for β-D-glucose. Therefore, complete oxidation of any sample containing both β-D-glucose and α-D-glucose requires conversion of the α form to the β

■ Chemstrip kit for blood glucose.
(Charles D. Winters)

acid is a strong acid, and, at the pH of cellular and intercellular fluids, both acidic protons of a phosphoric ester are ionized giving the ester group a charge of 2−.

D-Glucose   D-Glucose 6-phosphate   α-D-Glucose 6-phosphate

19.6 Disaccharides and Oligosaccharides

Most carbohydrates in nature contain more than one monosaccharide unit. Those that contain two units are called **disaccharides,** those that contain three units are called **trisaccharides,** and so forth. The general term **oligosaccharide** is often used for carbohydrates that contain from four to ten monosaccharide units. Carbohydrates containing larger numbers of monosaccharide units are called **polysaccharides.**

In a disaccharide, two monosaccharide units are joined together by a glycosidic bond between the anomeric carbon of one unit and an —OH of the other. Three important disaccharides are maltose, lactose, and sucrose.

> **Disaccharide** A carbohydrate containing two monosaccharide units joined by a glycosidic bond

> **Oligosaccharide** A carbohydrate containing from four to ten monosaccharide units, each joined to the next by a glycosidic bond

Maltose

Maltose derives its name from its presence in malt, the juice from sprouted barley and other cereal grains. Maltose consists of two molecules of D-glucopyranose joined by a glycosidic bond between carbon 1 (the anomeric carbon) of one unit and carbon 4 of the other unit. Because the oxygen atom on the anomeric carbon of the first glucopyranose unit is alpha, the bond joining the two is called an α-1,4-glycosidic bond. Shown in Figure 19.5 are Haworth and chair representations for β-maltose, so

Maltose is an ingredient in most syrups.

FIGURE 19.5 β-Maltose.

FIGURE 19.6   Lactose (milk sugar).

named because the —OH on the anomeric carbon of the glucose unit on the right is beta.

Maltose is a reducing sugar because the hemiacetal group on the right unit of D-glucopyranose is in equilibrium with the free aldehyde and can be oxidized to a carboxylic acid.

Lactose

Lactose is the principal sugar present in milk. It makes up about 5 to 8 percent of human milk and 4 to 6 percent of cow's milk. It consists of D-galactopyranose bonded by a β-1,4-glycosidic bond to carbon-4 of D-glucopyranose (Figure 19.6). Lactose is a reducing sugar.

Sucrose

Sucrose (table sugar) is the most abundant disaccharide in the biological world (Figure 19.7). It is obtained principally from the juice of sugar cane and sugar beets. In sucrose, carbon 1 of α-D-glucopyranose is joined to carbon 2 of D-fructofuranose by an α-1,2-glycosidic bond. Note that glucose is a six-membered (pyranose) ring, whereas fructose is a five-membered (furanose) ring. Because the anomeric carbons of both the glucopyranose and fructofuranose units are involved in formation of the glycosidic bond, sucrose is a nonreducing sugar.

FIGURE 19.7   Sucrose (table sugar).

A, B, AB, and O Blood Types

Membranes of animal plasma cells have large numbers of relatively small carbohydrates bound to them. In fact, it appears that the outsides of most plasma cell membranes are literally "sugar-coated." These membrane-bound carbohydrates are part of the mechanism by which cell types recognize one another and, in effect, act as biochemical markers. Typically, these carbohydrates contain from 4 to 17 monosaccharide units consisting primarily of relatively few monosaccharides, including D-galactose, D-mannose, L-fucose, N-acetyl-D-glucosamine, and N-acetyl-D-galactosamine. L-Fucose is a 6-deoxyaldohexose.

An L-monosaccharide because this —OH is on the left in the Fischer projection

CHO

HO —— H

H —— OH

H —— OH

HO —— H

CH₃ ← Carbon 6 is —CH₃ rather than —CH₂OH

L-Fucose

■ Bag of blood showing blood type. (© Larry Mulvehill/Photo Researchers, Inc.)

An example of the importance of these membrane-bound carbohydrates is the ABO blood group system, discovered in 1900 by Karl Landsteiner (1868–1943). Whether an individual belongs to type A, B, AB, or O is genetically determined and depends on the type of tri- or tetrasaccharide bound to the surface of the red blood cells. These surface-bound carbohydrates, designated A, B, and O, act as antigens. The type of glycosidic bond joining each monosaccharide is shown in the figure.

The blood carries antibodies against foreign substances. When a person receives a blood transfusion, the antibodies clump (aggregate) the foreign blood cells. A-Type blood has A antigens (N-acetyl-D-galactosamine) on the surfaces of its red blood cells and carries anti-B antibodies (against B antigen). B-Type blood carries B antigen (D-galactose) and has anti-A antibodies (against A antigens). Transfusion of A-type blood into a person with B-type blood can be fatal, and vice versa.

O-Type blood has neither A nor B antigens on its red blood cells but carries both anti-A and anti-B antibodies. The AB-type individual has both A and B antigens, but no anti-A or anti-B antibodies. These blood transfusion possibilities are summarized in this chart.

Thus, people with O-type blood are universal donors, and those with AB-type blood are universal acceptors. People with A-type blood can accept from A- or O-type donors only. Those with B-type blood can accept from B- or O-type donors only. AB-type people can accept from all blood types. O-type people can accept only from O-type donors.

EXAMPLE 19.5

Draw a chair conformation for the β anomer of a disaccharide in which two units of D-glucopyranose are joined by an α-1,6-glycosidic bond.

Solution

First draw a chair conformation of α-D-glucopyranose. Then connect the anomeric carbon of this monosaccharide to carbon 6 of a second D-glucopyranose unit by an α-glycosidic bond. The resulting molecule is either α or β depending on the orientation of the —OH group on the reducing end of the disaccharide. The disaccharide shown here is β.

Problem 19.5

Draw a chair conformation for the α form of a disaccharide in which two units of D-glucopyranose are joined by a β-1,3-glycosidic bond. ∎

19.7 Polysaccharides

Polysaccharides consist of large numbers of monosaccharide units joined together by glycosidic bonds. Three important polysaccharides, all made up of glucose units, are starch, glycogen, and cellulose.

Starch. Amylose and Amylopectin

Starch is used for energy storage in plants. It is found in all plant seeds and tubers and is the form in which glucose is stored for later use. Starch can be separated into two principal polysaccharides: amylose and amylopectin. Although the starch from each plant is unique, most starches contain 20 to 25 percent amylose and 75 to 80 percent amylopectin.

Complete hydrolysis of both amylose and amylopectin yields only D-glucose. Amylose is composed of continuous, unbranched chains of up to 4000 D-glucose units joined by α-1,4-glycosidic bonds. Amylopectin contains chains of up to 10,000 D-glucose joined by α-1,4-glycosidic bonds. In addition, there is considerable branching from this linear network. At branch points, new chains of 24 to 30 units are started by α-1,6-glycosidic bonds (Figure 19.8).

Glycogen

Glycogen is the energy-reserve carbohydrate for animals. Like amylopectin, glycogen is a branched polysaccharide of approximately 10^6 D-glucose units joined by α-1,4- and α-1,6-glycosidic bonds. The total amount of glycogen in the body of a well-nourished adult human is about 350 g, divided almost equally between liver and muscle.

FIGURE 19.8 Amylopectin is a branched polymer of approximately 10,000 D-glucose units joined by α-1,4-glycosidic bonds. Branches consist of 24 to 30 D-glucose units started by α-1,4-glycosidic bonds.

Cellulose

Cellulose, the most widely distributed plant skeletal polysaccharide, constitutes almost half of the cell wall material of wood. Cotton is almost pure cellulose. Cellulose is a linear polysaccharide of D-glucose units joined by β-1,4-glycosidic bonds (Figure 19.9). It has an average molecular weight of 400 000 g/mol, corresponding to approximately 2200 glucose units per molecule. Cellulose molecules act very much like stiff rods, a feature that enables them to align themselves side by side into well-organized water-insoluble fibers in which the OH groups form numerous intermolecular hydrogen bonds. This arrangement of parallel chains in bundles gives cellulose fibers their high mechanical strength. It is also the reason cellulose is insoluble in water. When a piece of cellulose-containing material is placed in water, there are not enough water molecules on the surface of the fiber to pull individual cellulose molecules away from the strongly hydrogen-bonded fiber.

Humans and other animals cannot use cellulose as food because our digestive systems do not contain β-glucosidases, enzymes that catalyze hydrolysis of β-glucosidic bonds. Instead, we have only α-glucosidases; hence, the polysaccharides we use as sources of glucose are starch and glycogen. On the other hand, many bacteria and microorganisms do con-

BOX 19E

Polysaccharides As Fillers in Drug Tablets

All drugs that have the same generic name contain the same active ingredient. In spite of this fact, many physicians prefer one trade–name drug over another. In many cases, the difference is in the inactive ingredients, called fillers, which are inactive but may affect the rate of drug delivery.

Polysaccharides have been used as fillers because of their solubility in water and their thickening and gelling properties. For example, starch is used as a binding agent in tablets and capsules for oral delivery. In water, the starch binding agent swells and dissolves and the tablet falls apart, allowing the drug to be delivered in the stomach and the intestines.

FIGURE 19.9 Cellulose is a linear polysaccharide of up to 2200 units of D-glucose joined by β-1,4-glycosidic bonds.

tain β-glucosidases and so can digest cellulose. Termites are fortunate (much to our regret) to have such bacteria in their intestines and can use wood as their principal food. Ruminants (cud-chewing animals) and horses can also digest grasses and hay because β-glucosidase-containing microorganisms are present in their digestive systems.

19.8 Acidic Polysaccharides

Acidic polysaccharides are a group of polysaccharides that contain carboxyl groups and/or sulfuric ester groups. These compounds play important roles in the structure and function of connective tissues, the matrix between organs and cells that provides mechanical strength and also filters the flow of molecular information between cells. Most connective tissues are usually made up of collagen, a structural protein, in combination with a variety of acidic polysaccharides that interact with collagen to form tight or loose networks.

There is no single general type of connective tissue. Rather, there are a large number of highly specialized forms, such as cartilage, bone, synovial fluid, skin, tendons, blood vessels, intervertebral disks, and cornea.

Hyaluronic Acid

Hyaluronic acid is the simplest acidic polysaccharide present in connective tissue. It has a molecular weight of between 10^5 and 10^7 g/mol and contains from 300 to 100 000 repeating units, depending on the organ in which it occurs. It is most abundant in embryonic tissues and in specialized connective tissues such as synovial fluid, the lubricant of joints in the body and also in the vitreous of the eye where it provides a clear, elastic gel that maintains the retina in its proper position.

Hyaluronic acid is composed of D-glucuronic acid linked by a β-1,3-glycosidic bond to N-acetyl-D-glucosamine, which is in turn linked to D-glucuronic acid by a β-1,4-glycosidic bond.

In rheumatoid arthritis, inflammation of the synovial tissue results in swelling of the joints.

The repeating unit of hyaluronic acid

BOX 19F

Retina Detachment

In a young person, the vitreous of the eye is a strong, clear, elastic gel located in the eyeball between the lens and the retina. In the vitreous, collagen fibers are entangled with random coil-like molecules of hyaluronic acid. As a person ages, the collagen and hyaluronic acid separate, and the hyaluronic acid forms liquid pockets in the otherwise collagen-rich gel. When these pockets become large, the vitreous is not sufficiently elastic to maintain the retina flattened on the back of the eyeball. Under these circumstances the retina can become detached, and the result is blindness.

One treatment for retina detachment is to remove the vitreous and replace it by a pure, noninflammatory hyaluronic acid gel.

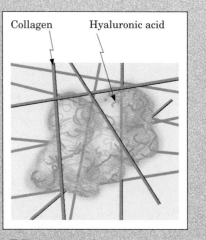

■ The structure of vitreous.

Heparin

Heparin is a heterogeneous mixture of variably sulfonated polysaccharide chains, ranging in molecular weight from 6000 to 30 000 g/mol. This acidic polysaccharide is synthesized and stored in mast cells of various tissues, particularly the liver, lungs, and gut. Heparin has many biological functions, the best known and understood of which is its anticoagulant activity. It binds strongly to antithrombin III, a plasma protein involved in terminating the clotting process.

The repeating monosaccharide units of heparin are D-glucosamine, D-glucuronic acid, and L-ioduronic acid bonded by a combination of α-1,4- and β-1,4-glycosidic bonds. Figure 19.10 shows a pentasaccharide unit of

FIGURE 19.10 A pentasaccharide unit of heparin.

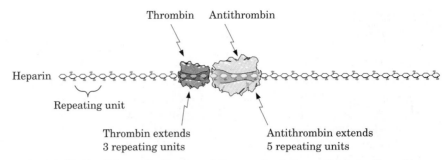

FIGURE 19.11 Schematic diagram of the complex between heparin, thrombin, and antithrombin.

heparin that binds to and inhibits the enzymatic activity of antithrombin III. Figure 19.11 shows a complex formed between heparin, thrombin, and antithrombin III. A heparin preparation with good anticoagulant activity has a minimum eight repeating units. The larger the molecule, the better its anticoagulant activity.

SUMMARY

Monosaccharides are polyhydroxyaldehydes or polyhydroxyketones. The most common have the general formula $C_nH_{2n}O_n$ where n varies from 3 to 8. Their names contain the suffix "-ose," and their prefixes "tri-," "tetr-," and so on indicate the number of carbon atoms in the chain. The prefix "aldo-" indicates an aldehyde, and the prefix "keto-" indicates a ketone.

In a **Fischer projection** of a carbohydrate, the carbon chain is written vertically with the most highly oxidized carbon toward the top. Horizontal lines show groups projecting above the plane of the page; vertical lines show groups projecting behind the plane of the page.

The **penultimate carbon** of a monosaccharide is the next to last carbon of a Fischer projection. A monosaccharide that has the same configuration at the penultimate carbon as D-glyceraldehyde is called a **D-monosaccharide;** one that has the same configuration at the penultimate carbon as L-glyceraldehyde is called an **L-monosaccharide.**

Monosaccharides exist primarily as cyclic hemiacetals. A six-membered cyclic hemiacetal is a **pyranose;** a five-membered cyclic hemiacetal is a **furanose.** The new stereocenter resulting from hemiacetal formation is referred to as an **anomeric carbon,** and the stereoisomers thus formed are **anomers.** The symbol β- indicates that the —OH on the anomeric carbon is on the same side of the ring as the terminal —CH$_2$OH. The symbol α- indicates that the —OH on the anomeric carbon is on the opposite side from the terminal —CH$_2$OH. Fura-

noses and pyranoses can be drawn as **Haworth projections.** Pyranoses can also be drawn as **chair conformations.**

Mutarotation is the change in specific rotation that accompanies formation of an equilibrium mixture of α and β anomers in aqueous solution.

A **glycoside** is a cyclic acetal derived from a monosaccharide. The name of the glycoside is composed of the name of the alkyl or aryl group bonded to the acetal oxygen atom followed by the name of the monosaccharide in which the terminal "-e" is replaced by "-ide."

An **alditol** is a polyhydroxy compound formed by reduction of the carbonyl group of a monosaccharide to a hydroxyl group. An **aldonic acid** is a carboxylic acid formed by oxidation of the aldehyde group of an aldose. Any carbohydrate that reacts with an oxidizing agent to form an aldonic acid is classified as a **reducing sugar** (it reduces the oxidizing agent).

A **disaccharide** contains two monosaccharide units joined by a glycosidic bond. Terms applied to carbohydrates containing larger numbers of monosaccharides are **trisaccharide, tetrasaccharide, oligosaccharide,** and **polysaccharide. Maltose** is a disaccharide of two molecules of D-glucose joined by an α-1,4-glycosidic bond. **Lactose** is a disaccharide consisting of D-galactose joined to D-glucose by a β-1,4-glycosidic bond. **Sucrose** is a disaccharide consisting of D-glucose joined to D-fructose by an α-1,2-glycosidic bond.

Starch can be separated into two fractions given the names amylose and amylopectin. **Amylose** is a linear polysaccharide of up to 4000 units of D-glucopyranose joined by α-1,4-glycosidic bonds. **Amylopectin** is a highly branched polysaccharide of D-glucose joined by α-1,4-glycosidic bonds and, at branch points, by α-1,6-glycosidic bonds. **Glycogen,** the reserve carbohydrate of animals, is a highly branched polysaccharide of D-glucopyranose joined by α-1,4-glycosidic bonds and, at branch points, by α-1,6-glycosidic bonds. **Cellulose,** the skeletal polysaccharide of plants, is a linear polysaccharide of D-glucopyranose joined by β-1,4-glycosidic bonds.

The carboxyl and sulfate groups of **acidic polysaccharides** are ionized to $-CO_2^-$ and $-SO_3^-$ at the pH of body fluids, which gives these polysaccharides net negative charges.

KEY TERMS

Alditol (Section 19.5)
Aldose (Section 19.2)
Anomeric carbon (Section 19.3)
Anomers (Section 19.3)
Blood type (Box 19D)
Carbohydrate (Section 19.1)
Chair conformation (Section 19.3)
Disaccharide (Section 19.6)

Fischer projection (Section 19.2)
Furanose (Section 19.3)
Glycoside (Section 19.5)
Glycosidic bond (Section 19.5)
Haworth projection (Section 19.3)
Hemiacetal (Section 19.3)
Ketose (Section 19.2)
Monosaccharide (Section 19.2)

Mutarotation (Section 19.3)
Oligosaccharide (Section 19.6)
Polysaccharide (Section 19.7)
Pyranose (Section 19.3)
Reducing sugar (Section 19.5)
Saccharide (Section 19.1)
Trisaccharide (Section 19.6)

KEY REACTIONS

1. Formation of Cyclic Hemiacetals (Section 19.3)

A monosaccharide existing as a five-membered ring is a furanose; one existing as a six-membered ring is a pyranose. A pyranose is most commonly drawn as either a Haworth projection or a chair conformation.

D-Glucose

β-D-Glucopyranose
(β-D-Glucose)

2. Mutarotation (Section 19.3)

Anomeric forms of a monosaccharide are in equilibrium in aqueous solution. Mutarotation is the change in specific rotation that accompanies this equilibration.

β-D-Glucopyranose
$[\alpha]_D^{25} + 18.7°$

Open-chain form

α-D-Glucopyranose
$[\alpha]_D^{25} + 112°$

3. Formation of Glycosides (Section 19.5)

Treatment of a monosaccharide with an alcohol in the presence of an acid catalyst forms a cyclic acetal called a glycoside. The bond to the new —OR group is called a glycosidic bond.

4. Reduction to Alditols (Section 19.5)

Reduction of the carbonyl group of an aldose or ketose to a hydroxyl group yields a polyhydroxy compound called an alditol.

D-Glucose

D-Glucitol
(D-Sorbitol)

5. Oxidation to an Aldonic Acid (Section 19.5)

Oxidation of the aldehyde group of an aldose to a carboxyl group by a mild oxidizing agent gives a polyhydroxycarboxylic acid called an aldonic acid.

D-Glucose

D-Gluconic acid

CONCEPTUAL PROBLEMS

19.A Of the eight D-aldohexoses, which is the most abundant in the biological world?

19.B When drawn in the β-D-pyranose form, which of the eight D-aldohexoses has the greatest number of substituents on its six-membered ring in equatorial positions.

19.C Both Haworth projections and chair conformations show the cis/trans configurations of groups on the six-membered rings of D-aldohexopyranoses. In what way are chair conformations a more accurate representation of molecular shape?

19.D Explain why all mono- and disaccharides are soluble in water.

PROBLEMS

Difficult problems are designated by an asterisk.

19.6 What percent of the dry weight of plants is carbohydrates? How does this percent compare with that in humans?

19.7 What gas is produced by plants during photosynthesis? By animals during metabolism?

Monosaccharides

19.8 Explain the meaning of the designations D and L as used to specify the configuration of carbohydrates.

19.9 How many stereocenters are present in D-glucose? In D-ribose?

19.10 Define each term. (a) Aldose (b) Ketose (c) Aldopentose (d) Ketohexose

19.11 Which carbon of an aldopentose determines whether the pentose has a D or L configuration?

19.12 Which compounds are D-monosaccharides and which are L-monosaccharides?

19.13 Write Fischer projections for L-ribose and L-arabinose.

The Cyclic Structure of Monosaccharides

19.14 Define the term anomeric carbon. In glucose, which carbon is the anomeric carbon?

19.15 Define. (a) Pyranose (b) Furanose

19.16 Which is the anomeric carbon in a 2-ketohexose?

19.17 Are α-D-glucose and β-D-glucose enantiomers? Explain.

19.18 Explain the conventions for using α and β to designate the configuration of cyclic forms of monosaccharides.

19.19 Are the hydroxyl groups on carbons 1, 2, 3, and 4 of α-D-glucose all in equatorial positions?

***19.20** Convert each Haworth projection to an open-chain form and then to a Fischer projection. Name the monosaccharide you have drawn.

***19.21** Convert each chair conformation to an open-chair form and then to a Fischer projection. Name the monosaccharide you have drawn.

19.22 Explain the phenomenon of mutarotation with reference to carbohydrates. By what means is it detected?

19.23 The specific rotation of α-D-glucose is +112.2°.
(a) What is the specific rotation of α-L-glucose?
(b) When α-D-glucose is dissolved in water, the specific rotation of the solution changes from +112.2° to +52.7°. Does the specific rotation of α-L-glucose also change when it is dissolved in water? If so, to what value does it change?

19.24 What is the most important monosaccharide in the human body?

Reactions of Monosaccharides

19.25 Draw Fischer projections for the product formed by treating each monosaccharide with H_2/Pt.
(a) D-Galactose (b) D-Ribose

19.26 Name the two alditols formed by treating D-fructose with H_2/Ni.

***19.27** One pathway for the metabolism of glucose 6-phosphate is its enzyme-catalyzed conversion to fructose 6-phosphate. Show that this transformation can be regarded as two enzyme-catalyzed keto-enol tautomerizations (Section 17.6).

D-Glucose 6-phosphate   D-Fructose 6-phosphate

19.28 Ribitol and β-D-ribose 1-phosphate are derivatives of D-ribose. Draw structural formulas for each compound.

19.29 Define the term glycosidic bond.

19.30 Do glycosides undergo mutarotation?

19.31 What is the difference in meaning between the terms glycosidic bond and glucosidic bond?

Disaccharides and Oligosaccharides

19.32 Following is the structural formula of a disaccharide.

(a) Name the two monosaccharide units in the disaccharide.

(b) Locate the acetal group.

(c) Locate the hemiacetal group.

(d) Is this disaccharide a reducing sugar?

19.33 Which disaccharides are reducing sugars? (a) Maltose (b) Lactose (c) Sucrose

*19.34 Trehalose is found in young mushrooms and is the chief carbohydrate in the blood of certain insects. Trehalose is a disaccharide consisting of two D-monosaccharide units, each joined to the other by an α-1,1-glycosidic bond.

Trehalose

(a) Is trehalose a reducing sugar?

(b) Does trehalose undergo mutarotation?

(c) Name the two monosaccharide units of which trehalose is composed.

Polysaccharides

19.35 What is the difference in structure between oligo- and polysaccharides?

19.36 Why is cellulose insoluble in water?

19.37 How is it possible that cows can digest grass but humans cannot?

*19.38 A Fischer projection of *N*-acetyl-D-glucosamine is given in Section 19.2.

(a) Draw Haworth and chair structures for the β-pyranose form of this monosaccharide.

(b) Draw Haworth and chair structures for the disaccharide formed by joining two units of the pyranose form of *N*-acetyl-D-glucosamine by a β-1,4-glycosidic bond. If you draw this correctly, you will have the structural formula for the repeating dimer of chitin, the structural polysaccharide component of the shells of lobsters and other crustaceans.

*19.39 Propose structural formulas for the repeating disaccharide unit in these polysaccharides.

(a) Alginic acid, isolated from seaweed, is used as a thickening agent in ice cream and other foods. Alginic acid is a polymer of D-mannuronic acid in the pyranose form joined by β-1,4-glycosidic bonds.

(b) Pectic acid is the main component of pectin, which is responsible for the formation of jellies from fruits and berries. Pectic acid is a polymer of D-galacturonic acid in the pyranose form joined by α-1,4-glycosidic bonds.

D-Mannuronic acid   D-Galacturonic acid

19.40 Hyaluronic acid acts as a lubricant in the synovial fluid of joints. In rheumatoid arthritis, inflammation breaks hyaluronic acid down to smaller molecules. Under these conditions, what happens to the lubricating power of the synovial fluid?

19.41 The anticlotting property of heparin is partly due to the negative charges it carries.

(a) Identify the functional groups that provide the negative charges.

(b) Which type of heparin is a better anticoagulant, one with a high or a low degree of polymerization?

Boxes

19.42 (Box 19A) Why does congenital galactosemia appear only in infants?

19.43 (Box 19A) Why can galactosemia be relieved by feeding an infant a formula containing sucrose as the only carbohydrate?

19.44 (Box 19B) What is the difference in structure between L-ascorbic acid and L-dehydroascorbic acid? What does the designation L indicate in these molecules?

19.45 (Box 19B) When L-ascorbic acid participates in redox reactions, it is converted to L-dehydroascorbic acid. In such redox reactions, is L-ascorbic acid oxidized or reduced?

19.46 (Box 19B) Is L-ascorbic acid a biological oxidizing agent or a biological reducing agent?

19.47 (Box 19C) Why is glucose assay one of the most common analytical tests performed in the clinical chemistry laboratory?

19.48 (Box 19C) What reaction is catalyzed by the enzyme glucose oxidase?

19.49 (Box 19D) The monosaccharide unit common to A, B, AB, and O blood types is L-fucose.

(a) Characterize this monosaccharide as an aldose or ketose; as a hexose or pentose.

(b) What is unusual about this monosaccharide.

(c) If L-fucose were to undergo a reaction in which its terminal —CH₃ group were converted to a —CH₂OH group, what monosaccharide would this be?

19.50 (Box 19D) Why can't a person with A-type blood donate to a person with B-type blood?

19.51 (Box 19E) How does a filler influence the effectiveness of a drug?

19.52 (Box 19F) What is the most important function of the vitreous and its component hyaluronic acid?

Additional Problems

19.53 How many aldooctoses are possible? How many D-aldooctoses are possible?

****19.54** 2,6-Dideoxy-D-altrose, known alternatively as D-digitoxose, is a monosaccharide obtained on hydrolysis of digitoxin, a natural product extracted from purple foxglove *(Digitalis purpurea).* Digitoxin has found wide use in cardiology because it reduces pulse rate, regularizes heart rhythm, and strengthens heart beat. Draw the structural formula of 2,6-dideoxy-D-altrose.

19.55 Is this compound (a) an aldose or a ketose, (b) a pyranose or a furanose, (c) a pentose or a hexose, (d) a D- or an L-monosaccharide?

19.56 Draw α-D-glucopyranose (α-D-glucose) as a Haworth projection. Now, using only the information given here, draw Haworth projections for these monosaccharides.
(a) α-D-mannopyranose (α-D-mannose). The configuration of D-mannose differs from the configuration of D-glucose only at carbon 2.
(b) α-D-gulopyranose (α-D-gulose). The configuration of D-gulose differs from the configuration of D-glucose at carbons 3 and 4.

19.57 In making candy or sugar syrups, sucrose is boiled in water with a little acid, such as lemon juice. Why

does the product mixture taste sweeter than the starting sucrose solution?

19.58 Hot water extracts of ground willow bark are an effective pain reliever (Box 18A). Unfortunately, the liquid is so bitter that most persons refuse it. The pain reliever in these infusions is salicin. Name the monosaccharide unit in salicin.

Salicin

19.59 Keratan sulfate is an important component of the cornea of the eye. Following is the repeating unit of this acidic polysaccharide.

(a) From what monosaccharides or derivatives of monosaccharides is keratan sulfate made?
(b) Describe the glycosidic bond in this repeating disaccharide unit.
(c) What is the net charge on this repeating disaccharide unit at pH 7.0?

Lipids

Note that, unlike the case of carbohydrates, we define lipids in terms of a property and not in terms of structure.

Above: Sea lions are marine mammals that require a heavy layer of fat in order to survive in cold waters. (Doug Perrine/TCL/Masterfile)

20.1 Introduction

Found in living organisms, **lipids** are substances that **are insoluble in water** but soluble in nonpolar solvents and solvents of low polarity, such as diethyl ether. This lack of solubility in water is an important property because our body chemistry is so firmly based on water. Most body constituents, including carbohydrates, are soluble in water. But the body also needs insoluble compounds for many purposes, including the separation of compartments containing aqueous solutions from each other, and that's where lipids come in.

The water-insolubility of lipids is due to the fact that the polar groups they contain are much smaller than their alkane-like (nonpolar) portions. These nonpolar portions provide the water-repellent, or *hydrophobic,* property.

An important use for lipids, especially in animals, is the storage of energy. As we saw in Section 19.8, plants store energy in the form of starch. Animals (including humans) find it more economical to use fats instead. Although our bodies do store some carbohydrates in the form of glycogen for quick energy when we need it, energy stored in the form of fats is much more important. The reason is simply that the burning of fats produces more than twice as much energy (about 9 kcal/g) as the burning of an equal weight of carbohydrates (about 4 kcal/g). Lipids also serve as chemical messengers.

For purposes of study, we can divide lipids into four groups: (1) fats and waxes; (2) complex lipids; (3) steroids; and (4) prostaglandins, thromboxanes, and leukotrienes.

20.2　The Structure of Triglycerides

Animal fats and vegetable oils are triglycerides. **Triglycerides are triesters of glycerol and long-chain carboxylic acids called fatty acids.** We saw in Section 18.4 that esters are made up of an alcohol part and an acid part.

$$
\begin{array}{l}
CH_2\!-\!OH \\
|\ \\
CH\!-\!OH \\
|\ \\
CH_2\!-\!OH \\
\end{array}
$$

Glycerol

$$
R\!-\!\overset{\overset{\textstyle O}{\|}}{C}\!-\!O\!-\!R'
$$

Acid part　　Alcohol part

In contrast to the alcohol part, the acid component of fats may be any number of acids, which do, however, have certain things in common:

1. They are practically all unbranched carboxylic acids.

2. They range in size from about 10 carbons to 20 carbons.

3. They have an even number of carbon atoms.

4. Apart from the —COOH group, they have no functional groups except that some do have double bonds.

5. In most unsaturated fatty acids, the cis isomers predominate.

The reason that only even-numbered acids are found in fats is that the body builds these acids entirely from acetic acid units and therefore puts the carbons in two at a time (Section 28.3). Table 20.1 shows the most important acids found in fats.

The fatty acids can be divided into two groups: saturated and unsaturated. **Saturated fatty acids** have only single bonds in the hydrocarbon chain. **Unsaturated fatty acids** have at least one C=C double bond in the chains. All the unsaturated fatty acids listed in Table 20.1 are the cis isomers. This explains their physical properties, reflected in their melting points. Saturated fatty acids are solids at room temperature because the regular nature of their aliphatic chains allows the molecules to be packed in a close, parallel alignment:

TABLE 20.1 **The Most Important Fatty Acids in Triglycerides**

Carbon Atoms: Double Bonds[a]	Structure	Common Name	Melting Point (°C)
Saturated Fatty Acids			
12:0	$CH_3(CH_2)_{10}COOH$	lauric acid	44
14:0	$CH_3(CH_2)_{12}COOH$	myristic acid	58
16:0	$CH_3(CH_2)_{14}COOH$	palmitic acid	63
18:0	$CH_3(CH_2)_{16}COOH$	stearic acid	70
20:0	$CH_3(CH_2)_{18}COOH$	arachidic acid	77
Unsaturated Fatty Acids			
16:1	$CH_3(CH_2)_5CH{=}CH(CH_2)_7COOH$	palmitoleic acid	1
18:1	$CH_3(CH_2)_7CH{=}CH(CH_2)_7COOH$	oleic acid	16
18:2	$CH_3(CH_2)_4(CH{=}CHCH_2)_2(CH_2)_6COOH$	linoleic acid	−5
18:3	$CH_3CH_2(CH{=}CHCH_2)_3(CH_2)_6COOH$	linolenic acid	−11
20:4	$CH_3(CH_2)_4(CH{=}CHCH_2)_4(CH_2)_2COOH$	arachidonic acid	−49

[a]The first number is the number of carbons in the fatty acid; the second is the number of carbon-carbon double bonds in its hydrocarbon chain.

The longer the aliphatic chain, the higher the melting point.

The interactions (London dispersion forces) between neighboring chains are weak. Nevertheless, the regular packing allows these forces to operate over a large portion of the chain so that a considerable amount of energy is needed to melt them.

Unsaturated fatty acids, in contrast, are all liquids at room temperature because the cis double bonds interrupt the regular packing of the chains:

The London dispersion forces act over shorter segments of the chain. Thus much less energy is required to melt them. The greater the degree of unsaturation, the lower the melting point because each double bond introduces more disorder into the packing of the molecules (Table 20.1).

Since glycerol has three —OH groups, a single molecule of glycerol can be esterified with three different acid molecules. Thus, a typical fat molecule might be

A triglyceride

Such compounds are called **triglycerides** or **triacylglycerols:** All three —OH groups of glycerol are esterified. Triglycerides are the most common lipid materials, although **mono-** and **diglycerides** are not infrequent. In the latter two types, only one or two —OH groups of the glycerol are esterified by fatty acids.

Triglycerides are complex mixtures. Although some of the molecules have three identical fatty acids, in most cases two or three different acids are present.

The hydrophobic character of fats is caused by the long hydrocarbon chains. The ester groups (—COO—), though polar, are buried in a nonpolar environment, and this makes the fats insoluble in water.

20.3 Properties of Fats

Physical State

With some exceptions, fats that come from animals are generally solids at room temperature, and those from plants or fish are usually liquids. Liquid fats are often called **oils,** even though they are esters of glycerol just like solid fats and should not be confused with petroleum, which is mostly alkanes.

What is the structural difference between solid fats and liquid oils? In most cases it is the degree of unsaturation. The physical properties of the fatty acids are carried over to the physical properties of the triglycerides. Solid animal fats contain mainly saturated fatty acids, and vegetable oils contain high amounts of unsaturated fatty acids. Table 20.2 shows the average fatty acid content of some common fats and oils. Note that even solid fats contain some unsaturated acids and that liquid fats contain some saturated acids. Some unsaturated fatty acids (linoleic and linolenic acids) are called *essential fatty acids* because the body cannot synthesize them from precursors; they must therefore be included in the diet.

Though most vegetable oils have high amounts of unsaturated fatty acids, there are exceptions. Note that coconut oil has only a small amount of unsaturated acids. This oil is a liquid not because it contains many double bonds but because it is rich in low-molecular-weight fatty acids (chiefly lauric).

Oils with an average of more than one double bond per fatty acid chain are called *polyunsaturated.* For some years there has been a controversy about whether a diet rich in unsaturated and polyunsaturated fats helps to prevent heart attacks (Box 20G).

> **Fat** A mixture of triglycerides containing a high proportion of long-chain, saturated fatty acids.

> **Oil** A mixture of triglycerides containing a high proportion of long-chain unsaturated fatty acids or short-chain saturated fatty acids.

Some vegetable oils. *(Charles D. Winters)*

BOX 20A

Rancidity

The double bonds in fats and oils are subject to oxidation by the air. When a fat or oil is allowed to stand out in the open, this reaction slowly turns some of the molecules into aldehydes and other compounds with bad tastes and odors. We say that the fat or oil has become rancid and is no longer edible. Vegetable oils, which generally contain more double bonds, are more susceptible to this than solid fats, but even fats contain some double bonds, so rancidity can be a problem here, too.

Another cause of unpleasant taste is hydrolysis. The hydrolysis of triglycerides may produce short-chain fatty acids, such as butyric acid, which have bad odors. In order to prevent rancidity, fats and oils should be kept refrigerated (these reactions are slower at low temperatures) and in dark bottles (the oxidation is catalyzed by ultraviolet light). In addition to this, antioxidants are often added to fats and oils to prevent rancidity.

TABLE 20.2 Average Percentage of Fatty Acids of Some Common Fats and Oils

	Lauric	Myristic	Palmitic	Stearic	Oleic	Linoleic	Linolenic	Other
Animal Fats								
Beef tallow	—	6.3	27.4	14.1	49.6	2.5	—	0.1
Butter	2.5	11.1	29.0	9.2	26.7	3.6	—	17.9
Human	—	2.7	24.0	8.4	46.9	10.2	—	7.8
Lard	—	1.3	28.3	11.9	47.5	6.0	—	5.0
Vegetable Oils								
Coconut	45.4	18.0	10.5	2.3	7.5	—	—	16.3
Corn	—	1.4	10.2	3.0	49.6	34.3	—	1.5
Cottonseed	—	1.4	23.4	1.1	22.9	47.8	—	3.4
Linseed	—	—	6.3	2.5	19.0	24.1	47.4	0.7
Olive	—	—	6.9	2.3	84.4	4.6	—	1.8
Palm	—	1.4	40.1	5.5	42.7	10.3	—	—
Peanut	—	—	8.3	3.1	56.0	26.0	—	6.6
Safflower	←——————— 6.8 ———————→				18.6	70.1	3.4	1.1
Soybean	0.2	0.1	9.8	2.4	28.9	52.3	3.6	2.7
Sunflower	—	—	5.6	2.2	25.1	66.2	—	—

Pure fats and oils are colorless, odorless, and tasteless. This statement may seem surprising because we all know the tastes and colors of such fats and oils as butter and olive oil. The tastes, odors, and colors are caused by substances dissolved in the fat or oil.

BOX 20B

Waxes

Waxes are simple esters. They are solids because of their high molecular weights. As in fats, the acid portions of the esters consist of a mixture of fatty acids, but the alcohol portions are not glycerol but simple long-chain alcohols. For example, a major component of beeswax is 1-triacontyl palmitate:

Palmitic acid portion   1-Triacontanol portion

$$CH_3(CH_2)_{13}CH_2\overset{\overset{\textstyle O}{\|}}{C}-OCH_2(CH_2)_{28}CH_3$$
Triacontyl palmitate

Waxes generally have higher melting points than fats (60 to 100 °C) and are harder. Animals and plants use them for protective coatings. The leaves of most plants are coated with wax, which helps to prevent microorganisms from attacking them and also allows them to conserve water. The feathers of birds and the fur of animals are also coated with wax. Ducks would not be able to swim without the protective coating.

Some important waxes are carnauba wax (from a Brazilian palm tree), lanolin (from lamb's wool), beeswax, and spermaceti (from whales). These are used to make cosmetics, polishes, candles, and ointments. Paraffin waxes are not esters.

■ Bees making beeswax. *(Charles D. Winters)*

They are mixtures of high-molecular-weight alkanes. Neither is ear wax a simple ester. It is a gland secretion and contains a mixture of fats (triglycerides) phospholipids, and esters of cholesterol.

Hydrogenation

In Section 12.8, we learned that we can reduce carbon-carbon double bonds to single bonds by treating them with hydrogen (H_2) and a catalyst. It is, therefore, not difficult to convert unsaturated liquid oils to solids, for example,

$$
\begin{array}{l}
\overset{O}{\underset{\|}{}}\quad \text{Oleic acid}\\
CH_2-O-\overset{O}{\overset{\|}{C}}-(CH_2)_7-CH=CH-(CH_2)_7-CH_3\\
\\
CH-O-\overset{O}{\overset{\|}{C}}-(CH_2)_7-CH=CHCH_2CH=CH-(CH_2)_4-CH_3 + 5H_2 \xrightarrow{Pt}\\
\\
CH_2-O-\overset{}{\underset{\|}{C}}-(CH_2)_7-CH=CHCH_2CH=CH-(CH_2)_4-CH_3\\
\quad\quad\quad\quad O
\end{array}
$$

Oleic acid, Linoleic acid, Linoleic acid →

Stearic acid product:
$$
\begin{array}{l}
CH_2-O-\overset{O}{\overset{\|}{C}}-(CH_2)_{16}-CH_3\\
\\
CH-O-\overset{O}{\overset{\|}{C}}-(CH_2)_{16}-CH_3\\
\\
CH_2-O-\overset{}{\underset{\|}{C}}-(CH_2)_{16}-CH_3\\
\quad\quad\quad\quad O
\end{array}
$$

This hydrogenation is carried out on a large scale to produce the solid shortening sold in stores under such brand names as Crisco, Spry, and Dexo. In making such products, manufacturers must be careful not to hydrogenate all the double bonds because a fat with no double bonds would be too solid. Partial, but not complete, hydrogenation results in a product with the right consistency for cooking.

Margarine is also made by partial hydrogenation of vegetable oils. Because less hydrogen is used, margarine contains more unsaturation than hydrogenated shortenings.

Saponification

Glycerides, being esters, are subject to hydrolysis, which can be carried out with either acids or bases. As we saw in Section 18.4, the use of bases is more practical. An example of the saponification of a typical fat is

$$
\begin{array}{l}
\overset{O}{}\ \overset{O}{}\\
\underset{\|}{O}\ CH_2OCR\\
RCOCH\ \ \underset{\|}{O}\\
\quad\ \ CH_2OCR
\end{array}
+ 3NaOH \xrightarrow{saponification}
\begin{array}{l}
CH_2OH\\
|\\
CHOH\\
|\\
CH_2OH
\end{array}
+ 3R\overset{O}{\overset{\|}{C}}O^-Na^+
$$

A triglyceride   1,2,3-Propanetriol   Sodium soaps
                   (Glycerol; glycerin)

The mixture of sodium salts of fatty acids produced by saponification of fats or oils is soap. **Soap** has been used for thousands of years, and saponification is one of the oldest known chemical reactions (Box 20C).

Many common products contain hydrogenated vegetable oils.
(Charles D. Winters)

20.4 Complex Lipids

The triglycerides discussed in the previous sections are significant components of fat storage cells. Other kinds of lipids, called complex lipids, are important in a different way. They constitute the main components of membranes (Section 20.5). Complex lipids can be divided into two groups: phospholipids and glycolipids.

Phospholipids contain an alcohol, two fatty acids, and a phosphate group. There are two types: **glycerophospholipids** and **sphingolipids.** In glycerophospholipids, the alcohol is glycerol (Section 20.6). In sphingolipids, the alcohol is sphingosine (Section 20.7).

Soaps and Detergents

Soaps clean because each soap molecule has a hydrophilic head and a hydrophobic tail. The —COOH end of the molecule (the hydrophilic end), being ionic, is soluble in water. The other end (the hydrophobic end) is a long-chain alkane-like portion, which is insoluble in water but soluble in organic compounds. Dirt may contain water-soluble and water-insoluble portions. The water-soluble portion dissolves in water; no soap is needed for it. The soap is needed for the water-insoluble portion. The hydrophilic end of the soap molecule dissolves in the water; the hydrophobic end dissolves in the dirt. This causes an emulsifying action in which the soap molecules surround the dirt particles in an orderly fashion, the hydrophobic tail interacting with the hydrophobic dirt particle and the hydrophilic head providing the attraction for water molecules (Figure 20C). The cluster of soap molecules in which the dirt is embedded is called a **micelle** [Figure 20C(d)].

The solid soaps and soap flakes with which we are all familiar are sodium salts of fatty acids. Liquid soaps are generally potassium salts of the same acids. Soap has the disadvantage of forming precipitates in hard water, which is water that contains relatively high amounts of Ca^{2+} and Mg^{2+} ions. Most community water supplies in the United States are hard water. Calcium and magnesium salts of fatty acids are insoluble in water. Therefore, when soaps are used in hard water, the Ca^{2+}

and Mg^{2+} ions precipitate the fatty acid anions, which deposit on clothes, dishes, sinks, and bathtubs (it is the "ring" around the bathtub). Because of these difficulties, much of the soap used in the United States has been replaced in the last 40 years by **detergents** for both household and industrial cleaning. Two typical detergent molecules are

$$CH_3(CH_2)_{11}-O-\overset{\overset{\displaystyle O}{\|}}{\underset{\underset{\displaystyle O}{\|}}{S}}-O^-\ Na^+$$

Sodium dodecyl sulfate
(Sodium lauryl sulfate)

$$CH_3(CH_2)_{10}CH_2-\langle\!\!\!\bigcirc\!\!\!\rangle-\overset{\overset{\displaystyle O}{\|}}{\underset{\underset{\displaystyle O}{\|}}{S}}-O^-\ Na^+$$

Sodium dodecylbenzenesulfonate

It can be seen immediately that these molecules also have hydrophobic tails and hydrophilic heads and so can function in the same way as soap molecules. However, they do not precipitate with Ca^{2+} and Mg^{2+} and so work very well in hard water.

■ **FIGURE 20C** The cleaning action of soap. (a) Without soap, water molecules cannot penetrate through molecules of hydrophobic dirt. (b and c) Soap molecules line up at the interface between dirt and water. (d) Soap molecules carry dirt particles away.

Glycolipids are complex lipids that contain carbohydrates (Section 20.8). Figure 20.1 shows schematic structures for all of these.

20.5 Membranes

These small structures inside the cell are called **organelles.**

The complex lipids mentioned in Section 20.4 form the **membranes** around body cells and around small structures inside the cells. Unsaturated fatty acids are important components of these lipids. The purpose of these membranes is to separate cells from the external environment and to

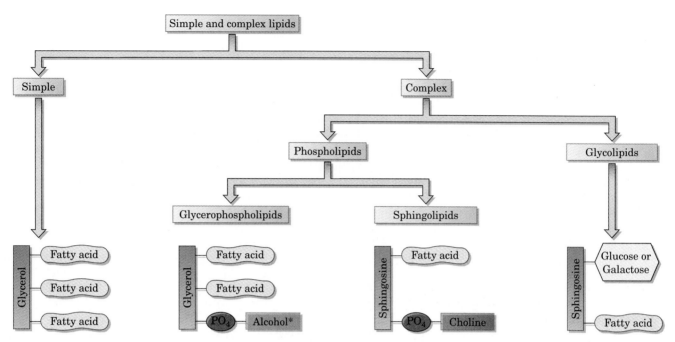

FIGURE 20.1 Schematic diagram of simple and complex lipids.
* The alcohol can be choline, serine, ethanolamine, inositol, or certain others.

provide selective transport for nutrients and waste products; that is, membranes allow the selective passage of substances into and out of cells.

These membranes are made of **lipid bilayers** (Figure 20.2). In a lipid bilayer, there are two rows (layers) of complex lipid molecules arranged tail to tail. The hydrophobic tails point toward each other because that

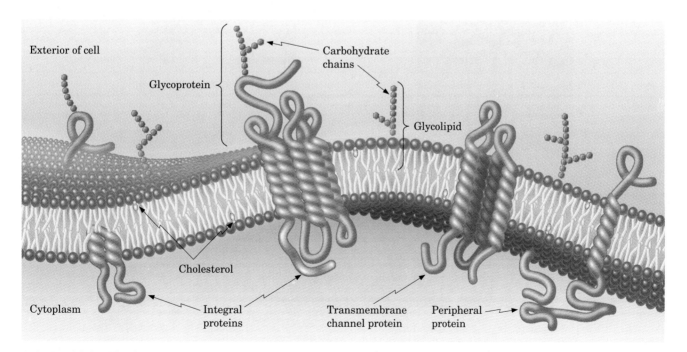

FIGURE 20.2 The fluid mosaic model of membranes. Note that proteins are embedded in the lipid matrix.

BOX 20D

Transport Across Cell Membranes

Membranes are not just random assemblies of complex lipids to provide a nondescript barrier. For example, in human red blood cells, the outer part of the bilayer is made largely of phosphatidylcholine and sphingomyelin, while the inner part is made mostly of phosphatidylethanolamine and phosphatidyl serine. In another example, in the membrane called sarcoplasmic reticulum in the heart muscles, phosphatidylethanolamine is found in the outer part of the membrane, phosphatidylserine is found in the inner part, and the phosphatidyl choline is equally distributed in the two layers of the membrane. Membranes are not static structures either. In many processes, membranes fuse with one another; in others they disintegrate, and their building blocks are used elsewhere. When membranes fuse, for example, in vacuole fusions inside of cells, certain restrictions prevent incompatible membranes from intermixing.

The protein molecules are not dispersed randomly in the bilayer. Sometimes they cluster in patches, but they also appear in regular geometric patterns. An example of the latter are **gap junctions,** channels made of six proteins that create a central pore (Figure 20D). This allows neighboring cells to communicate. Gap junctions are an example of **passive transport.** Small polar molecules, which include such essential nutrients as inorganic ions, sugars, amino acids, and nucleotides, can readily pass through gap junctions. Large molecules such as proteins, polysaccharides, and nucleic acid cannot.

In the **facilitated transport** there is a specific interaction between the transporter and the transported molecule. An example is the **anion transporter** of the red blood cells through which chloride and bicarbonate ions are exchanged in a one-for-one ratio. The transporter is a protein with 14 helical structures that span the membrane. One side of the helices contains the hydrophobic parts of the protein. These can interact with the lipid membrane. The other side of the helices forms a channel. The channel contains the hydrophilic portions of the protein, which can interact with the hydrated ions. In this manner, the anion passes through the erythrocyte membrane.

Active transport involves the passage of ions against concentration gradient. For example, the K^+ has a higher concentration inside the cells than outside in the surrounding environment. Still, potassium ion can be transported from the outside into a cell, but at the expense of energy. The transporter, a membrane protein called Na^+, K^+ ATPase, uses the energy from the hydrolysis of ATP molecule to change the conformation of the transporter which brings in K^+ and exports Na^+.

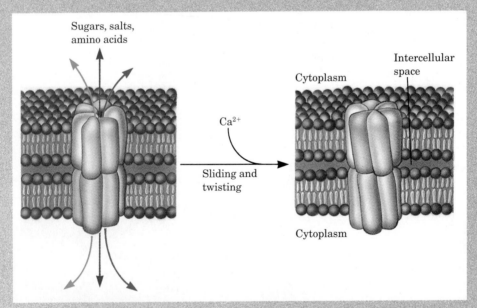

■ *FIGURE 20D* Gap junctions are made of six cylindrical protein subunits. They are lined up in two plasma membranes parallel to each other forming a pore. The pores of gap junctions are closed by a sliding and twisting motion of the cylindrical subunits.

enables them to get as far away from the water as possible. This leaves the hydrophilic heads projecting to the inner and outer surfaces of the membrane.

The unsaturated fatty acids prevent the tight packing of the hydrophobic chains in the lipid bilayer and thereby provide a liquid-like character to the membranes. This is of extreme importance because many products of the body's biochemical processes must cross the membrane, and the liquid nature of the lipid bilayer allows such transport.

The lipid part of the membrane serves as a barrier against any movement of ions or polar compounds into and out of the cells. In the lipid bilayer, protein molecules are either suspended on the surface or partly or fully embedded in the bilayer. These proteins stick out either on the inside or on the outside of the membrane; others are thoroughly embedded, going through the bilayer and projecting from both sides. The model shown in Figure 20.2, called the **fluid mosaic model** of membranes, allows the passage of nonpolar compounds by diffusion, since these compounds are soluble in the lipid membranes. Polar compounds are transported either through specific channels through the protein regions or by another mechanism called active transport (Box 20D). For any transport process, the membrane must behave like a nonrigid liquid so that the proteins can move sideways within the membrane.

> "Hydrophilic" means water-loving.

> Most lipid molecules in the bilayer contain at least one unsaturated fatty acid.

> This effect is similar to that which causes unsaturated fatty acids to have lower melting points than saturated fatty acids.

20.6 Glycerophospholipids

The structure of glycerophospholipids (also called phosphoglycerides) is very similar to that of simple fats. The alcohol is glycerol. Two of the three —OH groups are esterified by fatty acids. As with the simple fats, these fatty acids may be any long-chain carboxylic acids with or without double bonds. The third —OH group is esterified not by a fatty acid but by a phosphate group, which is also esterified to another alcohol. If the other alcohol is choline, a quaternary ammonium compound, the glycerophospholipids are called **phosphatidylcholines** (common name **lecithin**):

> Glycerophospholipids are membrane components of cells throughout the body.

$$HO-CH_2CH_2{}^+NCH_3$$
(Choline)

Structure labeled: Stearic acid, Linolenic acid, Choline, Phosphoric diester — A lecithin

$$CH_2-O-\overset{O}{\overset{\|}{C}}-(CH_2)_{16}CH_3$$

$$CH-O-\overset{O}{\overset{\|}{C}}-(CH_2)_7CH=CHCH_2CH=CH(CH_2)_4CH_3$$

$$CH_2-O-\overset{O}{\underset{O^-}{\overset{\|}{P}}}-O-CH_2CH_2{}^+NCH_3$$

> This typical lecithin molecule has stearic acid on one end and linoleic acid in the middle. Other lecithin molecules contain other fatty acids, but the one on the end is always saturated and the one in the middle is unsaturated.

Note that lecithin has a negatively charged phosphate group and a positively charged quaternary nitrogen from the choline. These charged parts of the molecule provide a strongly hydrophilic head, whereas the rest of the molecule is hydrophobic. Thus, when a phospholipid such as lecithin is part of a lipid bilayer, the hydrophobic tail points toward the middle of the

FIGURE 20.3 Space-filling molecular models of complex lipids in a bilayer. *(From L. Stryer,* Biochemistry, *2nd ed. New York: W. H. Freeman, 1981. ©1981 by W. H. Freeman and Co. All rights reserved.)*

R = glycerol + fatty acid portions

bilayer, and the hydrophilic heads line both the inner and outer surfaces of the membranes (Figures 20.2 and 20.3).

Lecithins are just one example of glycerophospholipids. Another is the **cephalins,** which are similar to the lecithins in every way except that, instead of choline, they contain other alcohols, such as ethanolamine or serine:

A phosphatidylethanolamine (a cephalin)

A phosphatidylserine (a cephalin)

Another important group of glycerophospholipids is the **phosphatidyl inositols, PI.** In PI, the alcohol is inositol linked to the rest of the molecule by a phosphate ester linkage. These compounds not only are integral structural parts of the biological membranes, but also, in their higher phosphorylated form, such as **phosphatidyl inositol 3,4 biphosphates (PIP$_2$), serve as signaling molecules** in chemical communication (see Chapter 23).

Phosphatidyl inositols, PI

20.7 Sphingolipids

Johann Thudichum, who discovered sphingolipids in 1874, named these brain lipids after the monster of Greek mythology, the sphinx. Part woman and part winged lion, the sphinx devoured all who could not provide the correct answer to her riddles. Sphingolipids appeared to Thudichum as part of a dangerous riddle of the brain.

The coating of nerve axons *(myelin)* contains a different kind of complex lipid called **sphingolipids.** In sphingolipids, the alcohol portion is sphingosine:

$$CH_3(CH_2)_{12}CH=CH-CH-CH-CH_2OH$$
$$\qquad\qquad\qquad\qquad\quad OH \quad NH_2$$

Sphingosine

BOX 20E

The Myelin Sheath and Multiple Sclerosis

The human brain and spinal cord can be divided into gray and white regions. Forty percent of the human brain is white matter. Microscopic examination reveals that the white matter is made up of nerve axons wrapped in a white lipid coating, the **myelin sheath,** which provides insulation and thereby allows the rapid conduction of electrical signals. The myelin sheath consists of 70 percent lipids and 30 percent proteins in the usual lipid bilayer structure.

Specialized cells, the **Schwann cells,** wrap themselves around the peripheral nerve axons to form numerous concentric layers (Figure 20E). In the brain, other cells do the wrapping in a similar manner.

Multiple sclerosis affects 250,000 people in the United States. In this disease, a gradual degradation of the myelin sheath can be observed. The symptoms are muscle weariness, lack of coordination, and loss of vision. The symptoms may vanish for a time and return with greater severity. Autopsy of multiple sclerotic brains shows scar-like plaques of white matter, with bare axons not covered by myelin sheaths. The symptoms are produced because the demyelinated axons cannot conduct nerve impulses. A secondary effect of the demyelination is damage to the axon itself.

Similar demyelination occurs in the Guillain-Barré syndrome that follows certain viral infections. In 1976 fears of a "swine flu" epidemic prompted a vaccination program that precipitated a number of cases of Guillain-Barré syndrome.

■ *FIGURE 20E* Myelination of a nerve axon outside the brain by a Schwann cell. The myelin sheath is produced by the Schwann cell and is rolled around the nerve axon for insulation.

The main result is a paralysis that can cause death unless artificial breathing is supplied. The U.S. government assumed legal responsibility for the few bad vaccines and paid compensation to the victims and their families.

A long-chain fatty acid is connected to the —NH$_2$ group by an amide bond, and the —OH group at the end of the chain is esterified by phosphorylcholine:

$$CH_3(CH_2)_{12}CH = CH - CH - CH - CH_2 - O - P - O - CH_2CH_2N^+CH_3$$

Ceramide ↗ portion

OH   NH$_2$

C=O

R

A sphingomyelin
(a sphingolipid)

Sphingomyelin (schematic diagram)

The combination of a fatty acid and sphingosine (shown in color) is often referred to as the **ceramide** part of the molecule because many of these compounds are also found in cerebrosides (Section 20.8). The ceramide part of complex lipids may contain different fatty acids; stearic acid occurs mainly in sphingomyelin.

The phospholipids are not randomly distributed in membranes. In human red blood cells, the sphingomyelin and phosphatidylcholine (lecithin) are on the outside of the membrane facing the blood plasma, whereas phosphatidylethanolamine and phosphatidylserine (cephalins) are on the inside of the membrane. In viral membranes, most of the sphingomyelin is on the inside of the membrane.

20.8 Glycolipids

Glycolipids are complex lipids that contain carbohydrates. Among the glycolipids are the **cerebrosides,** which are ceramide mono- or oligosaccharides.

In cerebrosides, the fatty acid of the ceramide part may contain either 18-carbon or 24-carbon chains; the latter is found only in these complex lipids. A glucose or galactose carbohydrate unit forms a beta glycosidic bond with the ceramide portion of the molecule. The cerebrosides occur primarily in the brain (7 percent of the dry weight) and at nerve synapses.

Glucocerebroside

EXAMPLE 20.1

A lipid isolated from the membrane of red blood cells had the following structure:

(a) To what group of complex lipids does this compound belong?
(b) What are the components?

Solution
The molecule is a triester of glycerol and contains a phosphate group; therefore, it is a glycerophospholipid. Besides glycerol and phosphate, it has a palmitic acid and an oleic acid component. The other alcohol is ethanolamine. Therefore, it belongs to the subgroup of cephalins.

Problem 20.1 ■■■■

A complex lipid had the following structure:

$$CH_2-O-\overset{\overset{\displaystyle O}{\|}}{C}-(CH_2)_{12}CH_3$$

$$CH-O-\overset{\overset{\displaystyle O}{\|}}{C}-(CH_2)_7CH=CHCH_2CH=CH(CH_2)_4CH_3$$

$$CH_2-O-\overset{\overset{\displaystyle O}{\|}}{\underset{\underset{\displaystyle O^-}{|}}{P}}-O-CH_2\underset{\underset{\displaystyle NH_3^+}{|}}{CH}COO^-$$

(a) To what group of complex lipids does this compound belong?
(b) What are the components? ■

20.9 Steroids

The third major class of lipids is the **steroids,** which are compounds containing this ring system:

There are three cyclohexane rings (A, B, and C) connected in the same way as in phenanthrene (Section 14.3) and a fused cyclopentane ring (D). Steroids are thus completely different in structure from the lipids already discussed. Note that they are not necessarily esters, though some of them are.

Cholesterol

The most abundant steroid in the human body, and the most important, is **cholesterol:**

Cholesterol

It serves as a membrane component, mostly in the plasma membranes of red blood cells and in the myelinated nerve cells. The second important function of cholesterol is to serve as a raw material for other steroids, such

Lipid Storage Diseases

Complex lipids are constantly being synthesized and decomposed in the body. Several genetic diseases are classified as lipid storage diseases. In these cases, some of the enzymes needed to decompose the complex lipids are defective or altogether missing from the body. As a consequence, the complex lipids accumulate and cause enlarged liver and spleen, mental retardation, blindness, and, in certain cases, early death. Table 20F summarizes some of these diseases and indicates the missing enzyme and the accumulating complex lipid (Figure 20F).

At present there is no treatment for these diseases. The best way to prevent them is by genetic counseling. Some of these diseases can be diagnosed during fetal development. For example, Tay-Sachs disease, which is carried by about 1 in every 30 Jewish Americans (versus 1 in 300 in the non-Jewish population), can be diagnosed from amniotic fluid obtained by amniocentesis.

However, there is now an experimental treatment available for patients with mild forms of certain lipid storage diseases such as Gaucher's disease. The human enzyme β-glucosidase can be extracted from placenta. It is encapsulated in a lipid bilayer and injected intravenously. The lipid bilayer capsule allows the drug to pass through the membranes of spleen cells and thus deliver the enzyme where it is needed. Unfortunately, because it is difficult to isolate the

■ **FIGURE 20F** The accumulation of glucocerebrosides in the cell of a patient with Gaucher's disease. These cells (Gaucher cells) infiltrate the bone marrow. *(Courtesy of Drs. P. G. Bullogh and V. J. Vigorita, and the Gower Medical Publishing Co., NY)*

human enzyme from placenta, each injection costs thousands of dollars. If and when the human gene for the enzyme is isolated, the enzyme can be manufactured by the recombinant DNA technique (Section 25.8), and the cost will be drastically lowered.

TABLE 20F Lipid Storage Diseases

Name	Accumulating Lipid	Missing or Defective Enzyme
Gaucher's disease	Glucocerebroside	β-Glucosidase
Krabbe's leukodystrophy	Galactocerebroside	β-Galactosidase
Fabry's disease	Ceramide trihexoside	α-Galactosidase

Table 20F *continued*

Tay-Sachs disease

Ceramide oligosaccharide (a ganglioside)

Hexosaminidase A

Niemann-Pick disease

Sphingomyelinase

Sphingomyelin

as the sex and adrenocorticoid hormones (Section 20.10) and bile salts (Section 20.11).

Cholesterol exists both in the free form and esterified with fatty acids. Gallstones contain free cholesterol (Figure 20.4).

Because the correlation between high serum cholesterol levels and such diseases as atherosclerosis has received so much publicity, many peo-

FIGURE 20.4 A human gallstone is almost pure cholesterol. This gallstone measures 5 mm in diameter. *(© Carolina Biological Supply Company / Phototake, NYC)*

Cholesterol crystals, taken from fluid in the elbow of a patient suffering from bursitis, as seen in a polarizing light microscopic photograph. *(Courtesy of Drs. P. A. Dieppe, P. A. Bacon, A. N. Bamji, I. Watt, and the Gower Medical Publishing Co. Ltd., London, England)*

ple are afraid of cholesterol and regard it as some kind of poison. It should be apparent from this discussion that, far from being poisonous, cholesterol is necessary for human life. Without it, we would die. Fortunately, there is no chance of that because, even if it were completely eliminated from the diet, our livers would make enough to satisfy our needs.

Cholesterol in the body is in a dynamic state. Most of the cholesterol ingested as well as that manufactured by the liver is used by the body to make other steroids, such as bile salts. The serum cholesterol level controls the amount of cholesterol synthesized by the liver.

Cholesterol, along with fat, is transported from the liver to the peripheral tissues by **lipoproteins.** There are two important kinds: **high-density lipoprotein (HDL),** which has a protein content of about 50 percent, and **low-density lipoprotein (LDL),** in which the protein content is only about 25 percent. LDL contains about 45 percent cholesterol; HDL contains only 18 percent. High-density lipoprotein transports cholesterol to the liver and also transfers cholesterol to LDL. Low-density lipoprotein plays an important role in cholesterol metabolism. The core of LDL (Figure 20.5) contains fats (triglycerides) and esters of cholesterol in which linoleic acid is an important constituent.

On the surface of the LDL cluster are phospholipids, free cholesterol, and proteins. These molecules contain many polar groups so that water molecules in the blood plasma can solvate them by forming hydrogen bonds, making the LDL soluble in the blood plasma. The LDL carries cholesterol to the cells, where specific LDL-receptor molecules line the cell surface in certain concentrated areas called **coated pits.** One of the proteins on the surface of the LDL binds specifically to the LDL-receptor molecules in the coated pits. After such binding, the LDL is taken inside the cell, where enzymes break it down, liberating the free cholesterol from cholesterol esters. In this manner, the cell can, for example, use cholesterol as a component of a membrane. This is the normal fate of LDL and the normal course of cholesterol transport.

Lipoproteins Spherically shaped clusters containing both lipid molecules and protein molecules.

When the cholesterol level goes above 150 mg/100 mL, cholesterol synthesis in the liver is reduced to half the normal rate of production.

Michael Brown and Joseph Goldstein of the University of Texas shared the Nobel Prize in Medicine in 1986 for the discovery of the LDL-receptor-mediated pathway.

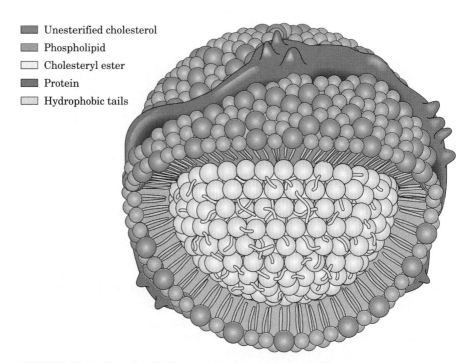

Unesterified cholesterol
Phospholipid
Cholesteryl ester
Protein
Hydrophobic tails

FIGURE 20.5 Low-density lipoprotein shown schematically.

BOX 20G

Cholesterol and Heart Attacks

Like all lipids, cholesterol is insoluble in water, and if its level is elevated in the blood serum, plaque-like deposits may form on the inner surfaces of the arteries. This leads to a decrease in the diameter of the blood vessels, which may lead to a decrease in the flow of blood. The result is atherosclerosis (Figure 20G), along with accompanying high blood pressure, which may lead to heart attack, stroke, or kidney dysfunction. Atherosclerosis enhances the possible complete blockage of some arteries by a clot at the point where the arteries are constricted by plaque. Furthermore, blockage may deprive cells of oxygen, and these may cease to function. The death of heart muscles due to lack of oxygen is called myocardial infarction.

The more general condition, arteriosclerosis, or hardening of the arteries with age, is also accompanied by increased levels of cholesterol in the blood serum. Young adults have, on the average, 1.6 g of cholesterol per liter of blood; however, in people older than 55, this almost doubles to 2.5 g/L because the rate of metabolism slows with age. Diets low in cholesterol and saturated fatty acids usually reduce the serum cholesterol level, and a number of drugs that inhibit the synthesis of cholesterol in the liver are available. The commonly used drug lovastatin and related compounds act to inhibit one of the key enzymes in cholesterol synthesis, HMG-CoA reductase. Thus, they block the synthesis of cholesterol inside the cells and stimulate the synthesis of LDL-receptor proteins. I this way, more LDL enters the cells, diminishing the amount of cholesterol that will be deposited on the inner walls of arteries. Although there is a good correlation between high serum cholesterol and various circulatory diseases, not everyone who suffers from hardening of the arteries has high serum cholesterol, nor do all patients with high serum cholesterol develop arteriosclerosis.

■ **FIGURE 20.G** Effect of atherosclerosis in arteries. *(© 1994 SIU/Photo Researchers, Inc.)*

In certain cases, however, there are not enough LDL receptors. In the disease called familial hypercholesterolemia, the cholesterol level in the plasma may be as high as 680 mg/100 mL as opposed to 175 mg/100 mL in normal subjects. These high levels of cholesterol can cause atherosclerosis and heart attacks (Box 20G). The high plasma cholesterol levels in these patients are caused because there are not enough functional LDL receptors or, if there are enough, they are not concentrated in the coated pits. Thus, high LDL content means high cholesterol content in the plasma because LDL cannot get into the cells and be metabolized. Therefore, a high LDL level together with a low HDL level is a symptom of faulty cholesterol transport and a warning for possible atherosclerosis. Today it is generally considered desirable to have high levels of HDL and low levels of LDL in the bloodstream. Premenopausal women have more HDL than men, which is why women have lower risk of coronary heart disease. HDL levels can be increased by exercise and weight loss.

At the present time, our knowledge of the role played by serum cholesterol in atherosclerosis is incomplete. The best we can say is that it probably makes good sense to reduce the amount of cholesterol and saturated fatty acids in the diet.

20.10 Steroid Hormones

Cholesterol is the starting material for the synthesis of steroid hormones. In this process, the aliphatic side chain on the D ring is shortened by the removal of a six-carbon unit, and the secondary alcohol group on C-3 is oxidized to a ketone. The resulting molecule, *progesterone,* serves as the starting compound for both the sex hormones and the adrenocorticoid hormones (Figure 20.6).

Adrenocorticoid Hormones

The adrenocorticoid hormones are products of the adrenal glands. We divide them into two groups according to function: *Mineralocorticoids* regulate the concentrations of ions (mainly Na^+ and K^+), and *glucocorticoids* control carbohydrate metabolism.

Aldosterone is one of the most important mineralocorticoids. Increased secretion of aldosterone enhances the reabsorption of Na^+ and Cl^- ions in the kidney tubules and increases the loss of K^+. Because Na^+ concentration controls water retention in the tissues, aldosterone also controls tissue swelling.

The term "adrenal" means adjacent to the renal (which refers to the kidney). The name "corticoid" indicates that the site of the secretion is the cortex (outer part) of the gland.

Sex hormones   **Adrenocorticoid hormones**

FIGURE 20.6 The biosynthesis of hormones from progesterone.

Anabolic Steroids

Testosterone, the principal male hormone, is responsible for the buildup of muscles in men. Because of this, many athletes have taken this drug in an effort to increase muscular development. This is especially common among athletes in sports in which strength and muscle mass are important, including weight lifting, shot put, and hammer throw, but participants in other sports, such as running, swimming, and cycling, would also like larger and stronger muscles.

Although used by many athletes, testosterone has two disadvantages: (1) Besides its effect on muscles, it also affects secondary sexual characteristics, and too much of it can result in undesired side effects. (2) It is not very effective when taken orally and must be injected for best results.

For these reasons, a large number of other anabolic steroids, all of them synthetic, have been developed. Some examples follow:

Methandienone

Methenolone

Nandrolone decanoate

Of the three steroids shown here, methandienone and methenolone can be taken orally, but nandrolone decanoate must be injected.

Some women athletes also use anabolic steroids. Because their bodies produce only small amounts of testosterone, women have much more to gain from anabolic steroids than men.

Another way to increase testosterone concentration is to use prohormones, which the body converts to testosterone. One such prohormone is 4-androstenedione, which is also known as "andro." Athletes have used it to enhance performance.

4-Androstene–3,17–dione

The use of anabolic steroids is forbidden in many sporting events, especially in international competition, largely for two reasons: (1) It gives some competitors an unfair advantage, and (2) these drugs can have many side effects, ranging from acne to liver tumors. Side effects can be especially disadvantageous for women; they can include growth of facial hair, baldness, deepening of the voice, and menstrual irregularities.

All athletes participating in the Olympic Games are required to pass a urine test for anabolic steroids. A number of winning athletes have had their victories taken away because they tested positive for steroid use. One tragic example is the Canadian Ben Johnson, a world-class sprinter, whose career was brought to an end at the 1988 Olympiad in Seoul, Korea. He had just won the 100-m race in world-record time when a urine analysis proved that he had taken steroid drugs. He was stripped of both his world record and his gold medal. A similar failed test for 'andro' resulted in banning the American shot put champion Randy Barnes from competition in 1998. Prohormones such as andro are not listed under the Anabolic Steroid Act of 1990; hence, their nonmedical use is not a federal offense as is the case with anabolic steroids. Mark McGwire hit his record-breaking home runs in 1998 while taking andro, as baseball rules did not prohibit its usage. Still the International Olympic Committee bans the use of prohormones together with anabolic steroids.

■ Home run champion Mark McGwire has used androstenedione, a muscle-building dietary supplement that is allowed in baseball, but is banned in professional football, college athletics, and the Olympics. (Jed Jacobson/AllSport USA)

Cortisol is the major glucocorticoid. Its function is to increase the glucose and glycogen concentrations in the body. This is done at the expense of other nutrients. Fatty acids from fat storage cells and amino acids from body proteins are transported to the liver, which, under the influence of cortisol, manufactures glucose and glycogen from these sources.

Cortisol and its ketone derivative, *cortisone,* have remarkable anti-inflammatory effects. These or similar synthetic derivatives, such as prednisolone, are used to treat inflammatory diseases of many organs, rheumatoid arthritis, and bronchial asthma.

Sex Hormones

The most important male sex hormone is testosterone (Figure 20.6). This hormone, which promotes the normal growth of the male genital organs, is synthesized in the testes from cholesterol. During puberty, increased testosterone production leads to such secondary male sexual characteristics as deep voice and facial and body hair.

Female sex hormones, the most important of which is estradiol (Figure 20.6), are synthesized from the corresponding male hormone (testosterone) by aromatization of the A ring.

Estradiol, together with its precursor progesterone, regulates the cyclic changes occurring in the uterus and ovaries known as the *menstrual cycle.* As the cycle begins, the level of estradiol in the body rises, and this causes the lining of the uterus to thicken. Then another hormone, called luteinizing hormone, triggers ovulation. If the ovum is fertilized, increased progesterone levels will inhibit any further ovulation. Both estradiol and progesterone then promote further preparation of the uterine lining to receive the fertilized ovum. If no fertilization takes place, progesterone production stops altogether, and estradiol production decreases. This decreases the thickening of the uterine lining, and it is then sloughed off with accompanying bleeding. This is menstruation (Figure 20.7).

Because progesterone is essential for the implantation of the fertilized ovum, blocking its action leads to termination of pregnancy (Box 20.I). Progesterone interacts with a receptor (a protein molecule) in the nucleus of cells. The receptor changes its shape when progesterone binds to it (see Section 23.7). A drug, widely used in France and China, called mifepristone, or RU486, acts as a competitor to progesterone.

RU486 also binds to the receptors of glucocorticoid hormones. Its use as an antiglucocorticoid is also recommended to alleviate a disease known as Cushing syndrome, the overproduction of cortisone.

Mifepristone
(RU486)

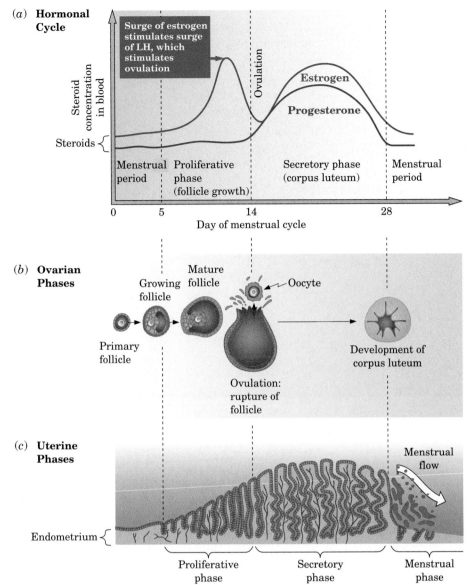

(a) **Hormonal Cycle**

Surge of estrogen stimulates surge of LH, which stimulates ovulation

Steroid concentration in blood

Ovulation

Estrogen

Progesterone

Steroids

Menstrual period | Proliferative phase (follicle growth) | Secretory phase (corpus luteum) | Menstrual period

0   5   14   28

Day of menstrual cycle

(b) **Ovarian Phases**

Mature follicle

Growing follicle

Oocyte

Primary follicle

Ovulation: rupture of follicle

Development of corpus luteum

(c) **Uterine Phases**

Menstrual flow

Endometrium

Proliferative phase | Secretory phase | Menstrual phase

FIGURE 20.7 Events of the menstrual cycle. (*a*) Levels of sex hormones in the bloodstream during the phases of one menstrual cycle in which pregnancy does not occur. (*b*) Development of an ovarian follicle during the cycle. (*c*) Phases of development of the endometrium, the lining of the uterus. The endometrium thickens during the proliferative phase. In the secretory phase which follows ovulation, the endometrium continues to thicken and the glands secrete a glycogen-rich nutritive material in preparation to receive an embryo. When no embryo implants, the new outer layers of the endometrium disintegrate and the blood vessels rupture, producing the menstrual flow.

Oral Contraception

Because progesterone prevents ovulation during pregnancy, it occurred to investigators that progesterone-like compounds might be used for birth control. Synthetic analogs of progesterone proved to be more effective than the natural compound. In "the Pill," a synthetic progesterone-like compound is supplied together with an estradiol-like compound (the latter prevents irregular menstrual flow). Triple-bond derivatives of testosterone, such as norethindrone, norethynodrel, and ethynodiol diacetate, are used most often in birth control pills:

Norethynodrel

Norethindone

Ethynodiol diacetate

It binds to the same receptor site. Because the progesterone molecule is prevented from reaching the receptor molecule, the uterus is not prepared for the implantation of the fertilized ovum, and the ovum is aborted. This is a "morning-after pill." If taken orally within 72 h after intercourse, it prevents pregnancy. However, the drug fails in 20 percent of cases unless followed 36 to 48 h later by an injection, by an intravaginal suppository, or by oral administration of a prostaglandin (Section 20.12). This chemical form of abortion has been in clinical trials in the United States as a supplement to surgical abortion for the last three years. The Food and Drug Administration has recently approved the use of this drug in the United States.

Estradiol and progesterone also regulate secondary female sex characteristics, such as the growth of breasts. Because of this, RU486, as an antiprogesterone, has also been reported to be effective against certain types of breast cancer.

Testosterone and estradiol are not exclusive to either males or females. A small amount of estradiol production occurs in males, and a small amount of testosterone production is normal in females. Only when the proportion of these two hormones (hormonal balance) is upset can one observe symptoms of abnormal sexual differentiation.

20.11 Bile Salts

Bile salts are oxidation products of cholesterol. First the cholesterol is converted to the trihydroxy derivative, and the end of the aliphatic chain is oxidized to the carboxylic acid. The latter in turn forms an amide linkage with an amino acid, either glycine or taurine:

Glycocholate

Taurocholate

Bile salts are powerful detergents. One end of the molecule is strongly hydrophilic because of the negative charge, and the rest of the molecule is largely hydrophobic. Thus, bile salts can disperse dietary lipids in the small intestine into fine emulsions and thereby help digestion.

Because they are eliminated in the feces, bile salts also remove excess cholesterol in two ways: They themselves are breakdown products of cholesterol (thus cholesterol is eliminated via bile salts), and they solubilize deposited cholesterol in the form of bile salt-cholesterol particles.

The dispersion of dietary lipids by bile salts is similar to the action of soap on dirt.

20.12 Prostaglandins, Thromboxanes, and Leukotrienes

Prostaglandins, a group of fatty-acid-like substances, were discovered by Ulf von Euler of Sweden in the 1930s, when he demonstrated that seminal fluid caused a hysterectomized uterus to contract. He isolated them from human semen, and, thinking that they had come from the prostate gland, named them **prostaglandin.** Even though the seminal gland secretes 0.1 mg of prostaglandin per day in mature males, small amounts of prostaglandins are present throughout the body in both sexes.

Prostaglandins are synthesized in the body from arachidonic acid by a ring closure at C-8 and C-12. The enzyme catalyzing this reaction is called **cyclooxygenase** or COX, for short. The product, known as PGG_2, is the common precursor of other prostaglandins, among them PGE and PGF.

Ulf S. von Euler won the Nobel Prize in Physiology and Medicine in 1970. *(Hulton Getty / Tony Stone Images)*

Other prostaglandins (PGAs and PGBs) are derived from PGE.

The prostaglandin E group (PGE) has a carbonyl group at C-9; the subscript indicates the number of double bonds in the hydrocarbon chain. The prostaglandin F group (PGF) has two hydroxyl groups on the ring at positions C-9 and C-11.

The COX enzyme comes in two forms in the body: COX-1 and COX-2. *COX-1* catalyzes the normal physiological production of prostaglandins, which are always present in the body. For example, PGE_2 and $PGF_{2\alpha}$ stimulate uterine contractions and induce labor. They are used in therapeutic abortion, for example, of a dead fetus. PGE_2 lowers blood pressure, but $PGF_{2\alpha}$ causes hypertension (increases blood pressure). PGE_2 in aerosol form is used to treat asthma; it opens up the bronchial tubes by relaxing the surrounding muscles. PGE_1 is used as a decongestant; it opens up nasal passages by constricting blood vessels.

The activity of the *COX-2 enzyme,* on the other hand, is responsible for the production of prostaglandins in inflammation. When a tissue is injured or damaged, special inflammatory cells invade the injured tissue and interact with resident cells, for example, smooth muscle cells. This interaction activates the COX-2, and prostaglandins are synthesized. Such tissue injury may occur in heart attack (myocardial infarction), rheumatic arthritis, and ulcerative colitis. Nonsteroidal anti-inflammatory drugs, for example, aspirin, inhibit COX enzymes (Box 20J).

PGH$_2$

Thromboxane A$_2$

Another class of arachidonic acid derivatives is the **thromboxanes.** Their structure also includes a ring closure. They are derived from PGH_2, but their ring is a cyclic acetal. Thromboxane A_2 is known to induce platelet aggregation. When a blood vessel is ruptured, the first line of defense is the platelets circulating in the blood, which form an incipient clot (see Box 31B). Thromboxane A_2 causes other platelets to clump and thereby increases the size of the blood clot. Aspirin and similar anti-inflammatory agents inhibit the COX enzyme. Consequently, PGH_2 and thromboxane A_2 synthesis is inhibited, and blood clotting is impaired. This is the reason that physicians recommend a daily supply of 81 mg aspirin for people at risk for heart attack or stroke. On the other hand, this is also the reason that physicians forbid patients to use aspirin and other anti-inflammatory agents for a week before a planned surgery because aspirin causes excessive bleeding.

Another group of substances that acts as mediators of hormonal responses are the **leukotrienes.** Like prostaglandins, leukotrienes are derived from arachidonic acid by an oxidative mechanism. However, in this case, there is no ring closure.

Arachidonic acid

Leukotriene B4

Action of Anti–inflammatory Drugs

Anti-inflammatory steroids (such as cortisone, Section 20.10) exert their function by inhibiting the enzyme phospholipase A_2, the enzyme that releases unsaturated fatty acids from complex lipids in the membranes. This is how arachidonic acid, one of the components of membranes, is made available to the cell. Because arachidonic acid is the precursor of prostaglandins, thromboxanes, and leukotrienes, inhibiting its release stops the synthesis of these compounds and prevents inflammation.

Steroids such as cortisone have many undesirable side effects (duodenal ulcer and cataract formation, among others). Therefore, their use must be controlled. Nonsteroidal anti-inflammatory agents (NSAIDS), among them aspirin, ibuprofen, ketoprofen, and indomethacin, are available.

Aspirin and other NSAIDS (Box 18A) inhibit the cyclooxygenase enzymes, which synthesize prostaglandins and thromboxanes. Aspirin (acetyl salicylic acid) acetylates the enzyme and thereby blocks the entrance of arachidonic acid to the active site. These agents inhibit both COX-1 and COX-2.

That is why aspirin and the other anti-inflammatory agents have undesirable side effects. NSAIDS also interfere with the COX-1 isoform of the enzyme, which is needed for normal physiological function. Such side effects include stomach and duodenal ulceration and renal (kidney) toxicity. Obviously it is desirable to have an anti-inflammatory agent without such side effects, the one that inhibits only the COX-2 isoform. The Food and Drug Administration has approved two new COX-2 inhibitor drugs so far: Celebrex, which became the most frequently prescribed drug, and VIOXX, a more recent drug. Many other COX-2 inhibitors are in the clinical trial stage. All anti-inflammatory agents reduce pain and relieve fever and swelling by reducing the prostaglandin production; however, they do not affect the leukotriene production. Thus, asthmatic patients must beware of using these anti-inflammatory agents. Even though they inhibit the prostaglandin synthesis, they may shift the available arachidonic acid to leukotriene production, which could precipitate a severe asthma reaction.

Leukotrienes occur mainly in leukocytes but also in other tissues of the body. They produce long-lasting muscle contractions, especially in the lungs, and they cause asthma-like attacks. They are a hundred times more potent than histamines. Both prostaglandins and leukotrienes cause inflammation and fever. The inhibition of their production in the body is a major pharmacological concern. One way to counteract the effects of leukotrienes is to inhibit their uptake by leukotriene receptors (LTR) in the body. A new drug, zafirlukast, marketed under the name Accolate, an antagonist of LTRs, is used to treat and control chronic asthma. Another antiasthmatic drug, zileuton, inhibits the enzyme, 5-lipoxygenase, which is the initial enzyme in leukotriene biosynthesis from arachidonic acid.

SUMMARY

Lipids are water-insoluble substances. They are divided into four groups: fats (glycerides) and waxes; complex lipids; steroids; and prostaglandins, thromboxanes, and leukotrienes. **Fats** are made up of fatty acids and glycerol. In saturated fatty acids, the hydrocarbon chains have only single bonds; unsaturated fatty acids have hydrocarbon chains with one or more double bonds, all in the cis configuration. Solid fats contain mostly saturated fatty acids, whereas **oils** contain substantial amounts of unsaturated fatty acids. The alkali salts of fatty acids are called **soaps.**

Complex lipids can be divided into two groups: phospholipids and glycolipids. **Phospholipids** are made of a central alcohol (glycerol or sphingosine), fatty acids, and a nitrogen-containing phosphate ester, such as phosphorylcholine or inositol phosphate. The **glycolipids** contain sphingosine and a fatty acid, which together are called the ceramide portion of the molecule, and a carbohydrate portion. Many phospholipids and glycolipids are important constituents of cell **membranes.**

Membranes are made of a **lipid bilayer** in which the hydrophobic parts of phospholipids (fatty acid residues) point to the middle of the bilayer, and the hydrophilic parts point toward the inner and outer surfaces of the membrane.

The third major group of lipids is the **steroids.** The characteristic feature of the steroid structure is a fused four-ring nucleus. The most common steroid is **cholesterol,** which also serves as a starting material for the synthesis of other steroids, such as bile salts and sex and other hormones. Cholesterol is also an integral part of membranes, occupying the hydrophobic region of the lipid bilayer. Because of its low solubility in water, cholesterol deposits are implicated in the formation of gallstones and the plaque-like deposits of atherosclerosis. Cholesterol is transported in the blood plasma by two kinds of lipoprotein: **HDL** and **LDL.** The LDL plays an important role in cholesterol metabolism. It is soluble in blood plasma

because of polar groups on its surface. An oxidation product of cholesterol is progesterone, which is a sex hormone and also gives rise to the synthesis of other **sex hormones,** such as testosterone and estradiol, as well as to the **adrenocorticoid hormones.** Among the latter, the best known are cortisol and cortisone for their anti-inflammatory action. **Bile salts** are also oxidation products of cholesterol. They emulsify all kinds of lipids, including cholesterol, and are essential in the digestion of fats.

Prostaglandins, thromboxanes, and **leukotrienes** are derived from arachidonic acid. They have a wide variety of effects on body chemistry; among other things, they can lower or raise blood pressure, cause inflammation and blood clotting, and induce labor. They act generally as mediators of hormone action.

KEY TERMS

Active transport (Box 20D)
Adrenocorticoid hormone
    (Section 20.10)
Anabolic steroid (Box 20H)
Anion transporter (Box 20D)
Bile salt (Section 20.11)
Cephalin (Section 20.6)
Ceramide (Section 20.7)
Cerebroside (Section 20.8)
Cholesterol (Section 20.9)
Coated pits (Section 20.9)
Complex lipid (Section 20.4)
Cyclooxygenase (COX)
    (Section 20.12)
Detergent (Box 20C)
Diglyceride (Section 20.2)
Facilitated transport (Box 20D)
Fat (Section 20.2)
Fluid mosaic model (Section 20.5)
Gap junction (Box 20D)

Glycerophospholipid (Section 20.6)
Glycolipid (Section 20.8)
HDL (Section 20.9)
Hydrophilic (Section 20.5)
Hydrophobic (Section 20.1)
LDL (Section 20.9)
Lecithin (Section 20.6)
Leukotriene (Section 20.12)
Lipid (Section 20.1)
Lipid bilayer (Section 20.5)
Lipoprotein (Section 20.9)
Membrane (Section 20.5)
Micelle (Box 20C)
Monoglyceride (Section 20.2)
Myelin (Section 20.7)
Myelin sheath (Box 20E)
Oil (Section 20.3)
Organelle (Section 20.5)
Passive transport (Box 20D)

Phosphatidylcholine (Section 20.6)
Phosphatidyl inositol, PI
    (Section 20.6)
Phosphatidyl inositol 3,4
    biphosphate (PIP_2) (Section 20.6)
Phospholipid (Section 20.4)
Polyunsaturated oils (Section 20.3)
Prostaglandin (Section 20.12)
Saturated fatty acid (Section 20.2)
Schwann cell (Box 20E)
Sex hormone (Section 20.10)
Soap (Section 20.3)
Sphingolipid (Section 20.7)
Steroid (Section 20.9)
Steroid hormone (Section 20.10)
Thromboxane (Section 20.12)
Triacylglycerol (Section 20.2)
Triglyceride (Section 20.2)
Unsaturated fatty acid (Section 20.2)

CONCEPTUAL PROBLEMS

20.A How many different triglycerides can you create using three different fatty acids (A, B, and C) in each case?

20.B Which portion of the phosphatidyl inositol molecule contributes to the (a) fluidity of the bilayer and (b) surface polarity of the bilayer?

20.C What are the common structural features of the oral contraceptive pills, including mifepristone?

20.D Prostaglandins have a five-membered ring closure; thromboxanes have a six-membered ring closure. Still the synthesis of both of these groups of compounds are prevented by COX inhibitors. The COX enzymes catalyze ring closure. Explain this fact.

PROBLEMS

Difficult problems are designated by an asterisk.

Structure and Properties of Fats

20.2 Why are fats a good source of energy for storage in the body?

20.3 What is the meaning of the term "hydrophobic"? Why is the hydrophobic nature of lipids important?

20.4 Draw the structural formula of a fat molecule (triglyceride) made of myristic acid, oleic acid, palmitic acid, and glycerol.

*20.5 Oleic acid has a melting point of 16°C. If you converted the cis double bond into a trans double bond, what would happen to the melting point? Explain.

*20.6 Draw schematically all possible diglycerides made up of glycerol, oleic acid, or stearic acid. How many are there all together? Draw the detailed structure of one of the diglycerides.

20.7 For the diglycerides in Problem 20.6, predict which two will have the highest and which two will have the lowest melting points.

20.8 Predict which acid in each pair has the higher melting point and explain why.
(a) Palmitic acid or stearic acid
(b) Arachidonic acid or arachidic acid

20.9 Which has the higher melting point: (a) a triglyceride containing only lauric acid and glycerol or (b) a triglyceride containing only stearic acid and glycerol?

20.10 Explain why the melting points of the saturated fatty acids increase as we move from lauric acid to stearic acid in Table 20.1.

20.11 Predict the order of the melting points of triglycerides containing fatty acids, as follows.
(a) Palmitic, palmitic, stearic
(b) Oleic, stearic, palmitic
(c) Oleic, linoleic, oleic

20.12 Look at Table 20.2. Which animal fat has the highest percentage of unsaturated fatty acids?

20.13 Rank in order of increasing solubility in water (assuming that all are made with the same fatty acids):
(a) triglycerides   (b) diglycerides
(c) monoglycerides.
Explain.

20.14 How many moles of H_2 are used up in the catalytic hydrogenation of 1 mole of a triglyceride containing glycerol, palmitic acid, oleic acid, and linoleic acid?

20.15 Name the products of the saponification of this triglyceride:

$$
\begin{array}{l}
\quad\quad\quad\quad O \\
\quad\quad\quad\quad \| \\
CH_2-O-C-(CH_2)_{14}CH_3 \\
|\quad\quad\quad O \\
\quad\quad\quad\quad \| \\
CH-O-C-(CH_2)_{16}CH_3 \\
|\quad\quad\quad O \\
\quad\quad\quad\quad \| \\
CH_2-O-C-(CH_2)_7(CH=CHCH_2)_3CH_3
\end{array}
$$

20.16 Using the equation in Section 20.3 as a guideline for stoichiometry, calculate the number of moles of NaOH we need to saponify 5 moles of (a) triglycerides (b) diglycerides (c) monoglycerides.

Membranes

20.17 (a) Where in the body are membranes found?
(b) What functions do they serve?

20.18 How do the unsaturated fatty acids of the complex lipids contribute to the fluidity of a membrane?

20.19 Which type of lipid molecules are most likely to be present in membranes?

20.20 What is the difference between an integral and a peripheral membrane protein?

Complex Lipids

*20.21 Which glycerophospholipid has the most polar groups capable of forming hydrogen bonds with water?

20.22 Draw the structure of a phosphatidyl inositol that contains oleic and arachidonic acid.

*20.23 Among the glycerophospholipids containing palmitic acid and linolenic acid, which will have the greatest solubility in water: (a) phosphatidylcholine, (b) phosphatidylethanolamine, or (c) phosphatidylserine? Explain.

20.24 Name all the groups of complex lipids that contain ceramides.

20.25 Are the various phospholipids randomly distributed in membranes? Give an example.

Steroids

*20.26 Cholesterol has a fused four-ring steroid nucleus and is a part of body membranes. The —OH group on C-3 is the polar head, and the rest of the molecule provides the hydrophobic tail that does not fit into the zig-zag packing of the hydrocarbon portion of the saturated fatty acids. Considering this structure, tell whether cholesterol contributes to the stiffening (rigidity) or to the fluidity of a membrane. Explain.

20.27 (a) Is cholesterol necessary for human life?
(b) Why do many people restrict cholesterol intake in their diet?

20.28 Where can pure cholesterol crystals be found in the body?

*20.29 (a) Find all the carbon stereocenters in a cholesterol molecule.
(b) How many total stereoisomers are possible?
(c) How many of these do you think are found in nature?

20.30 Look at the structures of cholesterol and the hormones shown in Figure 20.6. Which ring of the steroid structure undergoes the most substitution?

20.31 What makes LDL soluble in blood plasma?

20.32 How does LDL deliver its cholesterol to the cells?

Steroid Hormones and Bile Salts

20.33 What physiological functions are associated with cortisol?

20.34 Estradiol in the body is synthesized starting from progesterone. What chemical modifications occur when estradiol is synthesized?

20.35 Describe the difference in structure between the male hormone testosterone and the female hormone estradiol.

*__20.36__ Considering that RU486 can bind to the receptors of progesterone as well as those of cortisone and cortisol, what can you say regarding the importance of the functional group on C-11 of the steroid ring in drug and receptor binding?

20.37 (a) How does the structure of RU486 resemble that of progesterone?

(b) How do the structures differ?

20.38 Explain how the constant elimination of bile salts through the feces can reduce the danger of plaque formation in atherosclerosis.

Prostaglandins and Leukotrienes

20.39 What is the basic structural difference between:
(a) Arachidonic acid and prostaglandin PGE_2?
(b) PGE_2 and $PGF_{2\alpha}$?

20.40 Find and name all the functional groups in (a) Glycocholate (b) Cortisone (c) Prostaglandin PGE_2 (d) Leukotriene B4.

20.41 What is the major structural difference between prostaglandins and leukotrienes?

20.42 How does aspirin, an anti-inflammatory drug, prevent strokes caused by blood clots in the brain?

Boxes

20.43 (Box 20A) What causes rancidity? How can it be prevented?

20.44 (Box 20B) What is the structural difference between esters of ear wax and those of beeswax?

20.45 (Box 20C) What is the major structural difference between soaps and detergents?

20.46 (Box 20D) How do the gap junctions prevent the passage of proteins from cell to cell?

20.47 (Box 20E) (a) What is the role of sphingomyelin in the conductance of nerve signals?

(b) What happens to this process in multiple sclerosis?

*__20.48__ (Box 20F) Compare the complex lipid structures listed for the lipid storage diseases (Table 20F) with the missing or defective enzymes. Explain why in Fabry's disease the missing enzyme is α-galactosidase and not β-galactosidase.

20.49 (Box 20G) How does lovastatin reduce the severity of atherosclerosis?

20.50 (Box 20H) How does the oral anabolic steroid methenolone differ structurally from testosterone?

20.51 (Box 20I) What is the role of progesterone and similar compounds in contraceptive pills?

20.52 (Box 20J) How does cortisone prevent inflammation?

20.53 (Box 20J) How does indomethacin act in the body to reduce inflammation?

20.54 (Box 20J) What kind of prostaglandins are synthesized each by COX-1 and COX-2 enzymes?

20.55 (Box 20J) Steroids prevent asthma-causing leukotriene as well as inflammation-causing prostaglandin synthesis. Nonsteroidal anti-inflammatory agents (NSAIDS) such as aspirin reduce only the prostaglandin production. Why do NSAIDS drugs not affect leukotriene production?

Additional Problems

20.56 What is the role of taurine in lipid digestion?

20.57 Draw a schematic diagram of a lipid bilayer. Show how the bilayer prevents the passage by diffusion of a polar molecule such as glucose. Show why nonpolar molecules, such as $CH_3CH_2-O-CH_2CH_3$, can diffuse through the membrane.

20.58 What is the role of progesterone in pregnancy?

20.59 Suggest a reason why free cholesterol forms gallstones but the various esters of cholesterol do not.

20.60 What are the constituents of sphingomyelin?

20.61 What structural feature do detergents (Box 20C) and the bile salt, taurocholate, have in common?

*__20.62__ (Box 20D) What is the difference between a facilitated and an active transporter?

20.63 (a) Classify aldosterone into as many chemical families as appropriate.

(b) In what functional groups does aldosterone differ from cortisone?

*__20.64__ What is the major difference between aldosterone and all the other hormones listed in Figure 20.6?

20.65 (Box 20J) The new anti-inflammatory drug Celebrex does not have the usual side effect of stomach upset or ulceration common to other NSAIDS. Why is that so?

*__20.66__ How many grams of H_2 are needed to saturate 100.0 g of a triglyceride made of glycerol and one unit each of lauric, oleic, and linoleic acids?

Proteins

21.1 Introduction

Proteins are by far the most important of all biological compounds. The very word "protein" is derived from the Greek *proteios,* meaning of first importance, and the scientists who named these compounds more than 100 years ago chose an appropriate term. There are many types of proteins, and they perform many functions, including the following roles:

1. Structure We saw in Section 19.7 that the main structural material for plants is cellulose. For animals, it is structural proteins, which are the chief constituents of skin, bones, hair, and fingernails. Two important structural proteins are collagen and keratin.

2. Catalysis Virtually all the reactions that take place in living organisms are catalyzed by proteins called enzymes. Without enzymes, the reactions would take place so slowly as to be useless. We will discuss enzymes in Chapter 22.

3. Movement Every time we crook a finger, climb stairs, or blink an eye, we use our muscles. Muscle expansion and contraction are involved in every movement we make. Muscles are made up of protein molecules called myosin and actin.

> **Protein** A large biological molecule made of numerous amino acids linked together by amide bonds

The heart itself is a muscle, expanding and contracting about 70 to 80 times a minute.

Above: Spider silk is a fibrous protein that exhibits unmatched strength and toughness. *(Hans Strand/Tony Stone Images)*

4. Transport A large number of proteins fall into this category. Hemoglobin, a protein in the blood, carries oxygen from the lungs to the cells in which it is used and carbon dioxide from the cells to the lungs. Other proteins transport molecules across cell membranes.

5. Hormones Many hormones are proteins, among them insulin, oxytocin, and human growth hormone.

6. Protection When a protein from an outside source of other foreign substance (called an antigen) enters the body, the body makes its own proteins (called antibodies) to counteract the foreign protein. This is the major mechanism the body uses to fight disease. Blood clotting is another protective device carried out by a protein; this one is called fibrinogen. Without blood clotting, we would bleed to death from any small wound.

7. Storage Some proteins are used to store materials in the way that starch and glycogen store energy. Examples are casein in milk and ovalbumin in eggs, which store nutrients for newborn mammals and birds. Ferritin, a protein in the liver, stores iron.

8. Regulation Some proteins not only control the expression of genes, thereby regulating the kind of proteins manufactured in a particular cell, but also control when such manufacture takes place.

These are not the only functions of proteins, but they are among the most important. It is very easy to see that any individual needs a great many proteins to carry out all these varied functions. A typical cell contains about 9000 different proteins; an individual human being has about 100 000 different proteins.

We can divide proteins into two major types: **fibrous proteins,** which are insoluble in water and are used mainly for structural purposes, and **globular proteins,** which are more or less soluble in water and are used mainly for nonstructural purposes.

The antibodies are produced in the gamma globulin fraction of blood plasma.

Collagen, actin, and keratin are some fibrous proteins. Albumin, hemoglobin, and immunoglobulins are some globular proteins.

21.2 Amino Acids

Although there are so many different proteins, they all have basically the same structure: They are chains of amino acids. As its name implies, **an amino acid is an organic compound containing an amino group and a carboxyl group.** Organic chemists can synthesize many thousands of amino acids, but nature is much more restrictive and uses only 20 different amino acids to make up proteins. Furthermore, all but one of the 20 fit the formula

$$R-\underset{\underset{NH_2}{|}}{\overset{\overset{H}{|}}{C}}-COOH$$

and even the one that doesn't fit the formula (proline) comes pretty close. Proline would fit except that it has a bond between the R and the N. The 20 amino acids found in proteins are called **alpha amino acids.** They are shown in Table 21.1, which also shows the one- and three-letter abbreviations that chemists and biochemists use for them.

The most important aspect of the R groups is their polarity, and on this basis we can classify amino acids into the four groups shown in Table 21.1: nonpolar, polar but neutral, acidic, and basic. Note that the nonpolar side

Alpha (α-)amino acid An amino acid in which the amino group is linked to the carbon atom next to the —COOH carbon

The one-letter abbreviations are more recent, but the three-letter abbreviations are still frequently used.

TABLE 21.1 The 20 Amino Acids Commonly Found in Proteins (both the three-letter and one-letter abbreviations are shown) and Their Isoelectric Points (pI)

Nonpolar Side Chains

alanine (Ala, A)
pI = 6.01

$$CH_3\overset{\underset{\displaystyle NH_3^+}{|}}{C}HCOO^-$$

phenylalanine (Phe, F)
pI = 5.48

glycine (Gly, G)
pI = 5.97

$$H\overset{\underset{\displaystyle NH_3^+}{|}}{C}HCOO^-$$

proline (Pro, P)
pI = 6.48

isoleucine (Ile, I)
pI = 6.02

$$CH_3CH_2\overset{\underset{\displaystyle}{|}CH_3}{C}H\overset{\underset{\displaystyle NH_3^+}{|}}{C}HCOO^-$$

tryptophan (Trp, W)
pI = 5.88

leucine (Leu, L)
pI = 5.98

$$(CH_3)_2CHCH_2\overset{\underset{\displaystyle NH_3^+}{|}}{C}HCOO^-$$

methionine (Met, M)
pI = 5.74

$$CH_3SCH_2CH_2\overset{\underset{\displaystyle NH_3^+}{|}}{C}HCOO^-$$

valine (Val, V)
pI = 5.97

$$(CH_3)_2CH\overset{\underset{\displaystyle NH_3^+}{|}}{C}HCOO^-$$

Polar but Neutral Side Chains

asparagine (Asn, N)
pI = 5.41

$$H_2N\overset{\overset{\displaystyle O}{\|}}{C}CH_2\overset{\underset{\displaystyle NH_3^+}{|}}{C}HCOO^-$$

serine (Ser, S)
pI = 5.68

$$HOCH_2\overset{\underset{\displaystyle NH_3^+}{|}}{C}HCOO^-$$

glutamine (Gln, Q)
pI = 5.65

$$H_2N\overset{\overset{\displaystyle O}{\|}}{C}CH_2CH_2\overset{\underset{\displaystyle NH_3^+}{|}}{C}HCOO^-$$

threonine (Thr, T)
pI = 5.87

$$CH_3\overset{\overset{\displaystyle OH}{|}}{C}H\overset{\underset{\displaystyle NH_3^+}{|}}{C}HCOO^-$$

Acidic Side Chains

Basic Side Chains

aspartic acid (Asp, D)
pI = 2.77

$$^-OOCCH_2\overset{\underset{\displaystyle NH_3^+}{|}}{C}HCOO^-$$

arginine (Arg, R)
pI = 10.76

$$H_2N\overset{\overset{\displaystyle NH_2^+}{\|}}{C}NHCH_2CH_2CH_2\overset{\underset{\displaystyle NH_3^+}{|}}{C}HCOO^-$$

glutamic acid (Glu, E)
pI = 3.22

$$^-OOCCH_2CH_2\overset{\underset{\displaystyle NH_3^+}{|}}{C}HCOO^-$$

histidine (His, H)
pI = 7.59

cysteine (Cys, C)
pI = 5.07

$$HSCH_2\overset{\underset{\displaystyle NH_3^+}{|}}{C}HCOO^-$$

lysine (Lys, K)
pI = 9.74

$$H_3\overset{+}{N}CH_2CH_2Ch_2CH_2\overset{\underset{\displaystyle NH_3^+}{|}}{C}HCOO^-$$

tyrosine (Tyr, Y)
pI = 5.66

$$HO\text{—}\underset{}{\bigcirc}\text{—}CH_2\overset{\underset{\displaystyle NH_3^+}{|}}{C}HCOO^-$$

* Each ionizable group is shown in the form present in highest concentration at pH 7.0.

Two of the amino acids in Table 21.1 have a second carbon stereocenter and exist as two pairs of enantiomers. Can you find them?

chains are *hydrophobic* (they repel water), whereas polar but neutral, acidic, and basic side chains are *hydrophilic* (attracted to water). This aspect of the R groups is very important in determining both the structure and the function of each protein molecule.

When we look at the general formula for the 20 amino acids, we see at once that all of them (except glycine, in which R = H) are chiral with (carbon) stereocenters, since R, H, COOH, and NH$_2$ are four different groups. This means, as we saw in Section 19.2, that each of the amino acids (except glycine) exists as two enantiomers. As is the case for most examples of this kind, nature makes only one of the two possible enantiomers for each amino acid, and it is virtually always the L form. Except for glycine, which exists in only one form, all the amino acids in all the proteins in your body are the L form. D amino acids are extremely rare in nature; some are found, for example, in the cell walls of a few types of bacteria.

In Section 19.2 we learned the systematic use of the D,L system. We used glyceraldehyde as a reference point for the assignment of relative configuration. Here again, with amino acids, we can use glyceraldehyde as a reference point.

$$
\begin{array}{cc}
\underset{\text{L-Glyceraldehyde}}{
\begin{array}{c}
O\!\!\diagdown\!\!C\!\!-\!\!H \\
|\\
HO\!\blacktriangleright\!C\!\blacktriangleleft\!H \\
|\\
CH_2OH
\end{array}}
&
\underset{\text{L-Alanine}}{
\begin{array}{c}
O\!\!\diagdown\!\!C\!\!-\!\!O^- \\
|\\
{}^+H_3N\!\blacktriangleright\!C\!\blacktriangleleft\!H \\
|\\
CH_3
\end{array}}
\end{array}
$$

The spatial relationship of the functional groups around the carbon stereocenter in L-amino acids, as in L-alanine, can be compared to that of L-glyceraldehyde. When we put the carbonyl groups of both compounds in the same position (top) the —OH of L-glyceraldehyde and the NH$_3^+$ of L-alanine lies to the left of the carbon stereocenter.

21.3 Zwitterions

Up to now, we have shown the structural formula for amino acids as

$$
\begin{array}{c}
H \\
| \\
R\!-\!C\!-\!COOH \\
| \\
NH_2
\end{array}
$$

But in Section 18.2 we learned that carboxylic acids, RCOOH, cannot exist in the presence of a moderately weak base (such as NH$_3$). They donate a proton to become carboxylate ions, RCOO$^-$. Likewise, amines, RNH$_2$ (Section 16.6), cannot exist as such in the presence of a moderately weak acid (such as acetic acid). They gain a proton to become substituted ammonium ions, RNH$_3^+$.

An amino acid has —COOH and —NH$_2$ groups in the same molecule. Therefore, in water solution, the —COOH donates a proton to the —NH$_2$, so that an amino acid actually has the structure

$$
\begin{array}{c}
H \\
| \\
R\!-\!C\!-\!COO^- \\
| \\
NH_3^+
\end{array}
$$

Compounds that have a positive charge on one atom and a negative charge on another are called **zwitterions.** Amino acids are zwitterions, not only in water solution but in the solid state as well. They are therefore ionic compounds, that is, internal salts. *Un-ionized RCH(NH₂)COOH molecules do not actually exist, in any form.*

The fact that amino acids are zwitterions explains their physical properties, which would otherwise be quite puzzling. All of them are solids with high melting points (for example, glycine melts at 262°C). This is just what we expect of ionic compounds. The 20 amino acids are also fairly soluble in water, as ionic compounds generally are; if they had no charges, only the smaller ones would be expected to be soluble.

If we add an amino acid to water, it dissolves and then has the same zwitterionic structure that it has in the solid state. Let us see what happens if we change the pH of the solution, as we can easily do by adding a source of H_3O^+, such as HCl (to lower the pH), or a strong base, such as NaOH (to raise the pH). Since H_3O^+ is a stronger acid than a typical carboxylic acid (Section 18.2), it donates a proton to the —COO^- group, turning the zwitterion into a positive ion. This happens to all amino acids if the pH is sufficiently lowered.

$$R-\underset{\underset{NH_3^+}{|}}{\overset{\overset{H}{|}}{C}}-COO^- + H_3O^+ \longrightarrow R-\underset{\underset{NH_3^+}{|}}{\overset{\overset{H}{|}}{C}}-COOH + H_2O$$

Addition of OH^- to the zwitterion causes the —NH_3^+ to donate its proton, turning the zwitterion into a negative ion. This happens to all amino acids if the pH is sufficiently raised.

$$R-\underset{\underset{NH_3^+}{|}}{\overset{\overset{H}{|}}{C}}-COO^- + OH^- \longrightarrow R-\underset{\underset{NH_2}{|}}{\overset{\overset{H}{|}}{C}}-COO^- + H_2O$$

Note that in both cases the amino acid is still an ion so that it is still soluble in water. There is no pH at which an amino acid has no ionic character at all. If the amino acid is a positive ion at low pH and a negative ion at high pH, there must be some **pH at which all the molecules have equal positive and negative charges.** This pH is called the **isoelectric point** (symbol **pI**).

Every amino acid has a different isoelectric point, although most of them are not very far apart (see the values in Table 21.1). Fifteen of the 20 have isoelectric points near 6. However, the three basic amino acids have higher isoelectric points, and the two acidic amino acids have lower values.

At or near the isoelectric point, amino acids exist in aqueous solution largely or entirely as zwitterions. As we have seen, they react with either a strong acid, by taking a proton (the —COO^- becomes —COOH), or a strong base, by giving a proton (the —NH_3^+ becomes —NH_2). To summarize:

$$R-\underset{\underset{NH_3^+}{|}}{\overset{\overset{H}{|}}{C}}-COOH \underset{H_3O^+}{\overset{OH^-}{\rightleftharpoons}} R-\underset{\underset{NH_3^+}{|}}{\overset{\overset{H}{|}}{C}}-COO^- \underset{H_3O^+}{\overset{OH^-}{\rightleftharpoons}} R-\underset{\underset{NH_2}{|}}{\overset{\overset{H}{|}}{C}}-COO^-$$

"Zwitterion" comes from the German *zwitter,* meaning hybrid.

Isoelectric point A pH at which a sample of amino acids or protein has an equal number of positive and negative charges

Acidic amino acids have two —COO^- groups, one on the α-carbon and one on the side chain. At the isoelectric point, one of these must be in the un-ionized form (—COOH). To achieve this, H_3O^+ must be added to lower the pH. Similar considerations explain why the isoelectric points of basic amino acids are higher than pH 6.

We learned in Section 8.3 that a compound that is both an acid and a base is called amphiprotic. We also learned, in Section 8.10, that a solution that neutralizes both acid and base is a buffer solution. Amino acids are therefore *amphiprotic* compounds, and aqueous solutions of them are *buffers*.

Proteins also have isoelectric points and act as buffers. This will be discussed in Section 21.6.

21.4 Cysteine. A Special Amino Acid

One of the 20 amino acids in Table 21.1 has a chemical property not shared by any of the others. This amino acid is cysteine. It can easily be dimerized by many mild oxidizing agents:

We met these reactions in Section 13.7.

[O] symbolizes oxidation, and [H] stands for reduction.

A dimer is a molecule made up of two units.

$$2\,HS-CH_2-\underset{\underset{NH_3^+}{|}}{CH}-COO^- \underset{[H]}{\overset{[O]}{\rightleftarrows}} {}^-OOC-\underset{\underset{NH_3}{|}}{CH}-CH_2-\boxed{S-S}-CH_2-\underset{\underset{NH_3^+}{|}}{CH}-COO^-$$

a disulfide bond

Cysteine   Cystine

The dimer of cysteine, which is called **cystine,** can in turn be fairly easily reduced to give two molecules of cysteine. As we shall see, the presence of cystine has important consequences for the chemical structure and shape of protein molecules it is part of. The **S—S bond** (shown in color) is also called a **disulfide bond.**

21.5 Peptides and Proteins

Each amino acid has a carboxyl group and an amino group. In Section 18.5 we saw that a carboxylic acid and an amine could be combined to form an amide:

To allow easy visualization of amide bond formation, we present the amino acids in the un-ionized form in Section 21.5.

$$R-\overset{\overset{\displaystyle O}{\|}}{C}-OH + R'-NH_2 \longrightarrow R-\overset{\overset{\displaystyle O}{\|}}{C}-NH-R' + H_2O$$

In the same way, it is possible for the COOH group of one amino acid molecule, say glycine, to combine with the amino group of a second molecule, say alanine:

$$H_2N-CH_2-\overset{\overset{\displaystyle O}{\|}}{C}-OH + H_2N-\underset{\underset{CH_3}{|}}{CH}-\overset{\overset{\displaystyle O}{\|}}{C}-OH \longrightarrow H_2N-CH_2-\overset{\overset{\displaystyle O}{\|}}{C}-NH-\underset{\underset{CH_3}{|}}{CH}-\overset{\overset{\displaystyle O}{\|}}{C}-OH + H_2O$$

Glycine   Alanine   Glycylalanine (Gly—Ala)

Peptide bond An amide bond that links two amino acids

The synthesis of peptide bonds in cells is catalyzed by enzymes.

This reaction takes place in the cells by a mechanism that we shall examine in Section 25.5. The product is an amide. The two amino acids are joined together by a **peptide bond** also called a **peptide linkage.** The product is a **dipeptide.**

It is important to realize that glycine and alanine could also be linked the other way:

$$H_2N-\underset{\underset{CH_3}{|}}{CH}-\overset{\overset{\displaystyle O}{\|}}{C}-OH + H_2N-CH_2-\overset{\overset{\displaystyle O}{\|}}{C}-OH \longrightarrow H_2N-\underset{\underset{CH_3}{|}}{CH}-\overset{\overset{\displaystyle O}{\|}}{C}-NH-CH_2-\overset{\overset{\displaystyle O}{\|}}{C}-OH + H_2O$$

Alanine   Glycine   Alanylglycine (Ala—Gly)

In this case we get a *different* dipeptide. The two dipeptides are isomers, of course, but they are different compounds in all respects, with different properties.

Note that in Ala—Gly the —NH$_2$ is connected to a —CHCH$_3$, while in Gly—Ala the —NH$_2$ is connected to a —CH$_2$.

EXAMPLE 21.1

Show how to form the dipeptide aspartylserine (Asp—Ser).

Solution

The name implies that this dipeptide is made of two amino acids, aspartic acid (Asp) and serine (Ser), with the amide being formed between the carboxyl group of aspartic acid and the amino group of serine. Therefore, we write the formula of aspartic acid with its amino group on the left side. Next we place the formula of serine to the right, with its amino group facing the carboxyl group of aspartic acid. Finally, we eliminate a water molecule between the —COOH and —NH$_2$ groups next to each other, forming the peptide bond:

$$\underset{\text{Asp}}{\overset{\displaystyle O}{H_2N-CH-\overset{\|}{C}-OH}} \quad + \quad \underset{\text{Ser}}{\overset{\displaystyle O}{H_2N-CH-\overset{\|}{C}-OH}} \quad \longrightarrow$$

with Asp having CH$_2$ and COOH below, and Ser having CH$_2$OH below.

$$\underset{\text{Asp—Ser}}{H_2N-CH-\overset{O}{\overset{\|}{C}}-NH-CH-\overset{O}{\overset{\|}{C}}-OH \; + \; H_2O}$$

with CH$_2$, COOH below the first CH, and CH$_2$OH below the second CH.

Problem 21.1 ▬▬▬

Show how to form the dipeptide valylphenylalanine (Val—Phe). ■

Any two amino acids, the same or different, can be linked together to form dipeptides in a similar manner. Nor does it end there. Each dipeptide still contains a COOH and an amino group. We can therefore add a third amino acid to alanylglycine, say lysine:

$$\underset{\text{Ala—Gly}}{H_2N-CH-\overset{O}{\overset{\|}{C}}-NH-CH_2-\overset{O}{\overset{\|}{C}}-OH} \; + \; \underset{\text{Lys}}{H_2N-CH-\overset{O}{\overset{\|}{C}}-OH} \; \xrightarrow{-H_2O}$$

with CH$_3$ below the Ala, and (CH$_2$)$_4$, NH$_2$ below the Lys.

$$\underset{\text{Ala—Gly—Lys}}{H_2N-CH-\overset{O}{\overset{\|}{C}}-NH-CH_2-\overset{O}{\overset{\|}{C}}-NH-CH-\overset{O}{\overset{\|}{C}}-OH}$$

with CH$_3$ below the first CH, and (CH$_2$)$_4$, NH$_2$ below the last CH.

Glutathione

A very important tripeptide present in high concentrations in all tissues is glutathione. It contains L-glutamic acid, L-cysteine, and glycine. Its structure is unusual because the glutamic acid is linked to the cysteine by its γ-carboxyl group rather than by the α-carboxyl group as is usual in most peptides and proteins:

$$NH_3^+ - CH - CH_2 - CH_2 - \overset{\text{O}}{\underset{\|}{C}} - NH - CH - \overset{\text{O}}{\underset{\|}{C}} - NH - CH_2 - COO^-$$

with COO$^-$ on the first CH, and CH$_2$—SH on the middle CH.

Glutathione
(Glu—Cys—Gly)

Glutathione functions in the cells as a general protective agent. Oxidizing agents that would damage the cells, such as peroxides, will oxidize glutathione instead (the cysteine portion; see also Section 21.4), thus protecting the proteins and nucleic acids. Many foreign chemicals also get attached to glutathione, and in this sense it acts as a detoxifying agent.

The product is a **tripeptide.** Since it too contains a COOH and an NH$_2$ group, we can continue the process to get a tetrapeptide, a pentapeptide, and so on until we have a chain of hundreds or even thousands of amino acids. These **chains of amino acids** are the **proteins** that serve so many important functions in living organisms.

A word must be said about the terms used to describe these compounds. The shortest chains are often simply called **peptides,** longer ones are **polypeptides,** and still longer ones are proteins, but chemists differ about where to draw the line. Many chemists use the terms "polypeptide" and "protein" almost interchangeably. We shall consider a protein to be a polypeptide chain that contains a minimum of 30 to 50 amino acids.

The amino acids in a chain are often called **residues.** It is customary to use either the one-letter or the three-letter abbreviations shown in Table 21.1 to represent peptides and proteins. For example, the tripeptide shown on the previous page, alanylglycyllysine, is AGK or Ala—Gly—Lys. **C-terminal amino acid is the residue with the free COOH group** (lysine in Ala—Gly—Lys), and **N-terminal amino acid is the residue with the free amino group** (alanine in Ala—Gly—Lys). It is the universal custom to write peptide and protein chains with the N-terminal residue on the left. No matter how long a protein chain gets—hundreds or thousands of units—it always has just two ends: one C-terminal and one N-terminal.

The naming of the peptides also begins with the N-terminal amino acids.

21.6 Some Properties of Peptides and Proteins

α-carbon

α-carbon

Six atoms of the peptide backbone lie in an imaginary (shaded) plane.

The continuing pattern of peptide bonds is the backbone of the peptide or protein molecule; the R groups are called the **side chains.** The six atoms of the peptide backbone are rigid and lie in the same plane, and two adjacent peptide bonds can rotate relative to one another about the C—N and C—C bonds.

$$\underset{R_1}{\sim\sim N} - \overset{H}{\underset{|}{C}}H - \overset{O}{\underset{\|}{C}} - \overset{H}{\underset{|}{N}} - \overset{}{\underset{R_2}{C}}H - \overset{O}{\underset{\|}{C}} \sim\sim$$

AGE and Aging

There is a reaction that can take place between a primary amine and an aldehyde or a ketone, linking the two molecules (shown for an aldehyde):

$$R-\overset{\overset{\displaystyle O}{\|}}{C}-H + H_2N-R' \longrightarrow$$

$$R-CH=N-R' + H_2O$$
An imine

Since proteins have NH_2 groups and carbohydrates have aldehyde or keto groups, they can undergo this reaction, establishing a link between a sugar and a protein molecule. When this is not catalyzed by enzymes, it is called glycation of proteins. The process, however, does not stop there. When these linked products are heated in a test tube, high-molecular-weight,

water-insoluble brownish complexes are formed. These complexes are called **Advanced Glycation End-products** or **AGE.** In the body they cannot be heated, but the same result happens over long periods of time.

The longer we live, the more AGE products accumulate in the body, and the higher the blood sugar concentration becomes. AGE products show up in all the afflicted organs of diabetic patients: in the lens of the eye (cataracts), in the capillary blood vessels of the retina (diabetic retinopathy), and in the glomeruli of the kidneys (kidney failure). For people who do not have diabetes, these harmful protein modifications become disturbing only in advanced years. In a young person, metabolism functions properly, and the AGE products are decomposed and eliminated from the body. In an older person, metabolism slows and the AGE products accumulate. AGE products themselves are thought to enhance oxidative damages.

The 20 different amino acid side chains supply variety and determine the physical and chemical properties of proteins. Among these properties, acid-base behavior is one of the most important. Like amino acids (Section 21.3), proteins behave as zwitterions. The side chains of glutamic and aspartic acids provide COOH groups, whereas lysine and arginine provide basic groups (histidine does, too, but this side chain is less basic than the other two). (See the structures of these amino acids in Table 21.1.)

The isoelectric point of a protein occurs at the pH at which there are an equal number of positive and negative charges (the protein has no *net* charges). At any pH above the isoelectric point, the protein molecules have a net negative charge; at any pH below the isoelectric point, they have a net positive charge. Some proteins, such as hemoglobin, have an almost equal number of acidic and basic groups; the isoelectric point of hemoglobin is at pH 6.8. Others, like serum albumin, have more acidic groups than basic groups; the isoelectric point of this protein is 4.9. In each case, however, because proteins behave like zwitterions, they act as buffers, for example, in the blood (Figure 21.1).

The water-solubility of large molecules such as proteins often depends on the repulsive forces between like charges on their surfaces. When protein molecules are at a pH at which they have a net positive or negative charge, the presence of these like charges causes the protein molecules to repel each other. These repulsive forces are smallest at the isoelectric point, when the net charges are close to zero. When there are no repulsive forces, the protein molecules tend to clump together to form aggregates of two or more molecules, reducing their solubility. Therefore, *proteins are least soluble in water at their isoelectric points and can be precipitated from their solutions.*

We pointed out in Section 21.1 that proteins have many functions. In order to understand these functions, we must look at four levels of organization in their structures. *Primary structure* describes the linear sequence of amino acids in the polypeptide chain. *Secondary structure* refers to certain repeating patterns, such as the α-helix conformation or the pleated

The terminal —COOH and —NH_2 groups also ionize, but these are only 2 out of 50 or more residues.

Carbonates and phosphates are the other blood buffers (Section 8.10).

FIGURE 21.1 Schematic diagram
of a protein (*a*) at its isoelectric point
and its buffering action when (*b*) H⁺
or (*c*) OH⁻ ions are added.

(*b*)

(*a*)

(*c*)

sheet (Figure 21.5), or the absence of a repeating pattern, as with the random coil (Figure 21.6). *Tertiary structure* describes the overall conformation of the polypeptide chain. A good analogy for all this is a coiled telephone cord (Figure 21.2). The primary structure is the stretched-out cord. The secondary structure is the coil in the form of a helix. We can take the entire coil and twist it into various shapes. Any structure made by doing this is a tertiary structure. As we shall see, protein molecules twist and curl in a very similar manner. *Quaternary structure* applies only to proteins with more than one polypeptide chain (subunit) and has to do with how the different chains are spatially related to each other.

21.7 The Primary Structure of Proteins

Very simply, the primary structure of a protein consists of the sequence of amino acids that makes up the chain. Each of the very large number of peptide and protein molecules in biological organisms has a different sequence of amino acids—and that sequence allows the protein to carry out its function, whatever it may be.

Is it possible that so many different proteins can arise from different sequences of only 20 amino acids? Let us look at a little arithmetic, starting with a dipeptide. How many different dipeptides can be made from 20 amino acids? There are 20 possibilities for the N-terminal amino acid, and for each of these 20 there are 20 possibilities for the C-terminal amino acid. This means that there are $20 \times 20 = 400$ different dipeptides possible from the 20 amino acids. What about tripeptides? We can form a tripeptide by taking any of the 400 dipeptides and adding any of the 20 amino acids. Thus, there are $20 \times 20 \times 20 = 8000$ tripeptides, all different. It is easy to

FIGURE 21.2 The "structure" of a telephone cord: (*a*) primary, (*b*) secondary, (*c*) tertiary.

(*a*)   (*b*)   (*c*)

see that we can calculate the total number of possible peptides or proteins for a chain of n amino acids simply by raising 20 to the nth power (20^n).

Taking a typical small protein to be one with 60 amino acid residues, the number of proteins that can be made from the 20 amino acids is $20^{60} = 10^{78}$. This is an enormous number, possibly greater than the total number of atoms in the universe. It is clear that only a tiny fraction of all possible protein molecules has ever been made by biological organisms.

Each peptide or protein in the body has its own sequence of amino acids. We mentioned that proteins also have secondary, tertiary, and in some cases also quaternary structures. We will deal with these in Sections 21.8 and 21.9, but here we can say that **the primary structure of a protein determines to a large extent the native** (most frequently occurring) **secondary and tertiary structures.** That is, it is the particular sequence of amino acids on the chain that enables the whole chain to fold and curl in such a way as to assume its final shape. As we shall see in Section 21.11, without its particular three-dimensional shape, a protein cannot function.

Just how important is the exact amino acid sequence? Can a protein perform the same function if its sequence is a little different? The answer to this question is that a change in amino acid sequence may or may not matter, depending on what kind of a change it is. As an example, cytochrome c is a protein chain consisting of 104 amino acid residues in terrestrial vertebrates. It performs the same function in humans, chimpanzees, sheep, and others. Humans and chimpanzees have exactly the same amino acid sequence of this protein. Sheep cytochrome c differs in 10 positions out of the 104. (You can find more about biochemical evolution in Box 25D.)

Another example is the hormone insulin. Human insulin consists of two chains having a total of 51 amino acids connected by disulfide bonds. The sequence of amino acids is shown in Figure 21.3. Insulin is necessary for proper utilization of carbohydrates (Section 27.2), and people with severe diabetes (Box 23G) must take insulin injections. The amount of human insulin available is far too small to meet the need, so bovine insulin (from cattle) or insulin from hogs or sheep is used instead. Insulin from these sources is similar, but not identical, to human insulin. The differences are entirely in the 8, 9, and 10 positions of the A chain and the C-terminal position (30) of the B chain:

| | **A Chain** | | | **B Chain** |
	8	*9*	*10*	*30*
Human	—Thr	—Ser	— Ile —	—Thr
Bovine	—Als	—Ser	—Val—	—Ala
Hog	—Thr	—Ser	— Ile —	—Ala
Sheep	—Ala	—Gly	—Val—	—Ala

The remainder of the molecule is the same in all four varieties of insulin. Despite the slight differences in structure, all these insulins perform the same function and even can be used by humans. However, none of the other three is quite as effective in humans as human insulin.

Another factor showing the effect of substituting one amino acid for another is that sometimes patients become allergic to, say, bovine insulin and can switch to hog or sheep insulin without causing allergies.

In contrast to the previous examples, there are small changes in amino acid sequence that make a great deal of difference. First, we can consider

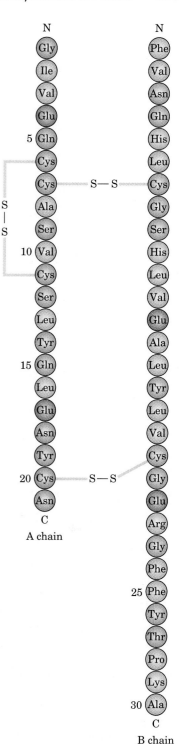

FIGURE 21.3 The hormone insulin consists of two polypeptide chains, A and B, held together by two disulfide cross-bridges (S—S). The sequence shown is for bovine insulin.

BOX 21C

The Use of Human Insulin

Although human insulin, manufactured by recombinant DNA techniques (see Section 25.8), is on the market, many diabetic patients continue to use hog or sheep insulin because it is cheaper. Changing from animal to human insulin creates an occasional problem for diabetics. All diabetics experience an insulin reaction (hypoglycemia) when the insulin level in the blood is too high relative to the blood sugar level. Hypoglycemia is preceded by symptoms of hunger, sweating, and poor coordination. These symptoms, called **hypoglycemic awareness,** are a signal to the patient that hypoglycemia is

coming and that it must be reversed, which the patient can do by eating sugar.

Some diabetics who changed from animal to human insulin reported that the hypoglycemic awareness from recombinant DNA human insulin is not as strong as that from animal insulin. This can create some hazards, and the effect is probably due to different rates of absorption in the body. The literature supplied with human insulin now incorporates a warning that hypoglycemic awareness may be altered.

BOX 21D

Sickle Cell Anemia

Normal adult human hemoglobin has two alpha chains and two beta chains (Figure 21.12). Some people, however, have a slightly different kind of hemoglobin in their blood. This hemoglobin (called HbS) differs from the normal type only in the beta chains and only in one position on these two chains: The glutamic acid in the sixth position of normal Hb is replaced by a valine residue in HbS.

	4	5	6	7	8	9
Normal Hb	—Thr	—Pro	—Glu	—Glu	—Lys	—Ala—
Sickle cell Hb	—Thr	—Pro	—Val	—Glu	—Lys	—Ala—

This change affects only two positions in a molecule containing 574 amino acid residues. Yet, it is enough to result in a very serious disease, **sickle cell anemia.**

Red blood cells carrying HbS behave normally when there is an ample oxygen supply. When the oxygen pressure decreases, the red blood cells become sickle-shaped (Figure 21D). This occurs in the capillaries. As a result of this change in shape, the cells may clog the capillaries. The body's defenses destroy the clogging cells, and the loss of the blood cells causes anemia.

This change at only a single position of a chain consisting of 146 amino acids is severe enough to cause a high death rate. A child who inherits two genes programmed to produce sickle cell hemoglobin (a homozygote) has an 80 percent smaller chance of surviving to adulthood than a child with only one such gene (a heterozygote) or a child with two normal genes. In spite of the high mortality of homozygotes, the genetic trait survives. In central Africa, 40 percent of the population in malaria-ridden areas carry the sickle cell gene, and 4 percent are homozygotes.

It seems that the sickle cell genes help to acquire immunity against malaria in early childhood so that in malaria-ridden areas the transmission of these genes is advantageous.

Although there is no known cure for the sickle cell anemia, recently the Food and Drug Administration approved

$$hydroxyurea, \quad H_2NC\overset{\overset{\displaystyle O}{\|}}{N}\overset{H}{\underset{OH}{\diagdown}}, \quad (sold \ under \ the \ name \ of \ Droxia)$$

to treat and control the symptom of the disease. Hydroxyurea promotes the bone marrow to manufacture fetal hemoglobin (HbF), which does not have beta chains where the sticky mutation occurs. Thus, red blood cells containing HbF do not sickle and do not clog the capillaries. Under hydroxyurea therapy, the bone marrow still manufactures mutated HbS, but the presence of cells with fetal hemoglobin dilutes the concentration of the sickling cells and thus relieves the symptom of the disease.

■ Blood cells from a patient with sickle cell anemia. Both normal cells (round) and sickle cells (shriveled) are visible. *(G. W. Willis, M.D./Biological Photo Service)*

Vasopressin

Oxytocin

FIGURE 21.4 The structures of vasopressin and oxytocin. Differences are shown in color.

two peptide hormones, oxytocin and vasopressin (Figure 21.4). These nonapeptides have identical structures, including a disulfide bond, except for different amino acids in positions 3 and 8. Yet their biological functions are quite different. Vasopressin is an antidiuretic hormone. It increases the amount of water reabsorbed by the kidneys and raises blood pressure. Oxytocin has no effect on water in the kidneys and slightly lowers blood pressure. It affects contractions of the uterus in childbirth and the muscles in the breast that aid in the secretion of milk. Vasopressin also stimulates uterine contractions, but much less so than oxytocin.

Another instance where a minor change makes a major difference is in the blood protein hemoglobin. A change in only one amino acid in a chain of 146 is enough to cause a fatal disease—sickle cell anemia (Box 21D).

Although slight changes in amino acid sequence make little or no difference in the functioning of peptides and proteins in some cases, it is clear that the sequence is highly important in most cases. The sequences of a large number of protein and peptide molecules have now been determined. The methods for doing it are complicated and will not be discussed in this book.

21.8 The Secondary Structure of Proteins

Proteins can fold or align themselves in such a manner that certain patterns repeat themselves. These **repeating patterns** are referred to as **secondary structures.** The two most common secondary structures encountered in proteins are the **α-helix** and the **β-pleated sheet** (Figure 21.5) originally proposed by Linus Pauling and Robert Corey. In contrast, those protein conformations that do not exhibit a repeated pattern are called **random coils** (Figure 21.6).

In the α-helix form, a single protein chain twists in such a manner that its shape resembles a right-handed coiled spring, that is, a helix. The shape of the helix is maintained by numerous **intramolecular hydrogen bonds** that exist between the backbone —C=O and H—N— groups. It can be seen from Figure 21.5 that there is a hydrogen bond between the —C=O oxygen atom of each peptide linkage and the —N—H hydrogen atom of another peptide linkage four amino acid residues farther along the chain. These hydrogen bonds are in just the right position to cause the molecule (or a portion of it) to maintain a helical shape. Each

Both oxytocin and vasopressin are secreted by the pituitary gland.

Both hormones are used as drugs, vasopressin to combat loss of blood pressure after surgery and oxytocin to induce labor.

The first sequence of an important protein, insulin, was obtained by Frederick Sanger (1918–   ) in England for which he received the Nobel Prize in 1958.

Linus Pauling (1901–1994), who won the 1954 Nobel Prize in Chemistry, determined these structures. He also discovered or contributed much to our understanding of certain fundamental concepts including chemical bonds and electronegativity (Section 3.2).

Beta (β-)pleated sheet A secondary protein structure in which the backbone of two protein chains in the same or different molecules is held together by hydrogen bonds

An intramolecular hydrogen bond goes from a hydrogen atom in a molecule to an O, N, or F atom in the same molecule.

Carbon atom ● Oxygen atom ● Nitrogen atom ● ☐Hydrogen atom ● R group

FIGURE 21.5 (*a*) The α-helix. (*b*) The β-pleated sheet structure.

FIGURE 21.6 A random coil.

—N—H points upward and each C=O points downward, roughly parallel to the axis of the helix. All the R-groups (the amino acid side chains) point outward from the helix.

The other important orderly structure in proteins is called the β-pleated sheet. In this case, the orderly alignment of protein chains is maintained by **intermolecular hydrogen bonds.** The β-pleated sheet structure can occur between molecules when polypeptide chains run parallel (all the N-terminal ends on one side) or antiparallel (neighboring N-terminal ends on opposite sides). β-Pleated sheets can also occur intramolecularly, when the polypeptide chain makes a U-turn, forming a hairpin structure, and the pleated sheet is antiparallel (Figure 21.5).

In all secondary structures, the hydrogen bonding is between backbone —C=O and H—N— groups. This is the distinction between secondary and tertiary structure. In the latter, as we shall see, the hydrogen bonding is between R groups on the side chains.

Few proteins have predominantly α-helix or pleated sheet structures. Most proteins, especially globular ones, have only certain portions of their

BOX 21E

Protein/Peptide Conformation – Dependent Diseases

There are a number of diseases in which a normal protein or peptide becomes pathological when its conformation changes. A common feature of these proteins is the property to self-assemble into β-sheet-forming amyloid (starch-like) plaques. These amyloid structures appear in a number of diseases. One example of this process is the prion protein, the discovery of which brought Stanley Prusiner of the University of California, San Francisco, the Nobel Prize in 1997. When prion undergoes conformational change, it can cause the Mad Cow Disease. During the conformational change, the α-helical content of the normal prion protein unfolds and reassembles in the β-pleated

sheet form. This new form supposedly has the potential to cause more normal prion proteins to undergo conformational change. In humans it causes spongiform encephalitis, and the Creutzfeld-Jakob disease is one variant that mainly afflicts the elderly. Although the transmission from diseased cows to humans is rare, fear of it caused the wholesale slaughter of British cattle in 1998 and for a while the embargo of the importation of such meat in most of Europe and America. Another such conformational change is found in the protein transthyretin, which causes a disease called senile systematic amyloidosis. β-Amyloid plaques also appear in Alzheimer's diseased brains (Box 23A).

molecules in these conformations. The rest of the molecule is random coil. Many globular proteins contain all three kinds of secondary structure in different parts of their molecules: α-helix, β-pleated sheet, and random coil. A schematic representation of such a structure is shown in Figure 21.7.

Keratin, a fibrous protein of hair, fingernails, horns, and wool, is one protein that does have a predominantly α-helix structure. Silk is made of fibroin, another fibrous protein, which exists mainly in the pleated sheet form. Silkworm silk and especially spider silk exhibit a combination of strength and toughness unmatched by high-performance synthetic fibers. In its primary structure, silk contains sections that consist of only alanine (25 percent) and glycine (42 percent). The formation of pleated sheets, largely by the alanine sections, allows microcrystals to orient along the fiber axis. This accounts for the superior mechanical strength.

Another repeating pattern classified as a secondary structure is the **triple helix of collagen** (Figure 21.8). It is quite different from the α-helix. Collagen is the structural protein of connective tissues (bone, cartilage, tendon, blood vessels, skin), where it provides strength and elasticity. It is the most abundant protein in the body, making up about 30 percent by weight of all the body's protein. The triple helix structure is made possible by the primary structure of collagen, which allows three polypeptide chains to come together. Each strand of collagen is made of repetitive units that can be symbolized as Gly—X—Y, that is, every third amino acid in the chain is glycine. Glycine, of course, has the shortest side chain (—H) of all amino acids, and this allows the three chains to come together. About one third of the X amino acid is proline, and the Y is often hydroxyproline. The triple helix units, called *tropocollagen,* constitute the soluble form of collagen; they are stabilized by hydrogen bonding between the backbones of the three chains. Collagen is made of many tropocollagen units.

FIGURE 21.7 Schematic structure of the enzyme carboxypeptidase. The β-pleated sheet portions are shown in blue, the green structures are the α-helix portions, and the orange strings are the random coil areas.

Hydroxyproline is not one of the 20 amino acids in Table 21.1, but the body makes it from proline and uses it in certain proteins of which collagen is but one example.

21.9 The Tertiary and Quaternary Structure of Proteins

Tertiary Structure

Tropocollagen is found only in fetal or young connective tissues. With aging, the triple helixes that organize themselves (Figure 21.8) into fibrils cross-link and form insoluble collagen. This cross-linking of collagen is an example of the **tertiary** structures that stabilize the three-dimensional conformations of protein molecules. In collagen, the **cross-linking** consists of covalent bonds that link together two lysine residues on adjacent chains of the helix.

In general, tertiary structures are stabilized four ways:

1. Covalent bonds Besides the covalent cross-linking in collagen, the covalent bond most often involved in stabilization of the tertiary structure of proteins is the disulfide bond. In Section 21.4, we noted that the amino acid cysteine is easily converted to the dimer cystine. When a cysteine residue is in one chain and another cysteine residue is in another chain (or in another part of the same chain), formation of a disulfide bond provides a covalent linkage that binds together the two chains or the two parts of the same chain:

$$\text{—SH HS—} \xrightarrow{\text{[O]}} \text{—S—S—}$$

Examples of both types are found in the structure of insulin (Figure 21.3).

FIGURE 21.8 The triple helix of collagen.

Primary, secondary, tertiary, and quaternary structures of protein.

Besides covalent bonds, three other interactions (Figure 21.9) can stabilize tertiary structures:

2. Hydrogen Bonding We saw in Section 21.8 that secondary structures are stabilized by hydrogen bonding between backbone —C=O and H—N— groups. **Tertiary structures are stabilized by hydrogen bonding between polar groups on side chains.**

FIGURE 21.9   Noncovalent interactions that stabilize the tertiary and quaternary structures of proteins: (*a*) hydrogen bonding, (*b*) salt bridge, (*c*) hydrophobic interaction.

3. Salt Bridges   These occur only between two amino acids with ionized side chains; that is, between an acidic amino acid and a basic amino acid, each in its ionized form. The two are held together by simple ion-ion attraction.

4. Hydrophobic Interactions   In aqueous solution, globular proteins usually turn their polar groups outward, toward the aqueous solvent, and their nonpolar groups inward, away from the water molecules. The nonpolar groups prefer to interact with each other, excluding water from these regions. This is called hydrophobic interaction. Although this type of interaction is weaker than hydrogen bonding or salt bridges, it usually acts over large surface areas so that cooperatively the interactions are strong enough to stabilize a loop or some other tertiary structure formation.

The four types of interaction that stabilize the tertiary structures of proteins are shown in Figure 21.10.

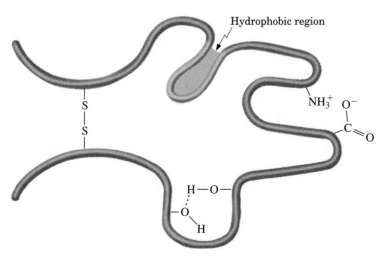

FIGURE 21.10   The four types of interaction that stabilize the tertiary structure of proteins, shown schematically.

The Role of Quaternary Structures in Mechanical Stress and Strain

Collagen is the main component of the extracellular matrix that exists between cells of higher organisms. The triple helix units of collagen (Figure 21.8) are further organized into a pattern called the **quarter-stagger arrangement** forming fibrils (Figure 21F.1). Fibers made of such fibrils provide elasticity to such tissues as skin and cornea of the eye. In the quarter-stagger arrangement, the units along a row are not spaced end to end. There is a gap between the end of one unit and the start of another. These gaps play an important role in bone formation. Collagen is one of the main constituents of bones and teeth, and the gaps in the quarter-stagger arrangement are essential for the deposition of inorganic crystals of calcium hydroxyapatite, $Ca_5(PO_4)_3OH$. The gaps serve as the nucleation sites for the growth of these crystals. The combination of hydroxyapatite crystals and collagen creates a hard material that still has some springiness, owing to the presence of collagen. Dentine, the main constituent of the internal part of a tooth, contains a higher percentage (about 75 percent) of inorganic crystals than does bone and is therefore harder. Enamel, the outer part of the tooth, has a still higher mineral content (about 95 percent) and is even harder.

The quaternary structures of other protein molecules residing inside of cells provide shape and strength. These fibrous filaments, commonly called cytoskeleton, come in different sizes. The microfilaments are 7 nm in diameter and are made of globular G-actin molecules (Figure 21F.2), which under physiological conditions form fibrous F-actin. In F-actin, the single globes are polymerized to form a double helical structure, which provides the elasticity required when cells move, change their shape, or divide. Some of these actin filaments

Packing of collagen molecules

Hole zone
0.6*d*

Overlap zone
0.4*d*

■ *FIGURE 21F.1* In the electron microscope, collagen fibers exhibit alternating light and dark bands. The dark bands correspond to the 40-nm gaps or "holes" between pairs of aligned collagen triple helices. *(J. Gross, Biozentrum/Science Photo Library/Photo Researchers, Inc.)*

■ *FIGURE 21F.2* The three-dimensional structure of an actin monomer from skeletal muscle. This view shows the two domains *(left and right)* of actin.

are anchored to cell membranes by other proteins. Through these proteins, they are connected to the collagen in the extracellular matrix. In this fashion, stresses and strains inside and outside the cells are coordinated. This same F-actin, in the specialized muscle cells, interacts with myosin (see Figure 20.11) to provide the contraction of muscles.

The cytoskeleton with the largest diameter is called a microtubule. These are 30-nm-wide cylindrical structures. Microtubules are formed by assembling two similar proteins, α- and β-tubulin. In the presence of Ca^{2+} ions, the dimer is quite stable. These dimers form a helical structure in which 13 units form a turn. The microtubules are not static. At the expense of energy, they constantly shed dimeric units at the "minus end" and add new dimeric units at the "plus end" at about the same rate (Figure 21F.3). Microtubules provide the internal scaffolding of cells in much the same way that bones are the scaffolding in our body. They also act as railroads along which molecules and organelles are transported from one part of the cell to another.

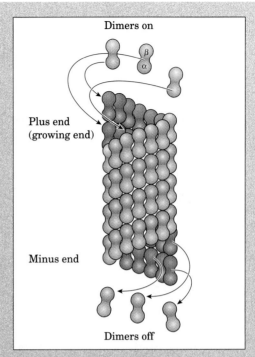

■ **FIGURE 21F.3** A model of the GTP-dependent treadmilling process.

EXAMPLE 21.2

What kind of noncovalent interaction occurs between the side chains of serine and glutamine?

Solution
The side chain of serine ends in an —OH group; that of glutamine ends in an —NH$_2$ group. The two groups can form a hydrogen bond.

Problem 21.2 ▬
What kind of noncovalent interaction occurs between the side chains of arginine and glutamic acid? ■

Section 21.7 pointed out that the primary structure of a protein largely determines its secondary and tertiary structures. We can now see the reason for this. When the particular R groups are in the proper positions, all the hydrogen bonds, salt bridges, disulfide linkages, and hydrophobic interactions that stabilize the three-dimensional structure of that molecule are allowed to form.

The side chains of some protein molecules allow them to fold (form a tertiary structure) in only one possible way, but other proteins, especially those with long polypeptide chains, can fold in a number of possible ways. Certain proteins in living cells, called **chaperones,** help a newly synthesized polypeptide chain to assume the proper secondary and tertiary structures that are necessary for the functioning of that molecule and prevent foldings that are not biologically active.

Quaternary Structure

The highest level of protein organization is quaternary structure, which applies to proteins with more than one polypeptide chain. **Quaternary structure determines how the different subunits of the protein fit into an organized whole.** The subunits are packed and held together by hydrogen bonds, salt bridges, and hydrophobic interactions, the same forces that operate within tertiary structures, but *no covalent bonds*. The hemoglobin molecule provides an important example. Hemoglobin in adult humans is made of four chains (called globins): two identical chains (called alpha) of 141 amino acid residues each and two other identical chains (beta) of 146 residues each. Figure 21.11 shows how the four chains fit together.

In hemoglobin each globin chain surrounds an iron-containing heme unit, the structure of which is shown in Figure 21.12. Proteins that contain non-amino acid portions are called conjugated proteins. The non-amino acid portion of a **conjugated protein** is called a **prosthetic group.** In hemoglobin, the globins are the amino acid portions and the heme units are the prosthetic groups.

Hemoglobin containing two alpha and two beta chains is not the only kind existing in the human body. In the early development stage of the fetus, the hemoglobin contains two alpha and two gamma chains. The fetal hemoglobin has greater affinity toward oxygen than the adult hemoglobin. In this way, the mother's red blood cells carrying oxygen can pass it to the fetus for its own use. Fetal hemoglobin also alleviates some of the symptoms of sickle cell anemia (Box 21D).

Another example of quaternary structure and higher organizations of subunits can be seen in collagen. The quarter-stagger alignment of tropocollagen forms collagen fibrils, and the twisting of these fibrils into a five-strand or six-strand helix builds the collagen fibers. These higher organizations provide the superior strength these fibers exhibit in tissues (Box 21F).

21.10 Glycoproteins

Although many proteins such as serum albumin consist exclusively of amino acids, others also contain covalently linked carbohydrates and are therefore classified as **glycoproteins.** These include most of the plasma

When a protein consists of more than one polypeptide chain, each is called a **subunit.**

The oxygen and carbon dioxide carrying functions of hemoglobin are discussed in Sections 31.3 and 31.4.

Heme molecules with iron ion

Hemoglobin

FIGURE 21.11 The quaternary structure of hemoglobin.

FIGURE 21.12 The structure of heme.

proteins (for example, fibrinogen), enzymes such as ribonuclease, hormones such as thyroglobulin, storage proteins such as casein and ovalbumins, and protective proteins such as immunoglobulins and interferon. The carbohydrate content of these proteins may vary from a few percent (immunoglobulins) up to 85 percent (blood group substances, Box 19D). Most of the proteins in membranes (lipid bilayers, Section 20.5) are glycoproteins.

Glycoproteins have predominantly two kinds of linkages between the protein and the carbohydrate parts. The **O-linked saccharides** (Section 19.5) are bonded to a —OH group on the serine, threonine, or hydroxylysine side chain of the protein. The linkage itself is a glycosidic bond. Mucins, which coat and protect the mucous membranes, are O-linked saccharides of glycoproteins. They line the respiratory and gastrointestinal tracts as well as that of the cervix. They are highly viscous and help to protect the underlying cells from harmful environmental agents.

The second kind of linkage between the carbohydrate and the protein is the N-glycosidic bond (—C—N—). This bond exists between the N of an aspargine residue of the protein chain and the C-1 (anomeric carbon) of N-acetylglucosamine (Section 19.2). The glycoprotein is called an **N-linked saccharide** (Section 19.5). N-Acetylglucosamine is the first in the fork-like branched oligosaccharidic chain. Immunoglobulins are an example of N-linked saccharides. In many cases, the function adding N-linked saccharides to the protein core is to make them more water-soluble and to facilitate their transport within the cell.

A special class of glycoproteins are the proteoglycans. These compounds have a protein core, and the side chains are long polysaccharide chains made of acidic polysaccharides (also called glycosaminoglycans) discussed in Section 19.9. Proteoglycans are essential parts of the extracellular matrix where they provide strength, flexibility, and elasticity in tissues such as skin, cartilage, and the cornea of the eye, among others.

21.11 Denaturation

Protein conformations are stabilized in their native states by secondary and tertiary structures and through the aggregation of subunits in quaternary structure. Any physical or chemical agent that destroys these stabilizing structures changes the conformation of the protein. We call this process **denaturation.** For example, heat cleaves hydrogen bonds, so boiling a protein solution destroys the α-helical structure (compare Figures 21.5 and 21.6). In collagen, the triple helixes disappear upon boiling, and the molecules have a largely random-coil conformation in the denatured state, which is gelatin. In other proteins, especially globular proteins, heat causes the unfolding of the polypeptide chains, and, because of subsequent intermolecular protein-protein interactions, precipitation or coagulation takes place. That is what happens when we boil an egg.

Similar conformational changes can be brought about by the addition of denaturing chemicals. Solutions such as 6 M aqueous urea break hydrogen bonds and cause the unfolding of globular proteins. Surface-active agents (detergents) change protein conformation by opening up the hydrophobic regions, whereas acids, bases, and salts affect the salt bridges as well as the hydrogen bonds.

Denaturation The loss of the secondary, tertiary, and quaternary structure of a protein by a chemical or physical agent that leaves the primary structure intact

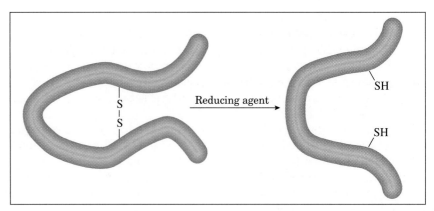

Reduction of disulfide bond.

Reducing agents, such as 2-mercaptoethanol ($HOCH_2CH_2SH$), can break the —S—S— disulfide bonds, reducing them to —SH groups. The processes of permanent waving and straightening of curly hair are examples of the latter. The protein keratin, which makes up human hair, has a high percentage of disulfide bonds. These are primarily responsible for the shape of the hair, straight or curly. In either permanent waving or straightening, the hair is first treated with a reducing agent that cleaves some of the —S—S— bonds.

This treatment allows the molecules to lose their rigid orientations and become more flexible. The hair is then set into the desired shape, using curlers or rollers, and an oxidizing agent is applied. The oxidizing agent reverses the preceding reaction, forming new disulfide bonds, which now hold the molecules together in the desired positions.

Heavy metal ions (for example, Pb^{2+}, Hg^{2+}, or Cd^{2+}) also denature protein by attacking the —SH groups. They form salt bridges, for example, —S^- Hg^{2+} ^-S—. This very feature is taken advantage of in the antidote for heavy metal poisoning: raw egg whites and milk. The egg and milk proteins are denatured by the metal ions, forming insoluble precipitates in the stomach. These must be pumped out or removed by inducing vomiting. In this way, the poisonous metal ions are removed from the body. If the antidote is not pumped out of the stomach, the digestive enzymes would degrade the proteins and release the poisonous heavy metal ions to be absorbed in the bloodstream.

Other chemical agents such as alcohol also denature proteins, coagulating them. This process is used in sterilizing the skin before injections. At a concentration of 70 percent, ethanol penetrates bacteria and kills them by coagulating their proteins, whereas 95 percent alcohol denatures only surface proteins.

Denaturation changes secondary, tertiary, and quaternary structures. It does not affect primary structures (that is, the sequence of amino acids that make up the chain). If these changes occur to a small extent, denaturation can be reversed. For example, in many cases, when we remove a denatured protein from a urea solution and put it back into water, it reassumes its secondary and tertiary structure. This is reversible denaturation. In living cells, some denaturation caused by heat can be reversed by chaperones (Section 21.9). These help a partially heat-denatured protein to regain its native secondary, tertiary, and quaternary structures. However, some denaturation is irreversible. We cannot unboil a hard-boiled egg.

Raw egg whites are an antidote to heavy metal poisoning. *(Charles D. Winters)*

BOX 21G

Laser Surgery and Protein Denaturation

Proteins can also be denatured by physical means, most notably by heat. Bacteria are killed and surgical instruments are sterilized by heat. A special method of heat denaturation that is seeing increasing use in medicine is the use of lasers. A laser beam (a highly coherent light beam of a single wavelength) is absorbed by tissues, and its energy is converted to heat energy. This process can be used to cauterize incisions so that a minimum amount of blood is lost during the operation. Laser beams can be delivered by an instrument called a **fiberscope.** The laser beam is guided through tiny fibers, thousands of which are fitted into a tube only 1 mm in diameter. In this way, the energy for denaturation is delivered where it is needed. It can, for example, seal wounds or join blood vessels without the necessity of cutting through healthy tissues. Fiberscopes have been used successfully to diagnose and treat many bleeding ulcers in the stomach, intestines, and colon.

A novel use of the laser fiberscope is in treating tumors that cannot be reached for surgical removal. A drug called Photofrin, which is activated by light, is given to patients intravenously. The drug in this form is inactive and harmless. The patient then waits 24 to 48 h during which the drug accumulates in the tumor but is removed and excreted from the healthy tissues. A laser fiberscope with 630 nm of red light is directed toward the tumor. An exposure between 10 and 30 min is applied.

The energy of the laser beam activates the Photofrin, which destroys the tumor. This is not a complete cure because the tumor may grow back, or it may have spread before the treatment. This treatment has only one side effect: The patient remains sensitive to exposure to strong light for approximately 30 days (i.e., sunlight must be avoided). But this is minor compared to the pain, nausea, hair loss, and the like that accompany radiation or chemotherapy of tumors. In the United States, Photofrin is only approved to treat esophageal cancer; however, in Europe, Japan, and Canada, it is also used to treat lung, bladder, gastric and cervical cancers. The light that acti-

vates Photofrin penetrates only a few mm, but the new drugs under development may use radiation in the near infrared that can penetrate tumors up to a few cm.

The most common use of laser in surgery is its application to correct near sightedness and astigmatism. In a computer-assisted laser surgery process, the curvature of the cornea is changed. Using the energy of the laser beam, physicians remove part of the cornea. In the procedure called photorefractive keratectomy (PRK), the outer layers of the cornea are denatured, burnt off. In the procedure called LASIK (*laser in situ keratomileusis*), the surgeon creates a flap or a hinge of the outer layers of the cornea and then with the laser beam burns off a computer programmed amount under the flap to change the shape of the cornea. After the 5- to 10-min procedure, the flap is put back, and it heals without stitches. The patient regains good vision one day after the surgery. There is no need for prescription lenses.

■ Argon krypton laser surgery. *(© Larry Mulvehill/Photo Researchers, Inc.)*

SUMMARY

Proteins are giant molecules made of amino acids linked together by **peptide bonds.** Proteins have many functions: structural (collagen), enzymatic (ribonuclease), carrier (hemoglobin), storage (casein), protective (immunoglobulin), and hormonal (insulin). **Amino acids** are organic compounds containing an amino ($-NH_2$) and a carboxylic ($-COOH$) acid group. The 20 amino acids found in proteins are classified by their side chains: nonpolar, polar but neu-

tral, acidic, and basic. All amino acids in human tissues are L amino acids. Amino acids in the solid state, as well as in water, carry both positive and negative charges; they are called **zwitterions.** The pH at which the number of positive charges is the same as the number of negative charges is the **isoelectric point** of an amino acid or protein.

When the amino group of one amino acid condenses with the carboxyl group of another, an amide (peptide) linkage is formed, with the elimination of water. The two amino acids form a dipeptide. Three amino acids form a tripeptide, and so forth. Many amino acids form a **polypeptide chain.** Proteins are made of one or more polypeptide chains.

The linear sequence of amino acids is the **primary structure** of proteins. The repeating short-range conformations (**α-helix, β-pleated sheet, triple helix of collagen,** or **random coil**) are the **secondary structures.** The tertiary structure is the three-dimensional conformation of the protein molecule. **Tertiary structures** are maintained by covalent cross-links such as **disulfide bonds** and by **salt bridges, hydrogen bonds,** and **hydrophobic interactions between the side chains.** The precise fit of polypeptide subunits into an aggregated whole is called the **quaternary structure.**

Secondary and tertiary structures stabilize the native conformation of proteins; physical and chemical agents, such as heat or urea, destroy these structures and **denature** the protein. Protein functions depend on native conformation; when a protein is denatured, it can no longer carry out its function. Some (but not all) denaturation is reversible; in some cases **chaperone** molecules reverse denaturation. Many proteins are classified as **glycoproteins** because they contain carbohydrate units.

KEY TERMS

AGE (Box 21B)
Alpha amino acid (Section 21.2)
Amino acid (Section 21.2)
Backbone (Section 21.6)
C-terminal amino acid (Section 21.5)
Chaperones (Section 21.9)
Conjugated protein (Section 21.9)
Cross-link (Section 21.9)
Cystine (Section 21.4)
Denaturation (Section 21.11)
Disulfide bond (Section 21.4)
Fiberscope (Box 21G)
Fibrous protein (Section 21.1)
Globular protein (Section 21.1)
Glycoprotein (Section 21.10)

α-Helix (Section 21.8)
Hydrophobic interaction
  (Section 21.9)
Hypoglycemic awareness (Box 21C)
Intramolecular hydrogen bond
  (Section 21.8)
Isoelectric point (Section 21.3)
N-linked saccharide (Section 21.10)
N-terminal amino acid (Section 21.5)
O-linked saccharide (Section 21.10)
Peptide (Section 21.5)
Peptide linkage (Section 21.5)
β-Pleated sheet (Section 21.8)
Polypeptide (Section 21.5)
Primary structure (Section 21.7)

Prosthetic group (Section 21.9)
Protein (Sections 21.1, 21.5)
Quarter-stagger arrangement
  (Box 21F)
Quaternary structure (Section 21.9)
Random coil (Section 21.8)
Residue (Section 21.5)
Salt bridge (Section 21.9)
Secondary structure (Section 21.8)
Sickle cell anemia (Box 21D)
Side chain (Section 21.6)
Subunit (Section 21.9)
Tertiary structure (Section 21.9)
Triple helix (Section 21.8)
Zwitterion (Section 21.3)

CONCEPTUAL PROBLEMS

21.A Why does glycine have no D or L form?

21.B Write the structure of the oxidation product of glutathione. (See Box 21A.)

21.C What kind of changes are necessary to transform a protein having predominantly α-helical structure into one having β-pleated sheet?

21.D Cytochrome c is an important protein in producing energy from food. It contains a heme surrounded by a polypeptide chain. What kind of structure do these two entities form? To what group of proteins does cytochrome c belong?

PROBLEMS

Difficult problems are designated by an asterisk.

21.3 The human body has about 100 000 different proteins. Why do we need so many?

21.4 The members of which class of proteins are insoluble in water and can serve as structural materials?

Amino Acids

21.5 What is the difference in structure between tyrosine and phenylalanine?

21.6 Classify the following amino acids as nonpolar, polar but neutral, acidic, or basic. (a) Arginine (b) Leucine (c) Glutamic acid (d) Asparagine (e) Tyrosine (f) Phenylalanine (g) Glycine

21.7 Which amino acid has the highest percent nitrogen (g N/100 g amino acid)?

21.8 Which amino acids have aromatic side chains?

21.9 Draw the structure of proline. Which class of heterocyclic compounds does this molecule belong to?

21.10 Which amino acid is also a thiol?

21.11 Why is it necessary to have proteins in our diets?

21.12 Which amino acids in Table 21.1 have more than one stereocenter?

21.13 What are the similarities and differences in the structures of alanine and phenylalanine?

21.14 What special structural feature or property does each of the following amino acids have that makes it different from all the others? (a) Glycine (b) Cystine (c) Tyrosine (d) Proline

21.15 Draw the structures of L- and D-valine.

Zwitterions

21.16 Why are all amino acids solids at room temperature?

21.17 Show how alanine, in solution at its isoelectric point, acts as a buffer (write equations to show why the pH does not change much if we add an acid or base).

21.18 Explain why an amino acid cannot exist in an un-ionized form [$RCH(NH_2)COOH$] at any pH.

21.19 Draw the structure of valine at pH 1 and at pH 12.

Peptides and Proteins

21.20 A tetrapeptide is abbreviated as DPKH. Which amino acid is at the N-terminal and which is at the C-terminal?

*__21.21__ Draw the structure of a tripeptide made of threonine, arginine, and methionine.

21.22 (a) Use the three-letter abbreviations to write the representation of the following tetrapeptide. (b) Which amino acid is the C-terminal end and which is in the N-terminal end?

21.23 Draw the structure of the dipeptide leucylproline.

21.24 (a) What is a protein backbone? (b) What is an N-terminal end of a protein?

21.25 Show by chemical equations how alanine and glutamine can be combined to give two different dipeptides.

Properties of Peptides and Proteins

21.26 (a) How many atoms of the peptide bond lie in the same plane? (b) Which atoms are they?

21.27 (a) Draw the structural formula of the tripeptide Met—Ser—Cys. (b) Draw the different ionic structures of this tripeptide at pH 2.0, 7.0, and 10.0.

21.28 How can a protein act as a buffer?

21.29 Proteins are least soluble at their isoelectric points. What would happen to a protein precipitated at its isoelectric point if a few drops of dilute HCl were added?

Primary Structure of Proteins

21.30 How many different tripeptides can be made (a) using only leucine, threonine, and valine (b) using all 20 amino acids?

21.31 How many different tetrapeptides can be made (a) if the peptides contain one residue each of asparagine, proline, serine, and methionine (b) if all 20 amino acids can be used?

21.32 How many amino acid residues in the A chain of insulin are the same in insulin from humans, cattle (bovine), hogs, and sheep?

21.33 Based on your knowledge of the chemical properties of amino acid side chains, suggest a substitution for leucine in the primary structure of a protein that would probably not change the character of the protein very much.

Secondary Structure of Proteins

21.34 Is a random coil a (a) primary, (b) secondary, (c) tertiary, or (d) quaternary structure? Explain.

21.35 Decide whether the following structures that exist in collagen are primary, secondary, tertiary, or quaternary.
(a) Tropocollagen
(b) Collagen fibril
(c) Collagen fiber
(d) The proline—hydroxyproline—glycine repeating sequence

21.36 Proline is often called an α-helix terminator; that is, it is usually in the random-coil secondary structure following an α-helix portion of a protein chain. Why does proline not fit easily into an α-helix structure?

Tertiary and Quaternary Structures of Proteins

21.37 Polyglutamic acid (a polypeptide chain made only of glutamic acid residues) has an α-helix conformation below pH 6.0 and a random-coil conformation above pH 6.0. What is the reason for this conformational change?

21.38 Distinguish between inter- and intramolecular hydrogen bonding between backbone groups. Where in protein structures do you find one and where the other?

21.39 Identify the different primary, secondary, and tertiary structures in the numbered boxes:

21.40 If both cysteine residues on the B chain of insulin were changed to alanine residues, how would the tertiary structure of insulin be affected?

21.41 Which amino acid side chains can form hydrophobic bonds?

21.42 (a) What is the difference in the quaternary structure between fetal hemoglobin and adult hemoglobin? (b) Which can carry more oxygen?

21.43 What is a conjugated protein?

Glycoprotein

*21.44 In many glycoproteins, the protein carbohydrate linkage occurs on a serine side chain between the —OH of the protein and the aldehyde chain, —HC=O, of the monosaccharide, for example D-glucose. Give the structure of such a protein-carbohydrate linkage in which a glycosidic bond has been established (Section 19.5).

21.45 Draw the structure of an N-linked saccharide where the amide nitrogen group on the side chain of asparagine is linked to C-1 of N-acetylglucosamine by a β-N-glycosidic bond.

Protein Denaturation

21.46 Which amino acid side chain is most frequently involved in denaturation by reduction?

21.47 What does the reducing agent do in straightening a curly hair?

21.48 Silver nitrate, $AgNO_3$, is sometimes put into the eyes of newborn infants as a disinfectant against gonorrhea. Silver is a heavy metal. Explain how this may work against the bacteria.

21.49 Why do nurses and physicians use 70 percent alcohol to wipe the skin before giving injections?

Boxes

21.50 (Box 21A) What is unusual about the peptide bond of glutathione?

21.51 (Box 21B) AGE products become disturbing only in elderly people, even though they also form in younger people. Why don't they harm younger people?

21.52 (Box 21C) Define hypoglycemic awareness.

21.53 (Box 21D) Why has the mutation in the beta chain of hemoglobin survived in many generations of heterozygotes, even though it causes sickle cell anemia?

21.54 (Box 21D) In sickle cell anemia, Val is substituted for Glu in the sixth position of the beta chain of hemoglobin. Some individuals have hemoglobin in which Asp is substituted for Glu. Would you expect this substitution also to be detrimental to health? Explain.

21.55 (Box 21D) How does the hydroxyurea therapy alleviate the symptoms of sickle cell anemia?

21.56 (Box 21E) What is the difference in the conformation between normal prion protein and the amyloid prion causing Mad Cow Disease?

21.57 (Box 12F) What constituent of bone determines its hardness?

21.58 (Box 21F) How does G-actin form F-actin?

21.59 (Box 21G) What is Photofrin? How does it destroy inoperable tumors?

21.60 (Box 21G) How does the fiberscope help to heal bleeding ulcers?

Additional Problems

21.61 Carbohydrates are integral parts of glycoproteins. Does the addition of carbohydrate to the protein core make the protein more or less soluble in water? Explain.

21.62 Can the hydrolysis of a peptide bond in a protein be considered a denaturation process?

21.63 How many different dipeptides can be made (a) using only alanine, tryptophan, glutamic acid, and arginine (b) using all 20 amino acids?

21.64 Denaturation is usually associated with transitions from helical structures to random coils. If an imaginary process were to transform the keratin in your

hair from an α-helix to a β-pleated sheet structure, would you call the process denaturation? Explain.

21.65 In diabetes, insulin is administered intravenously. Explain why this hormone protein cannot be taken orally.

21.66 Draw the structure of lysine (a) above, (b) below, and (c) at its isoelectric point.

21.67 Describe the important role cysteine plays in forming the tertiary structures of proteins.

21.68 Considering the vast number of animal and plant species on Earth (including those now extinct) and the large number of different protein molecules in each organism, have all possible protein molecules been used already by some species or other? Explain.

21.69 What kind of noncovalent interaction occurs between the following amino acids?
(a) Valine and isoleucine
(b) Glutamic acid and lysine
(c) Tyrosine and threonine
(d) Alanine and alanine

21.70 How many different decapeptides (peptides containing 10 amino acids each) can be made from the 20 amino acids?

21.71 Which amino acid does not rotate the plane of polarized light?

21.72 Write the expected products of the acid hydrolysis of the following tripeptide.

21.73 What charges are there on aspartic acid at pH 2.0?

For the discovery of ribozymes,
Sidney Altman of Yale University
and Thomas R. Cech of the University of Colorado were awarded the
1989 Nobel Prize in Chemistry.

Above: A hydrothermal vent on the
ocean floor along with marine life,
including tube worms. (© NSF Oasis
Project/Norbert Wu Photography)

Enzymes

22.1 Introduction

The cells in your body are chemical factories. Only a few of the thousands
of compounds necessary for the operation of the human organisms are
obtained from the diet. Most of them are synthesized within the cells,
which means that hundreds of chemical reactions are taking place in your
cells every minute of your life.

Nearly all these reactions are catalyzed by **enzymes,** which are large
**protein molecules that increase the rates of chemical reactions
without themselves undergoing any change.** Without enzymes, life as
we know it would not be possible.

Proteins are not the only biological catalysts. **Ribozymes** are enzymes
made of ribonucleic acids. They catalyze the self-cleavage of certain portions of their molecules (introns). Many biochemists believe that during
evolution RNA catalysts were first and protein enzymes arrived on the
scene later.

Like all catalysts, enzymes do not change the position of equilibrium.
That is, enzymes cannot make a reaction take place that would not take
place without them. What they do is increase the rate; they cause reactions

BOX 22A

Muscle Relaxants and Enzyme Specificity

Acetylcholine is a neurotransmitter (Section 23.1) that operates between the nerve endings and muscles. It attaches itself to a specific receptor in the muscle end plate. This transmits a signal to the muscle to contract; shortly thereafter, the muscles relax. A specific enzyme, acetylcholinesterase, then catalyzes the hydrolysis of the acetylcholine, removing it from the receptor site and thus preparing it for the next signal transmission: the next contraction.

Succinylcholine is sufficiently similar to acetylcholine so that it too attaches itself to the receptor of the muscle end plate. However, acetylcholinesterase can hydrolyze succinylcholine only very slowly. While it stays attached to the receptor, no new signal can reach the muscle to allow it to contract again. Thus, the muscle stays relaxed for a long time.

This feature makes succinylcholine a good muscle relaxant during minor surgery, especially when a tube must be inserted

into the bronchus (bronchoscopy). For example, after intravenous administration of 50 mg of succinylcholine, paralysis and respiratory arrest are observed within 30 s. While respiration is carried on artificially, the bronchoscopy can be performed within minutes.

to take place faster by lowering the activation energy. As catalysts, enzymes are remarkable in two respects: (1) They are extremely effective, increasing reaction rates by anywhere from 10^9 to 10^{20} times, and (2) most of them are extremely specific.

As an example of their effectiveness, consider the oxidation of glucose. A lump of glucose or even a glucose solution exposed to oxygen under sterile conditions would show no appreciable change for months. In the human body the same glucose is oxidized within seconds (see also Box 19C).

Every organism has many enzymes—many more than 3000 in a single cell. Most enzymes are very specific, each of them speeding up only one particular reaction or class of reactions. For example, the enzyme urease catalyzes only the hydrolysis of urea and not that of other amides, even closely related ones.

$$(NH_2)_2C{=}O + H_2O \xrightarrow{\text{urease}} 2\,NH_3 + CO_2$$
Urea

Another type of specificity can be seen with trypsin, an enzyme that cleaves the peptide bonds of protein molecules—but not every peptide bond, only those on the carboxyl side of lysine and arginine residues (Fig-

FIGURE 22.1 A typical amino acid sequence. The enzyme trypsin catalyzes the hydrolysis of this chain only at the points marked with an arrow (the —COOH side of lysine and arginine).

> **Enzyme specificity** The limitation of an enzyme to catalyze one specific reaction with one specific substrate

The enzyme that oxidizes D-glucose does not work on L-glucose. It is specific for β-D-glucose.

ure 22.1). The enzyme carboxypeptidase specifically cleaves only the last amino acid on a protein chain—the one at the C-terminal end. Lipases are less specific: They cleave any triglyceride, but they still don't affect carbohydrates or proteins.

The specificity of enzymes also extends to stereospecificity. The enzyme arginase hydrolyzes the amino acid L-arginine (the naturally occurring form) to a compound called L-ornithine and urea (Section 27.8) but has no effect on its mirror image, D-arginine.

Enzymes are distributed according to the body's need to catalyze specific reactions. A large number of protein-splitting enzymes are in the blood, ready to promote clotting. Digestive enzymes, which also catalyze the hydrolysis of proteins, are located in the secretions of the stomach and pancreas. Even within the cells themselves, some enzymes are localized according to the need for specific reactions. The enzymes that catalyze the oxidation of compounds that are part of the citric acid cycle (Section 26.4) are located in the mitochondria, and special organelles such as those called lysosomes contain an enzyme (lysozyme) that catalyzes the dissolution of bacterial cell walls.

22.2 Naming and Classifying Enzymes

Enzymes are commonly given names derived from the reaction they catalyze and/or the compound or type of compound they act on. For example, lactate dehydrogenase is the enzyme that speeds up the removal of hydrogen from lactate (a redox reaction). Acid phosphatase helps to cleave phosphate ester bonds under acidic conditions. As can be seen from these examples, the names of most enzymes end in "-ase." Some enzymes, however, have older names, ones that were assigned before their actions were clearly understood. Among these are pepsin, trypsin, and chymotrypsin—all enzymes of the digestive tract.

Enzymes can be classified into six major groups according to the type of reaction they catalyze (see also Table 22.1):

1. **Oxidoreductases** catalyze oxidations and reductions.

2. **Transferases** catalyze the transfer of a group of atoms, such as CH_3, CH_3CO, or NH_2, from one molecule to another.

3. **Hydrolases** catalyze hydrolysis reactions.

4. **Lyases** catalyze the addition of a group to a double bond or the removal of two groups from adjacent atoms to create a double bond.

5. **Isomerases** catalyze isomerization reactions.

6. **Ligases,** or synthetases, catalyze the joining of two molecules.

TABLE 22.1 Classifications of Enzymes

Class	Typical Example	Reaction Catalyzed	Section Number in This Book
1. Oxidoreductases	Lactate dehydrogenase	$CH_3—\overset{\underset{\|}{O}}{\overset{\|}{C}}—COO^- \longrightarrow CH_3—\overset{\underset{\|}{OH}}{\overset{\|}{CH}}—COO^-$ Pyruvate L-(+)-Lactate	27.2
2. Transferases	Aspartate amino transferase or Aspartate transaminase	Aspartate + α-Ketoglutarate ⟶ Oxaloacetate + Glutamate	27.8
3. Hydrolases	Acetylcholinesterase	$CH_3—\overset{\underset{\|}{O}}{\overset{\|}{C}}—OCH_2CH_2\overset{+}{N}(CH_3)_3 + H_2O \longrightarrow$ Acetylcholine $CH_3COOH + HOCH_2CH_2\overset{+}{N}(CH_3)_3$ Acetic acid Choline	23.3
4. Lyases	Aconitase	cis-Aconitate + $H_2O \longrightarrow$ Isocitrate	26.4
5. Isomerases	Phosphohexose isomerase	Glucose-6-phosphate ⟶ Fructose-6-phosphate	27.2
6. Ligases	Tyrosine-tRNA synthetase	$ATP + \text{L-tyrosine} + tRNA \longrightarrow$ $\text{L-tyrosyltRNA} + AMP + PP_i$	25.6

BOX 22B

Meat Tenderizers

Pepsin and trypsin are proteases, a class of hydrolases that catalyze the hydrolysis of proteins. Some of the meat that we eat is tough and difficult to chew. Since meat contains a lot of protein, coating the meat with a protease before cooking hydrolyzes some of the long protein chains, breaking them into shorter chains and making the meat easier to chew. Meat tenderizers do just this. They contain proteases such as papain. Papain is isolated from the latex of the green fruits and leaves of the plant *Carica papaya*. Proteases are also used to increase the yield of meat from bones and meat scraps in the manufacture of processed meats, such as bologna and frankfurters.

22.3 Common Terms in Enzyme Chemistry

Cofactor The nonprotein part of an enzyme necessary for its catalytic function

Coenzyme An organic molecule, frequently a B vitamin, which acts as a cofactor

Some enzymes have two kinds of cofactor: a coenzyme and a metallic ion.

Active site A three-dimensional cavity of the enzyme with specific chemical properties to accommodate the substrate

Some enzymes, such as pepsin and trypsin, consist of polypeptide chains only. Other enzymes contain nonprotein portions called **cofactors. The protein (polypeptide) portion of the enzyme is** called an **apoenzyme.** The cofactors may be metallic ions, such as Zn^{2+} or Mg^{2+}, or they may be organic compounds. **Organic cofactors are** called **coenzymes.** An important group of coenzymes are the B vitamins, which are essential to the action of many enzymes (Section 26.3). Another important coenzyme is heme (Figure 21.12), which is part of a number of oxidoreductases as well as of hemoglobin. In any case, an apoenzyme cannot catalyze a reaction without its cofactor, nor can the cofactor function without the apoenzyme. When a metal ion is a cofactor, it can be bound directly to the protein or to the coenzyme, if the enzyme contains one.

The compound on which the enzyme works, the one whose reaction it speeds up, is called the **substrate.** The substrate usually binds to the enzyme surface while it undergoes the reaction. There is a specific portion of the enzyme to which the substrate binds during the reaction. This part is called the **active site.** If the enzyme has coenzymes, they are located at the active site. Therefore, the substrate is simultaneously surrounded by parts of the apoenzyme, coenzyme, and metal ion cofactor (if any), as shown in Figure 22.2.

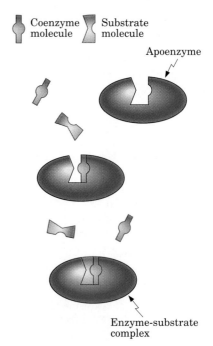

FIGURE 22.2 Schematic diagram of the active site of an enzyme and the participating components.

Activation is any process that initiates or increases the action of an enzyme. This can be the simple addition of a cofactor to an apoenzyme or the cleavage of a polypeptide chain of a proenzyme (Section 22.6). **Inhibition** is the opposite—**any process that makes an active enzyme less active or inactive** (Section 22.5). Inhibitors are compounds that accomplish this. Some inhibitors bind to the active site of the enzyme surface, thus preventing the binding of substrate. These are **competitive inhibitors.** Others, which bind to some other portion of the enzyme surface, may sufficiently alter the tertiary structure of the enzyme so that its catalytic effectiveness is slowed down. These are called **noncompetitive inhibitors.** Both competitive and noncompetitive inhibition are *reversible,* but some compounds alter the structure of the enzyme *permanently* and thus make it *irreversibly* inactive.

Competitive inhibition An enzyme regulation in which an inhibitor competes with the substrate for the active site

Noncompetitive inhibition An enzyme regulation in which an inhibitor binds to the enzyme outside of the active site, thereby changing the shape of the active site and reducing its catalytic activity

22.4 Factors Affecting Enzyme Activity

Enzyme activity is a measure of how much reaction rates are increased. In this section, we examine the effects of concentration, temperature, and pH on enzyme activity.

Enzyme and Substrate Concentration

If we keep the concentration of substrate constant and increase the concentration of enzyme, the rate increases linearly (Figure 22.3). That is, if the enzyme concentration is doubled, the rate also doubles; if the enzyme concentration is tripled, the rate also triples. This is the case in practically all enzyme reactions because the molar concentration of enzyme is almost always much lower than that of substrate (that is, there are almost always many more molecules of substrate present than molecules of enzyme).

On the other hand, if we keep the concentration of enzyme constant and increase the concentration of substrate, we get an entirely different type of curve, called a saturation curve (Figure 22.4). In this case the rate does not increase continuously. Instead, a point is reached after which the rate stays the same even if we increase the substrate concentration further. This happens because, at the saturation point, substrate molecules are bound to all the available active sites of the enzymes. Since the active sites are where the reactions take place, once they are all occupied the reaction is going at its maximum rate. Increasing the substrate concentration can no longer increase the rate because the excess substrate cannot find any active sites to attach to.

Temperature

Temperature affects enzyme activity because it changes the three-dimensional structure of the enzyme. In uncatalyzed reactions, the rate usually increases as the temperature increases (Section 7.4). The effect of temperature on enzyme-catalyzed reactions is different. When we start at a low temperature (Figure 22.5), an increase in temperature first causes an increase in rate. However, protein conformations are very sensitive to temperature changes. Once an optimum temperature is reached, any further increase in temperature causes changes in enzyme conformation. The substrate may then not fit properly onto the changed enzyme surface. Therefore the rate of reaction *decreases.*

After a *small* temperature increase above optimum, the decreased rate could still be increased again by lowering the temperature because, over a

FIGURE 22.3 The effect of enzyme concentration on the rate of an enzyme-catalyzed reaction. Substrate concentration, temperature, and pH are constant.

FIGURE 22.4 The effect of substrate concentration on the rate of an enzyme-catalyzed reaction. Enzyme concentration, temperature, and pH are constant.

FIGURE 22.5 The effect of temperature on the rate of an enzyme-catalyzed reaction. Substrate and enzyme concentrations and pH are constant.

Acidic Environment and Helicobacter

Certain organisms can live in an extreme pH environment. An example is the bacterium *Helicobacter pylori,* which survives happily in the acidic medium of the stomach wall. These bacteria live under the mucus coating of the stomach and release a toxin that causes ulcers. The role of *H. pylori* in ulceration was discovered only in the last decade because people did not think that enzymes of a bacterium can function in such a harsh acidic environment with pH 1.4. However, the enzymes in the cytoplasm of *H. pylori* are not really exposed to low pH. The bacteria contain a special enzyme, **urease,** which converts urea to the basic ammonia; therefore, in spite of the acidic environment, *H. pylori* lives in a protective, neutralizing basic cloud. After the destructive role of *H. pylori* was established, new treatment for gastric disorders was designed. Antibiotics such as amoxillin or tetracyclin can eradicate the bacteria, the ulcers are healed, and the recurrence is also prevented.

A black smoker vent pours out a sulfurous (as indicated by the yellow) mineral-rich fluid from its mound or chimney. *(B. Murton/Southampton Oceanography Centre/Science Photo Library/Photo Researchers, Inc.)*

FIGURE 22.6 The effect of pH on the rate of an enzyme-catalyzed reaction. Substrate and enzyme concentrations and temperature are constant.

> **Lock-and-key model** A model explaining the specificity of enzyme action by comparing the active site to a lock and the substrate to a key

narrow temperature range, changes in conformation are reversible. However, at some higher temperature above optimum, we reach a point where the protein denatures (Section 21.11); the conformation is altered irreversibly, and the polypeptide chain cannot refold. At this point, the enzyme is completely inactivated. The inactivation of enzymes at low temperatures is used in the preservation of food by refrigeration.

Most enzymes from bacteria and higher organisms have an optimal temperature around 37°C. However, the enzymes of organisms that live in the ocean floor at 2°C have an optimal temperature in that range. Other organisms live in ocean vents under extreme conditions, and their enzymes have optimal conditions at ranges up to 90 to 105°C. The enzymes of these hyperthermophile organisms also have other extreme requirements such as pressures up to 100 atm, and some of them have optimal pH in the range of 1 to 4. Enzymes from these hyperthermophiles, especially polymerases that catalyze the polymerization of DNA, have gained commercial importance (Section 24.6).

pH

Since the pH of its environment changes the conformation of a protein (Section 21.11), we expect effects similar to those observed when the temperature is changed. Each enzyme operates best at a certain pH (Figure 22.6). Once again, within a narrow pH range, changes in enzyme activity are reversible. However, at extreme pH values (either acidic or basic), enzymes are denatured irreversibly, and enzyme activity cannot be restored by changing back to the optimal pH.

22.5 Mechanism of Enzyme Action

We have seen that the action of enzymes is highly specific for a substrate. What kind of mechanism can account for such specificity? About 100 years ago, Arrhenius suggested that catalysts speed up reactions by combining with the substrate to form some kind of intermediate compound. In an enzyme-catalyzed reaction, the intermediate is the **enzyme-substrate complex.**

To account for the high specificity of most enzyme-catalyzed reactions, a number of models have been proposed. The simplest and most frequently quoted is the **lock-and-key model** (Figure 22.7). This model assumes that

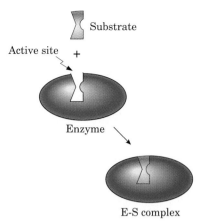

FIGURE 22.7 The lock-and-key model of enzyme mechanism.

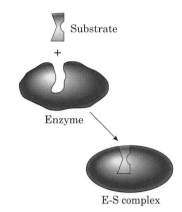

FIGURE 22.8 The induced-fit model of enzyme mechanism.

the enzyme is a rigid three-dimensional body. The surface that contains the active site has a restricted opening into which only one kind of substrate can fit, just as only the proper key can fit exactly into a lock and turn it open.

According to the lock-and-key mechanism, an enzyme molecule has its own particular shape because that shape is necessary to maintain the active site in exactly the geometric alignment required for that particular reaction. An enzyme molecule is very large (typically 100 to 200 amino acid residues), but the active site is usually composed of only two or a few amino acid residues, which may well be located at different places in the chain. The other amino acids—those not part of the active site—are located in the sequence in which we find them because that sequence causes the whole molecule to fold up in exactly the required way. This arrangement emphasizes that the shape and the functional groups on the surface of the active site are of utmost importance in recognizing a substrate.

The lock-and-key model was the first to explain the action of enzymes. But for most enzymes, there is evidence that this model is too restrictive. Enzyme molecules are in a dynamic state, not a static one. There are constant motions within them, so that the active site has some flexibility.

From x-ray diffraction, we know that the size and shape of the active site cavity change when the substrate enters. The American biochemist Daniel Koshland introduced the **induced-fit model** (Figure 22.8), in which he compared the changes occurring in the shape of the cavity upon substrate binding to the changes in the shape of a glove when a hand is inserted. That is, the enzyme modifies the shape of the active site to accommodate the substrate.

The induced-fit model explains the phenomenon of competitive inhibition (Section 22.3). The inhibitor molecule fits into the active site cavity in the same way the substrate does (Figure 22.9), preventing the substrate from entering. The result is that whatever reaction is supposed to take place on the substrate does not take place.

Many cases of noncompetitive inhibition can also be explained by the induced-fit model. In this case, the inhibitor does not bind to the active site but to a different part of the enzyme. Nevertheless, the binding causes a change in the three-dimensional shape of the enzyme molecule, and this so alters the shape of the active site that the substrate can no longer fit (Figure 22.10).

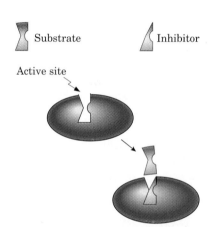

FIGURE 22.9 The mechanism of competitive inhibition. When a competitive inhibitor enters the active site, the substrate cannot get in.

Induced-fit model A model explaining the specificity of enzyme action by comparing the active site to a glove and the substrate to a hand

BOX 22D

Active Sites

The perception of the active site as either a rigid (lock-and-key model) or a partly flexible template (induced-fit model) is an oversimplification. Not only is the geometry of the active site impor tant but so are the specific interactions that take place between enzyme surface and substrate. To illus-trate, we take a closer look at the active site of the enzyme pyruvate kinase. This enzyme catalyzes the transfer of the phosphate group from phosphoenol pyruvate (PEP) to ADP, an important step in glycolysis (Section 27.2).

■ *FIGURE 22D.1* The active site and the substrates of pyruvate kinase.

The active site of the enzyme binds both substrates, PEP and ADP (Figure 22D.1). The rabbit muscle pyruvate kinase has two cofactors, K^+ and Mn^{2+} or Mg^{2+}. The divalent cation is coordinated to the carbonyl and carboxylate oxygen of pyruvate substrate and to a glutamyl 271 and aspartyl 295 residue of the enzyme. (The numbers indicate the position of the amino acid in the sequence.) The nonpolar $=CH_2$ group lies in a hydrophobic pocket formed by an alanyl 292, glycyl 294, and threonyl 327 residue. The K^+ on the other side of the active site is coordinated with the phosphate of the substrate and the seryl 76 and aspargyl 74 of the enzyme. Lysyl 269 and arginyl 72 are also part of the catalytic apparatus anchoring the ADP. This arrangement of the active site illustrates that specific folding into secondary and tertiary structures is required to bring important functional groups together. The residues of amino acids participating in the active sites are sometimes close in the sequence (aspargyl 74 and seryl 76) but mostly far apart (glutamyl 271 and aspartyl 295). The secondary and tertiary structures providing such a stable active site are illustrated in Figure 22D.2.

■ *FIGURE 22D.2* Ribbon cartoon of the pyruvate kinase monomer. Pyruvate, Mg^{2+}, and K^+ are depicted as space-filling models.

If we compare enzyme activity in the presence and absence of an inhibitor, we can tell whether competitive or noncompetitive inhibition is taking place (Figure 22.11). The maximum reaction rate is the same without an inhibitor and in the presence of a competitive inhibitor. The only difference is that this maximum rate is achieved at a low substrate concentration with no inhibitor but at a high substrate concentration when an inhibitor is present. This is the true sign of competitive inhibition because

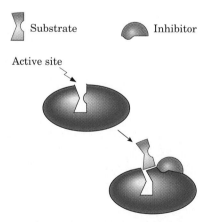

FIGURE 22.10 Mechanism of noncompetitive inhibition. The inhibitor attaches itself to a site other than the active site (allosterism) and thereby changes the conformation of the active site.

FIGURE 22.11 Enzyme kinetics in the presence and absence of inhibitors.

Sulfa Drugs As Competitive Inhibitors

The vitamin folic acid in its reduced form as tetrahydrofolate is a coenzyme in a number of biosynthetic processes, such as the synthesis of amino acids and of nucleotides. Humans obtain folic acid from the diet or from microorganisms in the intestinal tract. These microorganisms can synthesize folic acid if para-aminobenzoic acid is available to them.

In the 1930s it was discovered that the compound sulfanilamide could kill many types of harmful bacteria and thus cure several diseases. Because sulfanilamide itself has an unacceptably high toxicity to humans, a number of derivatives of this compound, such as sulfapyridine and sulfathia-

zole, which act in a similar way, are used instead. These drugs work by "tricking" the bacteria, which normally use para-aminobenzoic acid as a raw material in the synthesis of folic acid. When the bacteria get a sulfa drug instead, they cannot tell the difference and use it to make a molecule that also has a folic acid type of structure but is not exactly the same. When they try to use this fake folic acid as a coenzyme, not only doesn't it work, but it is also now a competitive inhibitor of the enzyme's action. Consequently, many of the bacteria's amino acids and nucleotides cannot be made, and the bacteria die.

Tetrahydrofolate

p-Aminobenzoic acid

Sulfapyridine

Sulfanilamide

Sulfathiazole

Noncompetitive Inhibition and Heavy Metal Poisoning

Many enzymes contain a number of —SH groups (cysteine residues) that are easily cross-linked by heavy metal ions (Section 13.3):

$$\text{SH} \quad \text{SH} + \text{Pb}^{2+} \longrightarrow \text{S}^- \quad \text{Pb}^{2+} \quad \text{S}^- + 2\text{H}^+$$

If these cross-linked cysteine residues are at or near active sites, the cross-linking can noncompetitively inhibit enzyme activity. This is the basis of lead and mercury poisoning.

Mercury poisoning has become an acute problem in recent years because the mercury and mercury compounds that factories dumped into streams and lakes for decades have entered the food chain. Mercury is also released into the environment from its use in mining. Certain microorganisms convert mercury metal to organic mercury compounds, mainly dimethylmercury, $(\text{CH}_3)_2\text{Hg}$. This compound enters the food chain when it is absorbed by algae. Later, it is concentrated in fish, and people who eat the contaminated fish can get mercury poisoning.

In Japan, where mercury was dumped in the Bay of Minamata in the 1950s, more than 100 people suffered mercury poisoning from a diet high in locally caught seafood, and 44 people died. Once such contaminated food was removed from the markets, the number of mercury-poisoning cases reported dropped drastically.

here the substrate and the inhibitor are competing for the same active site. If the substrate concentration is sufficiently increased, the inhibitor will be displaced from the active site by Le Chatelier's principle.

If, on the other hand, the inhibitor is noncompetitive, it cannot be displaced by addition of excess substrate because it is bound to a different site. In this case, the enzyme cannot be restored to its maximum activity, and the maximum rate of the reaction is lower than it would be in the absence of the inhibitor. An example of noncompetitive inhibition is the heavy metal poisoning (Box 22F).

Enzymes can also be inhibited irreversibly if a compound is bound covalently and permanently to or near the active site. Such an inhibition occurs when penicillin, which inhibits the enzyme transpeptidase necessary for cross-linking bacterial cell walls. Without cross-linking, the bacterial cytoplasm spills out, and the bacteria die (Box 18D).

22.6 Enzyme Regulation

Feedback Control

Enzymes are often regulated by environmental conditions. **Feedback control is an enzyme regulation process in which formation of a product inhibits an earlier reaction in the sequence.**

The reaction product of one enzyme may control the activity of another, especially in a complex system in which enzymes work cooperatively. For example, in a system each step is catalyzed by a different enzyme.

$$A \xrightarrow{E_1} B \xrightarrow{E_2} C \xrightarrow{E_3} D$$

The last product in the chain, D, may inhibit the activity of enzyme E_1 (by competitive or noncompetitive inhibition). When the concentration of D is low, all three reactions proceed rapidly, but as the concentration of D increases, the action of E_1 becomes inhibited and eventually stops. In this manner, the accumulation of D is a message that tells enzyme E_1 to shut down because the cell has enough D for its present needs. Shutting down E_1 stops the whole process.

Proenzymes

Some enzymes are manufactured by the body in an inactive form. In order to make them active, a small part of their polypeptide chain must be removed. These inactive forms of enzymes are called **proenzymes** or **zymogens.** After the excess polypeptide chain is removed, the enzyme becomes active. For example, trypsin is manufactured as the inactive molecule trypsinogen (a zymogen). When a fragment containing six-amino-acid residues is removed from the N-terminal end, the molecule becomes a fully active trypsin molecule. Removal of the fragment not only shortens the chain but also changes the three-dimensional structure (the tertiary structure), allowing the molecule to achieve its active form.

Why does the body go to all this trouble? Why not just make the fully active trypsin to begin with? The reason is very simple. As we have seen, trypsin is a protease—it catalyzes the hydrolysis of proteins (Figure 22.1)—and is, therefore, an important catalyst for the digestion of the proteins we eat. But it would not be good if it cleaved the proteins our own bodies are made of! Therefore, the body makes trypsin in an inactive form, and only after it has entered the digestive tract is it allowed to become active.

> **Proenzyme or zymogen** A protein that becomes an active enzyme after undergoing a chemical change

The removal of the six-amino-acid fragment is, of course, also catalyzed by an enzyme.

> **Allosterism** An enzyme regulation in which the binding of a regulator on one site on the enzyme modifies the enzyme ability to bind the substrate in the active site

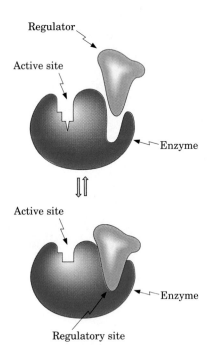

FIGURE 22.12 The allosteric effect. Binding of a regulator to a site other than the active site changes the shape of the active site.

Allosterism

Sometimes regulation takes place by means of an event that occurs at a site other than the active site but that eventually affects the active site. This type of interaction is called **allosterism,** and any enzyme regulated by this mechanism is called an **allosteric enzyme.** If a substance binds noncovalently and reversibly to a site *other than the active site,* it may affect the enzyme in either of two ways: It may inhibit enzyme action **(negative modulation)** or it may stimulate it **(positive modulation).**

The substance that binds to the allosteric enzyme is called a **regulator,** and the site it attaches to is called the **regulatory site.** In most cases, allosteric enzymes contain more than one polypeptide chain (subunits); the regulatory site is on one polypeptide chain and the active site on another.

Specific regulators can bind reversibly to the regulatory sites. For example, the enzyme protein kinase is an allosteric enzyme. In this case, the enzyme has only one polypeptide chain, so it carries both the active site and the regulatory site at different parts of this chain (Figure 22.12). The regulator is another protein molecule, one that binds reversibly to the regulatory site. As long as the regulator is bound to the regulatory site, the total enzyme-regulator complex is inactive. When the regulator is removed from the regulatory site, the protein kinase becomes active. Thus, the allosteric enzyme action is controlled by the regulator.

Allosteric regulation may occur with proteins other than enzymes. Section 31.3 explains how the oxygen-carrying ability of hemoglobin, an allosteric protein, is affected by modulators.

Protein Modification

The activity of an enzyme may also be controlled by **protein modification.** The modification is usually a change in the primary structure mostly by addition of a functional group covalently bound to the apoenzyme. The best known example of protein modification is the activation/inhibition of enzymes by phosphorylation. A phosphate group is bonded frequently to a serine or tyrosine residue. In some enzymes, such as glycogen phosphorylate (Section 28.1), the phosphorylated form is the active form of the enzyme. Without it, the enzyme is less active or not active at all. The opposite example is the enzyme pyruvate kinase, PK, discussed in Box 22D. Pyruvate kinase from liver is inactive when it is phosphorylated. Enzymes that catalyze such phosphorylation go under the common name of kinases. When the activity of PK is not needed, it is phosphorylated (to PKP) by a protein kinase using ATP as a substrate as well as a source of energy (Section 26.3). When the system wants to turn on PK activity, the phosphate group, P_i, is removed by another enzyme, phosphatase, which renders PK active.

Isoenzymes

Another form of regulation of enzyme activity occurs when the same enzyme appears in different forms in different tissues. Lactate dehydrogenase catalyzes the oxidation of lactate to pyruvate, and vice versa (Figure

One Enzyme. Two Functions

The enzyme, prostaglandin enderoperoxide synthase, PGHS, catalyzes the conversion of arachidonic acid to prostaglandin PGH_2 (Section 20.11):

Arachidonic acid

PGG$_2$

PGH$_2$

PGHS inserts two molecules of oxygen into arachidonic acid. The enzyme is a single protein molecule. It is associated with cell membrane and has a heme coenzyme. The first activity of this enzyme is to close a substituted cyclopentane ring; this is the cyclooxygenase activity. The peroxidase activity of the same enzyme yields the 15-hydroxy derivative of prostaglandin, PGH_2. The cyclooxygenase activity of the enzyme is inhibited by aspirin and related NSAIDS (Box 20J). Aspirin inhibits PGHS by acetylating serine in the active site; thus, the active site is no longer able to accommodate arachidonic acid. Other NSAIDS also inhibit the cyclooxygenase activity by competitive inhibition, but they do not inhibit the peroxidase activity. A second class of inhibitor has antioxidant activity. For example, acetaminophen (Tylenol) acts as a pain killer by inhibiting the peroxidase activity of PGHS.

21.3, step 11). The enzyme has four subunits. Two kinds of subunits, called H and M, exist. The enzyme that dominates in the heart is an H_4 enzyme, meaning that all four subunits are of the H type, although some M type subunits are also present. In the liver and skeletal muscles the M type dominates. Other types of tetramer (four units) combinations exist in different tissues: H_3M, H_2M_2, and HM_3. These different forms of the same enzyme are called **isozymes** or **isoenzymes.** The H_4 enzyme is allosterically inhibited by high levels of pyruvate, while the M_4 is not. In diagnosing the severity of heart attacks (Section 22.7), the release of H_4 isoenzyme is monitored in the serum.

Isozymes Enzymes that perform the same function but have different combinations of subunits, different quaternary structures

22.7 Enzymes in Medical Diagnosis and Treatment

Most enzymes are confined within the cells of the body. However, small amounts of enzymes can also be found in body fluids such as blood, urine, and cerebrospinal fluid. The level of enzyme activity in these fluids can easily be monitored. It has been found that abnormal activity (either high or low) of particular enzymes in various body fluids signals either the onset of certain diseases or their progression. Table 22.2 lists some enzymes used in medical diagnosis and their activities in normal body fluids.

A number of enzymes are assayed (measured) during myocardial infarction in order to diagnose the severity of the heart attack. Dead heart muscle cells spill their enzyme contents into the serum. Thus, the level of

TABLE 22.2 Enzyme Assays Useful in Medical Diagnosis

Enzyme	Normal Activity	Body Fluid	Disease Diagnosed
Alanine aminotransferase (ALT)	3–17 U/L[a]	Serum	Hepatitis
Acid phosphatase	2.5–12 U/L	Serum	Prostate cancer
Alkaline phosphatase (ALP)	13–38 U/L	Serum	Liver or bone disease
Amylase	19–80 U/L	Serum	Pancreatic disease or mumps
Aspartate aminotransferase (AST)	7–19 U/L	Serum	Heart attack or hepatitis
	7–49 U/L	Cerebrospinal fluid	
Lactate dehydrogenase (LD-P)	100–350 WU/mL	Serum	Heart attack
Creatine kinase (CK)	7–60 U/L	Serum	
Phosphohexose isomerase (PHI)	15–75 U/L	Serum	

[a]U/L = International units per liter; WU/mL = Wrobleski units per milliliter.

aspartate aminotransferase (AST) (formerly called glutamate-oxaloacetate transaminase, or GOT) in the serum rises rapidly after a heart attack. In addition to AST, lactate dehydrogenase (LD-P) and creatine kinase (CK) levels are also monitored. In infectious hepatitis, the alanine aminotransferase (ALT) (formerly called glutamate-pyruvate transaminase, or GPT) level in the serum can rise to ten times normal. There is also a concurrent increase in AST activity in the serum.

In some cases, the administration of an enzyme is part of therapy. After duodenal or stomach ulcer operations, patients are advised to take tablets containing digestive enzymes that are in short supply in the stomach after surgery. Such enzyme preparations contain lipases, either alone or combined with proteolytic enzymes, and are sold under such names as Pancreatin, Acro-lase, and Ku-zyme.

SUMMARY

Enzymes are proteins that catalyze chemical reactions in the body. Most enzymes are very specific—they catalyze only one particular reaction. The compound whose reaction is catalyzed by an enzyme is called the **substrate.** Enzymes are classified into six major groups according to the type of reaction they catalyze. Most enzymes are named after the substrate and the type of reaction they catalyze, adding the ending "-ase."

Some enzymes are made of polypeptide chains only. Others have, besides the polypeptide chain (the **apoenzyme**), nonprotein **cofactors** and either organic compounds **(coenzymes)** or inorganic ions. Only a small part of the enzyme surface participates in the actual catalysis of chemical reactions. This part is called the **active site.** Cofactors, if any, are part of the active site.

Compounds that slow enzyme action are called **inhibitors.** A **competitive inhibitor** attaches itself to the active site. A **noncompetitive inhibitor** binds to other parts of the enzyme surface. The higher the enzyme and substrate concentrations, the higher the

enzyme activity, except that, at sufficiently high substrate concentrations, a saturation point is reached. After this, increasing substrate concentration no longer increases the rate. Each enzyme has an optimum temperature and pH at which it has its greatest activity.

Two closely related mechanisms by which enzyme activity and specificity are explained are the **lock-and-key model** and the **induced-fit model.**

Enzyme activity is regulated by four mechanisms. In **feedback control,** the concentration of products influences the rate of the reaction. In **allosterism,** an interaction takes place at a position other than the active site but affects the active site, either positively or negatively. Enzymes can be activated or inhibited by **protein modification.** Some enzymes, called **proenzymes** or **zymogens,** must be activated by removing a small portion of the polypeptide chain. Finally, enzyme activity is also regulated by **isozymes,** which are different forms of the same enzyme.

Abnormal enzyme activity can be used to diagnose certain diseases.

KEY TERMS

Activation (Section 22.3)
Active site (Section 22.3)
Allosteric enzyme (Section 22.6)
Allosterism (Section 22.6)
Apoenzyme (Section 22.3)
Coenzyme (Section 22.3)
Cofactor (Section 22.3)
Competitive inhibition
  (Section 22.3)
Enzyme (Section 22.1)

Enzyme activity (Section 22.4)
Enzyme-substrate complex
  (Section 22.5)
Feedback control (Section 22.6)
Induced-fit model (Section 22.5)
Inhibition (Section 22.3)
Isoenzyme (Section 22.6)
Isozyme (Section 22.6)
Lock-and-key model (Section 22.5)
Negative modulation (Section 22.6)

Noncompetitive inhibition
  (Section 22.3)
Positive modulation (Section 22.6)
Proenzyme (Section 22.6)
Protein modification (Section 22.6)
Regulator (Section 22.6)
Regulatory site (Section 22.6)
Ribozyme (Section 22.1)
Substrate (Section 22.3)
Zymogen (Section 22.6)

CONCEPTUAL PROBLEMS

22.A In most enzyme-catalyzed reactions, the rate of reaction reaches a constant value with increasing substrate concentration. This is described as a saturation curve diagram (Figure 22.4). If the enzyme concentration, on a molar basis, would be twice as much as the maximum substrate concentration, would you obtain a saturation curve?

22.B A bacterial enzyme has the following temperature-dependent activity.

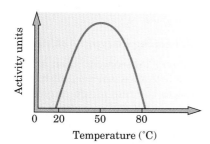

(a) Is this enzyme more or less active at normal body temperature than when a person has a fever?

(b) What happens to the enzyme activity if the patient's temperature is lowered to 35°C?

22.C The following reaction may be represented by the cartoon figures:

Glucose + ATP $\longrightarrow$ glucose-6-phosphate + ADP

In this enzyme-catalyzed reaction, Mg^{2+} is a cofactor, fluoroglucose is a competitive inhibitor, and Cd^{2+} is a noncompetitive inhibitor. Identify each component of the reaction by a cartoon figure and assemble them to show (a) normal enzyme reaction, (b) a competitive inhibition, and (c) a noncompetitive inhibition.

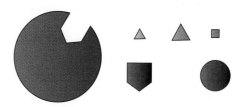

22.D The enzyme glycogen phosphorylase initiates the decomposition of glycogen. It comes in two forms: Phosphorylase b is less active, and phosphorylase a is more active. The difference between b and a is the modification of the apoenzyme. Phosphorylase a has two phosphate groups added to the polypeptide chain. In analogy with the pyruvate kinase discussed in the text, give a scheme indicating the transition between b and a. What enzymes and what cofactors control this reaction?

PROBLEMS

Difficult problems are designated by an asterisk.

Enzymes

22.1 What is the difference between the terms catalyst and enzyme?

22.2 What are ribozymes made of?

22.3 Approximately how many enzymes are in one single cell?

22.4 Compare the energy of activation of a reaction in uncatalyzed and in enzyme catalyzed reactions?

22.5 Why does the body need so many different enzymes?

22.6 Trypsin cleaves polypeptide chains at the carboxyl side of a lysine or arginine residue (Figure 22.1). Chymotrypsin cleaves polypeptide chains on the carboxyl side of an aromatic amino acid residue or any other nonpolar, bulky side chain. Which enzyme is more specific? Explain.

Naming and Classifying Enzymes

22.7 The text says lipases are less specific than trypsin. Explain why we can make such a statement.

22.8 Monoamine oxidases are important enzymes in brain chemistry. Judging from the name, which of these would be a suitable substrate for this class of enzymes:

(a) HO—⟨benzene ring⟩—CH(OH)—CH$_2$NH$_2$

(b) CH$_3$—C(=O)—N(CH$_3$)$_2$

(c) ⟨benzene ring⟩—NO$_2$

22.9 On the basis of the classification given in Section 22.2, decide to which group each of the following enzymes belongs:

(a) Phosphoglyceromutase

$$^-OOC—CH(OH)—CH_2—OPO_3^{2-} \rightleftharpoons$$

3-Phosphoglycerate

$$^-OOC—CH(OPO_3^{2-})—CH_2—OH$$

2-Phosphoglycerate

(b) Urease

$$H_2N—C(=O)—NH_2 + H_2O \rightleftharpoons 2NH_3 + CO_2$$

Urea

(c) Succinate dehydrogenase

$$^-OOC—CH_2—CH_2—COO^- + FAD \rightleftharpoons$$

Succinate   Coenzyme (oxidized form)

$$\text{C=C (Fumarate)} + FADH_2$$

Fumarate   Coenzyme (reduced form)

(d) Aspartase

$$\text{Fumarate} + NH_4^+ \rightleftharpoons$$

$$^-OOC—CH_2—CH(NH_3^+)—COO^-$$

L-Asparate

22.10 What kind of reaction does each of the following enzymes catalyze?
(a) Deaminases
(b) Hydrolases
(c) Dehydrogenases
(d) Isomerases

22.11 What is the difference between a coenzyme and a cofactor?

22.12 In the citric acid cycle, an enzyme converts succinate to fumarate (see reaction in Problem 22.9c). The enzyme consists of a protein portion and an organic molecule portion called FAD. What term do we use to refer to (a) the protein portion and (b) the organic molecule portion?

22.13 What is the difference between reversible and irreversible noncompetitive inhibition?

22.14 How does a heavy metal ion such as Hg^{2+} inactivate an enzyme?

Factors Affecting Enzyme Activity

22.15 At a very low concentration of a certain substrate, we find that, when the substrate concentration is doubled, the rate of the enzyme-catalyzed reaction is also doubled. Would you expect to find the same at a very high substrate concentration? Explain.

22.16 If we wish to double the rate of an enzyme-catalyzed reaction, can we do this by increasing the temperature 10°C? Explain.

22.17 Where can one find enzymes that are stable and active at 90°C?

22.18 The optimum temperature for the action of lactate dehydrogenase is 36°C. It is irreversibly inactivated at 85°C, but a yeast containing this enzyme can survive for months at −10°C. Explain how this can happen.

Mechanism of Enzyme Action

22.19 Urease can catalyze the hydrolysis of urea but not the hydrolysis of diethylurea. Explain why diethyl urea is not hydrolyzed.

$$H_2N—C(=O)—NH_2 \quad CH_3CH_2—NH—C(=O)—NH—CH_2CH_3$$

Urea   Diethylurea

22.20 Which is a correct statement describing the induced-fit model of enzyme action? Substrates fit into the active site

(a) Because both are exactly the same size and shape.

(b) By changing their size and shape to match those of the active site.

(c) By a change in the size and shape of the active site upon binding.

22.21 What is the maximum rate that can be achieved in competitive inhibition compared with noncompetitive inhibition?

22.22 Enzymes are long protein chains, usually containing more than 100 amino acid residues. Yet the active site contains only a few amino acids. Explain why all the other amino acids of the chain are present and what would happen to the enzyme activity if significant changes were made in the structure.

Enzyme Regulation

22.23 The hydrolysis of glycogen to yield glucose is catalyzed by the enzyme phosphorylase. Caffeine, which is not a carbohydrate and not a substrate for the enzyme, inhibits phosphorylase. What kind of regulatory mechanism is at work?

22.24 Can the product of a reaction in a sequence act as an inhibitor for another reaction of the sequence? Explain.

22.25 What is the difference between a zymogen and a proenzyme?

22.26 The enzyme trypsin is synthesized by the body in the form of a long polypeptide chain of 235 amino acids (trypsinogen) from which a piece must be cut before the trypsin can be active. Why does the body not synthesize the trypsin directly?

22.27 Give the structure of a tyrosyl residue of an enzyme modified by a protein kinase.

22.28 What is an isozyme?

Enzymes in Medical Diagnosis and Treatment

22.29 The enzyme formerly known as GPT (glutamate-pyruvate transaminase) has a new name: ALT (alanine aminotransferase). Looking at the equation in Section 27.9, which is catalyzed by this enzyme, what prompted this change of name?

22.30 If an examination of a patient indicated elevated levels of AST but normal levels of ALT, what would be your tentative diagnosis?

22.31 Which LD-P isozyme is monitored in the case of a heart attack?

22.32 Chemists who have been exposed for years to organic vapors usually show higher-than-normal activity when given the alkaline phosphatase test. What organ in the body is affected by organic vapors?

22.33 What enzyme preparation is given to patients after duodenal ulcer surgery?

22.34 Chymotrypsin is secreted by the pancreas and passed into the intestine. The optimal pH for this enzyme is 7.8. If a patient's pancreas cannot manufacture chymotrypsin, would it be possible to supply it orally? What happens to the chymotrypsin activity during its passage through the gastrointestinal tract?

Boxes

22.35 (Box 22A) Acetylcholine causes muscles to contract. Succinylcholine, a close relative, is a muscle relaxant. Explain the different effects of these related compounds.

22.36 (Box 22A) An operating team usually administers succinylcholine before bronchoscopy. What is achieved by this procedure?

22.37 (Box 22B) Which enzyme resembles papain in its action: trypsin or carboxypeptidase? Explain.

22.38 (Box 22C) How does *Helicobacter pylori* protect its enzymes against stomach acid?

22.39 (Box 22C) How do we treat ulcers caused by *Helicobacter pylori?*

22.40 (Box 22D) What is the role of Mn^{2+} in anchoring the substrate in the active site of protein kinase?

*__22.41__ (Box 22D) Which amino acids of the active site interact with the $=CH_2$ group of the phosphoenol pyruvate? Do both provide the same surface environment? What is the nature of the interaction?

22.42 (Box 22E) What part of folic acid is the key structure that sulfa drugs mimic?

*__22.43__ (Box 22E) Sulfa drugs kill certain bacteria by preventing the synthesis of folic acid, a vitamin. Why don't these drugs kill people?

22.44 (Box 22G) Which activity of the prostaglandin enderoperoxide synthase, PGHS, is inhibited by Tylenol and which is inhibited by aspirin?

Additional Problems

22.45 Food can be preserved by inactivation of enzymes that would cause spoilage, for example, by refrigeration. Give an example of food preservation in which the enzymes are inactivated by (a) heat and (b) lowering the pH.

22.46 Why is enzyme activity during myocardial infarction measured in the serum of patients rather than in the urine?

22.47 The activity of pepsin was measured at various pH values. The temperature and the concentrations of pepsin and substrate were held constant. The following activities were obtained:

pH	Activity
1.0	0.5
1.5	2.6
2.0	4.8
3.0	2.0
4.0	0.4
5.0	0.0

(a) Plot the pH dependence of pepsin activity. (b) What is the optimum pH? (c) Predict the activity of pepsin in the blood at pH 7.4.

22.48 What is the common characteristic of the amino acids of which the carboxyl groups of the peptide bonds can be hydrolyzed by trypsin?

22.49 Many enzymes are active only in the presence of Zn^{2+}. What common term is used for ions like this when discussing enzyme activity?

22.50 An enzyme has the following pH dependence:

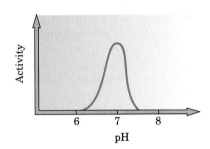

At what pH do you think this enzyme works best?

22.51 What enzyme is monitored in the diagnosis of infectious hepatitis?

22.52 The enzyme chymotrypsin catalyzes the following type of reaction:

$$R-CH-\overset{\overset{\displaystyle O}{\|}}{C}-NH-CH-R + H_2O$$

with CH_2 (phenyl) and CH_3 substituents

↓

$$R-CH-\overset{\overset{\displaystyle O}{\|}}{C}-OH + H_2N-CH-R$$

with CH (phenyl) and CH_3 substituents

On the basis of the classification given in Section 22.2, which group of enzymes does chymotrypsin belong to?

22.53 Nerve gases operate by forming covalent bonds at the active site of cholinesterase. Is this an example of competitive inhibition? Can the nerve gas molecules be removed by simply adding more substrate (acetylcholine) to the enzyme?

22.54 What would be the appropriate name for an enzyme that catalyzes each of the following reactions.

(a) $CH_3CH_2OH \longrightarrow CH_3\overset{\overset{\displaystyle O}{\|}}{C}-H$

(b) $CH_3\overset{\overset{\displaystyle O}{\|}}{C}-O-CH_2CH_3 + H_2O \longrightarrow$

$CH_3\overset{\overset{\displaystyle O}{\|}}{C}-OH + CH_3CH_2OH$

22.55 In Section 28.5, a reaction between pyruvate and glutamate, to form alanine and α-ketoglutarate, is given. How would you classify the enzyme that catalyzes this reaction?

22.56 A liver enzyme is made of four subunits: 2A and 2B. The same enzyme isolated from the brain has the following subunits 3A and 1B. What would you call these two enzymes?

22.57 What is the function of a ribozyme?

Chemical Communications: Neurotransmitters and Hormones

23.1   Introduction

Each cell in the body is an isolated entity enclosed in its own membrane. Furthermore, within each cell of higher organisms, organelles, such as the nucleus or the mitochondrion, are also enclosed by membranes separating them for the rest of the cell. If cells could not communicate with each other, the thousands of reactions in each cell would be uncoordinated. The same is true for organelles within a cell. Such communication allows the activity of a cell in one part of the body to be coordinated with the activity of cells in a different part of the body. There are three principal types of molecules for communications.

1. Receptors, which are protein molecules on the surface of cells embedded in the membrane;

Above: Scanning electron micrograph of neuron touching a muscle cell. (© Don Fawcett/Science Source/Photo Researchers, Inc.)

2. Chemical messengers, also called ligands, which interact with the receptors; and

3. Secondary messengers, which in many cases, carry the message from the receptor to the inside of the cell and amplify the message.

When your house is on fire and the fire is threatening your life, external signals—light, smoke, and heat—register alarm at specific receptors in your eyes, nose, and skin. From there the signals are transmitted by specific compounds to nerve cells, or **neurons.** In the neurons, the signals travel as electric impulses along the axons (Figure 23.1). When they reach the end of the neuron, the signals are transmitted to adjacent neurons by specific compounds called **neurotransmitters.** Communication between the eyes and the brain, for example, is by neural transmission.

As soon as the danger signals are processed in the brain, other neurons carry messages to the muscles and to the endocrine glands. The message to the muscles is to run away or to take some other action in response to the fire (save the baby or run to the fire extinguisher, for example). In order to do one of these things, the muscles must be activated. Again, neurotransmitters carry the messages from the neurons to the muscle cells and to endocrine glands. The endocrine glands are stimulated, and a different chemical signal, called a **hormone,** is secreted into your blood stream. "The adrenaline begins to flow." The danger signal carried by adrenaline makes quick energy available so that the muscles can contract and relax rapidly, allowing your body to take quick action to avoid the danger.

Without these chemical communicators, the whole organism—you—would not survive because there is a constant need for coordinated efforts to face a complex outside world. The chemical communication between different cells and different organs plays a role in the proper functioning of our bodies. Its significance is illustrated by the fact that **a large percentage of the drugs we encounter in medical practice try to influence this communication.** The scope of these drugs cover all fields from prescriptions against hypertension, heart disease, through antidepressants to pain killers, just to mention a few. Principally, there are four ways these drugs act in the body. A drug may affect either the receptor or the messenger.

1. An **antagonist** drug **blocks the receptor and prevents its stimulation.**

2. An **agonist** drug competes with the natural messenger for the receptor site. Once it is there, it **stimulates the receptor.**

3. Other drugs may **decrease the concentration of the messenger** by controlling the release of messengers from their storage.

4. Still others may **increase the concentration of the messenger** by inhibiting their removal from the receptors. A sampling of some drugs and their modes of action that affect neurotransmission is presented in Table 23.1.

23.2 Chemical Messengers. Neurotransmitters and Hormones

Neurotransmitters are compounds that communicate between two nerve cells or between a nerve cell and another cell (such as a muscle cell). If we look at a nerve cell (Figure 23.1), we see that it consists of a main cell body from which projects a long, fiber-like part called an **axon.** Coming off the other side of the main body are hair-like structures called **dendrites.**

Nerve cells are present throughout the body and, together with the brain, constitute the nervous system.

Neurotransmitter Chemical messengers between a neuron and another target cell: neuron, muscle cell, or cell of a gland

Hormone A chemical messenger released by an endocrine gland into the blood stream and transported there to reach its target cell

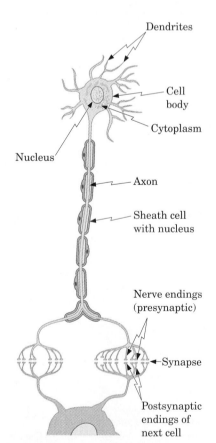

FIGURE 23.1 Neuron and synapse.

TABLE 23.1 Drugs That Affect Nerve Transmission

| Neurotransmitter | Drugs That Affect Receptor Sites | | Drugs That Affect Available Concentration of Neurotransmitters or Their Removal from Receptors | |
	Agonists (Activate Receptor Sites)	*Antagonists (Block Receptor Sites)*	*Increase Concentration*	*Decrease Concentration*
Acetylcholine (cholinergic)	Nicotine Pilocarpine Carbachol Succinylcholine	Curare Atropine Propantheline (Pro-Banthine)	Malathion Nerve gases Succinylcholine Donepezil (Aricept) Tacrine (Cognex)	*Clostridium botulinum* toxin
Norepinephrine (adrenergic)	Phenylephrine (Neo-Synephrine) Epinephrine (adrenalin)	Methyldopa (Aldomet) Propranolol (Inderal) Metoprolol (Lopressor) Selegiline (Deprenyl)	Amphetamines Iproniazide Antidepressants (Tofranil, Elavil)	Reserpine
Dopamine (adrenergic)	Methylphenidate (Ritalin)	Clozapine (Clorazil) Metoclopramide (Reglan) Promethazine (Phenergan)	(Deprenyl)	
Serotonin (adrenergic)		Ondansetron (Zofran)	Antidepressant Fluoxetin (Prozac)	
Histamine (adrenergic)	2-Methylhistamine Betazole Pentagastin	Fexofenadine (Allegra) Promethazine (Phenergan) Diphenylhydramine (Benadryl) Ranitidine (Zantac) Cimetidine (Tagamet)	Histidine	Hydrazino histidine
Glutamic acid (amino acid)	N-methyl D-aspartate	Phencyclidine		
Enkephalin (peptidergic)	Opiate	Morphine Heroin Demerol		Naloxone (Narcan)

Neurons do not typically touch each other. Between the axon end of one neuron and the cell body or dendrite end of the next, there is a space filled with an aqueous fluid. This fluid-filled space is called a **synapse.** If the chemical signal travels, say, from axon to dendrite, we call the nerve ends on the axon the **presynaptic** site. The neurotransmitters are stored at the presynaptic site in **vesicles,** which are small, membrane-enclosed packages. Receptors are located on the **postsynaptic** site of the cell body or the dendrite.

Synapse An aqueous small space between the tip of a neuron and its target cell

The neurotransmitters fit into the receptor sites in a manner reminiscent of the lock-and-key model mentioned in Section 22.5.

TABLE 23.2 The Principal Hormones and Their Action

Gland	Hormone	Action	Structures Shown in
Parathyroid	Parathyroid hormone	Increases blood calcium Excretion of phosphate by kidney	
Thyroid	Thyroxine (T$_4$) Triiodothyronine (T$_3$)	Growth, maturation, and metabolic rate Metamorphosis	Box 14C
Pancreatic islets Beta cells	Insulin	Hypoglycemic factor Regulation of carbohydrates, fats, and proteins	Section 21.7 and Box 23G
Alpha cells	Glucagon	Liver glycogenolysis	
Adrenal medulla	Epinephrine Norepinephrine	Liver and muscle glycogenolysis	Section 23.5
Adrenal cortex	Cortisol Aldosterone Adrenal androgens	Carbohydrate metabolism Mineral metabolism Androgenic activity (esp. females)	Section 20.20 Section 20.20
Kidney	Renin	Hydrolysis of blood precursor protein to yield angiotensin	
Anterior pituitary	Luteinizing hormone Interstitial cell-stimulating hormone Prolactin Mammotropin	Causes ovulation Formation of testosterone and progesterone in interstitial cells Growth of mammary gland Lactation Corpus luteum function	
Posterior pituitary	Vasopressin Oxytocin	Contraction of blood vessels Kidney reabsorption of water Stimulates uterine contraction and milk ejection	Section 21.7 Section 21.7
Ovaries	Estradiol Progesterone	Estrous cycle Female sex characteristics	Section 20.20 Section 20.20
Testes	Testosterone Androgens	Male sex characteristics Spermatogenesis	Section 20.20

Hormones are diverse compounds secreted by specific tissues (the endocrine glands), released into the blood stream, and then adsorbed onto specific receptor sites, usually relatively far from their source. This is the physiological definition of a hormone. Some of the principal hormones are listed in Table 23.2. Figure 23.2 shows the target organs of hormones secreted by the pituitary gland.

The distinction between hormones and neurotransmitters is physiological and not chemical. Whether a certain compound is considered to be a neurotransmitter or a hormone depends on whether it acts over a short distance across a synapse (2×10^{-6} cm), in which case it is a neurotransmitter, or over a long distance (20 cm) from secretory gland through the blood stream to target cell, in which case it is a hormone. For example, epinephrine and norepinephrine are neurotransmitters as well as hormones.

There are, broadly speaking, five classes of chemical messengers: *cholinergic, amino acid, adrenergic, peptidergic,* and *steroid* messengers. This classification is based on the chemical nature of the important messenger (ligand) in each group. Neurotransmitters can belong to the first four classes, and hormones can belong to the last three classes.

Messengers can also be classified according to how they work. Some of them—epinephrine, for example—*activate enzymes.* Others affect the *synthesis of enzymes and proteins* by working on the transcription of genes (Section 25.2); steroid hormones (Section 20.10) work in this manner. Finally, some affect the *permeability of membranes;* acetylcholine, insulin, and glucagon belong to this class.

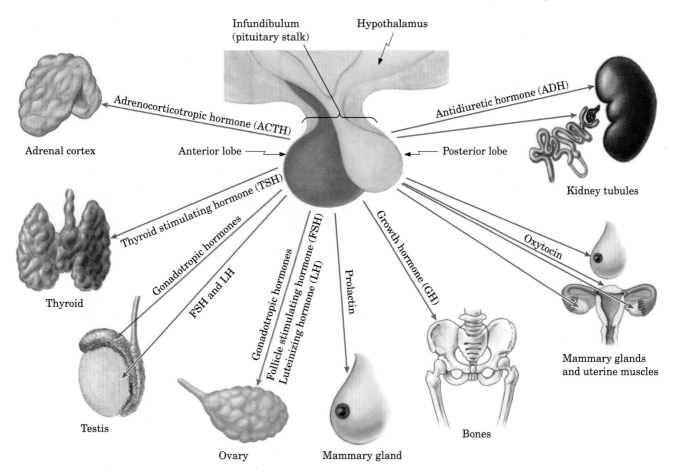

FIGURE 23.2 The pituitary gland is suspended from the hypothalamus by a stalk of neural tissue. The hormones secreted by the anterior and posterior lobes of the pituitary gland and the target tissues they act upon are shown. *(Modified from P. W. Davis and E. P. Solomon,* The World of Biology. *Philadelphia: Saunders College Publishing, 1986)*

Still another way of classifying messengers is according to their potential to *act directly* or through a *secondary messenger*. The steroid hormones act directly. They can penetrate the cell membrane and also pass through the membrane of the nucleus. For example, estradiol stimulates uterine growth.

Other chemical messengers act through secondary messengers. For example, epinephrine, glucagon, luteinizing hormone, norepinephrine, and vasopressin use cAMP as a secondary messenger.

In the following sections, we will sample the mode of communication within each of the five chemical categories of messengers.

23.3 Cholinergic Messenger. Acetylcholine

The main **cholinergic neurotransmitter** is acetylcholine:

$$CH_3-\overset{\overset{\displaystyle O}{\|}}{C}-O-CH_2-CH_2-\overset{\overset{\displaystyle CH_3}{|}}{\underset{\underset{\displaystyle CH_3}{|}}{N^+}}-CH_3$$

Acetylcholine

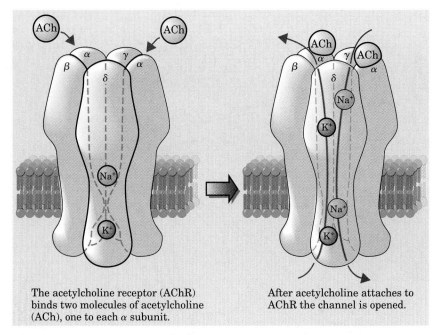

The acetylcholine receptor (AChR) binds two molecules of acetylcholine (ACh), one to each α subunit.

After acetylcholine attaches to AChR the channel is opened.

FIGURE 23.3 Acetylcholine in action. The receptor protein has five subunits. When two molecules of acetylcholine bind to the two α subunits, a channel opens to allow the passage of Na^+ and K^+ ions. *(Courtesy of Anthony Tu, Colorado State University)*

There are two kinds of receptors for this messenger. We shall look at one that exists on the motor end plates of skeletal muscles or in the sympathetic ganglia. The nerve cells that bring messages contain stored acetylcholine in the vesicles in their axons. The receptor on the muscle cells or neurons is also known as nicotinic receptor because nicotine (see Box 16B) inhibits the neurotransmission of these nerves. The receptor itself is a *transmembrane protein* (Figure 20.2) made of five different subunits. The central core of the receptor is an ion channel through which, when open, Na^+ and K^+ ions can pass (Figure 23.3). When ion channels are closed, the K^+ ion concentration is higher inside the cell than outside; the reverse is true for Na^+ ion concentration.

Events begin when a message is transmitted from one neuron to the next by neurotransmitters. The message is initiated by **calcium ions** (see Box 23A). When the Ca^{2+} concentration in a neuron reaches a certain level (more than $10^{-4}\,M$), the vesicles containing acetylcholine fuse with the presynaptic membrane of the nerve cells. Then they empty the neurotransmitters into the synapse. The messenger molecules travel across the synapse and are adsorbed onto specific receptor sites.

The presence of the acetylcholine molecules at the postsynaptic receptor site then triggers a conformational change (Section 21.11) in the receptor protein. This opens the *ion channel* and allows ions to cross membranes freely. Na^+ ions have higher concentration outside the neuron than K^+ ions; thus, more Na^+ enters the cell than K^+ leaves. Because it involves ions, which carry electric charges, this process is translated into an electric signal. After a few milliseconds, the channel closes again. The acetylcholine still occupies the receptor. In order that the channel should reopen and transmit a new signal, the acetylcholine must be removed, and the neuron must be reactivated. The acetylcholine

BOX 23A

Calcium As a Signaling Agent (Secondary Messenger)

The message delivered to the receptors on the cell membranes by neurotransmitters or hormones must be delivered intracellularly to the different locations within the cell. One mode of intracellular signaling is depicted in Figure 23.4 where cAMP acts as a second messenger or signaling agent. The most universal and, at the same time, the most versatile signaling agent is the cation Ca^{2+}.

Calcium ions in the cells come from either extracellular sources or intracellular stores, such as the endoplasmic reticulum. If it comes from the outside, it enters the cell through specific calcium channels. Calcium ion controls our heart beats, our movements through the action of skeletal muscles, and through the release of neurotransmitters in our neurons, learning, and memory. It is also involved in signaling the beginning of life at fertilization and its end at death. Calcium ion signaling controls these different functions by two mechanisms: (1) increased concentration (forming sparks and puffs) and (2) duration of the signals.

An increase in calcium ion concentration may take the form of sparks or puffs. The source of calcium ions may be external (calcium influx caused by the electric signal of nerve transmission) or internal (calcium released from the stores of endoplasmic reticulum). At the signal of calcium puffs, the vesicles storing acetylcholine travel to the membrane of the presynaptic cleft. There they fuse with the membrane and empty their contents into the synapse.

Calcium ions can also control signaling by the duration of the signal. The signal in arterial smooth muscle lasts for 0.1 to 0.5 s. The global wave of Ca^{2+} in the liver lasts 10 to 60 s. The calcium wave in the human egg lasts 1 to 35 min after fertilization. Thus, combining the concentration, localization, and duration of the signal, calcium ion can deliver messages to perform a variety of functions.

The effect of Ca^{2+} is modulated through specific calcium-binding proteins. In all nonmuscle cells and in the smooth muscle, **calmodulin** is the **calcium-binding protein.** Calmodulin-bound calcium activates an enzyme, protein kinase II, which then phosphorylates an appropriate protein substrate. Thus, the signal is translated into metabolic activity.

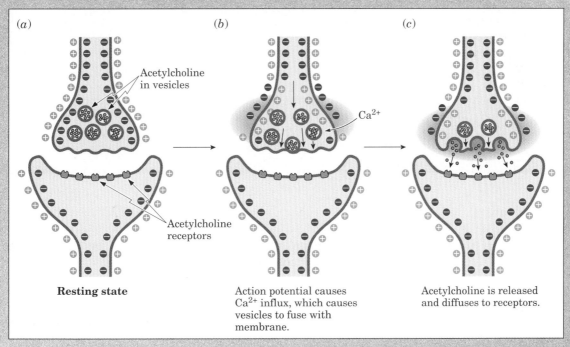

■ Calcium signaling to release acetylcholine from its vesicles.

Nerve Gases and Antidotes

Most nerve gases in the military arsenal exert their lethal effect by binding to acetylcholinesterase. Under normal conditions, this enzyme hydrolyzes the synaptic neurotransmitter acetylcholine within a few milliseconds after it is released at the nerve endings. Nerve gases such as Sarin (agent GB, also called Tabun), Soman (agent GD), and agent VX are organic phosphonates related to such pesticides as parathion (the latter being much less lethal, of course):

Sarin
(Agent GB)

Soman
(Agent GD)

Agent VX

Parathion

If any of these phosphonates bind to acetylcholinesterase, the enzyme is irreversibly inactivated, and the transmission of nerve signals stops. The result is a cascade of symptoms: sweating, bronchial constriction due to mucus buildup, dimming of vision, vomiting, choking, convulsions, paralysis, and respiratory failure. Direct inhalation of as little as 0.5 mg can cause death within a few min. If the dosage is less or if the nerve gas is absorbed through the skin, the lethal effect may take several hours. In warfare, protective clothing and gas masks are effective countermeasures. An enzyme of the bacteria *Pseudomonas diminuta*, organophosphorous hydroxylase, is capable of detoxifying nerve gases. Their inclusion in cotton towelettes may help to remove nerve gases from contaminated skin. Also, first-aid kits containing antidotes that can be injected are available. These antidotes contain the alkaloid atropine and pralidoxime chloride. These substances must be used in tandem and very quickly after exposure to nerve gases.

The nations of the world, with the exception of a few Arab countries, recognizing the dangers of proliferating these and other chemicals used for warfare, signed a treaty in 1993. The treaty obliges all the signatory governments to destroy their stockpiles of chemical weapons and to dismantle the plants that manufacture them within 10 years.

is removed rapidly from the receptor site by the enzyme *acetylcholinesterase*, which hydrolyzes it.

Acetylcholine

Acetate   Choline

Botulism and Acetylcholine Release

When meat or fish is improperly cooked or preserved, a deadly food poisoning, called botulism, may result. The culprit is the bacterium *Clostridium botulinum*, whose toxin prevents the release of acetylcholine from the presynaptic vesicles. Therefore, no neurotransmitter reaches the receptors on the surface of muscle cells, and the muscles no longer contract. If untreated, the person may die. Surprisingly, the botu-

lin toxin actually has a medical use. It is used in the treatment of involuntary muscle spasms, for example, in facial tics. These tics are caused by the uncontrolled release of acetylcholine. Therefore, controlled administration of the toxin, applied locally to the facial muscles, stops the uncontrolled contractions and relieves the facial distortions.

■ *Clostridium botulinum* is a food-poisoning bacterium. *(Alfred Pasieka/Science Photo Library/Photo Researchers, Inc.)*

This rapid removal enables the nerves to transmit more than 100 signals per second. By this means, the message moves from neuron to neuron until finally it gets transmitted, again by acetylcholine molecules, to the muscles or endocrine glands that are the ultimate target of the message.

The action of the acetylcholinesterase enzyme is obviously essential to the whole process. When this enzyme is inhibited, the removal of acetylcholine is incomplete, and nerve transmission ceases. Acetylcholinesterase is inhibited irreversibly by phosphonates in nerve gases and pesticides (Box 23B) or reversibly by succinylcholine (Box 22A) and decamethonium bromide.

$$CH_3 \overset{+}{-}\underset{\underset{CH_3}{|}}{\overset{\overset{CH_3}{|}}{N}}-CH_2(CH_2)_8CH_2 \overset{+}{-}\underset{\underset{CH_3}{|}}{\overset{\overset{CH_3}{|}}{N}}-CH_3$$

$$Br^- \qquad\qquad Br^-$$

Decamethonium bromide

Succinylcholine and decamethonium bromide resemble the choline end of acetylcholine and therefore act as a competitive inhibitor of acetylcholinesterase. In small doses, these reversible inhibitors relax the muscles temporarily and are used as a muscle relaxant in surgery. In large doses, they are just as deadly as the irreversible inhibitors.

BOX 23D

Alzheimer's Disease and Acetylcholine Transferase

Alzheimer's disease is the name given to the symptoms of senile behavior (second childhood) that afflict about 1.5 million older people in the United States. One of the post mortem indications for this disease is the presence of a neurofibrillary tangle. The tangle disrupts nerve cell functions, creating distrophied neurons. These neurons congregate, and plaques that are filled with a protein called β-amyloid form between them. This protein is water-soluble in its original form (β-amyloid precursor protein, β-APP). A 40 to 42 amino-acid-containing section is cut from β-APP. After changing its conformation to a β-pleated sheet, it self-assembles into water-insoluble aggregates and forms plaques.

The result is that these nerve cells in the cerebral cortex die, the brain becomes smaller, and part of the cortex atrophies. People with Alzheimer's disease are forgetful, especially about recent events. As the disease advances, they become confused and, in severe cases, lose their ability to speak; then they need total care. There is as yet no cure for this disease.

It has been found that patients with Alzheimer's disease have significantly diminished acetylcholine transferase activity in their brains. This enzyme synthesizes acetylcholine by transferring the acetyl group from acetyl-CoA to choline:

The diminished concentration of acetylcholine can be partially compensated for by inhibiting the enzyme acetylcholinesterase, which decomposes acetylcholine. Certain drugs, which act as acetylcholinesterase inhibitors, have been shown to improve memory and other cognitive functions in some people with the disease. Two such drugs, Cognex (tacrine) and Aricept (donepezil), are already on the market. The alkaloid huperzine A, an active ingredient of Chinese herb tea, used for centuries to improve memory, is also a potent inhibitor of acetylcholinesterase.

(a)

(b)

■ (a) Normal brain. (b) Brain of a person with Alzheimer's disease. *(Courtesy of Anatomical Collections, National Museum of Health and Medicine, Armed Forces Institute of Pathology, Washington, D.C.)* (c) PET scans comparing a normal brain with the brain of a person with Alzheimer's disease. *(Science VU/Visuals Unlimited)*

The inhibition of acetylcholinesterase is but one way in which cholinergic neurotransmission is controlled. Another way is to modulate the action of the receptor. Since acetylcholine enables the ion channels to open and propagate signals, this mode of action is called *ligand-gated ion channels*. The attachment of the ligand to the receptor is critical in signaling. Nicotine in low doses is a stimulant; it is an agonist because it prolongs the receptor's biochemical response. In large doses, however, nicotine becomes an antagonist and blocks the action on the receptor. As such, it may cause convulsion and respiratory paralysis. Succinylcholine, besides being a reversible inhibitor of acetylcholinesterase, also has this concentration-dependent agonist/antagonist effect on the receptor. A strong antagonist, which blocks the receptor completely, can interrupt the communication between neuron and muscle cell. The venom of a number of snakes, especially that of the cobra, cobratoxin, exerts its deadly influence in this manner. This is also how the plant extract curare works, which was used in poisoned arrows by the Amazon Indians. In small doses, curare is also used as a muscle relaxant.

Finally, the supply of the acetyl choline messenger can also influence the proper nerve transmission. If the acetyl choline messenger is not released from its storage as in botulism (Box 23C) or if its synthesis is impaired as in Alzheimer's disease (Box 23D), the concentration of acetylcholine is reduced, and nerve transmission is impaired.

23.4 Amino Acids as Messengers. Neurotransmitters

Amino acids are distributed throughout the neurons individually or as parts of peptides and proteins. They can also act as neurotransmitters. Some of them, such as glutamic acid, aspartic acid, and cysteine, act as **excitatory neurotransmitters** similar to acetylcholine and norepinephrine. Others, such as glycine, β-alanine, taurine, and mainly γ-aminobutyric acid (GABA), are **inhibitory neurotransmitters;** they reduce neurotransmission. Note that some of these neurotransmitter amino acids are not found in proteins.

$$^+H_3NCH_2CH_2SO_3^-$$
Taurine

$$^+H_3NCH_2CH_2COO^-$$
β-Alanine

$$^+H_3NCH_2CH_2CH_2COO^-$$
γ-Aminobutyric acid
(GABA)
(IUPAC name:
4-aminobutanoic acid)

Each of these amino acids has its own receptor; as a matter of fact, glutamic acid has at least five subclasses of receptors. The best known among them is the *N*-methyl-D-aspartate (NMDA) receptor. It is a *ligand-gated ion channel* similar to the nicotinic cholinergic receptor we discussed in Section 23.3:

$$
\begin{array}{c}
CH_3 \\
| \\
NH_2^+ \\
| \\
CHCH_2—COOH \\
| \\
COO^-
\end{array}
$$
N-methyl-D-aspartate

When glutamic acid binds to this receptor, the ion channel opens, Na^+ and Ca^{2+} flow in to the neuron, and K^+ ion flows out of the neuron. The same thing happens when NMDA, being an agonist, stimulates the receptor. The gate of this channel is closed by a Mg^{2+} ion.

Phencyclidine (PCP) is an antagonist of this receptor, and it induces hallucination. PCP, known by the street name of angel dust, is a controlled substance; it causes bizarre psychotic behavior and long-term psychological problems.

In contrast to acetylcholine, there is no enzyme that would degrade glutamic acid and thereby remove it from its receptor once the signaling has been done. Glutamic acid is removed by **transporter** molecules bringing it back through the presynaptic membrane into the neuron. This process is called **reuptake.**

> **Transporter** A protein molecule carrying small molecules, such as glucose or glutamic acid, across a membrane

23.5 Adrenergic Messengers (Neurotransmitters/Hormones). Monoamines

The second class of neurotransmitters/hormones, the adrenergic messengers, includes such monoamines as epinephrine, serotonin, dopamine, and histamine. (Structures of these compounds can be found later in this section and in Box 23E.) These monoamines transmit signals by a mechanism whose beginning is similar to the action of acetylcholine. But once the monoamine neurotransmitter/hormone (for example, norepinephrine) is adsorbed onto the receptor site, the signal will be amplified inside the cell. In the example of Figure 23.4, the receptor has an associated protein called

Martin Rodbell (1925–1998) of the National Institutes of Health was one of the pioneers in discovering and naming this process. For this discovery, he was awarded the Nobel Prize in Physiology and Medicine in 1994.

FIGURE 23.4 The sequence of events in the postsynaptic membrane when norepinephrine is absorbed onto the receptor site. (*a*) The active G-protein hydrolyzes GTP. The energy of hydrolysis of GTP to GDP activates adenylate cyclase. A molecule of cAMP is formed when adenylate cyclase cleaves ATP into cAMP and pyrophosphate. (*b*) Cyclic AMP activates protein kinase by dissociating the regulatory (R) unit from the catalytic unit (C). A second molecule of ATP, shown in (*b*), has phosphorylated the catalytic unit and has been converted to ADP. (*c*) The catalytic unit phosphorylates the ion-translocating protein that blocked the channel for ion flow. The phosphorylated ion-translocating protein changes its shape and position and opens the ion gates.

G-protein. This protein is the key to the cascade that produces many signals inside the cell (amplification). The whole process that occurs after the messenger binds to the receptor is called **signal transduction.** The active G-protein has an associated nucleotide, guanosine triphosphate, GTP. This is an analog of adenosine triphosphate, ATP, in which the aromatic base adenine is substituted by guanidine (Section 24.2). The G-protein becomes inactive when its associated nucleotide is hydrolyzed to GDP (guanosine diphosphate). Signal transduction starts with the active G-protein. It activates the enzyme adenylate cyclase.

Adenylate cyclase produces a secondary messenger inside the cell, called cyclic AMP (cAMP). The manufacture of cAMP activates processes that result in the transmission of an electrical signal. The cAMP is manufactured by adenylate cyclase from ATP:

> **Signal transduction** A cascade of events through which the signal of a neurotransmitter or hormone delivered to its receptor is carried inside the target cell and amplified into many signals that can cause protein modifications, enzyme activation, and the opening of membrane channels

Adenosine triphosphate
(ATP)

Cyclic-Adenosine monophosphate   Pyrophosphate
(cAMP)

G-protein also participates in another signal transduction cascade. This involves inositol-based compounds (Section 20.6) as signaling molecules. Phosphoinositoldiphosphate, PIP_2, mediates the action of hormones and neurotransmitters. They can phosphorylate enzymes similarly to the c-AMP cascade. They also play an important role in the release of calcium ions from their storage areas in the endoplasmic reticulum (ER) or sarcoplasmic reticulum (SR).

The activation of adenylate cyclase accomplishes two important goals: (1) It converts an event occurring at the outer surface of the target cell (adsorption onto receptor site) to a change inside the target cell (release of cAMP). Thus, the primary messenger (neurotransmitter or hormone) does not have to cross the membrane. (2) It amplifies the signal. One molecule adsorbed on the receptor triggers the adenylate cyclase to make many cAMP molecules. Thus, the signal is amplified many thousands of times.

How does this signal amplification stop? When the neurotransmitter or hormone dissociates from the receptor, the adenylate cyclase stops the manufacture of cAMP. The cAMP already produced is destroyed by the enzyme phosphodiesterase, which catalyzes the hydrolases of the phosphoric ester bond, yielding AMP.

The amplification through the secondary messenger (cAMP) is a relatively slow process. It may take from 0.1 s to a few minutes. Therefore, in cases where the transmission of signals must be fast, in ms or μs, a neurotransmitter such as acetylcholine acts on membrane permeability directly, without the mediation of a secondary messenger.

The G-protein-adenylate cyclase cascade in transduction signaling is not limited to monoamine messengers. A wide variety of peptide hormones and neurotransmitters (Section 23.6) use this signaling pathway. Among those are glucagon, vasopressin, luteinizing hormone, enkephalins, P-protein. Neither is the opening of ion channels, depicted in Figure 23.4, the only target of these signaling. A number of enzymes can be phosphorylated by protein kinases, and the phosphorylation controls whether these enzymes will be active or inactive (Box 27F).

The fine control of the G-protein-adenylate cyclase cascade is essential for health. The toxin of the bacteria, *Vibrio cholerae,* permanently **activates G-protein.** The result is the symptoms of cholera, namely severe dehydration as a result of diarrhea. This happens because the activated G-proteins overproduce cAMP. This, in turn, opens the ion channels, which produces a large outflow of Na^+ ions and accompanying water from the epithelial cells to the intestines. Therefore, the first aid for cholera victims is to replace the lost water and salt.

The **inactivation of the adrenergic neurotransmitters** is somewhat different from that of the cholinergic transmitters. While acetylcholine is decomposed by acetylcholinesterase, most of the adrenergic neurotransmitters are inactivated in a different way. *The body inactivates monoamines by oxidizing them to aldehydes.* Enzymes that catalyze these reactions are very common in the body. They are called monoamine oxidases (MAOs). For example, there is an MAO that converts both epinephrine and norepinephrine to the corresponding aldehyde:

Epinephrine (salt form)   MAO   MAO   Norepinephrine (salt form)

Many drugs that are used as antidepressants or antihypertensive agents are inhibitors of monoamine oxidases, for example, Tofranil and Elavil. These inhibitors prevent MAOs from converting monoamines to aldehydes, thus increasing the concentration of the active adrenergic neurotransmitters.

Shortly after adsorption onto the postsynaptic membrane, the neurotransmitter comes off the receptor site and is reabsorbed through the presynaptic membrane and stored again in the vesicles.

The neurotransmitter histamine is present in mammalian brains.

Histamine cannot readily pass the blood-brain barrier (Section 31.1) and must be synthesized in the brain neurons by a one-step decarboxylation of the amino acid histidine.

Histidine H^+ →   Histamine $+ CO_2$

Parkinson's Disease. Depletion of Dopamine

Parkinson's disease is characterized by spastic motion of the eyelids as well as rhythmic tremors of the hands and other parts of the body, often when the patient is at rest. The posture of the patient changes to a forward, bent-over position; walking becomes slow, with shuffling footsteps. This is a degenerative nerve disease. The neurons affected contain, under normal conditions, mostly dopamine as a neurotransmitter. People with Parkinson's disease have depleted amounts of dopamine in their brains. However, the dopamine receptors are not affected. Thus, the first line of remedy is to *increase the concentration of dopamine.* Dopamine cannot be administered directly because it cannot penetrate the blood-brain barrier and therefore does not reach the tissue where its action is needed. L-Dopa, on the other hand, is transported through the arterial wall and is converted to dopamine in the brain:

HO—⬡—CH₂—CH(COO⁻)(NH₃⁺) ... (S)-3,4-Dihydroxyphenylalanine (L-Dopa)

$$\text{(S)-3,4-Dihydroxyphenylalanine (L-Dopa)} \xrightarrow[\text{decarboxylation}]{\text{enzyme-catalyzed}} \text{Dopamine} + CO_2$$

When L-dopa is administered, many of these patients are able to synthesize dopamine and resume normal nerve transmission. In these patients, L-dopa reverses the symptoms of Parkinson's disease, although the respite is only temporary. In other patients, the L-dopa regimen provides little benefit.

Recently, a new drug treatment has shown promise as a possible route for slowing neuron degeneration. (R)-Selegiline (Deprenyl), a monoamine oxidase (MAO) inhibitor, given together with L-dopa, reduced the symptoms of Parkinson's disease and even increased the life span of patients. It appears that Deprenyl not only increases the level of dopamine by *preventing its oxidation by MAOs* but also reduces the degeneration of neurons.

Certain drugs designed to affect one neurotransmitter may also affect another. An example is the drug methylphenidate, Ritalin. This drug in higher doses promotes the dopamine concentration in the brain and acts as a stimulant. On the other hand, the same drug is prescribed in small doses to calm hyperactive children or to minimize ADD (attention deficit disorder). It seems that in smaller doses Ritalin raises the concentration of serotonin. This neurotransmitter decreases hyperactivity without affecting the dopamine levels of the brain.

Serotonin

The close connection between two monoamine neurotransmitters, dopamine and serotonin, is also evident in controlling nausea and vomiting which often follow general anesthesia and chemotherapy. Blockers of dopamine receptors in the brain, such as Reglan or Phenergan, can and do alleviate the symptoms after anesthesia. However, a blocker of serotonin receptors in the brain as well as on the terminals of the vagus nerve in the stomach, such as Zofran, is the drug of choice against vomiting induced by chemotherapy.

Synthesis and degradation of dopamine is not the only way the brain keeps its concentration at a steady state. The *concentration is also controlled by* specific proteins, called *transporters,* that ferry the used dopamine from the receptor back across the synapse into the original neuron for reuptake. Cocaine addiction (Box 18E) works through such a transporter. Cocaine binds to the dopamine transporter, like a reversible inhibitor, preventing the reuptake of dopamine. Thus, the dopamine is not transported back to the original neuron and stays in the synapse, increasing the continuous firing of signals, which is the psychostimulatory effect associated with a cocaine "high."

The action of histamine as a neurotransmitter is very similar to that of other monoamines. There are two kinds of receptors for histamine. One receptor, H_1, can be blocked by antihistamines such as dimenhydrinate (Dramamine) or diphenhydramine (Benadryl). The other receptor, H_2, can be blocked by ranitidine (Zantac) and cimetidine (Tagamet). H_1 receptors are found in the respiratory tract. They affect the vascular, muscular, and secretory changes associated with hay fever and asthma. Therefore, the

Antihistamines block the H_1 receptor for histamine. *(George Semple)*

Sir James W. Black of England received the 1988 Nobel Prize in Medicine for the invention of cimetidine and such other drugs as propranolol (Table 24.1).

antihistamines blocking H_1 receptors relieve these symptoms. The H_2 receptors are mainly in the stomach and affect the secretion of HCl. Cimetidine and ranitidine, both H_2 blockers, reduce acid secretion and, thus, are effective drugs for ulcer patients. However, the real cure is to kill the bacteria *Helicobacter pylori* (Box 22C) by antibiotics or by a combination of antibiotics and H_2 blockers.

23.6 Peptidergic Messengers. Neurotransmitters/Hormones

In the last few years, scientists have isolated a number of brain peptides that have affinity for certain receptors and, therefore, act as if they were neurotransmitters. Some 25 or 30 such peptides are now known.

The first brain peptides isolated were the **enkephalins.** These pentapeptides are present in certain nerve cell terminals. They bind to specific pain receptors and seem to control pain perception. Since they bind to the receptor site that also binds the pain-killing alkaloid morphine (Box 16D), it is assumed that it is the N-terminal end of the pentapeptide that fits the receptor (Figure 23.5). Even though morphine remains the most effective in reducing pain, its clinical use is limited because of its side effects. These include respiratory depression, constipation, and mainly addiction. The clinical use of enkephalins yielded only modest relief. The challenge is to develop analgesic drugs that do not involve the opiate receptors in the brain.

Another brain peptide, **neuropeptide Y,** affects the hypothalamus, a region that integrates the body's hormonal and nervous systems. Neuropeptide Y is a potent orexic (appetite-stimulating) agent. When its receptors are blocked, for example by leptin, the "thin" protein, appetite is suppressed. Leptin is an anorexic agent.

Still another **peptidergic neurotransmitter** is **substance P** (P for pain). This 11-amino-acid peptide is involved in transmission of pain signals. In injury or inflammation, sensory nerve fibers transmit signals from the peripheral nervous system (where the injury occurred) to the spinal cord, which processes the pain. The peripheral neurons synthesize and release substance P, which bonds to receptors on the surface of the spinal cord. The substance P in its turn removes the magnesium block at the *N*-methyl-D-aspartate (NMDA) receptor. Glutamic acid, an excitatory amino acid, can now bind to this receptor. In doing so, it amplifies the pain signal going to the brain.

Morphine

Methionine enkephalin

FIGURE 23.5 Similarities between the structure of morphine and that of the brain's own pain regulators, the enkephalins.

Nitric Oxide As a Secondary Messenger

The simple gaseous molecule NO has been long known for its toxic effect (see Box 3C). Therefore, it came as a surprise to find that it plays a major role in chemical communications. This simple molecule is synthesized in the cells when arginine is converted to citrulline. (These two compounds appear in the urea cycle, Section 27.8). Nitric oxide is a relatively non-polar molecule, and shortly after it has been produced in the nerve cell, it quickly diffuses across the lipid bilayer membrane. Within its short half-life, 4 to 6 s, it can reach a neighboring cell. Because NO passes through membranes, it does not need extracellular receptors to deliver its message. NO is very unstable; thus, there is no need for special mechanism for its destruction.

NO acts as an intercellular messenger between the endothelial cells surrounding the blood vessels and the smooth muscles surrounding these cells. Nitric oxide relaxes the muscle cells and, thereby, dilates the blood vessels. The consequence of this is less-restricted blood flow and a drop in the blood pressure. This is also the reason that nitroglycerin (Box 13A) is effective against angina since it produces NO in the body.

Another role of NO in dilating blood vessels is remedying impotence. The new impotence-relieving drug, Viagra, enhances the activity of NO by inhibiting an enzyme (phosphodiesterase) that otherwise would reduce the effect of NO on the smooth muscles. When the NO concentration is suffi-ciently high, the blood vessels dilate allowing enough blood to flow to provide an erection. In most cases, this happens within an hour after taking the pill.

Sometimes the dilation of blood vessels is not so beneficial. Headaches are caused by dilated arteries in the head. NO-producing compounds in the food—nitrites in smoked and cured meats and sodium glutamate in seasoning (Box 27D)—can cause headache. Nitroglycerin itself often induces headaches.

Nitric oxide is toxic, as we have seen in Box 3C. This toxicity is used by our immune system (Section 30.10) to fight infections by viruses.

The toxic effect of NO is also evident in strokes. A blocked artery restricts the blood flow to certain parts of the brain; this causes the oxygen-starved neurons to die. Then, neurons in the surrounding area, ten times larger than the place of the initial attack, release glutamic acid, which stimulates other cells. They, in turn, release NO, which kills all the cells in the area. Thus, the damage to the brain is spread tenfold. A concentrated effort is under way to find inhibitors against the NO-producing enzyme, nitric oxide synthase, that can be used as an antistroke drug. For the discovery of NO and its role in blood pressure control, three pharmacologists—Robert Furchgott, Louis Ignarro, and Ferid Murad—received the 1998 Nobel Prize in Physiology.

Diabetes

The disease diabetes mellitus affects about 10 million people in the United States. In a normal person, the pancreas, a large gland behind the stomach, secretes the hormone insulin, as well as other hormones. Diabetes usually results from low insulin secretion. Insulin is necessary for glucose molecules to penetrate such cells as brain, muscle, and fat cells, where they can be used. Insulin accomplishes this task by being adsorbed onto the receptors in the target cells. This adsorption triggers the manufacture of cyclic GMP (not cAMP), and this secondary messenger increases the transport of glucose molecules into the target cells.

In diabetic patients, the glucose level rises to 600 mg/100 mL of blood or higher (normal is 80 to 100 mg/100 mL). There are two kinds of diabetes. In insulin-dependent diabetes, patients do not manufacture enough of this hormone in the pancreas. This disease develops early, before the age of 20, and must be treated with daily injections of insulin. Even with daily injections of insulin, the blood sugar level fluctuates, and the fluctuation may cause other disorders, such as cataracts, blindness, kidney disease, heart attack, and nervous disorders.

In non-insulin-dependent diabetes, the patient has enough insulin in the blood but cannot utilize it properly because there is an insufficient number of receptors in the target cells. These patients usually develop the disease after age 40 and are likely to be obese. Overweight people usually have a lower-than-normal number of insulin receptors in their adipose (fat) cells.

Several oral drugs help this second type of diabetic patient. These are mostly sulfonyl urea compounds, such as the compound shown below.

These drugs seem to control the symptoms of diabetes, but it is not known exactly how.

Tolbutamine
(Orinase)

Chlorpropamide
(Diabinese)

All the peptidergic messengers, hormones, and neurotransmitters act through secondary messengers. Glucagon, luteinizing hormone, antidiuretic hormone, angiotensin, enkaphalin, and P-substance use the G-protein-adenylate cyclase cascade. Others such as vasopressin use membrane-derived phosphatidylinositol, PI, derivates (Section 20.6). Still others use calcium (Box 24A) as secondary messengers.

23.7 Steroid Messengers. Hormones

We saw in Section 20.10 that a large number of hormones possess steroid ring structures. These hormones, the sex hormones among them, are hydrophobic; therefore, they can cross plasma membranes of the cell by diffusion.

Progesterone

There is no need for special receptors embedded in the membrane for these hormones. It has been shown, however, that **steroid hormones** interact inside the cell with protein receptors. Most of these receptors are localized in the nucleus of the cell, but small amounts exist also in the cytoplasm. When they interact with steroids, they facilitate their migration through the aqueous cytoplasm; the proteins themselves are hydrophilic.

Once inside the nucleus, the steroid-receptor complex can either bind directly to the DNA or combine with a transcription factor (Section 25.4), influencing the synthesis of a certain key protein. Thyroid hormones, (see

BOX 23H

Tamoxifen and Breast Cancer

Large-scale clinical studies have shown that tamoxifen can reduce the recurrence of breast cancer in women who had been successfully treated for the disease. Even more importantly, a study on 13,000 women showed that tamoxifen reduced the occurrence of breast cancer by 45 percent in women who were at high risk because of family history, previous breast abnormality, or age. Tamoxifen is a nonsteroidal compound that binds to the estradiol receptors situated on the nuclei of cells. By binding to the receptor, it causes conformational changes. This then prevents estradiol's binding to its receptor and thus the stimulation of the growth of breast cancer. Tamoxifen also showed promise of treating refractory ovarian cancer.

Although it reduces the risk of breast cancer, tamoxifen also increases the risk of uterine cancer as well as clot formation. The latter can be life threatening if clots migrate to the lungs, causing pulmonary embolism. However, these last two diseases are much rarer than breast cancer.

Estradiol

Tamoxifen

Box 14E) having also large hydrophobic domains also have protein receptors, which facilitates their transport through aqueous media.

The steroid hormonal response through protein synthesis is not fast. It involves hours. Steroids can also act at the cell membrane, influencing ligand-gate ion channels. Such a response would involve only seconds. An example of such a fast response occurs in fertilization. The sperm head contains proteolytic enzymes, which are needed to act on the egg to facilitate its penetration. These enzymes are stored in acrosomes, an organelle on the sperm head. During fertilization, progesterone originating from the follicle cells surrounding the egg acts on the acrosome outer membrane, which disintegrates within seconds, releasing the proteolytic enzymes.

SUMMARY

Cell-to-cell communications are carried out by three different kinds of molecules. (1) **Receptors** are protein molecules embedded in the membranes of cells. (2) **Chemical messengers,** ligands, interact with the receptors. (3) **Secondary messengers** carry and amplify the signals from the receptor to inside the cell. **Neurotransmitters** send chemical messengers across a short distance—the **synapse** between two neurons or between a neuron and a muscle or endocrine gland cell. This communication occurs in milliseconds. **Hormones** transmit their signals more slowly and over a longer distance, from the source of their secretion (endocrine gland) through the blood stream into target cells. Many drugs affect chemical communications. **Antagonists** are drugs that block receptors. **Agonists** are drugs that stimulate receptors.

There are five kinds of chemical messengers: **cholinergic, amino acid, adrenergic, peptidergic,** and **steroid.** Neurotransmitters may belong to the first four classes; hormones to the last three classes. Acetylcholine is cholinergic, glutamic acid is an amino acid, epinephrine (adrenaline) and norepinephrine are adrenergic, and enkephalins are peptidergic. Progesterone is a steroid. Nerve transmission starts with the neurotransmitters packaged in

vesicles in the **presynaptic end** of neurons. When these neurotransmitters are released, they cross the membrane and the synapse and are adsorbed onto receptor sites on the **postsynaptic** membranes. This adsorption triggers an electrical response. Some neurotransmitters act directly, whereas others act through a secondary messenger, **cyclic AMP.** After the electrical signal is triggered, the neurotransmitter molecules must be removed from the postsynaptic end. In the case of acetylcholine, this is done by an enzyme called acetylcholinesterase; in the case of monoamines, by enzymes (MAOs) that oxidize them to aldehydes.

Amino acids bind to their receptors, which are ligand-gated ion channels, Peptides and proteins bind to receptors on the target cell membrane and use secondary messengers to exert their influence. **Signal transduction** is the process that occurs after a ligand binds to its receptor: The signal is carried inside the cell and is amplified. Steroids penetrate the cell membrane, and their receptors are in the cytoplasm. Together with their receptors, they penetrate the cell nucleus. Hormones can act in three ways: (1) They activate enzymes, (2) they affect the gene transcription of an enzyme or protein, and (3) they change membrane permeability.

KEY TERMS

Adrenergic messengers
   (Section 23.4)
Agonist (Section 23.1)
Amino acid neurotransmitter
   (Section 23.4)
Antagonist (Section 23.1)
Axon (Section 23.2)
Calmodulin (Box 23A)
Chemical messenger (Section 23.1)
Cholinergic neurotransmitter
   (Section 23.3)

Dendrite (Section 23.2)
Endocrine gland (Section 23.2)
Enkephalins (Section 23.6)
Excitatory neurotransmitters
   (Section 23.4)
Hormone (Sections 23.1, 23.2)
Inhibitory neurotransmitters
   (Section 23.4)
Neuron (Section 23.1)
Neurotransmitter
   (Sections 23.1, 23.2)

Peptidergic neurotransmitter
   (Section 23.6)
Postsynaptic (Section 23.2)
Presynaptic (Section 23.2)
Receptor (Section 23.1)
Secondary messenger (Section 23.1)
Signal transduction (Section 23.5)
Steroid hormones (Section 23.7)
Synapse (Section 23.2)
Transporter (Section 23.4)
Vesicle (Section 23.2)

CONCEPTUAL PROBLEMS

23.A What is the difference between a chemical messenger and a secondary messenger?

23.B Insulin is a hormone which, when it binds to a receptor, enables glucose molecules to enter the cell and be metabolized. If you have a drug that is an agonist, how would the glucose level in the serum change upon administering the drug?

23.C The action of protein kinase is the next to the last step in the G-protein-adenylate cyclase cascade signal transduction. What kind of effects can elicit the phosphorylation by this enzyme?

23.D The pituitary gland releases luteinizing hormone (LH), which enhances the production of progesterone in the uterus. Classify these two messengers and discuss how each delivers its message.

PROBLEMS

Difficult problems are designated by an asterisk.

Chemical Communication

23.1 What kind of compounds are receptors?

23.2 What kind of signal travels along the axon of a neuron?

Neurotransmitters, Hormones

23.3 What is the difference between the presynaptic and the postsynaptic ends of a neuron?

23.4 Define.
(a) Synapse
(b) Receptor
(c) Presynaptic
(d) Postsynaptic
(e) Mediator
(f) Vesicle

23.5 What is the role of Ca^{2+} in releasing neurotransmitters into the synapse?

23.6 Which signal takes longer (a) neurotransmitter or (b) hormone? Explain.

23.7 What gland controls lactation?

23.8 To which of the three groups of chemical messengers do these hormones belong?
(a) Norepinephrine
(b) Thyroxine
(c) Oxytocin
(d) Progesterone

Cholinergic Neurotransmitters

23.9 How does acetylcholine transmit an electric signal from neuron to neuron?

23.10 Which end of the acetylcholine molecule fits into the receptor site?

23.11 Explain how succinylcholine acts as a muscle relaxant (see Box 22A).

23.12 Which carbons of cAMP contain the phosphoric ester bonds?

Amino Acid Neurotransmitters

23.13 List two features by which taurine is different from the amino acids found in proteins.

23.14 How is glutamic acid removed from its receptor?

23.15 What is unique in the structure of GABA that distinguishes it from all the amino acids that are present in proteins?

23.16 What is the structural difference between NMDA, an agonist of a glutamic acid receptor, and L-aspartic acid?

Adrenergic Neurotransmitters

23.17 (a) Find two monoamine neurotransmitters in Table 23.1. (b) Explain how they act. (c) What medication controls the particular diseases caused by the lack of monoamine neurotransmitters?

23.18 What bond is hydrolyzed and what bond is formed in the synthesis of cAMP?

*23.19 How is the catalytic unit of protein kinase activated in adrenergic neurotransmission?

23.20 The formation of cyclic AMP is described in Section 23.5. Show by analogy how cyclic GMP is formed from GTP.

23.21 By analogy to the action of MAO on epinephrine, write the structural formula of the product of the corresponding oxidation of dopamine.

23.22 What happens to the cAMP in the cell after the neurotransmitter is removed from the receptor?

23.23 Explain how adrenergic neurotransmission is affected by (a) amphetamines (b) reserpine. (See Table 23.1.)

*23.24 Which step in the events depicted in Figure 23.3 provides an electrical signal?

23.25 What kind of product is the MAO-catalyzed oxidation of epinephrine?

23.26 How is histamine removed from the receptor site?

23.27 Cyclic AMP affects the permeability of membranes for ion flow. (a) What blocks the ion channel? (b) How is this blockage removed? (c) What is the direct role of cAMP in this process?

23.28 Dramamine and cimetidine are both antihistamines. Would you expect Dramamine to cure ulcers and cimetidine to relieve the symptoms of asthma? Explain.

Peptidergic Neurotransmitters

23.29 What is the chemical nature of enkephalins?

23.30 What is the mode of action of Demerol as a pain killer? (See Table 23.1.)

Boxes

23.31 (Box 23A) What is the difference between calcium sparks (puffs) and calcium waves?

23.32 (Box 23A) What is the role of calmodulin in signaling by Ca^{2+} ions?

23.33 (Box 23B) (a) What is the effect of nerve gases? (b) What molecular substance do they affect?

23.34 (Box 23C) What is the mode of action of botulinum toxin?

23.35 (Box 23C) How can a deadly botulinum toxin cure facial tics?

23.36 (Box 23D) What is the difference in the β-amyloid protein of a brain from a normal patient compared with that from a patient with Alzheimer's disease?

23.37 (Box 23D) Alzheimer's disease causes loss of memory. What kind of drugs may provide some relief for, if not cure, this disease? How do they act?

23.38 (Box 23E) Why would a dopamine pill be ineffective in treating Parkinson's disease?

23.39 (Box 23E) What is the mechanism by which cocaine stimulates the continuous firing of signals between neurons?

23.40 (Box 23F) How can NO cause headaches?

23.41 (Box 23F) How is nitric oxide synthesized in the cells?

23.42 (Box 23F) How is the toxicity of NO detrimental in strokes?

23.43 (Box 23G) What is the common chemical feature of oral antidiabetic drugs?

23.44 (Box 23G) What is the difference between insulin-dependent and non-insulin-dependent diabetes?

23.45 (Box 23H) Tamoxifen is an antagonist. How does it work in reducing the reoccurrence of breast cancer?

Additional Problems

23.46 Considering its chemical nature, how does aldosterone (Section 20.10) affect mineral metabolism (Table 23.2)?

23.47 What is the function of the ion-translocating protein in adrenergic neurotransmission?

*23.48 Decamethonium acts as a muscle relaxant. If an overdose of decamethonium occurs, can paralysis be prevented by administering large doses of acetylcholine? Explain.

23.49 Endorphin, a potent pain killer, is a peptide containing 22 amino acids; among them are the same five N-terminal amino acids found in the enkephalins. Does this explain its pain-killing action?

23.50 How do alpha and beta alanine differ in structure?

23.51 Where is a G-protein located in adrenergic neurotransmission?

23.52 (Box 23E) List a number of effects that are caused when NO as a secondary messenger relaxes smooth muscles.

23.53 (a) In terms of their action, what do the hormone vasopressin and the neurotransmitter dopamine have in common? (b) What is the difference in their mode of action?

23.54 What is the difference in the mode of action between acetylcholinesterase and acetylcholine transferase?

23.55 How does cholera toxin exert its effect?

23.56 Give the formulas for the reaction: $GTP \rightarrow GDP + P_i$.

23.57 How does the bite of a cobra exert its deadly effect?

23.58 (Box 23E) The drug ritalin is used to alleviate hyperactivity in attention deficit disorder of children. How does this drug work?

Chapter 24

Nucleotides, Nucleic Acids, and Heredity

There are two parents if reproduction is sexual but one parent if it is asexual.

Above: The world's first sheep clones. These Welsh Mountain sheep are the product of research by Dr. Ian Wilmut and colleagues at the Roslin Institute in Edinburgh, Scotland. *(James King-Holmes/Science Photo Library/ Photo Researchers, Inc.)*

24.1 Introduction

Each cell of our bodies contains thousands of different protein molecules. Recall from Chapter 21 that all these molecules are made up of the same 20 amino acids, but in different sequences. The hormone insulin has a different amino acid sequence than the globin of the red blood cell. Even the same protein, for example, insulin, has different sequence in different species (Section 21.7). Within a species also, different individuals may have some differences in their proteins, though the differences are much less than those between species. This shows up graphically in cases where people have such conditions as hemophilia, albinism, or color-blindness because they lack certain proteins that normal people have or the sequence of their amino acids is somewhat different (see Box 21D).

After scientists understood this, the next question was, how do the cells know which proteins to synthesize out of the extremely large number of possible amino acid sequences? The answer to this question is that an individual gets the information from its parents; this is called *heredity*. We all know that a pig gives birth to a pig and a mouse gives birth to a mouse. When an egg cell joins a sperm cell, a new cell called a *zygote* is produced. A pig zygote looks very much like a mouse zygote. Yet the pig zygote will produce the proteins necessary to cause the zygote to grow into a pig and not a mouse.

It was easy to determine that the information is obtained from the parent or parents, but what form does this information take? During the last 50 years, revolutionary developments have led to an answer to this question—the transmission of heredity on the molecular level.

From about the end of the nineteenth century, biologists suspected that the transmission of hereditary information from one generation to another took place in the nucleus of the cell. More precisely, they believed that structures within the nucleus, called **chromosomes,** have something to do with heredity. Different species have different numbers of chromosomes in the nucleus. The information that determines external characteristics (red hair, blue eyes) and internal characteristics (blood group, hereditary diseases) was thought to reside in **genes** located inside the chromosomes.

Studies on the nuclei of fruit flies in the early 1930s revealed that the genes that carry the different traits lie in sequences along the chromosomes. Chemical analysis of nuclei showed that they are largely made up of special basic proteins called **histones** and a type of compound called **nucleic acids.** By 1940, it became clear through the work of Oswald Avery (1877–1955) that, of all the material in the nucleus, only a nucleic acid called deoxyribonucleic acid (DNA) carries the hereditary information. That is, the genes are located in the DNA. Other work in the 1940s by George Beadle (1903–1989) and Edward Tatum (1909–1975) demonstrated that each gene controls the manufacture of one protein, and that external and internal characteristics are expressed through this gene. Thus, the expression of the gene (DNA) in terms of an enzyme (protein) led to the study of protein synthesis and its control. **The information that tells the cell which proteins to manufacture is carried in the molecules of DNA.**

In the following sections, we provide some of the highlights of this rapidly developing field.

24.2 Components of Nucleic Acids

Two kinds of nucleic acids are found in cells, each has its own role in the transmission of hereditary information. The two types are **ribonucleic acid (RNA)** and **deoxyribonucleic acid (DNA).** As we just saw, DNA is present in the chromosomes of the nucleus. RNA is not found in the chromosomes. It is located elsewhere in the nucleus and even outside the nucleus, in the cytoplasm. As we will see in Section 24.5, there are four types of RNA, all with similar structures.

Both DNA and RNA are polymers. Just as proteins consist of chains of amino acids, and polysaccharides consist of chains of monosaccharides, nucleic acids are also chains. The building blocks (monomers) of nucleic acid chains are **nucleotides.** Nucleotides themselves, however, are composed of three simpler units: a base, a monosaccharide, and phosphate. We will look at each of these in turn.

Bases

The **bases** found in DNA and RNA are chiefly those shown in Figure 24.1. All of them are basic because they are heterocyclic amines (Section 16.2). Two of these bases—adenine and guanine—are purines, and the other three—cytosine, thymine, and uracil—are pyrimidines. The two purines (A and G) and one of the pyrimidines (C) are found in both DNA and RNA, but uracil (U) is found only in RNA, and thymine (T) is found only in DNA. Note that thymine

Some simple organisms, such as bacteria, do not have a nucleus. Their chromosomes are condensed in a central region.

Gene The unit of heredity; a DNA segment that codes for one protein

Human chromosomes magnified about 8000 times. *(Biophoto Associates/Photo Researchers)*

Beadle and Tatum shared the 1958 Nobel Prize in physiology with Joshua Lederberg.

Bases Purines and pyrimidines, which are components of nucleotides, DNA, and RNA

The three pyrimidines and guanine are in their keto rather than their enol forms (see Section 17.6).

Purines   Pyrimidines

FIGURE 24.1 The five principal bases of DNA and RNA. Note how the rings are numbered. The hydrogens shown in blue are lost when the bases are bonded to monosaccharides.

differs from uracil only in the methyl group in the 5 position. Thus, both DNA and RNA contain four bases: two pyrimidines and two purines. For DNA, the bases are A, G, C, and T; for RNA, the bases are A, G, C, and U.

The initial letter of each base is used as an abbreviation for that base.

Sugars

The sugar component of RNA is D-ribose (Section 19.2). In DNA it is 2-deoxy-D-ribose (hence the name deoxyribonucleic acid):

The only difference between these molecules is that ribose has an OH group in the 2 position not found in deoxyribose.

β-D-Ribose β-2-Deoxy-D-ribose

> **Nucleoside** A compound composed of ribose or deoxyribose and a base

The combination of sugar and base is known as a **nucleoside.** The purine bases are linked to the C-1 of the monosaccharide through N-9 (the nitrogen at position 9 of the five-membered ring). This is a β-*N*-glycosidic bond:

The nucleoside made of guanine and ribose is called **guanosine.** The names of the other nucleosides are given in Table 24.1.

The pyrimidine bases are linked to the C-1 of the monosaccharide through their N-1 nitrogen; this is a β-*N*-glycosidic bond.

TABLE 24.1 The Names of the Eight Nucleosides and Eight Nucleotides in DNA and RNA

Base	Nucleoside	Nucleotide
		DNA
Adenine (A)	Deoxyadenosine	Deoxyadenosine 5'-monophosphate (dAMP)[a]
Guanine (G)	Deoxyguanosine	Deoxyguanosine 5'-monophosphate (dGMP)[a]
Thymine (T)	Deoxythymidine	Deoxythymidine 5'-monophosphate (dTMP)[a]
Cytosine (C)	Deoxycytidine	Deoxycytidine 5'-monophosphate (dCMP)[a]
		RNA
Adenine (A)	Adenosine	Adenosine 5'-monophosphate (AMP)
Guanine (G)	Guanosine	Guanosine 5'-monophosphate (GMP)
Uracil (U)	Uridine	Uridine 5'-monophosphate (UMP)
Cytosine (C)	Cytidine	Cytidine 5'-monophosphate (CMP)

[a]The d indicates that the sugar is deoxyribose.

Primed numbers are used for the ribose and deoxyribose portions of nucleosides, nucleotides, and nucleic acids. Unprimed numbers are used for the bases.

β-N-glycosidic bond

Uridine

Phosphate

The third component of nucleic acids is phosphoric acid. When this group forms a phosphate ester (Section 18.7) bond with the —CH$_2$OH group of a nucleoside, the result is a compound known as a **nucleotide.** For example, adenosine combines with phosphate to form the nucleotide adenosine 5'-monophosphate, AMP:

Nucleotides A nucleoside bonded to one, two, or three phosphate groups

AMP

Anticancer Drugs

A major difference between cancer cells and most normal cells is that the cancer cells divide much more rapidly. Rapidly dividing cells require a constant new supply of DNA. One component of DNA is the nucleoside deoxythymidine, which is synthesized in the cell by the methylation of uridine. If fluorouracil is administered to a cancer patient as part of chemotherapy, the body converts it to fluorouridine, a compound that irreversibly inhibits the enzyme that manufactures thymidine from uridine, greatly decreasing DNA synthesis.

Because this affects the rapidly dividing cancer cells more than the healthy cells, the growth of the tumor and the spread of the cancer are arrested. However, chemotherapy with fluorouracil or other anticancer drugs weakens the body because it also interferes with DNA synthesis in normal cells. Therefore, chemotherapy is used intermittently to give the body time to recover from the side effects of the drug. During the period after chemotherapy, special precautions must be taken so that bacterial infections do not debilitate the already weakened body.

Fluorouracil

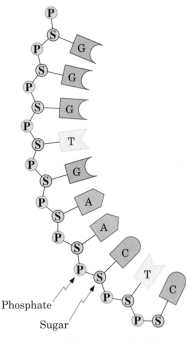

Phosphate

Sugar

FIGURE 24.2 Schematic diagram of a nucleic acid molecule. The four bases of each nucleic acid are arranged in various specific sequences.

Nucleic acid A polymer composed of nucleotides

The names of the other nucleotides are given in Table 24.1. Some of these nucleotides play an important role in metabolism. They are part of the structure of key coenzymes, cofactors, and activators (Section 26.3 and 28.2). Most notably, we shall see how adenosine 5′-triphosphate, ATP, serves as a common currency into which the energy gained from food is converted and stored. In ATP, two more phosphate groups are joined to AMP in phosphate anhydride bonds (Section 18.7). In adenosine 5′-diphosphate, ADP, only one phosphate group is attached to the AMP. Guanosine also has important multiphosphorylated forms: GMP, GDP and GTP. In Section 24.3, we will see how DNA and RNA are chains of nucleotides. In summary,

A nucleoside = Base + Sugar

A nucleotide = Base + Sugar + Phosphoric acid

A nucleic acid = A chain of nucleotides

24.3 Structure of DNA and RNA

In Chapter 21, we saw that proteins have primary, secondary, and higher structures. Nucleic acids, which are chains of monomers, also have primary and secondary structures.

Primary Structure

Nucleic acids are polymers of nucleotides, as shown schematically in Figure 24.2. Note that the structure can be divided into two parts: (1) the backbone of the molecule and (2) the bases that are the side-chain groups. The backbone in DNA consists of alternating deoxyribose and phosphate groups. Each phosphate group is linked to the 3′ carbon of one deoxyribose unit and simultaneously to the 5′ carbon of the next deoxyribose unit (Figure 24.3). Similarly, each monosaccharide unit forms a phosphate ester at the 3′ position and another at the 5′ position. The primary structure of

FIGURE 24.3 Primary structure of the DNA backbone. The hydrogens shown in blue cause the acidity of nucleic acids. In the body, at neutral pH, the phosphate groups carry a charge of −1, and the hydrogens are replaced by Na^+ and K^+.

RNA is the same except that each sugar is ribose (so there is an —OH group in the 2′ position) rather than deoxyribose and U is present instead of T.

Thus, the backbone of the DNA and RNA chains has two ends: a 3′ —OH end and a 5′ —OH end. These two ends have roles similar to those of the C-terminal and N-terminal ends in proteins. This backbone provides the structural stability of the DNA and RNA molecules.

The bases that are linked, one to each sugar unit, are the side chains, and they carry all the information necessary for protein synthesis. Analysis of the base composition of DNA molecules from many different species was done by Erwin Chargaff (1905–), who showed that in DNA taken from many different species, the quantity of adenine (in moles) is always approximately equal to that of thymine, and the quantity of guanine is always approximately equal to that of cytosine, though the adenine/guanine ratio varies widely from species to species. This important information helped to establish the secondary structure of DNA, as we shall soon see.

Just as the order of the amino acid residues of protein side chains determines the primary structure of the protein (for example, —Ala—Gly—Glu—Met—), **the order of the bases** (for example, —ATTGAC—) **provides the primary structure of DNA.** As with proteins, we need a convention to tell us which end to start with when we are writing the sequence of bases. For nucleic acids, the convention is to begin the sequence with the nucleotide that has the free 5′ —OH terminal. Thus, the sequence AGT means that adenine is the base at the 5′ terminal and thymine is the base at the 3′ terminal.

The sequence TGA is not the same as AGT, but rather its opposite.

Secondary Structure of DNA

In 1953 James Watson (1928–) and Francis Crick (1916–) established the three-dimensional structure of DNA. Their work is a cornerstone in the history of biochemistry. The model of DNA established by Watson and Crick was based on two important pieces of information obtained by other workers: (1) the Chargaff rule that (A and T) and (G and C) are present in equimolar quantities and (2) x-ray diffraction photographs obtained by Rosalind Franklin (1920–1958) and Maurice Wilkins (1916–). By the clever use of these facts, Watson and Crick concluded that DNA is composed of two strands entwined around each other in a **double helix,** as shown in Figure 24.4.

Watson, Crick, and Wilkins were awarded the 1962 Nobel Prize in Medicine for their discovery. Franklin died in 1958. The Nobel Committee does not award the Nobel Prize posthumously.

Double helix The arrangement of which two strands of DNA are coiled around each other in a screw-like fashion

Recall from Section 21.8 that a helix has a shape like a coiled spring or a spiral staircase. It was Pauling's discovery that human hair protein is helical that led Watson and Crick to look for helixes in DNA.

In the DNA double helix, the two polynucleotide chains run in opposite directions. This means that at each end of the double helix there is one 5′ —OH and one 3′ —OH terminal. The sugar-phosphate backbone is on the outside, exposed to the aqueous environment, and the bases point inward. They are hydrophobic, and thus try to avoid contact with water. Through their hydrophobic interactions, they stabilize the double helix. The bases are paired according to Chargaff's rule: For each adenine on one chain, a thymine is aligned opposite it on the other chain; each guanine on one chain has a cytosine aligned with it on the other chain. **The bases so paired form hydrogen bonds with each other, thereby stabilizing the double helix** (Figure 24.5). They are called **complementary base pairs.**

Rosalind Franklin (1920–1958).
(Chemical Heritage Foundation)

Watson and Crick with their model of the DNA molecule.

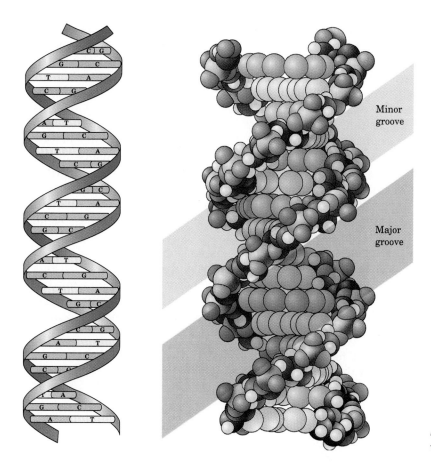

FIGURE 24.4 Three-dimensional structure of the DNA double helix.

Thymine   Adenine

AT pair

Cytosine   Guanine

GC pair

FIGURE 24.5 A and T pair up by forming two hydrogen bonds.

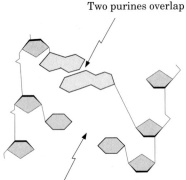

Two purines overlap

Gap between two pyrimidines

FIGURE 24.6 The bases of DNA cannot stack properly in the double helix if a purine is opposite a purine or a pyrimidine is opposite a pyrimidine.

The important thing here, as Watson and Crick realized, is that only adenine could fit with thymine and only guanine could fit with cytosine. Let us consider the other possibilities. Can two purines (AA, GG, or AG) fit opposite each other? Figure 24.6 shows that they would overlap. How about two pyrimidines (TT, CC, or CT)? As shown in Figure 24.6, they would be too far apart. *There must be a pyrimidine opposite a purine.* But could A fit opposite C, or G opposite T? Figure 24.7 shows that the hydrogen bonding would be much weaker.

The entire action of DNA—and of the heredity mechanism—depends on the fact that, **wherever there is an adenine on one strand of the helix, there must be a thymine on the other strand because that is the only base that fits and forms strong hydrogen bonds, and similarly for G and C.** The entire heredity mechanism rests on these slender hydrogen bonds (Figure 24.7), as we shall see in Section 24.4.

The beauty of establishing the three-dimensional structure of the DNA molecule was that the knowledge of this structure immediately led to the explanation for the transmission of heredity: how the genes transmit traits from one generation to another. Before we look at the mechanism of DNA replication (in the next section), let us summarize the three differences in structure between DNA and RNA:

1. DNA has the four bases A, G, C, and T. RNA has three of these—A, G, and C—but the fourth base is uracil, not thymine.

No hydrogen bonding

Thymine Guanine

Cytosine Adenine

No hydrogen bonding

FIGURE 24.7 Only one hydrogen bond is possible for GT or CA. These combinations are not found in DNA. Compare this with Figure 24.5.

2. In DNA, the sugar is 2-deoxy-D-ribose. In RNA, it is D-ribose.

3. DNA is almost always double-stranded, with the helical structure shown in Figure 24.4.

There are several kinds of RNA (as we shall see in Section 24.5); all of them are single-stranded, though base-pairing can occur within a chain (see, for example, Figure 24.9). When it does, adenine pairs with uracil because thymine is not present.

24.4 DNA Replication

The DNA in the chromosomes carries out two functions: (1) It reproduces itself, and (2) it supplies the information necessary to make all the proteins in the body, including enzymes. The second function is covered in Chapter 25. Here we are concerned with the first, **replication.**

Each gene is a section of a DNA molecule that contains a specific sequence of the four bases A, G, T, and C, typically containing about 1000 to 2000 nucleotides. The base sequence of the gene carries the information necessary to produce one protein molecule. If the sequence is changed (for example, if one A is replaced by a G, or if an extra T is inserted), a different protein is produced, which might mean that the individual would have brown eyes instead of blue or perhaps would not have some vital metabolic protein such as insulin.

But consider the task that must be accomplished by the organism. When an individual is conceived, the egg and sperm cells unite to form the zygote. This cell, which is very tiny in most mammals, contains a small amount of DNA, but this DNA contains all the genetic information the individual will ever have. A fully grown large mammal, such as a human being or a horse, may contain more than a trillion cells. Each cell (except the egg and sperm cells) contains the same amount of DNA as the original single cell. Furthermore, cells are constantly dying and being replaced. Thus, there must be a mechanism by which DNA molecules can be copied (just as we can copy a letter on a photocopying machine) over and over again, millions of times, without error. In Section 25.7, we shall see that such errors sometimes do happen and can have serious consequences, but here we want to examine this remarkable mechanism that takes place every day in billions of organisms, from microbes to whales, and has been taking place for billions of years—with only a tiny percentage of errors.

The DNA double helix contains millions of bases. One DNA strand may carry many inheritable genes, each of which is a stretch of DNA a few hundred or thousand bases long. Genetic information is transmitted from one cell to the next when cell division occurs. The two new cells carry all the information that the original cell possessed. Where originally there was one set of DNA molecules, there will now be two sets of DNA molecules, one set in each new cell.

The replication of DNA molecules starts with the unwinding of the double helix. This can occur at either end or in the middle. Special **unwinding protein molecules,** called **helicases,** attach themselves to one DNA strand (Figure 24.8) and cause the separation of the double helix. All four kinds of free DNA nucleotide molecules are present in the vicinity. These nucleotides constantly move into the area and try to fit themselves into new chains. The key to the process is that, as we saw in Section 24.3, **only thymine can fit opposite adenine, and only cytosine can fit opposite guanine.** Wherever a cytosine, for example, is present on one of the strands of an unwound portion of the helix, all four nucleotides may

Replication A process by which copies of DNA are made during cell division

The pigments responsible for the blue or brown color of eyes are synthesized with the help of specific enzymes. If one of these enzymes is lacking, the eye color may be different.

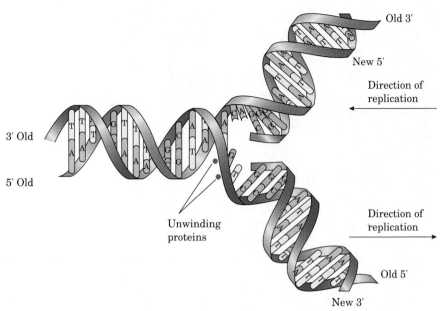

FIGURE 24.8   The replication of DNA. The two strands of the DNA double helix are shown separating. *(Modified from P. W. Davis and E. P. Solomon,* The World of Biology. *Philadelphia: Saunders College Publishing, 1986)*

The base sequence of each newly synthesized DNA chain is **complementary** to the chain already there.

Tuneko and Reiji Okazaki provided biochemical verification for the semicontinuous replication of DNA in 1968.

Okazaki fragment A short DNA segment made of about 200 nucleotides in higher organisms (eukaryotes) and 2000 nucleotides in prokaryotes

approach, but three of them are turned away because they do not fit. Only the nucleotide of guanine fits.

While the bases of the newly arrived nucleotides are being hydrogen bonded to their partners, enzymes called polymerases join the nucleotide backbones. At the end of the process, there are two double-stranded DNA molecules, each exactly the same as the original one because only T fits opposite A and only G fits against C. The process is called **semiconservative** because only two of the four strands are new; the other two were present in the original molecule.

If the unwinding begins in the middle, the synthesis of new DNA molecules on the old templates continues in both directions until the whole molecule is duplicated. This is the more common pathway. The unwinding can also start at one end and proceed in one direction until the whole double helix is unwound.

An interesting detail of DNA replication is that the two daughter strands are synthesized in different ways. One of the syntheses is continuous in the $5' \rightarrow 3'$ direction (see Section 24.3), and the polymerase enzymes involved are capable of linking millions of phosphate ester bonds continuously in this direction. However, since the strands in the double helix run in opposite directions, only one strand of the double helix runs $5' \rightarrow 3'$; the other runs $3' \rightarrow 5'$. Along this second strand, there is no possibility of continuous synthesis. What happens is that along the $3' \rightarrow 5'$ strand, the enzymes can synthesize only short fragments because the only way they can work is from $5'$ to $3'$. These short fragments consist of about 200 nucleotides each, named **Okazaki fragments** after their discoverer. They are eventually joined together by the enzyme DNA ligase. The two newly formed strands run in opposite directions. The new strand synthesized continuously is called the **leading strand;** the one assembled from Okazaki fragments is called the **lagging strand.**

The model of replication discussed earlier visualizes an immobile DNA template along which mobile enzymes, the polymerases, move or slide. This mode of action is similar to a train proceeding along a track. Recently, more

Telomers, Telomerase, and Immortality

Telomers are specialized structures at the ends of chromosomes. In vertebrates, telomers are TTAGGG sequences that are repeated hundreds to thousands of times. In normal somatic cells that divide in a cyclic fashion throughout the life of the organism (mitosis), chromosomes lose about 50 to 200 nucleotides from their telomers at each cell division. This loss is due to the lagging strand, which is synthesized in Okazaki fragments. Okazaki fragments themselves require a template. DNA polymerase, the enzyme that links the fragments, does not work at the end of linear DNA. This results in the shortening of the telomers at each replication. The shortening of the telomers acts as a clock by which the cells count the number of times they have divided. After a certain number of divisions, the cells stop dividing, reaching the limit of the aging process.

In contrast to **somatic cells** that do not give rise to reproductive cells in all immortal cells (germ cells in proliferative stem cells, in normal fetal cells, and in cancer cells), there is an enzyme, telomerase, which is able to extend the shortened telomers by synthesizing new chromosomal ends. This telomerase is a ribonucleoprotein; that is, it is made of RNA and protein. The activity of this enzyme seems to confer, among others, immortality to the cells where it is present.

and more evidence is accumulated; it indicates that the polymerase enzymes are immobilized in "factories" through which the DNA moves. Such factory replication centers or foci may contain many polymerases. These factories may be bound to membranes in bacteria. In higher organisms, the replication centers are not permanent structures. They may be disassembled and their part reassembled in larger and larger factories.

24.5 RNA

We previously noted that there are three types of RNA.

1. Messenger RNA (mRNA) mRNA carry the genetic information from the DNA in the nucleus directly to the cytoplasm, where the protein is synthesized. It consists of a chain of nucleotides whose sequence is exactly complementary to that of one of the strands of the DNA. This type of RNA is not very stable. It is synthesized as needed and then degraded. Thus, its concentration at any time is rather low. The size of mRNA varies widely, the average size may contain 750 nucleotides.

2. Transfer RNA (tRNA) Containing from 73 to 93 nucleotides per chain, tRNA are relatively small molecules. There is at least one different tRNA molecule for each of the 20 amino acids from which the body makes its proteins. The three-dimensional tRNA molecules are L-shaped, but they are conventionally represented as a cloverleaf in two dimensions. A typical one is shown in Figure 24.9. Transfer RNA molecules contain not only cytosine, guanine, adenine, and uracil, but also several other modified nucleotides.

3. Ribosomal RNA (rRNA) Ribosomes, which are small spherical bodies located in the cells but outside the nuclei, contain rRNA. They consist of about 35 percent protein and 65 percent ribosomal RNA (rRNA). These are large molecules with molecular weights up to 1 million. As we shall see in Section 25.5, protein synthesis takes place on the ribosomes.

Ribozymes (catalytic RNA) are either mRNAs or tRNAs with special enzyme function. In messenger RNA, the precursor molecule is shortened to mature mRNA to be able to perform its function. These splicings are catalyzed by the **ribozymes,** which are segments of mRNA themselves. Similar splicing is catalyzed by ribozymes in forming mature tRNA from its precursors. Even though ribozymes have sim-

Messenger RNA (mRNA) The RNA that carries genetic information from DNA to the ribosome and acts as a template for protein synthesis

In DNA–RNA interactions, the complementary bases are

DNA RNA
A—U
G—C
C—G
T—A

Transfer RNA (tRNA) The RNA that transports amino acids to the site of protein synthesis in ribosomes

Ribosomal RNA (rRNA) The RNA complexed with proteins in ribosomes

Ribosome Small spherical bodies in the cell made of protein and RNA; the site of protein synthesis

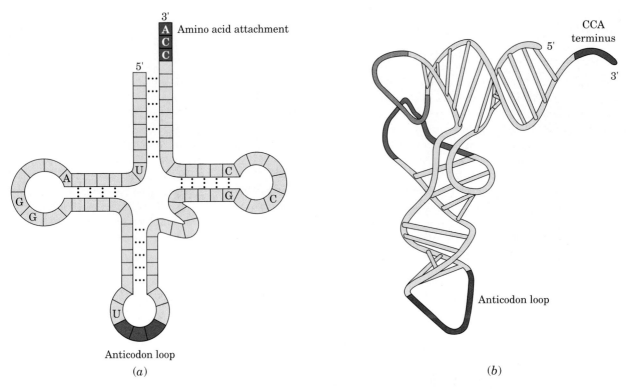

FIGURE 24.9   Structure of tRNA. (*a*) Two-dimensional simplified cloverleaf structure. (*b*) Three-dimensional structure. *[From* Biochemistry, *2nd ed., by Lubert Stryer. Copyright 1981 by W. H. Freeman and Co. All rights reserved. Part (b) also courtesy of Dr. Sung-Hou Kim.]*

pler primary structures that proteins (only 4 building blocks versus 20), they form complicated secondary structures by hydrogen bonding. These are twisted into different tertiary structures. Ribozymes also have an active site just like protein enzymes where the substrate binds.

24.6 Genes, Exons, Introns, and Cloning

A **gene** is a stretch of DNA that carries one particular message; for example, "make a globin molecule." In bacteria, this message is continuous. In higher organisms, the message is not continuous. Stretches of DNA that spell out (code for) the amino acid sequence to be assembled are interrupted by long stretches that seemingly do not code for anything. The coding sequences are called **exons,** and the noncoding sequences, **introns.** For example, the globin gene has three exons broken up by two introns. Because DNA contains exons and introns, the mRNA transcribed from it also contains exons and introns. The introns are cut out by enzymes, and the exons are spliced together before the mRNA is actually used to synthesize a protein (Figure 24.10). In other words, the introns function as spacers as well as enzymes (ribozymes), catalyzing the splicing of exons into "mature mRNA."

Exon Nucleotide sequence in mRNA that codes for a protein

Intron A nucleotide sequence in mRNA that does not code for a protein

Philip A. Sharp (1944–   ) and Richard J. Roberts (1943–   ) were awarded the 1993 Nobel Prize in Medicine for the discovery of the noncoding nature of introns.

FIGURE 24.10   Introns are cut out of mRNA before the protein is synthesized.

BOX 24C

DNA Fingerprinting

Every person has a genetic makeup consisting of about 3 billion pairs of nucleotides, distributed over 46 chromosomes. The base sequence in the nucleus of every one of our billions of cells is identical. However, except for people who have an identical twin, the base sequence in the total DNA of one person is different from that of every other person. This makes it possible to identify suspects in criminal cases from a bit of skin or a trace of blood left at the scene of the crime and to prove the identity of a child's father in paternity cases. The nuclei of these cells are extracted, and restriction enzymes are used to cut the DNA molecules at specific points. The resulting DNA fragments are put on a gel and undergo a process called **electrophoresis.** In this process, the DNA fragments move with different velocities; the smaller fragments move faster and the larger fragments slower. After a sufficient amount of time the fragments separate, and when they are made visible in the form of an auto radiogram, one can see bands in a lane. This is called a **DNA fingerprint.**

When the DNA fingerprint made from a sample taken from a suspect matches that from a sample obtained at the scene of the crime, the police have a positive identification. Figure 24C shows DNA fingerprints derived by using one particular restriction enzyme. A total of nine lanes can be seen. Three (numbers 1, 5, and 9) are control lanes. They contain the DNA fingerprint of a virus, using one particular restriction enzyme. Three other lanes (2, 3, and 4) were used in a paternity suit: These contain the DNA fingerprints of the mother, the child, and the alleged father. The child's DNA fingerprint (lane 3) contains six bands. The mother's DNA fingerprint (lane 2) has five bands, all of which match those of the child. The alleged father's DNA fingerprint (lane 4) also contains six bands, of which three match those of the child. This is a positive identification. In such cases, one cannot expect a perfect match even if the man is the actual father because the child has inherited only half of its genes from the father. Thus, of six bands, only three are expected to match. In the case just described, the paternity suit was won on the basis of the DNA fingerprint matching.

In the left area of the radiogram are three more lanes (6, 7, and 8). These DNA fingerprints were used in an attempt to identify a rapist. In lanes 7 and 8 are the DNA fingerprints of semen obtained from the rape victim. In lane 6 is the DNA fingerprint of the suspect. The DNA fingerprints of the semen do not match those of the suspect. This is a negative identifi-

■ FIGURE 24C DNA fingerprint. *(Courtesy of Dr. Lawrence Kobilinsky)*

cation and excluded the suspect from the case. When positive identification occurs, the probability that a positive match is due to chance is 1 in 100 billion.

However, a certain caution has developed lately in court cases involving DNA fingerprinting. Scientific committees and expert witnesses now demand that, in addition to matching or not matching auto radiograms, some internal controls should appear on the gel. This is demanded to prove that the two lanes to be compared were run and processed under identical conditions. Furthermore, identifications in courts do not rely solely on DNA fingerprinting. Additional evidence is provided by analysis of blood group substances (Box 16B), either directly from blood or from secretions such as saliva or cervical mucus samples, which also contain these substances.

In humans, only 3 percent of the DNA codes for proteins or RNA with clear functions. Introns are not the only noncoding DNA sequences. **Satellites** are DNAs in which short nucleotide sequences are repeated hundreds or thousands of times. Large satellite stretches appear at the ends and centers of chromosomes and are necessary for the stability of the chromosomes. Smaller repetitive sequences, **mini-** or **microsatellites,** when they mutate, are associated with cancer.

The Human Genome Project

The complete DNA sequence of any organism is called a **genome.** The genome of an average human being contains approximately 3 billion base pairs. These are distributed among 22 pairs of chromosomes plus two sex chromosomes. Each chromosome consists of a single DNA molecule. Among the 3 billion pairs that constitute the human genome are some 300 million base pairs, which represent 100 000 genes. It is the task of the Human Genome Project to determine the total sequences of the genome and in the process to identify the sequence and location of the genes.

A subsidiary but also an essential goal is to determine the total DNA sequence of model systems of simpler organisms: the bacterium *Escherichia coli* (3 million base pairs); the yeast *Saccharomyces cerevisiae* (14 million); the nematode *Caenorhabditis elegans* (97 million); the fruit fly *Drosophila melanogaster* (165 million); and the mouse *Mus musculus* (3 billion). The first organism of which the complete genome was established was the bacterium, *Hemophilus influenzae*, in 1995. This bacterium has only 1.8 million base pairs, which code for 1743 genes. Shortly thereafter the even smaller genome of *Mycobacterium genitalium* (0.6 million base pairs) was obtained. Many pharmaceutical companies are keenly interested in the genomes of pathogenic organisms in order to make better drug designs to combat them. The bacteria that causes tuberculosis, *Myobacterium tuberculosis*, has 4.4 million base pairs in its genome which was sequenced recently.

At the end of 1998, the genome of the worm *C. Elegans* was completed. The 97 million base pairs code for 19 000 genes. In 2000, the genome of the fruit fly was elucidated. It is hoped that the whole Human Genome Project will be finished by 2001. Knowing the genomes of the model systems will help the workers studying the human genome to elucidate how the different genes are linked. Many laboratories around the world are participating in this project.

In the course of the project, it is hoped that many of the genes that govern human diseases, such as cystic fibrosis and breast cancer, just to name a few, will be identified. With such identification, it is expected that strategies can be developed for the diagnosis, prevention, and perhaps therapy of these diseases.

■ *Hemophilus influenzae* bacteria resting on nasal tissue. *(© Dr. Tony Brain/SPL/Photo Researchers)*

> **Genome** A complete DNA sequence of an organism

One DNA molecule may have between 1 million and 100 million bases. Therefore, there are many genes in one DNA molecule. If a human DNA molecule were fully stretched out, its length would be perhaps 1 m. However, the DNA molecules in the nuclei are not stretched out. They are coiled around basic protein molecules called **histones.** The acidic DNA and the basic histones attract each other by electrostatic (ionic) forces. The DNA and histones combine to form units called **nucleosomes.** A nucleosome is a core of eight histone molecules around which the DNA double helix is wrapped (Figure 24.11). Nucleosomes are further condensed into chromatin, the DNA thread linking the nucleosomes into three stranded fibers with a width of 30 nm.

Cloning is the exact copying segments of DNA. Millions of copies of selected DNA fragments can be made within a few hours with high precision by a technique called **polymerase chain reaction (PCR),** discovered by Kary B. Mullis (1945–   ), who shared the 1993 Nobel Prize in chemistry for this discovery.

PCR technique can be used only if the sequence of a gene to be copied is known. In such a case, one can synthesize two primers that are complimentary to the 3′ ends of the gene. The primers are polynucleotides of 12 to 16 nucleotide length. When added to a target DNA segment, they hybridize with the 3′ end of the gene.

5′CATAGGACAGC—OH   Primer

3′TACGTATCCTGTCGTAGG—   Gene

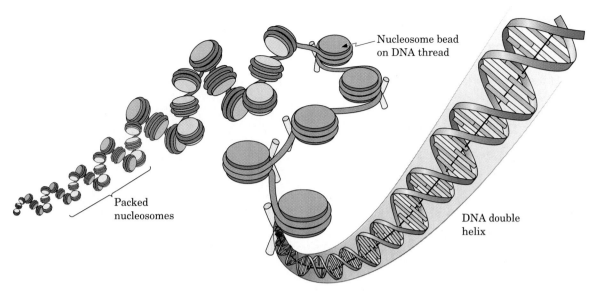

FIGURE 24.11 Schematic diagram of a nucleosome. The bandlike DNA double helix winds around cores consisting of eight histones.

In cycle 1 (Figure 24.12), the polymerase extends the primers in each direction as individual nucleotides are assembled and connected on the template DNA. Thus, two new copies are created. The two-step process is repeated (cycle 2) when the primers are hybridized with the new strands, and the primers are extended again. At that point, four new copies have

FIGURE 24.12 Schematic diagram of PCR. *(Adapted from* The Unusual Origin of the Polymerase Chain Reaction, *by Kary B. Mullis, illustrated by Michael Goodman.* Scientific American, *April 1990)*

Apoptosis, Programmed Cell Death

Cells of an organism die in large clusters when exposed to toxins or deprived of oxygen. Such a typical necrosis occurs in heart attack and stroke. *Apoptosis* or programmed cell death is clearly distinguishable from necrosis. Only scattered cells die at a time, and no inflammation or scar results. In apoptosis, the cells cleave into apoptopic bodies containing intact organelles and large nuclear fragments. These bodies are swallowed up by roving scavenger cells.

At the heart of this self-immolation process are a family of protein-cleaving enzymes, called **caspases.** The name is an acronym for cysteine-containing aspartate-specific proteases, which describes the mode of action: cysteine is in the active site of the enzyme, and the target protein is cleaved next to the amino acid **aspartate.** Caspases are zymogens, and they are activated by self-clipping. Once they are active, they attack targeted proteins. For example, they cleave gelsolin, a protein that normally binds to actin filaments and that maintains the shape of the cell. As a result of actin degradation, the apoptotic cells acquire one of their signatures: They lose their normal shape, become more rounded, and develop blister-like bumps on their surfaces (Figure 24E).

Caspases also attack another protein, a companion of the enzyme, **endonuclease.** When this enzyme is paired with its companion, it is unable to enter the cell nucleus. After it is freed from the companion, endonuclease can enter the nucleus and cleave DNA. The result is the second distinguishing fingerprint of death by apoptosis: the appearance of DNA fragments in multiples of 180 base pairs (180, 360, or 540 base pairs). These appear in the membrane-enclosed apoptotic bodies, the remnants of the dying cells.

■ *FIGURE 24E* Signs of apoptosis. (a) Blisters. (b) Apoptotic bodies. *(© Dr. Larry Schwartz, University of Massachusetts, Amherst)*

been created. The process continues, and in 25 cycles, 2^{25} or some 33 million copies can be made. In practice, only a few million are produced, which is sufficient for the isolation of a gene.

This fast process is possible because of the discovery of heat-resistant polymerases isolated from bacteria that live in hot thermal vents in the sea floor (Section 22.4). The high temperatures are needed because the double helix must be annealed (unwound) to hybridize the primer to the target DNA. Once single strands of DNA have been exposed, the mixture is cooled to 70°C. The primers are hybridized and subsequent extensions take place. The 95°C and 70°C cycles are repeated over and over. No new enzyme is required because the polymerase is stable even at the annealing temperature.

PCR techniques are routinely used when a gene or a segment of DNA must be amplified from a few molecules. It is used in the genome study (Box 24D), in obtaining evidence from a crime scene (Box 24C), or even in obtaining the gene of a long-extinguished species found fossilized in amber.

SUMMARY

Nucleic acids are composed of sugars, phosphates, and organic bases. There are two kinds: **ribonucleic acid (RNA)** and **deoxyribonucleic acid (DNA).** In DNA, the sugar is the monosaccharide 2-deoxy-D-ribose; in RNA, it is D-ribose. In DNA the bases are adenine (A), guanine (G), cytosine (C), and thymine (T). In RNA, they are A, G, C, and uracil (U). Nucleic acids are giant molecules with backbones made of alternating units of sugar and phosphate. The bases are side chains joined by β-N-glycosidic bonds to the sugar units.

DNA is made of two strands that form a double helix. The sugar-phosphate backbone runs on the outside of the double helix, and the hydrophobic bases point inward. There is **complementary pairing** of the bases in the double helix. Each A on one strand is hydrogen-bonded to a T on the other, and each G is hydrogen-bonded to a C. No other pairs fit. The DNA molecule carries, in the sequence of its bases, all the information necessary to maintain life.

When cell division occurs and this information is passed from parent cell to daughter cells, the sequence of the parent DNA is copied.

A **gene** is a segment of a DNA molecule that carries the sequence of bases that directs the synthesis of one particular protein. There are four kinds of RNA: **messenger RNA (mRNA), transfer RNA (tRNA), ribosomal RNA (rRNA),** and **ribozyme (catalytic RNA).** DNA in higher organisms contains sequences, called **introns,** that do not code for proteins. The sequences that do code for proteins are called **exons.** DNA is coiled around basic protein molecules called **histones.** Together they form **nucleosomes,** which are further condensed into the chromatins of chromosomes.

Cloning is copying genes or segments of DNA. The **polymerase chain reaction (PCR)** technique can make millions of copies with high precision in a few hours.

KEY TERMS

Apoptosis (Box 24E)
Base (Section 24.2)
Chromosome (Section 24.1)
Cloning (Section 24.6)
Complementary base pairs
  (Section 24.3)
Deoxyribonucleic acid (Section 24.2)
DNA (Section 24.2)
DNA fingerprinting (Box 24C)
Double helix (Section 24.3)
Exon (Section 24.6)
Gene (Sections 24.1, 24.6)
Genome (Box 24D)
Helicases (Section 24.4)

Histone (Sections 24.1, 24.6)
Intron (Section 24.6)
Lagging strand (Section 24.4)
Leading strand (Section 24.4)
Messenger RNA (mRNA)
  (Section 24.5)
Nucleic acid (Sections 24.1, 24.3)
Nucleoside (Section 24.2)
Nucleosome (Section 24.6)
Nucleotide (Section 24.2)
Okazaki fragment (Section 24.4)
Polymerase chain reaction (PCR)
  (Section 24.6)
Replication (Section 24.4)

Ribonucleic acid (Section 24.2)
Ribosomal RNA (rRNA)
  (Section 24.5)
Ribosome (Section 24.5)
Ribozymes (Section 24.5)
RNA (Section 24.2)
Satellites (Section 24.6)
Transfer RNA (tRNA)
  (Section 24.5)
Unwinding protein molecules
  (Section 24.4)

CONCEPTUAL PROBLEMS

24.A How would you classify the functional groups that bond together the three different components of a nucleotide?

24.B Why do we call DNA or RNA nucleic acids?

24.C Which nucleic acid molecule is the largest?

24.D What kind of bonds are broken during replication? Does the primary structure of DNA change during replication?

PROBLEMS

Difficult problems are designated by an asterisk.

Nucleic Acids and Heredity

24.1 What are chromosomes made of?

24.2 What structures of the cell, visible in a microscope, contain hereditary information?

24.3 Name one hereditary disease.

Components of Nucleic Acids

24.4 (a) Where in a cell is the DNA located? (b) Where in a cell is the RNA located?

24.5 What are the components of (a) a nucleotide (b) a nucleoside?

*24.6 Draw the structures of ADP and GDP. Are these structures parts of nucleic acids?

24.7 What is the difference in structure between thymine and uracil?

24.8 Which DNA and RNA bases contain a carbonyl group?

24.9 Draw the structures of (a) cytidine (b) deoxycytidine

24.10 Which DNA and RNA bases are primary amines?

24.11 What is the difference in structure between D-ribose and 2-deoxy-D-ribose?

24.12 What is the difference between a nucleoside and a nucleotide?

Structure of DNA and RNA

24.13 In RNA, which carbons of the ribose are linked to the phosphate group and which to the base?

24.14 What constitutes the backbone of DNA?

24.15 Draw the structures of (a) UMP (b) dAMP.

24.16 In DNA, which carbon atoms of the 2-deoxy-D-ribose are bonded to the phosphate groups?

24.17 The sequence of a short DNA segment is ATGGCAATAC.
(a) What name do we give to the two ends (terminals) of a DNA molecule?
(b) In this segment, which end is which?

*24.18 Chargaff showed that, in samples of DNA from many different species, the molar quantity of A was always approximately equal to the molar quantity of T, and the same for C and G. How did this information help to establish the structure of DNA?

*24.19 How many hydrogen bonds can be formed between uracil and adenine?

24.20 What interactions stabilize the three-dimensional structure of DNA?

24.21 Which nucleic acid is single-stranded?

DNA Replication

24.22 A DNA molecule normally replicates itself millions of times, with almost no errors. What single fact about the structure is most responsible for this?

24.23 What functional groups on the bases form hydrogen bonds in the DNA double helix?

24.24 Draw the structures of adenine and thymine, and show with a diagram the two hydrogen bonds that stabilize A-T pairing in DNA.

24.25 Draw the structures of cytosine and guanine, and show with a diagram the three hydrogen bonds that stabilize C-G pairing in nucleic acids.

24.26 How many different bases are there in a DNA double helix?

24.27 Where does the unwinding of DNA occur?

24.28 What are helicases? What is their function?

24.29 What do we call the enzymes that join nucleotides into a DNA strand?

24.30 In which direction is the DNA molecule synthesized continuously?

24.31 Which two models have been proposed to function in replication that explain the interaction of DNA with polymerases?

RNA

24.32 Which RNA has enzyme activity? Where does it function mostly?

24.33 Which has the longest chains: tRNA, mRNA, or rRNA?

24.34 What is the central dogma of molecular biology?

24.35 Which kind of RNA has a sequence complementary to that of DNA?

24.36 Where is rRNA located in the cell?

Genes, Exons, Introns, and Cloning

24.37 What is the nature of the interaction between histones and DNA in nucleosomes?

24.38 Define.
(a) Intron
(b) Exon

24.39 Does mRNA also have introns and exons? Explain.

24.40 (a) What percentage of human DNA codes for proteins? (b) What is the function of the rest of the DNA?

24.41 How many histones are in a nucleosome?

*24.42 What 12-nucleotide primer would you use in the PCR technique when you want to clone a gene whose 3′ end is as follows:
3′ TACCGTCATCCGGTG—?

Boxes

24.43 (Box 24A) Draw the structure of the fluorouridine nucleoside that inhibits DNA synthesis.

24.44 (Box 24A) Give an example of how anticancer drugs work in chemotherapy.

24.45 (Box 24B) What sequence of nucleotide is repeated many times in telomers?

24.46 (Box 24B) Why are up to 200 nucleotides lost at each replication?

24.47 (Box 24B) How does telomerase make a cancer cell immortal?

24.48 (Box 24C) After having been cut by restriction enzymes, how are the DNA fragments separated from each other?

24.49 (Box 24C) How is DNA fingerprinting used in paternity suits?

24.50 (Box 24D) What was the first organism to have its genome completely sequenced?

*24.51 (Box 24D) Does the human genome contain only nucleotides representing genes?

*24.52 (Box 24E) How do caspase enzymes, which operate in the cytoplasm of the cell, cause the cleavage of DNA in the nucleus?

24.53 (Box 24E) What are the distinguishing fingerprints of cell death by apoptosis?

Additional Problems

24.54 What is the active site of a ribozyme?

24.55 Why is it important that a DNA molecule be able to replicate itself millions of times without error?

24.56 Why is DNA replication called semiconservative?

*24.57 Which nuclear superstructure is stabilized by an acid-base interaction?

24.58 Draw the structures of (a) uracil (b) uridine.

Chapter 25

Although this statement is correct in the vast majority of cases, in certain viruses the flow of information goes from RNA to DNA.

Above: Ribosomes along mRNA: emerging nascent polypeptides from ribosomes during translation. (© E. Kiseleva, D. Fawcett/Visuals Unlimited)

Gene Expression and Protein Synthesis

25.1 Introduction

We have seen that the DNA molecule is a storehouse of information. We can compare it to a loose-leaf cookbook, each page of which contains one recipe. The pages are the genes. In order to prepare a meal, we use a number of recipes. Similarly, to provide a certain inheritable trait, a number of genes, segments of DNA, are needed.

However, the recipe itself is not the meal. The information in the recipe must be expressed in the proper combination of food ingredients. Similarly, the information stored in DNA must be expressed in the proper combination of amino acids representing a particular protein. The way this works is now so well established that it is called the **central dogma of molecular biology.** The dogma states that **the information contained in DNA molecules is transferred to RNA molecules, and then from the RNA molecules the information is expressed in the structure of proteins. Gene expression is the turning on or activation of a gene.** Transmission of information occurs in two steps.

1. Transcription. Since the information (that is, the DNA) is in the nucleus of the cell and the amino acids are assembled outside the nucleus, the information must first be carried out of the nucleus. This is analogous to copying the recipe from the cookbook. All the necessary information is copied, though in a slightly different format, as if we were converting the printed page into handwriting. On the molecular level, this is accomplished by transcribing the information from the DNA molecule onto a molecule of messenger RNA (Section 24.4), so named because it carries the message from the nucleus to the site of protein synthesis. The transcribed information on the mRNA molecule is then carried out of the nucleus.

2. Translation. The mRNA serves as a template on which the amino acids are assembled in the proper sequence. In order for this to happen, the information that is written in the language of nucleotides must be translated into the language of amino acids. The translation is done by the second type of RNA, transfer RNA (Section 24.4). There is an exact word-to-word translation. Each amino acid in the protein language has a corresponding word in the RNA language. Each word in the RNA language is a sequence of three bases. This correspondence between three bases and one amino acid is called the genetic code (we will discuss the code in Section 25.4).

A summary of the process follows:

$$\text{DNA} \xrightarrow{\text{replication}} \text{DNA} \xrightarrow{\text{transcription}} \text{mRNA} \xrightarrow{\text{translation}} \text{protein}$$

> **Transcription** The process in which information encoded in a DNA molecule is copied into an mRNA molecule

> **Translation** The process in which information encoded in an mRNA molecule is used to assemble a specific protein

25.2 Transcription

An enzyme called RNA polymerase, which catalyzes the synthesis of mRNA, helps to copy the information (the recipe from the cookbook). First, the DNA double helix begins to unwind at a point near the gene that is to be transcribed. Only one strand of the DNA molecule is transcribed. Ribonucleotides assemble along the unwound DNA strand in the complementary sequence. Opposite each C on the DNA there is a G on the growing mRNA, and the other complementary bases follow the patterns $G \rightarrow C$, $A \rightarrow U$, and $T \rightarrow A$.

On the DNA strand, there is always a sequence of bases that the RNA polymerase recognizes as an **initiation signal,** saying, in essence, "Start here." At the end of the gene, there is a **termination sequence** that tells the enzyme, "Stop the synthesis." Between these two signals, the enzyme zips up the complementary bases by forming a phosphate ester bond (Section 18.7) between each ribose and the next phosphate group. In higher organisms (eukaryotes), the RNA polymerase has little affinity for binding to DNA. Certain binding proteins called **transcription factors** facilitate the DNA-RNA polymerase interaction. The enzyme synthesizes the mRNA molecule from the 5′ to the 3′ end (the zipper can move only in one direction). But because the complementary chains (RNA and DNA) run in opposite directions, the enzyme must move along the DNA template in the $3' \rightarrow 5'$ direction of the DNA (Figure 25.1). After the mRNA molecule has been synthesized, it moves away from the DNA template, which then rewinds to the original double-helix form. Transfer RNA and ribosomal RNA are also synthesized on DNA templates in this manner.

Transcription does not occur at the same rate throughout the process. It is accelerated or slowed down as the need arises. The signal to activate transcription may originate from outside of the cell. One such signal, the

These are not the deoxyribonucleotides used in DNA replication.

Note again that RNA contains no thymine but has uracil instead.

After RNA molecules are synthesized, they move out of the nucleus and into the cytoplasm.

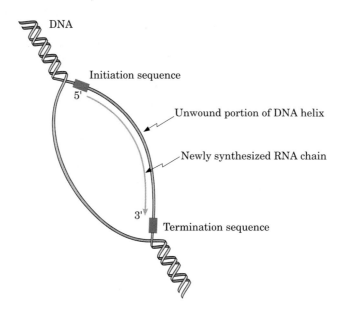

GTP-adenylate cyclase-cAMP pathway (Section 23.4) produces phosphory-lated protein kinase. The catalytic unit of this enzyme enters the nucleus. There it phosphorylates transcription factors, which initiate a transcription cascade.

25.3   The Role of RNA in Translation

Translation is the process by which the genetic information preserved in the DNA and transcribed into the mRNA is converted to the language of proteins, that is, the amino acid sequence. All three types of RNA partici-pate in the process.

The synthesis of proteins takes place on the ribosomes (Section 24.5). These spheres dissociate into two parts—a larger and a smaller body. Each of these bodies contains rRNA and some polypeptide chains that act as enzymes, speeding up the synthesis. In higher organisms, including humans, the larger ribosomal fragment is called the 60S ribosome, and the smaller one is called the 40S ribosome. As the mRNA is being made on the DNA template (Section 25.2), the 5′ end of the mRNA coming off this assembly is first attached to the smaller ribosomal body and later joined by the larger body. Together they form a unit on which the mRNA is stretched out. After the mRNA is attached to the ribosome in this way, the 20 amino acids are brought to the site, each carried by its own particular tRNA molecule.

The most important segments of the tRNA molecule are (1) the site to which enzymes attach the amino acids and (2) the recognition site. Figure 24.9 shows that the 3′ terminal of the tRNA molecule carries the amino acid.

As we have said, each tRNA is specific for one amino acid only. How does the body make sure that alanine, for example, attaches only to the one tRNA molecule that is specific for alanine? The answer is that each cell carries at least 20 specific enzymes for this purpose. Each of these enzymes recognizes only one amino acid and only one tRNA. The enzyme attaches the activated amino acid to the 3′ terminal —OH group of the tRNA, forming an ester bond.

The second important segment of the tRNA molecule carries the **recognition site, which is a sequence of three bases called an anti-codon** located at the opposite end of the molecule in the three-dimensional structure of tRNA. This triplet of bases can align itself in a complementary fashion to another triplet on mRNA. The triplets of bases on the mRNA are called **codons.**

The S, or Svedberg unit, is a meas-ure of the size of these bodies.

The 20 amino acids are always available in the cytoplasm, near the site of protein synthesis.

Anticodon A sequence of three nucleotides on tRNA compli-mentary to the codon in mRNA

Codon The sequence of three nucleotides in messenger RNA that codes for a specific amino acid

25.4 The Genetic Code

By 1961 it was apparent that the order of bases in a DNA molecule corresponds to the order of amino acids in a particular protein. But the code was unknown. Obviously, it could not be a one-to-one code. There are only four bases, so if for example, A coded for glycine, G for alanine, C for valine, and T for serine, there would be 16 amino acids that could not be coded.

In 1961 Marshall Nirenberg (1927–   ) and his co-workers attempted to break the code in a very ingenious way. They made a synthetic molecule of mRNA consisting of uracil bases only. They put this into a cellular system that synthesized proteins and then supplied the system with all 20 amino acids. The only polypeptide produced was a chain consisting solely of the amino acid phenylalanine. This showed that the code for phenylalanine must be UUU or some other multiple of U.

A series of similar experiments by Nirenberg and other workers followed, and by 1967 the entire genetic code had been broken. **Each amino acid is coded for by a sequence of three bases,** called a **codon.** The complete code is shown in Table 25.1.

The first important aspect of **the genetic code is** that it is almost **universal.** In virtually every organism, from a bacterium to an elephant to a human, the same sequence of three bases codes for the same amino acid. The universality of the genetic code implies that all living matter on earth arose from the same primordial organisms. This is perhaps the strongest evidence for Darwin's theory of evolution.

There are 20 amino acids in proteins, but there are 64 possible combinations of four bases into triplets. All 64 codons (triplets) have been deciphered. Three of them—UAA, UAG, and UGA—are stop signs. They terminate protein synthesis. The remaining 61 codons all code for amino acids. Since there

Three-dimensional structure of tRNA.

Genetic code The sequence of triplets of nucleotides (codons) that determines the sequence of amino acids in a protein

TABLE 25.1 The Genetic Code

First Position (5'-end)	Second Position								Third Position (3'-end)
	U		*C*		*A*		*G*		
U	UUU	Phe	UCU	Ser	UAU	Tyr	UGU	Cys	U
	UUC	Phe	UCC	Ser	UAC	Tyr	UGC	Cys	C
	UUA	Leu	UCA	Ser	UAA	Stop	UGA	Stop	A
	UUG	Leu	UCG	Ser	UAG	Stop	UGG	Trp	C
C	CUU	Leu	CCU	Pro	CAU	His	CGU	Arg	U
	CUC	Leu	CCC	Pro	CAC	His	CGC	Arg	C
	CUA	Leu	CCA	Pro	CAA	Gln	CGA	Arg	A
	CUG	Leu	CCG	Pro	CAG	Gln	CGG	Arg	G
A	AUU	Ile	ACU	Thr	AAU	Asn	AGU	Ser	U
	AUC	Ile	ACC	Thr	AAC	Asn	AGC	Ser	C
	AUA	Ile	ACA	Thr	AAA	Lys	AGA	Arg	A
	AUG*	Met	ACG	Thr	AAG	Lys	AGG	Arg	G
G	GUU	Val	GCU	Ala	GAU	Asp	GGU	Gly	U
	GUC	Val	GCC	Ala	GAC	Asp	GGC	Gly	C
	GUA	Val	GCA	Ala	GAA	Glu	GGA	Gly	A
	GUG	Val	GCG	Ala	GAG	Glu	GGG	Gly	G

* AUG also serves as the principal initiation codon.

Some exceptions to the genetic code in Table 23.2 occur in mitochondrial RNA. Because of that and other evidences, it is thought that the mitochondrion may have been an ancient entity. During evolution it developed a symbiotic relationship with eukaryotic cells. For example, some of the respiratory enzymes located on the cristae of the mitochondrion (see Section 26.2) are encoded in the mitochondrial DNA, and other members of the same respiratory chain are encoded in the nucleus of the eukaryotic cell.

Because of this multiple coding, the genetic code is called a multiple code or a degenerate code.

are only 20 amino acids, there must be more than one codon for each amino acid. Indeed, some amino acids are coded for by as many as six codons. Leucine, for example, is coded for UUA, UUG, CUU, CUC, CUA, and CUG.

Just as there are three stop signs in the code, there is also an initiation sign. The initiation sign is AUG, which is also the codon for the amino acid methionine. This means that, in all protein synthesis, the first amino acid is always methionine. Methionine can also be put into the middle of the chain because there are two kinds of tRNA for it.

Although all protein synthesis starts with methionine, most proteins in the body do not have a methionine residue at the N-terminal of the chain. In most cases, the initial methionine is removed by an enzyme before the polypeptide chain is completed. The code on the mRNA is always read in the $5' \rightarrow 3'$ direction, and the first amino acid to be linked to the initial methionine is the N-terminal end of the translated polypeptide chain.

25.5 Translation and Protein Synthesis

So far we have met the molecules that participate in protein synthesis (Section 25.3) and the dictionary of the translation, the genetic code. Now let us look at the actual mechanism by which the polypeptide chain is assembled.

All protein synthesis takes place outside the nucleus, in the cytosol.

There are four major stages in protein synthesis: activation, initiation, elongation, and termination.

Activation

In the activation step, energy is used up. Two high-energy phosphate bonds are broken for each amino acid added to the chain: $ATP \rightarrow AMP + PP_i$ and $PP_i \rightarrow 2P_i$.

Each amino acid is first activated by reacting with a molecule of ATP:

$$\text{Adenosine} - O - \underset{\underset{O^-}{|}}{\overset{\overset{O}{\|}}{P}} - O - \underset{\underset{O^-}{|}}{\overset{\overset{O}{\|}}{P}} - O - \underset{\underset{O^-}{|}}{\overset{\overset{O}{\|}}{P}} - O^- \quad + \quad {}^-O - \overset{\overset{O}{\|}}{C} - \underset{\underset{R}{|}}{CH} - NH_3{}^+ \longrightarrow$$

ATP   An amino acid

$$\text{Adenosine} - O - \underset{\underset{O^-}{|}}{\overset{\overset{O}{\|}}{P}} - O - \overset{\overset{O}{\|}}{C} - \underset{\underset{R}{|}}{CH} - NH_3{}^+ \quad + \quad {}^-O - \underset{\underset{O^-}{|}}{\overset{\overset{O}{\|}}{P}} - O - \underset{\underset{O^-}{|}}{\overset{\overset{O}{\|}}{P}} - O^-$$

An amino acid-AMP   Pyrophosphate

The activated amino acid is then bonded to its own particular tRNA molecule with the aid of an enzyme (a synthetase) that is specific for that particular amino acid and that particular tRNA molecule:

tRNA   Amino acid–tRNA

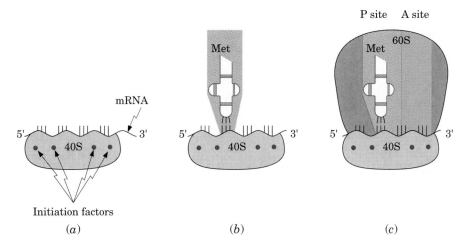

FIGURE 25.2 The initiation of protein synthesis. (*a*) The mRNA attaches to the 40S ribosomal body. (*b*) The first tRNA anticodon binds to the initiation mRNA codon. (*c*) The 60S ribosomal body joins the unit.

Initiation

This stage consists of three steps:

1. The mRNA molecule, which carries the information necessary to synthesize one protein molecule, attaches itself to the 40S ribosome (the smaller body; see Section 25.3). This is shown in Figure 25.2(a).

2. To initiate the protein synthesis, a unique tRNA molecule, methionine tRNA$_i$, is used (Figure 25.2[b]). (In the elongation process, a completely different methionine tRNA participates in incorporating methionine into the growing chain.) The anticodon of the methionine tRNA, UAC, binds to the codon of the mRNA, AUG, that represents the initiation signal.

3. The 60S ribosome (the larger portion) now combines with the 40S body, as shown in Figure 25.2(c).

The 60S body carries two binding sites. The one shown on the left in Figure 25.2(c) is called the **P site** because that is where the growing peptide chain will bind. The one right next to it is called the **A site** because that is where the incoming tRNA will bring the next amino acid. When the 60S ribosome attaches itself to the 40S one, it does so in such a way that the P site is right where the methionine tRNA$_i$ already is.

Elongation

At this point, the A site is vacant, and each of the 20 tRNA molecules can come in and try to fit itself in. But only one of the 20 carries exactly the right anticodon that corresponds to the next codon on the mRNA. (This is an alanine tRNA in Figures 25.3 and 25.4.) The binding of this tRNA to the A site takes place with the aid of proteins called **elongation factors.** At the A site, the new amino acid, alanine (Ala), is linked to the Met in a peptide bond by the enzyme peptidyl **transferase.** The empty tRNA remains on the P site.

In the next phase of elongation, the whole ribosome moves one codon along the mRNA. Simultaneously with this move, the dipeptide is **translocated** from the A site to the P site, as shown in Figure 25.4(d), while the empty tRNA dissociates and goes back to the tRNA pool to pick up another

FIGURE 25.3 Alanine tRNA aligning on the ribosome with its complementary codon.

More recently a third site was identified on the ribosome. It is the site from which the tRNA exits when it is no longer needed. It is called the E site.

In prokaryotes, the corresponding ribosomes are smaller, 30S and 50S.

This enzyme, which is part of the 60S ribosome unit, is not a protein but a ribozyme.

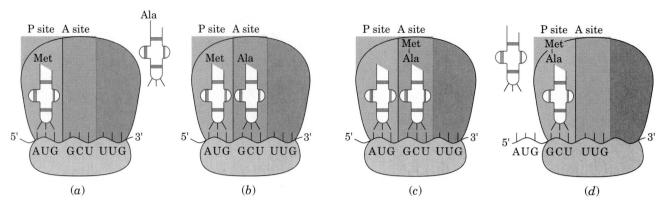

(a)　　　　(b)　　　　(c)　　　　(d)

FIGURE 25.4 Phases of elongation in protein synthesis. (*a*) After initiation, the tRNA of the second amino acid (alanine in this case) approaches the ribosome. (*b*) The tRNA binds to the A site. (*c*) The enzyme transferase connects methionine to alanine, forming a peptide bond. (*d*) The peptide tRNA is moved (translocated) from the A to the P site while the ribosome moves to the right and simultaneously releases the empty tRNA. In this diagram, the alanine tRNA is shown in blue.

amino acid. After the translocation, the A site is associated with the next codon on the mRNA, which is UUG in Figure 25.4(d). Once again, each tRNA can try to fit itself in, but only the one whose anticodon is AAC can align itself with UUG. This one, the tRNA that carries leucine (Leu), now comes in. The transferase establishes a new peptide bond between Leu and Ala, moving the dipeptide from the P site to the A site and forming a tripeptide. These elongation steps are repeated until the last amino acid is attached.

Figure 25.5 shows a three-dimensional model of the translational process, which has been constructed on the basis of recent cryoelectron microscopy and x-ray diffraction studies. This model clearly shows how the

FIGURE 25.5 Ribosome in action. The lower yellow half represents the 30S ribosome; the upper blue half represents the 50S ribosome. The yellow and green twisted cones are tRNAs and the chain of beads stand for mRNA. (*Courtesy of Dr. J. Frank, Wadsworth Center, Albany, New York*)

(continued on page 596)

Viruses

Nucleic acids are essential for life as we know it. No living thing can exist without them because they carry the information necessary to make protein molecules. The smallest form of life, the viruses, consist only of a molecule of nucleic acid surrounded by a "coat" of protein molecules. In some viruses, the nucleic acid is DNA; in others, it is RNA. No virus has both. The shapes and sizes of viruses vary greatly. Some of them are shown in Figure 25A.1.

Because their structures are so simple, viruses are unable to reproduce themselves in the absence of other organisms. They carry DNA or RNA but do not have the nucleotides, enzymes, amino acids, and other molecules necessary to replicate their nucleic acid (Section 24.4) or to synthesize proteins (Section 25.5). Instead, viruses invade the cells of other organisms and cause those cells (the hosts) to do these tasks for them. Typically, the protein coat of a virus remains outside the host cell, attached to the cell wall, while the DNA or RNA is pushed inside. Once the viral nucleic acid is inside the cell, the cell stops replicating its own DNA and making its own proteins and now replicates the viral nucleic acid and synthesizes the viral protein, according to the instructions on the viral nucleic acid. One host cell can make many copies of the virus.

In many cases, the cell bursts when a large number of new viruses have been synthesized, sending the new viruses out into the intercellular material, where they can infect other cells. This kind of process causes the host organisms to get sick, per-haps to die. Among the many human diseases caused by viruses are measles, hepatitis, mumps, influenza, the common cold, rabies, and smallpox. There is no cure for most viral diseases. Antibiotics, which can kill bacteria, have no effect on viruses. So far, the best defense against these diseases has been immunization (Box 30C), which under the proper circumstances can work spectacularly well. Smallpox, once one of the most dreaded diseases, has been totally eradicated from this planet by many years of vaccination, and comprehensive programs of vaccination against such diseases as polio and measles have greatly reduced the incidence of these diseases.

Lately, a number of antiviral agents have been developed. They completely stop the reproduction of viral nucleic acids (DNA or RNA) inside infected cells without preventing the DNA of normal cells from replicating. One such drug is called vidarabine, or Ara-A, and is sold under the trade name Vira-A. Antiviral agents often act like the anticancer drugs (Box 24A), in that they have structures similar to one of the nucleotides necessary for the synthesis of nucleic acids. Vidarabine is the same as adenosine, except that the sugar is arabinose instead of ribose (Figure 25A.2). Vidarabine is used to fight a life-threatening viral illness, herpes encephalitis. It is also effective in neonatal herpes infection and chicken pox. However, as with many other anticancer and antiviral drugs, vidarabine is toxic, causing nausea and diarrhea. In some cases, it has caused chromosomal damage.

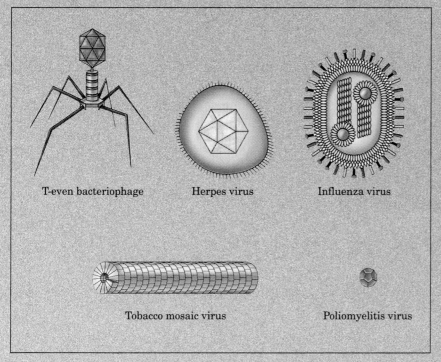

T-even bacteriophage   Herpes virus   Influenza virus

Tobacco mosaic virus   Poliomyelitis virus

Vidarabine

■ *FIGURE 25A.1* The shape of various viruses. *(From* The Structure of Viruses *by R. W. Horne. Copyright 1963 by Scientific American, Inc. All rights reserved.)*

■ *FIGURE 25A.2* Vidarabine.

AIDS

In recent years a deadly virus, commonly called the AIDS virus (for Acquired Immune Deficiency Syndrome) or HIV (for Human Immunodeficiency Virus), has spread alarmingly fast all over the world. Perhaps 100 million people are infected, and this number is increasing daily, especially in underdeveloped countries. The AIDS virus invades the human immune system, and especially enters the T (lymphocyte) cells and kills them (Figure 25B.1). The AIDS virus, an RNA-containing virus or **retrovirus,** decreases the population of T cells (blood cells that fight invading foreign cells; see Section 30.2), and thus allows other opportunistic invaders, such as the protozoan *Pneumocystis carinii,* to proliferate, causing pneumonia and eventual death.

The progressive paralysis of the immune system makes the AIDS virus a lethal threat. Those who are ill with AIDS or have been exposed to the virus carry antibodies against the virus in their blood. One can detect these antibodies through patented tests, and all blood banks are now testing donated blood for AIDS infection. In the United States, currently two to three out of 10 000 blood samples are found to carry antibodies against AIDS. With such screening, infections through blood transfusions are minimized.

At the time of writing, there is no known cure for AIDS. At least two approaches have been tried. One involves an antiviral agent, azidothymidine (AZT). AIDS patients with pneumo-

■ *FIGURE 25B.1* HIV particles (pink) on the surface of a lymphocyte (green). *(© NIBSC/Science Photo Library/Photo Researchers, Inc.)*

nia caused by *Pneumocystis carinii* show some improvement after taking this drug. However, AZT is toxic, and most patients cannot tolerate therapeutic doses. Also, AZT provides only transient improvement—after about 2 years the virus develops resistance to the drug. Other antiviral agents, such as dideoxyinosine (ddI), dideoxycytidine (ddCyd), and lamivudine, also known as (−)-2-deoxy-3'-thiacytidine (3TC), have been found beneficial in patients who cannot tolerate AZT or whose HIV becomes resistant to AZT. None of these drugs provides a cure. They only slow the development of the disease. All four are nucleoside-analog inhibitors, which inhibit a key viral enzyme, reverse transcriptase.

Another group of drugs provides a second approach acting against a different viral enzyme, protease. These drugs—abacavir, indinavir, ritonavir, and saquinavir (Figure 25B.2)—are protease inhibitors, entering into the active site of the enzyme. By slowing down the action of protease, they suppress viral replication. These drugs have proved to be effective, even in AIDS patients who are very ill, and have few of the side effects of AZT and the other drugs of its type. However, to suppress the viral infection of lymphocytes, large doses are required. Furthermore, as with the first class of drugs, HIV quickly mutates, and after two to three months of treatment, protease-inhibitor-drug-resistant strains of HIV appear. To slow down the mutation of the virus, even larger doses of protease inhibitors are required. The present strategy is to treat AIDS patients with a battery of different drugs (for example, AZT and indinavir, or ritonavir, ddCyd, and 3TC) to minimize the viral mutations. It has been shown that employment of such two- or three-drug cocktails greatly reduces the presence of HIV in the patients' blood for short periods, up to eight weeks. (See Figure 25B.2.) Whether the virus can be eradicated with long-term treatment is a question yet to be answered, though at least temporarily this treatment reduces the death rate.

The strategy of treating a patient with a battery of drugs developed in fighting AIDS paid unexpected dividends in treating other diseases. Hepatitis viruses B and C are present in 1.5 percent of the U.S. population. A combination of interferon, which bolsters the immune system, and ribavirin, another viral protease inhibitor, is the cocktail of treatment for hepatitis C.

A vaccine would be the ideal remedy. But because of the deadly nature of the virus, one cannot use attenuated or even "killed" virus for immunization as was done with the polio vaccine. Even if only one virus out of a million survived, it could kill the vaccinated person.

2′,3′-Dideoxyinosine (ddI)
(Didanosine)

2′,3′-Dideoxycytidine (ddCyd)
(Zalcitabine)

(−)-2′-Deoxy-3′-thiacytidine (3TC)
(Lamivudine)

Indinavir

Ritonavir

Saquinavir

■ **FIGURE 25B.2** Structural formulas for some AIDS drugs.

elongation factor proteins (in blue) are fitting in a cleft between the 50S (pale blue) and the 30S (pale yellow) of prokaryotic ribosomes. The tRNAs on the P site (green) and on the A site (yellow) occupy a central cavity in the ribosomal complex. The orange beads represent the mRNA. Similar arrangement exists in eukaryotes, which have 60S and 40S ribosomes.

Termination

After the last translocation, the next codon reads "Stop" (UAA, UGA, or UAG). No more amino acids can be added. Releasing factors then cleave the polypeptide chain from the last tRNA in a mechanism not yet fully understood. The tRNA itself is released from the P site. While the mRNA is attached to the ribosomes, many polypeptide chains are synthesized on it simultaneously. This can be seen in the electron micrograph shown in the chapter opening photo, where a number of polypeptides in the making are attached to the mRNA. At the end, the whole mRNA is released from the ribosome.

> Their work done, the two parts of the ribosome separate.

25.6 Gene Regulation

> **Gene regulation** The control process by which the expression of a gene is turned on or off

Every embryo that is formed by sexual reproduction inherits its genes from the parent sperm and egg cells. But the genes in its chromosomal DNA are not active all the time. They are switched on and off during development and growth of the organism. Soon after formation of the embryo, the cells begin to differentiate. Some cells become neurons, some cells become muscle cells, some become liver cells, and so on. Each cell is a specialized unit that uses only some of the many genes it carries in its DNA. This means that each cell must switch some of its genes on and off—either permanently or temporarily. How this is done is the subject of gene regulation. We know less about gene regulation in eukaryotes than in the simpler prokaryotes. Even with our limited knowledge, however, we can state that organisms do not have a single, unique way of controlling genes. Many gene regulations occur at the transcriptional level (DNA → RNA). Some operate at the translational level (mRNA → protein). A few of these processes are listed here as examples.

> Since RNA synthesis proceeds in one direction (5′ → 3′) (see Figure 25.2), the gene (DNA) to be transcribed runs from 3′ → 5′. Thus, the control sites are in front of, or upstream of, the 3′ end of the structural gene.

1. Operon, a case of negative regulation This process operates only in prokaryotes. In the common species of bacteria *Escherichia coli*, the genes for a number of enzymes are organized in a unit. This unit contains DNA sequences that code for proteins as well as sequences that play a part in gene regulation. Sequences that code for proteins are called **structural genes.** Preceding the structural genes is a DNA sequence called the **control sites,** made up of two parts called the **operator site** and the **promoter site.** Preceding the control sites is a DNA sequence called the **regulatory gene.** The structural genes, the control sites, and the regulatory gene together form a unit called the **operon.**

One of the most studied and understood operons in *E. coli* is called the *lac* operon. It contains three structural genes that code for three enzymes, called β-galactosidase, lactose permease, and thiogalactosidase transacetylase (Figure 25.6). These enzymes are associated with the breakdown of milk sugar, lactose (Section 19.6).

There is an allosteric protein, called a repressor protein, that prevents the transcription of the structural genes. This binds to the operator site of the control sites. The RNA polymerase that actually synthesizes the mRNA must be bound to the promoter site in order to

BOX 25C

The Promoter. A Case of Targeted Expression

Although the major part of a gene contains the region that codes for a particular protein, a portion of the gene does not code for any part of the protein. Instead it acts as a switch that turns the transcription process on or off. This regulatory part of the gene is called the **promoter.** By using the gene insertion technique discussed in Section 25.8, one can combine the coding sequence of one gene with the promoter of another gene, thereby targeting the expression of a protein to a certain organ.

A recent development in AIDS research (Box 25B) involved an enzyme called HIV-1 protease. This enzyme is essential if the virus that causes AIDS is to multiply inside the host cell. It has been argued that if suitable drugs that can inhibit HIV-1 protease could be found, the development of AIDS could be arrested, and perhaps the disease cured. But how can we test the hundreds of organic compounds that give promise of being HIV-1 protease inhibitors?

This is where the concept of the promoter comes in. For example, if the coding region of the HIV-1 protease is linked to the promoter of αA-crystallin, a protein molecule that appears only in the lens of the eye, the expression of the HIV-1 protease could be targeted to the eye only. True to the design, a DNA sequence containing the gene for HIV-1 protease and the promoter of αA-crystallin were transfected into a mouse. The result was transgenic mice that became blind 24 days after birth. Otherwise, the mice were not affected at all—they ate, ran, and bred as normal mice do. The blindness was the result of the action of HIV-1 protease in the lens, causing cataract formation (Figure 25C).

These newly created transgenic mice could now be used to test drugs to find out if they would be good inhibitors of HIV-1 protease. If injection of a potential drug delayed cataract formation, that would be proof that the drug truly acts as an inhibitor of HIV-1 protease in a living organism and not just in test tubes. Many of the new drugs now used in treatment of AIDS patients (see Box 25B), for example, indinavir, have first been tested in this manner. This is clearly more efficient than testing all such potential inhibitors directly in humans.

■ *FIGURE 25C* Eye with cataractous lens of a transgenic mouse containing the HIV-1 protease linked to αA-crystallin promoter, 27 days after birth. *(Photo courtesy of Dr. Paul Russell, National Eye Institute, NIH)*

transcribe the gene. When the repressor occupies the operator site, the RNA polymerase cannot bind to the promoter site. Thus, the gene will be inactive. This is negative regulation. However, when the proper inducer is present—in this case, lactose—it will form a complex with the repressor. As a consequence, the operator site will be empty, and the RNA polymerase can bind to the promoter site to perform the transcription.

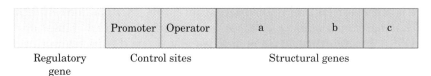

Regulatory gene	Promoter	Operator	a	b	c
	Control sites		Structural genes		

FIGURE 25.6 A schematic view of the *lac* operon in *E. coli*. The a, b, and c correspond, respectively, to the genes that code for the proteins β-galactosidase, lactose permease, and thiogalactosidase transacetylase.

2. Response elements   We have seen that steroid hormones exert their influence in the cell nucleus (Section 23.7) of eukaryotes. Such a hormone binds to a steroid receptor protein. Specific promoter modules on the DNA molecule, called **response elements,** activate the transcription of a gene. The response element of steroids is in front of and is 250 base pairs upstream from the starting point of transcription. Only the receptor with the bound steroid hormone can interact with its response element and thus initiate the transcription.

3. Transcription factors   In eukaryotes, the enzyme RNA-polymerase has little affinity for binding to DNA. In these cells, selective binding proteins bind to a promoter site. These binding proteins are called **transcription factors.** The binding one after another of these factors to the promoter site controls the rate of initiation of transcription. This control is very selective. Although a prokaryotic operon can vary the rate by a factor of perhaps one thousand, a eukaryotic assembly of transcription factors may allow the synthesis of mRNA (and from there the target protein) to vary by a factor of one million. An example of such a wide variation in eukaryotic cells is that a specific gene, for example, the α-A-crystallin gene, can be expressed in the lens of the eye, at a rate one-million-fold higher than the same protein gene in the liver cell of the same organism.

How do these transcription factors find the specific gene control sequences into which they fit, and how do they bind to them? The interaction between the protein and DNA is by nonspecific electrostatic interactions (positive ions attracting negative ones and repelling other positive ions) as well as by more specific hydrogen bonding. They find their targeted sites by twisting their protein chains so that a certain amino acid sequence is present at the surface. One such conformational twist is provided by what are called **metal-binding fingers** (Figure 25.7).

These finger shapes are created by Zn^{2+} ions, which form covalent bonds with the amino acid side chains of the protein. The zinc fingers interact with specific DNA (or sometimes RNA) sequences. The recognition comes by hydrogen bonding between a nucleotide (for example, guanine) and the side chain of a specific amino acid (for example, arginine). Besides metal-binding fingers, at least two other prominent transcription factors exist, called helix-turn-helix and leucine zipper. Transcription factors can also be repressors, reducing the rate of transcription. Some can serve as both activators and repressors.

FIGURE 25.7   Schematic view of a metal-binding finger (a common motif in many transcription factors). The zinc ion forms coordinate-covalent bonds to two cysteines and two histidines.

Protein chain

25.7   Mutations, Mutagens, and Genetic Diseases

In Section 24.4, we saw that the base-pairing mechanism provides an almost perfect way to copy a DNA molecule during replication. The key word here is "almost." No machine, not even the copying mechanism of DNA replication, is totally without error. It has been estimated that, on average, there is one error for every 10^{10} bases (that is, one in 10 billion). An error in the copying of a sequence of bases is called a **mutation.** Mutations can occur during replication. Base errors can also occur during transcription in protein synthesis (a nonheritable error). These errors may have widely varying consequences. For example, the codon for valine in mRNA can be GUA, GUG, GUC, or GUU. In DNA, these correspond to CAT, CAC, CAG, and CAA, respectively. Let us assume that the original codon in the DNA is CAT. If during replication a mistake is made and the CAT was spelled as

Mutations and Biochemical Evolution

We can trace the genetic relationship of different species through the variability of their amino acid sequences in different proteins. For example, the blood of all mammals contains hemoglobin, but the amino acid sequences of the hemoglobins are not identical. In Table 25D, we see that the first ten amino acids in the β-globin of humans and gorillas are exactly the same. As a matter of fact, there is only one amino acid difference, at position 104, between us and apes. The β-globin of the pig differs from ours at 10 positions, of which 2 are in the N-terminal decapeptide. That of the horse differs from ours in 26 positions, of which 4 are in this decapeptide. β-Globin seems to have gone through many mutations during the evolutionary process because only 26 of the 146 sites are invariant, that is, exactly the same in all species studied so far.

The relationship between different species can also be established by similarities in mRNA primary structures. Since the mutations actually occurred on the original DNA molecule, and it was perpetuated in the progeny by the mutant DNA, it is instructive to learn how a point mutation may occur in different species. Looking at position 4 of the β-globin molecule (Table 25D), we see a change from serine to threonine. The code for serine is AGU or AGC, whereas that for threonine is ACU or ACC (Table 25.1). Thus, a change from G to C in the second position of the codon brought the divergence between the β-globins of humans and horses. The genes of closely related species, such as humans and apes, have very similar primary structures, presumably because these two species diverged on the evolutionary tree only recently. On the other hand, species far removed from each other diverged long ago and have undergone more mutations, which show up in differences in primary structures of their DNA, mRNA, and consequently proteins.

The number of amino acid substitutions is significant in the evolutionary process caused by mutation, but the kind of substitution is even more important. If the substitution is by an amino acid with physicochemical properties similar to those of the amino acid in the ancestor protein, the mutation is most probably viable. For example, in human and gorilla β-globin, position 4 is occupied by threonine, but it is occupied by serine in the pig and horse. Both amino acids provide an —OH-carrying side chain.

TABLE 25D **Amino Acid Sequence of the N-Terminal Decapeptides of β-Globin in Different Species**

Species	Position									
	1	*2*	*3*	*4*	*5*	*6*	*7*	*8*	*9*	*10*
Human	Val	His	Leu	Thr	Pro	Glu	Glu	Lys	Ser	Ala
Gorilla	Val	His	Leu	Thr	Pro	Glu	Glu	Lys	Ser	Ala
Pig	Val	His	Leu	Ser	Ala	Glu	Glu	Lys	Ser	Ala
Horse	Val	Glu	Leu	Ser	Gly	Glu	Glu	Lys	Ala	Ala

CAG in the copy, there will be no harmful mutation because, when a protein is synthesized, the CAG will be transcribed onto the mRNA as GUC, which also codes for valine. Therefore, although a mutation occurred, the same protein is manufactured.

On the other hand, assume that the original sequence in the gene's DNA is CTT, which transcribes onto mRNA as GAA and codes for glutamic acid. If, during replication, a mutation occurs and CTT becomes ATT, the new cells will probably die. The reason is that ATT transcribes to UAA, which does not code for any amino acid but rather is a stop signal. Thus, instead of continuing to build a protein chain with glutamic acid, the synthesis stops altogether. An important protein is not manufactured, and the organism may die. In this way, very harmful mutations are not carried over from one generation to the next.

Ionizing radiation (x-rays, ultraviolet light, gamma rays) can cause mutations. Furthermore, a large number of chemicals can induce mutation by reacting with DNA. Such chemicals are called **mutagens.** Many

Sickle cell anemia (Box 21D) is caused by a single amino acid change: valine for glutamic acid. In this case, CTT is mutated to CAT.

This is discussed in Section 9.5.

Examples of mutagens are benzene, carbon tetrachloride, and vinyl chloride.

BOX 25E

Oncogenes

An **oncogene** is a gene that in some way or other participates in the development of cancer. Cancer cells differ from normal cells in a variety of structural and metabolic ways; however, the most important difference is their uncontrolled proliferation. Most ordinary cells are quiescent. When they lose their controls, they give rise to tumors—benign or malignant. The uncontrolled proliferation allows these cells to spread, invade other tissues and colonize them, a process called metastasis. Malignant tumors can be caused by a variety of agents: chemical carcinogens such as benzo(e)pyrene (Box 14B), ionizing radiation such as x-rays, and viruses. However, for a normal cell to become a cancer cell, certain transformations must occur. Many of these transformations take place on the level of the genes. Certain genes in normal cells have counterparts in viruses, especially in retroviruses (which carry their genes in the form of RNA rather than DNA). This means that a normal cell in a human being and in a virus both have genes that code for similar proteins. Usually these genes code for proteins that control cell growth. Such a protein is epidermal growth factor (EGF). When a retrovirus gets into the cell it takes control of the cell's own EGF gene. At that point, the EGF gene becomes an oncogene, a cancer-causing gene, because it produces large amounts of the EGF protein, which in turn allows the epidermal cells to grow in an uncontrolled manner.

Because the normal EGF gene in the normal cell was turned into an oncogene, we call it a **proto-oncogene.** Such conversions of a proto-oncogene to an oncogene can occur not just by viral invasion but also by a mutation of the gene. For example, a form of cancer called human bladder carcinoma is caused by a single mutation of a gene—a change from guanine (G) to thymine (T). The resulting protein has a valine in place of a glycine. This change is sufficient to transform the cell to a cancerous state because the protein for which the gene is coded is a protein that amplifies signals (Section 23.4). Thus, because of the mutation that caused the proto-oncogene to become an oncogene, the cell stays "on"—its metabolic process is not shut off—and it is now a cancer cell that proliferates without control. There are a number of other mechanisms by which a proto-oncogene is transformed to an oncogene, but in each case the gene product has something to do with growth, hormonal action, or some other kind of cell regulation. About 100 viral and cellular oncogenes have been identified.

changes caused by radiation and mutagens do not become mutations because the cell has its own repair mechanism, called nucleotide excision repair (NER), which can prevent mutations by cutting out damaged areas and resynthesizing them. In spite of this defense mechanism, certain errors in copying that result in mutations do slip by. Many chemicals (both synthetic and natural) are mutagens, and some can cause cancer when introduced into the body. These are called **carcinogens** (Box 14B). One of

BOX 25F

A p53 Central Tumor Suppressor Protein

Not all gene mutations causing cancer have their origin in oncogenes. There are some 36 known **tumor suppressor genes,** the products of which are proteins controlling cell growth. None of them are more important than a protein with a molecular mass of 53 000 simply named **p53.** In about 40 percent of all cancer cases, the tumor contains a p53 that underwent mutation. Mutated p53 protein can be found in 55 percent of lung cancers, about half of the colon and rectal cancers, and some 40 percent of lymphomas, stomach and pancreatic cancers. On top of this, in one third of the soft tissue sarcomas, p53 is inactive, even though it did not undergo mutation.

All these indicate that the normal function of p53 protein is to suppress tumor growth. p53 protein binds to specific sequences of double-stranded DNA. When x-rays or γ-rays damage DNA, an increase in p53 protein concentration is observed. The increased binding of p53 controls the cell cycle; it holds it between cell division and DNA replication. The time gained in this arrested cell cycle allows the DNA to repair its damage. If that fails, the p53 protein triggers apoptosis, the programmed cell death of the injured cell (see also Box 24E). This happens in normal cells with normal p53. In any case, the damaged DNA would be repaired or eliminated.

The mutated p53 in cancer cells can no longer bind to the DNA in the normal fashion. Therefore, it loses its tumor suppressor ability.

the main tasks of the U.S. Food and Drug Administration and the Environmental Protection Agency is to identify these chemicals and eliminate them from our food, drugs, and environment. Even though most carcinogens are mutagens, the reverse is not true.

Not all mutations are harmful. Certain ones are beneficial because they enhance the survival rate of the species. For example, mutation is used to develop new strains of plants that can withstand pests.

If a mutation is harmful, it results in an inborn genetic disease. This may be carried as a recessive gene from generation to generation with no individual demonstrating the symptoms of the disease. Only when both parents carry recessive genes does an offspring have a 25 percent chance of inheriting the disease. If the defective gene is dominant, on the other hand, every carrier will develop symptoms.

A recessive gene is one that expresses its message only if both parents transmit it to the offspring. Specific diseases that result from defective or missing genes are discussed in Boxes 20F, 21D, 27B, and 27F.

25.8 Recombinant DNA

There are no cures for the inborn genetic diseases that we discussed in the preceding section. The best we can do is detect the carriers and, through genetic counseling of prospective parents, try not to perpetuate the defective genes. However, a technique called the recombinant DNA technique gives some hope for the future. At this time, recombinant DNA techniques are used mostly in bacteria, plants, and test animals (such as mice), but it is possible that they may someday be successfully extended to human cells as well.

One example of the **recombinant DNA technique** begins with certain circular DNA molecules found in the cells of the bacteria *Escherichia coli*. These molecules, called **plasmids** (Figure 25.8), consist of double-stranded DNA arranged in a ring. Certain highly specific enzymes called **restriction endonucleases** cleave DNA molecules at specific locations (a

FIGURE 25.8   Plasmids from a bacterium used in the recombinant DNA technique. *(Thomas Broker / Cold Spring Harbor Laboratory)*

different location for each enzyme). For example, one of these enzymes may split a double-stranded DNA as follows:

$$\text{\~\~\~GAATTC\~\~\~ } \xrightarrow{\text{enzyme}} \text{\~\~\~G} \qquad + \qquad \text{AATTC\~\~\~}$$
$$\text{\~\~\~CTTAAG\~\~\~ } \qquad\qquad \text{\~\~\~CTTAA} \qquad \text{G\~\~\~}$$

The enzyme is so programmed that whenever it finds this specific sequence of bases in a DNA molecule, it cleaves it as shown. Since a plasmid is circular, cleaving it in this way produces a double-stranded chain with two ends (Figure 25.9). These are called "sticky ends" because on one strand each has several free bases that are ready to pair up with a complementary section if they can find one.

The next step is to give them one. This is done by adding a gene from some other species. The gene is a strip of double-stranded DNA that has the necessary base sequence. For example, we can put in the human gene that manufactures proinsulin, which we can get in two ways:

1. It can be made in a laboratory by chemical synthesis; that is, chemists can combine the nucleotides in the proper sequence to make the gene.

2. We can cut a human chromosome with the same restriction enzyme.

Since it is the same enzyme, it cuts the human gene so as to leave the same sticky ends:

We use "H" to indicate a human gene.

$$\text{H—GAATTC—H} \xrightarrow{\text{enzyme}} \text{H—G} \qquad + \qquad \text{AATTC—H}$$
$$\text{H—CTTAAG—H} \qquad\qquad \text{H—CTTAA} \qquad \text{G—H}$$

The human gene must be cut at two places so that a piece of DNA that carries two sticky ends is freed. To splice the human gene into the plasmid, the two are mixed in the presence of DNA ligase, and the sticky ends come together:

We use "B" to indicate a bacterial plasmid.

$$\text{H—G} \qquad + \qquad \text{AATTC—B} \xrightarrow{\text{enzyme}} \text{H—GAATTC—B}$$
$$\text{H—CTTAA} \qquad \text{G—B} \qquad\qquad \text{H—CTTAAG—B}$$

This reaction takes place at both ends of the human gene, and the plasmid is a circle once again (Figure 25.9).

The modified plasmid is then put back into a bacterial cell where it replicates naturally every time the cell divides. Bacteria multiply quickly, and soon we have a large number of bacteria, all containing the modified plasmid. All these cells now manufacture human insulin by transcription and translation. Thus, we can use bacteria as a factory to manufacture specific proteins. This new industry has tremendous potential for lowering the price of drugs that are now manufactured by isolation from human or animal tissues (for example, human interferon, a molecule that fights infection). Not only bacteria but also plant cells can be used (Figure 25.10). Ultimately, if recombinant DNA techniques can be applied to humans and not just to bacteria, it is possible that genetic diseases might someday be cured by this powerful technique. An infant or fetus missing a gene might be given that gene. Once in the cells, the gene would reproduce itself for an entire lifetime.

Human insulin is now marketed by the Lilly Corporation in two forms, called Humulin R and Humulin N (Box 18D). The R form has a faster onset action. Both of them are manufactured by the recombinant DNA technique.

(a)

(b)

FIGURE 25.10 (a) Injection of an aqueous DNA solution into the nucleus of a proto-plast. *(Courtesy of Dr. Anne Crossway, Calgene, Inc., Davis, California, and* Biotechniques, *Eaton Publishing, Natick, Massachusetts)* (b) A luminescent tobacco plant. The gene of the enzyme luciferase (from a firefly) has been incorporated into the genetic material of the tobacco plant. *(Courtesy of Dr. Marlene DeLuca, University of California at San Diego)*

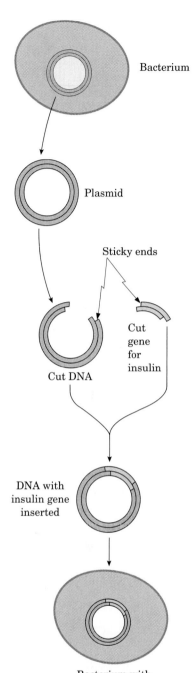

FIGURE 25.9 The recombinant DNA technique can be used to turn a bacterium into an insulin "factory." *(From P. B. Berlow et al.,* Introduction to the Chemistry of Life. *Philadelphia: Saunders College Publishing, 1982)*

At the time of writing (mid 2000), no such gene therapy has been suc-cessfully achieved in humans. This is not for lack of trying. A tremendous amount of research is extended trying not just to cure genetic diseases but also to deliver a fatal bullet to cancer cells by recombinant DNA technique. The difficulties are numerous; however, finding the proper vector (a harm-less virus, for example) is one of the bigger challenges. The virus must be proficient enough to deliver the selected gene without eliciting an immune reaction from the organ itself. The hope is that, in the future, genes can be inserted in human organs and they will express themselves in sufficient quantity. Even then the product of the healthy gene will compete with the product of the faulty gene, hopefully alleviating the symptoms of the dis-ease or curing it altogether.

By far the greatest potential of **genetic engineering,** that is, insert-ing new genes into cells, is in the field of agriculture. Genetic engineering has produced tomatoes that can be picked in their naturally ripened states and yet do not spoil quickly on the shelves of supermarkets. Other geneti-cally engineered plants, corn and cotton, for example, can resist damages by insects or fungi or herbicides. Most of the soybeans grown in the United States today are genetically engineered. In addition to herbicide- and insect-resistant plants, genetic engineering offers opportunities to improve crop yields and resist freezing temperatures. All these will contribute to the increase of the food supply necessary to feed our ever-growing world population.

Some people feel that genetically engineered organisms may create havoc with the ecology of the planet by introducing new, dominant species. But plant breeding has been going on for centuries trying to achieve exactly the same goals. Which is better, a genetically engineered ripe tomato or one picked green and ripened in the warehouse by ethylene gas? Obviously, care must be taken and controls exercised so as not to release harmful newly engineered life-forms into the environment.

SUMMARY

A **gene** is a segment of a DNA molecule that carries the sequence of bases that directs the synthesis of one particular protein. The information stored in the DNA is transcribed onto RNA and then expressed in the synthesis of a protein molecule. This is done in two steps: **transcription** and **translation.** In transcription, the information is copied from DNA onto mRNA by complementary base-pairing ($A \rightarrow U$, $T \rightarrow A$, $G \rightarrow C$, $C \rightarrow G$). There are also start and stop signals. The mRNA is strung out along the ribosomes. Transfer RNA carries the individual amino acids. Each tRNA goes to a specific site on the mRNA. A sequence of three bases (a triplet) on mRNA constitutes a **codon.** It spells out the particular amino acid that the tRNA brings to this site. Each tRNA has a recognition site, the **anticodon,** that pairs up with the codon. When two tRNA molecules are aligned at adjacent sites, the amino acids that they carry are linked by an enzyme forming a peptide bond. The process continues until the whole protein is synthesized. The **genetic code** provides the correspondence between a codon and an amino acid. In most cases, there is more than one codon for each amino acid.

There are a number of mechanisms for gene regulation, the switching on and off of the action of genes. In *E. coli,* portions of DNA called **operons** contain not only structural genes but also control sites and a regulatory gene. In higher organisms, there are different ways to regulate genes. Among them, **response elements** and **transcription factors** regulate gene expression by positive enhancement. Most of human DNA (96 to 98 percent) does not code for proteins.

A change in the sequence of bases is called a **mutation.** Mutations can be caused by an internal mistake or induced by chemicals or radiation. A change in just one base can cause a mutation. This may be harmful or beneficial or may cause no change whatsoever in the amino acid sequence. If a mutation is very harmful, the organism may die. Chemicals that cause mutations are called **mutagens.** Chemicals that cause cancer are called **carcinogens.** Many carcinogens are mutagens, but the reverse is not true.

With the discovery of restriction enzymes that can cut DNA molecules at specific points, scientists have found ways to splice DNA segments together. In this manner, a human gene (for example, the one that codes for insulin) can be spliced into a bacterial plasmid. Then the bacteria, when multiplied, transmit this new information to the daughter cells. Therefore, the ensuing generations of bacteria are able to manufacture human insulin. This powerful method is called the **recombinant DNA technique.** Genetic engineering is the process by which genes are inserted into cells.

KEY TERMS

A site (Section 25.5)
Activation (Section 25.5)
Anticodon (Section 25.3)
Central dogma (Section 25.1)
Codon (Sections 25.3, 25.4)
Control sites (Section 25.6)
Elongation (Section 25.5)
Elongation factor (Section 25.5)
Eukaryote (Section 25.6)
Exon (Section 25.5)
Gene expression (Section 25.1)
Gene regulation (Section 25.6)
Genetic code (Section 25.4)
Genetic engineering (Section 25.8)
Initiation (Section 25.5)
Initiation signal (Section 25.2)

Metal-binding fingers
  (Section 25.6)
Metastasis (Box 25E)
Mutagen (Section 25.7)
Mutation (Section 25.7)
Oncogene (Box 25E)
Operator site (Section 25.6)
Operon (Section 25.6)
P site (Section 25.5)
Plasmid (Section 25.8)
Prokaryote (Section 25.6)
Promoter (Box 25C)
Promoter site (Section 25.6)
Proto-oncogene (Box 25E)
Recognition site (Section 25.3)
Recombinant DNA (Section 25.8)

Regulatory gene (Section 25.6)
Response elements (Section 25.6)
Restriction endonuclease
  (Section 25.8)
Retrovirus (Box 25B)
Structural genes (Section 25.6)
Termination (Section 25.5)
Termination sequence (Section 25.2)
Transcription (Section 25.1)
Transcription factors
  (Sections 25.2, 25.6)
Transfection (Box 25C)
Transgenic (Box 25C)
Translation (Section 25.1)
Translocation (Section 25.5)
Tumor suppressor protein (Box 25F)

CONCEPTUAL PROBLEMS

25.A In both the transcription and the translation steps of protein synthesis, a number of different molecules come together to act as a factor unit. What are these units of (a) transcription and (b) translation?

25.B In what sense does the universality of genetic code support the theory of evolution?

25.C Where are the codons located? Where are the anticodons located?

25.D A new endonuclease is found. It cleaves double-stranded DNA at every location where C and G are paired on opposite strands. Could this enzyme be used in producing human insulin by the recombinant DNA technique? Explain.

PROBLEMS

Difficult problems are designated by an asterisk.

Transcription and Translation

25.1 What initiates a transcription cascade?

25.2 Where is an initiation signal located?

25.3 Which end of the DNA contains the initiation signal?

25.4 In what part of the cell does transcription occur?

25.5 (a) Which ribosome portion has specific A and P sites? (b) What happens at each site during translation?

25.6 What are the two most important sites on tRNA molecules?

The Genetic Code

25.7 (a) If a codon is GCU, what is the anticodon? (b) What amino acid does this codon code for?

*25.8 If a segment of DNA is 981 units long, how many amino acids appear in the protein this DNA segment codes for? (Assume that the entire segment is used to code for the protein and that there is no methionine at the N-terminal end of the protein.)

25.9 Does the same codon signal for both initiation and methionine insertion in the middle of a chain? What is the difference between these two processes?

Translation and Protein Synthesis

25.10 To which end of the tRNA is the amino acid bonded? Where does the energy come from to form the tRNA-amino acid bond?

25.11 There are three sites on the ribosome, each participating in the translation. Identify them and describe what is happening at each site.

25.12 What is the main role of (a) the 40S ribosome (b) the 60S ribosome?

25.13 What is the function of elongation proteins?

Gene Regulation

25.14 When gene regulation occurs at the transcriptional level, what molecules are involved?

25.15 Name the two kinds of control sites.

25.16 What is an operon?

*25.17 Compare the interactions of RNA polymerase with the DNA chain in prokaryotes as opposed to eukaryotes.

25.18 What kind of interactions exist between metal-binding fingers and DNA?

Mutations

25.19 Using Table 25.1, give an example of a mutation that (a) does not change anything in a protein molecule (b) might cause fatal changes in a protein.

25.20 How do cells repair mutations caused by x-rays?

25.21 Can a harmful mutation-causing genetic disease exist from generation to generation without exhibiting the symptoms of the disease? Explain.

25.22 Are all mutagens also carcinogens?

Recombinant DNA

25.23 How do restriction endonucleases operate?

25.24 What are sticky ends?

25.25 A new genetically engineered corn has been approved by the Food and Drug Administration. This new corn shows increased resistance to a destructive insect called a corn borer. What is the difference, in principle, between this genetically engineered corn and one that developed insect resistance by mutation (natural selection)?

25.26 Can a human genetic disease be cured by recombinant DNA technique?

Boxes

25.27 (Box 25A) Why are viruses considered to be parasites?

*25.28 (Box 25A) What is a viral "coat"? Where do the ingredients—amino acids, enzymes, and so forth—necessary to synthesize the coat come from?

25.29 (Box 25B) What are the target cells for the AIDS virus?

25.30 (Box 25B) What HIV-1 viral enzymes are inhibited by (a) AZT (b) indinavir?

*25.31 (Box 25C) What was the role of the αA-crystallin promoter in the development of protease-inhibitor drugs as a treatment for AIDS?

25.32 (Box 25D) What is an invariant site?

25.33 (Box 25E) What kind of mutation transforms the EGF proto-oncogene to an oncogene in human bladder carcinoma?

*__25.34__ (Box 25F) What is p53? Why is its mutated form associated with cancer?

25.35 (Box 25F) How does p53 promote DNA repair?

Additional Problems

25.36 How does an inducer such as lactose enhance the production of the enzyme β-galactosidase?

*__25.37__ In the tRNA structure, there are stretches where complementary base-pairing is necessary and other areas where it is absent. Describe two functionally critical areas (a) where base-pairing is mandatory and (b) where it is absent.

25.38 Is there any way to prevent a hereditary disease? Explain.

25.39 How does the cell make sure that a specific amino acid (say valine) attaches itself only to the one tRNA molecule that is specific for valine?

25.40 (a) What is a plasmid? (b) How does it differ from a gene?

25.41 Why do we call the genetic code degenerate?

*__25.42__ Glycine, alanine, and valine are classified as nonpolar amino acids. Compare their codons. What similarities do you find? What differences do you find?

*__25.43__ Looking at the multiplicity (degeneracy) of the genetic code, you may get the impression that the third base of the codon is irrelevant. Point out where this is not the case. Out of the 16 possible combinations of the first and second bases, in how many cases is the third base irrelevant?

25.44 What polypeptide is coded for by the mRNA sequence 5'-GCU-GAA-GUC-GAG-GUG-UGG-3'?

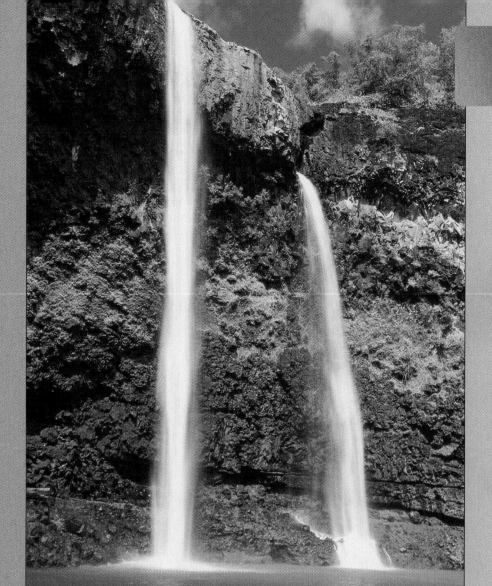

Bioenergetics. How the Body Converts Food to Energy

26.1   Introduction

Living cells are in a dynamic state, which means that compounds are constantly being synthesized and then broken down into smaller fragments. Thousands of different reactions are taking place at the same time (see chapter opener).

> **The sum total of all the chemical reactions involved in maintaining the dynamic state of the cell is called metabolism.**

In general, we can divide metabolic reactions into two broad groups: (1) those in which molecules are broken down to provide the energy needed by cells and (2) those that synthesize the compounds needed by cells—both simple and complex.

The same compounds may be synthesized in one part of a cell and broken down in a different part of the cell.

Above: Wailua Falls, Hawaii, is a natural demonstration of the conversion of potential to kinetic energy. *(A & L Sinibaldi/Tony Stone Images)*

> The process of breaking down molecules to supply energy is catabolism. The process of synthesizing, building up molecules is anabolism.

In spite of the large number of chemical reactions, there are only a few that dominate cell metabolism. In this chapter and the next, we focus our attention on the catabolic pathways that yield energy. A **biochemical pathway** is **a series of consecutive biochemical reactions.** We will see the actual reactions by means of which the chemical energy stored in our food is converted to the energy we use every minute of our lives—to think,

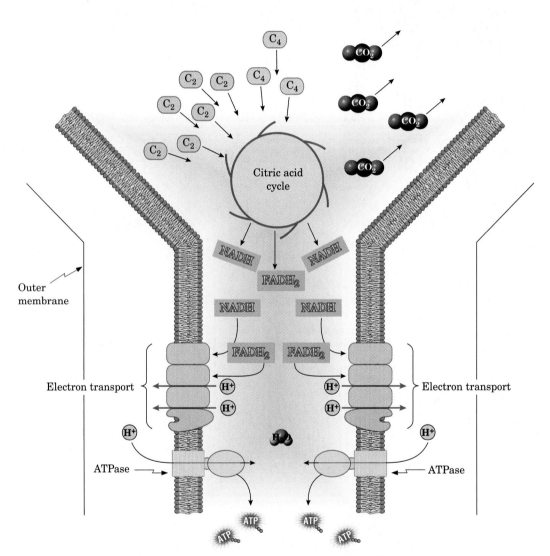

FIGURE 26.1 In this schematic diagram of the common catabolic pathway, an imaginary funnel represents what happens in the cell. The diverse catabolic pathways drop their products into the funnel of the common catabolic pathway, mostly in the form of C_2 fragments (Section 20.4). (The source of the C_4 fragments will be shown in Section 21.9.) The spinning wheel of the citric acid cycle breaks these molecules down further. The carbon atoms are released in the form of CO_2, and the hydrogen atoms and electrons are picked up by special compounds such as NAD^+ and FAD. Then the reduced NADH and $FADH_2$ cascade down into the stem of the funnel, where the electrons are transported inside the walls of the stem and the H^+ ions (represented by dots) are expelled to the outside. In their drive to get back, the H^+ ions form the energy carrier ATP. Once back inside, they combine with the oxygen that picked up body electrons and produce water.

to breathe, and to use our muscles to walk, write, eat, and everything else. In Chapter 28, we will look at some synthetic (anabolic) pathways.

The food we eat consists of many types of compounds, largely the ones we discussed: carbohydrates, lipids, and proteins. All of them can serve as fuel, and we derive our energy from them. To convert those compounds to energy, the body uses a different pathway for each type of compound. *However, all these diverse pathways converge to one* **common catabolic pathway,** which is highlighted in the chapter opener and illustrated in Figure 26.1. The diverse pathways are shown as different food streams. The small C_2 and C_4 molecules produced from the original large molecules in food drop into an imaginary collecting funnel that represents the common catabolic pathway. At the end of the funnel appears the energy carrier molecule adenosine triphosphate (ATP).

The whole purpose of catabolic pathways is to convert the chemical energy in foods to molecules of ATP. In the process, the food also yields metabolic intermediates, which the body can use for synthesis. In this chapter, we deal with the common catabolic pathway only. In Chapter 27, we will discuss the ways in which the different types of food (carbohydrates, lipids, and proteins) feed molecules into the common catabolic pathway.

> **Common catabolic pathway**
> A series of chemical reactions through which most degraded food goes through to yield energy in the form of ATP; the common catabolic pathway consists of (1) the citric acid cycle (Section 26.4) and (2) oxidative phosphorylation (Sections 26.5 and 26.6)

26.2 Cells and Mitochondria

A typical animal cell has many components, as shown in Figure 26.2. Each serves a different function. For example, the replication of DNA (Section 24.4) takes place in the **nucleus; lysosomes** remove damaged cellular components and some unwanted foreign materials; and Golgi bodies package and process proteins for secretion and delivery to other cellular compartments. The specialized structures within cells are called **organelles.**

The **mitochondria,** which possess two membranes (Figure 26.3), are the organelles in which the common catabolic pathway takes place in higher organisms. The enzymes that catalyze the common pathway are all located in these organelles. Because the enzymes are located inside the inner membrane of mitochondria, the starting material of the reactions in the common pathway must get through the two membranes to enter the mitochondria, and products must leave the same way.

The inner membrane of a mitochondrion is quite resistant to the penetration of any ions and of most uncharged molecules. However, ions and molecules can still get through the membrane—they are transported across it by the numerous protein molecules embedded in it (Figure 20.2). The outer membrane, on the other hand, is quite permeable to small molecules and ions and does not need many different kinds of transporting membrane proteins.

The inner membrane is highly corrugated and folded, and the folds are called **cristae.** One can compare the organization of a mitochondrion to that of the galley of an ancient ship. The mitochondrion as a whole is the ship. The cristae are the benches to which the enzymes of the oxidative phosphorylation cycle are chained like ancient slaves, who provide the driving power. The space between the inner and outer membranes is like the space within the double hull of a ship.

The enzymes of the citric acid cycle are located in the matrix, which is the inner nonmembranous portion of a mitochondrion (Figure 26.3). We shall soon see in detail how the specific sequence of the enzymes causes the chain of events in the common catabolic pathway. Beyond that, we must also discuss the ways in which nutrients and reaction products move into and out of the mitochondria.

Mitochondria is the plural form, and mitochondrion is the singular.

An animal cell

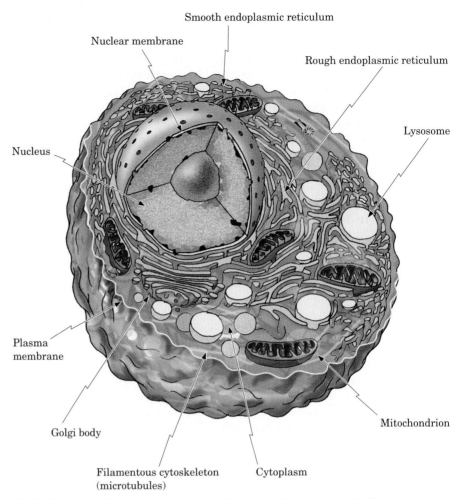

FIGURE 26.2   Diagram of a rat liver cell, a typical higher animal cell. *(Adapted from R. H. Garrett and C. M. Grisham,* Biochemistry, *Philadelphia: Saunders College Publishing, 1995)*

FIGURE 26.3   (a) Schematic of a mitochondrion cut to reveal the internal organization. (b) Colorized transmission electron micrograph of a mitochondrion in a heart muscle cell. *(b, © Dennis Kunkel / Phototake)*

26.3 The Principal Compounds of the Common Catabolic Pathway

There are two parts to the common catabolic pathway. The first is the **citric acid cycle** (also called tricarboxylic acid cycle or the Krebs cycle), and the second is the oxidative phosphorylation pathway, also called the **electron transport chain** or the **respiratory chain.**

In order to understand what is actually happening in these reactions, we must first introduce the principal compounds participating in the common catabolic pathway. The most important of these are three rather complex compounds: **adenosine monophosphate (AMP), adenosine diphosphate (ADP),** and **adenosine triphosphate (ATP)** (Figures 26.4 and 26.5). All three of these molecules contain the heterocyclic amine adenine (Section 24.2) and the sugar D-ribose (Section 19.8) joined together by a β-N-glycosidic bond, forming adenosine (Section 24.2).

AMP, ADP, and ATP all contain adenosine connected to phosphate groups. The only difference between AMP, ADP, and ATP is the number of phosphate groups. As you can see from Figure 26.5, each phosphate is attached to the next by an anhydride bond (Section 18.7). ATP contains three phosphates—one phosphoric ester and two anhydride bonds. In all three molecules, the first phosphate is attached to the ribose by a phosphoric ester bond (Section 18.7).

A phosphoric anhydride linkage, P—O—P, contains more chemical energy (7.3 kcal/mol) than a phosphate-ester linkage, C—O—P (3.4 kcal/mol). This means that when ATP and ADP are hydrolyzed to yield inorganic phosphate (Figure 26.5), they release more energy per phosphate group than does AMP. Conversely, when inorganic phosphate bonds to AMP or ADP, greater amounts of energy are added to the chemical bond than when it bonds to adenosine. ADP and ATP contain *high-energy* phosphate bonds.

Of the three phosphate groups, the one closest to the adenosine, the phosphate ester bond, is the least energetic; the other two possess a good deal more energy. Consequently, ATP releases the most energy and AMP releases the least, when each gives up one phosphate group. This makes ATP a very useful compound for energy storage and release. The energy gained in the oxidation of food is stored in the form of ATP. However, this is only a short-term storage. ATP molecules in the cells normally do not last longer than about 1 min. They are hydrolyzed to ADP and inorganic phosphate to yield energy for other processes, such as muscle contraction and nerve signal conduction. This means that ATP is constantly being formed

The PO_4^{3-} ion is generally called inorganic phosphate.

When one phosphate group is hydrolyzed from each, the following energy yields are obtained:

ATP = 7.3 kcal/mol
ADP = 7.3 kcal/mol
AMP = 3.4 kcal/mol

FIGURE 26.4 Adenosine 5′-monophosphate (AMP).

FIGURE 26.5 Hydrolysis of ATP produces ADP plus inorganic phosphate plus energy.

and decomposed. Its turnover rate is very high. Estimates suggest that during strenuous exercise the human body manufactures and degrades as much as 1 kg (approximately 2.2 lb) of ATP every 2 min.

In summary, when the body takes in food, some of it goes to produce energy, and some is used for building molecules and for other purposes. The energy extracted from the food is converted to ATP. This is the form in which the body stores its energy. In order to release this energy, the body hydrolyzes the ATP to ADP (sometimes to AMP). Exactly how these things happen is what we will be discussing in the rest of this chapter and in the next.

Two other actors in this drama are the coenzymes (Section 22.3) NAD^+ (nicotinamide adenine dinucleotide) and FAD (flavin adenine dinucleotide) (Figure 26.6), both of which contain an ADP core. In NAD^+, the operative part of the coenzyme is the nicotinamide part. In FAD, the operative part is the flavin. In both molecules, the ADP is the handle by which the apoenzyme holds onto the coenzyme; the other end of the molecule carries out the actual chemical reaction. For example, when NAD^+ is reduced, the nicotinamide part of the molecule gets reduced:

> The body is able to extract only 40 to 60 percent of the total caloric content of food.

> The + in NAD^+ refers to the positive charge on the nitrogen.

> Nicotinamide and riboflavin are both members of the vitamin B group.

> The R stands for the rest of the NAD^+ molecule; R′ stands for the rest of the FAD molecule.

The reduced form of NAD^+ is called NADH. The same reduction happens on two nitrogens of the flavin portion of FAD:

The reduced form of FAD is called $FADH_2$.

FIGURE 26.6 The structures of NAD⁺ and FAD.

We view **NAD⁺** and **FAD** coenzymes as the **hydrogen ion** and **electron-transporting molecules.**

The final principal compound in the common catabolic pathway is **coenzyme A** (CoA, Figure 26.7), which is the **acetyl-carrying group.** Coenzyme A also contains ADP, but here the next structural unit is pantothenic acid, another B vitamin. Just as ATP can be looked upon as an ADP molecule to which a —PO₃ is attached by a high-energy bond, so can **acetyl coenzyme A** be considered a CoA molecule linked to an acetyl group by another high-energy bond. The active part of coenzyme A is the

Acetyl group The CH_3CO- group

The hydrolysis of acetyl coenzyme A yields 7.51 kcal/mol.

FIGURE 26.7 The structure of coenzyme A.

The bond-linking acetyl group to CoA is a thioester bond.

mercaptoethylamine. The acetyl group of acetyl coenzyme A is attached to the SH group:

$$CoA-S-\overset{\overset{\displaystyle O}{\|}}{C}-CH_3$$

Acetyl coenzyme A

26.4 The Citric Acid Cycle

The common catabolism of carbohydrates and lipids begins when they have been broken down into pieces of two-carbon atoms each. The two-carbon fragments are the acetyl (CH_3CO-) portions of acetyl coenzyme A. The acetyl is now fragmented further in the citric acid cycle (named after the main component of the cycle), which is also called the **Krebs cycle** and, sometimes, the **tricarboxylic acid cycle.**

Hans Krebs (1900–1981), Nobel laureate in 1953, established the relationships among the different components of the cycle.

The details of the citric acid cycle are given in Figure 26.8. A good way to gain an insight is to use Figure 26.8 in connection with the simplified schematic diagram shown in Figure 26.9, which shows only the carbon balance.

We will now follow the two carbons of the acetyl group (C_2) through each step in the citric acid cycle. The circled numbers correspond to those in Figure 26.8.

In step ⑧ we will see where the oxaloacetate comes from.

Step 1 Acetyl coenzyme A enters the cycle by combining with a C_4 compound called oxaloacetate:

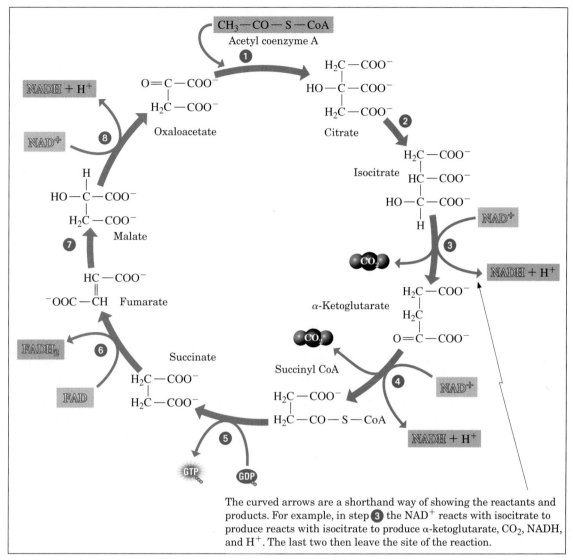

The curved arrows are a shorthand way of showing the reactants and products. For example, in step ❸ the NAD^+ reacts with isocitrate to produce reacts with isocitrate to produce α-ketoglutarate, CO_2, NADH, and H^+. The last two then leave the site of the reaction.

FIGURE 26.8 The citric acid (Krebs) cycle. The numbered steps are explained in detail in the text.

The first thing that happens is the addition of the $-CH_3$ group of the acetyl CoA to the $C=O$ of the oxaloacetate, catalyzed by the enzyme citrate synthase. This is followed by hydrolysis to produce the C_6 compound citrate ion and CoA. Therefore, step ① is a building up rather than a breaking down process.

Step 2 The citrate ion is dehydrated to *cis*-aconitate after which the *cis*-aconitate is hydrated, but this time to isocitrate instead of citrate:

$$\begin{array}{c}
COO^- \\
| \\
CH_2 \\
| \\
C-COO^- \\
\| \\
C-H \\
| \\
COO^-
\end{array}
\xrightarrow[\text{aconitase}]{H_2O}
\begin{array}{c}
COO^- \\
| \\
CH_2 \\
| \\
CH-COO^- \\
| \\
HO-CH \\
| \\
COO^-
\end{array}$$

cis-Aconitate   Isocitrate

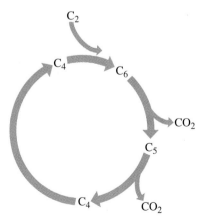

FIGURE 26.9 A simplified view of the citric acid cycle, showing only the carbon balance.

In citrate, the alcohol is a tertiary alcohol. We learned in Section 13.5 that tertiary alcohols cannot be oxidized. The alcohol in the isocitrate is a secondary alcohol, which upon oxidation yields a ketone.

Decarboxylation The process that leads to the loss of CO_2 from a —COOH group

Step 3 The isocitrate is now oxidized and decarboxylated at the same time:

$$\begin{array}{c} COO^- \\ | \\ CH_2 \\ | \\ CH-COO^- \\ | \\ HO-CH \\ | \\ COO^- \end{array} + NAD^+ \xrightarrow[\text{dehydrogenase}]{\text{isocitrate}} \begin{array}{c} COO^- \\ | \\ CH_2 \\ | \\ CH_2 \\ | \\ C=O \\ | \\ COO^- \end{array} + NADH + CO_2$$

Isocitrate α-Ketoglutarate

In oxidizing the secondary alcohol to a ketone the oxidizing agent NAD^+ has removed two hydrogens. One of them has been used to reduce NAD^+ to NADH. The other has replaced the COO^- that goes into making CO_2. Note that the CO_2 given off comes from the original oxaloacetate and not from the two carbons of the acetyl CoA. Both of these carbons are still present in the α-ketoglutarate. Also note that we are now down to a C_5 compound, α-ketoglutarate.

The CO_2 molecules given off in steps ③ and ④ are the ones we exhale.

Steps 4 and 5 Next a complex system removes another CO_2, once again from the original oxaloacetate portion rather than from the acetyl CoA portion:

The P_i is inorganic phosphate.

$$\begin{array}{c} COO^- \\ | \\ CH_2 \\ | \\ CH_2 \\ | \\ C=O \\ | \\ COO^- \end{array} + NAD^+ + GDP + P_i + H_2O \xrightarrow[\text{system}]{\text{complex} \atop \text{enzyme}}$$

α-Ketoglutarate

$$\begin{array}{c} COO^- \\ | \\ CH_2 \\ | \\ CH_2 \\ | \\ COO^- \end{array} + CO_2 + NADH + H^+ + GTP$$

Succinate

We are now down to a C_4 compound, succinate. This oxidative decarboxylation is more complex than the first. It occurs in many steps and requires a number of cofactors. For our purpose, it is sufficient to know that, during this second oxidative decarboxylation, a high-energy compound called **guanosine triphosphate (GTP)** is also formed.

GTP is similar to ATP except that guanine replaces adenine. Otherwise, the linkages of the base to ribose and the phosphates are exactly the same as in ATP. The function of GTP is also similar to that of ATP, namely, to store energy in the form of high-energy phosphate bonds (chemical energy). The energy from the hydrolysis of GTP drives many important biochemical reactions—for example, the signal transduction in neurotransmission (Section 23.5).

Step 6   In this step, the succinate is oxidized by FAD, which removes two hydrogens to give fumarate (the double bond in this molecule is trans):

Succinate   Fumarate

This reaction cannot be carried out in the laboratory, but with the aid of an enzyme catalyst, the body does it easily.

Step 7   The fumarate is now hydrated to give the malate ion:

Fumarate   Malate

Step 8   In the final step of the cycle, malate is oxidized by NAD^+ to give oxaloacetate:

Malate   Oxaloacetate

Thus, the final product of the Krebs cycle is oxaloacetate, which is the compound that we started with in step ①.

What has happened in the entire process is that the original two acetyl carbons of acetyl CoA were added to the C_4 oxaloacetate to produce a C_6 unit, which then lost two carbons in the form of CO_2, to produce, at the end of the process, the C_4 unit oxaloacetate. The net effect is the conversion of the two acetyl carbons of acetyl CoA to two molecules of carbon dioxide.

How does this produce energy? We have already learned that one step in the process produces a high-energy molecule of GTP. But other steps contribute also. In several of the steps, the citric acid cycle converts NAD^+ to NADH and FAD to $FADH_2$. These reduced coenzymes carry the H^+ and electrons that eventually will provide the energy for the synthesis of ATP (discussed in detail in Sections 26.5 and 26.6).

This stepwise degradation and oxidation of acetate in the citric acid cycle results in the most efficient extraction of energy. Rather than in one burst, the energy is released in small packets carried away step by step in the form of NADH and $FADH_2$.

But the cyclic nature of this acetate degradation has other advantages besides maximizing energy yield:

1. The citric acid cycle components also provide raw materials for amino acid synthesis as the need arises (Chapter 28).

2. The many-component cycle provides an excellent method for regulating the speed of catabolic reactions.

For example, α-ketoglutaric acid is used to synthesize glutamic acid.

The regulation can occur at many different parts of the cycle, so that feedback information can be used at many points to speed up or slow down the process as necessary.

The following equation represents the overall reactions in the citric acid cycle.

$$CH_3COOH + 2H_2O + 3NAD^+ + FAD \longrightarrow$$
$$2CO_2 + 3NADH + FADH_2 + 3H^+ \quad (20.1)$$

The citric acid cycle is controlled by a feedback mechanism. When the essential products of the cycle, such as ATP and NADH + H$^+$, accumulate, they inhibit some of the enzymes in the cycle. Citrate synthase (step ①), isocitrate dehydrogenase (step ③), and α-ketoglutarate dehydrogenase (part of the complex enzyme system in step ④) are inhibited by ATP and/or by NADH + H$^+$. This slows down or shuts off the cycle. On the other hand, when the feed material, acetyl CoA, is in abundance, the cycle is speeded up. The enzyme isocitrate dehydrogenase (step ③) is stimulated by ADP and NAD$^+$.

26.5 Electron and H$^+$ Transport (Oxidative Phosphorylation)

The reduced coenzymes NADH and FADH$_2$ are end products of the citric acid cycle. They carry hydrogen ions and electrons and, thus, the potential to yield energy when these combine with oxygen to form water:

The oxygen in this reaction is the oxygen we breathe.

$$4H^+ + 4e^- + O_2 \longrightarrow 2H_2O$$

This simple exothermic reaction is carried out in many steps.

A number of enzymes are involved, all embedded in the inner membrane of the mitochondria. These enzymes are situated in a particular *sequence* in the membrane so that the product from one enzyme can be passed on to the next enzyme, in a kind of assembly line. The enzymes are arranged in order of increasing affinity for electrons, so electrons flow through the enzyme system (Figure 26.10).

The sequence of the electron carrying enzyme systems starts with complex I. This is the largest complex containing some 40 subunits, among them a flavoprotein and several FeS clusters, and **coenzyme Q,** or CoQ (also called ubiquinone). Complex I oxidizes the NADH produced in the citric acid cycle and reduces the CoQ:

The letters used to designate the cytochromes were given in order of their discovery.

$$NADH + H^+ + CoQ \longrightarrow NAD^+ + CoQH_2$$

Some of the energy released in this reaction is used to move 2H$^+$ across the membrane, from the matrix to the intermembrane space. The CoQ is soluble in lipids and can move laterally in the membrane.

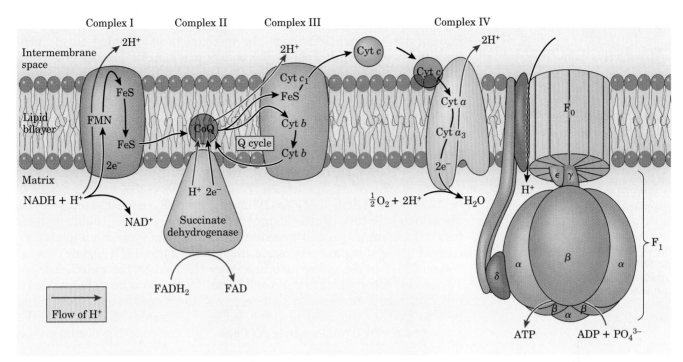

FIGURE 26.10   Schematic diagram of oxidative phosphorylation (also called the electron transport chain).

Complex II also catalyzes the transfer of electrons to CoQ. The source of the electrons is the oxidation of succinate in the citric acid cycle producing $FADH_2$. The final reaction is

$$FADH_2 + CoQ \longrightarrow FAD^+ + CoQH_2$$

The energy of this reaction is not sufficient to pump two protons across the membrane, nor is there an appropriate channel for such a transfer.

Complex III delivers the electrons from $CoQH_2$ to **cytochrome** c. This integral membrane complex contains 11 subunits, among them cytochrome b, cytochrome c_1, and FeS clusters. This complex has two channels through which two H$^+$ are pumped from the $CoQH_2$ into the intermembrane space. The process is very complicated. Simplified, we can imagine that it occurs in two distinct steps and so is the electron transfer. Since each cytochrome c can pick up only one electron, two cytochrome c's are needed:

$$CoQH_2 + 2 \text{ cytochrome c (reduced)} \longrightarrow$$
$$CoQ + 2H^+ + 2 \text{ cytochrome c (oxidized)}$$

Cytochrome c is also a mobile carrier of electrons, it can move laterally in the membrane.

Complex IV is known as cytochrome oxidase. It contains 13 subunits, most importantly cytochrome a_3, which has a copper center. It is an integral membrane protein complex. The electron flows from cytochrome c to cytochrome a to cytochrome a_3. There, the electrons are transferred to the oxygen molecule, and the O—O bond is cleaved. The oxidized form of the enzyme takes up 2H$^+$ from the matrix for each oxygen atom. The water molecule thus formed is released into the matrix:

$$\tfrac{1}{2}O_2 + 2H^+ + 2e^- \longrightarrow H_2O$$

During this process, two more H^+ are pumped out of the matrix into the intermembrane space. Although the mechanism of pumping out protons from the matrix is not known, the energy driving this process is derived from breaking the covalent bond of O_2. This final pumping into the intermembrane space makes a total of six H^+ per $NADH + H^+$ and four per $FADH_2$ molecules.

Mitchell received the Nobel Prize in Chemistry in 1978.

Chemiosmotic theory Mitchell proposed that the electron transport is accompanied by an accumulation of protons in the intermembrane space of the mitochondrion, which in turn creates an osmotic pressure; the protons driven back to the mitochondrion under this pressure generate ATP

26.6 Phosphorylation and the Chemiosmotic Pump

In 1961 Peter Mitchell (1920–1992), an English chemist, proposed the **chemiosmotic theory:** The energy in the electron transfer chain creates a proton gradient. A **proton gradient** is a continuous variation in the H^+ concentration along a given region. In this case, there is a higher concentration of H^+ in the intermembrane space than inside the mitochondrion. The driving force, which is the result of the spontaneous flow of ions from a region of high concentration to a region of low concentration, propels the protons back to the mitochondrion through a complex that is given the name of **proton translocating ATPase.** This compound is located on the

BOX 26A

2,4-Dinitrophenol As an Uncoupling Agent

Nitrated aromatic compounds are highly explosive. Trinitrotoluene (TNT) is the best known. During World War I, many ammunition workers were exposed to 2,4-dinitrophenol (DNP), a compound used to prepare the explosive picric acid. It was observed that these workers lost weight. As a consequence, DNP was used as a weight-reducing drug during the 1920s. Unfortunately, DNP eliminated not only the fat but sometimes also the patient, and its use as a diet pill was discontinued after 1929.

Today we know why DNP works as a weight-reducing drug. It is an effective protonophore, which is a compound that transports H^+ ions through a membrane passively, without the expenditure of energy. We have seen that H^+ ions accumulate in the intermembrane space of mitochondria and, under normal conditions, drive the synthesis of ATP while they are going back to the inside. This is Mitchell's chemiosmotic principle. When DNP is ingested, it transfers the H^+ back to the mitochondrion easily, and no ATP is manufactured. The energy of the electron separation

is dissipated as heat and is not built in as chemical energy in ATP. The loss of this energy-storing compound makes the utilization of food much less efficient, resulting in weight loss.

There exists in the medical literature a case history of a woman whose muscles contained mitochondria in which electron transport was not coupled to oxidative phosphorylation. This unfortunate woman could not utilize electron transport to generate ATP. As a consequence, she was severely incapacitated and bedridden. When the uncoupling worsened, the heat generated by the uncontrolled oxidation (no ATP production) was so severe that the patient required continuous cooling.

A similar mechanism is used to provide heat in hibernating bears. The bears have brown fat; its color is derived from the numerous mitochondria in the tissue. The brown fat also contains an uncoupling protein, a protonophore, which allows the H^+ ions to stream back into the mitochondrial matrix without manufacturing ATP. The heat generated in this manner keeps the animal alive during cold winter days.

DNP   TNT   Picric acid

BOX 26B

Protection Against Oxidative Damage

As we have seen, most of the oxygen we breathe in is used in the final step of the common pathway, namely, in the conversion of our food supply to ATP. This reaction occurs in the mitochondria of cells. However, some 10 percent of the oxygen we consume is used elsewhere, in specialized oxidative reactions. One such reaction is the hydroxylation of compounds (for example, steroids). This is catalyzed by the enzyme cytochrome P-450, which usually resides outside the mitochondria in the endoplasmic reticulum of cells. The reaction catalyzed by P-450 is the following, where RH represents a steroid or other molecule:

$$RH + O_2 + NADPH + H^+ \longrightarrow$$
$$ROH + H_2O + NADP^+$$

This reaction is also used by our body to detoxify foreign substances (for example, barbiturates). But the action of P-450 is not always beneficial. Some of the most powerful carcinogens are converted to their active forms by P-450.

Another use of oxygen for detoxification is in the fight against infection by bacteria. Specialized white blood cells called leukocytes (Section 30.2) engulf bacteria and kill them by generating superoxide, O_2^-, a highly reactive form of oxygen. But superoxide and other highly reactive forms of oxygen, such as $\cdot OH$ radicals and hydrogen peroxide (H_2O_2), not only destroy foreign cells but damage our own cells as well. They especially attack unsaturated fatty acids and thereby damage cell membranes (Section 20.5). Our bodies have

their own defenses against these very reactive compounds. One example is glutathione (Box 21B), which protects us against oxidative damages. Furthermore, we have in our cells two powerful enzymes that decompose these highly reactive oxidizing agents:

$$2O_2^- + 2H^+ \xrightarrow{\text{superoxide dismutase}} H_2O_2 + O_2$$
$$2H_2O_2 \xrightarrow{\text{catalase}} 2H_2O + O_2$$

These enzymes also play an important defensive role in heart attacks and strokes. As a result of oxygen deprivation of the heart muscles or the brain when arteries are blocked, a condition known as ischemia-reperfusion may occur. This means that when these tissues suddenly become reoxygenated, superoxide anions (O_2^-) can form. This, and the $\cdot OH$ radicals derived from it, can further damage the tissues. Superoxide dismutase and catalase prevent reperfusion injuries.

One theory that has been advanced to explain why humans and animals show symptoms of old age is that these oxidizing agents damage healthy cells over the years. All the enzymes involved in reactions with highly reactive forms of oxygen contain a heavy metal: iron in the case of P-450 and catalase and copper and/or zinc in the case of superoxide dismutase. If the aging theory mentioned earlier is correct, then these protecting agents keep us alive for many years, but eventually the oxidizing processes do so much damage that we exhibit symptoms of aging and eventually die.

---

inner membrane of the mitochondrion (Figure 26.10) and is the active enzyme that catalyzes the conversion of ADP and inorganic phosphate to ATP (the reverse of the reaction shown in Figure 26.5):

$$ADP + P_i \underset{\text{ATPase}}{\rightleftharpoons} ATP + H_2O$$

The proton translocating ATPase is a complex rotor engine made of 16 different proteins. The F_0 sector, which is embedded in the membrane, contains the **proton channel.** The 12 subunits that form this channel rotate every time a proton passes from the cytoplasmic side (intermembrane) to the matrix side of the mitochondrion. This rotation is transmitted to a "rotor" in the F_1 sector. F_1 contains five different kinds of polypeptides. The rotor (γ and ϵ subunits) is surrounded by the catalytic unit (made of α and β subunits) that synthesizes the ATP. The F_1 catalytic unit converts the mechanical energy of the rotor into chemical energy of the ATP molecule. The last unit, the "stator," containing the δ subunit, stabilizes the whole complex.

The proton translocating ATPase can catalyze the reaction in both directions. When protons that have accumulated on the outer surface of the mitochondrion stream inward, the enzyme manufactures ATP and stores the electrical energy in the form of chemical energy. On the other hand, the enzyme can also hydrolyze ATP and as a consequence pump out H^+ from the mitochondrion. Each pair of protons that is translocated gives rise to the formation of one ATP molecule.

The protons that enter a mitochondrion combine with the electrons transported through the electron transport chain and with oxygen to form water. The net result of the two processes (electron/H^+ transport and ATP formation) is that the oxygen we breathe in combines with four H^+ ions and four electrons to give two water molecules. The four H^+ ions and four electrons come from the NADH and $FADH_2$ molecules produced in the citric acid cycle. The functions of the oxygen, therefore, are

1. To oxidize NADH to NAD^+ and $FADH_2$ to FAD so that all these molecules can go back and participate in the citric acid cycle

2. To provide energy for the conversion of ADP to ATP

The latter function is accomplished indirectly. The entrance of the H^+ ions into the mitochondrion drives the ATP formation, but the H^+ ions enter the mitochondrion because the O_2 depleted the H^+ ion concentration when water was formed. It is a rather complex process involving the transport of electrons along a whole series of enzyme molecules (which catalyze all these reactions); however, the cell cannot utilize the O_2 molecules without it and eventually will die. The following equations represent the overall reactions in oxidative phosphorylation:

$$NADH + 3ADP + \tfrac{1}{2}O_2 + 3P_i + H^+ \longrightarrow NAD^+ + 3ATP + H_2O \quad (20.2)$$

$$FADH_2 + 2ADP + \tfrac{1}{2}O_2 + 2P_i \longrightarrow FAD + 2ATP + H_2O \quad (20.3)$$

26.7 The Energy Yield

The energy released during electron transport is now finally built into the ATP molecule. Therefore, it is instructive to look at the energy yield in the universal biochemical currency: the number of ATP molecules.

Each pair of protons entering a mitochondrion results in the production of one ATP molecule. For each NADH molecule, three pairs of protons are pumped into the intermembrane space in the electron transport process. Therefore, for each NADH molecule, we get three ATP molecules, as can be seen in Equation 26.2. For each $FADH_2$ molecule, we have seen that only four protons are pumped out of the mitochondrion. Therefore, only two ATP molecules are produced for each $FADH_2$, as seen in Equation 26.3.

Now we can produce the energy balance for the whole common catabolic pathway (citric acid cycle and oxidative phosphorylation combined). For each C_2 fragment entering the citric acid cycle, we obtain three NADH and one $FADH_2$ (Equation 26.1) plus one GTP, which is equivalent in energy to one ATP. Thus, the total number of ATP molecules produced per C_2 fragment is

$$
\begin{aligned}
3\ \text{NADH} \times 3\ \text{ATP/NADH} &= 9\ \text{ATP} \\
1\ \text{FADH}_2 \times 2\ \text{ATP/FADH}_2 &= 2\ \text{ATP} \\
1\ \text{GTP} \phantom{\times 2\ \text{ATP/FADH}_2} &= \underline{1\ \text{ATP}} \\
&= 12\ \text{ATP}
\end{aligned}
$$

Each C_2 fragment that enters the cycle produces 12 ATP molecules and uses up two O_2 molecules. The total effect of the energy-production chain of reactions we have discussed in this chapter (the common catabolic pathway) is to oxidize one C_2 fragment with two molecules of O_2 to produce two molecules of CO_2 and 12 molecules of ATP:

$$C_2 + 2O_2 + 12\,ADP + 12P_i \longrightarrow 12\,ATP + 2CO_2$$

This is the net reaction for the whole common pathway.

The important thing is not the waste product, CO_2, but the 12 ATP molecules because these will now release their energy when they are converted to ADP.

26.8 Conversion of Chemical Energy to Other Forms of Energy

As mentioned in Section 26.3, the storage of chemical energy in the form of ATP lasts only a short time. Usually, within a minute, the ATP is hydrolyzed (an exergonic reaction) and thus releases its chemical energy. How does the body utilize this chemical energy? To answer this question, let us look at the different forms in which energy is needed in the body.

Conversion to Other Forms of Chemical Energy

The activity of many enzymes is controlled and regulated by phosphorylation. For example, the enzyme phosphorylase, which catalyzes the breakdown of glycogen (Box 27B), occurs in an inactive form, phosphorylase b. When ATP transfers a phosphate group to a serine residue, the enzyme becomes active. Thus, the chemical energy of ATP is used in the form of chemical energy to activate phosphorylase b so that glycogen can be utilized. We shall see several other examples of this in Chapters 27 and 28.

Electrical Energy

The body maintains a high concentration of K^+ ions inside the cells despite the fact that outside the cells the K^+ concentration is low. The reverse is true for Na^+. So that K^+ does not diffuse out of the cells and Na^+ does not penetrate in, special transport proteins in the cell membranes constantly pump K^+ into the cells and Na^+ out. This pumping requires energy, which is supplied by the hydrolysis of ATP to ADP. With this pumping, the charges inside and outside the cell are unequal, and this generates an electric potential. Thus, the chemical energy of ATP is transformed into electrical energy, which operates in neurotransmission (Section 23.2).

Mechanical Energy

ATP is the immediate source of energy in muscle contraction. In essence, muscle contraction takes place when thick and thin filaments slide past each other (Figure 26.11). The thick filament is myosin, an ATPase enzyme (that is, one that hydrolyzes ATP). The thin filament, actin, binds strongly to myosin in the contracted state. However, when ATP binds to myosin, the actin-myosin dissociates, and the muscle relaxes. When myosin hydrolyzes ATP, it interacts with actin once more, and a new contraction occurs. Therefore, the hydrolysis of ATP drives the alternating association and dis-

(*a*) Relaxed muscle

(*b*) Contracted muscle

FIGURE 26.11 Schematic diagram of muscle contraction.

sociation of actin and myosin and, consequently, the contraction and relaxation of the muscle.

Heat Energy

One molecule of ATP upon hydrolysis to ADP yields 7.3 kcal/mole. Some of this energy is released as heat and is used by the body to maintain body temperature. If we estimate that the specific heat of the body is about the same as that of water, a person weighing 60 kg would need to hydrolyze approximately 99 moles of ATP to raise the temperature of the body from room temperature, 25°C, to 37°C. Not all body heat is derived from ATP hydrolysis; some other exothermic reactions in the body also contribute.

SUMMARY

The sum total of all the chemical reactions involved in maintaining the dynamic state of cells is called **metabolism.** The breaking down of molecules is **catabolism;** the building up of molecules is **anabolism.** The **common metabolic pathway** uses a two-carbon C_2 fragment (acetyl) from different foods. Through the **citric acid cycle** and **oxidative phosphorylation (electron transport chain),** the C_2 fragment is oxidized. The products formed are water and carbon dioxide. The energy from oxidation is built into the high-chemical-energy-storing molecule ATP.

Both the citric acid cycle and oxidative phosphorylation take place in the **mitochondria.** The enzymes of the citric acid cycle are located in the matrix, whereas the enzymes of the oxidative phosphorylation chain are on the inner mitochondrial membrane. Some of them project into the intermembrane space.

The principal carriers in the common pathway are as follows: **ATP** is the phosphate carrier, **CoA** is the

C_2 fragment carrier, and **NAD$^+$** and **FAD** carry the hydrogen ions and electrons. The unit common to all these carriers is **ADP.** This is the nonactive end of the carriers, which acts as a handle that fits into the active sites of the enzymes.

In the citric acid cycle, the C_2 fragment first combines with a C_4 fragment (oxaloacetate) to yield a C_6 fragment (citrate). An oxidative decarboxylation yields a C_5 fragment. One CO_2 is released, and one $NADH + H^+$ is passed on to the electron transport chain. Another oxidative decarboxylation provides a C_4 fragment. Once again, a CO_2 is released, and another $NADH + H^+$ is passed on to the electron transport chain. Subsequently, two dehydrogenation (oxidation) steps yield one $FADH_2$ and one additional $NADH + H^+$, along with an analog of ATP called GTP. The cycle is controlled by a feedback mechanism.

The NADH enters the electron transport chain at the complex I stage. The coenzyme Q (CoQ) of this complex picks up the electrons and the H^+ and becomes $CoQH_2$. Two H^+ ions are expelled into the intermembrane space of the mitochondrion from the matrix. Complex II also has CoQ. Electrons and H^+ are passed to this complex. Complex II also catalyzes the transfer of electrons from $FADH_2$. However, no H^+ ions are pumped into the intermembrane space at this point. The electrons are passed along by $CoQH_2$ to complex III of the electron transport chain. Two H^+ from the $CoQH_2$ are expelled into the intermembrane space. Cytochrome c of complex III transfers electrons to the next in the series through redox reactions. As the electrons are transported to complex IV, the cytochrome oxidase additional two H^+ ions are expelled from the matrix of the mitochondrion to the intermembrane space. For each NADH, six H^+ ions are expelled. For each $FADH_2$, four H^+ are expelled. Finally, the electrons inside the mitochondrion combine with oxygen and H^+ to form water. When the expelled H^+ ions stream back into the mitochondrion, they drive a complex enzyme called **proton translocating ATPase,** which makes one ATP molecule for each two H^+ ions that enter the mitochondrion. Therefore, for each NADH + H^+ coming from the citric acid cycle, three ATP molecules are formed. For each $FADH_2$, only two ATP molecules are formed. The overall result is that for each C_2 fragment entering the citric acid cycle, 12 ATP molecules are produced. The proton-translocating ATPase is a complex rotor engine. The proton channel part (F_0) is embedded in the membrane and the catalytic unit (F_1) converts mechanical energy to chemical energy of the ATP molecule.

The chemical energy is stored in ATP only for a short time. ATP is hydrolyzed, usually within a minute. This chemical energy is used to do chemical, mechanical, and electrical work in the body and to maintain body temperature.

KEY TERMS

Acetyl coenzyme A (Section 26.3)
ADP (Section 26.3)
AMP (Section 26.3)
Anabolism (Section 26.1)
ATP (Section 26.3)
Biochemical pathway (Section 26.1)
Catabolism (Section 26.1)
Chemiosmotic hypothesis
   (Section 26.6)
Citric acid cycle (Section 26.3)
Coenzyme A (Section 26.3)
Coenzyme Q (Section 26.5)
Cytochrome (Section 26.5)

Common catabolic pathway
   (Section 26.1)
Cristae (Section 26.2)
Electron transport chain
   (Section 26.3)
Energy yield (Section 26.7)
FAD (Section 26.3)
Guanosine triphosphate (GTP)
   (Section 26.4)
Inorganic phosphate
   (Section 26.3)
Krebs cycle (Section 26.4)
Metabolism (Section 26.1)

Mitochondria (Section 26.2)
NAD^+ (Section 26.3)
Organelles (Section 26.2)
Oxidative phosphorylation
   (Section 26.5)
Pathway (Section 26.1)
Proton channel (Section 26.6)
Proton gradient (Section 26.6)
Proton translocating ATPase
   (Section 26.6)
Respiratory chain (Section 26.3)
Tricarboxylic acid cycle
   (Section 26.4)

CONCEPTUAL PROBLEMS

26.A The reaction

$$NADH \rightleftharpoons NAD^+ + H^+ + 2e^-$$

is a reversible reaction. (a) Where does the forward reaction occur in the common catabolic pathway? (b) Where does the reverse reaction occur?

26.B What substrate of the citric acid cycle is oxidized by FAD? What is the oxidation product?

26.C What are the mobile electron carriers of the oxidative phosphorylation?

26.D The passage of H^+ from the cytoplasmic side into the matrix generates mechanical energy. Where is this energy of motion exhibited first?

PROBLEMS

Difficult problems are designated by an asterisk.

Metabolism

26.1 To what end product is the energy of foods converted in the catabolic pathways?

26.2 (a) How many sequences are there in the common catabolic pathway? (b) Name them.

Cells and Mitochondria

26.3 (a) How many membranes do mitochondria have? (b) Which membrane is permeable to ions and small molecules?

26.4 How do ions and molecules enter the mitochondria?

26.5 What is the name of the nonmembranous portion of a mitochondrion?

26.6 (a) Where are the enzymes of the citric acid cycle located? (b) Where are the enzymes of oxidative phosphorylation located?

Principal Compounds of the Common Catabolic Pathway

26.7 How many high-energy phosphate bonds are in the ATP molecule?

26.8 What are the products of the following reaction?

$$AMP + H_2O \xrightarrow{\text{H}^+}$$

26.9 Which yields more energy, the hydrolysis (a) of ATP to ADP or (b) of ADP to AMP?

26.10 ATP is often called the molecule for storing chemical energy. Does "storage" mean a long-term preservation of energy? Explain.

26.11 What kind of chemical bond exists between the ribitol and the phosphate group in FAD?

*26.12 When NAD^+ is reduced, two electrons enter the molecule, together with one H^+. Where in the product will the two electrons be located?

26.13 Which atoms in the flavin portion of FAD are reduced to yield $FADH_2$?

26.14 Which part of the FAD molecule is (a) the operative part and (b) the handle?

26.15 In the common catabolic pathway, a number of important molecules act as carriers (transfer agents). (a) What is the carrier of phosphate groups? (b) Which are the coenzymes transferring hydrogen ions and electrons? (c) What kind of groups does coenzyme A carry?

*26.16 The sugar units in FAD and NAD^+ are different. Explain the difference.

26.17 What kind of chemical bond exists between the pantothenic acid and mercaptoethylamine in the structure of CoA?

26.18 Name the vitamin B molecules that are a part of the structure of (a) NAD^+ (b) FAD (c) coenzyme A.

*26.19 In both NAD^+ and FAD, the vitamin B portion of the molecule is the active part. Is this also true for CoA?

26.20 What type of compound is formed when coenzyme A reacts with acetate?

26.21 The fats and carbohydrates metabolized by our bodies are eventually converted to a single compound. What is it?

The Citric Acid Cycle

26.22 The first step in the citric acid cycle is abbreviated as

$$C_2 + C_4 = C_6$$

(a) What do these symbols stand for?
(b) What are the common names of the three compounds involved in this reaction?

26.23 What is the only C_5 compound in the citric acid cycle?

26.24 Identify by number those steps of the citric acid cycle that are not redox reactions.

26.25 What kind of reaction occurs in the citric acid cycle when a C_6 compound is converted to a C_5 compound?

26.26 In steps ③ and ④ of the citric acid cycle, the compounds are shortened by one carbon each time. What is the form of this one-carbon compound? What happens to it in the body?

26.27 What is the role of succinate dehydrogenase in the citric acid cycle?

26.28 (a) In the citric acid cycle, how many steps can be classified as decarboxylation reactions? (b) In each case, what is the concurrent oxidizing agent?

26.29 Is ATP directly produced during any step of the citric acid cycle? Explain.

26.30 There are four dicarboxylic acid compounds, each containing four carbons, in the citric acid cycle. Which is (a) the least oxidized (b) the most oxidized?

26.31 Why is a many-step cyclic process more efficient in utilizing energy from food than a single-step combustion?

26.32 Did the two CO_2 molecules given off in one citric acid cycle originate from the entering acetyl group?

26.33 Which intermediates of the citric acid cycle contain $C=C$ double bonds?

26.34 The citric acid cycle can be regulated by the body. It can be slowed down or speeded up. By what mechanism is this process controlled?

26.35 Oxidation is defined as loss of electrons. When oxidative decarboxylation occurs, as in step ④, where do the electrons of the α-ketoglutarate go?

Oxidative Phosphorylation

26.36 What is the main function of oxidative phosphorylation (the electron transport chain)?

26.37 How many pairs of H^+ ions are expelled for each (a) NADH molecule (b) $FADH_2$ molecule?

26.38 What kind of motion is set up in the proton translocating ATPase by the passage of H^+ from the intermembrane space into the matrix?

26.39 In oxidative phosphorylation, water is formed from H^+, e^-, and O_2. Where does this take place?

26.40 At what points in oxidative phosphorylation are the H^+ ions and the electrons separated from each other?

26.41 How many ATP molecules are generated (a) for each H^+ translocated through the ATPase complex (b) for each C_2 fragment that goes through the complete common catabolic pathway?

26.42 Since H^+ is pumped out into the intermembrane space, is the pH there increased, decreased, or unchanged compared with that in the matrix?

The Chemiosmotic Pump

26.43 What is the channel through which H^+ ions reenter the matrix of mitochondria?

*26.44 The proton gradient accumulated in the intermembrane area of a mitochondrion drives the ATP-manufacturing enzyme ATPase. Why do you think Mitchell called this the "chemiosmotic theory"?

The Energy Yield

26.45 If each mole of ATP yields 7.3 kcal of energy upon hydrolysis, how many kcal of energy would you get from 1 g of $CH_3COO^-(C_2)$ entering the citric acid cycle?

26.46 A hexose sugar (C_6) enters the common metabolic pathway in the form of two C_2 fragments. (a) How many molecules of ATP are produced from one hexose molecule? (b) How many O_2 molecules are used up in the process?

Conversion of Chemical Energy to Other Forms

26.47 (a) How do muscles contract? (b) Where does the energy in muscle contraction come from?

26.48 Give an example of the conversion of the chemical energy of ATP to electrical energy.

26.49 How is the enzyme phosphorylase activated?

Boxes

26.50 (Box 26A) What is a protonophore?

*26.51 (Box 26A) Oligomycin is an antibiotic that allows electron transport to continue but stops phosphorylation in bacteria as well as in humans. Would you use this as an antibacterial drug for people? Explain.

26.52 (Box 26B) How does superoxide dismutase prevent damage to the brain during a stroke?

26.53 (Box 26B) (a) What kind of reaction is catalyzed by cytochrome P-450? (b) Where does the oxygen reactant come from?

Additional Problems

26.54 (a) What is the difference in structure between ATP and GTP? (b) Compared with ATP, would you expect GTP to carry more, less, or about the same amount of energy?

*26.55 How many g of CH_3COOH molecules must be metabolized in the common metabolic pathway to yield 87.6 kcal of energy?

*26.56 What is the basic difference in the functional groups between citrate and isocitrate?

26.57 What structural characteristics do citric acid and malic acid have in common?

*26.58 Two keto acids are important in the citric acid cycle. Identify the two keto acids, and tell how they are manufactured.

26.59 Which filament of muscles is an enzyme, catalyzing the reaction that converts ATP to ADP?

26.60 One of the end products of food metabolism is water. How many molecules of H_2O are formed from the entry of each molecule of (a) $NADH + H^+$ (b) $FADH_2$? (Use Figure 26.10.)

26.61 What is the difference between NADH and $FADH_2$ in terms of the atoms that are reduced?

*26.62 Acetyl CoA is labeled with radioactive carbon as shown: $CH_3{}^*CO-S-CoA$. This enters the citric acid cycle. If the cycle is allowed to progress only to the α-ketoglutarate level, will the CO_2 expelled by the cell be radioactive?

26.63 Where is the H^+ ion channel located in the proton translocating ATPase complex?

*26.64 Is the passage of H^+ ion through the channel converted directly into chemical energy?

26.65 Does all the energy used in ATP synthesis come from the mechanical energy of rotation?

Above: Ballet dancer leaping.
(© Dennis Degnan/Corbis)

Specific Catabolic Pathways. Carbohydrate, Lipid, and Protein Metabolism

27.1 Introduction

The food we eat serves two main purposes: (1) It fulfills our energy needs, and (2) it provides the raw materials to build the compounds our bodies need. Before either of these processes can take place, the food—carbohydrates, fats, and proteins—must be broken down into small molecules that can be absorbed through the intestinal walls.

Carbohydrates

Complex carbohydrates (di- and polysaccharides) are broken down by stomach acid and enzymes to produce monosaccharides (Section 29.8). Monosaccharides, the most important of which is glucose, may also come from the enzymatic breakdown of glycogen. As you may recall, this highly

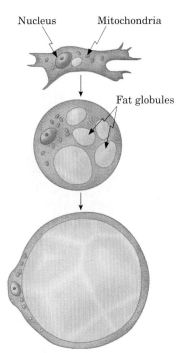

Nucleus   Mitochondria

Fat globules

FIGURE 27.1 Storage of fat in a fat cell. As more and more fat droplets accumulate in the cytoplasm, they coalesce to form a very large globule of fat. Such a fat globule may occupy most of the cell, pushing the cytoplasm and the organelles to the periphery. *(Modified from C. A. Villee, E. P. Solomon, and P. W. Davis,* Biology, *Philadelphia, Saunders College Publishing, 1985.)*

branched polymer stores carbohydrates in the liver and muscles until needed. Once monosaccharides are produced, they can be used either to build new oligo- and polysaccharides or to provide energy. The specific pathway by which energy is extracted from monosaccharides is called **glycolysis** (Sections 27.2 and 27.3).

Lipids

Ingested fats are broken down by lipases to glycerol and fatty acids or to monoglycerides, which are absorbed through the intestine (Section 29.9). In a similar way, complex lipids are also hydrolyzed to smaller units before absorption. As with carbohydrates, these smaller molecules (fatty acids, glycerol, and so on) can be used to build complex molecules needed in membranes, or they can be oxidized to provide energy, or they can be stored in **fat storage depots** (Figure 27.1). The stored fats can then be hydrolyzed later to glycerol and fatty acids whenever they are needed as fuel.

Specialized cells that store fats are called fat depots or adipocytes.

The specific pathway by which energy is extracted from glycerol involves the same glycolysis pathway as that used for carbohydrates (Section 27.4). The specific pathway used by the cells to obtain energy from fatty acids is called **β-oxidation** (Section 27.5).

Proteins

As you might expect from a knowledge of their structures, proteins are broken down by HCl in the stomach and by digestive enzymes in the stomach and intestines (pepsin, trypsin, chymotrypsin, and carboxypeptidases) to produce their constituent amino acids. The amino acids absorbed through the intestinal wall enter the **amino acid pool.** They serve as building blocks for proteins as needed and, to a smaller extent (especially during starvation), as a fuel for energy. In the latter case, the nitrogen of the amino acids is catabolized through **oxidative deamination** and the **urea cycle** and is expelled from the body as urea in the urine (Section 27.8). The

Amino acid pool The name used for the free amino acids found both inside and outside cells throughout the body

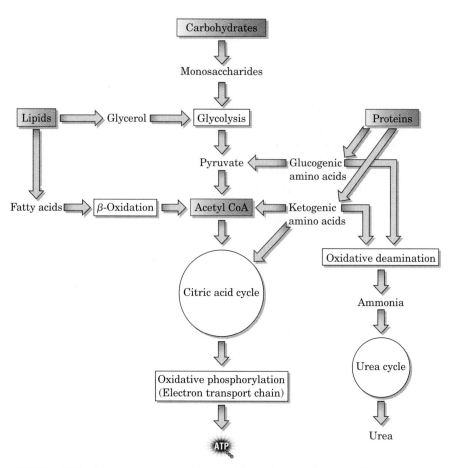

FIGURE 27.2 The convergence of the specific pathways of carbohydrate, fat, and protein catabolism into the common catabolic pathway, which is made up of the citric acid cycle and oxidative phosphorylation.

carbon skeletons of the amino acids enter the common catabolic pathway (Chapter 26) either as α-keto acids (pyruvic, oxaloacetic, and α-ketoglutaric acids) or as acetyl coenzyme A (Section 27.9).

In all cases, **the specific pathways of carbohydrate, fat, and protein catabolism converge to the common catabolic pathway** (Figure 27.2). In this way, the body needs fewer enzymes to get energy from diverse food materials. Efficiency is thus achieved because a minimal number of chemical steps are required and also because the energy-producing factories of the body are localized in the mitochondria.

27.2 Glycolysis

> **Glycolysis** The biochemical pathway that breaks down glucose to pyruvate, which yields chemical energy in the form of ATP and reduced coenzymes

Glycolysis is the specific pathway by which the body gets energy from monosaccharides. The detailed steps are shown in Figures 27.3, and the most important features are shown schematically in Figure 27.4.

In the first steps of glucose metabolism, energy is consumed rather than released. At the expense of two molecules of ATP (which are converted to ADP), glucose (C_6) is phosphorylated; first glucose 6-phosphate is formed in step ①, and then, after isomerization to fructose 6-phosphate in step ②, a second phosphate group is attached to yield fructose 1,6-bisphosphate in step ③. We can consider these steps as the activation process.

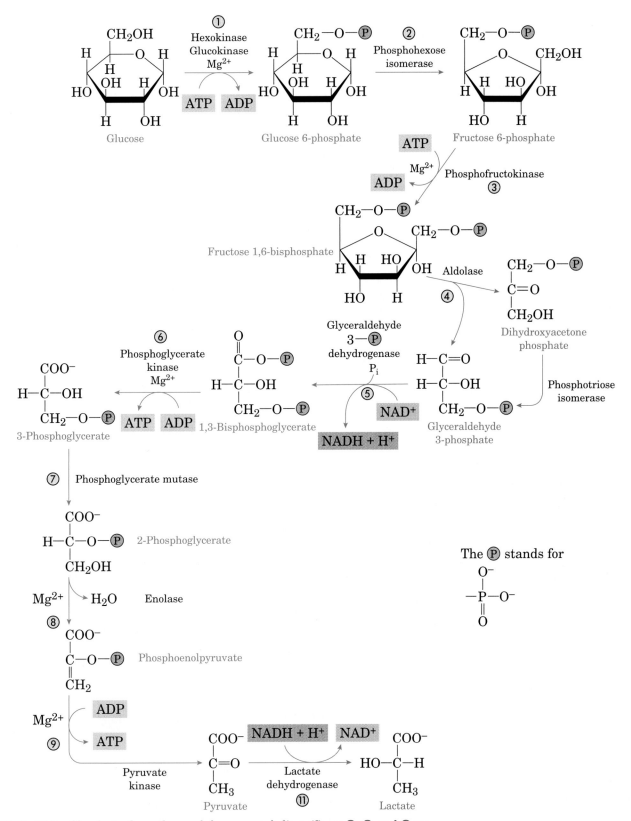

FIGURE 27.3 Glycolysis, the pathway of glucose metabolism. (Steps ⑩, ⑫, and ⑬ are shown in Figure 27.4.) Some of the steps are reversible, but equilibrium arrows are not shown. They are shown in Figure 27.4.

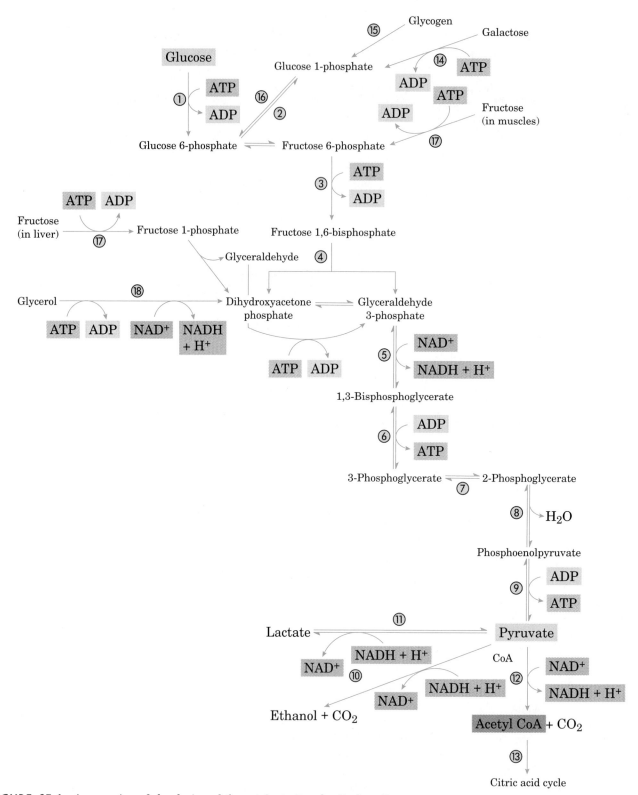

FIGURE 27.4 An overview of glycolysis and the entries to it and exits from it.

In the second stage, the C_6 compound fructose 1,6-bisphosphate is broken in step ④ into two C_3 fragments. The two C_3 fragments, glyceraldehyde 3-phosphate and dihydroxyacetone phosphate, are in equilibrium (they can be converted to each other). Only the glyceraldehyde 3-phosphate is oxidized in glycolysis, but as this species is removed from the equilibrium mix-

BOX 27A

Lactate Accumulation

Many athletes suffer muscle cramps when they undergo strenuous exercise (see Chapter 8 opener). This is the result of a shift from normal glucose catabolism (glycolysis → citric acid cycle → oxidative phosphorylation) to that of lactate production in step ⑪ (Figure 27.4). During exercise, oxygen is used up rapidly, and this slows down the rate of the common pathway. The demand for energy makes anaerobic glycolysis proceed at a high rate, but because the aerobic (oxygen-demanding) pathways are slowed down, not all the pyruvate produced in glycolysis can enter the citric acid cycle. The excess pyruvate ends up as lactate, which causes painful muscle contractions.

The same shift in catabolism also occurs in heart muscle when coronary thrombosis leads to cardiac arrest. The oxygen supply is cut off by the blockage of the artery to the heart muscles. The common pathway and its ATP production are shut off. Glycolysis proceeds at an accelerated rate, accumulating lactate. The heart muscle contracts, producing a cramp. Just as in skeletal muscle, massage of heart muscles can relieve the cramp and start the heart beating. Even if heartbeat is restored within 3 min (the amount of time the brain can survive without being damaged), acidosis develops as a result of the cardiac arrest. Therefore, at the same time that efforts are under way to start the heart beating by chemical, physical, or electrical means, an intravenous infusion of 8.4 percent bicarbonate solution is given to combat acidosis.

---

ture, Le Chatelier's principle ensures that the dihydroxyacetone phosphate is converted to glyceraldehyde 3-phosphate.

In the third stage, the glyceraldehyde 3-phosphate is oxidized to 1,3-bisphosphoglycerate in step ⑤. The hydrogen of the aldehyde group is removed by the NAD^+ coenzyme. In step ⑥, the phosphate from the carboxyl group is transferred to ADP, yielding ATP and 3-phosphoglycerate. The latter, after isomerization in step ⑦ and dehydration in step ⑧, is converted to phosphoenolpyruvate, which loses its remaining phosphate in step ⑨ and yields pyruvate ion and another ATP molecule.

All these glycolysis reactions occur in the cytosol outside the mitochondria. They occur in the absence of O_2 and, therefore, are also called reactions of the **anaerobic pathway.** As indicated in Figure 27.4, the end product of glycolysis, pyruvate, does not accumulate in the body. In certain bacteria and yeast, pyruvate undergoes reductive decarboxylation in step ⑩ to produce ethanol (Section 12.4). In some bacteria, and also in mammals in the absence of oxygen, pyruvate is reduced to lactate in step ⑪. But most importantly, pyruvate goes through an oxidative decarboxylation in the presence of coenzyme A in step ⑫ to produce acetyl CoA:

$$CH_3-\overset{\overset{\displaystyle O}{\|}}{C}-CO^- + CoA-SH \longrightarrow CH_3-\overset{\overset{\displaystyle O}{\|}}{C}-S-CoA + CO_2$$
$$\text{Pyruvate} \qquad\qquad\qquad \text{Acetyl coenzyme A}$$

This reaction is catalyzed by a complex enzyme system, pyruvate dehydrogenase, that sits on the inner membrane of the mitochondrion. The reaction produces acetyl CoA, CO_2, and $NADH + H^+$. The acetyl CoA then enters the citric acid cycle in step ⑬ and goes through the common pathway.

In summary, after converting complex carbohydrates to glucose, the body gets energy from glucose by converting it to acetyl CoA (by way of pyruvate) and then using the acetyl CoA as a raw material for the common pathway.

As we saw in Figure 27.4, glucose 6-phosphate plays a central role in different entries into the glycolytic pathway. However, glucose 6-phosphate can also be utilized by the body for other purposes, not just for the production of energy in the form of ATP. Most importantly, glucose 6-phosphate can be shunted to the **pentose phosphate pathway** ⑲. This pathway has the capacity to produce NADPH, ribose ⑳ as well as energy.

In step ⑨, after hydrolysis of the phosphate, the resulting enol of pyruvic acid tautomerizes to the more stable keto form (Section 17.7).

The **cytosol** is the fluid inside the cell that surrounds the nucleus and organelles such as mitochondria.

Pentose phosphate pathway The biochemical pathway that produces ribose and NADPH form glucose 6-phosphate or, alternatively, energy

NADPH is badly needed in red blood cells (RBC) for defense against oxidative damages. Glutathione (Box 21A) is the main agent to keep the hemoglobin in its reduced form. On the other hand, glutathione is regenerated by NADPH. Thus, the lack of proper supply of NADPH leads to the destruction of RBC, causing severe anemia.

α-D-Glucose 6-phosphate

Ribulose 5-phosphate

Ribose 5-phosphate

Glucose 6-phosphate + 2NADP$^+$ $\xrightarrow{\text{⑲}}$ Ribulose 5-phosphate + NADPH + CO$_2$

$\downarrow$ ⑳

Ribose 5-phosphate

$\updownarrow$ ㉑

Glyceraldehyde 3-phosphate

Simplified schematic representation of the pentose phosphate pathway also called shunt.

The structure of NADP$^+$.

Muscles attached to bones are called skeletal muscles.

NADPH is needed in many biosynthetic processes, among them synthesis of unsaturated fatty acid (Section 28.3), photosynthesis (Box 28A), cholesterol synthesis, amino acid synthesis, and the reduction of ribose to deoxyribose for DNA. Ribose is needed for the synthesis of RNA (Section 24.3). Therefore, when the body needs more of these synthetic ingredients than energy, the glucose 6-phosphate is utilized through the pentose phosphate pathway. When energy is needed, the glucose 6-phosphate remains in the glycolytic pathway and even ribose 5-phosphate can be channeled back to glycolysis through glyceraldehyde 3-phosphate ㉑. Through this reversible reaction, the cells can obtain ribose also directly from the glycolytic intermediates.

27.3 The Energy Yield from Glucose

In conjunction with Figure 27.4, we can sum up the energy derived from glucose catabolism in terms of ATP production. Before we begin, however, we must take into account that glycolysis takes place in the cytosol, whereas oxidative phosphorylation occurs in the mitochondria. Therefore, the NADH + H$^+$ produced in glycolysis must penetrate the mitochondrial membrane in order to be utilized in oxidative phosphorylation.

The NADH is too large to cross the mitochondrial membrane. There are two routes available to get the electrons of NADH + H$^+$ into the mitochondria; these have different efficiencies. In one, which operates in muscle and nerve cells, only two ATP molecules are produced for each NADH + H$^+$. In the other, which operates in the heart and the liver, three ATP molecules are produced for each NADH + H$^+$ produced in the cytosol, just as is the case in the mitochondria (Section 26.7). Since most energy production takes place in skeletal muscle cells, when we construct the energy balance sheet, we use two ATP molecules for each NADH + H$^+$ produced in the cytosol.

TABLE 27.1 ATP Yield from Complete Glucose Metabolism

Step Numbers in Figure 27.4	Chemical Steps	Number of ATP Molecules Produced
① ② ③	Activation (glucose → 1,6-fructose bisphosphate)	−2
⑤	Oxidative phosphorylation 2(glyceraldehyde 3-phosphate → 1,3-bisphosphoglycerate), producing 2NADH + H⁺ in cytosol	4
⑥ ⑨	Dephosphorylation 2(1,3-bisphosphoglycerate → pyruvate)	4
⑫	Oxidative decarboxylation 2(pyruvate → acetyl CoA), producing 2NADH + H⁺ in mitochondrion	6
⑬	Oxidation of two C_2 fragments in citric acid cycle and oxidative phosphorylation common pathways, producing 12 ATP for each C_2 fragment	24
	Total	36

With this knowledge, we are ready to calculate the energy yield of glucose in terms of ATP molecules produced. This is shown in Table 27.1. In the first stage of glycolysis (steps ①, ②, and ③), two ATP molecules are used up, but this is more than compensated for by the production of 14 ATP molecules in step ⑤, ⑥, ⑨, and ⑫ and in the conversion of pyruvate to acetyl CoA. The net yield of these steps is 12 ATP molecules. As we saw in Section 20.7, the oxidation of one acetyl CoA produces 12 ATP molecules, and one glucose molecule provides two acetyl CoA molecules. Therefore, the total net yield from metabolism of one glucose molecule in skeletal muscle is 36 molecules of ATP.

$$C_6H_{12}O_6 + 6O_2 \longrightarrow 6CO_2 + 6H_2O$$

This calculation applies to glucose metabolism in the skeletal muscle cells, which is what happens most frequently. The two NADH produced in glycolysis yield only a total of four ATP because of the efficiency of the glycerol 3-phosphate transport. Thus, a total of 36 ATP molecules are produced from one glucose molecule, which is 6 APT molecules per carbon atom. If, however, the same glucose is metabolized in the heart or liver, the electrons of the two NADH produced in the glycolysis are transported into the mitochondrion by the malate-aspartate shuttle. Through this shuttle, the two NADH yield a total of six ATP molecules, so that, in this case, there are 38 ATP molecules produced for each glucose molecule. It is instructive to note that most of the energy (in the form of ATP) from the glucose is produced in the common metabolic pathway.

Glucose is not the only monosaccharide that can be used as an energy source. Other hexoses, such as galactose (step ⑭), and fructose (step ⑰), enter the glycolysis pathway at the stages indicated in Figure 27.4. They also yield 36 molecules of ATP per hexose molecule. Furthermore, the glycogen stored in the liver and muscle cells and elsewhere can also be converted by enzymatic breakdown and phosphorylation to glucose 1-phosphate (step ⑮). This in turn isomerizes to glucose 6-phosphate, providing

Recent investigations suggest that only about 30 to 32 ATP molecules are produced per glucose molecule (2.5 ATP/NADH and 1.5 ATP/$FADH_2$). Futher research into the complexity of the oxidative phosphorylation pathway is needed to verify these numbers.

BOX 27B

Effects of Signal Transduction on Metabolism

The ligand-binding/G-protein/adenylate cyclase cascade, which activates proteins by phosphorylation, has a wide range of effects other than opening or closing ion-gated channels (Figure 23.2). An example of such a target is the glycogen phosphorylase enzyme that participates in the breakdown of glycogen stored in muscles. The enzyme cleaves units of glucose 1-phosphate from glycogen, which enters into the glycolytic pathway yielding quick energy (Section 27.3). The active form of the enzyme, phosphorylase a, is phosphorylated. When it is dephosphorylated (phosphorylase b), it is inactive. When epinephrine signals danger to a muscle cell, phosphorylase is activated through the cascade, and quick energy is produced. Thus, the signal is converted to a metabolic event, allowing the muscles to contract rapidly and enabling the person in danger to fight or run away.

Not all phosphorylation of enzymes results in activation. An example is the glycogen synthase. In this case, the phosphorylated form of the enzyme is inactive, and the dephosphorylated one is active. This enzyme participates in glycogenesis, the conversion of glucose to, and its storage in the form of, glycogen. The action of glycogen synthase is the opposite of phosphorylase. Nature provides a beautiful balance, inasmuch as the danger signal of the epinephrine hormone has a dual target. It activates the phosphorylase to get quick energy, but at the same time it inactivates the glycogen synthase so that the available glucose will be used solely for energy and will not be stored away.

Glycogenolysis The biochemical pathway for breakdown of glycogen to glucose

an entry to the glycolytic pathway (step ⑯). The pathway in which glycogen breaks down to glucose is called **glycogenolysis.**

27.4 Glycerol Catabolism

Glycerol 1-phosphate is the same as glycerol 3-phosphate.

The glycerol hydrolyzed from fats or complex lipids (Chapter 20) can also be a rich energy source. The first step in glycerol utilization is an activation step. The body uses one ATP molecule to form glycerol 1-phosphate:

The glycerol phosphate is oxidized by NAD^+ to dihydroxyacetone phosphate, yielding $NADH + H^+$ in the process. Dihydroxyacetone phosphate then enters the glycolysis pathway (step ⑱) and is isomerized to glyceraldehyde phosphate, as shown in Figure 27.4. A net yield of 20 ATP molecules is produced from each glycerol molecule, which is 6.7 ATP molecules per carbon atom.

27.5 β-Oxidation of Fatty Acids

The β-carbon is the second carbon from the COOH group.

As early as 1904, Franz Knoop in Germany proposed that the body utilizes fatty acids as an energy source by breaking them down into C_2 fragments. Prior to fragmentation, the β carbon is oxidized:

β-oxidation The biochemical pathway that degrades fatty acids to acetyl CoA by removing two carbons at a time and yielding energy

$$-\underset{}{C}-\underset{}{C}-\underset{}{C}-\overset{\beta}{\underset{}{C}}-\overset{\alpha}{\underset{}{C}}-COOH$$

Thus, the name **β-oxidation** has its origin in Knoop's prediction. It took about 50 years to establish the mechanism by which fatty acids are utilized as an energy source.

FIGURE 27.5 The β-oxidation spiral of fatty acids. Each loop in the spiral contains two dehydrogenations, one hydration, and one fragmentation. At the end of each loop, one acetyl CoA is released.

The overall process of fatty acid metabolism is shown in Figure 27.5. As is the case with the other foods we have seen, the first step is an activation step. This occurs in the cytosol, where the fat was previously hydrolyzed to glycerol and fatty acids. The activation (step ①) converts ATP to AMP and inorganic phosphate. This is equivalent to the cleavage of two high-energy phosphate bonds. The chemical energy derived from the

Transfer into the mitochondrion is accomplished by an enzyme system called carnitine acyltransferase.

splitting of ATP is built into the compound acyl CoA, which is formed when the fatty acid combines with coenzyme A. The fatty acid oxidation occurs inside the mitochondrion, so the acyl group of acyl CoA must pass through the mitochondrial membrane. Carnitine is the acyl group transporter.

Once the fatty acid in the form of acyl CoA is inside the mitochondrion, the β-oxidation starts. In the first oxidation (dehydrogenation) (step ②), two hydrogens are removed, creating a trans double bond between the alpha and beta carbons of the acyl chain. The hydrogens and electrons are picked up by FAD.

In step ③, the double bond is hydrated. An enzyme specifically places the hydroxy group on C-3. The second oxidation (dehydrogenation), (step ④) requires NAD^+ as a coenzyme. The two hydrogens and electrons removed are transferred to the NAD^+ to form $NADH + H^+$. In the process, a secondary alcohol group is oxidized to a keto group at the beta carbon. In step ⑤, the enzyme thiolase cleaves the terminal C_2 fragment from the chain, and the rest of the molecule is attached to a new molecule of coenzyme A.

The cycle now starts again with the remaining acyl CoA, which is now two carbon atoms shorter. At each turn of the cycle, one acetyl CoA is produced. Most fatty acids contain an even number of carbon atoms. The cyclic spiral is continued until we reach the last four carbon atoms. When this fragment enters the cycle at the end, two acetyl CoA molecules are produced in the fragmentation step.

The β-oxidation of unsaturated fatty acids proceeds in the same way. There is an extra step involved, in which the cis double bond is isomerized to a trans bond, but otherwise the spiral is the same.

27.6 The Energy Yield from Stearic Acid

In order to compare the energy yield from fatty acids with that of other foods, let us select a typical and quite abundant fatty acid—stearic acid, the C_{18} saturated fatty acid.

We start with the initial step, in which energy is used up rather than produced. The reaction breaks two high-energy phosphate bonds.

$$ATP \longrightarrow AMP + 2P_i$$

This is equivalent to hydrolyzing two molecules of ATP to ADP. In each cycle of the spiral, we obtain one $FADH_2$, one $NADH + H^+$, and one acetyl CoA. Stearic acid (C_{18}) goes through seven cycles in the spiral before it reaches the final C_4 stage. In the last (eighth) cycle, one $FADH_2$, one $NADH + H^+$, and two acetyl CoA molecules are produced. Now we can add up the energy. Table 27.2 shows that, for a C_{18} compound, we obtain a total of 146 ATP molecules.

It is instructive to compare the energy yield from fats with that from carbohydrates, since both are important constituents of the diet. In Section 27.3 we saw that glucose, $C_6H_{12}O_6$, produces 36 ATP molecules, that is, 6 for each carbon atom. For stearic acid, there are 146 ATP molecules and 18 carbons, or $146/18 = 8.1$ ATP molecules per carbon atom. Since additional ATP molecules are produced from the glycerol portion of fats (Section 27.4), fats have a higher caloric value than carbohydrates.

27.7 Ketone Bodies

In spite of the high caloric value of fats, the body preferentially uses glucose as an energy supply. When an animal is well fed (plenty of sugar intake), fatty acid oxidation is inhibited, and fatty acids are stored in the

TABLE 27.2 **ATP Yield from Complete Stearic Acid Metabolism**

Step Number in Figure 27.5	Chemical Steps	Happens	Number of ATP Molecules Produced
①	Activation (stearic acid → stearyl CoA)	Once	−2
②	Dehydrogenation (acyl CoA → transenoyl CoA), producing FADH$_2$	8 times	16
④	Dehydrogenation (hydroxy-acyl CoA → ketoacyl CoA), producing NADH + H$^+$	8 times	24
	C$_2$ fragment (acetyl CoA → common catabolic pathway), producing 12 ATP for each C$_2$ fragment	9 times	108
		Total	146

form of neutral fat in fat depots. When physical exercise demands energy, or when the glucose supply dwindles, as in fasting or starvation, or when glucose cannot be utilized, as in the case of diabetes, the β-oxidation pathway of fatty acid metabolism is mobilized.

Unfortunately, low glucose supply also slows down the citric acid cycle. This happens because oxaloacetate is produced from the carboxylation of pyruvate. This oxaloacetate normally enters the citric acid cycle (Figure 20.8) where it is essential for the continuous operation of the cycle. But if there is no glucose, there will be no glycolysis, no pyruvate formation, and therefore no oxaloacetate production.

Thus, even though the fatty acids are oxidized, not all the resulting C$_2$ fragments (acetyl CoA) can enter the citric acid cycle because there is not enough oxaloacetate. Therefore, acetyl CoA builds up in the body, with the following consequences.

The liver is able to condense two acetyl CoA molecules to produce acetoacetyl CoA:

In some pathological conditions, it is possible for glucose not to be available at all.

$$2CH_3-\overset{\overset{\displaystyle O}{\|}}{C}-SCoA \longrightarrow CH_3-\overset{\overset{\displaystyle O}{\|}}{C}-CH_2-\overset{\overset{\displaystyle O}{\|}}{C}-SCoA + CoASH$$

Acetyl CoA   Acetoacetyl CoA

When the acetoacetyl CoA is hydrolyzed, it yields acetoacetate, which can be reduced to form β-hydroxybutyrate:

$$CH_3-\overset{\overset{\displaystyle O}{\|}}{C}-CH_2-\overset{\overset{\displaystyle O}{\|}}{C}-SCoA \xrightarrow{H_2O} CH_3-\overset{\overset{\displaystyle O}{\|}}{C}-CH_2-\overset{\overset{\displaystyle O}{\|}}{C}-O^- + CoASH + H^+$$

Acetoacetyl CoA   Acetoacetate

NADH + H$^+$ → NAD$^+$

H$^+$ → CO$_2$

$$CH_3-\overset{\overset{\displaystyle H}{|}}{\underset{\underset{\displaystyle OH}{|}}{C}}-CH_2-\overset{\overset{\displaystyle O}{\|}}{C}-O^-$$

β-Hydroxybutyrate

$$CH_3-\overset{\underset{\displaystyle O}{\|}}{C}-CH_3$$

Acetone

Ketoacidosis in Diabetes

In untreated diabetes, the glucose concentration in the blood is high because the lack of insulin prevents utilization of glucose by the cells. Regular injections of insulin remedy this situation. However, in some stressful conditions, **ketoacidosis** can still develop. A typical case was a diabetic patient admitted to the hospital in semicoma. He showed signs of dehydration, his skin was inelastic and wrinkled, his urine showed high concentrations of glucose and ketone bodies, and his blood contained excess glucose and had a pH of 7.0, a drop of 0.4 pH units from normal, which is an indication of severe acidosis. The urine also contained the bacterium *Escherichia coli*. This indication of urinary tract infection explained why the normal doses of insulin were insufficient to prevent ketoacidosis.

The stress of infection can upset the normal control of diabetes by changing the balance between administered insulin and other hormones produced in the body. This happened during the infection, and his body started to produce ketone bod-

ies (Section 27.7) in large quantities. Both glucose and ketone bodies appear in the blood before they show up in the urine.

The acidic nature of ketone bodies (acetoacetic acid and β-hydroxybutyric acid) lowers the blood pH. A large drop in pH is prevented by the bicarbonate/carbonic acid buffer (Section 8.10), but even a drop of 0.3 to 0.5 pH units is sufficient to decrease the Na^+ concentration. The decrease of Na^+ in the interstitial tissues draws out K^+ ions from the cells. This in turn impairs brain function and leads to coma. During the secretion of ketone bodies and glucose in the urine, a lot of water is lost, the body becomes dehydrated, and the blood volume shrinks. Thus, the blood pressure drops, and the pulse rate increases to compensate. Smaller quantities of nutrients reach the brain cells, and this too can cause coma.

The patient mentioned earlier was infused with physiological saline solution to remedy the dehydration. Extra doses of insulin restored his glucose level to normal, and antibiotics cured the urinary infection.

Ketone bodies A collective name for acetone, acetoacetate and β-hydroxybutyrate; compounds produced from acetyl CoA in the liver that are used as a fuel for energy production by muscle cells and neurons

These two compounds, together with smaller amounts of acetone, are collectively called **ketone bodies.** Under normal conditions, the liver sends these compounds into the blood stream to be carried to the tissues and utilized there via the common catabolic pathway. Normally the concentration of ketone bodies in the blood is low. But during starvation and in untreated diabetes mellitus, ketone bodies accumulate in the blood and can reach high concentrations. When this occurs, the excess is secreted in the urine. A check of urine for ketone bodies is used in the diagnosis of diabetes.

Test kit for the presence of ketone bodies in the urine. *(Charles D. Winters)*

27.8 Catabolism of the Nitrogen of Amino Acids

The proteins of our foods are hydrolyzed to amino acids in digestion. These amino acids are primarily used to synthesize new proteins. However, unlike carbohydrates and fats, they cannot be stored, so excess amino acids are catabolized for energy production. What happens to the carbon skeleton of the amino acids is dealt with in the next section. Here we discuss the catabolic fate of the nitrogen. An overview of the whole process of protein catabolism is shown in Figure 27.6.

In the tissues, amino (—NH_2) groups freely move from one amino acid to another. The enzymes that catalyze these reactions are the transaminases.

In essence, there are three stages is nitrogen catabolism in the liver. The final product of the three stages is urea, which is excreted in the urine of mammals. The first stage is a **transamination.** Amino acids transfer their amino groups to α-ketoglutarate:

> **Transamination** The exchange of the amino group of an amino acid and a keto group of an α-ketoacid

$$R\text{—}\underset{\underset{NH_3^+}{|}}{CH}\text{—}COO^- + \underset{\underset{COO^-}{\overset{\overset{COO^-}{|}}{\underset{CH_2}{\overset{CH_2}{\overset{C=O}{|}}}}}}{} \xrightarrow{\text{transaminase}} R\text{—}\underset{\overset{O}{\|}}{C}\text{—}COO^- + \underset{COO^-}{\overset{CH\text{—}NH_3^+}{\overset{|}{\underset{CH_2}{\overset{CH_2}{|}}}}}{}$$

α-Amino acid α-Ketoglutarate α-Ketoacid   Glutamate
(zwitterion form)

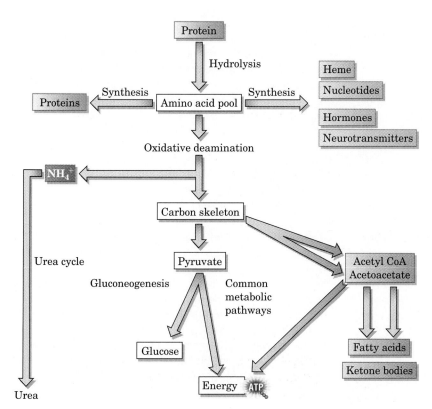

FIGURE 27.6 An overview of pathways in protein catabolism.

MSG

Glutamic acid is one of the 20 amino acids found in proteins (Section 18.2). It is also an excitatory neurotransmitter (Section 23.4). Its monosodium salt, **monosodium glutamate (MSG),** enhances the flavor of many foods without itself contributing any significant taste. It makes such foods as meats, vegetables, soups, and stews taste better. Because of this, MSG is added to many canned and frozen foods. In pure form, it is sold in supermarkets under such brand names as Accent. The spice racks of many kitchens contain a bottle labeled MSG, next to the nutmeg, thyme, and basil. In par-

ticular, Chinese restaurants in this country used copious quantities of MSG in the past.

Although most people can consume MSG with no problems, there have been claims that some are allergic to it and get symptoms that include dizziness, numbness spreading from the jaw to the back of the neck, hot flashes, and sweating. Collectively, these symptoms are known as Chinese restaurant syndrome. However, recent studies cast doubt on the theory that these symptoms are caused by MSG.

Ubiquitin and Protein Targeting

In previous sections, we dealt with the catabolism of dietary proteins, especially with their use as an energy source. The body's own cellular proteins are also broken down and degraded. This occurs sometimes as a response to a stress such as starvation, but more often it is done to maintain a steady-state level of regulatory proteins or to eliminate damaged proteins. But how does the cell know which proteins to degrade and which to leave alone? One pathway that targets proteins for destruction depends on an ancient protein, called ubiquitin. (As the name implies, it is present in all cells belonging to higher organisms.) Its sequence of 76 amino acids is essentially identical in yeast and in humans. The C-terminal amino acid of ubiquitin is glycine, the carboxyl group of which forms an amide linkage with the side chain of a lysine residue in the protein that is targeted for destruction.

This process, known as **ubiquitinylation,** requires three enzymes and costs energy by using up one ATP molecule per ubiquitin molecule. In many cases, a polymer of ubiquitin molecules (polyubiquitin) is attached to the target protein molecule. In attaching one ubiquitin molecule to the next, the same linkage is used as with the targeted protein: The glycine at the C-terminal of one ubiquitin forms an amide linkage with the amino group on the side chain of a lysyl residue of a second ubiquitin. Once a protein molecule becomes "flagged" by several ubiquitin molecules, it is delivered to an organelle containing the machinery for proteolysis. Part of the proteolytic system is proteasome, a large water-soluble complex assembly of proteases and regulatory proteins. In this assembly, the polyubiquitin is removed, and the protein is fully or partially degraded. Cleavage of a protein is called **proteolysis.**

One role of the ubiquitin targeting is that of "garbage

collector." Damaged proteins that the chaperones (Section 21.9) cannot salvage by proper refolding must be removed. For example, oxidation of hemoglobin causes unfolding, and rapid degradation by ubiquitin targeting follows. A second purpose is to control the concentration of some regulatory proteins. The protein products of some oncogenes (Box 25E) are also removed by ubiquitin targeting.

Both of these cases result in complete degradation of the targeted proteins. A third case involves only a partial degradation of the target protein. For example, a virus invades a cell. The infected cell recognizes a foreign protein from the virus and targets it for destruction through the ubiquitin system. The targeted protein is partially degraded. A peptide segment of the viral protein is incorporated into a system that rises to the surface of the infected cell and displays the viral peptide as an **antigen.** An antigen is a foreign body recognized by the immune system (Section 30.3). The T cells (Section 30.3) can now recognize the infected cell as a foreign cell; they attack and kill it.

The excessive zeal of ubiquitinylation as a garbage collector lies in the heart of cystic fibrosis, a genetic disease. Cystic fibrosis causes severe bronchopulmonary disorders and pancreatic insufficiency. The molecular culprit is a protein in the membranes that serves as a chloride channel providing the transport of chloride ions in and out of cells. In cystic fibrosis, the protein has a mutation that does not allow it to fold properly. The system perceives these chloride channel proteins as faulty and targets them for destruction by ubiquitinylation. The result is that there are not enough functioning chloride channels. This produces obstructions in the respiratory as well as in the intestinal tracts by clogging up the secretion pathways with highly viscous mucins.

The carbon skeleton of the amino acid remains behind as an α-ketoacid. The second stage of the nitrogen catabolism is the **oxidative deamination** of glutamate, which occurs in the mitochondrion:

The catabolism of the α-ketoacid is discussed in the next section.

> **Oxidative deamination** The reaction in which the amino group of an amino acid is removed and an α-ketoacid is formed

$$
\begin{array}{ccc}
\text{COO}^- & & \text{COO}^- \\
| & & | \\
\text{H}-\text{C}-\text{NH}_3{}^+ & & \text{C}=\text{O} \\
| & +\ \boxed{\text{NAD}^+}\ +\ \text{H}_2\text{O} \rightleftharpoons \text{NH}_4{}^+\ +\ & | \\
\text{CH}_2 & & \text{CH}_2 \quad +\quad \boxed{\text{NADH} + \text{H}^+} \\
| & & | \\
\text{CH}_2 & & \text{CH}_2 \\
| & & | \\
\text{COO}^- & & \text{COO}^- \\
\text{Glutamate} & & \alpha\text{-Ketoglutarate}
\end{array}
$$

The oxidative deamination yields $NH_4{}^+$ and regenerates α-ketoglutarate, which can again participate in the first stage (transamination). The $NADH + H^+$ produced in the second stage enters the oxidative phosphorylation pathway and eventually produces three ATP molecules.

In the third stage, the $NH_4{}^+$ is converted to urea through the **urea cycle** (Figure 27.7). In step ①, $NH_4{}^+$ is condensed with CO_2 in the mitochondrion to form an unstable compound, carbamoyl phosphate. This condensation occurs at the expense of two ATP molecules. In step ②, car-

The body must get rid of $NH_4{}^+$ because both it and NH_3 are toxic.

> **Urea cycle** A cyclic pathway that produces urea from ammonia and carbon dioxide

FIGURE 27.7 The urea cycle.

Not all organisms dispose of the metabolic nitrogen in the form of urea. Bacteria and fish release ammonia in the surrounding water as such. Birds and reptiles secrete nitrogen in the form of uric acid, the concentrated white solid so familiar in bird droppings.

Ornithine does not occur in proteins.

bamoyl phosphate is condensed with ornithine, a basic amino acid similar in structure to lysine:

Ornithine   Carbamoyl phosphate   Citrulline

The result is citrulline, which diffuses out of the mitochondrion into the cytosol.

A second condensation reaction in the cytosol takes place between citrulline and aspartate, forming argininosuccinate (step ④):

Citrulline   Aspartate   Argininosuccinate

The PP$_i$ stands for pyrophosphate.

The energy for this reaction comes from the hydrolysis of ATP to AMP and pyrophosphate (PP$_i$).

In step ④, the argininosuccinate is split into arginine and fumarate:

Argininosuccinate   Arginine   Fumarate

In step ⑤, the final step, arginine is hydrolyzed to urea and ornithine:

$$
\begin{array}{ccc}
\underset{\text{Arginine}}{
\begin{array}{l}
\text{NH} \\
\parallel \\
\text{C}-\text{NH}_2 \\
\mid \\
\text{NH} \\
\mid \\
\text{CH}_2 \\
\mid \\
\text{CH}_2 \\
\mid \\
\text{CH}_2 \\
\mid \\
\text{CH}-\text{NH}_3{}^+ \\
\mid \\
\text{COO}^-
\end{array}}
&
\xrightarrow{\text{H}_2\text{O}}
&
\underset{\text{Ornithine}}{
\begin{array}{l}
\text{NH}_3{}^+ \\
\mid \\
\text{CH}_2 \\
\mid \\
\text{CH}_2 \\
\mid \\
\text{CH}_2 \\
\mid \\
\text{CH}-\text{NH}_3{}^+ \\
\mid \\
\text{COO}^-
\end{array}}
\quad + \quad
\underset{\text{Urea}}{
\begin{array}{c}
\text{O} \\
\parallel \\
\text{H}_2\text{N}-\text{C}-\text{NH}_2
\end{array}}
\end{array}
$$

The urea is excreted in the urine. The ornithine re-enters the mitochondrion and, thus, completes the cycle. It is now ready to pick up another carbamoyl phosphate.

An important aspect of carbamoyl phosphate as an intermediate is that it can be used for synthesis of nucleotide bases (Chapter 24). Furthermore, the urea cycle is linked to the citric acid cycle because both involve fumarate.

Hans Krebs, who elucidated the citric acid cycle, was also instrumental in establishing the urea cycle.

27.9 Catabolism of the Carbon Skeleton of Amino Acids

After transamination of amino acids (Section 27.8) to glutamate, the alpha amino group is removed from glutamate by oxidative deamination (Section 27.8). The remaining carbon skeleton is used as an energy source (Figure 27.8).

We are not going to study the pathways involved except to point out the eventual fate of the skeleton. Not all the carbon skeletons of amino acids are used as fuel. Some of them may be degraded up to a certain point, and the resulting intermediate then used as a building block to construct another needed molecule. For example, if the carbon skeleton of an amino acid is catabolized to pyruvate, there are two possible choices for the body: (1) to use the pyruvate as an energy supply via the common pathway or (2) to use it as a building block to synthesize glucose (Section 28.2). Those amino acids that yield a carbon skeleton that is degraded to pyruvate or another intermediate capable of conversion to glucose (such as oxaloacetate) are called **glucogenic.** One example is alanine. When alanine reacts with α-ketoglutaric acid, the transamination produces pyruvate directly, as shown in Figure 27.8:

$$
\begin{array}{ccccc}
\underset{\text{Alanine}}{
\begin{array}{l}
\text{COO}^- \\
\mid \\
\text{CH}-\text{NH}_3{}^+ \\
\mid \\
\text{CH}_3
\end{array}}
&+&
\underset{\alpha\text{-Ketoglutarate}}{
\begin{array}{l}
\text{COO}^- \\
\mid \\
\text{C}=\text{O} \\
\mid \\
\text{CH}_2 \\
\mid \\
\text{CH}_2 \\
\mid \\
\text{COO}^-
\end{array}}
&\longrightarrow&
\underset{\text{Pyruvate}}{
\begin{array}{l}
\text{COO}^- \\
\mid \\
\text{C}=\text{O} \\
\mid \\
\text{CH}_3
\end{array}}
\quad +\quad
\underset{\text{Glutamate}}{
\begin{array}{l}
\text{COO}^- \\
\mid \\
\text{CH}-\text{NH}_3{}^+ \\
\mid \\
\text{CH}_2 \\
\mid \\
\text{CH}_2 \\
\mid \\
\text{COO}^-
\end{array}}
\end{array}
$$

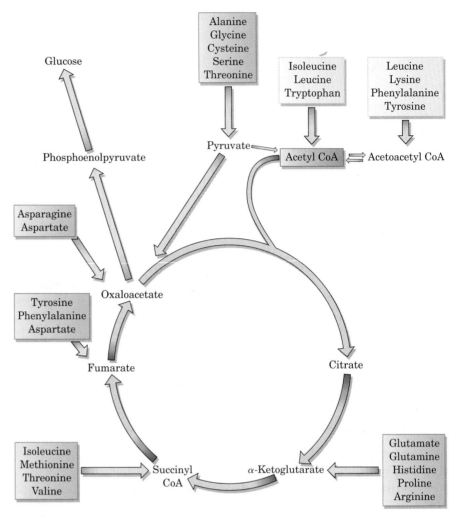

FIGURE 27.8 Catabolism of the carbon skeletons of amino acids. The glucogenic amino acids are in the blue boxes; the ketogenic ones in the brown boxes.

BOX 27F

Hereditary Defects in Amino Acid Catabolism. PKU

Many hereditary diseases involve missing or malfunctioning enzymes that catalyze the breakdown of amino acids. The oldest known of such diseases is cystinuria, which was described as early as 1810. In this disease, cystine shows up as flat hexagonal crystals in the urine. Stones form because of the low solubility of cystine in water. This leads to blockage in the kidneys or the ureters and requires surgery. One way to reduce the amount of cystine secreted is to remove as much methionine as possible from the diet. Beyond that, an increased fluid intake increases the volume of the urine, reducing the solubility problem. It has been found that penicillamine can also prevent cystinuria.

An even more important genetic defect is the absence of the enzyme phenylalanine hydroxylase, causing a disease called phenylketonuria (PKU). In normal catabolism, this enzyme helps degrade phenylalanine by converting it to tyro-

sine. If the enzyme is defective, phenylalanine is converted to phenylpyruvate (see the conversion of alanine to pyruvate in Section 27.9). Phenylpyruvate (a ketoacid) accumulates in the body and inhibits the conversion of pyruvate to acetyl CoA, thus depriving the cells of energy via the common pathway. This is most important in the brain, which gets its energy from the utilization of glucose. The result is mental retardation. This genetic defect can be detected early because phenylpyruvic acid appears in the urine. A federal regulation requires that all infants be tested for this disease. When PKU is detected, mental retardation can be prevented by restricting the intake of phenylalanine in the diet. Patients with PKU should avoid the artificial sweetener aspartame (Box 29C) because it yields phenylalanine when hydrolyzed in the stomach.

Jaundice

In normal individuals there is a balance between the destruction of heme in the spleen and the removal of bilirubin from the blood by the liver. When this balance is upset, the bilirubin concentration in the blood increases. If this condition continues, the skin and whites of the eyes both become yellow in a condition known as **jaundice.** Therefore, jaundice can signal a malfunctioning of the liver, the spleen, or the gallbladder, where bilirubin is stored before excretion. Hemolytic jaundice is the acceleration of the destruction of red blood cells in the spleen to such a level that the liver cannot cope with the bilirubin production. When gallstones obstruct the bile ducts and the bilirubin cannot be excreted in the feces, the backup of bilirubin causes jaundice. In infectious hepatitis, the liver is incapacitated, and bilirubin is not removed from the blood, causing jaundice.

■ Yellowing of the sclera (white outer coat of eyeball) in jaundice. The yellowing is caused by an excess of bilirubin, a bile pigment, in the blood. *(Science Photo Library/Photo Researchers, Inc.)*

On the other hand, many amino acids are degraded to acetyl CoA and acetoacetic acid. These compounds cannot form glucose but are capable of yielding ketone bodies; they are called **ketogenic.** Leucine is an example of a ketogenic amino acid.

Both glucogenic and ketogenic amino acids, when used as an energy supply, enter the citric acid cycle at some point (Figure 27.8) and are eventually oxidized to CO_2 and H_2O. The oxaloacetate (a C_4 compound) produced in this manner enters the citric acid cycle, adding to the oxaloacetate produced in the cycle itself.

27.10 Catabolism of Heme

Red blood cells are continuously being manufactured in the bone marrow. Their life span is relatively short—about four months. Aged red blood cells are destroyed in the phagocytic cells. When a red blood cell is destroyed, its hemoglobin is metabolized: The globin (Section 21.9) is hydrolyzed to amino acids, the heme is converted to *bilirubin,* and the iron is preserved in ferritin, an iron-carrying protein, and is reused. The bilirubin enters the liver via the blood and is then transferred to the gallbladder, where it is stored in the bile and finally excreted via the small intestine. The color of feces is provided by *urobilin,* an oxidation product of bilirubin.

Phagocytes are specialized blood cells that destroy foreign bodies.

SUMMARY

The foods we eat are broken down into small molecules in the stomach and intestines before being absorbed. These small molecules serve two purposes: They can be the building blocks of new materials the body needs to synthesize (anabolism), or they can be used for energy supply (catabolism). Each group of compounds—carbohydrates, fats, and proteins—has its own catabolic pathway. All the different catabolic pathways converge to the common pathway.

The specific pathway of carbohydrate catabolism is **glycolysis.** In this process, hexose monosaccharides are activated by ATP and eventually converted to two C_3 fragments, dihydroxyacetone phosphate and glyceraldehyde phosphate. The glyceraldehyde phosphate is further oxidized and eventually ends up as pyruvate. All these reactions occur in the cytosol. Pyruvate is converted to acetyl CoA, which is further catabolized in the common pathway. When completely metabolized, a hexose yields the energy of 36 ATP molecules.

When the body needs intermediates for synthesis rather than energy, the glycolytic pathway can be shunted to the **pentose phosphate pathway.** By doing so, NADPH is obtained, which is necessary for reduction. The pentose phosphate pathway also yields ribose, which is necessary for synthesis of RNA.

Fats are broken down to glycerol and fatty acids. Glycerol is catabolized in the glycolysis pathway and yields 20 ATP molecules.

Fatty acids are broken down into C_2 fragments in the **β-oxidation** spiral. At each turn of the spiral, one acetyl CoA is released together with one $FADH_2$ and one $NADH + H^+$. These products go through the common pathway. Stearic acid, a C_{18} compound,

yields 146 molecules of ATP. In starvation and under certain pathological conditions, not all the acetyl CoA produced in the β-oxidation of fatty acids enters the common pathway. Some of it forms acetoacetate, β-hydroxybutyrate, and acetone, commonly called **ketone bodies.** Excess ketone bodies in the blood are secreted in the urine.

Proteins are broken down to amino acids. The nitrogen of the amino acids is first transferred to glutamate. This in turn is **oxidatively deaminated** to yield ammonia. Mammals get rid of the toxic ammonia by converting it to urea in the **urea cycle.** Urea is secreted in the urine. The carbon skeletons of amino acids are catabolized via the citric acid cycle. Some of these enter as pyruvate or other intermediates of the citric acid cycle; these are **glucogenic amino acids.** Others are incorporated into acetyl CoA or ketone bodies and are called **ketogenic amino acids.** Heme is catabolized to bilirubin, which is excreted in the feces.

KEY TERMS

Amino acid pool (Section 27.1)
Anaerobic pathway (Section 27.2)
Cytosol (Section 27.2)
Fat depot (Section 27.1)
Glucogenic amino acid (Section 27.9

Glycogenolysis (Section 27.3)
Glycolysis (Sections 27.1, 27.2)
Ketoacidosis (Box 21C)
Ketogenic amino acid (Section 27.9)
Ketone bodies (Section 27.7)
β-Oxidation (Sections 27.1, 27.5)

Oxidative deamination (Sections 27.2, 27.8)
Pentose phosphate pathway (Section 27.2)
Transamination (Section 27.8)
Ubiquitinylation (Box 27E)
Urea cycle (Sections 27.2, 27.8)

CONCEPTUAL PROBLEMS

27.A At which intermediate of the glycolytic pathway does the oxidation and hence the energy production begin? In what form is the energy produced?

27.B In Figure 27.3, step ⑤ yields one NADH. Yet in Table 27.1, the same step indicates the yield of 2NADH + H. Is there a discrepancy between these two statements? Explain.

27.C Is the β-oxidation of fatty acid (without the subse-

quent metabolism of C_2 fragments via common metabolic pathway) more efficient with short-chain fatty acid than with long-chain fatty acid? Is there more ATP produced per C atoms in short-chain fatty acid than in long-chain fatty acid during β-oxidation?

27.D Is the urea cycle an energy-producing or -consuming pathway?

PROBLEMS

Difficult problems are designated by an asterisk.

Specific Pathways

27.1 In what form is carbohydrate stored in the body?

27.2 What are the products of the lipase-catalyzed hydrolysis of fats?

27.3 What is the main use of amino acids in the body?

Glycolysis

27.4 Although catabolism of a glucose molecule eventually produces a lot of energy, the first step uses up energy. Explain why this step is necessary.

27.5 In one step of the glycolysis pathway, a C_6 chain is broken into two C_3 fragments, only one of which can be further degraded in the glycolysis pathway. What happens to the other C_3 fragment?

27.6 Kinases are enzymes that catalyze the addition or removal of a phosphate group to (from) a substance. ATP is also involved. How many kinases are in glycolysis? Name them.

27.7 (a) Which steps in glycolysis of glucose need ATP? (b) Which steps in glycolysis yield ATP directly?

27.8 What two C_3 fragments are obtained from splitting fructose 1,6-bisphosphate?

27.9 Find all the oxidation steps in the glycolysis pathway shown in Figure 27.3

27.10 The end product of glycolysis, pyruvate, cannot enter as such into the citric acid cycle. What is the name of the process that converts this C_3 compound to a C_2 compound?

27.11 What essential compound is produced in the pentose phosphate pathway that is needed for synthesis as well as for defense against oxidative damages?

*27.12 Which of these steps yields energy and which consumes energy?
(a) Pyruvate $\rightarrow$ lactate
(b) Pyruvate $\rightarrow$ acetyl CoA + CO_2

27.13 How many moles of lactate are produced from 3 moles of glucose?

*27.14 How many moles of net NADH + H^+ are produced from 1 mole of glucose going to (a) acetyl CoA? (b) lactate?

Energy Yield from Glucose

27.15 Of the 36 molecules of ATP produced by the complete metabolism of glucose, how many are produced in glycolysis alone, that is, before the common pathway?

27.16 How many net ATP molecules are produced in the skeletal muscles for each glucose molecule
(a) In glycolysis alone (up to pyruvate)?
(b) In converting pyruvate to acetyl CoA?
(c) In the total oxidation of glucose to CO_2 and H_2O?

*27.17 (a) If fructose is metabolized in the liver, how many moles of net ATP are produced from each mole during glycolysis? (b) How many moles are produced if the same thing occurs in a muscle cell?

27.18 What is the difference between glycolysis and glycogenolysis?

Glycerol Catabolism

*27.19 Based on the names of the enzymes participating in glycolysis, what would be the name of the enzyme catalyzing the activation of glycerol?

27.20 Which yields more energy upon hydrolysis, ATP or glycerol 1-phosphate? Why?

β-Oxidation of Fatty Acids

*27.21 Two enzymes participating in the β-oxidation have the word "thio" in their names. (a) Name the two enzymes. (b) To what chemical group does this name

refer? (c) What is the common feature in the action of these two enzymes?

27.22 (a) Which part of the cells contains the enzymes needed for the β-oxidation of fatty acids? (b) How does the activated fatty acid get there?

27.23 Assume that lauric acid is metabolized through β-oxidation. What are the products of the reaction after three turns of the spiral?

27.24 How many turns of the spiral are there in the β-oxidation of (a) lauric acid (b) palmitic acid?

Energy Yield from Fatty Acids

27.25 Calculate the number of ATP molecules obtained in the β-oxidation of myristic acid, $CH_3(CH_2)_{12}COOH$.

*27.26 Assume that the cis-trans isomerization in the β-oxidation of unsaturated fatty acids does not require energy. Which C_{18} fatty acid yields the greater amount of energy, saturated (stearic acid) or mono-unsaturated (oleic acid)? Explain.

27.27 Assuming that both fats and carbohydrates are available, which does the body preferentially use as an energy source?

27.28 If equal weights of fats and carbohydrates are eaten, which will give more calories? Explain.

Ketone Bodies

27.29 Why do starving people have ketone bodies in their urine?

27.30 Do ketone bodies have nutritional value?

27.31 What happens to the oxaloacetate produced from carboxylation of pyruvate?

Catabolism of Amino Acids

27.32 What enzymes facilitate the movement of an amino group from one amino acid to another?

27.33 Write an equation for the oxidative deamination of alanine.

27.34 Ammonia, NH_3, and ammonium ion, NH_4^+, are both soluble in water and could easily be excreted in the urine. Why does the body convert them to urea rather than excreting them directly?

27.35 The metabolism of the carbon skeleton of tyrosine yields pyruvate. Why is tyrosine a glucogenic amino acid?

*27.36 What are the sources of the nitrogen in urea?

27.37 What compound is common to both the urea and citric acid cycles?

27.38 (a) What is the toxic product of the oxidative deamination of glutamate? (b) How does the body get rid of it?

27.39 If leucine is the only carbon source in the diet of an experimental animal, what would you expect to find in the urine of such an animal?

Catabolism of Heme

*27.40 Why is a high bilirubin content in the blood an indication of liver disease?

27.41 When hemoglobin is fully metabolized, what happens to the iron in it?

27.42 Which component of the metabolized hemoglobin molecule ends up in the amino acid pool?

Boxes

27.43 (Box 27A) What causes cramps of the muscles when fatigued?

*27.44** (Box 27B) How does the signal of epinephrine result in depletion of glycogen in the muscle?

27.45 (Box 27C) What system counteracts the acidic effect of ketone bodies in the blood?

27.46 (Box 27C) The patient whose condition is described in Box 27C was transferred to a hospital in an ambulance. Could a nurse in the ambulance tentatively diagnose the diabetic condition without running blood and urine tests? Explain.

27.47 (Box 27D) Draw the structure of monosodium glutamate.

27.48 (Box 27E) What is meant by the expression: "the amino acid sequence of ubiquitin is highly conserved?

27.49 (Box 27E) How does ubiquitin attach itself to a target protein?

27.50 (Box 27E) How does the body eliminate damaged (oxidized) hemoglobin from the blood?

*27.51** (Box 27F) Draw structural formulas for each reaction component, and complete the following equation.

$$\text{phenylalanine} \longrightarrow \text{phenylpyruvate} + ?$$

27.52 (Box 27G) What compound causes the yellow coloration of jaundice?

Additional Problems

27.53 If you have received a laboratory report showing the presence of a high concentration of ketone bodies in the urine of a patient, what disease would you suspect?

27.54 Ornithine is a basic amino acid similar to lysine. (a) What is the structural difference between the two? (b) Is ornithine a constituent of proteins?

*27.55** (a) At which step of the glycolysis pathway does NAD^+ participate (Figures 27.3 and 27.4)? (b) At which step does $NADH + H^+$ participate? (c) As a result of the overall pathway, is there a net increase of NAD^+, of $NADH + H^+$, or of neither?

*27.56** What is the net energy yield in moles of ATP produced when yeast converts 1 mole of glucose to ethanol?

*27.57** Can the intake of alanine, glycine, and serine relieve hypoglycemia caused by starvation? Explain.

*27.58** How can glucose be utilized to produce ribose for RNA synthesis?

27.59 Write the products of the transamination reaction between alanine and oxaloacetate:

$$\begin{array}{ccc} & & \text{COO}^- \\ \text{COO}^- & & | \\ | & & \text{C}{=}\text{O} \\ \text{CH}{-}\text{NH}_3^+ & + & | \\ | & & \text{CH}_2 \\ \text{CH}_3 & & | \\ & & \text{COO}^- \end{array} \longrightarrow$$

*27.60** Phosphoenolpyruvate (PEP) has a high-energy phosphate bond that has more energy than the anhydride bonds in ATP. What step in glycolysis suggests that this is so?

*27.61** Suppose that a fatty acid labeled with radioactive carbon-14 is fed to an experimental animal. Where would you look for the radioactivity?

27.62 What functional groups are in carbamoyl phosphate?

27.63 What compound causes jaundice?

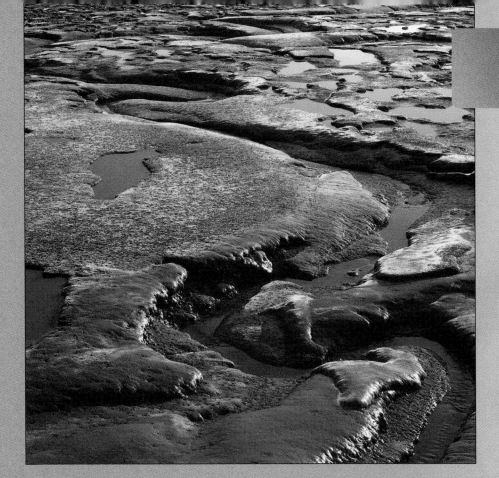

Biosynthetic Pathways

28.1 Introduction

In the human body, and in most other living tissues, the pathways by which a compound is synthesized (anabolism) are usually different from the pathways by which it is degraded (catabolism). There are several reasons why it is biologically advantageous for anabolic and catabolic pathways to be different. We will give two of them.

1. Flexibility If the normal **biosynthetic pathway** is blocked, the body can often use the reverse of the degradation pathway instead (remember that most steps in degradation are reversible), thus providing another way to make the necessary compounds.

2. Overcoming the effect of Le Chatelier's principle This can be illustrated by the cleavage of a glucose unit from a glycogen molecule, an equilibrium process:

$$(\text{Glucose})_n + \text{P}_i \xrightleftharpoons{\text{phosphorylase}} (\text{Glucose})_{n-1}$$

Glycogen   Glycogen
(one unit smaller)   (28.1)

$$+ \text{ Glucose 1-phosphate}$$

Phosphorylase catalyzes not only glycogen degradation (the forward reaction) but also glycogen synthesis (the reverse reaction). However,

Anabolic pathways are also called biosynthetic pathways.

Above: Algae on mudflats. *(© Bruce M. Herman/Photo Researchers, Inc.)*

in the body there is a large excess of inorganic phosphate, P_i. This would drive the reaction, on the basis of Le Chatelier's principle, to the right, which represents glycogen degradation. In order to provide a method for the synthesis of glycogen even in the presence of excess inorganic phosphate, a different pathway is needed in which P_i is not a reactant. Thus, the body uses the following synthetic pathway:

The structure of UDP-glucose is shown in Section 28.2.

$$\text{(Glucose)}_{n-1} + \text{UDP-glucose} \longrightarrow \text{(Glucose)}_n + \text{UDP} \quad (28.2)$$

Glycogen   Glycogen
(one unit larger)

Not only are the synthetic pathways different from the catabolic pathways, but the energy requirements are also different, as is the location. Most catabolic reactions occur in the mitochondria, whereas anabolic reactions generally take place in the cytosol. We shall not go into the energy balances of the biosynthetic processes as we did for catabolism. However, it must be kept in mind that, while energy (in the form of ATP) is *obtained* in the degradative processes, biosynthetic processes *consume* energy.

28.2 Biosynthesis of Carbohydrates

We discuss the biosynthesis of carbohydrates under three headings: (a) conversion of atmospheric CO_2 to glucose in plants, (b) synthesis of glucose in animals and humans, and (c) conversion of glucose to other carbohydrate molecules in animals and humans.

Conversion of Atmospheric CO_2

Photosynthesis The process in which plants synthesize carbohydrates from CO_2 and H_2O with the help of sunlight and chlorophyll

The most important biosynthesis of carbohydrates takes place in plants. This is **photosynthesis.** In this process, the energy of the sun is built into chemical bonds of glucose; the overall reaction is

Although the primary product of photosynthesis is glucose, it is largely converted to other carbohydrates, mainly cellulose and starch.

$$n\text{H}_2\text{O} + n\text{CO}_2 \xrightarrow[\text{chlorophyll}]{\substack{\text{energy in} \\ \text{the form of} \\ \text{sunlight}}} (\text{CH}_2\text{O})_n + n\text{O}_2 \quad (28.3)$$

where $(\text{CH}_2\text{O})_n$ is a general formula for carbohydrates. This is a very complicated process and takes place only in plants, not in animals (Box 28A). We shall not discuss it further here except to note that the carbohydrates of plants—starch, cellulose, and other mono- and polysaccharides—serve as the basic carbohydrate supply of all animals, including humans.

Synthesis of Glucose

Gluconeogenesis The process by which glucose is synthesized in the body

We saw in Chapter 27 that when the body needs energy, carbohydrates are broken down via glycolysis. When energy is not needed, glucose can be synthesized from the intermediates of the glycolytic and citric acid pathways. This process is called **gluconeogenesis.** As shown in Figure 28.1, a large number of intermediates—pyruvate, lactate, oxaloacetate, malate, and several amino acids (the glucogenic amino acids we met in Section 27.9)—can serve as starting compounds. Gluconeogenesis proceeds in reverse order from glycolysis, and many of the enzymes of glycolysis also catalyze gluconeogenesis. However, at four points there are unique enzymes (marked in Figure 28.1) that catalyze only gluconeogenesis and not the breakdown reactions. These four enzymes make gluconeogenesis a pathway that is distinct from glycolysis.

ATP is used up in gluconeogenesis and produced in glycolysis.

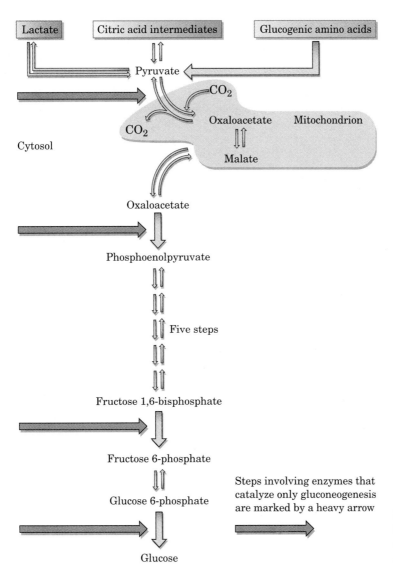

FIGURE 28.1 Gluconeogenesis. All reactions take place in the cytosol, except for those shown in the mitochondria.

Conversion of Glucose to Other Carbohydrates

The third important biosynthetic pathway for carbohydrates is the conversion of glucose to other hexoses and to hexose derivatives and the synthesis of di-, oligo-, and polysaccharides. The common step in all of these is the activation of glucose by uridine diphosphate (UDP) to form UDP-glucose:

Photosynthesis

Photosynthesis requires sunlight, water, CO_2, and pigments found in plants, mainly chlorophyll. The overall reaction shown in Equation 28.3 in the text actually occurs in two distinct steps. First, the light interacts with the pigments that are located in highly membranous organelles of plants, called **chloroplasts.** Chloroplasts resemble mitochondria (Section 26.2) in many respects: They contain a whole chain of oxidation–reduction enzymes similar to the cytochrome and iron–sulfur complexes of mitochondrial membranes, and they also contain a proton translocating ATPase. In a manner similar to mitochondria, in chloroplasts too the proton gradient accumulated in the intermembrane region drives the synthesis of ATP (see the chemiosmotic pump, Section 20.6). The chlorophylls themselves, buried in a complex protein that traverses the chloroplast membranes, are molecules similar to the heme we have already encountered in hemoglobin (Figure 21.12). In contrast to heme, the chlorophylls contain Mg^{2+} instead of Fe^{2+}:

Chlorophyll a

The first reaction that takes place in photosynthesis is called the light reaction because chlorophyll captures the energy of sunlight and with its aid strips the electrons and protons from water to form oxygen, ATP, and NADPH + H$^+$ (see Figure 28.3):

$$H_2O + ADP + P_i + NADP^+ + \text{sunlight} \longrightarrow \\ O_2 + ATP + NADPH + H^+$$

The second reaction, called the dark reaction because it does not need light, in essence converts CO_2 to carbohydrates:

$$CO_2 + ATP + NADPH + H^+ \longrightarrow \\ \text{carbohydrates} + ADP + P_i + NADP^+$$

In this step, the energy, now in the form of ATP, is used to help NADPH + H$^+$ reduce carbon dioxide to carbohydrates. Thus, the protons and electrons stripped in the light reaction are added to the carbon dioxide in the dark reaction. This reduction takes place in a multistep cyclic process called the **Calvin cycle,** named after its discoverer, Melvin Calvin (1911–1997), who was awarded the 1961 Nobel Prize in Chemistry for this work. In this cycle, the CO_2 is first attached to a C_5 fragment and breaks down to two C_3 fragments (triose phosphates) that are, through a complex series of steps, converted to a C_6 compound and, eventually, to glucose.

$$CO_2 + C_5 = 2C_3 = C_6$$

The critical step in the dark reaction (Calvin cycle) is the attachment of CO_2 to the ribulose (Table 19.2), a C_5 biphosphate intermediate. The enzyme that catalyzes this reaction,

Glycogenesis The conversion of glucose to glycogen

UDP is similar to ADP except that the base is uracil instead of adenine. UTP, an analog of ATP, contains two high-energy phosphate bonds. For example, when the body has excess glucose and wants to store it as glycogen (this process is called **glycogenesis**), the glucose is first converted to glucose 1-phosphate, but then a special enzyme catalyzes the reaction:

$$\text{glucose 1-phosphate} + \boxed{\text{UTP}} \longrightarrow \text{UDP-glucose} + {}^-O-\overset{\overset{\displaystyle O}{\|}}{\underset{\underset{\displaystyle O^-}{|}}{P}}-O-\overset{\overset{\displaystyle O}{\|}}{\underset{\underset{\displaystyle O^-}{|}}{P}}-O^-$$

$$\underset{\text{Glycogen}}{\text{UDP-glucose} + (\text{glucose})_n} \longrightarrow \boxed{\text{UDP}} + \underset{\substack{\text{Glycogen} \\ \text{(one unit larger)}}}{(\text{glucose})_{n+1}}$$

ribulose-1,5-bisphosphate carboxylase-oxygenase, nicknamed RuBisCO, is one of the slowest in nature. As in traffic, the slowest moving vehicle determines the overall flow, so RuBisCO is the main factor in the low efficiency of the Calvin cycle. Because of this enzyme, most plants convert less than 1

percent of the energy into carbohydrates. To overcome the enzyme inefficiency, plants must synthesize large quantities of this enzyme. More than half of the soluble proteins in plant leaves are RuBisCO enzymes, the synthesis of which requires a lot of energy expenditure.

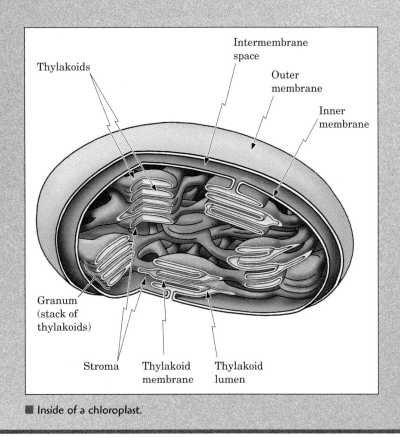

■ Inside of a chloroplast.

The biosynthesis of many other di- and polysaccharides and their derivatives also uses the common activation step: forming the appropriate UDP compound.

28.3 Biosynthesis of Fatty Acids

The body can synthesize all the fatty acids it needs except for linoleic and linolenic acids (essential fatty acids, Section 20.3). The source of carbon in this synthesis is acetyl CoA. Since acetyl CoA is also a degradation product of the β-oxidation spiral of fatty acids (Section 27.5), we might expect that the synthesis is the reverse of the degradation. This is not the case. For one thing, the majority of fatty acid synthesis occurs in the cytosol, while degradation takes place in the mitochondria. Fatty acid synthesis is catalyzed by a multienzyme system.

However, there is one aspect of fatty acid synthesis that is the same as in fatty acid degradation: Both processes involve acetyl CoA; therefore, both proceed in units of two carbons. Fatty acids are built up two carbons at a time, just as they are broken down two carbons at a time (Section 27.5).

Most of the time, fatty acids are synthesized when excess food is available. That is, when we eat more food than we need for energy, our bodies turn the excess acetyl CoA (produced by catabolism of carbohydrates—Section 27.2) into fatty acids and then to fats. The fats are stored in the fat depots, which are specialized fat-carrying cells (see Figure 27.1).

The key to fatty acid synthesis is an **acyl carrier protein** called **ACP.** This can be looked upon as a merry-go-round—a rotating protein molecule to which the growing chain of fatty acids is bonded. As the growing chain rotates with the ACP, it sweeps over the multienzyme complex, and at each enzyme one reaction of the chain is catalyzed (Figure 28.2).

At the beginning of this cycle, the ACP picks up an acetyl group from acetyl CoA and delivers it to the first enzyme, fatty acid synthase, here called synthase for short:

$$CH_3\overset{\overset{\displaystyle O}{\|}}{C}-S-CoA + HS-ACP \longrightarrow HS-CoA + CH_3\overset{\overset{\displaystyle O}{\|}}{C}-S-ACP$$

$$CH_3\overset{\overset{\displaystyle O}{\|}}{C}-S-ACP + synthase-SH \longrightarrow$$

$$CH_3\overset{\overset{\displaystyle O}{\|}}{C}-S-synthase + HS-ACP$$

The C_2 fragment on the synthase is condensed with a C_3 fragment attached to the ACP in a process in which CO_2 is given off:

$$CH_3\overset{\overset{\displaystyle O}{\|}}{C}-S-synthase + \underset{\underset{\displaystyle COO^-}{|}}{CH_2}-\overset{\overset{\displaystyle O}{\|}}{C}-S-ACP \longrightarrow$$

Malonyl-ACP

$$CH_3\overset{\overset{\displaystyle O}{\|}}{C}-CH_2-\overset{\overset{\displaystyle O}{\|}}{C}-S-ACP + CO_2 + synthase-SH$$

Acetoacetyl-ACP

The result is a C_4 fragment, which is reduced twice and dehydrated before it becomes a fully saturated C_4 group. This is the end of one cycle of the merry-go-round.

In the next cycle, the C_4 fragment is transferred to the synthase, and another malonyl-ACP (C_3 fragment) is added. When CO_2 splits out, a C_6 fragment is obtained. The merry-go-round continues to turn. At each turn,

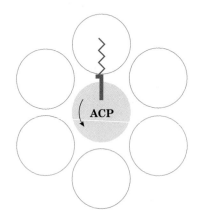

FIGURE 28.2 The biosynthesis of fatty acids. The ACP (central dark sphere) has a long side chain (⌐) that carries the growing fatty acid (⌁). The ACP rotates counterclockwise, and its side chain sweeps over a multienzyme system (empty spheres). As each cycle is completed, a C_2 fragment is added to the growing fatty acid chain.

another C_2 fragment is added to the growing chain. Chains up to C_{16} (palmitic acid) can be obtained in this process. If the body needs longer fatty acids—for example, stearic (C_{18})—another C_2 fragment is added to palmitic acid by a different enzyme system.

Unsaturated fatty acids are obtained from saturated fatty acids by an oxidation step in which hydrogen is removed and combined with O_2 to form water:

$$R—CH_2—CH_2—(CH_2)_n COOH + O_2 + \boxed{NADPH} + H^+ \xrightarrow{\text{enzyme}}$$

$$\underset{R}{\overset{H}{\underset{(CH_2)_n COOH}{\overset{H}{C=C}}}} \quad + 2H_2O + \boxed{NADP^+}$$

The structure of $NADP^+$ is the same as that of NAD^+, except that an additional phosphate group is attached to one of the ribose units (Figure 28.3).

Nicotinamide adenine dinucleotide phosphate
($NADP^+$)

FIGURE 28.3 The structure of $NADP^+$.

28.4 Biosynthesis of Membrane Lipids

The various membrane lipids (Sections 20.6 to 20.8) are assembled from their constituents. We just saw how fatty acids are synthesized in the body. These fatty acids are activated by CoA, forming acyl CoA. The compound glycerol 3-phosphate, which is obtained from the reduction of dihydroxy-acetone (a C_3 fragment of glycolysis, Figure 27.4), is the second building block of glycerophospholipids. This compound combines with two acyl CoA molecules, which can be the same or different:

We encountered glycerol 3-phosphate as a vehicle for transporting electrons in and out of mitochondria (Section 21.3).

$$
\begin{array}{ccc}
\text{CH}_2\text{—OH} & & \text{CH}_2\text{—O—}\overset{\displaystyle O}{\overset{\|}{\text{C}}}\text{—R} \\[2ex]
| & \overset{\displaystyle O}{\overset{\|}{\text{R—C—S—CoA}}} & | \quad \overset{\displaystyle O}{\overset{\|}{}} \\
\text{CH—OH} \quad + & & \text{CH—O—C—R}' \;+\; 2\text{CoA—SH} \\[2ex]
| & \overset{\displaystyle O}{\overset{\|}{\text{R}'\text{—C—S—CoA}}} & | \quad \overset{\displaystyle O}{\overset{\|}{}} \\
\text{CH}_2\text{—O—}\overset{\displaystyle O}{\overset{\|}{\text{P}}}\text{—O}^- & & \text{CH}_2\text{—O—}\overset{\displaystyle O}{\overset{\|}{\text{P}}}\text{—O}^- \\
\quad | & & \quad | \\
\quad \text{O}^- & & \quad \text{O}^- \\[1ex]
\text{Glycerol} & \text{Acyl CoA} & \text{A phosphatidate} \\
\text{3-phosphate} & &
\end{array}
$$

To complete the molecule, an activated serine or choline or ethanolamine is added to the $-\text{OPO}_3^{2-}$ group (see the structures in Section 20.6; a model of phosphatidylcholine is shown in Figure 28.4). Choline is activated by cytidine diphosphate (CDP), yielding CDP-choline. This is similar to the activation of glucose by UDP (Section 28.2) except that the base is cytosine rather than uracil (Section 24.2).

Sphingolipids (Section 20.7) are similarly built up from smaller molecules. An activated phosphocholine is added to the sphingosine part of ceramide (Section 20.7) to make sphingomyelin.

Table 22.1 shows the essential and nonessential amino acids.

The glycolipids are made in a similar fashion. Ceramide is assembled as above, and the carbohydrate is added one unit at a time in the form of activated monosaccharides (UDP glucose and so on).

Cholesterol, the molecule that controls the fluidity of membranes and is a precursor of all steroid hormones, is also synthesized by the human body. It is assembled in the liver from C_2 fragments in the form of acetyl CoA. All the carbon atoms of cholesterol come from the carbons of acetyl

FIGURE 28.4 A model of phosphatidylcholine, commonly called lecithin.

Phospholipid Synthesis and Hyaline Membrane Disease

Hyaline membrane disease, which affects the breathing of infants and has caused numerous sudden infant deaths, is not completely understood at this time. Analysis has shown that there is a great difference in the phospholipid composition of the lung between normal and affected individuals. The phospholipid of the lung is mainly phosphatidylcholine, which in normal lungs contains large quantities of unsaturated fatty acids. In contrast, the phosphatidylcholine in the lungs of patients with hyaline membrane disease is largely saturated. Furthermore, these patients have less phosphatidylcholine in their lung tissue than normal individuals. This indicates a faulty biosynthetic apparatus, but the exact cause has not yet been determined.

CoA molecules (Figure 28.5). During the synthesis, geranyl, C_{10}, and farnesyl, C_{15}, intermediates are formed.

Geranyl pyrophosphate

Farnesyl pyrophosphate

The drug lovastatin inhibits the biosynthesis of cholesterol and is frequently prescribed to control the cholesterol level in the blood in order to prevent atherosclerosis (Box 20G).

FIGURE 28.5 Biosynthesis of cholesterol. The highlighted carbon atoms come from the —CH_3 group, and the others come from the —CO— group of the acetate.

28.5 Biosynthesis of Amino Acids

The human body needs 20 different amino acids to make its protein chains—all 20 are found in a normal diet. Some of the amino acids can be synthesized from other compounds: These are the nonessential amino acids. Others cannot be synthesized by the human body and must be supplied in the diet. These are the **essential amino acids** (see Section 29.4). Most nonessential amino acids are synthesized from some intermediate of either glycolysis (Section 27.2) or the citric acid cycle (Section 26.4). Glutamic acid plays a central role in the synthesis of five nonessential amino acids. Glutamic acid itself is synthesized from α-ketoglutaric acid, one of the intermediates in the citric acid cycle:

$$
\mathrm{NADH} + \mathrm{H}^+ + \mathrm{NH_4}^+ +
\begin{array}{c}
\mathrm{COO}^- \\
| \\
\mathrm{C{=}O} \\
| \\
\mathrm{CH_2} \\
| \\
\mathrm{CH_2} \\
| \\
\mathrm{COO}^-
\end{array}
\rightleftharpoons
\begin{array}{c}
\mathrm{COO}^- \\
| \\
\mathrm{CH{-}NH_3}^+ \\
| \\
\mathrm{CH_2} \\
| \\
\mathrm{CH_2} \\
| \\
\mathrm{COO}^-
\end{array}
+ \mathrm{NAD}^+ + \mathrm{H_2O}
$$

α-Ketoglutaric acid   Glutamic acid

The forward reaction is the synthesis, and the reverse reaction is the oxidative deamination (degradation) reaction we encountered in the catabolism of amino acids (Section 27.8). This is one case in which the synthetic and degradative pathways are exactly the reverse of each other.

BOX 28C

Amino Acid Transport and Blue Diaper Syndrome

Amino acids, both essential and nonessential, are usually more concentrated inside the cells than in the surroundings. Because of this, special transport mechanisms are available to carry the amino acids into the cells. Sometimes these transport mechanisms are faulty, and the result is a lack of an amino acid inside the cells despite the fact that the particular amino acid is being provided by the diet.

An interesting example of this is blue diaper syndrome, in which infants excrete blue urine. Indigo blue, a dye, is an oxidation product of the amino acid tryptophan. But how does this oxidation product get into the system to be excreted in the urine? The patient's tryptophan transport mechanism is faulty. Although enough tryptophan is being supplied by the diet, most of it is not absorbed through the intestine. It accumulates there and is oxidized by bacteria in the gut. The oxidation product is moved into the cells, but because the cells cannot use it, it is excreted in the urine. Much of the blue diaper syndrome can be eliminated with antibiotics. This treatment kills many of the gut's bacteria, and the nonabsorbed tryptophan is then excreted in the feces because the bacteria cannot now oxidize it to indigo blue.

L-Tryptophan   Indigo

■ Structures of tryptophan and indigo.

Glutamic acid can serve as an intermediate in the synthesis of alanine, serine, aspartic acid, asparagine, and glutamine. For example, the transamination reaction we saw in Section 27.8 leads to alanine formation:

$$
\begin{array}{ccc}
\underset{\text{Pyruvate}}{\begin{array}{c} COO^- \\ | \\ C=O \\ | \\ CH_3 \end{array}} +
\underset{\text{Glutamic acid}}{\begin{array}{c} COO^- \\ | \\ CH-NH_3^+ \\ | \\ CH_2 \\ | \\ CH_2 \\ | \\ COO^- \end{array}}
\rightleftharpoons
\underset{\text{Alanine}}{\begin{array}{c} COO^- \\ | \\ CH-NH_3^+ \\ | \\ CH_3 \end{array}} +
\underset{\substack{\alpha\text{-Ketoglutaric} \\ \text{acid}}}{\begin{array}{c} COO^- \\ | \\ C=O \\ | \\ CH_2 \\ | \\ CH_2 \\ | \\ COO^- \end{array}}
\end{array}
$$

Besides being the building blocks of proteins, amino acids also serve as intermediates for a large number of biological molecules. We have already seen that serine is needed in the synthesis of membrane lipids (Section 28.4). Certain amino acids are also intermediates in the synthesis of heme and of the purines and pyrimidines that are the raw materials for DNA and RNA (Chapter 24).

SUMMARY

For most biochemical compounds, the biosynthetic pathways are different from the degradation pathways. Carbohydrates are synthesized in plants from CO_2 and H_2O, using sunlight as an energy source (**photosynthesis**). Glucose can be synthesized by animals from the intermediates of glycolysis, from those of the citric acid cycle, and from glucogenic amino acids. This process is called **gluconeogenesis.** When glucose or other monosaccharides are built into di-, oligo-, and polysaccharides, each monosaccharide unit in its activated form is added to a growing chain.

Fatty acid biosynthesis is accomplished by a multienzyme system. The key to this process is the **acyl carrier protein, ACP,** which acts as a merry-go-round transport system; it carries the growing fatty acid chain over a number of enzymes, each of which catalyzes a specific reaction. With each complete turn of the merry-go-round, a C_2 fragment is added to the growing fatty acid chain. The source of the C_2 fragment is malonyl ACP, a C_3 compound attached to the ACP. This becomes C_2 by loss of CO_2. Membrane lipids are synthesized in the body by assembling the constituent parts. The fatty acids are activated by conversion to acyl CoA.

Many nonessential amino acids are synthesized in the body from the intermediates of glycolysis or of the citric acid cycle. In half of these, glutamic acid is the donor of the amino group in transamination. Amino acids serve as building blocks for proteins.

KEY TERMS

Acyl carrier protein (Section 28.3)
Biosynthetic pathway (Section 28.1)
Calvin cycle (Box 28A)

Chloroplasts (Box 28A)
Essential amino acid (Section 28.5)
Gluconeogenesis (Section 28.2)

Glycogenesis (Section 28.2)
Photosynthesis (Section 28.2)

CONCEPTUAL PROBLEMS

28.A In most biosynthetic processes, the reactant is reduced to obtain the desired product. Verify if this statement holds for the overall reaction of photosynthesis.

28.B Glucose is the only carbohydrate compound the brain can use for energy. Which pathway is mobilized to supply the need of the brain during starvation: (a) glycolysis, (b) gluconeogenesis, or (c) glycogenesis? Explain.

28.C Are fatty acids for energy, in the form of fat, synthesized the same way as fatty acids for the lipid bilayer of membrane?

28.D Consider the fact that the deamination of glutamic acid and its synthesis from α-ketoglutaric acid are equilibrium reactions. Which way will the equilibrium shift when we are exposed to cold temperature?

PROBLEMS

Difficult problems are designated by an asterisk.

Biosynthesis

28.1 Why are the pathways the body uses for anabolism and catabolism mostly different?

28.2 Why is it advantageous to have more than one pathway for the biosynthesis of a compound?

28.3 Glycogen can be synthesized in the body by the same enzymes that degrade it. Why is this process utilized in glycogen synthesis only to a small extent, while most glycogen biosynthesis occurs via a different synthetic pathway?

28.4 Do most anabolic and catabolic reactions take place in the same location?

Carbohydrate Biosynthesis

28.5 What is the difference in the overall chemical equations between photosynthesis and respiration?

28.6 In photosynthesis, what is the source of (a) carbon (b) hydrogen (c) energy?

28.7 Name a compound that can serve as a raw material for gluconeogenesis and is (a) from the glycolytic pathway (b) from the citric acid cycle (c) an amino acid.

28.8 How is glucose activated for glycogen synthesis?

28.9 Which steps in gluconeogenesis are not reversible?

28.10 Are the enzymes that combine two C_3 compounds to a C_6 compound in gluconeogenesis the same as or different from those that cleave the C_6 compound into two C_3 compounds in glycolysis?

28.11 Which part of gluconeogenesis occurs in the mitochondria?

28.12 Glycogen is written as $(glucose)_n$. (a) What does the n stand for? (b) What is the approximate value of n?

28.13 What are the constituents of UTP?

Fatty Acid Biosynthesis

28.14 What is the source of carbon in fatty acid synthesis?

28.15 (a) Where in the body does fatty acid synthesis occur? (b) Does fatty acid degradation occur in the same location?

28.16 Is ACP an enzyme?

28.17 In fatty acid biosynthesis, which compound is added repeatedly to the synthase?

28.18 (a) What is the name of the first enzyme in fatty acid synthesis? (b) What does it do?

28.19 From what compound is the CO_2 released in fatty acid synthesis?

28.20 What is the functional group of the synthase to which the C_2 fragment is attached?

*28.21 In the synthesis of unsaturated fatty acids, $NADPH + H^+$ is converted to $NADP^+$. Yet this is an oxidation step and not a reduction step. Explain.

28.22 Which of these fatty acids can be synthesized by the multienzyme fatty acid synthesis complex alone?
(a) Oleic
(b) Stearic
(c) Myristic
(d) Arachidonic
(e) Lauric

*28.23 Some enzymes can use NADH as well as NADPH as a coenzyme. Other enzymes use one or the other exclusively. What features would prevent NADPH from fitting into the active site of an enzyme that otherwise can accommodate NADH?

28.24 Under what conditions does the body synthesize fatty acids?

*28.25 Linoleic and linolenic acids cannot be synthesized in the human body. Does this mean that the human body cannot make an unsaturated fatty acid from a saturated one?

Biosynthesis of Membrane Lipids

28.26 When the body synthesizes this membrane lipid, from what building blocks is it assembled?

$$
\begin{array}{l}
\quad\quad\quad\quad\quad O \\
\quad\quad\quad\quad\quad \parallel \\
CH_2-O-C-(CH_2)_{14}CH_3 \\
| \\
\quad\quad\quad\quad\quad O \\
\quad\quad\quad\quad\quad \parallel \\
CH-O-C-(CH_2)_{10}CH_3 \\
| \\
\quad\quad\quad\quad\quad O \\
\quad\quad\quad\quad\quad \parallel \\
CH_2-O-P-O-CH_2-CH-COO^- \\
\quad\quad\quad\quad\quad | \quad\quad\quad\quad\quad\quad\quad | \\
\quad\quad\quad\quad\quad O^- \quad\quad\quad\quad\quad NH_3^+
\end{array}
$$

*28.27 Name the activated constituents necessary to form the glycolipid glucoceramide.

28.28 What are the names of the C_{10} and C_{15} intermediates in cholesterol biosynthesis?

Biosynthesis of Amino Acids

*28.29 What compound reacts with glutamate in a transamination process to yield serine?

28.30 What reaction is the reverse of the synthesis of glutamate from α-ketoglutarate, ammonia, and $NADH + H^+$?

*28.31 Which amino acid will be synthesized by this process?

$$\begin{array}{c} COO^- \\ | \\ C{=}O \\ | \\ CH_2 \\ | \\ COO^- \end{array} + NADH + H^+ + NH_4{}^+ \longrightarrow$$

*28.32 Draw the structure of the compound needed to synthesize asparagine from glutamate by transamination.

28.33 Name the products of the transamination reaction

$$(CH_3)_2CH{-}\overset{\overset{\displaystyle O}{\|}}{C}{-}COO^-$$
$$+\ {}^-OOC{-}CH_2{-}CH_2{-}\underset{\underset{\displaystyle NH_3{}^+}{|}}{CH}{-}COO^- \longrightarrow$$

Boxes

28.34 (Box 28A) What is the major difference in structure between the chlorophylls and heme?

28.35 (Box 28A) What is the coenzyme that reduces CO_2 in the Calvin cycle?

28.36 (Box 28C) (a) What compound produces the colored urine of blue diaper syndrome? (b) Where does this compound come from?

Additional Problems

28.37 How are unsaturated fatty acids synthesized in the body?

28.38 What feature is common to the structures of ACP and CoA?

28.39 What C_3 fragment carried by ACP is used in fatty acid synthesis?

28.40 When glutamate transaminates phenylpyruvate, what amino acid is produced?

$$C_6H_5{-}CH_2{-}\overset{\overset{\displaystyle O}{\|}}{C}{-}COO^-$$
$$+\ {}^-OOC{-}CH_2{-}CH_2{-}\underset{\underset{\displaystyle NH_3{}^+}{|}}{CH}{-}COO^- \longrightarrow$$

28.41 Can the complex enzyme system participating in every fatty acid synthesis manufacture fatty acids of any length?

*28.42 Each activation step in the synthesis of complex lipids occurs at the expense of one ATP molecule. How many ATP molecules are used up in the synthesis of one molecule of lecithin?

28.43 Would a person on a completely cholesterol-free diet totally lack the cholesterol necessary to synthesize steroid hormones?

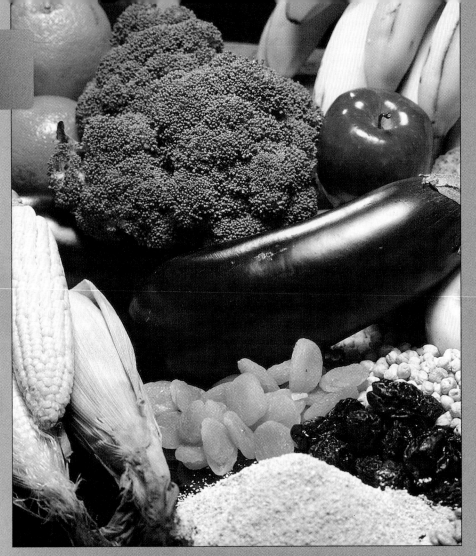

Nutrition and Digestion

29.1 Introduction

In Chapters 26 and 27, we saw what happens to the food that we eat in its final stages—after the proteins, lipids, and carbohydrates have been broken down into their components. In this chapter we discuss the earlier stages—nutrition and diet—and then the digestive processes that break down the large molecules to the small ones that undergo metabolism. We have seen that the purpose of food is to provide energy and new molecules to replace those that the body uses. The synthesis of new molecules is especially important for the period during which a child becomes an adult.

29.2 Nutrition

The components of food and drink that provide growth, replacement, and energy are called nutrients. Not all components of food are nutrients. Some components of food and drinks, such as those that provide

Above: Foods high in fiber include whole grains, legumes, fruits, and vegetables. *(Charles D. Winters)*

flavor, color, or aroma, enhance our pleasure in the food but are not themselves nutrients.

Nutritionists classify nutrients into six groups:

1. carbohydrates
2. lipids
3. proteins
4. vitamins
5. minerals
6. water

A healthy body needs the proper intake of all nutrients. However, nutrient requirements vary from one person to another. For example, more energy is needed to maintain the body temperature of an adult than that of a child. For this reason, nutritional requirements are usually given per kg of body weight. Furthermore, the energy requirements of a physically active body are greater than those of people in sedentary occupations. Therefore, when average values are given, as in **recommended daily allowances (RDA),** one should be aware of the wide range that these average values represent.

The public interest in nutrition and diet changes with time and geography. Seventy or eighty years ago, the main nutritional interest of most Americans was getting enough food to eat and avoiding diseases caused by vitamin deficiency, such as scurvy or beriberi. This is still the main concern of the large majority of the world's population. Today, in affluent societies such as ours, the nutritional message is no longer "eat more" but rather "eat less and discriminate more in your selection of food." For example, a sizable percentage of the American population avoids foods containing substantial amounts of cholesterol (Box 20G) and saturated fatty acids to reduce the risk of heart attacks.

Along with **discriminatory curtailment diets** came many different faddish diets. Diet faddism is an exaggerated belief in the effects of nutrition upon health and disease. This is not new; it has been prevalent for many years. A recommended food is rarely as good and a condemned food is never as bad as faddists claim. Scientific studies prove, for instance, that food grown with chemical fertilizers is just as healthy as food grown with organic (natural) fertilizers.

Each food contains a large variety of nutrients. For example, a typical breakfast cereal lists as its ingredients: milled corn, sugar, salt, malt flavoring, vitamins A, B, C, and D, plus flavorings and preservatives. Consumer laws require that most packaged food be labeled in a uniform manner to show the nutritional values of the food. Figure 29.1 shows a typical label of the type found on almost every can, bottle, or box of food that we buy.

These labels must list the percentages of daily values for four key vitamins and minerals: vitamins A and C, calcium, and iron. If other vitamins or minerals have been added, or if the product makes a nutritional claim about other nutrients, their values must be shown as well. The percent daily values on the labels are based on a daily intake of 2000 Cal. For anyone who eats more than that, the actual percentage figures would be lower (and higher for those who eat less). Note that each label specifies the serving size; the percentages are based on that, not on the entire contents of the package. The section at the bottom of the label is exactly the same on all labels, no matter what the food, and shows the daily amounts of nutrients recommended by the government, based on a consumption of either 2000 or 2500 Cal. Some food packages are allowed to carry shorter labels, either because they have only a few nutrients or because the package has

In the days of sailing ships, many seamen contracted scurvy because they had no source of fresh fruit or vegetables on voyages that could last many weeks away from land. In response to this, British ships carried limes, which could be stored for long periods. Limes are an excellent source of vitamin C. British sailors acquired the nickname "limey" from this practice, a name that was often extended to any Englishman.

> **Discriminatory curtailment diet**
> A diet that avoids certain food ingredients that are considered harmful to the health of an individual; for example, low-sodium diets for people with high blood pressure

A late nineteenth-century advertisement for a cure-all tonic. *(Courtesy of the National Library of Medicine)*

Nutrition Facts

Serving size 1 Bar (28g)
Servings per Container 6

Amount Per Serving

Calories 120	Calories from Fat 35

	% Daily Value*
Total Fat 4g	6%
Saturated Fat 2g	10%
Cholesterol 0mg	0%
Sodium 45mg	2%
Potassium 100mg	3%
Total Carbohydrate 19g	6%
Dietary fiber 2g	8%
Sugars 13g	
Protein 2g	

Vitamin A 15%	•	Vitamin C 15%
Calcium 15%	•	Iron 15%
Vitamin D 15%	•	Vitamin E 15%
Thiamin 15%	•	Riboflavin 15%
Niacin 15%	•	Vitamin B$_6$ 15%
Folate 15%	•	Vitamin B$_{12}$ 15%
Biotin 10%	•	Pantothenic Acid 10%
Phosphorus 15%	•	Iodine 2%
Magnesium 4%	•	Zinc 4%

*Percent Daily Values are based on a 2,000 calorie diet. Your daily values may be higher or lower depending on your calorie needs.

	Calories:	2,000	2,500
Total Fat	Less than	65g	80g
Sat Fat	Less than	20g	25g
Cholesterol	Less than	300mg	300mg
Sodium	Less than	2,400mg	2,400mg
Potassium		3,500mg	3,500mg
Total Carbohydrate		300g	375g
Dietary Fiber		25g	30g

FIGURE 29.1 A food label for a peanut butter crunch bar. The portion at the bottom (following the asterisk) is the same on all labels that carry it.

limited label space. The labels make it much easier for the consumer to know exactly what he or she is eating.

The U.S. Department of Agriculture issues occasional guidelines as to what constitutes a healthy diet. The latest, issued in 1992, depicts this in the form of a pyramid (Figure 29.2). As shown, they consider the basis of a healthy diet to be foods richest in starch (bread, rice, and so on), plus lots of fruits and vegetables (which are rich in vitamins and minerals). Protein-rich foods (meat, fish, dairy products) are to be consumed more sparingly, and fats, oils, and sweets are not considered necessary at all.

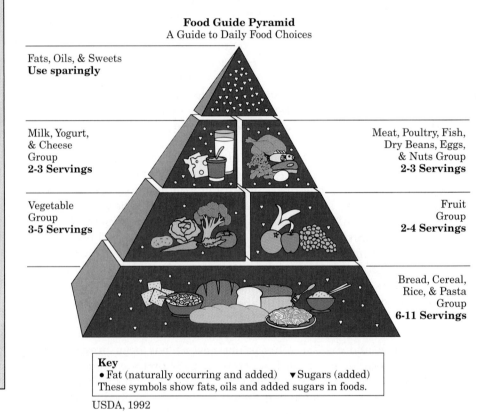

Food Guide Pyramid
A Guide to Daily Food Choices

Fats, Oils, & Sweets
Use sparingly

Milk, Yogurt, & Cheese Group
2-3 Servings

Meat, Poultry, Fish, Dry Beans, Eggs, & Nuts Group
2-3 Servings

Vegetable Group
3-5 Servings

Fruit Group
2-4 Servings

Bread, Cereal, Rice, & Pasta Group
6-11 Servings

Key
• Fat (naturally occurring and added) ▼ Sugars (added)
These symbols show fats, oils and added sugars in foods.

USDA, 1992

FOOD GROUP SERVING SIZES

Bread, Cereal, Rice, and Pasta
  1/2 cup cooked cereal, rice, pasta
  1 ounce dry cereal
  1 slice bread
  2 cookies
  1/2 medium doughnut

Vegetables
  1/2 cup cooked or raw chopped
    vegetables
  1 cup raw leafy vegetables
  3/4 cup vegetable juice
  10 french fries

Fruit
  1 medium apple, banana, or orange
  1/2 cup chopped, cooked, or canned
    fruit
  3/4 cup fruit juice
  1/4 cup dried fruit

Milk, Yogurt, and Cheese
  1 cup milk or yogurt
  1½ ounces natural cheese
  2 ounces process cheese
  2 cups cottage cheese
  1½ cups ice cream
  1 cup frozen yogurt

Meat, Poultry, Fish, Dry Beans, Eggs, and Nuts
  2–3 ounces cooked lean meat, fish,
    or poultry
  2–3 eggs
  4–6 tablespoons peanut butter
  1½ cups cooked dry beans
  1 cup nuts

Fats, Oils, and Sweets
  Butter, mayonnaise, salad dressing,
    cream cheese, sour cream, jam, jelly

FIGURE 29.2 The Food Guide Pyramid developed by the U.S. Department of Agriculture as a general guide to a healthful diet.

Parenteral Nutrition

In clinical practice, it occasionally happens that a patient cannot be fed through the gastrointestinal tract. Under these conditions, intravenous feeding or parenteral nutrition is required. In the past, patients could be maintained on parenteral nutrition only for short periods. Calories were provided in the form of glucose, and amino acids were supplied in the form of protein hydrolysates. Over long periods, however, the use of parenteral nutrition resulted in deterioration and progressive emaciation of the patient. More recently, the solutions used in parenteral nutrition are being manufactured to contain both essential and nonessential amino acids in fixed proportions. Calories now do not come only from glucose. Glucose supplies only about 20 percent of the caloric needs. Fat emulsions provide the rest of the caloric requirements and the essential fatty acids. Finally, the full vitamin and mineral requirements are also supplied in the parenteral solution.

An important non-nutrient in some foods is **fiber.** This generally consists of the indigestible portion of vegetables such as lettuce, cabbage, and celery. Chemically, it is made of cellulose, which, we saw in Section 20.11, cannot be digested by humans. But though we cannot digest it, fiber is necessary for proper operation of the digestive system; without it constipation may result. More seriously, a diet lacking sufficient fiber may also lead to colon cancer. Among other foods high in fiber are whole wheat, brown rice, peas, and beans.

Fiber The cellulosic non-nutrient component in our food

29.3 Calories

The largest part of our food supply goes to provide energy for our bodies. As we saw in Chapters 20 and 21, the energy comes from the oxidation of carbohydrates, fats, and proteins. The energy derived from food is usually measured in calories. One nutritional calorie (Cal) equals 1000 cal or 1 kcal. Thus, when we say that the average daily nutritional requirement for a young adult male is 2900 Cal, we mean the same amount of energy needed to raise the temperature of 2900 kg of water by 1°C or 29 kg (64 lb) of water by 100°C. A young adult female needs 2100 Cal/day. These are peak requirements. Children and older people, on the average, require less energy. Keep in mind that these energy requirements are for active people. For bodies completely at rest, the corresponding energy requirement for young adult males is 1800 Cal/day, and that for females is 1300 Cal/day. The requirement for a resting body is called the **basal caloric requirement.**

One calorie is the amount of heat necessary to raise the temperature of 1 g of liquid water by 1°C (Section 1.9).

Basal caloric requirement The caloric requirement for a resting body

An imbalance between the caloric requirement of the body and the caloric intake creates health problems. Chronic caloric starvation exists in many parts of the world where people simply do not have enough food to eat because of prolonged drought, devastation of war, natural disasters, or simply overpopulation. Famine particularly affects infants and children. Chronic starvation, called **marasmus,** increases infant mortality up to 50 percent. It results in arrested growth, muscle wasting, anemia, and general weakness. Even if starvation is later alleviated, it leaves permanent damage, insufficient body growth, and lowered resistance to disease.

At the other end of the caloric spectrum is excessive caloric intake. This results in obesity, or the accumulation of body fat. **Obesity** increases the risk of hypertension, cardiovascular disease, and diabetes. Reducing diets aim at decreased caloric intake without sacrificing any essential nutrients. A combination of exercise and lower caloric intake can eliminate obesity, but usually these diets must achieve their goal over an extended period.

Excessive caloric intake results from all sources of food: carbohydrates and proteins as well as fats.

Different activity levels require different caloric needs. *(Charles D. Winters)*

Essential amino acid An amino acid that the body cannot synthesize in the required amounts and so must be obtained in the diet

---

Crash diets give the illusion of quick weight loss, but most of this is due to loss of water, which can be regained very quickly. To reduce obesity, we must lose body fat and not water; this takes a lot of effort, because fats contain so much energy. A pound of body fat is equivalent to 3500 Cal. Thus, to lose 10 pounds, it is necessary either to consume 3500 fewer Cal, which can be achieved if one reduces caloric intake by 350 Cal every day for 100 days (or by 700 Cal daily for 50 days), or to use up, through exercise, the same numbers.

29.4 Carbohydrates, Fats, and Proteins in the Diet

Carbohydrates are the major source of energy in the diet. They also furnish important compounds for the synthesis of cell components (Chapter 28). The main dietary carbohydrates are the polysaccharide starch, the disaccharides lactose and sucrose, and the monosaccharides glucose and fructose. Nutritionists recommend a minimum carbohydrate intake equivalent to 500 Cal/day. In addition to these digestible carbohydrates, cellulose in plant materials provides dietary fiber. Artificial sweeteners (Box 29C) can be used to reduce mono- and disaccharide intake.

Fats are the most concentrated source of energy. About 98 percent of the lipids in our diet are fats (triglycerides); the remaining 2 percent consist of complex lipids and cholesterol. Only two fatty acids are essential in higher animals, including humans: linolenic and linoleic acids (Section 20.3). These fatty acids are needed in the diet because our bodies cannot synthesize them. Nutritionists occasionally list arachidonic acid as an **essential fatty acid.** However, our bodies can synthesize arachidonic acid from linoleic acid.

Although the proteins in our diet can be used for energy (Section 27.9), there main use is to furnish amino acids from which the body synthesizes its own proteins (Section 25.6). The human body is incapable of synthesizing ten of the amino acids needed to make proteins. These ten (the **essential amino acids**) must be obtained from our food; they are shown in Table 21.1. The body hydrolyzes food proteins into their amino acid constituents and then puts the amino acids together again to make body proteins. For proper nutrition, our diet should contain about 20 percent protein.

A dietary protein that contains all the essential amino acids is called a complete protein. Casein, the protein of milk, is a complete protein, as are most other animal proteins—those found in meat, fish, and eggs. People who eat adequate quantities of meat, fish, eggs, and dairy products get all the amino acids they need to keep healthy. About 50 g/day of complete proteins constitutes an adequate quantity.

An important animal protein that is not complete is gelatin, which is made by denaturing collagen (Section 21.11). Gelatin lacks tryptophan and is low in several other amino acids, including isoleucine and methionine. Many people on quick reducing diets take "liquid protein." This is nothing but denatured and partially hydrolyzed collagen (gelatin). Therefore, if this is the only protein source in the diet, some essential amino acids will be lacking.

Most plant proteins are incomplete. For example, corn protein lacks lysine and tryptophan; rice protein lacks lysine and threonine; wheat protein lacks lysine; legumes are low in methionine and cystein, and even soy protein, one of the best plant proteins, is very low in methionine. Adequate amino acid nutrition is possible with a vegetarian diet, but only if a range of different vegetables is eaten. **Protein complementation** is such a diet. In protein complementation, two or more foods complement the others' deficiencies. For example, grains and legumes complement each other, with grains being low in lysine but high in methionine. Such protein complementation in vegetarian diets became the staple in different parts of the world—corn tortillas and beans in Central and South America, rice and lentils in India, and rice and tofu in China and Japan.

In many underdeveloped countries, protein deficiency diseases are widespread because the people get their protein mostly from plants. Among these is a disease called **kwashiorkor,** the symptoms of which are a swollen stomach, skin discoloration, and retarded growth.

29.5 Vitamins, Minerals, and Water

Vitamins and **minerals** are essential for good nutrition. Animals maintained on diets that contain sufficient carbohydrates, fats, and proteins and provided with an ample water supply cannot survive on these alone. They also need the essential organic components called vitamins and inorganic ions called minerals. Deficiencies in vitamins and minerals lead to many nutritionally controllable diseases (one example is shown in Figure 29.3). Vitamins and minerals, together with their sources, **recommended dietary allowances (RDA),** and the disease their deficiency may cause are listed in Table 29.1. Although the concept of RDA has been used since

The name "vitamin" comes from a mistaken generalization. Casimir Funk (1884–1967) discovered that certain diseases such as beriberi, scurvy, and pellagra are caused by lack of certain nutrients. He found that the "antiberiberi factor" compound is an amine and mistakenly thought that all such nutrients are amines. Hence, he coined the name "Vitamine."

(a)

(b)

(c)

FIGURE 29.3 (a) Symptoms of rickets, a vitamin D deficiency in children. The nonmineralization of the bones of the radius and the ulna results in prominence of the wrist. *(Courtesy of Drs. P. G. Bullogh and V. J. Vigorita and the Gower Medical Publishing Co., New York.)* Histology of (b) normal and (c) osteomalacic bone. The latter shows the accumulation of osteoids (red stain) due to vitamin D deficiency. *(Courtesy of Drs. P. A. Dieppe, P. A. Bacon, A. N. Bamji, and I. Watt and Gower Medical Publishing Ltd., London, England)*

(text continued on page 674)

TABLE 29.1 Vitamins and Minerals: Sources, Functions, Deficiency Diseases, and Daily Requirements

Name	Structure	Best Food Source	Function	Deficiency Symptoms and Diseases	Recommended Dietary Allowance[a]
Fat-soluble Vitamins					
A		Liver, butter, egg yolk, carrots, spinach, sweet potatoes	Vision; to heal eye and skin injuries	Night blindness; blindness; keratinization of epithelium and cornea	800 μg (1500 μg)[b]
D		Salmon, sardines, cod liver oil, cheese, eggs, milk	Promotes calcium and phosphate absorption and mobilization	Rickets (in children): pliable bones; osteomalacia (in adults); fragile bones	5–10 μg; exposure to sunlight
E		Vegetable oils, nuts, potato chips, spinach	Antioxidant	In cases of malabsorption such as in cystic fibrosis: anemia. In premature infants: anemia	8–10 mg
K		Spinach, potatoes, cauliflower; beef liver	Blood clotting	Uncontrolled bleeding (mostly in newborn infants)	65–80 μg
Water-soluble Vitamins					
B_1 (thiamine)		Beans, soybeans, cereals, ham, liver	Coenzyme in oxidative decarboxylation and in pentose phosphate shunt	Beriberi. In alcoholics: heart failure; pulmonary congestion	1.1 mg
B_2 (riboflavin)		Kidney, liver, yeast, almonds, mushrooms, beans	Coenzyme of oxidative processes	Invasion of cornea by capillaries; cheilosis; dermatitis	1.4 mg

Name	Food sources	Function	Deficiency disease	Symptoms	Amount
Nicotinic acid (niacin)	Chickpeas, lentils, prunes, peaches, avocados, figs, meat, mushrooms, peanuts, bread, rice, beans, berries	Coenzyme of oxidative processes	Pellagra		15–18 mg
B_6 (pyridoxal)	Meat, fish, nuts, oats, wheat germ, potato chips	Coenzyme in transamination; heme synthesis	Convulsions; chronic anemia; peripheral neuropathy		1.6–2.2 mg
Folic acid	Liver, kidney, eggs, spinach, beets, orange juice, avocados, cantaloupe	Coenzyme in methylation and in DNA synthesis	Anemia		200 μg
B_{12}	Oysters, salmon, liver, kidney	Part of methyl-removing enzyme in folate metabolism	Patchy demyelination; degradation of nerves, spinal cord, and brain		1–3 μg

(continued on page 672)

Name	Structure	Best Food Source	Function	Deficiency Symptoms and Diseases	Recommended Dietary Allowance[a]
Pantothenic acid	$\text{HOCH}_2\text{C} \overset{\underset{\text{CH}_3}{\mid}}{\underset{}{\overset{\text{CH}_3}{\mid}}} \text{—CHCNHCH}_2\text{CH}_2\text{COOH}$ with C=O and OH	Peanuts, buckwheat, soybeans, broccoli, lima beans, liver kidney, brain, heart	Part of CoA; fat and carbohydrate metabolism	Gastrointestinal disturbances; depression	4–7 mg
Biotin	(biotin structure with $(CH_2)_4COOH$, HN, NH, O, S)	Yeast, liver, kidney, nuts, egg yolk	Synthesis of fatty acids	Dermatitis; nausea; depression	30–100 μg
C (ascorbic acid)	(ascorbic acid structure with OH, OH, CH—OH, CH₂OH, O)	Citrus fruit, berries, broccoli, cabbage, peppers, tomatoes	Hydroxylation of collagen; wound healing; bond formation; antioxidant	Scurvy; capillary fragility	60 mg
Minerals					
Potassium		Apricots, bananas, dates, figs, nuts, raisins, beans, chickpeas, cress, lentils	Provides membrane potential	Muscle weakness	3500 mg
Sodium		Meat, cheese, cold cuts, smoked fish, table salt	Osmotic pressure	None	2000–2400 mg
Calcium		Milk, cheese, sardines, caviar	Bone formation; hormonal function; blood coagulation; muscle contraction	Muscle cramps; osteoporosis; fragile bones	800–1200 mg
Chloride		Meat, cheese, cold cuts, smoked fish, table salt	Osmotic pressure	None	1700–5100 mg

Element	Sources	Function	Deficiency/Excess	RDA[a]
Phosphorus	Lentils, nuts, oats, grain flours, cocoa, egg yolk, cheese, meat (brain, sweetbreads)	Balancing calcium in diet	Excess causes structural weakness in bones	800–1200 mg
Magnesium	Cheese, cocoa, chocolate, nuts, soybeans, beans	Cofactor in enzymes	Hypocalcemia	280–350 mg
Iron	Raisins, beans, chickpeas, parsley, smoked fish, liver, kidney, spleen, heart, clams, oysters	Oxidative phosphorylation; hemoglobin	Anemia	15 mg
Zinc	Yeast, soybeans, nuts, corn, cheese, meat, poultry	Cofactor in enzymes, insulin	Retarded growth; enlarged liver	12–15 mg
Copper	Oysters, sardines, lamb, liver	Oxidative enzymes cofactor	Loss of hair pigmentation; anemia	1.5–3 mg
Manganese	Nuts, fruits, vegetables, whole-grain cereals	Bone formation	Low serum cholesterol levels; retarded growth of hair and nails	2.0–5.0 mg
Chromium	Meat, beer, whole wheat and rye flours	Glucose metabolism	Glucose not available to cells	0.05–0.2 mg
Molybdenum	Liver, kidney, spinach, beans, peas	Protein synthesis	Retarded growth	0.075–0.250 mg
Cobalt	Meat, dairy products	Component of vitamin B_{12}	Pernicious anemia	0.05 mg (20–30 mg)[b]
Selenium	Meat, seafood	Fat metabolism	Muscular disorders	0.05–0.07 mg (2.4–3.0 mg)[b]
Iodine	Seafood, vegetables, meat	Thyroid glands	Goiter	150–170 μg (1000 μg)[b]
Fluorine	Fluoridated water; fluoridated toothpaste	Enamel formation	Tooth decay	1.5–4.0 mg (8–20 mg)[b]

[a]The U.S. RDAs are set by the Food and Nutrition Board of the National Research Council. The numbers given here are based on the latest recommendations (National Research Council Recommended Dietary Allowances, 10th ed., 1989, National Academy Press, Washington, D.C.). The RDA varies with age, sex, and level of activity; the numbers given are average values for both sexes between the ages of 18 and 54.

[b]Toxic if doses above the level shown in parentheses are taken.

BOX 29B

Too Much Vitamin Intake?

One might suppose that, since vitamins are essential nutrients, the more you take the better off you are. Therefore, some people consider that the RDAs listed for each vitamin and mineral in Table 29.1 are only a minimum, and larger quantities will be even more beneficial. It has become fashionable to take large doses of vitamins, especially those with antioxidant properties, vitamins A, C, and E. No scientific proof exists for such claims. Larger doses are usually harmless because the body excretes the excess, but not in all cases.

High dosages can do unexpected harm in certain instances, and the RDA can prevent harm not usually associated with the vitamin. The case of folic acid illustrates this point. Table 29.1 shows an RDA of 400 mg for this vitamin, as the amount necessary to prevent anemia. Later it has been shown that this amount of folic acid is especially important during pregnancy. It can prevent birth defects that result in spina bifida, the birth of a child with an open spine, and anencephaly, the birth of a child without most of the brain. Since vitamins are not stored in the body, it is important that the RDA be taken every day. But not more. An intake of 1000 mg or more of folic acid daily is harmful because it can hide the symptoms of certain anemias without actually preventing these diseases. The Food and Drug Administration now requires that food labeled as enriched must carry certain amounts of folic acid. The amount for a slice of enriched bread is 40 to 70 mg, and that for enriched cereals is 100 mg per serving.

Similarly, high dosages for niacin and of vitamin A have been shown to cause health problems.

1940 and periodically updated as new knowledge dictated, a new concept is being developed in the field of nutrition. The **Dietary Reference Intake (DRI),** when completed, is designed to replace the RDA and it is tailor made to different ages and genders. For example, the RDA for vitamin D is 10 μg, the adequate intake listed in DRI for vitamin D between ages 9 and 50 years is 5 μg.

BOX 29C

Dieting

Obesity is a serious problem in the United States. Many people would like to lose weight but are unable to control their appetites enough to do so, which is why artificial sweeteners are popular. They have a sweet taste but do not add calories. Many people restrict their sugar intake. Some are forced to do so by diseases such as diabetes; others, by the desire to lose weight. Since most of us like to eat sweet foods, artificial sweeteners are added to many foods and drinks for those who must (or want to) restrict their sugar intake. Four noncaloric artificial sweeteners are now approved by the Food and Drug Administration. The oldest of these, saccharin, is 450 times sweeter than sucrose, yet it has very little caloric content. Saccharin has been in use for about a hundred years. Unfortunately, some tests have shown that saccharin, when fed in massive quantities to rats, caused some of these rats to develop cancer. Other tests have given negative results. Recent research has shown that this cancer in rats is not relevant to human consumphtion. Saccharin continues to be sold on the market.

Saccharin   Aspartame   Acesulfame-K

A newer artificial sweetener, aspartame, does not have the slight aftertaste that saccharin does. Aspartame is the methyl ester of a simple dipeptide, aspartylphenylalanine (Asp-Phe). Its sweetness was discovered in 1969. After extensive biological testing, it was approved by the U.S. Food and Drug Administration in 1981 for use in cold cereals, drink mixes, and gelatins and as tablets or powder to be used as a sugar substitute. Aspartame is 100 to 150 times sweeter than sucrose. It is made from natural amino acids so that both the aspartic acid and the phenylalanine have the L configuration. The other possibilities have also been synthesized: the L-D, the D-L, and the D-D. They are all bitter rather than sweet. This is another example of the principle that, when a biological organism uses or makes a compound that is chiral, in most cases only one stereoisomer is used or made. Aspartame is sold under such brand names as Equal and NutraSweet.

A third artificial sweetener, acesulfame-K, is 200 times sweeter than sucrose and is used (under the name Sunette) mostly in dry mixes. The most recent addition is Sucralose, approved by the Food and Drug Administration in 1998 for use in sodas, baked goods, and tabletop packets. It is six hundred times sweeter than sucrose and has no aftertaste. It is a trichloro derivative of a disaccharide. Neither Sucralose nor acesulfame-K is metabolized in the body; that is, they pass through unchanged.

Sucralose

But calories in the diet do not come only from sugar. Dietary fat is an even more important source (Section 27.6), and it has long been hoped that some kind of artificial fat, which would taste the same but would have no (or only a few) calories, would help in losing weight. Procter & Gamble Company has developed such a product, called olestra. Like the real fats, this molecule is a carboxylic ester, but instead of glycerol, the alcohol

component is sucrose (Section 19.7). All eight of sucrose's OH groups are converted to ester groups; the carboxylic acids are the fatty acids containing eight to ten carbons. The one illustrated here is made with the C_{10} decanoic (capric) acid.

Olestra

Although olestra has a chemical structure similar to a fat, the human body cannot digest it because the enzymes that digest ordinary fats are not designed for the particular size and shape of this molecule. Therefore, it passes through the digestive system unchanged, and we derive no calories from it. Olestra can be used instead of an ordinary fat in cooking such items as cookies and potato chips. People who eat these will then be consuming fewer calories.

However, olestra does cause diarrhea, cramps, and nausea in some people, effects which increase with the amount of olestra consumed, and those that are susceptible to such side effects will have to decide if the potential to lose weight is worth the discomfort involved. Furthermore, olestra dissolves and sweeps away some vitamins and nutrients in other foods being digested at the same time. To counter this, the manufacturers add some of these nutrients (vitamins A, D, E, and K) to the product. The Food and Drug Administration requires that all packages of olestra-containing foods carry a warning label about these side effects.

Water makes up 60 percent of our body weight. Most of the compounds in our body are dissolved in water, which also serves as a transporting medium to carry nutrients and waste materials. These functions are discussed in Chapter 31. We must maintain a proper balance between water intake and water excretion via urine, feces, sweat, and exhalation of breath. A normal diet requires about 1200 to 1500 mL of water per day. This is in addition to the water content of our foods.

29.6 Digestion

In order for food to be used in our bodies, it must be absorbed through the intestinal walls into the bloodstream or lymph system. Some nutrients, such as vitamins, minerals, glucose, and amino acids, can be absorbed directly. Others, such as starch, fats, and proteins, must first be broken down into smaller components before they can be absorbed. This breakdown process is called **digestion.**

> **Digestion** The process in which the body breaks down large molecules into smaller ones that can then be absorbed and metabolized

29.7 Digestion of Carbohydrates

Before the body can absorb carbohydrates, it must break down di-, oligo-, and polysaccharides into monosaccharides because only monosaccharides can pass into the blood stream.

The monosaccharide units are connected to each other by glycosidic (acetal) bonds. As we saw in Section 17.5, the cleavage of acetals by water is called hydrolysis. In the body, this hydrolysis is catalyzed by acids and by enzymes. When the metabolic need arises, storage polysaccharides are hydrolyzed to yield glucose and maltose.

The hydrolysis is aided by a number of enzymes: **α-amylase** attacks all three storage polysaccharides at random, hydrolyzing the $\alpha(1 \rightarrow 4)$ glycosidic bonds, and **β-amylase** also hydrolyzes the $\alpha(1 \rightarrow 4)$ glycosidic bonds but in an orderly fashion, cutting disaccharidic maltose units one by one from the nonreducing end of a chain. A third enzyme, called the **debranching enzyme,** attacks and hydrolyzes the $\alpha(1 \rightarrow 6)$ glycosidic bonds (Figure 29.4). In acid-catalyzed hydrolysis, storage polysaccharides are attacked at random points, although acid catalysis is slower than enzyme catalyzed hydrolysis at body temperature.

The digestion (hydrolysis) of starch and glycogen in our food supply starts in the mouth, where α-amylase is one of the main components of saliva. The hydrochloric acid in the stomach and the other hydrolytic enzymes in the intestinal tract decompose starch and glycogen to produce mono- and disaccharides (D-glucose and maltose).

The D-glucose produced by hydrolysis of the di-, oligo-, and polysaccharides enters the bloodstream and is carried to the cells to be utilized (Section 27.2). For this reason, D-glucose is often called blood sugar (Box 29D). In healthy people, little or none of this sugar ends up in the urine except for short periods of time. In the condition known as diabetes mellitus, how-

The three storage polysaccharides are amylose, amylopectin, and glycogen.

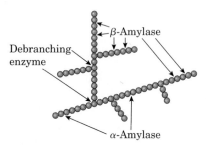

FIGURE 29.4 The action of different enzymes on glycogen and starch.

---

ever, glucose is not completely metabolized and does appear in the urine (Box 31C). Because of this, it is necessary to test the urine of diabetic patients for the presence of glucose (Box 19C).

29.8 Digestion of Lipids

The lipids in the food we eat must be hydrolyzed into smaller components before they can be absorbed into the blood or lymph system through the intestinal walls. The enzymes that promote this hydrolysis are located in the small intestine and are called *lipases*. However, since lipids are insoluble in the aqueous environment of the gastrointestinal tract, they must be dispersed into fine colloidal particles before the enzymes can act on them.

The *bile salts* (Section 20.11) perform this important function. Bile salts are manufactured in the liver from cholesterol and stored in the gallbladder. From there, they are secreted through the bile ducts into the intestine. The emulsion produced by the bile salts and dietary fats is acted upon by the lipases. These break fats into glycerol and fatty acids and complex lipids into fatty acids, alcohols (glycerol, choline, ethanolamine, sphingosine), and carbohydrates. All these hydrolysis products are absorbed through the intestinal walls.

29.9 Digestion of Proteins

The digestion of dietary proteins begins with cooking, which denatures proteins. (Denatured proteins are broken down more easily by the hydrochloric acid in the stomach and by digestive enzymes than are native proteins.) *Stomach acid* contains about 0.5 percent HCl. The HCl both denatures the proteins and hydrolyzes the peptide bonds randomly. *Pepsin,* the proteolytic enzyme of stomach juice, breaks peptide bonds on the C=O side of three amino acids only: tryptophan, phenylalanine, and tyrosine (Figure 29.5).

Most protein digestion occurs in the small intestine. There, the enzyme *chymotrypsin* breaks internal peptide bonds at the same positions as does pepsin, while another enzyme, *trypsin,* hydrolyzes them only on the C=O side of arginine and lysine. Other enzymes, such as *carboxypeptidase,* hydrolyze amino acids one by one from the C-terminal end of the protein. The amino acids and small peptides are then absorbed through the intestinal walls.

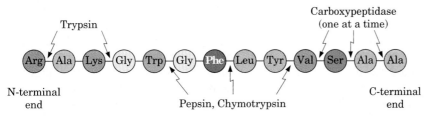

FIGURE 29.5 Different enzymes hydrolyze peptide chains in different but specific ways.

Food for Performance Enhancement

Athletes do whatever they can legally to enhance their performance. After a vigorous exercise of 15 minutes, the performance declines because the glycogen stores in the muscle are depleted. If athletes can maintain their blood glucose levels, the performance improves. This is why many athletes load up on carbohydrates in the form of energy bars and sport drinks.

In recent years a new performance enhancer food appeared on the market, creatine, and it quickly became a best-seller. It is sold over the counter. Creatine is a naturally occurring amino acid in the muscles, which store energy in the form of high energy phosphocreatine. During a short and strenuous exercise, such as a 100-meter sprint, the muscles first use up the ATP obtainable from the reaction of phosphocreatine with ADP and only after that do they rely on the glycogen stores.

Creatine together with carbohydrates are natural food and body components and therefore cannot be considered banned performance enhancers such as anabolic steroids or "andro" (Box 20H). It also has no known hazards even in long-term use. Creatine is beneficial in improving performance where short bursts of energy are needed, weight lifting, jumping, or sprinting. Lately it has been used experimentally to preserve muscle neurons in degenerative diseases such as Parkinson's or Huntington's disease and muscular dystrophy.

Phosphocreatine   Creatine

■ Creatine is sold over the counter. *(George Semple)*

SUMMARY

Nutrients are components of foods that provide growth, replacement, and energy. Nutrients are classified into six groups: carbohydrates, lipids, proteins, vitamins, minerals, and water. Each food contains a variety of nutrients. The largest part of our food intake is used to provide energy for our bodies. A typical young adult needs 2900 Cal (male) or 2100 Cal (female) as an average daily caloric intake. **Basal caloric requirements** are the energy needs when the body is completely at rest. These are less than the normal requirements. Imbalance between need and caloric intake may create health problems; chronic starvation increases infant mortality, whereas obesity leads to hypertension, cardiovascular diseases, and diabetes.

Carbohydrates are the major source of energy in our diet. Fats are the most concentrated source of energy. Essential fatty and amino acids are needed as building blocks because our bodies cannot synthesize them. Vitamins and minerals are essential constituents of diets that are needed in small quantities. The fat-soluble vitamins are A, D, E, and K. Vitamins C and the B group are water-soluble vitamins. Most of the B vitamins are essential coenzymes. The most important dietary minerals are Na^+, Cl^-, K^+, PO_4^{3-}, Ca^{2+}, and Mg^{2+}, but trace minerals are also necessary. Water makes up 60 percent of body weight.

The digestion of carbohydrates, fats, and proteins is aided by the stomach acid, HCl, and by enzymes that reside in the mouth, stomach, and intestines.

KEY TERMS

α-Amylase (Section 29.7)
Basal caloric requirement
  (Section 29.3)
β-Amylase (Section 29.7)
Complete protein (Section 29.4)
Debranching enzyme (Section 29.7)
Dietary Reference Intake (DRI)
  (Section 29.5)
Digestion (Section 29.6)

Discriminatory curtailment diets
  (Section 29.2)
Essential amino acid
  (Section 29.4)
Essential fatty acid (Section 29.4)
Fiber (Section 29.2)
Hyperglycemia (Box 29D)
Hypoglycemia (Box 29D)
Kwashiorkor (Section 29.4)

Lipase (Section 29.8)
Marasmus (Section 29.3)
Mineral (Sections 29.5)
Nutrient (Section 29.2)
Obesity (Section 29.3)
Parenteral nutrition (Box 29A)
RDA (Sections 29.2, 29.5)
Vitamin (Section 29.5)
Water (Section 29.5)

CONCEPTUAL PROBLEMS

29.A Can a chemical that, in essence, goes through the body unchanged be an essential nutrient? Explain.

29.B Humans cannot digest wood; termites do. Is there a basic difference in the digestive enzymes synthesized by humans and termites?

PROBLEMS

Difficult problems are designated by an asterisk.

Nutrition

29.1 Are nutrient requirements uniform for everyone?

29.2 Is banana flavoring, isopentyl acetate, a nutrient?

29.3 If sodium benzoate, a food preservative, is excreted as such, and if calcium propionate, another food preservative, is hydrolyzed in the body and metabolized to CO_2 and H_2O, would you consider either of these preservatives as nutrients? If so, why?

29.4 Is corn grown solely with organic fertilizers more nutritious than that grown with chemical fertilizers?

29.5 Which part of the Nutrition Facts label found on food packages is the same for all labels carrying it?

29.6 Which kinds of food does the government recommend that we have the most servings of each day?

29.7 What is the importance of fiber in the diet?

Calories

29.8 Define basal caloric requirement.

29.9 A young adult female needs a caloric intake of 2100 Cal/day. Her basal caloric requirement is only 1300 Cal/day. Why is the extra 800 Cal needed?

29.10 What ill effects may obesity bring?

29.11 Assume that you want to lose 20 lb of body fat in 60 days. Your present dietary intake is 3000 Cal/day. What should your caloric intake be, in Cal/day, in order to achieve this goal, assuming no change in exercise habits?

Carbohydrates, Fats, and Proteins in the Diet

29.12 Which nutrient provides energy in its most concentrated form?

29.13 Is it possible to get a sufficient supply of nutritionally adequate proteins by eating only vegetables?

29.14 How many (a) essential fatty acids and (b) essential amino acids do humans need in their diets?

29.15 What is the precursor of arachidonic acid in the body?

29.16 Suggest a way to cure kwashiorkor.

Vitamins, Minerals, and Water

29.17 In a prison camp during a war, the prisoners are fed plenty of rice and water but nothing else. What would be the result of such a diet in the long run?

29.18 (a) How many mL of water per day does a normal diet require? (b) How many calories does this contribute?

29.19 Why did British sailing ships carry a supply of limes?

29.20 What are the symptoms of vitamin A deficiency?

29.21 What is the function of vitamin K?

29.22 (a) Which vitamin contains cobalt? (b) What is the function of this vitamin?

29.23 Vitamin C is recommended in megadoses by some people for all kinds of symptoms from colds to cancer. What disease has been scientifically proven to be prevented when sufficient daily doses of vitamin C are in the diet?

29.24 What are the best dietary sources of calcium, phosphorus, and cobalt?

29.25 Which vitamin contains a sulfur atom?

29.26 What are the symptoms of vitamin B_{12} deficiency?

Digestion

29.27 What chemical processes take place during digestion?

29.28 What is the product of the reaction when β-amylase acts on amylose?

29.29 Do lipases degrade (a) cholesterol (b) fatty acids?

29.30 Does the debranching enzyme help in digesting amylose?

29.31 Does HCl in the stomach hydrolyze both the $(1 \rightarrow 4)$ and $(1 \rightarrow 6)$ glycosidic bonds?

29.32 What is the difference between protein digestion by trypsin and by HCl?

Boxes

29.33 (Box 29A) Why cannot glucose supply all the caloric needs in parenteral nutrition over a long period?

29.34 (Box 29B) The RDA for niacin is 18 mg. Would it be still better to take a 100-mg tablet of this vitamin every day?

29.35 (Box 29C) (a) Describe the difference between the structure of aspartame and the methyl ester of phenylalanylaspartic acid. (b) Do you expect this second compound to be as sweet as aspartame?

*29.36 (Box 29C) (a) Which artificial sweeteners are not metabolized in the body at all? (b) What could be products of the digested aspartame?

29.37 (Box 29C) What is common in the structures of olestra and sucralose?

29.38 (Box 29D) A patient complains of dizziness. Blood analysis shows that he has 30 mg of glucose in 100 mL of blood. What would you recommend to alleviate the dizziness?

29.39 (Box 29E) Looking in Table 21.1 listing the common amino acids found in proteins, which amino acid most resembles creatine?

Additional Problems

29.40 Which vitamin is part of coenzyme A (CoA)?

29.41 (a) Name the essential fatty acids. (b) Are they saturated or unsaturated?

29.42 Why is it necessary to have proteins in our diets?

29.43 Which vitamin is part of the coenzyme NAD^+?

29.44 According to the government's food pyramid, are there any foods that we can completely omit from our diets and still be healthy?

29.45 What kind of carbohydrate provides dietary fiber?

29.46 As an employee of a company that markets walnuts, you are asked to provide information for an ad that would stress the nutritional value of walnuts. What information would you provide?

29.47 In diabetes, insulin is administered intravenously. Explain why this hormone protein cannot be taken orally.

29.48 What kind of supplemental enzyme would you recommend for a patient after a peptic ulcer operation?

29.49 Name three essential amino acids.

29.50 Why is zinc needed in the body?

29.51 In a trial, a woman was accused of poisoning her husband by adding arsenic to his meals. Her attorney stated that this was done to promote her husband's health, arsenic being an essential nutrient. Would you accept this argument?

Immunochemistry

30.1 Introduction

When you were in elementary school, you may have had chickenpox along with many other children. The viral disease passed from one person to the next and ran its course, but after the children recovered, they never had chicken pox again. Those who were infected became immune to this disease.

Humans and other vertebrates possess a highly developed, complex immune system that defends the body against foreign invaders. An overview of this complex system is presented in Figure 30.1. This figure could serve as a road map as you read the different sections. Referring back to Figure 30.1, you will find the precise location of the topic under discussion and its relationship to the whole immune system. The first line of defense is the natural resistance of the body: the **innate immunity.** It comes in two forms: external and internal. The salient characteristics of innate immunity are that it is nonspecific and without memory. One example of **external innate immunity** is the skin, which provides a barrier against penetration of pathogens. The skin also secretes lactic acid and fatty acids, both creating a low pH, uncomfortable environment for bacteria. Tears in the eyes and mucus in the respiratory and gastrointestinal tracts perform the same function.

Innate immunity The natural nonspecific resistance of the body against foreign invaders, which has no memory

Above: Two natural killer (NK) cells, shown in yellow-orange, attacking a leukemia cell, shown in red. (© Meckes/Ottawa/Photo Researchers, Inc.)

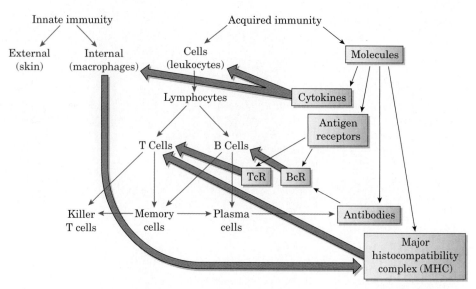

FIGURE 30.1   Overview of the immune system: Its components and their interactions.

Acute inflammation and its symptoms were described two millennia ago by one of the founding fathers of systematic medicine, the Roman physician, Celsus. These are redness, swelling, heat, and pain.

This is why, for example, if we have measles once, we will not get this disease a second time.

Antigen A substance foreign to the body that triggers an immune response

Antibody A glycoprotein molecule that interacts with an antigen

Immunoglobulin superfamily Glycoproteins that are composed of constant and variable protein segments having significant homologies to suggest that they evolved from a common ancestry

After a pathogen penetrates a tissue, **internal innate immunity** takes over. This system creates physiological barriers such as high oxygen pressures that kill anaerobic bacteria. Other nonspecific responses include the proliferation of macrophages, which engulf and digest bacteria and reduce inflammation. In inflammatory response to injury or infection, the capillaries dilate to allow greater flow of blood to the site of the injury.

Vertebrates have a second line of defense called **acquired or adaptive immunity,** and we are referring to this type of immunity when we talk about the **immune system.** The key features of the immune system are *specificity* and *memory.* The immune system uses antibodies designed specifically for each type of invader. In a second encounter with the same invader, the response is more rapid, more vigorous, and more prolonged than it was in the first case because the immune system remembers the nature of the invader from the first encounter.

The invaders may be bacteria, viruses, molds, or pollen grains. A body with no defense against such invaders could not survive. There is a rare genetic disease in which a person is born without a functioning immune system. Attempts have been made to bring such children up in an enclosure totally sealed from the environment. While in this environment, they can survive, but when the environment is removed, such people always die within a short time. The disease AIDS (Box 23D) slowly destroys the immune system, leaving its victims to die from some invading organism that a person without AIDS would easily be able to fight.

As we shall see, the beauty of the body's immune system is that it is flexible. The system is capable of making millions of potential defenders, so that it can almost always find just the right one to counter the invader, even when it has never seen that particular organism before. Foreign substances that invade the body are called **antigens.** The immune system is made of **cells** and **molecules.** Two types of **white blood cells,** called lymphocytes, fight against the invaders: (1) **T cells** kill the invader by contact and (2) **B cells** manufacture **antibodies** which are soluble **immunoglobulin molecules** that immobilize antigens.

The basic **molecules** of the immune system belong to the **immunoglobulin superfamily.** All molecules of this class have a certain portion of the molecule that can interact with antigens. They are all glycoproteins. The polypeptide chains in this superfamily all have two domains: a constant region and a variable region. The constant region has the same

Myasthenia Gravis: An Autoimmune Disease

Myasthenia gravis manifests itself as muscular weakness and excessive fatigue. In its severe form the patient has difficulty chewing, swallowing, and even breathing. Death follows from respiratory failure. Patients with myasthenia gravis develop antibodies against their own acetylcholine receptors (Figure 23.3). This is an example of an autoimmune disease, since the body mistakes its own protein for a foreign body. The anti-bodies produced interact with the acetylcholine receptor in the neuromuscular junctions. This interaction blocks the reception of acetylcholine molecules as well as the signaling nerve impulses. As a result, the muscles do not respond, and paralysis sets in. Furthermore, it also leads to the degradation and lysis of the blocked receptors.

amino acid sequence in each of the same class molecules. The variable region is antigen specific. This means that the amino acid sequence in this region is unique for each antigen. They are designed to recognize only one specific antigen.

There are three different representatives of the immunoglobulin superfamily in our immune system: (1) **Antibodies** are soluble immunoglobulins secreted by the plasma cells. (2) **Receptors** on the surface of T cells (TcR) recognize and bind antigens presented to them. (3) Molecules that present antigens also belong to this superfamily. They reside inside the cells. These protein molecules are known as **major histocompatibility complex** (MHC). When a cell is infected by an antigen, MHC molecules interact with it and bring a characteristic portion of the antigen to the surface of the cell. Such a surface presentation then marks the diseased cell for destruction. This can happen in a cell that was infected by a virus, and it can happen in macrophages that engulf and digest bacteria and virus.

The process of finding and making just the right immunoglobulin to fight a particular invader is relatively slow, compared to the chemical messengers discussed in Chapter 23. While neurotransmitters act within a millisecond and hormones within seconds, minutes, or hours, immunoglobulins respond to an antigen over a longer span of time: weeks and months.

Although the immune system can be considered as another form of chemical communication (Chapter 23), it is much more complex than neurotransmission because it involves molecular signals and interplay between various cells. In these constant interactions, the major elements are: (1) the cells of the immune system, (2) the antigens and their perception by the immune system, (3) the antibodies that are immunoglobulin molecules designed to immobilize antigens, (4) the receptor molecules on the surface of the cells that recognize antigens, and (5) the cytokine molecules that control these interactions. Because the immune system is the cornerstone of the body defenses, its importance for students in health-related sciences is undeniable.

Other glycoproteins belonging to the immunoglobulin superfamily are the cell-adhesion molecules that operate among cells and cause loose cells to adhere to each other to form layers and then tissues such as epithelium. They also play a significant part in the development of embryos.

30.2 Cells and Organs of the Immune System

The blood plasma circulates in the body and is in contact with the other body fluids through the semipermeable membranes of the blood vessels. Therefore, blood can exchange chemical compounds with other body fluids and, through them, with the cells and organs of the body (Figure 30.2).

The lymphatic capillary vessels that drain the **interstitial fluids** enter certain organs, called **lymphoid organs,** such as the thymus, the spleen, and the lymph nodes. The cells primarily responsible for the functioning of the immune system are the specialized white blood cells called **lymphocytes.** As their names imply, these cells are mostly found in the lymphoid organs (Figure 30.3).

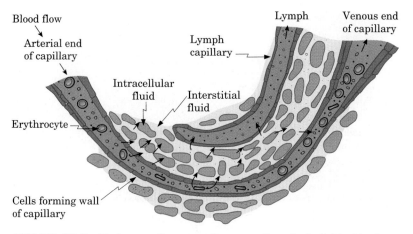

FIGURE 30.2 Exchange of compounds among three body fluids: blood, interstitial fluid, and lymph. *(After J. R. Holum,* Fundamentals of General, Organic and Biological Chemistry. *New York: John Wiley & Sons, 1978, p. 569)*

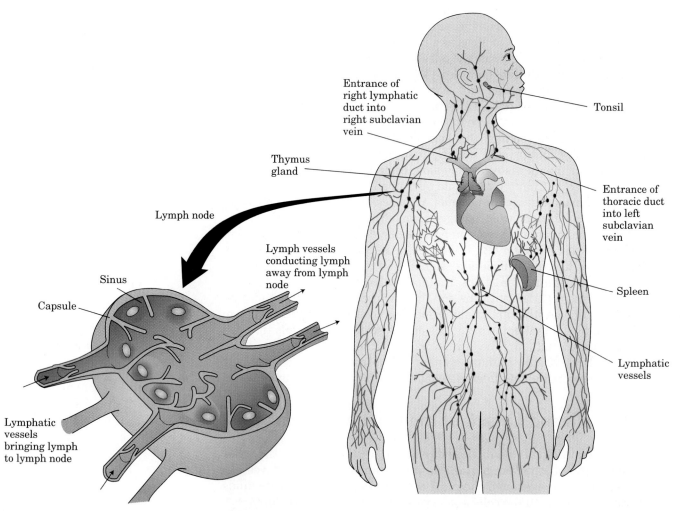

FIGURE 30.3 The lymphatic system is a system of lymphatic vessels containing a clear fluid, called lymph, and various lymphatic tissues and organs located throughout the body. Lymph nodes are masses of lymphatic tissue covered with a fibrous capsule. Lymph nodes filter the lymph. In addition, they are packed with macrophages and lymphocytes.

TABLE 30.1	Interactions Among the Different Cells of the Immune System

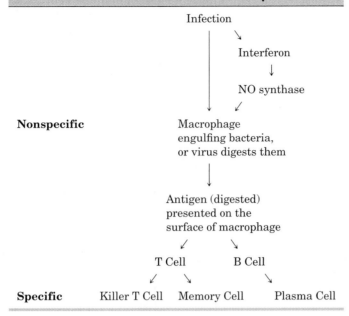

T cells are lymphocytes that originate in the bone marrow but
mature in the thymus gland. B cells are lymphocytes that originate
and mature in the bone marrow. Both B and T cells are found mostly in
the lymph, where they circulate looking for invaders, but small numbers of
lymphocytes are also found in the blood. In order to get there, they must
squeeze through tiny openings between the endothelial cells. This process
is aided by signaling molecules called **cytokines** (Section 30.5).

The response of the body to a foreign invader is depicted schematically
in Table 30.1. **Macrophages** are the first cells in the blood that encounter
an antigen. Macrophages belong to the internal innate immunity system;
inasmuch as they are nonspecific, they attack virtually anything that is not
recognized as part of the body. These include pathogens, cancer cells, and
damaged tissues. Macrophages engulf an invading bacterium or virus and
kill it. The magic bullet is the NO molecule, which we have seen is both
toxic (Box 3C) and can act as a secondary messenger (Box 23F).

The NO molecule is short lived and must be constantly manufactured
anew. When an infection begins, the immune system manufactures the protein interferon. This in turn activates a gene that produces the enzyme,
nitric oxide synthase. With the aid of this enzyme, the macrophages,
endowed with NO, kill the invading organisms. Macrophages then digest
the engulfed antigen and display a small portion of it on their surface.

T cells interact with the antigen presented by the macrophage, and
produce other T cells that are now highly specific to the antigen. When
these T cells differentiate, some of them become **killer T cells,** which kill
the invading foreign cells by cell-to-cell contact. Killer T cells release a protein, aptly named perforin, which attaches itself to the target cell, in effect
punching holes in its membranes. Through these holes water rushes into
the target cell, and it swells and eventually bursts. Other T cells will
become **memory cells.** These will remain in the bloodstream, so that if the
same antigen enters the body again, even years after the primary infection,
the body will not need to build up its defenses anew but is ready to kill
them instantly.

A macrophage-ingesting bacteria
(the rod-shaped structures). The
bacteria will be pulled inside the
cell within a membrane-bound vesicle and quickly killed. (© *David M.
Phillips/Photo Researchers, Inc.*)

The interferon-induced, enzyme-
manufactured NO is critical for the
immune system. But this enzyme is
also implicated in the overproduction of NO. Such an overproduction
may result in pathologies, like septic
shock, multiple sclerosis, rheumatoid arthritis, and cancer formation.

There are rare exceptions: In an
autoimmune disease, the body
mistakes its own protein as foreign
(see Box 30A).

The production of antibodies is the task of **plasma cells.** These cells are derived from B cells after the B cells have been exposed to an antigen.

The lymphatic vessels, in which most of the antigen attacking takes place, flow through a number of lymph nodes (Figure 30.3). These nodes are essentially filters. Most plasma cells reside in lymph nodes. Therefore most antibodies are produced there. Each lymph node is also packed with millions of other lymphocytes. More than 99 percent of all invading bacteria and foreign particles are filtered out in the lymph nodes. As a consequence, the outflowing lymph is almost free of invaders and is packed with antibodies produced by the plasma cells.

30.3 Antigens and Their Presentation by Major Histocompatibility Complex

Antigens are any foreign substances that elicit an immune response; therefore, they are also called immunogens. Three features characterize an antigen: The first is foreignness. Molecules of your own body will not elicit an immune response. The second condition is that the antigen must be of molecular weight greater than 6000. The third condition is that the molecule must have sufficient complexity. A polypeptide made of lysine only is not immunogenic.

Antigens can be proteins, polysaccharides, or nucleic acids, all being large molecules. Antigens may be soluble in the cytoplasm, or they may be at the surface of cells, either embedded in the membrane or just absorbed on it. An example of polysaccharidic antigenicity is ABO blood groups (Box 19D). In protein antigens, only part of the primary structure is needed to cause an immune response. Between 5 and 7 amino acids are needed to interact with an antibody, and between 10 and 15 amino acids are necessary to bind to a receptor on a T cell. The smallest unit of an antigen capable of binding with an antibody is called the **epitope.** The amino acids in an epitope do not have to be in sequence in the primary structure. Folding and secondary structures may bring amino acids that are not in sequence into each other's proximity. For example, amino acids in positions 20 and 28 may form part of an epitope. *Antibodies can recognize all antigens, but T cell receptors recognize only peptide antigens.*

Antigens may be in the interior of an infected cell or on the surface of a virus or bacteria that penetrated the cell. To elicit an immune reaction, the antigen or its epitope must be brought to the surface of the infected cell. Similarly, after a macrophage swallows up and partially digests an antigen, the macrophage must bring the epitope back to its surface to elicit an immune response from T cells (Table 30.1). The task of bringing an antigen's epitope to the infected cell's surface is performed by a protein complex called **major histocompatibility complex (MHC).** As the name implies, its role in immune response was first discovered in organ transplants.

MHC molecules are transmembrane proteins belonging to the immunoglobulin superfamily. There are two classes of MHC molecules (Figure 30.4), both having peptide-binding variable domains. Class I MHC is made of a single polypeptide chain, while class II MHC is a dimer. Class I MHC molecules seek out antigens that have been synthesized *inside a virus-infected cell,* and class II MHC molecules pick up *"dead"* antigens. For example, if a macrophage engulfed and digested a virus the result would be a dead antigen; the epitope of it will be recognized by MHC II. In each case, the epitope attached to the MHC is brought to the cell surface to be presented to T cells.

Epitope The smallest number of amino acids on an antigen that elicits an immune response

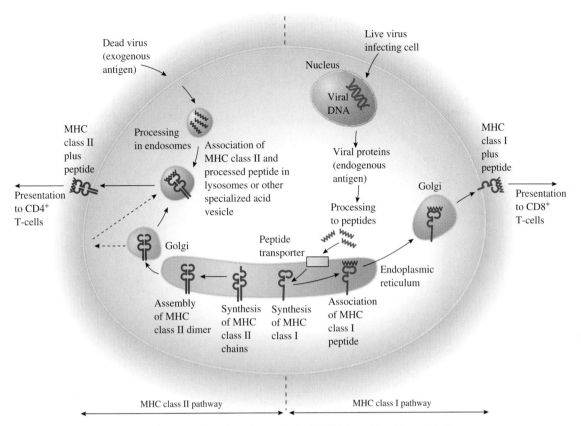

FIGURE 30.4 Differential processing of antigens in the MHC class II pathway (left) or MHC class I pathway (right).

Immunoglobulins are glycoproteins, that is, carbohydrate-carrying protein molecules. Not only do the different classes of immunoglobulins vary in molecular weight and carbohydrate content, but there is considerable variation in their concentration in the blood (Table 30.2). The IgG and IgM antibodies are the most important antibodies in the blood. They interact with antigens and trigger the swallowing up (phagocytosis) of these cells by such specialized cells as the macrophages.

The IgA molecules are found mostly in secretions: tears, milk, and mucus. Therefore, these immunoglobulins attack the invading material before it gets into the bloodstream. IgE immunoglobulins play a part in such allergic reactions as asthma and hay fever.

Each immunoglobulin molecule is made of four polypeptide chains: two identical light chains and two identical heavy chains. The four polypeptide chains are arranged symmetrically, forming a Y shape (Figure

TABLE 30.2 Immunoglobulin Classes

Class	Molecular Weight (MW)	Carbohydrate Content (%)	Concentration in Serum (mg/100 mL)
IgA	200 000–700 000	7–12	90–420
IgD	160 000	<1	1–40
IgE	190 000	10–12	0.01–0.1
IgG	150 000	2–3	600–1800
IgM	950 000	10–12	50–190

In immunoglobulin IgG, the light chains have a molecular weight of 25 000 and the heavy chains have a molecular weight of 50 000.

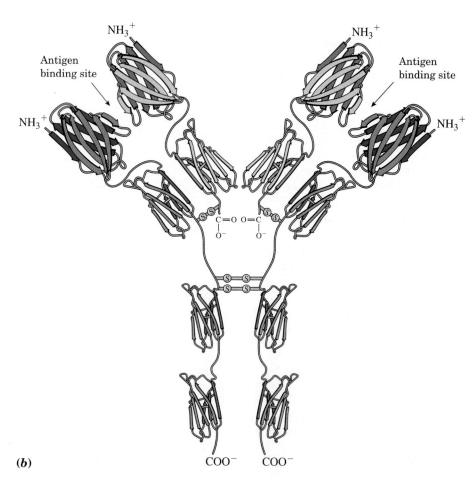

FIGURE 30.5 (*a*) Schematic diagram of an IgG-type antibody consisting of two heavy chains and two light chains connected by disulfide linkages. The amino terminal end of each chain has the variable portion. (*b*) The same in the ribbon model. (*c*) Model showing how an antibody bonds to an antigen.

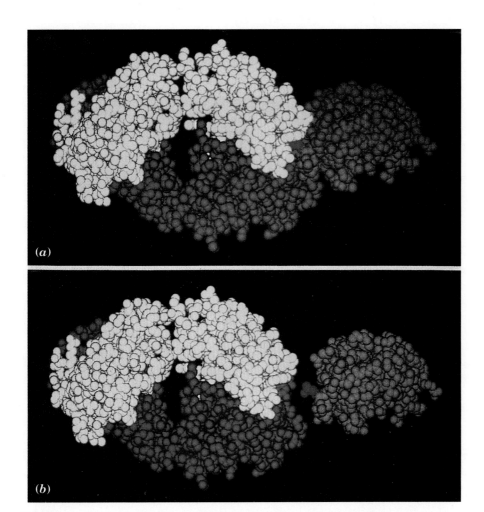

FIGURE 30.6   (*a*) An antigen-antibody complex. The antigen (shown in green) is
lysozyme. The heavy chain of the antibody is shown in blue; the light chain in yellow.
The most important amino acid residue (glutamine in the 121 position) on the antigen
is the one that fits into the antibody groove (shown in red). (*b*) The antigen-antibody
complex has been pulled apart. Note how they fit each other. *(Courtesy of Dr. R. J. Poijak,
Pasteur Institute, Paris, France)*

30.5). Three disulfide bridges link the four chains into a unit. Both light
and heavy chains have constant and variable regions. The constant
regions have the same amino acid sequences in different antibodies, and
the variable regions have different amino acid sequences in different antibodies.

The variable regions of the antibody recognize the foreign substance
(the antigen) and bind to it (Figure 30.6). Since each antibody contains two
variable regions, it can bind two antigens, and this results in a large aggregate (Figure 30.5).

The binding of the antigen to the variable region of the antibody is
not by covalent bonds but by much weaker intermolecular forces such as
London dispersion forces, dipole-dipole interactions, and hydrogen bonds
(Section 5.7). This binding is similar to the way neurotransmitters and hormones bind to a receptor site. The antigen must fit into the antibody surface. Since there are a large number of different antigens against which

Studies have shown that many
other molecules that function as
communication signals between
cells also have the same basic
structural design as the
immunoglobulins.

Antibodies and Cancer Therapy

Antibodies come in a variety of forms. As a response to a specific antigen, the body may develop a number of closely related antibodies, the so-called **multiclonal antibodies.** Monoclonal antibodies, on the other hand, are identical copies of a single antibody molecule that bind to a specific antigen. Georg Kohler and Cesar Milstein developed techniques for producing monoclonal antibodies in large quantities. For their pioneering work, they received the Nobel Prize in 1984.

Monoclonal antibody therapy is the latest advance in fighting cancer. The drug Rituxan is a monoclonal antibody against the antigen, CD20, which appears in high concentration on the surface of B cells in patients suffering from non-Hodgkin lymphoma. Rituxan kills the tumor cells in two ways. First, by binding to the diseased B cells, it enlists the whole immune system to attack the marked cells and to destroy them. Second, the monoclonal antibody causes the tumor cells to stop growing and die through apoptosis, the programmed cell death (Section 23.11). Another monoclonal antibody is Herceptin, recommended in treatment of metastatic breast cancers that have the specific antigen, c-erb-B2, on the surface of cancer cells.

One advantage monoclonal antibody treatment has over chemotherapy is its relatively minimal toxicity.

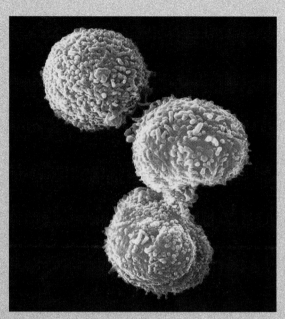

■ False-color scanning electron micrograph of hybridoma cells producing a monoclonal antibody to cytoskeleton protein. *(Dr. Jeremy Burgess / Science Photo Library / Photo Researchers, Inc.)*

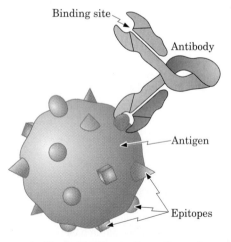

Antibody binding to the epitope of an antigen.

the human body must fight, there are more than 10 000 different antibodies in our systems.

Each B cell synthesizes only one unique immunoglobulin antibody, and that antibody contains a unique antigen-binding site to one epitope. Before encountering an antigen, these antibodies are inserted in the plasma membrane where they serve as receptors. When an antigen interacts with its receptor, it stimulates the B cell to divide and differentiate into plasma cells. These daughter cells secrete soluble antibodies that have the same antigen-binding sites as the original antibody/receptor. The soluble secreted antibodies appear in the serum (the noncellular part of blood) and can react with the antigen. Thus, an immunoglobulin produced in B cells acts both as a receptor to be stimulated by the antigen and as a secreted messenger ready to neutralize and eventually destroy the antigen.

When an antigen is injected in an organism, for example, human lysozyme into a rabbit, the initial response is quite slow. It may take from 1 to 2 weeks before the anti-lysozyme antibody may show up in the rabbit's serum. Those antibodies, however, are not uniform. The antigen may have many epitopes, and the antisera contains a mixture of immunoglobulins with varying specificity for all the epitopes. Even antibodies to a single epitope usually have a variety of specificities. **Monoclonal antibodies,** in contrast, are the product of cells cloned from a single B cell and have a single specificity.

Immunization

Smallpox was a scourge over the centuries, each outbreak leaving many dead and others maimed by deep pits on the face and body. A form of immunization was practiced in ancient China and the Middle East by intentionally exposing people to scabs and fluids of lesions of victims of smallpox. This practice was known as variolation in the Western world, where the disease was called variola. Variolation was introduced to England and the American Colonies in 1721. Edward Jenner, an English physician, noted that milkmaids who had contracted cowpox from infected cows seemed to be immune to smallpox. Cowpox was a mild disease, whereas smallpox could be lethal. In 1796, Jenner performed a daring experiment: He dipped a needle into the pus of a cowpox-infected milkmaid and then scratched a boy's hand with the needle. Two months later, Jenner injected the boy with a lethal dose of smallpox-carrying agent. The boy survived, and did not develop any symptoms of the disease. The word spread, and Jenner was soon established in the immunization business. When the news reached France, the skeptics there coined the derogatory term: vaccination, which literally means "encowment." The derision did not last long, and the practice was soon adopted worldwide. A century later in 1879, Pasteur found that storage of tissue infected with rabies has much weakened virus in it. When injected into patients, it elicits an immune response that protects against rabies. Pasteur named these attenuated, protective antigens as vaccines in honor of Jenner's work. Today, immunization and vaccination are synonymous.

Vaccines are available for several diseases including polio, measles, and smallpox, to name a few. A vaccine may be made up of either dead viruses or bacteria or weakened ones. For example, the Salk polio vaccine is a polio virus that has been made harmless by treatment with formaldehyde; it is given by intramuscular injection. In contrast, the Sabin polio vaccine is a mutated form of the wild virus; the mutation makes the virus sensitive to temperature. The mutated live virus is taken orally. The body temperature and the gastric juices render it harmless before it penetrates the bloodstream.

Vaccines change lymphocytes into plasma cells that produce large quantities of antibodies to fight any invading antigens. This is, however, only the immediate, short-term response. Some lymphocytes become memory cells rather than plasma cells. These memory cells do not secrete antibodies; they store them to serve as a detecting device for future invasion of the same foreign cells. In this way, long-term immunity is conferred. If a second invasion occurs, these memory cells divide directly into antibody-secreting plasma cells as well as into more memory cells. This time the response is faster because it does not have to go through the process of activation and differentiation into plasma cells, which usually takes two weeks.

Smallpox, which was once one of humanity's worst scourges, has been totally wiped out, and smallpox vaccination is no longer required.

30.5   T Cells and Their Receptors

As with B cells, T cells carry on their surface unique receptors that interact with antigens. We noted earlier that T cells respond only to protein antigens. An individual has millions of different T cells, each of which carries on its surface a unique T cell receptor (TcR) that is specific for one antigen only. The TcR is a glycoprotein made of two different subunits cross-linked by disulfide bridges. Like immunoglobulins, TcRs also have constant (C) and variable (V) regions. The antigen binding occurs on the variable region. The similarity in amino acid sequence between immunoglobulins (Ig) and TcR as well as the organization of the polypeptide chains makes TcR molecules members of the immunoglobulin superfamily. There are, however, basic differences between immunoglobulins (Ig) and TcRs.

Immunoglobulins have four polypeptide chains, TcRs contain only two subunits. Immunoglobulins can interact directly with antigens, TcRs can interact only when the epitope of an antigen is presented by an MHC molecule. Lastly, Ig can undergo somatic mutation. This mutation can be found in all body cells except the ones involved in sexual reproduction. Thus, Ig molecules can increase their diversity by somatic mutation; TcRs cannot.

A TcR is anchored in the membrane by hydrophobic transmembrane segments (Figure 30.7). TcR alone, however, is not sufficient for antigen binding. Also needed are additional protein molecules that act as corecep-

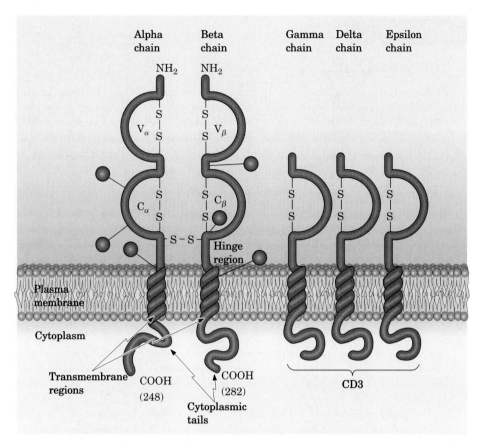

FIGURE 30.7 Schematic structure of TcR complex. TcR consists of two chains, α and β. Each has two extracellular domains, an amino-terminal V domain and a carboxyl-terminal C domain. Domains are stabilized by intrachain disulfide bonds between cysteine residues. The α and β chains are linked by an interchain disulfide bridge near the cell membrane (hinge region). Each chain is anchored on the membrane by a hydrophobic transmembrane segment and ends in the cytoplasm with a carboxyl-terminal segment rich in cationic residues. Both chains are glycosylated. (red spheres). The cluster determinant (CD) coreceptor consists of three chains, γ, δ, and ϵ. Each is anchored in the plasma membrane by a hydrophobic transmembrane segment. Each is cross-linked by a disulfide bridge, and the carboxylic terminal is located in the cytoplasm.

tors and/or signal transducers. These molecules go under the name of CD3, CD4, and CD8, the CD standing for **cluster determinant.** TcR and CD together form the **T cell receptor complex.**

The CD3 molecule adheres to the TcR in the complex not through covalent bonds but by intermolecular forces (Section 5.7). It is a signal transducer because, upon antigen binding, CD3 becomes phosphorylated. This sets up a signaling cascade inside the cell, which is carried out by different kinases. We have seen a similar signaling cascade in neurotransmission (Section 23.5).

The CD4 and CD8 molecules act also as **adhesion molecules** as well as signal transducers. A T cell has either a CD4 or a CD8 molecule to help bind the antigen to the receptor (Figure 30.4). A unique characteristic of the CD4 molecule is that it binds strongly to a glycoprotein that has a molecular weight of 120 000 (gp120). This glycoprotein exists on the surface of the human immunodeficiency virus (HIV). Through this binding to CD4, an HIV virus can enter and infect T cells and cause AIDS (Box 23C). T cells die as a result of HIV infection, and this depletes the T cell population to

Adhesion molecules Various protein molecules that help to bind an antigen to the T cell receptor the extent that the immune system cannot function anymore. As a consequence, the body succumbs to opportunistic pathogen infections.

30.6 Control of Immune Response: Cytokines

Cytokines are glycoprotein molecules that are produced by one cell but alter the function of another cell. They have no antigen specificity. Cytokines transmit intercellular communications between different types of cells at diverse sites in the body. They are short lived and are not stored in the cells.

> **Cytokine** A glycoprotein that traffics between cells and alters the function of a target cell

Cytokines facilitate a coordinated and appropriate inflammatory response by controlling many aspects of the immune reaction. Cytokines are released in bursts, in response to all manner of insult or injury (real or perceived). They travel and bind to specific cytokine receptors on the surface of macrophages and B and T cells and induce cell proliferation.

One set of cytokines are called **interleukins** (IL) because they communicate between and coordinate the actions of leukocytes (all kinds of white blood cells). Macrophages secrete IL-1 upon a bacterial infection. The presence of IL-1 then induces shivering and fever. The elevated body temperature both reduces the bacterial growth and speeds up the mobilization of the immune system (Box 7A).

One leukocyte may make many different cytokines, and one cell may be the target of many cytokines. Cytokines can be classified according to their mode of action, origin, or targets. The best way to classify them is by their structure, the secondary structure of their polypeptide chains.

One class of cytokines is made of *four α-helical segments*. A typical example is interleukin-2 (IL-2). It is a 15 000 MW polypeptide chain. A prominent source of IL-2 is T cells. IL-2 activates other B and T cells and macrophages. Its action is to enhance proliferation and differentiation of the target cells.

Another class of cytokines has only *β-pleated sheets* in its secondary structure. Tumor necrosis factor (TNF) is produced mostly by T cells and macrophages. Its name comes from its ability to lyse (dissolve) susceptible tumor cells after it binds to receptors on the tumor cell.

A third class of cytokines has both *α-helix and β-sheet* secondary structure. A representative of this class is the epidermal growth factor (EGF) which is a cysteine-rich protein. As its name indicates, EGF stimulates the growth of epidermal cells and its main function is in wound healing.

TNF targets other cells as well. For example, it plays a role in rheumatoid arthritis. When the immune system goes awry, the TNF attacks the patient's own cartilage. It binds to the receptors of the cartilage cells, which lyse. In this manner, it eats away even the bones. A new drug, called Enebrel, interacts with TNF in the blood. It immobilizes the cytokine and prevents its migration to the joints where they would cause damage.

A subgroup of cytokines, the chemotactic cytokines, are also called **chemokines.** In humans, there are some 40 chemokines, all low-molecular-weight proteins with distinct structural characteristics. They attract leukocytes to the site of infection or inflammation. All chemokines have four cysteine residues, which form two disulfide bridges: Cys1-Cys3 and Cys2-Cys4. Chemokines have a variety of names such as interleukin (IL-8) and monocyte chemotactic proteins (MCP-1 to MCP-4). Chemokines interact with specific receptors, which consist of seven helical segments coupled to GTP-binding proteins.

> **Chemokine** Low-molecular-weight polypeptides that interact with special receptors on target cells and alter their functions

Two subgroups or chemokines exist in which the first two Cys residues (a) are adjacent (CC) or (b) are separated by one amino acid (CXC).

Special receptors are for the CC and for the CXC chemokines on the surface of leukocytes and lymphocytes. The crucial parts of the chemokines in these interactions are the N-terminal domain and the loop between Cys-2 and Cys-4.

When a tissue is injured, leukocytes are rushed to the inflamed area. Chemokines help leukocytes migrate out of the blood vessels to the site of injury (Box 30D). There **leukocytes,** in all their forms—neutrophils, monocytes, lymphocytes—accumulate and attack the invaders by engulfing them **(phagocytosis)** and later killing them. Other phagocytic cells, the macrophages, which reside in the tissues and thus do not have to migrate, do the same. These activated phagocytic cells kill their prey by releasing endotoxins that kill bacteria, and/or by producing such highly

Mobilization of Leukocytes: Tight Squeeze

When injury occurs, leukocytes in the bloodstream must migrate into the lymph or inflammatory sites to provide protection. There are a number of adhesion and signaling molecules that guide this migration. Two major protein molecules on the surface of leukocytes facilitate their movement.

Selectins are protein molecules belonging to the CD (cluster determinant) class. When activated, they interact with carbohydrate portions of glycoproteins on the surface of endothelial cells. These endothelial cells line the inner wall of blood vessels and, with tight junctions, seal the vessel against leakiness. Endothelial cells have their own selectin molecules that interact with the carbohydrates of the leukocytes. These alternate interactions create a rolling motion of leukocytes along the endothelial wall. When the leukocyte arrives at a point near the injury or infection, another surface molecule is activated. This molecule, called integrin, exists in almost every cell type of the animal kingdom. Activated integrin on endothelial cells strongly interacts with surface adhesion molecules of the leukocyte, which stops rolling. This attachment and the force of the streaming blood make the leukocyte flatten out.

Chemokines, such as IL-8, also play a role in this leukocyte trafficking. When a chemokine is bound to its receptor on the leukocyte, the cell undergoes a remarkable transformation. Inside the cell, the cytoskeleton, actin, is first broken down and then repolymerized, resulting in armlike and leglike projections (lamellipodia) that enable the cells to migrate. In addition, chemokines signal the endothelial cells to change their shape too. This results in the opening of gaps between the cells. The flattened cells then manage to squeeze through the gaps between the tight junctions and enter the tissue or lymph to migrate to the point of injury. After accumulating in the inflamed tissue, the leukocytes may kill the invading bacteria by engulfing and digesting them.

reactive oxygen intermediates as superoxide, singlet oxygen, hydrogen peroxide and hydroxyl radicals (Box 20C).

Chemokines are also major players in chronic inflammations, autoimmune diseases, asthma, and other forms of allergic inflammation and even in transplant rejection.

SUMMARY

The human immune system protects us against foreign invaders. It consists of two parts: (a) the natural resistance of the body, called innate immunity, and (b) acquired or adaptive immunity. **Innate immunity** is nonspecific. **Acquired immunity** is highly specific directed against one particular invader. Acquired immunity also has memory, which innate immunity does not possess. The acquired immunity is referred to as the immune system. Antigen may be a bacterium, a virus, or a toxin. The principal cellular components of the immune system are the white blood cells, or **leukocytes.** The specialized leukocytes in the lymph system are called **lymphocytes.** They circulate mostly in the **lymphoid organs.** The lymph is a collection of vessels extending throughout the body and connected to the interstitial fluid on one hand, and to the blood vessels on the other. Lymphocytes that mature in the bone marrow and produce soluble immunoglobulins are **B cells.** Lymphocytes that mature in the thymus gland are **T cells. Antigens** are large, complex molecules of foreign origin. They interact with antibodies, T cell receptors (TcR), or major histocompatibility complex (MHC) molecules. All three types of molecules belong to the **immunoglobulin superfamily.** An **epitope** is the smallest part or parts of an antigen that binds to antibodies, TcRs and MHCs. Antibodies are **immunoglobulins.** They are water-soluble glycoproteins. They are made of two heavy chains and two light chains. All four are linked together by disulfide bridges. Immunoglobulins contain variable regions in which the amino acid composition of each antibody is different. These regions interact with antigens. Immunoglobulins interact with antigens to form insoluble large aggregates.

All antigens whether proteins, polysaccharides, or nucleic acids interact with soluble immunoglobulins produced by B cells. Protein antigens also interact with T cells. The binding of the epitope to the TcR is facilitated by MHC, which carries the epitope to the T cell surface where it is presented to the receptor. Upon epitope binding to the receptor, the T cell is stimulated. It proliferates and can differentiate into (a) killer T cells or (b) memory cells. The T cell receptor has a number of helper molecules, such as CD4 or CD8, that enable it to bind the epitope tightly. CD (**cluster determinant**) molecules also belong to the immunoglobulin superfamily.

The control and coordination of the immune response are handled by **cytokines,** which are small protein molecules. Chemotactic cytokines, the **chemokines,** such as interleukin-8, facilitate the migration of leukocyte from the blood vessel into the site of injury or inflammation. Other cytokines activate B and T cells and macrophages, enabling them to engulf the foreign body, digest them, or destroy them by releasing special toxins. Some cytokines, such as tumor necrosis factor (TNF), can lyse tumor cells.

KEY TERMS

Acquired immunity (Section 30.1)
Antibody (Section 30.1)
Antigen (Sections 30.1, 30.3)
B cell (Sections 30.1, 30.2)
Chemokine (Section 30.6)
Cluster determinant (Section 30.5)
Cytokine (Sections 30.2, 30.6)
Epitope (Section 30.3)
External innate immunity
 (Section 30.1)
Immunoglobulin (Section 30.4)

Immune system (Section 30.1)
Immunoglobulin superfamily
 (Section 30.1)
Innate immunity (Section 30.1)
Interleukins (Section 30.6)
Internal innate immunity
 (Section 30.1)
Interstitial fluid (Section 30.2)
Killer T cells (Section 30.2)
Leukocyte (Section 30.6)
Lymphocyte (Section 30.2)

Lymphoid organs (Section 30.2)
Macrophage (Section 30.2)
Major histocompatibility complex
 (MHC) (Section 30.3)
Memory cells (Section 30.2)
Monoclonal antibody (Section 30.5,
 Box 30B)
Phagocytosis (Section 30.6)
Plasma cells (Section 30.2)
T cell (Sections 30.1, 30.2)
T cell receptor complex (Section 30.5)

CONCEPTUAL PROBLEMS

30.A Where in the body do you find the largest concentration of antibodies as well as T cells?

30.B If you could isolate two monoclonal antibodies from a certain population of lymphocytes, in what sense would they be similar to each other and in what sense would they differ?

30.C To what class of compounds do MHCs belong? Where would you find them?

30.D T cell receptor molecules are made of two polypeptide chains? Which part of the chain acts as a binding site and what binds to it?

30.E Which compound or complex of compounds of the immune system is mostly responsible for the proliferation of leukocytes?

PROBLEMS

Difficult problems are designated by an asterisk.

Introduction

30.1 Give two examples for the external innate immunity in humans.

30.2 Which form of immunity is characteristic of vertebrates only?

30.3 Which way does the skin fight bacterial invasion?

30.4 What are the classic symptoms of inflammation?

30.5 What differentiates innate immunity from acquired (adaptive) immunity?

Cells and Organs of the Immune System

30.6 Where do T cells and B cells originate?

30.7 Where do T and B cells mature and differentiate?

30.8 What are memory cells? What is their function?

30.9 What causes a B cell to become a plasma cell?

Antigen and Its Presentation

30.10 Would a foreign substance, such as aspirin (MW = 180) be considered an antigen by our body?

30.11 What kind of antigen does a T cell recognize?

30.12 What is the smallest unit of an antigen that is capable of binding to an antibody?

30.13 What is the function of the major histocompatibility complex in the immune response?

***30.14** What role does the MHC play in the immune response of the ABO blood groups?

30.15 What is common between an immunoglobulin antibody and an MHC?

Immunoglobulins

30.16 When a foreign substance is injected in a rabbit, how long does it take to find antibodies against the foreign substance in the rabbit serum?

30.17 Distinguish among the roles of IgA, IgE, and IgG immunoglobulins.

30.18 (a) Which immunoglobulin has the highest carbohydrate content and the lowest concentration in the serum? (b) What is its main function?

30.19 Box 16B states that the antigen in the red blood cells of a person with B-type blood is a galactose unit. Show schematically how the antibody of a person with A-type blood would aggregate the red blood cells of a B-type person if such a transfusion were made by mistake.

30.20 In the immunoglobulin structure, there is a region called the "hinge region" that joins the stem of the Y to the arms. The hinge region can be cleaved by a specific enzyme to yield one Fc fragment (the stem of the Y) and two Fab fragments (the two arms). Which of these two kinds of fragments can interact with an antigen? Explain.

30.21 How are the light and heavy chains of an antibody held together?

30.22 What do we mean by the term immunoglobulin superfamily?

30.23 What are monoclonal antibodies?

30.24 What kind of interaction takes place between an antigen and an antibody?

30.25 How does the action of B cells differ from that of T cells?

T Cells and Their Receptors

30.26 What is common between immunoglobulin and T cell receptor?

*__30.27__ What is the difference between T cell receptor (TcR) and TcR complex?

*__30.28__ What kind of tertiary structure characterizes TcR?

30.29 What are the components of the T cell receptor complex?

30.30 What is needed for an antigen to stimulate the TcR complex?

30.31 Which adhesion molecule in the TcR complex helps the HIV virus to infect a leukocyte?

*__30.32__ There are three different kinds of molecules in the T cell that belong to the immunoglobulin superfamily. List them and indicate briefly their function.

Control of Immune Response: Cytokines

30.33 What kind of molecules are cytokines?

30.34 What do cytokines interact with? Do they bind to antigens?

30.35 As in most biochemistry literature, one encounters a veritable alphabet soup when reading about cytokines. Identify by their full name (a) TNF (b) IL (c) EGF.

*__30.36__ What are chemokines? How do they deliver their message?

30.37 What is the characteristic chemical signature in the structure of chemokines?

30.38 What are the chemical characteristics of cytokines that allows their classification?

30.39 What specific amino acid appears in all chemokines?

Boxes

30.40 (Box 30A) What causes the weakening and eventual paralysis of the muscles in Myasthenia gravis?

*__30.41__ (Box 30A) One finds elevated antibodies in the serum of Myasthenia gravis patients. Where does this antibody come from? Which cells of the immune system manufacture them?

30.42 (Box 30B) How does Rituxan, a monoclonal antibody, find its way to the surface of non-Hodgkins lymphoma cells?

30.43 (Box 30B) Why is monoclonal antibody treatment better than chemotherapy?

30.44 (Box 30C) What made Edward Jenner the father of immunization? In your opinion, could one do such an experiment today legally?

30.45 (Box 30C) What is the difference between memory cells and plasma cells?

30.46 (Box 30D) In leukocyte trafficking, what keeps the leukocytes rolling along the endothelial wall of the blood vessels?

30.47 (Box 30D) What is the role of a chemokine in facilitating the passage of leukocytes from the blood vessel to the site of injury?

Additional Problems

30.48 Which immunoglobulins form the first line of defense against invading bacteria?

30.49 Where do all the lymphatic vessels come together?

30.50 (Box 30B) Why is a monoclonal antibody treatment preferable to a multiclonal antibody treatment?

30.51 What feature of the immunoglobulins enables them to interact with thousands of different antigens?

30.52 Is the light chain of an immunoglobulin the same as the V region?

30.53 Where are TNF receptors located?

30.54 The variable regions of immunoglobulins bind the antigens. How many polypeptide chains carry variable regions in one immunoglobulin molecule?

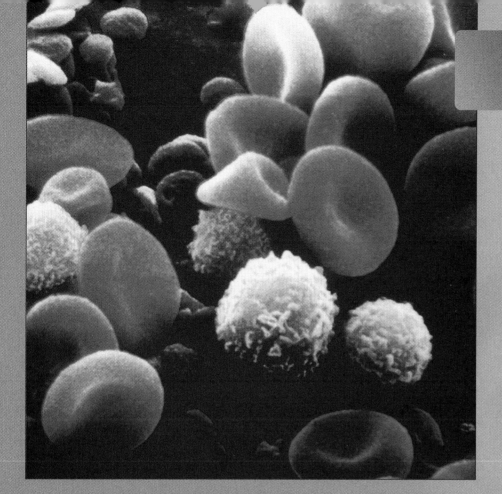

Body Fluids

31.1 Introduction

Single-cell organisms receive their nutrients directly from the environment and discard waste products directly into it. In multicellular organisms, the situation is not so simple. There, too, each cell needs nutrients and produces wastes, but most of the cells are not directly in contact with the environment.

Body fluids serve as a medium for carrying nutrients to and waste products from the cells and also for carrying the chemical communicators (Chapter 23) that coordinate activities among cells.

All the body fluids that are not inside the cells are collectively known as **extracellular fluids.** These fluids make up about one quarter of a person's body weight. The most abundant is **interstitial fluid,** which directly surrounds most cells and fills the spaces between them. It makes up about 17 percent of body weight. Another body fluid is **blood plasma,** which flows in the arteries and veins. This makes up about 5 percent of body weight. Other body fluids that occur in lesser amounts are urine, lymph, cerebrospinal fluid, aqueous humor, and synovial fluid. All body fluids are aqueous solutions. Water is the only solvent in the body.

The blood plasma circulates in the body and is in contact with the other body fluids through the semipermeable membranes of the blood

The fluid inside the cells is called **intracellular fluid.**

> **Interstitial fluid** The fluid that surrounds the cells and fills the spaces between them

> **Blood plasma** The noncellular portion of blood

Above: Human circulatory cells: red blood cells, platelets, and white blood cells. (© Ken Eward/Biografx/Photo Researchers, Inc.)

Using the Blood-Brain Barrier to Eliminate Undesirable Side Effects of Drugs

Many drugs have undesirable side effects. For example, many antihistamines, such as Dramamine and Benadryl (Section 23.5), cause drowsiness. These antihistamines are supposed to act on the peripheral H_1 histamine receptors to relieve seasickness, hay fever, or asthma. But because they penetrate the blood-brain barrier, they also act as antagonists to the H_1 receptors in the brain, causing sleepiness. A drug that also acts on the peripheral H_1 receptors, fexofenadine (sold under the trade name Allegra), cannot penetrate the blood-brain barrier. This antihistamine alleviates seasickness and asthma in the same way as the old antihistamines, but it does not cause drowsiness as a side effect.

Fexofenadine

vessels (Figure 30.1). Therefore, blood can exchange chemical compounds with other body fluids, such as lymph and interstitial fluid and, through them, with the cells and organs of the body.

There is only a limited exchange between blood and cerebrospinal fluid, on the one hand, and blood and the interstitial fluid of the brain, on the other. The limited exchange between blood and interstitial fluid is referred to as the **blood-brain barrier.** This is permeable to water, oxygen, carbon dioxide, glucose (Figure 9E.4), alcohols, and most anesthetics but only slightly permeable to electrolytes such as Na^+, K^+, and Cl^- ions. Many higher-molecular-weight compounds are also excluded.

The blood-brain barrier protects the cerebral tissue from detrimental substances in the blood and allows it to maintain low K^+ concentration, which is needed to generate the high electrical potential essential for neurotransmission.

It is vital that the body maintain in the blood a proper balance of levels of salts, proteins, and all other components. **Homeostasis is the process of maintaining the levels of nutrients of the blood as well as the temperature.**

Body fluids have special importance for the health care professions. Samples of body fluid can be taken with relative ease. The chemical analysis of blood plasma, blood serum, urine, and occasionally cerebrospinal fluid is of major importance in diagnosing disease.

> **Blood-brain barrier** A barrier limiting the exchange of blood components between blood and the cerebrospinal and brain interstitial fluids to water, carbon dioxide, glucose, and other small molecules but excluding electrolytes and large molecules

31.2 Functions and Composition of Blood

It has been known for centuries that "life's blood" is essential to human life. The blood has many functions including the following:

1. It carries O_2 from the lungs to the tissues.
2. It carries CO_2 from the tissues to the lungs.
3. It carries nutrients from the digestive system to the tissues.

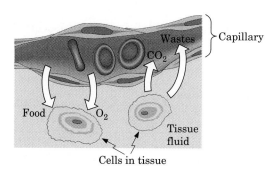

FIGURE 31.1 Nutrients, oxygen, and other materials diffuse out of the blood and through the tissue fluid that bathes the cells. Carbon dioxide and other waste products diffuse out of the cells and enter the blood through the capillary wall.

4. It carries waste products from the tissues to the excretory organs.

5. With its buffer systems, it maintains the pH of the body (with the help of the kidneys).

6. It maintains a constant body temperature.

7. It carries hormones from the endocrine glands to wherever they are needed.

8. It fights infection.

Figure 31.1 shows some of these functions. The rest of this chapter describes how the blood carries out some of these functions.

Whole blood is a complicated mixture. It contains several types of cells (Figure 31.2) and a liquid, noncellular portion called plasma, in which many substances are dissolved (Table 31.1). There are three main types of **cellular elements** of blood: erythrocytes, leukocytes, and platelets.

A 150-lb (68-kg) man has about 6 L of whole blood, 50 to 60 percent of which is plasma.

FIGURE 31.2 Some cellular elements of blood. The dimensions are shown in microns (μm; also called micrometers).

TABLE 31.1   **Blood Components and Some Diseases Associated with Their Abnormal Presence in the Blood**

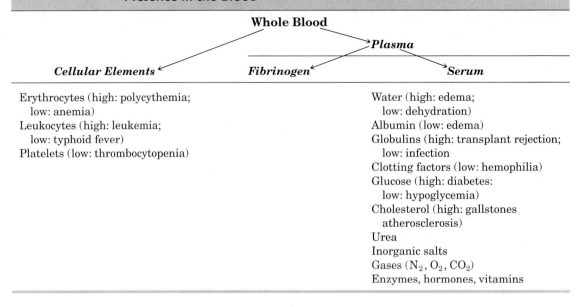

Whole Blood		
		Plasma
Cellular Elements	*Fibrinogen*	*Serum*

Erythrocytes (high: polycythemia; low: anemia)		Water (high: edema; low: dehydration)
Leukocytes (high: leukemia; low: typhoid fever)		Albumin (low: edema)
Platelets (low: thrombocytopenia)		Globulins (high: transplant rejection; low: infection
		Clotting factors (low: hemophilia)
		Glucose (high: diabetes: low: hypoglycemia)
		Cholesterol (high: gallstones atherosclerosis)
		Urea
		Inorganic salts
		Gases (N_2, O_2, CO_2)
		Enzymes, hormones, vitamins

Erythrocytes

> **Erythrocytes** Red blood cells; transporters of blood gases

In an adult male, there are approximately 30 trillion erythrocytes.

The most prevalent cells of blood are the red blood cells, which are called **erythrocytes.** There are about 5 million red blood cells in every cubic millimeter of blood, or roughly 100 million in every drop. Erythrocytes are very specialized cells. They have no nuclei and, hence, no DNA. Their main function is to carry oxygen to the cells and carbon dioxide away from the cells. Erythrocytes are formed in the bone marrow. They stay in the bloodstream for about 120 days. Old erythrocytes are removed by the liver and spleen and destroyed. The constant formation and destruction of red blood cells maintain a steady number of erythrocytes.

Leukocytes

> **Leukocytes** White blood cells; part of the immune system

Leukocytes (white blood cells) are relatively minor cellular components (minor in numbers, not in function). For each 1000 red blood cells, there are only one or two white blood cells. Most of the different leukocytes destroy invading bacteria or other foreign substances by devouring them (**phagocytosis**). Like erythrocytes, leukocytes are manufactured in the bone marrow. Specialized white blood cells formed in the lymph nodes and in the spleen are called lymphocytes. They synthesize immunoglobulins (antibodies, Section 30.4) and store them.

Platelets

> **Platelets** Cellular element of blood; essential to clot formation

There are about 300 000 platelets in each mm³ of blood, or one for every 10 or 20 red blood cells.

When a blood vessel is cut or injured, the bleeding is controlled by a third type of cellular element: the **platelets** or **thrombocytes.** They are formed in the bone marrow and spleen and are more numerous than leukocytes but less numerous than erythrocytes.

Plasma

If all the cellular elements from whole blood are removed by centrifugation, the resulting liquid is the plasma. The cellular elements—mainly red

BOX 31B

Blood Clotting

When body tissues are damaged, the blood flow must be stopped or else enough of it will pour out to cause death. The mechanism used by the body to stem leaks in the blood vessels is clotting. It is a complicated process involving many factors. Here we mention only a few important steps.

When a blood vessel is injured, the first line of defense is the platelets, which are constantly circulating in the blood. These rush to the site of injury, adhere to the collagen molecules in the capillary wall that are exposed by the cut, and form a gel-like plug. At the signal of thromboxane A_2, more platelets enlarge the size of the clot (Section 20.12). This plug is porous, however, and, to seal the site, a firmer gel (a clot) is needed. The clot is a three-dimensional network of fibrin molecules that also contains the platelets. The fibrin network is formed from the blood fibrinogen by the enzyme thrombin. Together with the embedded platelets, this constitutes the blood clot.

The question arises: Why doesn't the blood clot in the blood vessels under normal conditions (with no injury or disease)? The reason is that the enzyme that starts clot formation, thrombin, exists in the blood only in its inactive form, called prothrombin. Prothrombin itself is manufactured in the liver, and vitamin K is needed for its production. Even when prothrombin is in sufficient supply, a number of proteins are needed to change it to thrombin. These proteins are given

the collective name thromboplastin. Any thromboplastic substance can activate prothrombin in the presence of Ca^{2+} ions. Thromboplastic substances exist in the platelets, in the plasma, and in the injured tissue itself.

Clotting is nature's way of protecting us from loss of blood. However, we don't want the blood to clot during blood transfusions because this would stop the flow. To prevent this, we add sodium citrate. Since sodium citrate interacts with Ca^{2+} ions and removes them from the solution, the thromboplastic substances cannot activate prothrombin in the absence of Ca^{2+} ions, and no clot forms. After surgery, **anticoagulant drugs** are occasionally administered to prevent clot formation. A clot is not dangerous if it stays near the injury because, once the body repairs the tissue, the clot is digested and removed. However, a clot formed in one part of the body may break loose and travel to other parts, where it may lodge in an artery. This is **thrombosis.** If the clot then blocks oxygen and nutrient supply to the heart and brain, it can result in paralysis and death. The most common anticoagulants are bishydroxycoumarin (Dicoumarol, Box 18B) and heparin. Heparin enhances the inhibition of thrombin by antithrombin (Section 19.8), and bishydroxycoumarin blocks the transport of vitamin K to the liver, preventing prothrombin formation.

■ Components of a blood clot. (*From* The Functioning of Blood Platelets *by M. B. Zucker. Copyright 1980 by Scientific American, Inc. All rights reserved.*)

blood cells, which settle at the bottom of the centrifuge tube—occupy between 40 and 50 percent of the blood volume.

Blood plasma is 92 percent water. The dissolved solids in the plasma are mainly proteins (7 percent). The remaining 1 percent contains glucose, lipids, enzymes, vitamins, hormones, and such waste products as urea and CO_2. Of the plasma proteins, 55 percent is albumin, 38.5 percent is globu-

A dried clot becomes a scab.

> **Serum** Blood plasma from which fibrinogen has been removed

See the discussion of osmotic pressure in Section 6.9.

A model of heme with central Fe^{2+} ion.

The O_2 picked up by the hemoglobin binds to the Fe^{2+}.

lin, and 6.5 percent is fibrinogen. If plasma is allowed to stand, it forms a clot, a gel-like substance. We can squeeze out a clear liquid from the clot. This fluid is the **serum.** It contains all the components of the plasma except the fibrinogen. This protein is involved in the complicated process of clot formation (Box 31B).

As for the other plasma proteins, most of the globulins take part in the immune reactions (Section 30.4), and the albumin provides proper osmotic pressure. If the albumin concentration drops (from malnutrition or from kidney disease, for example), the water from the blood oozes into the interstitial fluid and creates the swelling of tissues called **edema.**

31.3 Blood As a Carrier of Oxygen

One of the most important functions of blood is to carry oxygen from the lungs to the tissues. This is done by hemoglobin molecules located inside the erythrocytes. As we saw in Section 21.9, hemoglobin is made up of two alpha and two beta protein chains, each attached to a molecule of heme.

The active sites are the hemes, and at the center of each heme is an iron(II) ion. The heme with its central Fe^{2+} ion forms a plane. Because each hemoglobin molecule has four hemes, it can hold a total of four O_2 molecules. However, the ability of the hemoglobin molecule to hold O_2 depends on how much oxygen is in the environment. To see this, look at Figure 31.3, which shows how the oxygen-carrying ability of hemoglobin depends on oxygen pressure. When oxygen enters the lungs, the pressure is high (100 mm Hg). At this pressure, all the Fe^{2+} ions of the hemes bind oxygen molecules; they are fully saturated. By the time the blood reaches the muscles through the capillaries, the oxygen pressure in the muscle is only 20 mm Hg. At this pressure, only 30 percent of the binding sites carry oxygen. Thus, 70 percent of the oxygen carried will be released to the tissues. The S shape of the

FIGURE 31.3 An oxygen dissociation curve. Saturation (%) means the percentage of Fe^{2+} ions that carry O_2 molecules.

binding (dissociation) curve (Figure 31.3) implies that the reaction is not a simple equilibrium reaction.

$$HbO_2 \rightleftharpoons Hb + O_2$$

Each heme has a cooperative effect on the other hemes. This cooperative action allows hemoglobin to deliver twice as much oxygen to the tissues as it would if each heme acted independently. The reason for this is as follows: When a hemoglobin molecule is carrying no oxygen, the four globin units coil into a certain shape (Section 21.9). When the first oxygen molecule attaches to one of the heme subunits, it changes the shape not only of that subunit but also of a second one, making it easier for the second one to bind an oxygen. When the second heme binds, the shape changes once again, and the oxygen-binding ability of the two remaining subunits is increased still more.

This is an allosteric effect (Section 22.6) and explains the S-shaped curve shown in Figure 25.4.

The oxygen-delivering capacity of hemoglobin is also affected by its environment. A slight change in the pH of the environment changes the oxygen-binding capacity. This is called the **Bohr effect.** An increase in CO_2 pressure also decreases the oxygen-binding capacity of hemoglobin. When a muscle contracts, both H^+ ions and CO_2 are produced. The H^+ ions, of course, lower the pH of the muscle. Thus, at the same pressure (20 mm Hg), more oxygen is released for an active muscle than for a muscle at rest. Similarly, an active contracting muscle also produces CO_2, which then accumulates and further enhances the release of oxygen.

Bohr effect The effect caused by a change in the pH on the oxygen-carrying capacity of hemoglobin

Lowering the pH decreases the oxygen-binding capacity of hemoglobin.

This is how the body delivers more oxygen to those tissues that need it.

31.4 Transport of Carbon Dioxide in the Blood

At the end of the previous section, we saw that the waste products of tissue cells, H^+ ions and CO_2, facilitate the release of oxygen from hemoglobin. This enables the cells to receive the oxygen they need. What happens to the CO_2 and H^+? They bind to the hemoglobin. There is an equilibrium in which, upon the release of oxygen, carbaminohemoglobin is formed:

$$HbO_2 + H^+ + CO_2 \rightleftharpoons Hb \overset{\displaystyle H^+}{\underset{\displaystyle CO_2}{}} + O_2$$

Carbaminohemoglobin

The CO_2 is bound to the terminal $-NH_2$ groups of the four polypeptide chains so that each hemoglobin can carry a maximum of four CO_2 molecules, one for each chain. How much CO_2 each hemoglobin actually carries depends on the pressure of CO_2. The higher the CO_2 pressure, the more carbaminohemoglobin is formed.

But only 25 percent of the total CO_2 produced by the cells is transported to the lungs in the form of carbaminohemoglobin. Another 70 percent is converted in the red blood cells to carbonic acid by the enzyme carbonic anhydrase:

The remaining 5 percent of the CO_2 is transported as CO_2 gas dissolved in the plasma.

$$CO_2 + H_2O \underset{\text{anhydrase}}{\overset{\text{carbonic}}{\rightleftharpoons}} H_2CO_3$$

A large part of this H_2CO_3 is carried as such to the lungs, where it is converted back to CO_2 by carbonic anhydrase and released.

The reaction proceeds to the left because loss of CO_2 from the lungs causes this equilibrium to shift to the left (Le Chatelier's principle).

31.5   Blood Cleansing. A Kidney Function

Hemodialysis is discussed in Box 6F.

The balance between filtration and reabsorption is controlled by a number of hormones.

Arteries are blood vessels that carry oxygenated blood away from the heart.

Glomeruli Part of the kidney filtration cell, nephron; a tangle of capillaries surrounded by a fluid-filled space

The singular is glomerulus.

We have seen that one of the functions of the blood is to maintain the pH at 7.4 (Section 8.10). Another is to carry waste products away from the cells. CO_2, one of the principal waste products of respiration, is carried by the blood to the lungs and exhaled. The other waste products are filtered out by the kidneys and eliminated in the urine. The kidney is a superfiltration machine. In the event of kidney failure, filtration can be done by hemodialysis (Figure 31.4). About 100 L of blood pass through a normal human kidney daily. Of this, only about 1.5 L of liquid are excreted as urine. Obviously, we do not have here just a simple filtration system in which small molecules are lost and large ones are retained. The kidneys also reabsorb from the urine those small molecules that are not waste products.

The biological units inside the kidneys that perform these functions are called **nephrons,** and each kidney contains about a million of them. A nephron is made up of a filtration head called a **Bowman's capsule** connected to a tiny, U-shaped tube called a tubule. The part of the tubule close to the Bowman's capsule is the **proximal tubule,** the U-shaped twist is called **Henle's loop,** and the part of the tubule farthest from the capsule is the **distal tubule** (Figure 31.5).

Blood vessels penetrate the kidney throughout. The arteries branch into capillaries, and one tiny capillary enters each Bowman's capsule. Inside the capsule, the capillary first branches into even smaller vessels, called **glomeruli,** and then leaves the capsule. The blood enters the glomeruli with every heartbeat, and the pressure forces the water, ions, and all small molecules (urea, sugars, salts, amino acids) through the walls of the glomeruli and the Bowman's capsule. These molecules and ions enter the proximal tubule. Blood cells and large molecules (proteins) are retained in the capillary and exit with the blood.

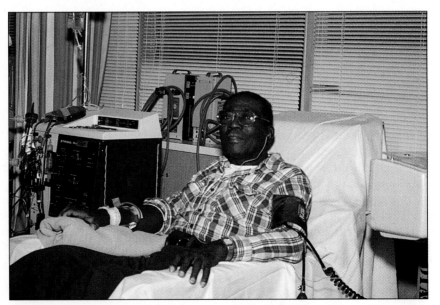

FIGURE 31.4   A patient undergoing hemodialysis for kidney disease. *(Beverly March; courtesy of Long Island Jewish Hospital)*

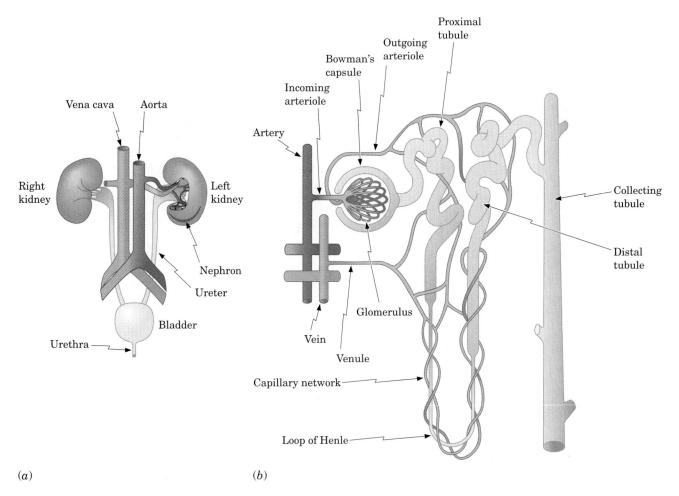

FIGURE 31.5 Excretion through the kidneys. (*a*) The human urinary system.
(*b*) A kidney nephron and its components, with the surrounding circulatory system.

As shown in Figure 31.5, the tubules and Henle's loop are surrounded by blood vessels; these blood vessels reabsorb vital nutrients. Eighty percent of the water is reabsorbed in the proximal tubule. Almost all the glucose and amino acids are reabsorbed here. When excess sugar is present in the blood (diabetes, Box 23G), some of it passes into the urine, and the measurement of glucose concentration in the urine is used in diagnosing diabetes.

By the time the glomerular filtrate reaches Henle's loop, the solids and most of the water have been reabsorbed, and only wastes (such as urea, creatinine, uric acid, ammonia, and some salts) pass into the collecting tubules that bring the urine to the ureter, from which it passes to the bladder, as shown in Figure 31.5(a).

Urine

Normal urine contains about 4 percent dissolved waste products. The rest is water. The daily amount of urine varies greatly but averages about 1.5 L per day. The pH varies from 5.5 to 7.5. The main solute is urea, the end product of protein metabolism (Section 27.8). Other nitrogenous waste

BOX 31C

Composition of Body Fluids

In Section 31.1, it was pointed out that there are three main body fluids: intracellular fluid, interstitial fluid, and blood plasma. Figure 31C.1 shows the average concentrations of the substances dissolved in these three fluids. Blood plasma and interstitial fluid have similar compositions (the main difference is that the blood has a higher protein concentration), but intracellular fluid is quite different. On the cation side, the intracellular fluid has a high concentration of K^+ and a low concentration of Na^+; in the other two fluids, these concentrations are reversed. Intracellular fluid also has a higher Mg^{2+} concentration. On the anion side, there is virtually no Cl^- inside cells, while this is the most abundant anion outside the cells. On the other hand, the concentrations of phosphate (mostly HPO_4^{2-}) and, to a lesser extent, of sulfate are much greater inside the cells. As might be expected, the protein concentration is much higher inside the cells.

■ **FIGURE 31C.1** The composition of body fluids.

31.5 Blood Cleansing. A Kidney Function **707**

BOX 31C

TABLE 31C	**The Range of Blood Constituent Concentrations Detectable by the Portable Digital Instrument**

$pO_2 = 20-699$ mm Hg	$pCO_2 = 10-99$ mm Hg
$Na^+ = 80-190$ mM	$K^+ = 0.2-20$ mM
$Ca^{2+} = 0.2-4.25$ mM	pH $= 6.80-7.70$

Hematocrit (percent of the red blood cells in the blood sample) = 12–70 percent.

In many emergency cases, a quick blood analysis is needed to provide a proper diagnosis. New automated portable instruments are available to read out blood gases and blood electrolyte concentration. The instrument is attached to the arm of the patient and a pump withdraws a small volume (1.5 mL) of arterial blood and pumps it across sensors before returning the sample to the patient's body. The results are displayed digitally.

■ A digital portable blood gas instrument. *(© James Prince/ Photo Researchers, Inc.)*

products, such as creatine, creatinine, ammonia, and hippuric acid, are also present, though in much smaller amounts.

Creatine

Creatinine

Hippuric acid

In addition, normal urine contains inorganic ions such as Na^+, Ca^{2+}, Mg^{2+}, Cl^-, PO_4^{3-}, SO_4^{2-}, and HCO_3^-.

Under certain pathological conditions, other substances can appear in the urine. It is because of this that urine analysis is such an important part of diagnostic medicine. We have already noted that the presence of glucose and ketone bodies is an indication of diabetes. Among other abnormal constituents may be proteins, which can indicate kidney diseases such as nephritis.

31.6 Buffer Production. Another Kidney Function

Among the waste products sent into the blood by the tissues are H^+ ions. These are neutralized by the HCO_3^- ions that are a part of the blood's buffer system:

$$H^+ + HCO_3^- \rightleftharpoons H_2CO_3$$

When the blood reaches the lungs, the H_2CO_3 is decomposed by carbonic anhydrase, and the CO_2 is exhaled. If the body had no mechanism for replacing HCO_3^-, we would lose most of the bicarbonate buffer from the blood, and the blood pH would eventually drop (acidosis). The lost HCO_3^- ions are continuously replaced by the kidneys—this is another principal kidney function. The replacement takes place in the distal tubules. The cells lining the walls of the distal tubules reabsorb the CO_2 that was lost in the glomeruli. With the aid of carbonic anhydrase, carbonic acid forms quickly and then dissociates to HCO_3^- and H^+ ions:

$$CO_2 + H_2O \rightleftharpoons H_2CO_3 \rightleftharpoons H^+ + HCO_3^-$$

The H^+ ions move from the cells into the urine in the tubule, where they are partially neutralized by a phosphate buffer. To compensate for the lost positive ions, Na^+ ions from the tubule enter the cells. When this happens, Na^+ and HCO_3^- ions move from the cells into the capillary. Thus, the H^+ ions picked up at the tissues and temporarily neutralized in the blood by HCO_3^- are finally pumped out into the urine. At the same time, the HCO_3^- ions lost in the lungs are regained by the blood in the distal tubules.

31.7 Water and Salt Balance in Blood and Kidneys

We mentioned in Section 31.5 that the balance in the kidneys between filtration and reabsorption is under hormonal control. The reabsorption of water is promoted by vasopressin, a small peptide hormone manufactured in the hypophysis (Section 21.7). In the absence of this hormone, only the proximal tubules—not the distal tubules or the collecting tubules—reabsorb water. As a consequence, too much water passes into the urine without vasopressin. In the presence of vasopressin, water is also reabsorbed in these parts of the nephrons; thus, vasopressin causes the blood to retain more water and produces a more concentrated urine. The production of urine is called **diuresis.** Any agent that reduces the volume of urine is called an **antidiuretic.**

Usually, the vasopressin level in the body is sufficient to maintain the proper amount of water in tissues under various levels of water intake. However, when severe dehydration occurs (as a result of diarrhea, excessive sweating, or insufficient water intake), another hormone also helps to maintain proper fluid level. This hormone is aldosterone (Section 20.10); it controls the Na⁺ ion concentration in the blood. In the presence of aldos-

A high concentration of solids in the urine is a symptom of diabetes insipidus.

> **Diuresis** Enhanced excretion of water in urine

For this reason vasopressin is also called ADH, for antidiuretic hormone.

BOX 31D

Sex Hormones and Old Age

The same sex hormones that provide a healthy reproductive life in the young can and do cause problems in the aging body. For male fertility, the prostate gland, among others, provides the semen. The prostate gland is situated at the exit of the bladder surrounding the ureter. Its growth is promoted by the conversion of the male sex hormone, testosterone, to 5-α-dihydrotestosterone by the enzyme 5-α-reductase. The majority of men over the age of 50 have enlarged prostate glands, a condition known as benign prostatic hyperplasia. This condition affects more than 90 percent of men over the age of 70. The enlarged prostate gland constricts the ureter and causes a restricted flow of urine.

This symptom can be relieved by oral intake of a drug called terazosin (trade name Hytrin). This drug was originally approved for high blood pressure, where its action results from blocking the α-1-adrenoreceptors, reducing the constriction around peripheral blood vessels. In a similar way, it reduces constriction around the ureter (the duct that conveys urine from the kidney to the bladder), allowing an unrestricted flow of urine. Terazosin, if taken as the sole antihypertensive, antiprostate drug, came under scrutiny lately. In a study released in 2000, it was concluded that this drug may be contributing to the occurrence of heart attacks.

There is a measure, the concentration of prostate-specific antigen (PSA) in the serum, that indicates the seriousness of the problem. A PSA of 4.0 ng/mL and less is considered to be normal, or an indicator of benign prostatic hyperplasia. A concentration above 10 ng/mL is an indicator for prostate cancer. Between 4.0 and 10 is a gray area deserved to be closely monitored. A more accurate measure of prostate cancer is the percent of free (not bound) PSA. When this level is above 15 percent, the likelihood of cancer diminishes with increasing percentage. When this number is 9 percent or below, the indication is that the prostate cancer has already spread.

Estradiol and progesterone control sexual characteristics in the female body (Section 20.10). With the onset of the menopause, smaller amounts of these and other sex hormones are secreted. This not only stops fertility but occasionally creates medical problems. The most serious is osteoporosis, which is the loss of bone tissue through absorption. This results in brittle and occasionally deformed bone structure. Estrogens (a general name for female sex hormones other than progesterone) in the proper amount help to absorb calcium from the diet and thus prevent osteoporosis. The lack of enough female sex hormones can also result in atrophic vaginitis and atrophic urethritis. A large number of postmenopausal women take pills that contain a synthetic progesterone analog and natural estrogens. However, continuous medication with these pills, for example, Premarin (mixed estrogens) and Provera (a progesterone analog), is not without danger. A slight but statistically significant increase in cancer of the uterus has been reported in postmenopausal women who use such medication. A physician's advice is needed in weighing the benefits against the risks.

terone, the reabsorption of Na^+ ions is increased. When more Na^+ ions enter the blood, more Cl^- ions follow (to maintain electroneutrality), as well as more water to solvate these ions. Thus, increased aldosterone production enables the body to retain more water. When the Na^+ ion and water levels in the blood return to normal, aldosterone production stops.

31.8 Blood Pressure

Blood pressure is produced by the pumping action of the heart. The blood pressure at the arterial capillary end is about 32 mm Hg. This pressure is higher than the osmotic pressure of the blood (18 mm Hg). The osmotic pressure of the blood is caused by the fact that more solutes are dissolved in the blood than in the interstitial fluid. At the capillary end, therefore, nutrient solutes flow from the capillary into the interstitial fluid and from there into the cell (see Figure 30.1). On the other hand, in the venous capillary the blood pressure is about 12 mm Hg. This is less than the osmotic pressure, so solutes (waste products) from the interstitial fluid flow into the capillaries.

The blood pressure is maintained by the total volume of blood, by the pumping of the heart, and by the muscles that surround the blood vessels and provide the proper resistance to blood flow. Blood pressure is controlled by several very complex systems, some of them acting within seconds and some that take days to respond after a change in blood pressure occurs. For example, if a patient hemorrhages, three different nervous control systems begin to function within seconds. The first is the **baroreceptors** in the neck, which detect the consequent drop in pressure and send appropriate

> **Baroreceptors** A feedback mechanism part in the neck that detects changes in the blood pressure and sends signals to the heart to counteract those changes by pumping harder or slower as the need arises

BOX 31E

Hypertension and Its Control

Nearly 60 million people in the United States have high blood pressure **(hypertension)**. Epidemiological studies have shown that even mild hypertension (diastolic pressure between 90 and 104 mm Hg) brings about the risk of cardiovascular disease. This can lead to heart attack, stroke, or kidney failure. Hypertension can be managed effectively with diet and drugs. As noted in the text, blood pressure is under a complex control system, so hypertension must be managed on more than one level. Recommended dietary practices are low Na^+ ion intake and abstention from caffeine and alcohol.

The most common drugs for lowering blood pressure are the **diuretics.** There are a number of synthetic organic compounds, (mostly thiazides such as Diuril and Enduron) that increase urine excretion. By doing so, they decrease the blood volume as well as the Na^+ ion concentration and thus lower blood pressure. Other drugs affect the nerve control of blood pressure. Propranolol hydrochloride blocks the adrenoreceptor sites of beta-adrenergic neurotransmitters (Section 23.5). By doing so, it reduces the flow of nerve signals and thereby reduces the blood output of the heart; it affects the responses of baroreceptors and chemoreceptors in the central nervous system and the adrenergic receptors in the smooth muscles surrounding the blood vessels. Propranolol and other beta blockers (such as atenolol) not only lower blood pressure but

also reduce the risk of myocardial infarction. Similar lowering of blood pressure is achieved by drugs that block calcium channels (for example, nifedipine/Adalat or diltiazem hydrochloride/Cardizem). For proper muscle contraction, calcium ions must pass through muscle cell membranes. This is facilitated by calcium channels. If these channels are blocked by the drug, the heart muscles will contract less frequently, pumping less blood and reducing blood pressure.

Other drugs that reduce hypertension are the vasodilators that relax the vascular smooth muscles. However, many of these have side effects causing fast heartbeat (tachycardia) and also promote Na^+ and water retention. Some drugs on the market act on one specific site rather than on many sites as the blockers do. This obviously reduces the possible side effects. An example is an enzyme inhibitor that prevents the production of angiotensin. Captopril (Box 15A) and Accupril are ACE (angiotensin-converting enzyme) inhibitors that lower blood pressure only if the sole cause of hypertension is angiotensin production. Prostaglandins PGE and PGA (Section 20.12) lower the blood pressure and, taken orally, have prolonged duration of action. Since hypertension is such a common problem, and it is manageable by drugs, there is great activity in drug research to come up with even more effective and safer antihypertension drugs.

signals to the heart to pump harder and to the muscles surrounding the blood vessels to contract and thus restore pressure. Chemical receptors on the cells that detect less O_2 delivery or CO_2 removal also send nerve signals. Finally, the central nervous system also reacts to oxygen deficiency with a feedback mechanism.

Hormonal controls act somewhat more slowly and may take minutes or even days. The kidneys secrete an enzyme called renin, which acts on an inactive blood protein called angiotensinogen, converting it to **angiotensin,** a potent vasoconstrictor. The action of this peptide increases blood pressure. Aldosterone (Section 31.7) also increases blood pressure by increasing Na^+ ion and water reabsorption in the kidneys.

Finally, there is a long-term renal blood control for volume and pressure. When blood pressure falls, the kidneys retain more water and salt, thus increasing blood volume and pressure.

SUMMARY

The most important body fluid is **blood plasma,** which is whole blood from which cellular elements have been removed. Other important body fluids are urine and the **interstitial fluids** directly surrounding the cells of the tissues. The **cellular elements** of the blood are the red blood cells **(erythrocytes),** which carry O_2 to the tissues and CO_2 away from the tissues; the white blood cells **(leukocytes),** which fight infection; and the **platelets,** which control bleeding. The plasma contains fibrinogen, which is necessary for blood clot formation. When fibrinogen is removed from the plasma, what remains is the **serum.** The serum contains albumins, globulins, nutrients, waste products, inorganic salts, enzymes, hormones, and vitamins dissolved in water. The body maintains **homeostasis.**

The red blood cells carry the O_2. The Fe(II) in the heme portion of hemoglobin binds the oxygen, which is then released at the tissues by a combination of factors: low O_2 pressure in the tissue cells, high H^+ ion concentration **(Bohr effect),** and high CO_2 concentration. Of the CO_2 in the venous blood, 25 percent binds to the terminal $—NH_2$ of the polypeptide chains of hemoglobin to form carbaminohemoglobin, 70 percent is carried in the plasma as H_2CO_3, and 5 percent is carried as dissolved gas.

Waste products are removed from the blood in the kidneys. The filtration units—**nephrons**—contain an entering blood vessel that branches out into fine vessels called **glomeruli.** Water and all the small molecules in the blood diffuse out of the glomeruli, enter the **Bowman's capsules** of the nephrons, and from there go into the **tubules.** The blood cells and large molecules remain in the blood vessels. The water, inorganic salts, and organic nutrients are then reabsorbed into the blood vessels. The waste products plus water form the urine, which goes from the tubule into the ureter and finally into the bladder.

The kidneys not only filter out waste products but also produce HCO_3^- ions for buffering, replacing those lost through the lungs. The water balance and salt balance of the blood and urine are under hormonal control. Vasopressin helps to retain water, and aldosterone increases the blood Na^+ ion concentration by reabsorption. Blood pressure is controlled by a large number of factors.

KEY TERMS

Angiotensin (Section 31.8)
Antidiuretic (Section 31.7)
Baroreceptor (Section 31.8)
Blood-brain barrier
  (Section 31.1)
Blood plasma (Section 31.1)
Bohr effect (Section 31.3)
Bowman's capsule (Section 31.5)
Cellular elements (Section 31.2)
Distal tubule (Section 31.5)

Diuresis (Section 31.7)
Diuretic (Box 31E)
Edema (Section 31.2)
Erythrocyte (Section 31.2)
Extracellular fluid (Section 31.1)
Glomeruli (Section 31.5)
Henle's loop (Section 31.5)
Homeostasis (Section 31.1)
Hypertension (Box 31E)
Interstitial fluid (Section 31.1)

Intracellular fluid (Section 31.1)
Leukocyte (Section 31.2)
Nephron (Section 31.5)
Phagocytosis (Section 31.2)
Plasma (Section 31.2)
Platelet (Section 31.2)
Proximal tubule (Section 31.5)
Serum (Section 31.2)
Thrombocyte (Section 31.2)
Thrombosis (Box 31B)

CONCEPTUAL PROBLEMS

31.A The pH of blood is maintained at pH 7.4. What would be the pH of an active muscle cell relative to that of blood? What component and process generates this pH?

31.B Caffeine is a diuretic. What effect will drinking a lot of coffee have on the specific gravity of your urine?

31.C A number of drugs combatting high blood pressure go under the group name ACE (angiotensin-converting enzyme) inhibitors. A patient takes diuretics and ACE inhibitors. How do these drugs lower blood pressure?

PROBLEMS

Difficult problems are designated by an asterisk.

Body Fluids

31.1 What is the common solvent in all body fluids?

31.2 How much of our body weight is made of blood plasma?

31.3 (a) Is the blood plasma in contact with the other body fluids? (b) If so, in what manner?

Functions and Composition of Blood

31.4 Name the three principal types of blood cells.

31.5 What is the function of erythrocytes?

31.6 Which blood cells are the largest?

31.7 Define. (a) Interstitial fluid (b) Plasma (c) Serum (d) Cellular elements

31.8 Where are the following manufactured?
(a) Erythrocytes
(b) Leukocytes
(c) Lymphocytes
(d) Platelets

31.9 How does a low albumin content of the serum cause edema?

31.10 State the function of (a) albumin (b) globulin (c) fibrinogen.

31.11 Which blood plasma component protects against infections?

31.12 How is serum prepared from blood plasma?

Blood As a Carrier of Oxygen

31.13 How many molecules of O_2 are bound to a hemoglobin molecule at full saturation?

31.14 A chemist analyzes contaminated air from an industrial accident and finds that the partial pressure of oxygen is only 38 mm Hg. How many O_2 molecules, on the average, will be transported by each hemoglobin molecule when someone breathes such air? (Consult Figure 31.3.)

31.15 What is the oxygen pressure in the muscles?

*31.16** From Figure 31.3, predict the O_2 pressure at which the hemoglobin molecule binds four, three, two, one, and zero O_2 molecules to its heme.

31.17 Explain how the absorption of O_2 on one binding site in the hemoglobin molecule increases the absorption capacity at other sites.

31.18 (a) What happens to the oxygen-carrying capacity of hemoglobin when the pH is lowered? (b) What is the name of this effect?

Transport of CO_2 in the Blood

31.19 (a) Where does CO_2 bind to hemoglobin? (b) What is the name of the complex formed?

31.20 How many CO_2 molecules can bind to a hemoglobin molecule at full saturation?

*31.21** If the plasma carries 70 percent of its carbon dioxide in the form of H_2CO_3 and 5 percent in the form of dissolved CO_2 gas, calculate the equilibrium constant of the reaction catalyzed by carbonic anhydrase:

$$H_2O + CO_2 \rightleftharpoons H_2CO_3$$

(Assume that the H_2O concentration is included in the equilibrium constant.)

Kidney Functions and Urine

31.22 What is the difference between glomeruli and nephron?

31.23 Where is most of the water reabsorbed in the kidney?

31.24 What nitrogenous waste products are eliminated in the urine?

31.25 Besides nitrogenous waste products, what other components are normal constituents of urine?

31.26 What happens to the H^+ ions sent into the blood as waste products by the cells of the tissues?

Water and Salt Balance in the Blood and Kidneys

31.27 In which part of the kidneys does vasopressin change the permeability of membranes?

31.28 How does aldosterone production counteract the excessive sweating that usually accompanies a high fever?

Blood Pressure

31.29 What would happen to the glucose in the arterial capillary if the blood pressure at the end of the capillary were 12 mm Hg rather than the normal 32 mm Hg?

31.30 Some drugs used as antihypertensive agents, for example, Captopril (Box 31E), inhibit the enzyme that activates angiotensinogen. What is the mode of action of such drugs in lowering blood pressure?

Boxes

31.31 (Box 31A) What side effect of antihistamines is avoided by using the drug fexofenadine?

31.32 (Box 31B) What are the functions of (a) thrombin, (b) vitamin K, and (c) thromboplastin in clot formation?

31.33 (Box 31B) What is the first line of defense when a blood vessel is cut?

31.34 (Box 31C) (a) In which body fluids are the negative ions mostly Cl^-? (b) In which body fluid is Cl^- essentially absent?

31.35 (Box 31D) What are the symptoms of enlarged prostate gland (benign prostatic hyperplasia)?

31.36 (Box 31D) What causes osteoporosis in post-menopausal women?

31.37 (Box 31E) How do calcium channel blockers act in lowering blood pressure?

31.38 (Box 31E) People with high blood pressure are advised to have a low-sodium diet. How does such a diet lower blood pressure?

Additional Problems

31.39 Albumin makes up 55 percent of the protein content of plasma. Is the albumin concentration of the serum higher, lower, or the same as that of the plasma? Explain.

31.40 Which nitrogenous waste product of the body can be classified as an amino acid?

31.41 What happens to the oxygen-carrying ability of hemoglobin when acidosis occurs in the blood?

31.42 Does the heme of hemoglobin participate in CO_2 transport as well as in O_2 transport?

31.43 What is a diuretic drug?

*31.44 When kidneys fail, the body swells, a condition called edema. Why and where does the water accumulate?

31.45 Which hormone controls water retention in the body?

31.46 Describe the action of renin.

*31.47 The oxygen dissociation curve of fetal hemoglobin is lower than that of adult hemoglobin. How does the fetus obtain its oxygen supply from the mother's blood when the two circulations meet?

Exponential Notation

The **exponential notation** system is based on powers of 10 (see table). For example, if we multiply $10 \times 10 \times 10 = 1000$, we express this as 10^3. The 3 in this expression is called the **exponent** or the **power,** and it indicates how many times we multiplied 10 by itself and how many zeros follow the 1.

There are also negative powers of 10. For example, 10^{-3} means 1 divided by 10^3:

$$10^{-3} = \frac{1}{10^3} = \frac{1}{1000} = 0.001$$

Numbers are frequently expressed like this: 6.4×10^3. In a number of this type, 6.4 is the **coefficient** and 3 is the exponent, or power of 10. This number means exactly what it says:

$$6.4 \times 10^3 = 6.4 \times 1000 = 6400$$

Similarly, we can have coefficients with negative exponents:

$$2.7 \times 10^{-5} = 2.7 \times \frac{1}{10^5} = 2.7 \times 0.00001 = 0.000027$$

For numbers greater than 10 in exponential notation, we proceed as follows: *Move the decimal point to the left,* to just after the first digit. The (positive) exponent is equal to the number of places we moved the decimal point.

Exponential notation is also called scientific notation.

For example, 10^6 means a one followed by six zeros, or 1 000 000, and 10^2 means 100.

APP. I.1	Examples of Exponential Notation
$10\ 000 = 10^4$	
$1000 = 10^3$	
$100 = 10^2$	
$10 = 10^1$	
$1 = 10^0$	
$0.1 = 10^{-1}$	
$0.01 = 10^{-2}$	
$0.001 = 10^{-3}$	

EXAMPLE

$$3\ 7\ 5\ 0\ 0 = 3.75 \times 10^4 \quad \text{4 because we went four places to the left}$$

Four places to the left — Coefficient

$$628 = 6.28 \times 10^2$$

Two places to the left — Coefficient

$$859\ 600\ 000\ 000 = 8.596 \times 10^{11}$$

Eleven places to the left — Coefficient

We don't really have to place the decimal point after the first digit, but by doing so we get a coefficient between 1 and 10, and that is the custom.

Using exponential notation, we can say that there are 2.95×10^{22} copper atoms in a copper penny. For large numbers, the exponent is always *positive*. Note that we do not usually write out the zeros at the end of the number.

For small numbers (less than 1), we move the decimal point *to the right,* to just after the first nonzero digit, and use a *negative exponent.*

> **EXAMPLE**
>
> $$0.00346 = 3.46 \times 10^{-3}$$
>
> Three places to
> the right
>
> $$0.000004213 = 4.213 \times 10^{-6}$$
>
> Six places to
> the right

In exponential notation, a copper atom weighs 2.3×10^{-25} pounds.

To convert exponential notation into fully written-out numbers, we do the same thing backward.

> **EXAMPLE**
>
> Write out in full: (a) 8.16×10^{7} (b) 3.44×10^{-4}.
>
> **Solution**
>
> (a) $8.16 \times 10^{7} = 81\ 600\ 000$
>
> Seven places to the right
> (add enough zeros)
>
> (b) $3.44 \times 10^{-4} = 0.000344$
>
> Four places to the left

When scientists add, subtract, multiply, and divide, they are always careful to express their answers with the proper number of digits, called significant figures. This method is described in Appendix II.

Adding and Subtracting Numbers in Exponential Notation

We are allowed to add or subtract numbers expressed in exponential notation *only if they have the same exponent.* All we do is add or subtract the coefficients and leave the exponent as it is.

> **EXAMPLE**
>
> Add 3.6×10^{-3} and 9.1×10^{-3}.
>
> **Solution**
>
> $$\begin{array}{r} 3.6 \times 10^{-3} \\ +\ \ 9.1 \times 10^{-3} \\ \hline 12.7 \times 10^{-3} \end{array}$$
>
> The answer could also be written in other, equally valid ways:
>
> $$12.7 \times 10^{-3} = 0.0127 = 1.27 \times 10^{-2}$$

When it is necessary to add or subtract two numbers that have different exponents, we first must change them so that the exponents are the same.

A calculator with exponential notation changes the exponent automatically.

EXAMPLE

Add 1.95×10^{-2} and 2.8×10^{-3}.

Solution
In order to add these two numbers, we make both exponents -2. Thus, $2.8 \times 10^{-3} = 0.28 \times 10^{-2}$. Now we can add:

$$
\begin{array}{r}
1.95 \times 10^{-2} \\
+\ 0.28 \times 10^{-2} \\
\hline
2.23 \times 10^{-2}
\end{array}
$$

Multiplying and Dividing Numbers in Exponential Notation

To multiply numbers in exponential notation, we first multiply the coefficients in the usual way and then algebraically *add* the exponents.

EXAMPLE

Multiply 7.40×10^5 by 3.12×10^9.

Solution

$$7.40 \times 3.12 = 23.1$$

Add exponents:

$$10^5 \times 10^9 = 10^{5+9} = 10^{14}$$

Answer:

$$23.1 \times 10^{14} = 2.31 \times 10^{15}$$

EXAMPLE

Multiply 4.6×10^{-7} by 9.2×10^4

Solution

$$4.6 \times 9.2 = 42$$

Add exponents:

$$10^{-7} \times 10^4 = 10^{-7+4} = 10^{-3}$$

Answer:

$$42 \times 10^{-3} = 4.2 \times 10^{-2}$$

To divide numbers expressed in exponential notation, the process is reversed. We first divide the coefficients and then algebraically *subtract* the exponents.

EXAMPLE

Divide: $\dfrac{6.4 \times 10^8}{2.57 \times 10^{10}}$

Solution

$$6.4 \div 2.57 = 2.5$$

Subtract exponents:

$$10^8 \div 10^{10} = 10^{8-10} = 10^{-2}$$

Answer:

$$2.5 \times 10^{-2}$$

EXAMPLE

Divide: $\dfrac{1.62 \times 10^{-4}}{7.94 \times 10^7}$

Solution

$$1.62 \div 7.94 = 0.204$$

Subtract exponents:

$$10^{-4} \div 10^7 = 10^{-4-7} = 10^{-11}$$

Answer:

$$0.204 \times 10^{-11} = 2.04 \times 10^{-12}$$

Scientific calculators do these calculations automatically. All that is necessary is to enter the first number, press $+$, $-$, $\times$, or $\div$, enter the second number, and press $=$. (The method for entering numbers of this form varies; consult the instructions that come with the calculator.) Many scientific calculators also have a key that will automatically convert a number such as 0.00047 to its scientific notation form (4.7×10^{-4}), and vice versa. For problems relating to exponential notation, see Chapter 1, Problems 1.16 through 1.23.

Significant Figures

If you measure the volume of a liquid in a graduated cylinder (Figure 1.6), you might find that it is 36 mL, to the nearest milliliter, but you cannot tell if it is 36.2, or 35.6, or 36.0 mL because this measuring instrument does not give the last digit with any certainty. A buret (Figure 1.6) gives more digits, and if you use one you should be able to say, for instance, that the volume is 36.3 mL and not 36.4 mL. But even with a buret, you could not say whether the volume is 36.32 or 36.33 mL. For that, you would need an instrument that gives still more digits. This example should show you that *no measured number can ever be known exactly.* No matter how good the measuring instrument, there is always a limit to the number of digits it can measure with certainty.

Scientists have found it useful to establish a method for telling to what degree of certainty any measured number is known. The method is very simple; it consists merely of writing down all the digits that are certain and not writing down any that are not certain. We define the number of **significant figures** in a measured number as **the number of digits that are known with certainty.**

What do we mean by this? Assume that you are weighing a small object on a laboratory balance that can weigh to the nearest 0.1 g, and you find that the object weighs 16 g. Because the balance weighs to the nearest 0.1 g, you can be sure that the object does not weigh 16.1 g or 15.9 g. In this case, you would write the weight as 16.0 g. To a scientist, there is a difference between 16 g and 16.0 g. Writing 16 g says that you don't know the digit after the 6. Writing 16.0 g says that you do know it: It is 0. However, you don't know the digit after that.

There are several rules governing the use of significant figures in reporting measured numbers.

Determining the Number of Significant Figures

All digits written down are significant except two types:

1. Zeros that come before the first nonzero digit are *not* significant.

EXAMPLE

How many significant figures are there in each of these numbers?

Number	Number of Significant Figures
23.742	5
332	3
0.023	2
0.230	3
0.000023	2
3.004	4
0.050008	5

2. Zeros that come after the last nonzero digit are significant if the number is a decimal, but if it is a whole number (no decimal point) they may or may not be significant. We cannot tell without knowing something about the number. This is the ambiguous case.

EXAMPLE

How many significant figures are there in each of these numbers?

Number	Number of Significant Figures
32.0400	6
0.0002300	4
1.02000	6
32500	3, 4, or 5

The ambiguous case is the only flaw in the system. If you know that a certain small business made a profit of $36,000 last year, you can be sure that the 3 and 6 are significant, but what about the rest? It might have been $36,126, or $35,786.53, or maybe even exactly $36,000. We just don't know because it is customary to round off such numbers. On the other hand, if the profit were reported as $36,000.00, then all seven digits are significant.

In science, we often get around the ambiguous case by using exponential notation (Section 1.3). Suppose a measurement comes out to be 2500 g. If we made the measurement, we of course know whether the two zeros are significant, but we need to tell others. If these digits are *not* significant, we write our number as 2.5×10^3. If one zero is significant, we write 2.50×10^3. If both zeros are significant, we write 2.500×10^3. Since we now have a decimal point, all the digits shown are significant.

Multiplying and Dividing

The rule in multiplication and division is that the final answer should have the *same* number of significant figures as there are in the number with the *fewest* significant figures.

This means that most of those beautiful digits on your calculator display are usually meaningless (insignificant).

EXAMPLE

Do the following multiplications and divisions:
(a) 3.6×4.27
(b) 0.004×217.38
(c) $\dfrac{42.1}{3.695}$
(d) $\dfrac{0.30652 \times 138}{2.1}$

Solution
(a) 15 (3.6 has two significant figures)
(b) 0.9 (0.004 has one significant figure)
(c) 11.4 (42.1 has three significant figures)
(d) 2.0×10^1 (2.1 has two significant figures)

Adding and Subtracting

In addition and subtraction, the rule is completely different. The number of significant figures in each number doesn't matter. The answer is given to the same number of *decimal places* as the term with the fewest decimal places.

EXAMPLE

Add or subtract:

(a) 320.084
      80.47
     200.23
      20.0
     ───────
     620.8

(b) 61.4532
     13.7
     22
      0.003
     ───────
     97

(c)   14.26
     − 1.05041
     ───────
      13.21

Solution
In each case, we add or subtract in the normal way but then round off so that the only digits that appear in the answer are those in the columns in which every digit is significant.

Rounding Off

When we have too many significant figures in our answer, it is necessary to round off. In this book we have used the rule that *if the first digit dropped is 5, 6, 7, 8, or 9, we raise the last digit kept to the next number; otherwise, we do not.*

EXAMPLE

In each case, drop the last two digits:

(a) 33.679   (b) 2.4715   (c) 1.1145   (d) 0.001309   (e) 3.52

Solution
(a) 33.679   = 33.7

(b) 2.4715   = 2.47

(c) 1.1145   = 1.11

(d) 0.001309 = 0.0013

(e) 3.52   = 4

Counted or Defined Numbers

All the preceding rules apply to *measured* numbers and **not** to any numbers that are *counted* or *defined*. Counted and defined numbers are known exactly. For example, a triangle has 3 sides, not 3.1 or 2.9. Here, we treat the number 3 as if it has an infinite number of zeros following the decimal point.

EXAMPLE

Multiply 53.692 (a measured number) × 6 (a counted number).

Solution

$$322.152$$

Because 6 is a counted number, we know it exactly, and 53.692 is the number with the fewest significant figures. All we really are doing is adding 53.692 six times.

For problems relating to significant figures, see Chapter 1, Problems 1.24 to 1.29.

ANSWERS *

Chapter 1 Matter, Energy, and Measurement

1.A The major contributor is the advance in medical sciences, among them chemistry. Advances in both the diagnosis and treatment of diseases, and especially in drug development, contribute to longer life expectancy.

1.B It is at best a hypothesis and at worst a hoax. It is not a fact, nor a theory proven by facts. It is a hypothesis if there is some evidence that carrots may increase the life expectancy of diabetic patients. It is definitely not a cure. It may be a hoax without any scientific foundation.

1.C You will arrive within an hour. The difference between kilometers and miles is much greater (1.6 km equals 1 mile) than the difference between 76 and 80.

1.D The motion of the wheels of the car generates kinetic energy; this is stored in your battery as potential energy.

1.1 (a) 147°F   (b) 8.3°C
1.2 109 kg
1.3 13.8 km
1.4 743 mph
1.5 78.5 g
1.6 2.43 g/mL
1.7 1.016 g/mL
1.8 48×10^3 cal = 48 kcal
1.9 46°C
1.10 0.0430 cal/g·deg
1.11 See Section 1.1.
1.13 A fact is a statement based on experience. A hypothesis is a statement offered to explain the facts.
1.15 Chemical changes: (a) (e) (f) (g)
Physical changes: (b) (c) (d)
1.17 (a) 403 000   (b) 3200   (c) 0.0000713
(d) 0.000000000555
1.19 (a) 2.00×10^{18} (b) 1.37×10^5
(c) 2.36×10^{-10} (d) 8.8×10^1
(e) 1.56×10^{-8}
1.21 (a) 7.74×10^3 (b) 8.81×10^{-2} (c) 1.3×10^2
*1.23 3.25×10^{-5}

1.25 (a) 3   (b) 2   (c) 1   (d) 4   (e) 5
1.27 (a) 25 000   (b) 4.1   (c) 15.5
1.29 (a) 10963.1   (b) 244   (c) 172.34
1.31 (a) 1000 g   (b) 0.001 g
1.33 (a) 100 cm   (b) 230 mL   (c) 75 kg   (d) 15 mL
(e) 50 mg   (f) 100 mm   (g) 40 g
1.35 (a) 160°C, 433 K   (b) 100°C, 373 K
(c) −18°C, 255 K   (d) −157°C, 116 K
1.37 (a) 93.9 lb   (b) 735 g   (c) 86 cm
(d) 23.1 miles   (e) 10.3 L   (f) 2.2 oz (avdp)
(g) 31.80 L   (h) 11.5 gal   (i) 1.8 km
(j) 1.18 fl oz
1.39 A student weighing 127 lb weighs 57.6 kg. A student 5 feet 5 inches tall (65.0 in) is 1.65 m tall.
1.41 10.6 km/L
1.43 bottom: manganese; top: sodium acetate; middle: calcium chloride
1.45 1.023 g/cc
1.47 water
*1.49 They are equally good buys. Butter A costs $5.00 for 1135 g, or 0.44 cents/g; butter B costs $4.00 for 908 g, or 0.44 cents/g.
1.51 151 kcal
*1.53 0.34 cal/g·C°
1.55 334 mg
1.57 The body shivers. Further temperature lowering results in unconsciousness and then death.
1.59 Methanol, because its higher specific heat allows it to retain the heat longer.
1.61 0.732
1.63 No. mL is a unit of volume, g a unit of mass.
1.65 the European car
1.67 kinetic energy
1.69 The largest is 41 g. The smallest is 4.1×10^{-8} kg.
1.71 10.9 h
1.73 5.9 lb/gal
1.75 (a) 1.57 g/mL   (b) 1.25 g/mL
1.77 two

Chapter 2 Atoms

2.A The last element in the Periodic Table is Uub, with atomic number 112. An element with 6 more protons will fall in the column 8A (18). Its properties will be that of noble gases, similar to radon, Rn.

2.B Si has four valence electrons. Two of them are in the $3s$ subshell; the other two are in the p_x and p_y subshells, respectively. The answer, then, is three.

2.C It belongs to the Group 2A.

2.D Xe
2.1 (a) $NaClO_3$ (b) AlF_3
2.2 (a) 31 amu and (b) 222 amu
2.3 (a) phosphorus; $^{31}_{15}P$ (b) radon; $^{222}_{86}Rn$
2.4 (a) 80 and 82   (b) 80 and 82   (c) 200 and 202
    (d) mercury-200; $^{200}_{80}Hg$; lead-202; $^{202}_{82}Pb$
2.5 (a) 72 and (b) 78; $^{125}_{53}I$ and $^{131}_{53}I$
2.6 lithium-7
2.7 (a) $[Ne]3s^23p$ (b) $[Ne]$
2.9 mixture
2.11 (a) K_2O (b) Na_3PO_4 (c) $LiNO_2$
2.13 the law of conservation of mass
2.15 (a) nucleus   (b) outside nucleus   (c) nucleus
2.17 (a) titanium ion   (b) osmium atom
    (c) selenium ion   (d) plutonium atom
2.19 the same, since the number of protons would not change
2.21 All contain 86 protons. They have 124, 132, and 136 neutrons, respectively.
2.23 (a) 10p, 12n   (b) 46p, 58n   (c) 17p, 18n
    (d) 52p, 76n   (e) 3p, 4n   (f) 92p, 146 n
2.25 tin-120, tin-121, tin-124
2.27 (a) F^- (b) K^+ (c) Ar   (d) H^+
    (e) Na^+ (f) S^{2-} (g) Br^- (h) Mn^{2+}
*2.29 6.926 amu; heavier
2.31 62.5% ^{121}Sb and 37.5% ^{123}Sb
2.33 (a) (c) (d) (f)
2.35 metals: (c) (e) (i); nonmetals: (a) (g) (j) (k);
    metalloids: (b) (d) (f) (h)
2.37 (a) $1s^22s^1$ (b) $1s^22s^22p^6$ (c) $1s^22s^2$
    (d) $1s^22s^22p^2$ (e) $1s^2$ (f) $1s^22s^22p^63s^23p^64s^23d^5$

(g) $1s^22s^22p^63s^23p^5$ (h) $1s^22s^22p^63s^23p^3$
2.39 (a) Mg: $1s^22s^22p^63s^2$; Mg^{2+}: $1s^22s^22p^6$
    (b) Na: $1s^22s^22p^63s^1$; Na^+: same as Mg^{2+}
    (c) Al: $1s^22s^22p^63s^23p^1$; Al^{3+}: same as Mg^{2+}
2.41 (a) $1s^22s^22p^63s^23p^63d^{10}4s^24p^64d^{15}5s^2$
    (b) $1s^22s^22p^63s^23p^63d^{10}4s^24p^64d^{10}5s^25p^66s^2$
    (c) $1s^22s^22p^63s^23p^63d^{10}4s^24p^64d^{10}5s^25p^5$
2.43 6s: 2; 6p: 6; 6d: 10; 6f: 14; 32 altogether
2.45 The properties are similar because all of them have the same outer-shell configurations. They are not identical because each has a different number of filled inner shells.
2.47 Calcium and phosphorus are the main components of the bones, and sulfur is an essential element in proteins.
2.49 heavier
2.51 copper and tin
2.53 (a) O, Si   (b) O, C
2.55 metals: (a) (c) (e); nonmetals: (b) (d)
2.57 (a) 17p, 17e, 18n   (b) 42p, 42e, 56n
    (c) 20p, 18e, 24n   (d) 53p, 54e, 74n
    (e) 64p, 64e, 94n   (f) 83p, 80e, 129n
2.59 more
2.61 (d)
*2.63 47.9 amu
2.65 (a) 4   (b) 7   (c) 5   (d) 1   (e) 2
    (f) 2   (g) 8   (h) 4   (i) 6   (j) 6
*2.67 (a) 18   (b) fifth
2.69 (a) electron: -1; proton: $+1$; neutron: no charge
    (b) electron 1/1835, proton 1, neutron 1
*2.71 76.5% rubidium-85 and 23.5% rubidium-87

Chapter 3 Chemical Bonds

3.A Electronegativity increases from left to right in the Periodic Table as the number of positive charges in the nucleus increases. It decreases from top to bottom in a column as the principal energy level in which the valence electrons are found increases

*3.B Predict a shape like that created by putting together sides of two triangular-based pyramids. This shape is called trigonal bipyramidal.

*3.C Predict a shape like that created by putting together sides of two square-based pyramids. This shape is called octahedral because it has eight sides.

3.D (a) $SiCl_4$ (b) PH_3 (c) H_2S
3.1 By losing two electrons, Mg acquires a complete octet. By gaining two electrons, sulfur acquires a complete octet.
    (a) Mg(12 electrons): $1s^22s^22p^63s^2 \longrightarrow$
            Mg^{2+} (10 electrons): $1s^22s^22p^6$

    (b) S (16 electrons): $1s^22s^22p^63s^23p^4 \longrightarrow$
            S^{2-} (18 electrons): $1s^22s^22p^63s^23p^6$
3.2 Each pair of elements is in the same column of the Periodic Table, and electronegativity increases from bottom to top within a column. Therefore,
    (a) Li > K   (b) N > P   (c) C > Si
3.3 (a) KCl   (b) CaF_2 (c) Fe_2O_3
3.4 (a) S—H $(2.5 - 2.1 = 0.4)$ nonpolar covalent
    (b) P—H $(2.1 - 2.1 = 0.0)$ nonpolar covalent
    (c) C—F $(4.0 - 2.5 = 1.5)$ polar covalent
    (d) C—Cl $(3.0 - 2.5 = 0.5)$ polar covalent

3.5 (a) $\overset{\delta+}{C}—\overset{\delta-}{N}$ (b) $\overset{\delta+}{N}—\overset{\delta-}{O}$ (c) $\overset{\delta+}{C}—\overset{\delta-}{Cl}$

3.6 (a)   (b)   (c)

3.7 Given are three-dimensional structures showing all unshared electron pairs.

(a)   (b)

(c)

$$109.5° \quad \overset{\cdot\cdot}{\underset{}{O}} \quad 109.5°$$

H—Ö—C—Ö—H

120°

(d) F—B—F with F below
(e) H—C=O with H above and H below
(f) Cl—C—C—C—Cl with Cl Cl Cl above and Cl Cl Cl below

3.8 (a) CH_2Cl_2 contains two polar bonds and is a polar molecule.
(b) HCN contains one polar bond and is a polar molecule.
(c) C_2H_6 contains no polar bonds and is not a polar molecule.

3.9 (a) Magnesium oxide   (b) Barium iodide
(c) Potassium chloride

3.10 (a) $MgCl_2$ (b) Al_2O_3 (c) LiI

3.11 (a) Iron(II) oxide
(b) Mercury(II) chloride

3.12 (a) Copper(I) bromide or cuprous bromide
(b) Iron(III) bromide or ferric bromide

3.13 (a) Potassium hydrogen phosphate
(b) Aluminum sulfate
(c) Iron(II) carbonate or ferrous carbonate

3.14 (a) Nitrogen dioxide   (b) Phosphorus tribromide
(c) Sulfur dichloride

3.15 The patient is given calcium ion, Ca^{2+}, in the form of a salt such as calcium carbonate, calcium gluconate, calcium lactate, or calcium dihydrogen phosphate.

3.17 (a) Mg^{2+} (b) F^- (c) Al^{3+} (d) S^{2-}
(e) K^+ (f) Br^-

3.19 Li^- has two electrons in its valence shell, which means that it does not have an electron configuration like that of helium, the noble gas nearest it in atomic number.

*3.21 No. Copper is a transition metal, almost all of which form two or more different ions, each with a different positive charge.

3.23 The 2 means that in the solid crystalline form of Na_2SO_4, there are two Na^+ ions for each SO_4^{2-} ion.

3.25 (a) NaBr   (b) Na_2O (c) $AlCl_3$
(d) $BaCl_2$ (e) MgO

3.27 (a) Ca^{2+} and S^{2-} (b) Mg^{2+} and F^-
(c) Cs^+ and O^{2-} (d) Sc^{3+} and Cl^-
(e) Al^{3+} and S^{2-}

3.29 (a) N forms three bonds and has one unshared pair of electrons.
(b) F forms one bond and has three unshared pairs of electrons.
(c) C forms four bonds and has no unshared pairs of electrons.
(d) Br forms one bond and has three unshared pairs of electrons.
(e) O forms two bonds and has two unshared pairs of electrons.

*3.31 In these structural formulas, only atoms and bonds are shown. Unshared pairs of electrons are not shown.

(a) H—C—H with H above and H below
(b) H—C≡C—H
(c) C=C with H H on left and H H on right

3.33 A molecular formula shows how many of each atom are present in the simplest unit of a compound. A structural formula shows how atoms are bonded in a compound. A Lewis structure shows how atoms are bonded in a compound, as well as the location of all unshared pairs of electrons.

3.35 (a) NH_3 has 8   (b) C_3H_6 has 18
(c) $C_2H_4O_2$ has 24   (d) C_2H_6O has 20
(e) CCl_4 has 32   (f) NO_3^- has 24
(g) PO_4^{3-} has 32   (h) O_2 has 12

*3.37 Each Lewis structure shows all valence electrons and the location of all charges.

(a) H—C≡N: (b) H—Ö—N̈=Ö
(c) H—Ö—S—Ö—H with O above and O below
(d) H—C—C—Ö: with H above and H below, O above
(e) :Ö—C—Ö: with O above

3.39 All three molecules contain a triple bond.
(a) H—C≡C—H   (b) H—C≡N: (c) :N≡N:

*3.41 Given is the shape of the molecule and approximate bond angles about its central atom.
(a) tetrahedral, 109.5°   (b) pyramidal, 109.5°
(c) tetrahedral, 109.5°   (d) bent, 120°
(e) trigonal planar, 120°   (f) tetrahedral, 109.5°
(g) bent, 120°   (h) pyramidal, 109.5°
(i) tetrahedral, 109.5°   (j) pyramidal, 109.5°

3.43 Given are differences in electronegativity between the bonded atoms. C—O is the most polar bond and C—C is the least polar (it is nonpolar).
(a) C—N $3.0 - 2.5 = 0.5$ (b) C—C $2.5 - 2.5 = 0.0$
(c) C—O $3.5 - 2.5 = 1.0$

*3.45 The differences are their shapes. Because CO_2 is a linear molecule, it is nonpolar. Because SO_2 is a bent molecule, it is polar.
(a) :Ö=C=Ö: (b) S with O O below (bent)

*3.47 For parts in which a bond will form, use differences in electronegativity to predict the polarity of the bond.
(a) Cl—I $3.0 - 2.5 = 0.5$; a polar covalent bond.
(b) No bond. Each loses electrons; neither gains electrons to become a negative ion.
(c) K—O $3.5 - 0.8 = 2.7$; an ionic bond.
(d) Al—N $3.0 - 1.5 = 1.5$; a polar covalent bond.
(e) No bond. Each loses electrons; neither gains electrons to become a negative ion.
(f) No bond. Neon is a noble gas, and neither gains, loses, or shares electrons.
(g) No bond. Each loses electrons; neither gains electrons to become a negative ion.
(h) No bond. Helium is a noble gas, and neither gains, loses, or shares electrons.

3.49 (a) Sulfite ion   (b) Nitrate ion
(c) Carbonate ion   (d) Ammonium ion
(e) Hydrogen phosphate ion

3.51 (a) Sodium fluoride   (b) Magnesium sulfide
(c) Aluminum oxide   (d) Barium chloride
(e) Calcium hydrogen sulfite   (f) Potassium iodide
(g) Strontium phosphate   (h) Iron(II) hydroxide or
ferrous hydroxide   (i) Sodium dihydrogen
phosphate   (j) Lead(II) acetate or plumbous
acetate   (k) Barium hydride   (l) Ammonium
hydrogen phosphate

3.53 Titanium dioxide, TiO_2.

3.55 Lead(IV) oxide, PbO_2, and lead(IV) carbonate,
$Pb(CO_3)_2$.

3.57 Fe^{2+}

3.59 (a) $CaSO_3$ (b) $Ca(HSO_3)_2$ (c) $Ca(OH)_2$
(d) $CaHPO_4$

3.61 To supply iodide ion for the synthesis of the thyroid
hormones thyroxin and triiodothyronine.

3.63 as a topical anti-infective

3.65 calcium dihydrogen phosphate, calcium phosphate,
and calcium carbonate

3.67 O^{6+} has a charge too concentrated for a small ion.

3.69 In each compound, the carbon in question must have
four bonds. All unshared electron pairs are shown in
these structures.

3.71 An argon atom has a complete octet so it (a) does not
lose or gain electrons and (b) has no need to share
electrons.

3.73 (a) S = 12   (b) O = 8   (c) P = 10
(d) N = 8   (e) C = 8

3.75 (a) CO   (b) BrCl   (c) PCl_5 (d) $SiBr_4$
(e) N_2O_3 (f) SF_6

Chapter 4 Chemical Reactions

4.A CH_4O; 32 amu

4.B (a)

4.C the carbon

4.1 (a) 180.0 amu   (b) 601.9 amu

4.2 83 moles of water

4.3 222 g

4.4 15 moles of C atoms, 30 moles of H atoms, 15 moles of
O atoms

4.5 4.9×10^{-4} moles of Cu^+ ions

4.6 7.86×10^{24} molecules of water

4.7 $P_4 + 6Cl_2 \longrightarrow 4PCl_3$

4.8 $2C_6H_{14} + 19O_2 \longrightarrow 14H_2O + 12CO_2$

4.9 $3K_2C_2O_4 + Ca_3(AsO_4)_2 \longrightarrow 2K_3AsO_4 + 3CaC_2O_4$

4.10 8.3 moles of CH_4

4.11 1.25×10^3 g of $C_2H_2Br_4$

4.12 (a) C is the limiting reagent and H_2 is in excess.
(b) 8.0 g CH_4 is produced

4.13 80.7%

4.14 $Cu^{2+} + S^{2-} \longrightarrow CuS$

4.15 (a) Cr is oxidized; Ni^{2+} is reduced. Ni^{2+} is the oxidizing agent; Cr the reducing agent.
(b) H_2 is oxidized; H_2CO is reduced. H_2CO is the oxidizing agent; H_2 the reducing agent.

4.17 (a) 74.6 amu   (b) 164.0 amu   (c) 89.8 amu
(d) 342.0 amu   (e) 210.2 amu   (f) 77.0 amu
(g) 342.3 amu   (h) 96.0 amu   (i) 354.5 amu

4.19 (a) 81.4 g NO_2 (b) 50 g 2-propanol
(c) 1.33×10^3 g WCl_5 (d) 62.6 g galactose
(e) 8.6 g vitamin C

4.21 (a) 6.56   (b) 24.96   (c) 0.86 moles

4.23 (a) 5.0×10^{-21} g   (b) 1.5×10^{-19} g
(c) 2.5×10^{-16} g   (d) 100 g

4.25 (a) 1.7×10^{24} molecules of TNT
(b) 1.67×10^{21} molecules of water

4.27 1.1×10^{-19} g

4.29 (a) $HI + NaOH \longrightarrow NaI + H_2O$

(b) $Ba(NO_3)_2 + H_2S \longrightarrow BaS + 2HNO_3$
(c) $CH_4 + 2O_2 \longrightarrow CO_2 + 2H_2O$
(d) $2C_4H_{10} + 13O_2 \longrightarrow 8CO_2 + 10H_2O$
(e) $2Fe + 3CO_2 \longrightarrow Fe_2O_3 + 3CO$
(f) $2Al + 6HBr \longrightarrow 2AlBr_3 + 3H_2$
(g) $P_4 + 5O_2 \longrightarrow P_4O_{10}$

4.31 (a) 2/3 mole N_2 (b) 2/3 mole N_2O_3
(c) 12 moles O_2

4.33 719 g Br_2

4.35 879 g O_2

4.37 4660 g CO_2

4.39 111 g of aspirin

*4.41 (a) H_2 (b) 14.3 g N_2 (c) 18.8 g NH_3

*4.43 94.0%

4.45 (a) Na^+, Cl^-
(b) Sr^{2+} (aq) + CO_3^{2-}(aq) $\longrightarrow SrCO_3$(s)

4.47 Pb^{2+}(aq) + $2Cl^-$(aq) $\longrightarrow PbCl_2$(s)

4.49 $2H^+$(aq) + SO_3^{2-}(aq) $\longrightarrow SO_2$(g) + H_2O(l)

4.51 Soluble: (a) (b) (d) (e)
Insoluble: (c) (f)

4.53 (a) $Ca_3(PO_4)_2$ (c) $BaCO_3$ (d) $Fe(OH)_2$
(f) Sb_2S_3 (g) $PbSO_4$
No precipitate will form in (b) and (e).

4.55 No, one gains electrons and the other loses electrons.
Electrons are not destroyed, but transferred.

4.57 (a) C_7H_{12} gets oxidized; O_2 gets reduced.
(b) O_2 is the oxidizing agent; C_7H_{12} the reducing agent.

4.59 exothermic: (b) (c) (e); endothermic: (a) (d)

4.61 160 000 cal

4.63 Fluoride ion replaces the hydroxide ion in the tooth
enamel, thereby making it more resistant to acids.

4.65 Fe(s) is the anode, and Zn^{2+} is the cathode.

4.67 In those parts of the film where the light hits, AgBr is
reduced to silver. These parts turn black. In the parts
where the light does not hit, no silver forms. This creates the image.

4.69 It was oxidized.

4.71 More than 90 percent of the energy needed to operate our cars, trucks, airplanes, ships, and farm and factory machinery and to light, heat, and cool our buildings comes from the combustion of oil, coal, and natural gas.

4.73 0.00271 mole aspirin; 1.63×10^{21} molecules aspirin

*4.75 (a) 5.68 moles CO_2
      (b) 7.00 moles CO_2

*4.77 (a) $AlCl_3$ is in excess   (b) 82.5%

*4.79 9.961 g

*4.81 4×10^{10} molecules

Chapter 5 Gases, Liquids, and Solids

5.A (c)

5.B the intramolecular covalent bonds

5.C 180 cal

5.D by reducing the pressure to 0.001 atm

5.1 0.4 atm

5.2 16.4 atm

5.3 0.053 atm

5.4 4.84 atm

5.5 0.422 mole

5.6 9.91 g

5.7 0.107 atm

5.8 (a) yes   (b) no

5.9 83.3 g

5.10 229 cal

5.11 The vapor first will condense to become liquid water and eventually solid ice.

5.13

*5.15 The barometric pressure drops because the air column above us weighs less than before.

5.17 1.5 atm

5.19 1.87 L

5.21 475°C

5.23 836 mL

5.25 61.0 mL

5.27 1.87 atm

5.29 1.17 atm

5.31 4 amu

5.33 8040 mm Hg

5.35 93.2 mL

5.37 1.15×10^{22} oxygen molecules

5.39 MW = 52

5.41 the same

5.43 Yes; the partial pressures added up to 785 mm Hg. There must be another gas with 5 mm Hg partial pressure.

5.45 (c)

5.47 Yes; the —S=O····H—O—H form hydrogen bonds.

5.49 Yes, a small dipole, as in NO, may provide weaker interaction forces than London dispersion forces acting over large surfaces.

5.51 See Section 5.9.

5.53 CH_4, because it has only London dispersion forces acting between them, while the stronger H-bonds act between water molecules.

*5.55 When the temperature decreases, the kinetic energy decreases; it is not lost but converted to potential energy or energy of interaction. More interaction means more attraction and, hence, condensation.

5.57 10^7 or ten million

5.59

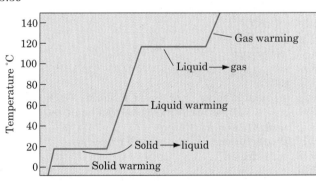

5.61 The gas at 100°C. At the lower temperature the molecules move slower, so there is more order, and hence a lower entropy than at the higher temperature.

5.63 First the teeth become loose, and then they fall out.

5.65 The volume of the chest cavity increases, lowering the pressure. Air flows in from outside.

5.67 highest in venous blood, lowest in arterial blood

5.69 The volume of water increases when it freezes. This expansion will crack the bottle.

5.71 112 mL

5.73 CO

*5.75 0.34 mole

*5.77 MW = 91.9

5.79 pentane

*5.81 0.633 g/L

5.83 Water molecules are held together with H-bonds; H_2S molecules possess only dipole-dipole interactions, which are much weaker.

Chapter 6 Solutions and Colloids

6.A The solubility of a gas increases as the temperature is lowered. Thus, all the CO_2 put in the bottle must have been dissolved in water in the refrigerator, while at room temperature the excess CO_2 escaped when the bottle was opened.

6.B In cooling, the protein molecules must have been aggregated to large colloidal size particles, which scattered light, causing a turbid appearance.

6.C The acetic acid did not dissociate to any extent, while the KF dissociated and yielded two ions.

6.D One may not observe any osmotic pressure with a dialysis membrane because small molecules such as glucose can easily cross the membrane.

6.1 To 11 g KBr add enough water to make 250 mL of solution.

6.2 1.9% w/v

6.3 To 158.2 g KCl (1.06 moles × 2) add enough water to make 2.0 L of solution.

6.4 0.013 M

6.5 185 mL

6.6 15 g

6.7 Take 16.5 mL of concentrated HCl solution and add enough water to make 300 mL solution.

6.8 To 0.13 mL of the concentrated solution add enough water to make 20.0 mL solution.

6.9 −12.5°C

6.10 (c)

6.11 0.80 osmol

6.12 (a)

6.13 water

6.15 (a) solid in solid   (b) solid in liquid   (c) gas in gas   (d) gas (carbon dioxide) and liquid (ethyl alcohol) in liquid (water)

6.17 True solutions. They mix in all proportions.

6.19 unsaturated

6.21 (c)

6.23 by cooling it without stirring

6.25 The warm water contains less oxygen than cold water. The fish die because of lack of oxygen.

6.27 (a)

6.29 (a) 76 mL ethyl alcohol   (b) 7.8 mL ethyl acetate   (c) 132 mL benzene

6.31 (a) 4.15% w/v casein   (b) 0.030% w/v vitamin C   (c) 1.75% w/v sucrose

6.33 In each case add enough water to the given weight of solute to make the total volume specified:

(a) 19.5 g of NH_4Br—175 mL of solution
(b) 167 g NaI—1.35 L of solution
(c) 2.4 g C_2H_5OH—330 mL of solution

6.35 0.17 M

6.37 0.405 M

6.39 0.568 M

*6.41 25.0 mL

6.43 To 2.08 mL of the stock solution add enough water to make 250 mL of solution.

6.45 (a) 3.85×10^4 ppm captopril   (b) 6.77×10^4 ppm Mg^{2+} and (c) 831 ppm Ca^{2+}

*6.47 No, the concentration is 0.01 ppb.

6.49 See Section 6.6

6.51 (b)

6.53 soluble: (b), (c), (d)

6.55 homogeneous: (a); heterogeneous: (b), (d); colloid: (c), (e), (f)

6.57 Water molecules in the solvation layer of a colloidal particle move together with the particle; water molecules in the bulk move randomly.

6.59 (a) −3.72°C   (b) −5.58°C   (c) −5.58°C   (d) −7.44°C

6.61 344 g methanol

6.63 5.4 moles

6.65 (a) B   (b) B   (c) A   (d) B   (e) A   (f) No rise. In each case, the side with the higher osmolarity rises.

6.67 (a) 1.2   (b) 2.5   (c) 8.4   (d) 0.036 osmol

*6.69 (b)

6.71 HNO_2, HNO_3

*6.73 1.768 g CO_2 lost

6.75 Nitrogen under high pressure creates a feeling similar to alcohol intoxification.

6.77 $CaCO_3$

6.79 the pressure that is larger than the osmotic pressure of the sea water

6.81 (a) 26.3% w/v   (b) 25.0% w/w

6.83 4 ppm

*6.85 Methanol is more efficient. It has a lower molecular weight; hence there are more moles in the same mass.

*6.87 $CO_2(g) + H_2O(l) \longrightarrow H_2CO_3(aq)$,
$SO_2(g) + H_2O(l) \longrightarrow H_2SO_3(aq)$

6.89 Take 38.6 mL of the concentrated solution, add enough water to bring to 300-mL volume.

6.91 0.006 g

6.93 354 ppm; 7.09 kg

Chapter 7 Reaction Rates and Equilibrium

7.A One must increase the pressure of the two gases entering the reaction chamber. The higher the pressure, the higher the velocity of the gases; this will provide the maximum effective collisions.

7.B The amount of product will be small, but the reaction will be fast.

7.C If one can find a way to remove the product continuously from the reaction vessel, then according to Le Chatelier's principle, the equilibrium will shift to the product side and more products will form.

7.D the use of a catalyst

7.1 0.022 L O_2/min

7.2 -3.6×10^{-3} moles H_2O_2/L min

7.3 $K = \dfrac{[H_2SO_4]}{[SO_3][H_2O]}$

7.4 $K = \dfrac{[N_2][H_2]^3}{[NH_3]^2}$

7.5 $K = 0.602$

7.6 The equilibrium shifts to the left.

7.7 The equilibrium shifts to the right because after saturation, bubbles form and the oxygen is removed from the reaction mixture.

7.8 endothermic

7.9 1.82×10^{-2} mole/L·min

7.11 Two are broken, and two are formed.

7.13 The activation energy is low because the breaking of covalent bonds is not required.

7.15 The spoiling is caused by chemical reactions initiated by bacteria. These reactions are much faster at room temperature than at refrigerator temperature.

7.17 50°C

7.19 (1) increase the concentrations (2) increase the temperature (3) add a catalyst

7.21 It provides an alternate pathway of lower activation energy.

7.23 digesting a piece of candy; explosion of TNT; rusting of iron; addition of Na or K metal to water

7.25 (a) $K = [H_2O]^2[O_2]/[H_2O_2]^2$
      (b) $K = [N_2O_4]^2[O_2]/[N_2O_5]^2$
      (c) $K = [C_6H_{12}O_6][O_2]^6/[H_2O]^6[CO_2]^6$

7.27 $K = 1.5$

7.29 $K = 9.9 \times 10^{-2}\, M$

7.31 products: (a), (b), (c); reactants; (d), (e)

7.33 (a) No, the speed of reaction depends on the energy of activation, not on the exothermic character.
      (b) Yes, endothermic reactions always have high energies of activation (Fig. 7.5) and so must be very slow.

7.35 (a) right (b) left (c) left (d) left
      (e) no shift

7.37 It shifts to the left.

*7.39 (a) no change (b) no change
      (c) K smaller

7.41 The higher temperature causes chemical reactions in the body to go faster. If the reactions go too fast, it is dangerous to the health of the patient. At low temperature, the reactions are too slow.

7.43 The capsule, because the beads have much larger surface area than the solid pill.

7.45 because the reaction is too slow at low temperature

*7.47

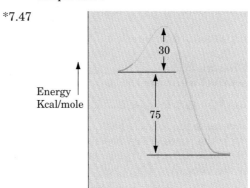

*7.49 -0.37/h

7.51 -2.0×10^{-3} mole of N_2O_4 per L per sec

*7.53 $0.4\, M$

7.55 $0.8\, M$

*7.57 A, because the orientation does not matter, but rod-like molecules need special orientations in order for effective collisions to occur.

*7.59 0.030 mole/L per min

Chapter 8 Acids and Bases

8.A the dissociation constant, or pK_a

8.B Yes, but it rarely happens. The H^+ ion concentration must be $1.0\, M$. In case of HCl, this means 36% and that all must be completely ionized.

8.C By adding acid or base, thereby changing the ratio of acid and the conjugated base. The buffer capacity can be changed by increasing or decreasing the concentrations of both the acid and its conjugated base.

8.D When we deal with a monoprotic acid.

8.1 (a) and (b) both go from left to right.

8.2 $1 \times 10^{-2}\, M$

8.3 3

8.4 $1 \times 10^{-4}\, M$

8.5 pOH is 4 and pH is 10.

8.5 3

8.6 8.2

8.7 $4 \times 10^{-8}\, M$

8.8 9.3

8.9 (a) 9.25 (b) 4.75

8.10 Use $CH_3CH(OH)COO^-/CH_3CH(OH)COOH$ in a ratio of 2 to 1.

8.11 4.27

8.12 (a) 29.17 (b) 127.93 (c) 36.64

8.13 (a) 1.55 equiv (b) 0.0430 equiv

8.14 $4.6 \times 10^{-5}\, N$

8.15 $0.0836\, N$

8.17 $LiOH(s) \xrightarrow{H_2O} Li^+(aq) + OH^-(aq)$

$(CH_3)_2NH + H_2O \rightleftharpoons (CH_3)_2NH_2^+ + OH^-$

8.19 (a) H_2O (b) NH_3 (c) NH_4^+
      (d) C_6H_5OH (e) HCO_3^- (f) H_2CO_3
      (g) H_3O^+ (h) $H_2PO_4^-$ (i) $CH_3{-}NH_3^+$
      (j) HPO_4^{2-}

8.21 No. The Brønsted-Lowry definitions are simply an extension of the Arrhenius definitions.

8.23 (a) HCOOH (b) HF (c) CH_3COOH

8.25 (a) $CO_3^{2-} + 2H_3O^+ \rightleftharpoons H_2CO_3 + 2H_2O$ or
      $Na_2CO_3 + 2HCl \rightleftharpoons H_2CO_3 + 2NaCl$
      (b) $Mg + 2H_3O^+ \rightleftharpoons Mg^{2+} + H_2 + 2H_2O$ or
      $Mg + 2HCl \rightleftharpoons MgCl_2 + H_2$
      (c) $OH^- + H_3O^+ \rightleftharpoons 2H_2O$ or
      $NaOH + HCl \rightleftharpoons H_2O + NaCl$

(d) $NH_3 + H_3O \rightleftharpoons NH_4^+ + H_2O$ or
$NH_3 + HCl \rightleftharpoons NH_4^+ + Cl^-$
(e) $CH_3NH_2 + H_3O^+ \rightleftharpoons CH_3NH_3^+ + H_2O$ or
$CH_3NH_2 + HCl \rightleftharpoons CH_3NH_3^+ + Cl^-$
(f) $HCO_3^- + H_3O^+ \rightleftharpoons H_2CO_3 + H_2O$ or
$NaHCO_3 + HCl \rightleftharpoons H_2CO_3 + NaCl$

8.27 (a) $10^{-4}\,M$ (b) $10^{-12}\,M$ (c) $10^{-7}\,M$
(d) $10^{-14}\,M$

8.29 pH = (a) 11 (b) 13 (c) 9 (d) 7
pOH = (a) 3 (b) 1 (c) 5 (d) 7

8.31 (a) $1 \times 10^{-12}\,M$ (b) $1 \times 10^{-7}\,M$
(c) $1 \times 10^{-2}\,M$ (d) $1 \times 10^{-13}\,M$
(e) $4 \times 10^{-9}\,M$ (f) $8 \times 10^{-13}\,M$
(g) $3 \times 10^{-4}\,M$; acidic: (c) (g); basic: (a) (d) (e) (f);
neutral: (b)

*8.33 (a) butanoic 1.5×10^{-5}; barbituric 1.0×10^{-5};
lactic 1.4×10^{-4} (b) strongest: lactic acid;
weakest: barbituric acid

8.35 Na^+ does not react with water, but CO_3^{2-} does:

$$CO_3^{2-} + H_2O \rightleftharpoons HCO_3^- + OH^-$$

The OH^- ions make the solution basic.

*8.37 Buffer action is the result of Le Chatelier's principle.
Adding acid or base disturbs an equilibrium, which
shifts in the direction that minimizes the effect of the
addition.

8.39 (a) $H_3O^+ + HPO_4^{2-} \rightleftharpoons H_2PO_4^- + H_2O$
(b) $OH^- + H_2PO_4^- \rightleftharpoons HPO_4^{2-} + H_2O$

8.41 The ability of the buffer to neutralize added amounts
of H_3O^+ or OH^-. The greater the ability, the greater
the capacity.

8.43 6.8
8.45 (a) 7.3 (b) 6.5 (c) 9.2 (d) 9.8
8.47 $0.348\,N$
8.49 $8 \times 10^{-4}\,N$
8.51 (a) $3.0\,N$ (b) $9.0\,N$ (c) $0.34\,N$
(d) $5.4 \times 10^{-3}\,N$
8.53 11.3 mL
8.55 0.00330 equiv
*8.57 To 2.0 g of $NaHCO_3$ add enough water to make a total
volume of 1 L.
8.59 immediate washing with a steady stream of cold
water
8.61 Alka-Seltzer, Brioschi, Chooz, Riopan, Tums
8.63 $NaHCO_3$ and an acid
8.65 (a) 3.6 (b) 8.6 (c) 4.0
*8.67 The $H_3BO_3/H_2BO_3^-$ ratio should be 5.5.
*8.69 The HCl solution is more acidic because its $[H_3O^+]$ is
higher.
8.71 $2 \times 10^{-9}\,M$
8.73 $0.00740\,N$
8.75 The concentrations, $1 \times 10^{-7}\,M$ for H_3O^+ and OH^- are
too low.
8.77 No. HF, being a stronger acid, reacts with water to a
greater extent, producing a higher $[H_3O^+]$, so the pH
of its solution is lower.
*8.79 EW = 172
*8.81 Yes. The ethoxide ion will not react with NH_3, but it
will react with water, since the latter is a stronger
acid than ethyl alcohol (Table 8.2):

$$C_2H_5O^- + H_2O \rightleftharpoons C_2H_5OH + OH^-$$

Chapter 9 Nuclear Chemistry

9.A Eight months have passed.
9.B The radiation is ionizing radiation, which knocks out
electrons when they interact with matter. The ionized
species react and cause great damage to the body.
9.C It is a fusion reaction.
9.D In the first step, alpha emission occurs; in the second
and third steps, beta emissions occur.
9.1 xenon-139
9.2 radium-219
9.3 germanium-74
9.4 0.31 g
9.5 3.3×10^{-3} mCi
9.6 1.5 mL
9.7 Yes; all electromagnetic radiations have the same
speed in vacuum. Gamma rays are one form of electromagnetic radiation.
*9.9 $5.2 \times 10^9\ s^{-1}$
9.11 red
9.13 (a) $^{19}_{9}F$ (b) $^{32}_{15}P$ (c) $^{87}_{37}Rb$
9.15 none
9.17 (a) $^{159}_{64}Gd$ (b) $^{141}_{57}La$ (c) $^{242}_{96}Cm$
9.19 (a) $^{206}_{81}Tl$ (b) $^{234}_{92}U$ (c) $^{170}_{70}Yb$
9.21 $^{29}_{15}P \longrightarrow {}^{29}_{14}Si + {}^{0}_{+1}e$

9.23 $^{214}_{84}P_0 \longrightarrow {}^{210}_{82}Pb + {}^{4}_{2}He$; $^{210}_{82}Pb \longrightarrow {}^{210}_{83}Bi + {}^{0}_{-1}e$
9.25 (a) $^{28}_{14}Si$ (b) $^{143}_{54}Xe$ (c) $^{1}_{0}n$ (d) $^{210}_{84}Po$ (e) $^{13}_{7}N$
9.27 12.5%
9.29 No, the conversion $Ra \longrightarrow Ra^{2+}$ involves electrons
from the electron cloud. Radioactivity is a nuclear
phenomenon.
9.31 15 min
9.33 We cannot see or feel it nor can our senses detect it in
any other way.
9.35 (a)
9.37 (a), (e) the amount of radiation absorbed by tissues
(b), (g) the effective dose absorbed by humans
(c) the effective dose delivered
(d), (f) intensity of radiation
*9.39 The penetrating power of alpha particles is so low that
most of them are stopped by the outer skin. But when
they get into the lung, the membranes offer little resistance so the particles can damage the cells of the lung.
9.41 alpha particles
9.43 iodine-131, because any iodine taken in by the body
goes to the thyroid gland
9.45 The half-life is too short. It would not last long enough
to reach a hospital.

9.47 $_1^1$H

9.49 curium-248

9.51 Neutrons. They are controlled by inserting or removing boron control rods.

9.53 The three rays have different charges: positive, negative, and neutral.

*9.55 one-tenth of a half-life

9.57 It gives off an alpha particle.

9.59 See Box 9E.

9.61 because they are destroyed by collisions with electrons, and electrons are in all matter

9.63 If ingested, iodine concentrates in the thyroid gland where radioactive iodine can cause cancer.

9.65 fluorine-19 and neon-20

9.67 the same

9.69 (a) 82%   (b) 11%   (c) 0.1%

9.71 x-rays

9.73 Neptunium-237. Approximately 25% of the original americium-241 will be around after 1000 years.

9.75 One sievert is 100 rem. This is sufficient to cause radiation sickness but not certain death.

9.77 (a) because a radioactive sample decays, and the decay products are always mixed with the starting isotope   (b) because a neutron in the nucleus is converted to a proton

*9.79 Oxygen-16 is stable because it has an equal number of protons and neutrons. The others are unstable because the numbers are unequal; in this case the less equal, the faster the isotope decays.

Chapter 10 Organic Chemistry

*10.A Living organisms use the substances they take in from their environment to produce organic compounds. They also use them to produce inorganic compounds, as for example the calcium-containing compounds of bones, teeth, and shells.

10.B The characteristic is the presence of functional groups which undergo the same types of chemical reactions in whatever compound they occur.

*10.C Among the textile fibers, think of natural fibers such as cotton, wool, and silk. Think also of synthetic textile fibers such as Nylon, Dacron polyester, and polypropylene.

10.D Carbon which forms four bonds, hydrogen which forms one bond, nitrogen which forms three bonds and has one unshared pair of electrons, and oxygen which forms two bonds and has two unshared pairs of electrons.

10.1 Following are Lewis structures showing all bond angles. Also shown are ball-and-stick models.

(a)

(b)

10.2 Of the four alcohols of molecular formula $C_4H_{10}O$, two are primary, one is secondary, and one is tertiary. Note that for the lower two alcohols, three C—C bonds are made longer to avoid crowding in the structural formula.

10.3 Following are structural formulas for the three secondary (2°) amines of molecular formula $C_4H_{11}N$.

$$CH_3—CH_2—CH_2—NH—CH_3 \qquad CH_3—\overset{\overset{\displaystyle CH_3}{|}}{CH}—NH—CH_3$$

$$CH_3—CH_2—NH—CH_2—CH_3$$

10.4 Following are structural formulas for the three ketones of molecular formula $C_5H_{10}O$.

$$CH_3-CH-\overset{\overset{\displaystyle O}{\|}}{C}-CH_3$$
$$|$$
$$CH_3$$

10.5 Following are structural formulas for the two carboxylic acids of molecular formula $C_4H_8O_2$.

$$CH_3-CH_2-CH_2-\overset{\overset{\displaystyle O}{\|}}{C}-OH \qquad CH_3-\underset{\underset{\displaystyle CH_3}{|}}{CH}-\overset{\overset{\displaystyle O}{\|}}{C}-OH$$

10.7 Under each Lewis structure is its name and the number of its valence electrons.

(a) $H-\ddot{O}-\overset{\overset{\displaystyle \cdot\cdot\ddot{O}\cdot}{\|}}{C}-\ddot{\ddot{O}}\!:\!^-$

Bicarbonate ion (24)

(b) $:\!\ddot{O}-\overset{\overset{\displaystyle \cdot\cdot\ddot{O}\cdot}{\|}}{C}-\ddot{\ddot{O}}\!:\!^-$

Carbonate ion (24)

(c) $H-\underset{\underset{\displaystyle H}{|}}{\overset{\overset{\displaystyle H}{|}}{C}}-\overset{\overset{\displaystyle \cdot\cdot\ddot{O}\cdot}{\|}}{C}-\ddot{O}\!:\!^-$

Acetate ion (24)

(d) $:\!\ddot{\underset{\cdot\cdot}{Cl}}\!:\!^-$

Chloride ion (8)

10.9 (a) Carbon 109.5° and oxygen 109.5°
(b) Carbon 120°   (c) Carbon 180°
(d) Carbon 120° and oxygen 109.5°
(e) Nitrogen 109.5°   (f) Nitrogen 120°

10.11 There is no difference.

10.13 Here are Lewis structures for these four functional groups.

(a) $-\overset{\overset{\displaystyle \cdot\cdot\ddot{O}\cdot}{\|}}{C}-$ (b) $-\overset{\overset{\displaystyle \cdot\cdot\ddot{O}\cdot}{\|}}{C}-\ddot{O}-H$ (c) $-\ddot{O}-H$

(d) $-\overset{\overset{\displaystyle \cdot\cdot}{\ddot{N}}}{\underset{\underset{\displaystyle H}{|}}{}}-H$

*10.15 (a) Incorrect. Nitrogen has four bonds; it should have only three. An uncharged nitrogen atom has only three bonds and one unshared pair of electrons.
(b) Incorrect. The carbon on the right has five bonds; it should have only four.
(c) Correct.
(d) Incorrect. Oxygen has one bond; it should have two. An uncharged oxygen atom has two bonds and two unshared pairs of electrons.
(e) Correct.
(f) Incorrect. The carbon on the right has five bonds; it should have only four.
(g) Correct.
(h) Correct.

10.17 The one tertiary alcohol of molecular formula $C_4H_{10}O$ has this structural formula.

$$CH_3-\underset{\underset{\displaystyle CH_3}{|}}{\overset{\overset{\displaystyle CH_3}{|}}{C}}-OH$$

10.19 The one tertiary amine of molecular formula $C_4H_{11}N$ has this structural formula.

$$CH_3-\underset{}{\overset{\overset{\displaystyle CH_3}{|}}{N}}-CH_2-CH_3$$

10.21 Drawn first are the two aldehydes and then the one ketone of molecular formula $C_4H_{10}O$.

$$CH_3-CH_2-CH_2-\overset{\overset{\displaystyle O}{\|}}{C}-H \qquad CH_3-\underset{\underset{\displaystyle CH_3}{|}}{CH}-\overset{\overset{\displaystyle O}{\|}}{C}-H$$

$$CH_3-\overset{\overset{\displaystyle O}{\|}}{C}-CH_2-CH_3$$

*10.23 Here are the six ketones of molecular formula $C_6H_{12}O$.

$$CH_3-\overset{\overset{\displaystyle O}{\|}}{C}-CH_2-CH_2-CH_2-CH_3$$

$$CH_3-CH_2-\overset{\overset{\displaystyle O}{\|}}{C}-CH_2-CH_2-CH_3$$

$$CH_3-CH_2-\overset{\overset{\displaystyle O}{\|}}{C}-\underset{\underset{\displaystyle CH_3}{|}}{CH}-CH_3$$

$$CH_3-\overset{\overset{\displaystyle O}{\|}}{C}-\underset{\underset{\displaystyle CH_3}{|}}{CH}-CH_2-CH_3$$

$$CH_3-\overset{\overset{\displaystyle O}{\|}}{C}-CH_2-\underset{\underset{\displaystyle CH_3}{|}}{CH}-CH_3$$

$$CH_3-\overset{\overset{\displaystyle O}{\|}}{C}-\underset{\underset{\displaystyle CH_3}{|}}{\overset{\overset{\displaystyle CH_3}{|}}{C}}-CH_3$$

*10.25 Here are structural formulas for each part.
(a) The four primary (1°) amines:

$$CH_3-CH_2-CH_2-CH_2-NH_2 \qquad CH_3-CH_2-\underset{\underset{\displaystyle NH_2}{}}{\overset{\overset{\displaystyle CH_3}{|}}{CH}}$$

$$CH_3-\underset{\underset{\displaystyle CH_3}{|}}{CH}-CH_2-NH_2 \qquad CH_3-\underset{\underset{\displaystyle CH_3}{|}}{\overset{\overset{\displaystyle CH_3}{|}}{C}}-NH_2$$

(b) The three secondary (2°) amines:

$$CH_3-CH_2-CH_2-NH-CH_3 \qquad CH_3-\underset{\underset{\displaystyle CH_3}{|}}{CH}-NH-CH_3$$

$$CH_3-CH_2-NH-CH_2-CH_3$$

(c) The one tertiary (3°) amine:

$$CH_3-CH_2-\underset{\underset{CH_3}{|}}{N}-CH_3$$

10.27 Taxol inhibits the disassembly of microtubules.

10.29 Silicon is surrounded by four regions of electron density. Therefore, predict each C—Si—C bond angle to be 109.5°.

*10.31 Here are the eight aldehydes of molecular formula $C_6H_{12}O$. Note that in two of them, a CH_3 group is turned and written H_3C to avoid crowding.

$$CH_3-CH_2-CH_2-CH_2-CH_2-\overset{\overset{O}{\|}}{C}-H$$

$$CH_3-CH_2-CH_2-\underset{\underset{CH_3}{|}}{CH}-\overset{\overset{O}{\|}}{C}-H$$

$$CH_3-CH_2-\underset{\underset{CH_3}{|}}{CH}-CH_2-\overset{\overset{O}{\|}}{C}-H$$

$$CH_3-\underset{\underset{CH_3}{|}}{CH}-CH_2-CH_2-\overset{\overset{O}{\|}}{C}-H \qquad CH_3-\underset{\underset{H_3C}{|}}{CH}-\underset{\underset{CH_3}{|}}{CH}-\overset{\overset{O}{\|}}{C}-H$$

$$CH_3-CH_2-\underset{\underset{CH_3}{|}}{\overset{\overset{H_3C}{|}}{C}}-\overset{\overset{O}{\|}}{C}-H$$

$$CH_3-\underset{\underset{CH_3}{|}}{\overset{\overset{CH_3}{|}}{C}}-CH_2-\overset{\overset{O}{\|}}{C}-H \qquad CH_3-CH_2-\underset{\underset{CH_2-CH_3}{|}}{CH}-\overset{\overset{O}{\|}}{C}-H$$

10.33 As a rule-of-thumb, if the difference in electronegativity between bonded atoms is less than 0.5 units, the bond is nonpolar; if it is between 0.5 and 1.9, the bond is polar covalent.
(a) C—C (2.5 − 2.5 = 0.0) nonpolar covalent
(b) C≡C (2.5 − 2.5 = 0.0) nonpolar covalent
(c) C—H (2.5 − 2.1 = 0.4) nonpolar covalent
(d) C—O (3.5 − 2.5 = 1.0) polar covalent
(e) O—H (3.5 − 2.1 = 1.4) polar covalent
(f) C—N (3.0 − 2.5 = 0.5) polar covalent
(g) N—H (3.0 − 2.1 = 0.9) polar covalent
(h) N—O (3.5 − 3.0 = 0.5) polar covalent

10.35 Under each formula is given the difference in electronegativity between the atoms of the most polar bond.

(a) O—H (3.5 − 2.1 = 1.4)   (b) N—H (3.0 − 2.1 = 0.9)

(c) C—O (3.5 − 2.5 = 1.0)   (d) C—O (3.5 − 2.5 = 1.0)

(e) N—H (3.0 − 2.1 = 0.9)   (f) O—H (3.5 − 2.1 = 1.4)

Chapter 11 Alkanes and Cycloalkanes

11.A Restricted rotation about a carbon-carbon single bond because of its presence in a ring.

11.B No. There is free rotation about all carbon-carbon bonds in an alkane.

11.C The carbon chain of an alkane is not straight; it is bent with all bond angles being 109.5°.

11.D In general, boiling points of hydrocarbons increase as molecular weight and size increase.

11.1 (a) The longest chain in each structural formula is six carbons with two one-carbon branches. In the formula on the left, the branches in carbons 3 and 4 of the chain. In the formula on the right, they are on carbons 2 and 4 of the chain. These formulas, therefore, represent constitutional isomers.
(b) The longest chain in each structural formula is five carbons with one-carbon branches on carbons 2 and 3. These structural formulas, therefore, represent the same compound.

11.2 There is one constitutional isomer with five carbons in the longest chain, one with four carbons in the longest chain, and one with three carbons in the longest chain.

$$CH_3CH_2CH_2CH_2CH_3 \qquad CH_3\underset{\underset{CH_3}{|}}{CH}CH_2CH_3 \qquad CH_3\underset{\underset{CH_3}{|}}{\overset{\overset{CH_3}{|}}{C}}CH_3$$

11.3 The longest carbon chain is numbered from the end toward the substituent encountered first. Substituents are numbered and listed in alphabetical order.

(a)
$$
\begin{array}{c}
\quad\quad CH_3 \quad\quad\quad\quad CH_3 \\
\quad\; 1 \;\; 2| \;\; 3 \;\; 4 \;\; 5\; | \\
CH_3CHCH_2CH_2CHCHCH_3 \\
\quad\quad\quad\quad\quad 6| \;\; 7 \;\; 8 \\
\quad\quad\quad\quad CH_2CH_2CH_3
\end{array}
$$

5-Isopropyl-2-methyloctane

(b)
$$
\begin{array}{c}
\quad\quad\quad\quad\quad CH_2CH_2CH_3 \\
\quad 1 \;\; 2 \;\; 3 \;\; 4| 5 \;\; 6 \;\; 7 \\
CH_3CH_2CH_2CCH_2CH_2CH_3 \\
\quad\quad\quad\quad | \\
\quad\quad\quad CH_3CHCH_3
\end{array}
$$

4-Isopropyl-4-propylheptane

11.4 (a) isobutylcyclopentane, C_9H_{18}
(b) *sec*-butylcycloheptane, $C_{11}H_{22}$
(c) 1-ethyl-1-methylcyclopropane, C_6H_{12}

11.5 (a) propanone   (b) pentanal
(c) cyclopentanone   (d) cycloheptene

11.6 Shown are equatorial methyl groups on carbons 1, 2, and 4 of the ring.

11.7 Cycloalkanes (a) and (c) show cis-trans isomerism.

(a) *cis*-1,3-Dimethylcyclopentane

trans-1,3-Dimethylcyclopentane

(c) *cis*-1,3-Dimethylcyclohexane

trans-1,3-Dimethylcyclohexane

11.8 In order of increasing boiling point, they are:
(a) 2,2-dimethylpropane, 2-methylbutane, pentane
(b) 2,2,4-trimethylhexane, 3,3-dimethylheptane, nonane

11.9 (a) Hydrocarbon: a compound composed entirely of carbon and hydrogen atoms.
(b) Alkane: a hydrocarbon containing only carbon-carbon single bonds and whose carbons are arranged in an open chain.
(c) Cycloalkane: a saturated hydrocarbon that contains carbon atoms joined to form a ring.
(d) Aliphatic hydrocarbon: an alternative name for alkane.

11.11 (a) Not constitutional isomers; they have different molecular formulas.
(b) Constitutional isomers.
(c) Constitutional isomers.
(d) Not constitutional isomers; they have different molecular formulas.
(e) Not constitutional isomers; they have different molecular formulas.
(f) Not constitutional isomers; they have different molecular formulas.

*11.13 The nine constitutional isomers of molecular formula C_7H_{16} along with their boiling points are

$CH_3CH_2CH_2CH_2CH_2CH_2CH_3$
Heptane
(bp 98°C)

$$
\begin{array}{c}
CH_3 \\
| \\
CH_3CHCH_2CH_2CH_2CH_3
\end{array}
$$
2-Methylhexane
(bp 90°C)

$$
\begin{array}{c}
CH_3 \\
| \\
CH_3CH_2CHCH_2CH_2CH_3
\end{array}
$$
3-Methylhexane
(bp 92°C)

$$
\begin{array}{c}
CH_3 \\
| \\
CH_3CCH_2CH_2CH_3 \\
| \\
CH_3
\end{array}
$$
2,2-Dimethylpentane
(bp 79°C)

$$
\begin{array}{c}
CH_3 \\
| \\
CH_3CHCHCH_2CH_3 \\
| \\
CH_3
\end{array}
$$
2,3-Dimethylpentane
(bp 90°C)

$$
\begin{array}{c}
CH_3 \\
| \\
CH_3CHCH_2CHCH_3 \\
| \\
CH_3
\end{array}
$$
2,4-Dimethylpentane
(bp 81°C)

$$
\begin{array}{c}
CH_3 \\
| \\
CH_3CH_2CCH_2CH_3 \\
| \\
CH_3
\end{array}
$$
3,3-Dimethylpentane
(bp 86°C)

$$
\begin{array}{c}
CH_3CH_2CHCH_2CH_3 \\
| \\
CH_2CH_3
\end{array}
$$
3-Ethylpentane
(bp 94°C)

$$
\begin{array}{c}
H_3C \quad CH_3 \\
| \quad\;\; | \\
CH_3CHCCH_3 \\
| \\
CH_3
\end{array}
$$
2,2,3-Trimethylbutane
(bp 81°C)

11.15 (a) ethyl   (b) isopropyl   (c) *sec*-butyl
(d) *tert*-butyl

11.17 (a) 2-methylpentane   (b) 2,5-dimethylhexane
(c) 3-ethyloctane   (d) 2,2,3-trimethylbutane
(e) isobutylcyclopentane
(f) 1-ethyl-2,4-dimethylcyclohexane

11.19 Following are structural formulas for each compound.

(a) CH_3CH_2OH

(b)
$$
\begin{array}{c}
\quad\;\; O \\
\quad\;\; \| \\
CH_3CH
\end{array}
$$

(c)
$$
\begin{array}{c}
\quad\;\;\;\; O \\
\quad\;\;\;\; \| \\
CH_3COH
\end{array}
$$

(d)
$$
\begin{array}{c}
\quad\;\; O \\
\quad\;\; \| \\
CH_3CCH_2CH_3
\end{array}
$$

(e)
$$
\begin{array}{c}
\quad\quad\quad\quad O \\
\quad\quad\quad\quad \| \\
CH_3CH_2CH_2CH
\end{array}
$$

(f) $CH_3CH_2CH_2\overset{\overset{\displaystyle O}{\|}}{C}OH$ (g) $CH_3CH_2\overset{\overset{\displaystyle O}{\|}}{C}H$ (h) ▷—OH

(i) ⬠—OH   (j) ⬠   (k) ⬠=O

11.21 A condensed structural formula does not show bond angles. Nor does it show conformations due to rotation about carbon-carbon single bonds.

11.23 Here are ball-and-stick models of two conformations of ethane. In the conformation on the top, the hydrogen atoms on one carbon are as far apart as possible from the hydrogens on the other carbon. In the conformation on the bottom, they are as close as possible.

Hydrogens are as far apart as possible

Hydrogens are as close as possible

11.25 Following are structural formulas for the six cycloalkanes of molecular formula C_5H_{10}.

Cyclopentane   Methylcyclobutane   1,1-Dimethyl-cyclopropane

trans-1,2-Dimethyl-cyclopropane   *cis*-1,2-Dimethyl-cyclopropane   Ethylcyclopropane

*11.27 The first two representations of menthol show the cyclohexane ring as a planar hexagon. The third shows it as a chair conformation. Notice that in the chair conformation, all three substituents are equatorial.

11.29 The boiling point of each isomer is given in the answer to Problem 11.14. The isomer with the highest boiling point is heptane, which is unbranched. The isomer with the lowest boiling point is 2,2-dimethylpentane, one of the most branched isomers.

11.31 Alkanes are less dense than water.

11.33 Both are colorless liquids, and you cannot tell the difference between them by looking at them. One way to tell which is which is to determine their boiling points. Hexane has the lower molecular weight, the smaller molecules, and the lower boiling point (69°C). Octane has the higher molecular weight, the larger molecules, and the higher boiling point (126°C).

11.35 In these balanced equations, only whole number coefficients are used. Notice that (a) hexane and (c) 2-methylpentane are constitutional isomers and that, in terms of numbers of atoms, their balanced equations are identical.

(a) $2CH_3(CH_2)_4CH_3 + 19O_2 \longrightarrow 12CO_2 + 14H_2O$
   Hexane

(b) ⬡ $+ 9O_2 \longrightarrow 6CO_2 + 6H_2O$

   Cyclohexane

(c) $2CH_3\overset{\overset{\displaystyle CH_3}{|}}{C}HCH_2CH_2CH_3 + 19O_2 \longrightarrow 12CO_2 + 14H_2O$
   2-Methylpentane

*11.37 (a) Only one ring contains just carbon atoms.
   (b) One ring contains nitrogen atoms.
   (c) One ring contains two oxygen atoms.

11.39 2,2,4-Trimethylpentane was assigned an octane rating of 100 and heptane an octane rating of 0.

11.41 (a) constitutional isomers   (b) constitutional isomers   (c) constitutional isomers   (d) different compounds   (e) different compounds   (f) constitutional isomers

11.43 Compounds (a) and (c) show cis-trans isomerism.

(a) [structure: cyclohexane with OH and CH₃]   [structure: cyclohexane with OH and CH₃]

   cis and trans isomers of
   3-methylcyclohexanol

(c)

cis and trans isomers of
4-methylcyclohexanol

*11.45 (a) The —OH group in ring A is equatorial; those on rings B and C are axial.

(b) The —CH$_3$ group at the junction of rings A/B is equatorial to ring A but axial to ring B.

(c) The —CH$_3$ group at the junction of rings C/D is axial to ring D.

11.47 Dodecane (a) does not dissolve in water, (b) dissolves in hexane, (c) burns when ignited, (d) is a liquid at room temperature, and (e) is less dense than water.

Chapter 12   Alkenes and Alkynes

12.A   The structural feature that makes cis-trans isomerism possible in alkenes is restricted rotation about the carbon-carbon double bond. In cycloalkanes, it is restricted rotation about a ring carbon-carbon bond. These two structural features have in common restricted rotation about a carbon-carbon bond.

12.B   A saturated fat is one in which all carbon-carbon bonds are single bonds. An unsaturated fat is one in which one or more carbon-carbon bonds are double bonds. A polyunsaturated fat has two or more carbon-carbon double bonds.

12.C   Reaction mechanisms focus on the change in location of electron pairs as some bonds are broken and others made. Because of this focus on pushing electron pairs from place to place, reaction mechanisms are often referred to as electron pushing.

12.D   When carbon bonds with other atoms, it shares one of its valence electrons in forming each new bond. In a carbocation, carbon has only three electrons to share with other atoms; it has lost one of its valence electrons and so has a positive charge. When nitrogen bonds with other atoms, it also shares one of its valence electrons in forming each new bond. Its five valence electrons are one from each covalent bond plus the two unshared electrons. When nitrogen forms four covalent bonds, it only has four valence electrons; one from each shaped pair. This nitrogen now has only four valence electrons, one less than an uncharged nitrogen atom and, therefore, has a positive charge.

12.1   (a) 3,3-dimethyl-1-pentene   (b) 2,3-dimethyl-2-butene   (c) 3,3-dimethyl-1-butyne

12.2   (a) *trans*-3,4-dimethyl-2-pentene
(b) *cis*-4-ethyl-3-heptene

12.3   (a) 1-isopropyl-4-methylcyclohexene
(b) cyclooctene   (c) 4-*tert*-butylcyclohexene

12.4   The other two isomers have cis-trans and cis-cis configurations.

cis-2-trans-4-heptadiene   cis-2-cis-4-heptadiene

12.5   The structural formula for this compound is drawn here in two alternative ways.

12.6   (a) CH$_3$CHCH$_3$ (b)

12.7   In order of increasing stability, they are (c) which is 1°, (b) which is 2°, and (a) which is 3°.

12.8   Propose a two-step mechanism similar to that proposed for the addition of HCl to propene.
Step 1: Reaction of H$^+$ with the carbon-carbon double bond gives a 3° carbocation intermediate.

A 3° carbocation intermediate

Step 2: Reaction of the 3° carbocation intermediate with bromide ion completes the valence shell of carbon and gives the product.

12.9   The product is the same from each reaction:

12.10 Propose a three-step mechanism similar to that for the acid-catalyzed hydration of propene.
Step 1: Reaction of the carbon-carbon double bond with H$^+$ gives a 3° carbocation intermediate.

Step 2: Reaction of the carbocation intermediate with water completes the valence shell of carbon and gives an oxonium ion.

An oxonium ion

Step 3: Loss of H⁺ from the oxonium ion completes the reaction and generates a new H⁺ catalyst.

$$+ \ H^+$$

12.11 (a) $CH_3\overset{\underset{\displaystyle CH_3}{|}}{C}\overset{\underset{\displaystyle H_3CBr}{|}}{C}H-\overset{\underset{\displaystyle Br}{|}}{C}H_2$

(b)

12.13 (a) 109.5°, 120°

(b) 120° CH₂OH

(c) 180° HC≡C—CH=CH₂ 120°

(d) 120°

12.15 (a) $CH_3\overset{\underset{\displaystyle CH_3}{|}}{C}H$, H, C=C, H, CH₂CH₃

(b) $CH_3\overset{\underset{\displaystyle CH_3}{|}}{C}HC≡CCH_2CH_3$

(c) $CH_2=\overset{\underset{\displaystyle CH_3}{|}}{C}CH_2CH_3$

(d) $HC≡C\overset{\underset{\displaystyle CH_2CH_3}{|}}{C}CH_2CH_3$, CH₃

(e) H₃C, CH₃, C=C, H₃C, CH₂CH₃

(f) H, H, C=C, H₃C, CH₂CH₂CH₃

(g) $CH_2=CHCH_2Cl$

(h) —CH₃

(i) CH₃, CH₃

(j) H₃C, CH₃, C=C, CH₃CH₂, CH₂CH₂CH₃

(k) ▷

(l) $CH_3CH_2C≡CCH_2CH_3$

*12.17 There are six alkenes of molecular formula C_5H_{10}.

$CH_2=CHCH_2CH_2CH_3$ $CH_2=\overset{\underset{\displaystyle CH_3}{|}}{C}CH_2CH_3$
1-Pentene   2-Methyl-1-butene

$CH_2=CH\overset{\underset{\displaystyle CH_3}{|}}{C}HCH_3$
3-Methyl-1-butene

H₃C, H, C=C, H, CH₂CH₃
trans-2-Pentene

H, H, C=C, H₃C, CH₂CH₃
cis-2-Pentene

H₃C, H, C=C, H₃C, CH₃
2-Methyl-2-butene

12.19 Only parts (b) and (d) show cis-trans isomerism.

(b) CH₃, CH₃

(d) CH₃, CH₃

*12.21 The structural formula of β-ocimene is drawn on the left showing all atoms and on the right as a line-angle drawing.

H₃C, H, C=C, CH₃, CH₂=CH, CH₂CH=CCH₃

12.23 The four isoprene units of vitamin A are shown in bold.

H₃C, CH₃, CH₃, CH₃, CH₂OH, CH₃

*12.25 (a) The isoprene units are shown as bold bonds, and the points of cross linking are shown by dashed bonds.

CH₂OH, HO, H₃C, CH₃, CH₃, H₃C, O, H
Zoapatanol

(b) Cis-trans isomerism is possible about the carbon-carbon double bond to the seven-membered ring, and with the —OH and —CH₃ groups bonded to the seven-membered ring. Four cis-trans isomers possible.

12.27 (a) HBr   (b) H_2O/H_2SO_4 (c) HI   (d) Br_2

12.29 (a) $CH_3CH_2CH_2^+$ or $\boxed{CH_3\overset{+}{C}HCH_3}$

(b) $CH_3\overset{CH_3}{\underset{\;}{CH}}CH\overset{+}{CH}CH_3$ or $\boxed{CH_3\overset{CH_3}{\underset{\;}{C}}\overset{+}{C}H_2CH_3}$

12.31 (a) $CH_3\overset{CH_3}{\underset{Cl}{C}}CH_2CH_2CH_3$ (b) $CH_3\overset{CH_3}{\underset{OH}{C}}CH_2CH_2CH_3$

*12.33 (a) $CH_2{=}\overset{CH_3}{\underset{\;}{C}}CH_2CH_3$ or $CH_3\overset{CH_3}{\underset{\;}{C}}{=}CHCH_3$

(b) $CH_3\overset{CH_3}{\underset{\;}{CH}}CH{=}CH_2$

(c) $=CH_2$ or $-CH_3$

*12.35 (a) $CH_3\overset{CH_3}{\underset{\;}{C}}{=}CHCH_3$ or $CH_2{=}\overset{CH_3}{\underset{\;}{C}}CH_2CH_3$

(b) $CH_3\overset{CH_3}{\underset{\;}{CH}}CH{=}CH_2$ (c) $CH_2{=}CHCH_2CH_2CH_3$

*12.37 (a) Terpin contains two —OH groups, each a tertiary alcohol. (b) Two cis-trans isomers are possible. (c) The isomer with methyl and isopropyl groups trans to each other are shown in three alternative representations.

12.39 The hydrocarbon is 2-methyl-1,3-butadiene.

12.41 Reagents are shown over each arrow.

12.43 Its molecular formula is $C_{16}H_{32}O_2$, and its molecular weight is 256 g/mol.

12.45 Rods are primarily responsible for black and white vision. Cones are responsible for color vision.

12.47 The most common consumer items made of high-density polyethylene (HDPE) are milk and water jugs, grocery bags, and squeezable bottles. The most common consumer items made of low-density polyethylene (LDPE) are packaging for baked goods, vegetables and other produce, and trash bags. Currently, only HDPE materials are recyclable.

12.49 An alkyne. An alkyne has four fewer hydrogens than an alkane. An alkene has only two fewer hydrogens.

12.51 (a) The carbon skeleton of lycopene can be divided into eight isoprene units.

(b) Of the 13 double bonds in lycopene, cis-trans isomerism is possible in about 11. Only the double bonds at either end of the carbon chain cannot show this type of isomerism.

12.53 (a) $-CH_3$ or $\boxed{\text{} -CH_3}$

(b) $\boxed{\text{}-CH_3}$ or $-CH_3{}^+$

*12.55 (a)   (b) $-CH_3$

(c)   or

(d) $CH_2{=}\overset{CH_3}{\underset{\;}{C}}CH_2CH_3$

Chapter 13 Alcohols, Ethers, and Thiols

13.A Low-molecular-weight alcohols dissolve in water because they contain polar —OH groups, which interact with water molecules by hydrogen bonding. Low-molecular-weight alkanes are nonpolar compounds and do not dissolve in water because they are not capable of interacting with water molecules.

13.B Expect an O—H···O hydrogen bond to be stronger.

13.C Low-molecular-weight alcohols form hydrogen bonds with water molecules through both the oxygen and hydrogen atoms of their —OH groups. Low-molecular-weight ethers form hydrogen bonds with water molecules only through the oxygen atom of their —O— group. The greater extent of hydrogen bonding between alcohol and water molecules makes the low-molecular-weight alcohols more soluble in water than low-molecular-weight ethers.

13.D A change in any of the factors determining an equilibrium will cause the system to adjust to reduce or minimize the effect of the change. In a mixture containing alkene, alcohol, and water, adding water will cause more alcohol to be formed; removing water will cause more alkene to be formed.

13.1 (a) *trans*-4-isopropylcyclohexanol
(b) 1-methylcyclopentanol
(c) 2,2-dimethyl-1-propanol

13.2 (a) primary   (b) secondary   (c) primary
(d) tertiary

13.3 (a) 3-buten-1-ol   (b) 2-cyclopentenol

13.4 ethyl isobutyl ether   (b) cyclopentyl methyl ether

13.5 (a) 3-methyl-1-butanethiol
(b) 3-methyl-2-butanethiol

13.6 In each case, the major product (circled) contains the more substituted double bond.

(a) $CH_3C{=}CHCH_3$ + $CH_2{=}CCH_2CH_3$ (each with CH_3 substituent)

(b) [cyclopentene with CH_3] + [methylenecyclopentane with CH_2]

13.7 Each secondary alcohol is oxidized to a ketone.

(a) [cyclohexanone structure] =O   (b) $CH_3\overset{O}{\overset{\|}{C}}CH_2CH_2CH_3$

13.9 Only (c) and (d) are secondary alcohols.

13.11 (a) $CH_3\overset{OH}{\underset{|}{C}}HCH_3$ (b) $CH_3\overset{HO\ OH}{\underset{|\ \ |}{C}}HCH_2$

(c) $CH_3\overset{CH_3}{\underset{|}{C}}HCH_2CH_2\overset{OH}{\underset{|}{C}}HCH_3$ (d) $HOCH_2\overset{CH_3}{\underset{|}{C}}CH_2OH$ (with $CH_2CH_2CH_3$)

(e) $CH_3(CH_2)_6CH_2OH$ (f) $CH_3\overset{CH_3}{\underset{|}{C}}HCH_2OH$

(g) $HOCH_2CH_2CH_2CH_2OH$ (h) [structure: $CH_3\overset{CH_3}{\underset{|}{C}}HCH_2$ and CH_2OH on a C=C with H, H]

(i) [H_3C and CH_2CH_2OH on C=C with H, H]   (j) $HO{\blacktriangleright}\!\!\langle\ \rangle\!\!{\cdots}OH$

13.13 (b) There are four cis-trans isomers possible. Only the one shown here is the active pheromone for the silk worm.

[structure with carbon positions labeled: 16 $CH_3CH_2CH_2$ ... $^{13}C{=}C$ ... ^{12} ... ^{11} ^{10} C=C ... 9 $CH_2CH_2CH_2CH_2CH_2CH_2CH_2CH_2CH_2OH$ 5 1, with H atoms]

trans-10-*cis*-12-Hexadecadien-1-ol

13.15 In order of increasing boiling point, they are:

$CH_3CH_2CH_2CH_3$ $CH_3CH_2OCH_2CH_3$
   0°C   35°C

$CH_3CH_2CH_2OH$
   97°C

13.17 The interaction between molecules of methanol in the liquid state is by hydrogen bonding. The interaction between molecules of methanethiol in the liquid state is only by the considerably weaker London dispersion forces. Because of the stronger intermolecular forces of attraction between its molecules, methanol has a higher boiling point than methanethiol.

13.19 The thickness (viscosity) of these three liquids is related to the degree of hydrogen bonding between their molecules in the liquid state. It is strongest between molecules of glycerol, weaker between molecules of ethylene glycol, and weakest between molecules of ethanol.

13.21 In order of decreasing solubility in water, they are:
(a) ethanol > diethyl ether > butane
(b) 1,2-hexanediol > 1-hexanol > hexane

13.23 Cyclohexene will react with bromine (a red-purple liquid) to give 1,2-dibromocyclohexane, which has no color. Cyclohexanol does not react with bromine. Therefore, the liquid that causes the red-purple of bromine to disappear must be cyclohexene.

13.25 The first reaction is an acid-catalyzed dehydration, the second is an oxidation.

(a) $CH_3CH_2\overset{OH}{\underset{|}{C}}HCH_3 \xrightarrow[\text{heat}]{H_2SO_4} CH_3CH{=}CHCH_3 + H_2O$

(b) $CH_3CH_2\overset{\underset{\displaystyle OH}{|}}{C}HCH_3 \xrightarrow[H_2SO_4]{K_2Cr_2O_7} CH_3CH_2\overset{\underset{\displaystyle}{\overset{\displaystyle O}{\|}}}{C}CH_3$

13.27 Reaction (1) is a dehydration of a tertiary alcohol to give an alkene. Reaction (2) is hydration of an alkene to give a secondary alcohol. Reaction (3) is the oxidation of a secondary alcohol to a ketone.

13.29 The two functional groups are a carboxyl group and a disulfide group. Its reduction product is dihydrolipoic acid.

Lipoic acid

Dihydrolipoic acid

*13.31 Acid-catalyzed dehydration of cyclohexanol (circled) gives cyclohexene. Reduction of cyclohexene gives cyclohexane, and oxidation of cyclohexanol gives cyclohexanol.

Cyclohexane   Cyclohexene   Cyclohexanol

Cyclohexanone

13.33 Oxidize the primary alcohol first to an aldehyde and then to a carboxylic acid.

13.35 Nobel discovered that diatomaceous earth absorbs nitroglycerin so that it will not explode without a fuse.

13.37 Normal bond angles about carbon and oxygen are 109.5°. In ethylene oxide, the C—C—O and C—O—C bond angles are compressed to 60°, which results in strain within the molecule.

13.39 Ethanol can cause brain damage to the developing fetus of a pregnant woman.

13.41 The relationship is that 2100 mL of breath contains the same amount of ethanol as 1.00 mL of blood.

13.43 Both enflurane and isoflurane have the molecular formula, $C_3H_2ClF_5O$.

*13.45 The eight alcohols of molecular formula $C_5H_{12}O$ and their IUPAC names are:

$$\overset{\underset{\displaystyle}{}}{CH_3CH_2CH_2CH_2CH_2}\overset{\underset{\displaystyle}{}}{}$$
OH
1-Pentanol

$CH_3CH_2CH_2CHCH_3$
OH
2-Pentanol

$CH_3CH_2CHCH_2CH_3$
OH
3-Pentanol

$CH_2CHCH_2CH_3$
OH CH_3
2-Methyl-1-butanol

$CH_3CCH_2CH_3$
OH CH_3
2-Methyl-2-butanol

$CH_3CHCHCH_3$
OH CH_3
3-Methyl-2-butanol

$CH_3CHCH_2CH_2$
OH CH_3
3-Methyl-1-butanol

CH_3CCH_2OH
CH_3 CH_3
2,2-Dimethyl-1-propanol

13.47 Ethylene glycol has two —OH groups by which each molecule participates in hydrogen bonding, whereas 1-propanol has only one. The stronger intermolecular forces of attraction between molecules of ethylene glycol give it the higher boiling point.

13.49 Arranged in order of increasing solubility in water, they are:

$CH_3CH_2CH_2CH_2CH_2CH_3$ $CH_3CH_2CH_2CH_2CH_2OH$
Hexane   1-Pentanol
(insoluble)   (2.3 g/mL water)

$HOCH_2CH_2CH_2CH_2OH$
1,4-Butanediol
(infinitely soluble)

*13.51 Each can be prepared from 2-methyl-1-propanol (circled) as shown in this flow chart.

Chapter 14 Benzene and Its Derivatives

14.A No. Aromaticity requires sites of unsaturation within a molecule. In the case of benzene, the simplest and best known aromatic compound, there are three sites of unsaturation; that is, there are three carbon-carbon double bonds that give rise to the so-called aromatic sextet.

14.B Benzene consists of carbons each surrounded by three regions of electron density, which gives 120° for all bond angles within the molecule. Bond angles of 120° in benzene can only be maintained if the molecule is flat. Cyclohexane, on the other hand, consists of carbons each surrounded by four regions of electron density, which gives 109.5° for all bond angles within the molecule. Angles of 109.5° in cyclohexane can be maintained only if the molecule is nonplanar.

14.C It works this way: Neither a unicorn, which has a horn like a rhinoceros, nor a dragon, which has a tough, leathery hide like a rhinoceros, exists. But if they did and you were to make a hybrid of them, you would have a rhinoceros. Furthermore, a rhinoceros is not a dragon part of the time and a unicorn the rest of the time; a rhinoceros is a rhinoceros all of the time. To carry this analogy to aromatic compounds, resonance-contributing structures for them do not exist; they are imaginary. But if they did exist and you made a hybrid of them, you would have the real aromatic compound.

14.D Polynuclear means that each contains two or more rings bonded in such a way that each ring shares two adjacent atoms with another ring. Aromatic means that each ring is six-membered and has three carbon-carbon double bonds; that is, each has an aromatic sextet. Hydrocarbon means that these compounds consist of only carbon and hydrogen.

14.1 (a) 1-phenylethanol   (b) 2,4-dichlorophenol
(c) 3-nitrobenzoic acid

14.3 A saturated compound contains only single covalent bonds. An unsaturated compound contains one or more double bonds. The most common double bonds are C=C, but they can also be C=N.

14.5 Yes, they have double bonds, at least in the contributing structures we normally use to represent them. Yes, they are unsaturated in the sense that they have fewer hydrogens than a cycloalkane of the same number of carbons.

14.7 (a) 1-chloro-4-nitrobenzene (*p*-chloronitrobenzene)
(b) 2-bromotoluene (*o*-bromotoluene)
(c) 3-phenyl-1-propanol
(d) 2-phenyl-2-butanol
(e) 2-nitroaniline (*o*-nitroaniline)
(f) 2-phenylphenol (*o*-phenylphenol)
(g) *trans*-1,2-diphenylethene
(*trans*-1,2-diphenylethylene)
(h) 2,4-dichlorotoluene

14.9 Phenol is a sufficiently strong acid that reacts with strong bases such as sodium hydroxide to form sodium phenoxide, a water-soluble salt. Cyclohexanol has no comparable acidity and does not react with sodium hydroxide.

14.11 (a) HNO_3/H_2SO_4 (b) $Br_2/FeCl_3$

14.13 The two sulfonated naphthalenes are:

SO$_3$H

SO$_3$H

1-Naphthalene-
sulfonic acid

2-Naphthalene-
sulfonic acid

14.15 Advantages of DDT are that it is extremely effective in killing the insect hosts that transmit diseases such as malaria and typhus. It is also effective in killing crop-destroying insect pests. Disadvantages are that it persists in the environment and is toxic to many higher organisms.

14.17 Loss of hydrogen and chlorine from adjacent carbons gives a carbon-carbon double bond.

Cl—〈 〉—C—〈 〉—Cl
          ‖
         CCl$_2$
         DDE

14.19 By definition, a carcinogen is a cancer-causing substance. The carcinogens present in cigarette smoke belong to a class called polynuclear aromatic hydrocarbons (PAHs).

14.21 Capsaicin is isolated from fruit of various species of *Capsicum,* otherwise known as chili peppers.

14.23 It is used in medicine as a topical anesthetic.

14.25 One oxygen-containing functional group in triclosan is a phenol; the second is a phenol ether.

14.27 Following are the three contributing structures for naphthalene.

(1)   (2)   (3)

*14.29 *Butylated* indicating the two *tert*-butyl group. *Hydroxyl* indicating the —OH group on the aromatic ring, and *toluene* indicating that toluene has been taken as the parent name for this compound.

Chapter 15 Chirality

15.A Share your findings with others. You will find it interesting to compare your results.

15.B The helical coil of a telephone cord, a spiral binding on a notebook, or any other helical coil has the same handedness viewed from either end.

15.C Chirality is the property of an object. Objects like a decorated cup, a baseball glove, or a helical coil are chiral but have no stereocenter. The only necessary condition for chirality is that an object and its mirror image not be superposable.

15.1 The enantiomers of each part are drawn with two groups in the plane of the paper, a third group toward you in front of the plane, and the fourth group away from you behind the plane.

(a)

(b)

15.2 The group of higher priority in each set is circled.

(a) $\boxed{-CH_2OH}$ and $-CH_2CH_2COOH$

(b) $\boxed{-CH_2NH_2}$ and $-CH_2CH_2COOH$

15.3 The order of priorities and the configuration of each is shown in the drawing.

(a) (b)

15.4 (a) Compounds (1) and (3) are one pair of enantiomers. Compounds (2) and (4) are a second pair of enantiomers. (b) Compounds 1 and 2, 1 and 4, 2 and 3, and 3 and 4 are diastereomers.

15.5 Four stereoisomers are possible for 3-methylcyclohexanol. The cis isomer is one pair of enantiomers; the trans isomer is a second pair of enantiomers.

15.6 Stereocenters are marked by an asterisk, and the number of possible stereoisomers is shown under the structural formula.

(a)

$2^1 = 2$

(b) $CH_2{=}CHCHCH_2CH_3$ (c)
  $\overset{*}{}$

$2^1 = 2$ $2^2 = 4$

15.7 (a) Chiral (b) Achiral (c) Achiral
(d) Chiral if it has markings. Achiral if it does not.
(e) Chiral, unless you are on the equator, in which case it goes straight down.

15.9 The stereocenter in each chiral compound is marked by an asterisk.

(a) $CH_3CH_2\overset{*}{C}HCH_2CH_2CH_3$ (b) $CH_3\overset{*}{C}HCH_2CH_3$
  $\quad\quad\quad |$ $\quad\quad\quad |$
  $\quad\quad CH_3$ $\quad\quad OH$

(c) (d)

(e)

15.11 Only three of these molecules have stereocenters. In each the stereocenter is marked by an asterisk.

(b) (c)

(f) $CH_3CH_2\overset{*}{C}HCH{=}CH_2$
  $\quad\quad\quad |$
  $\quad\quad OH$

15.13 Molecules (b) and (d) have R configurations.

(a) (b)

(c) (d)

15.15 The optical rotation of its enantiomer is +41°.

15.17 (a) To say that a drug is chiral means that it has handedness: that it has one or more stereocenters and has the possibility for two or more stereoisomers. (b) Just because it is chiral does not mean that it will be optically active. It may be chiral and present as a racemic mixture, in which case it will have no effect on the plane of polarized light. If, however, it is present as a single enantiomer, it will rotate the plane of polarized light.

15.19 There is one stereocenter in ibuprofen. Two stereoisomers are possible.

15.21 Of the eight alcohols of molecular formula $C_5H_{12}O$, only three are chiral.

$CH_3CH_2CH_2CHCH_3$ | $CH_2CHCH_2CH_3$ | $CH_3CHCHCH_3$
with OH and * markings

2-Pentanol   2-Methyl-1-butanol   3-Methyl-2-butanol

15.23 The two stereocenters are marked by asterisks.

OH
| *
$CH_3CHCHCH_2CH_3$
* |
    CH_3

15.25 The four stereocenters are shown by asterisks.

15.27 (a) The only possible compound that does not show cis,trans isomerism and has no stereocenter is 1-methylcyclohexanol.

1-Methylcyclohexanol

(b) Only 4-methylcyclohexanol shows cis,trans isomerism but has no stereocenter. Drawn here are the cis and trans isomers.

trans-4-Methylcyclohexanol   *cis*-4-Methylcyclohexanol

(c) Both 2-methylcyclohexanol and 3-methylcyclohexanol have two stereocenters and can exist as cis and trans isomers. The stereocenters in each are marked by asterisks.

2-Methylcyclohexanol   3-Methylcyclohexanol

15.29 Parts (b), (c), (d), (e), and (f) show cis-trans isomerism

Chapter 16 Amines

16.A When used to classify amines, secondary refers to the presence of two carbon groups on nitrogen. When used to classify alcohols, secondary refers to the presence of two carbon groups on the carbon bearing the —OH group.

16.B Nitrogen is less electronegative than oxygen and more willing to donate its unshared pair of electrons to H^+ in an acid-base reaction to form a salt.

16.C In order of decreasing ability to form intermolecular hydrogen bonds, they are $CH_3OH > (CH_3)_2NH > CH_3SH$. An O—H bond is more polar than an N—H bond, which is more polar than an S—H bond.

16.1 Pyrollidine has nine hydrogens; its molecular formula is C_4H_9N. Pyrrole has five hydrogens; its molecular formula is C_4H_5N.

16.2 Following are structural formulas for each compound.

CH_3
|
(a) $CH_3CHCH_2NH_2$ (b) cyclohexyl—NH_2

(c) $H_2NCH_2CH_2CH_2CH_2NH_2$

16.3 Following are structural formulas for each compound.

CH_3
|
(a) $CH_3CHCH_2NH_2$ (b) phenyl—NH—phenyl

(c) $CH_3CHNHCHCH_3$
         | |
        CH_3 CH_3

16.4 The stronger base is circled.

(a) pyridine N or [cyclohexyl—NH_2]

(b) [NH_3] or phenyl—NH_2

16.5 The product of each reaction is an amine salt.

(a) $(CH_3CH_2)_3\overset{+}{N}H$ Cl^- (b) piperidine $\overset{+}{N}$ with H, H and $CH_3\overset{O}{\overset{\|}{C}}O^-$

16.7 In an aliphatic amine, all carbon groups bonded to nitrogen are alkyl groups. In an aromatic amine, one or more of the carbon groups bonded to nitrogen are aryl (aromatic) groups.

16.9 Following is a structural formula for each amine.

NH_2
|
(a) $CH_3CHCH_2CH_3$ (b) $CH_3(CH_2)_6CH_2NH_2$

(c) CH$_3$CCH$_2$NH$_2$ (with CH$_3$ groups above and below)   (d) H$_2$N(CH$_2$)$_5$NH$_2$

(e) [benzene ring with NH$_2$ and Br substituents]   (f) (CH$_3$CH$_2$CH$_2$CH$_2$)$_3$N

(g) [benzene ring with —N(CH$_3$)$_2$]   (h) [dicyclohexylamine, two cyclohexyl rings joined by N—H]

(i) CH$_3$CHCH$_2$CH$_3$ (with NH$_2$ above)   (j) [benzene ring with NH$_2$ and two CH$_3$ groups]

16.11 (a) Each is a secondary aliphatic amine. (b) Each contains a benzene ring with a phenolic —OH group, a stereocenter with an R configuration, and a secondary aliphatic amine. One difference is that one phenolic OH of adrenaline is replaced by a CH$_2$OH group in albuterol. The other difference is that the N-methyl group in adrenaline is replaced by a N-*tert*-butyl group in albuterol.

16.13 Both propylamine (a 1° amine) and ethylmethylamine (a 2° amine) have an N—H group and show hydrogen bonding between their molecules in the liquid state. Because of this intermolecular force of attraction, these two amines have higher boiling points than trimethylamine, which has no N—H bond and therefore cannot participate in intermolecular hydrogen bonding.

16.15 Following are structural formulas for each amine salt.

(a) (CH$_3$)$_2$NH$_2^+$ I$^-$ (b) (CH$_3$)$_3$NCH$_2$CH$_3$ OH$^-$

(c) (CH$_3$)$_4$N$^+$ Cl$^-$ (d) [benzene ring]—NH$_3^+$ Br$^-$

16.17 Morphine is insoluble in water. Its salt with sulfuric acid, however, is water-soluble.

16.19 Structural formula A contains both an acid (the carboxyl group) and a base (the 1° amino group). The acid-base reaction between them gives structural formula B, which is the better representation of alanine.

16.21 The form of amphetamine present at both pH 1.0 and pH 7.4 is the ammonium ion shown in the answer to Problem 16.20(b).

16.23 (a) The secondary aliphatic amine is the more basic nitrogen. This molecule has three stereocenters, each marked with an asterisk.

Epibatidine [structure with Cl on pyridine ring, bicyclic amine, N—H labeled "This is the more basic nitrogen", stereocenters marked with *]

16.25 (a) Epinephrine has two phenolic —OH groups, whereas amphetamine has none. Epinephrine has a secondary alcohol on its carbon side chain, whereas amphetamine has none. Epinephrine is a secondary amine, whereas amphetamine is a primary amine. Finally, both epinephrine and amphetamine are chiral, but their stereocenters are in different locations within each molecule.
(b) Methamphetamine is a secondary aliphatic amine, whereas amphetamine is a primary aliphatic amine. Both compounds are chiral at the same carbon.

16.27 Alkaloids are basic nitrogen-containing compounds found in the roots, bark, leaves, berries, or fruits of plants. In almost all alkaloids, the nitrogen atom is present as a member of a ring; that is, it is present in a heterocyclic ring. By definition, alkaloids are nitrogen-containing bases and, therefore, turn red litmus blue.

16.29 The tertiary aliphatic amine in the five-membered ring is the stronger base.

[structure of nicotine-like molecule with pyridine ring and N-methyl pyrrolidinium, Cl$^-$]

16.31 The common structural feature is a benzene ring fused to a seven-membered ring containing two nitrogen atoms. This parent structural feature is named benzodiazepine.

16.33 The structural formula of 4-aminobutanoic acid is drawn on the left showing the primary amino group (a base) and the carboxyl group (an acid). It is drawn on the right as an internal salt.

a primary amine   a carboxylic acid

H$_2$NCH$_2$CH$_2$CH$_2$COH (with O double bond)

un-ionized amino and carboxyl groups

H$_3$NCH$_2$CH$_2$CH$_2$CO$^-$ (with + on N and O double bond)

internal salt

16.35 Each structural feature is shown here.

3° amine

4° carbon

Morphine   Dextromethorphan

3° amine   4° carbon

Meperidine

16.37 No. There is no unreacted HCl present.

16.39 Following are structural formula for each part.

(a) ⟨benzene⟩—NHCH$_3$ (b) ⟨benzene⟩—N(CH$_3$)CH$_3$

(c) ⟨benzene⟩—CH$_2$NH$_2$

(d) CH$_3$CH(NH$_2$)CH$_2$CH$_3$ (e) ⟨N-methylpyrrolidine⟩

(f) H$_3$C—⟨ring with two CH$_3$⟩—NH$_2$ (g) CH$_3$N$^+$(CH$_2$CH$_3$)(CH$_3$CHCH$_3$)CH$_2$CH$_2$CH$_3$ Cl$^-$CH$_2$CH$_3$

16.41 Both alcohols and amines can interact with water by hydrogen bonding. Because amines and alcohols of the same molecular weight have about the same solubility in water, the strength of hydrogen bonding interactions between their molecules must be comparable.

16.43 (a) Following is its structural formula. (b) It has two stereocenters marked with asterisks, and $2^2 = 4$ stereoisomers are possible.

16.45 (a) Atropine contains a tertiary heterocyclic amine. (b) Its one stereocenter is marked with an asterisk. (c) Following is the structural formula of the salt it forms with one mole of sulfuric acid. (d) Because it is an ammonium salt, atropine sulfate is more soluble in water than atropine. (e) Dilute aqueous solutions of atropine are basic because of the reaction in its tertiary aliphatic amine with water to produce hydroxide ions.

Chapter 17 Aldehydes and Ketones

17.A The VSEPR model treats both single and double bonds as single regions of electron density. The experimental evidence that supports this assumption is that the assumption leads to predictions that are identical or almost identical to experimentally measured bond angles.

17.B The aldehyde group of an aromatic aldehyde is bonded to a benzene or other aromatic ring. The aldehyde group of an aliphatic aldehyde is bonded to a carbon that is not part of an aromatic ring.

17.C No. To be a carbon stereocenter, the carbon atom must have four different groups bonded to it. The carbon atom of a carbonyl group has only three groups bonded to it.

17.D Hydration means adding water, as for example hydration of an alkene to give an alcohol. Hydrolysis means splitting or breaking apart by water.

Hydration:

Hydrolysis:

17.1 (a) 3,3-dimethylbutanal (b) cyclopentanone
(c) (S)-2-phenylpropanal

17.2 Following are line-angle drawings for each aldehyde of molecular formula $C_6H_{12}O$. In the three that are chiral, the stereocenter is marked by an asterisk.

Hexanal   4-Methylpentanal

3-Methylpentanal   2-Methylpentanal

2,3-Dimethylbutanal   3,3-Dimethylbutanal

2,2-Dimethylbutanal

17.3 (a) 2-hydroxypropanal
(b) 2-aminobenzaldehyde
(c) 5-amino-2-pentanone

17.4 Shown first is the hemiacetal and then the acetal.

(a)
Hemiacetal   Acetal

(b)
Hemiacetal   Acetal

17.5 (a) A hemiacetal formed from 3-pentanone, a ketone.
(b) Neither a hemiacetal nor an acetal. This compound is the dimethyl ether of ethylene glycol.
(c) An acetal derived from 5-hydroxypentanal.

17.6 Following is the keto form of each enol.

(a)   (b)   (c)

17.7 The product of oxidation of each aldehyde group is a carboxylic acid.

(a) $HOCCH_2CH_2CH_2CH_2COH$
Hexanedioic acid
(Adipic acid)

(b)
3-Phenylpropanoic acid

17.8 Each primary alcohol comes from reduction of an aldehyde. Each secondary alcohol comes from reduction of a ketone.

(a)   (b) CH_3O——CH_2CH

(c) $CH_3C(CH_2)_3CCH_3$

17.9 Oxidation of a primary alcohol gives either an aldehyde or a carboxylic acid, depending on the experimental conditions. Oxidation of a secondary alcohol gives a ketone.

(a)   (b) or

17.11 The carbonyl carbon of an aldehyde is bonded to at least one hydrogen. The carbonyl carbon of a ketone is bonded to two carbon groups.

17.13 Of these four aldehydes, only one is chiral. Its stereocenter is marked by an asterisk.

$CH_3CH_2CH_2CH_2CH$ CH_3CHCH_2CH
                                              CH_3

$CH_3CH_2\overset{*}{C}HCH$ CH_3CCH
         CH_3 CH_3 (with H_3CO)

17.15 Following are structural formulas for each ketone.

(a) $CH_3CH_2CCHCH_3$ (b)
                CH_3

(c)   (d) $CH_3CHCCHCH_3$
                                                CH_3 CH_3

(e) CH_3CCH_3 (f)

17.17 The compound with the higher boiling point is circled.

(a) CH_3CHO or $\boxed{CH_3CH_2OH}$

(b) CH_3CCH_3 or $\boxed{CH_3CH_2CCH_2CH_3}$

(c) $\boxed{CH_3(CH_2)_3CHO}$ or $CH_3(CH_2)_3CH_3$

(d) $CH_3CH_2CCH_3$ or $\boxed{CH_3CH_2CHCH_3}$ (with OH)

17.19 Acetone has the higher boiling point because of the intermolecular attraction between the carbonyl groups of acetone molecules.

17.21 They have no —OH group through which to form intermolecular hydrogen bonds.

17.23 (a) an acetal   (b) a hemiacetal
(c) an acetal   (d) an acetal   (e) an acetal
(f) neither

17.25 Drawn first is the hemiacetal and then the acetal.

(a) $CH_3CH_2\overset{OH}{\underset{H}{C}}$—$OCH_2CH_3$ ⟶ $CH_3CH_2\overset{OCH_2CH_3}{\underset{H}{C}}$—$OCH_2CH_3$

(b) ⟶

(c) ⟶

(d) [chemical structure: cyclohexane with OH and OCH₂CH₂OH groups] → [chemical structure: bicyclic dioxolane]

17.27 Compound (a), (b), (d), and (f) undergo keto-enol tautomerism.

17.29 For each enol, there is only one possible keto form.

(a) CH_3CH with O double bond (b) $CH_3CCH_2CH_2CH_2CH_3$ with O double bond

(c) [benzene ring]—CH_2CCH_3 with O double bond   (d) [cyclopentanone ring with O]

17.31 Only aldehydes are oxidized by Tollens' reagent.

(a) $CH_3CH_2CH_2COH$ with O double bond   (b) [benzene ring]—COH with O double bond

(c) No reaction   (d) No reaction

17.33 The white solid is benzoic acid, formed by air oxidation of benzaldehyde.

17.35 These experimental conditions reduce aldehydes to primary alcohols and ketones to secondary alcohols.

(a) $CH_3CHCH_2CH_3$ with OH   (b) $CH_3(CH_2)_4CH_2OH$

(c) [cyclopentane ring with OH and CH₃]   (d) [benzene ring with CH₂OH and OH]

17.37 The first reaction is a reduction, the second two are oxidations.

(a) $CH_3CH_2CH_2CH_2OH$ (b), (c) $CH_3CH_2CH_2COH$ with O double bond

17.39 The preparation of each compound is shown in this flow chart.

[flow chart: cyclohexanone →(H₂/Metal) cyclohexanol →(H₂SO₄ heat) cyclohexene; cyclohexene →(HCl) chlorocyclohexane; cyclohexene →(Br₂) trans-1,2-dibromocyclohexane; cyclohexene →(H₂/Metal) cyclohexane]

17.41 The alkene of three carbons is propene. Acid-catalyzed hydration of this alkene gives 2-propanol as the major product, not 1-propanol.

17.43 Each conversion can be brought about by acid-catalyzed hydration of the alkene to a secondary alcohol followed by oxidation of the secondary alcohol to a ketone.

(a) $CH_2{=}CHCH_2CH_2CH_3 \xrightarrow[H_2SO_4]{H_2O}$

[chemical structure: $CH_3CHCH_2CH_2CH_3$ with OH] $\xrightarrow[H_2SO_4]{K_2Cr_2O_7}$ $CH_3CCH_2CH_2CH_3$ with O double bond

(b) [cyclohexene] $\xrightarrow[H_2SO_4]{H_2O}$ [cyclohexanol with OH] $\xrightarrow[H_2SO_4]{K_2Cr_2O_7}$ [cyclohexanone with O]

17.45 Oral, intravenous, and interperitoneal.

17.47 LD_{50} measures only the lethal dose; it does not measure the ability of less-than-lethal doses to cause serious injury and the nature of those serious injuries.

17.49 Formulas for this one ketone and two aldehydes of molecular formula C_4H_8O are:

(a) $CH_3CCH_2CH_3$ with O double bond (b) $CH_3CH_2CH_2CH$ with O double bond and CH_3CHCH with O double bond and CH_3

17.51 2-Propanol has the higher boiling point because of the greater attraction between its molecules due to hydrogen bonding through its hydroxyl group.

17.53 (a) The hydroxyaldehyde is first redrawn to show the OH group nearer the CHO group. Closing the ring in hemiacetal formation gives the cyclic hemiacetal.
(b) 5-Hydroxyhexanal has one stereocenter, and 2 stereoisomers (one pair of enantiomers) are possible.
(c) The cyclic hemiacetal has two stereocenters, and four stereoisomers (two pairs of enantiomers) are possible.

[chemical structure: $CH_3\overset{*}{C}HCH_2CH_2CH_2CH$ with O double bond and OH] redrawn to show OH near CHO →

5-Hydroxyhexanal

[chemical structures showing open-chain hydroxyaldehyde $\overset{H^+}{\rightleftharpoons}$ the cyclic hemiacetal]

The cyclic hemiacetal

17.55 During any attempts to isolate the enol form, it undergoes keto-enol tautomerism, and only the more stable keto form is isolated.

17.57 First is given the alkene that undergoes acid-catalyzed hydration to give the desired alcohol, then the aldehyde or ketone that gives it upon reduction.

(a) $CH_2{=}CH_2$ CH_3CH with O double bond   (b) [cyclohexene] [cyclohexanone with O]

(c) $CH_2{=}CHCH_3$ CH_3CCH_3 with O double bond

(d) [benzene ring]—$CH{=}CH_2$ [benzene ring]—CCH_3 with O double bond

Chapter 18 Carboxylic Acids, Anhydrides, Esters, and Amides

18.A Of these three classes of compounds containing the —OH group, carboxylic acids are the strongest acids, phenols are intermediate, and alcohols are the weakest.

18.B They have in common an atom, either carbon or phosphorus, with a double bond to one oxygen and a single bond to a second oxygen.

18.C Following are sections of two parallel Nylon 66 chains, with hydrogen bonds between N—H and C=O groups indicated by dashed lines.

18.D An atom of high electronegativity adjacent to a carboxyl group pulls electron density toward it and away from the oxygen of the O—H bond, thus weakening the bond between oxygen and hydrogen and making the carboxyl group a stronger acid. Because fluorine is more electronegative than chlorine, predict that trifluoroacetic is a stronger acid than trichloroacetic acid. Trifluoroacetic acid is, in fact, a considerably stronger acid.

18.1 (a) 2,3-dihydroxypropanoic acid
(b) *cis*-2-butenedioic acid
(c) 3,5-dihydroxy-3-methylpentanoic acid

18.2 Each acid is converted to its ammonium salt.
(a) $CH_3(CH_2)_2COOH + NH_3 \longrightarrow CH_3(CH_2)_2COO^-NH_4^+$

    Butanoic acid   Ammonium butanoate
    (Butyric acid)   (Ammonium butyrate)

(b)
$$\overset{\;\;OH}{CH_3CHCOOH} + NH_3 \longrightarrow \overset{\;\;OH}{CH_3CHCOO^-NH_4^+}$$

    2-Hydroxypro-   Ammonium
    panoic acid   2-hydroxypropanoate
    (Lactic acid)   (Ammonium lactate)

18.3 Following is the structural formula of each ester.

18.4 (a) ethyl benzoate   (b) *sec*-butyl acetate

18.5 Under basic conditions, as in part (a), each carboxyl group is present as the carboxylic anion. Under acidic conditions, as in part (b), each carboxyl group is present in its un-ionized form.

(a)

(b) $CH_3\overset{O}{\overset{\|}{C}}(CH_2)_2\overset{O}{\overset{\|}{C}}OCH_2CH_3 + H_2O \xrightarrow{HCl}$

$$CH_3\overset{O}{\overset{\|}{C}}(CH_2)_2\overset{O}{\overset{\|}{C}}OH + CH_3CH_2OH$$

18.6 Each ester is converted to its amide.

$$2CH_3\overset{O}{\overset{\|}{C}}NH_2 + HO\!-\!\!\!\bigcirc\!\!\!-\!OH$$

(b)

18.7 Following is a structural formula for each amide.

18.8 In aqueous NaOH, each carboxyl group is present as a carboxylic anion, and each amine is present in its unprotonated form.

(a) $CH_3\overset{O}{\overset{\|}{C}}N(CH_3)_2 + NaOH \xrightarrow[\text{heat}]{H_2O} CH_3\overset{O}{\overset{\|}{C}}O^-Na^+ + (CH_3)_2NH$

(b)

18.9 Of these four carboxylic acids, only one is chiral. Its stereocenter is marked by an asterisk.

$$CH_3CH_2CH_2CH_2COOH \qquad \overset{\quad\quad}{CH_3CHCH_2COOH}$$
$$\qquad\qquad\qquad\qquad\qquad\qquad\;\; \overset{|}{CH_3}$$

      Pentanoic acid   3-Methylbutanoic
                             acid

$$CH_3CH_2\overset{*}{C}HCOOH$$
$$|$$
$$CH_3$$

2-Methylbutanoic
acid

$$\overset{\displaystyle CH_3}{\underset{\displaystyle CH_3}{CH_3\overset{|}{\underset{|}{C}}COOH}}$$

2,2-Dimethylpropanoic
acid

18.11 Following are structural formulas for each carboxylic acid.

(a) $O_2N-\!\!\!\!\bigcirc\!\!\!\!-CH_2\overset{O}{\overset{\|}{C}}OH$ (b) $H_2NCH_2CH_2CH_2\overset{O}{\overset{\|}{C}}OH$

(c) $\bigcirc\!\!\!-CH_2CH_2CH_2\overset{O}{\overset{\|}{C}}OH$

(d)
$$\underset{H}{\overset{H}{\Large>}}C\!\!=\!\!C\underset{CH_2\overset{O}{\overset{\|}{C}}OH}{\overset{CH_2\overset{O}{\overset{\|}{C}}OH}{}}$$

(e) $HOCH_2\overset{O}{\overset{\|}{C}}HCOH$
$$|$$
$$HO$$

(f) $CH_3CH_2CH_2\overset{O}{\overset{\|}{C}}CH_2\overset{O}{\overset{\|}{C}}OH$

18.13 Oxalic acid (IUPAC name: ethanedioic acid) is a dicarboxylic acid. In calcium oxalate, each of the carboxyl groups is present as its carboxylic anion, giving a charge of -2. The structural formula is drawn here to show Ca^{2+} forming ionic bonds with each carboxylic anion.

$$\overset{O}{\underset{O}{\Large\nwarrow}}C\!\!-\!\!O^-$$
$$Ca^{2+}$$
$$\overset{O}{\underset{O^-}{\Large\swarrow}}C$$

18.15 Hydrogen bonding is between the carbonyl oxygen of one carboxyl group and the hydrogen of the hydroxyl group of the other.

hydrogen bonding
between two
molecules

$$\overset{\delta^-\quad\delta^+}{O\cdots H-O}$$
$$H-C\qquad\qquad C-H$$
$$\underset{\delta^+\quad\delta^-}{O-H\cdots O}$$

18.17 The carboxyl group contributes to water solubility; the hydrocarbon chain prevents water solubility.

18.19 Following are structural formulas for the organic product of each reaction.

(a) $CH_3(CH_2)_4\overset{O}{\overset{\|}{C}}OH$

(b)
$$\overset{\overset{O}{\overset{\|}{C}}OH}{\bigcirc\!\!\!\!}\underset{OH}{\overset{OCH_3}{}}$$

(c) $O\!\!=\!\!\bigcirc\!\!\!-\overset{O}{\overset{\|}{C}}OH$

18.21 Following are completed acid-base reactions.

(a) $\bigcirc\!\!\!-CH_2COOH + NaOH \longrightarrow$

$\bigcirc\!\!\!-CH_2COO^-Na^+ + H_2O$

(b) $CH_3CH\!\!=\!\!CHCH_2COOH + NaHCO_3 \longrightarrow$
$CH_3CH\!\!=\!\!CHCH_2COO^-Na^+ + H_2O + CO_2$

(c)
$$\underset{OCH_3}{\overset{COH}{\bigcirc\!\!\!\!}} + NaHCO_3 \longrightarrow$$

$$\underset{OCH_3}{\overset{COO^-Na^+}{\bigcirc\!\!\!\!}} + H_2O + CO_2$$

(d) $CH_3\overset{OH}{\overset{|}{C}}HCOOH + H_2NCH_2CH_2OH \longrightarrow$

$CH_3\overset{OH}{\overset{|}{C}}HCOO^-\ H_3\overset{+}{N}CH_2CH_2OH$

(e) $CH_3CH\!\!=\!\!CHCH_2COO^-\ Na^+ + HCl \longrightarrow$
$CH_3CH\!\!=\!\!CHCH_2COOH + NaCl$

18.23 The pK_a of lactic acid is 4.07. In a solution whose pH is 4.07, 50% of this acid is present as un-ionized lactic acid and 50% as lactate anion. When the pH of the solution is 7.35–7.45, which is slightly on the basic side, lactic acid is present primarily as lactate anion.

18.25 Treat the mixture with aqueous sodium bicarbonate. Octanoic acid (an acid) will react with sodium bicarbonate (a base) to form sodium octanoate, a water-soluble salt. 1-Octanol is insoluble in water and can be separated from the aqueous solution. Now treat the aqueous solution with concentrated HCl or H_2SO_4 to convert sodium octanoate back to octanoic acid, which is insoluble in water and can be separated from water.

18.27 Following are structural formulas for the ester formed in each reaction.

(a) $CH_3\overset{O}{\overset{\|}{C}}OCH_2CH_2CH(CH_3)_2$ (b)

(c) $CH_3CH_2O\overset{O}{\overset{\|}{C}}CH_2CH_2\overset{O}{\overset{\|}{C}}OCH_2CH_3$

18.29 Following is the alcohol and carboxylic acid from which each ester is derived.

$$\text{(a) } 2CH_3\overset{\overset{\displaystyle O}{\|}}{C}OH + HO-\!\!\!\bigcirc\!\!\!-OH$$

$$\text{(b) } 2CH_3OH + HO\overset{\overset{\displaystyle O}{\|}}{C}CH_2CH_2\overset{\overset{\displaystyle O}{\|}}{C}OH$$

$$\text{(c) } \bigcirc\!\!-\overset{\overset{\displaystyle O}{\|}}{C}OH + CH_3OH$$

$$\text{(d) } CH_3CH_2CH\!=\!CH\overset{\overset{\displaystyle O}{\|}}{C}OH + CH_3\overset{\overset{\displaystyle OH}{|}}{C}HCH_3$$

18.31 Following are structural formulas for each compound.

$$\text{(a) } CH_3O\overset{\overset{\displaystyle O}{\|}}{C}OCH_3 \qquad \text{(b) } O_2N-\!\!\!\bigcirc\!\!\!-\overset{\overset{\displaystyle O}{\|}}{C}NH_2$$

$$\text{(c) } CH_3\overset{\overset{\displaystyle OH}{|}}{C}HCH_2\overset{\overset{\displaystyle O}{\|}}{C}OCH_2CH_3$$

$$\text{(d) } CH_3CH_2O\overset{\overset{\displaystyle OO}{\|\|}}{C}COCH_2CH_3$$

$$\text{(e) } \begin{array}{c} CH_3CH_2 \\ \diagdown \\ \end{array} C\!=\!C \begin{array}{c} \overset{\overset{\displaystyle O}{\|}}{C}OCH_2CH_3 \\ \diagup \\ \end{array} $$
$$\qquad\qquad H \qquad\quad H$$

$$\text{(f) } CH_3CH_2CH_2\overset{\overset{\displaystyle OO}{\|\|}}{C}CCH_2CH_2CH_3$$

18.33 Following is a structural formula for hexadecyl hexadecanoate.

$$CH_3(CH_2)_{14}\overset{\overset{\displaystyle O}{\|}}{C}OCH_2(CH_2)_{14}CH_3$$

18.35 Following is the structural formula for each synthetic flavoring agent.

$$\text{(a) } H\overset{\overset{\displaystyle O}{\|}}{C}OCH_2CH_3 \qquad \text{(b) } CH_3\overset{\overset{\displaystyle O}{\|}}{C}OCH_2CH_2CH(CH_3)_2$$

$$\text{(c) } CH_3\overset{\overset{\displaystyle O}{\|}}{C}OCH_2(CH_2)_6CH_3$$

$$\text{(d) } CH_3CH_2CH_2\overset{\overset{\displaystyle O}{\|}}{C}OCH_3$$

$$\text{(e) } CH_3CH_2CH_2\overset{\overset{\displaystyle O}{\|}}{C}OCH_2CH_3 \qquad \text{(f) } \begin{array}{c}\bigcirc\!\!-\overset{\overset{\displaystyle O}{\|}}{C}OCH_3 \\ \!\!\!\!-NH_2\end{array}$$

18.37 Propanoic acid, which has the possibility of intermolecular association by hydrogen bonding, has the higher boiling point.

18.39 Following are equations for the synthesis of each amide from a methyl ester.

$$\text{(a) } \bigcirc\!\!-NH_2 + CH_3O\overset{\overset{\displaystyle O}{\|}}{C}(CH_2)_4CH_3 \longrightarrow$$

$$\bigcirc\!\!-NH\overset{\overset{\displaystyle O}{\|}}{C}(CH_2)_4CH_3 + CH_3OH$$

$$\text{(b) } (CH_3)_2CH\overset{\overset{\displaystyle O}{\|}}{C}OCH_3 + HN(CH_3)_2 \longrightarrow$$

$$(CH_3)_2CH\overset{\overset{\displaystyle O}{\|}}{C}N(CH_3)_2 + CH_3OH$$

$$\text{(c) } CH_3O\overset{\overset{\displaystyle O}{\|}}{C}(CH_2)_4\overset{\overset{\displaystyle O}{\|}}{C}OCH_3 + 2NH_3 \longrightarrow$$

$$H_2N\overset{\overset{\displaystyle O}{\|}}{C}(CH_2)_4\overset{\overset{\displaystyle O}{\|}}{C}NH_2 + 2CH_3OH$$

18.41 Following is a balanced equation for each reaction.

$$\text{(a) } CH_3O-\!\!\!\bigcirc\!\!\!-NH_2 + CH_3\overset{\overset{\displaystyle OO}{\|\|}}{C}CCH_3 \longrightarrow$$

$$CH_3O-\!\!\!\bigcirc\!\!\!-NH\overset{\overset{\displaystyle O}{\|}}{C}CH_3 + CH_3\overset{\overset{\displaystyle O}{\|}}{C}OH$$

$$\text{(b) } CH_3\overset{\overset{\displaystyle O}{\|}}{C}OCH_3 + HN\!\!\bigcirc \longrightarrow$$

$$CH_3\overset{\overset{\displaystyle O}{\|}}{C}-N\!\!\bigcirc + CH_3OH$$

18.43 Convert nicotinic acid to its ethyl ester by Fischer esterification with ethanol. Then treat the ethyl ester with ammonia.

18.45 Reaction (a) is saponification. Reaction (b) is conversion of an ester to an amide.

$$\text{(a) } \bigcirc\!\!-\overset{\overset{\displaystyle O}{\|}}{C}OCH_2CH_3 + NaOH \overset{H_2O}{\longrightarrow}$$

$$\bigcirc\!\!-\overset{\overset{\displaystyle O}{\|}}{C}O^-Na^+ + CH_3CH_2OH$$

$$\text{(b) } \bigcirc\!\!-\overset{\overset{\displaystyle O}{\|}}{C}OCH_2CH_3 + H_2N(CH_2)_3CH_3 \longrightarrow$$

$$\bigcirc\!\!-\overset{\overset{\displaystyle O}{\|}}{C}NH(CH_2)_3CH_3 + CH_3CH_2OH$$

18.47 Each carbonyl group in meprobamate is part of an amide group in one direction and an ester group in the other direction. The three carbonyl groups in phenobarbital are all part of amide groups.

$$H_2NCOCH_2\overset{CH_3}{\underset{CH_2CH_2CH_3}{C}}CH_2OCNH_2 \xrightarrow[H_2O]{NaOH} 2NH_3$$

Meprobamate   Ammonia

$$+ \; Na^+{}^-OCO^-Na^+ \; + \; HOCH_2\overset{CH_3}{\underset{CH_3}{C}}CH_2OH$$

Sodium carbonate   2,2-Dimethyl-1,3-propanediol

$$\text{Phenobarbital} \xrightarrow[H_2O]{NaOH} \; C_6H_5\overset{CO^-Na^+}{\underset{CO^-Na^+}{C}}CH_2CH_3$$

Phenobarbital

$$+ \; 2NH_3 \; + \; Na^+{}^-OCO^-Na^+$$

18.49 (a) Both lidocaine and mepivacaine contain amide groups. In addition, both contain a tertiary aliphatic amine. (b) Both are derived from 2,6-dimethylaniline. In addition, in each the aliphatic amine nitrogen is separated by one carbon from the carbonyl group of the amide.

18.51 Following is the structural formula of dihydroxyacetone phosphate shown as it would be ionized at pH 7.40.

$$^-O-\overset{O}{\underset{O^-}{P}}-OCH_2\overset{O}{C}CH_2OH$$

Dihydroxyacetone phosphate

18.53 Hydrolysis requires one mole of water per mole of ester.

$$CH_3O-\overset{O}{\underset{OCH_3}{P}}-OCH_3 + H_2O \longrightarrow$$

Trimethyl phosphate

$$CH_3O-\overset{O}{\underset{OCH_3}{P}}-OH + CH_3OH$$

Dimethyl phosphate

18.55 The two functional groups in aspirin are a carboxyl group and an ester.

18.57 The oxygen-containing functional group in coumarin is a lactone (a cyclic ester). The two types of oxygen-containing functional groups in dicoumarol are a lactone (a cyclic ester) and an enol (an —OH group on a carbon-carbon double bond).

18.59 Warfarin is an anticoagulant; it inhibits the clotting of blood. It is often referred to as a "blood thinner."

18.61 The portion derived from urea is the same in all three compounds. It is shown here and circled for two of the three compounds.

Secobarbital
(Seconal)

Thiopental
(Pentothal)

18.63 The two esters in cocaine are circled. Hydrolysis requires two moles of water per mole of cocaine.

$$+ \; 2H_2O \longrightarrow$$

$$+ \; CH_3OH \; +$$

18.65 Of the two amines in procaine, the tertiary aliphatic amine is the stronger base, and the one that reacts with HCl to form a salt.

18.67 Lactomer stitches dissolve as the ester groups in the polymer chain are hydrolyzed until only glycolic and lactic acid remain. These small molecules are metabolized and excreted by existing biochemical pathways.

18.69 Note that ibuprofen has one stereocenter, here marked by an asterisk.

18.71 The parent alkane, undecane, has 11 carbons in its chain. There is a double bond in its chain between carbons 10-11. Zinc ion is Zn^{2+}, which means that this compound is a salt consisting of one zinc ion and two undecenoate anions.

18.73 In a solution whose pH is 8.4, which is basic, ascorbic acid is present primarily as ascorbate anion.

18.75 Following is the structural formula of benzocaine.

18.77 Following is a balanced equation for this synthesis of acetaminophen.

Acetic anhydride   4-Aminophenol

Acetaminophen

18.79 Following are structural formulas for each form.

Dimethyl phosphate   Sodium dimethyl phosphate

Chapter 19 Carbohydrates

19.A D-Glucose is by far the most abundant D-aldohexose in the biological world.

19.B D-glucose. When drawn in a chair conformation, all substituents on the ring are equatorial.

19.C Chair conformations are a more accurate representation of the actual shape and bond angles of the six-membered pyranose ring.

19.D Each carbon in a mono- and disaccharide has on it an oxygen group able to participate in hydrogen bonding with water molecules.

19.1 Following are Fischer projections for the four 2-ketopentoses.

one pair of enantiomers

D-Ribulose   L-Ribulose

a second pair of enantiomers

D-Xylulose   L-Xylulose

19.2 D-Mannose differs in configuration from D-glucose only at carbon 2. One way to arrive at the structures of the α and β forms of D-mannopyranose is to draw the corresponding α and β forms of D-glucopyranose and then invert the configuration in each at carbon 2.

β-D-Mannopyranose
(β-D-Mannose)

α-D-Mannopyranose
(α-D-Mannose)

19.3 D-Mannose differs in configuration from D-glucose only at carbon-2.

β-D-Mannopyranose   α-D-Mannopyranose
(β-D-Mannose)   (α-D-Mannose)

19.4 Following are structural formulas for these glycosides.

Methyl β-D-fructofuranoside
(Methyl β-D-fructoside)

Methyl α-D-Mannopyranoside
(Methyl α-D-Mannoside)

19.5 The β-glycosidic bond is between carbon 1 of the left unit and carbon 3 of the right unit.

β-1,3-glucosidic bond

19.7 During photosynthesis, plants produce oxygen, O_2. During metabolism, animals produce carbon dioxide, CO_2.

19.9 In D-glucose, carbons 2, 3, 4, and 5 are stereocenters. In D-ribose, carbons 2, 3, and 4 are stereocenters.

19.11 The D or L configuration in an aldopentose is determined by its configuration at carbon 4.

19.13 First draw the D form of each, and then its mirror image.

D-Ribose   L-Ribose

D-Arabinose   L-Arabinose

19.15 (a) A pyranose is a six-membered cyclic hemiacetal form of a monosaccharide.
(b) A furanose is a five-membered cyclic hemiacetal form of a monosaccharide.

19.17 They are not enantiomers; that is, they are not mirror images. They differ in configuration only at carbon 1 and, therefore, they are diastereomers.

19.19 No. The hydroxyl groups on carbon 2, 3, and 4 of α-D-glucose are equatorial, but the hydroxyl group on carbon 1 is axial.

19.21 Compound (a) differs from D-glucose only in the configuration at carbon 4. Compound, only at carbon 3.

D-Galactose

D-Allose

19.23 (a) The specific rotation of α-L-glucose is −112.2°.
(b) Yes, the specific rotation of α-L-glucose also changes to −52.7°.

19.25 Each aldehyde group is reduced to a primary alcohol.

(a)   (b)

19.27 Review Section 17.6 and your answer to Problem 17.54. The intermediate in this conversion is an enediol; that is, it contains a carbon-carbon double bond with two OH groups on it.

D-Glucose 6-phosphate   An enediol intermediate

D-Fructose 6-phosphate

19.29 A glycosidic bond is the bond from the anomeric carbon of a monosaccharide furanoside or pyranoside to an —OR group.

19.31 Glycosidic bond refers to the bond from the anomeric carbon to an —OR group of any monosaccharide unit. Glucosidic bond refers specifically to a glycosidic bond from the anomeric carbon of glucose.

19.33 Maltose and lactose are reducing sugars because one of the monosaccharide units is in its hemiacetal form and, thus, in equilibrium with its aldehyde form. Sucrose is not a reducing sugar because both anomeric carbons participate in formation of the glycosidic bond.

19.35 An oligosaccharide contains approximately four to ten monosaccharide units. A polysaccharide contains more, generally many more, than ten monosaccharide units.

19.37 Cows have within their digestive systems enzymes that catalyze the hydrolysis of the glycosidic bonds of cellulose. We do not have these enzymes in our digestive systems.

19.39 (a) One way to construct each repeating disaccharide is to draw a disaccharide of two units of D-glucose

and then modify each glucose unit to make it the appropriate unit in alginic acid or pectic acid.

(a)

units of D-mannuronic acid

(b)

units of D-galacturonic acid

19.41 (a) The negative charges are provided by —OSO$_3^-$ and —COO$^-$ groups. (b) The higher the degree of polymerization, the better the anticoagulant activity.

19.43 Feeding an infant a formula containing sucrose eliminates galactose from its diet.

19.45 L-Ascorbic acid is oxidized (there is loss of two hydrogen atoms) when it is converted to L-dehydroascorbic acid.

19.47 It is used to monitor blood glucose levels in diabetics and prediabetics.

19.49 (a) L-Fucose is a L-aldohexose. (b) What is unusual about it in human biochemistry is that it belongs to the L series of monosaccharides, and that it has no oxygen atom at carbon 6. (a) If its terminal —CH$_3$ group were converted to a —CH$_2$OH group, the monosaccharide formed would be L-galactose.

19.51 It has none.

19.53 In an aldooctose, carbons 2, 3, 4, 5, 6, and 7 are stereocenters. The number of stereoisomers possible, therefore, is $2^6 = 64$. Of these, 32 are D-aldooctoses and 32 are L-aldooctoses.

19.55 It is the furanose form of a D-aldopentose. If you compare the configuration of this furanose with that of D-ribose (a good reference form to remember), you see that this furanose differs in configuration from D-ribose only at carbon 3. Now consult Table 19.2 and discover that the aldopentose that differs in configuration from D-ribose only at carbon 3 is D-xylose. Therefore, the compound drawn in this problem is β-D-xylofuranose (β-D-xylose).

19.57 During boiling in water with an acid catalyst, some of the sucrose is hydrolyzed to glucose and fructose, and fructose is sweeter than sucrose.

19.59 (a) The left unit is derived from D-galactose. The right unit is derived from D-glucosamine. (b) The monosaccharide units are joined by a β-1,4-glycosidic bond. (c) The net charge on this disaccharide unit is −1.

Chapter 20 Lipids

20.A   Three: ABC, ACB, and BAC

20.B   (a) the unsaturated fatty acid   (b) inositol and the phosphate group

20.C   They all contain the steroid ring structure and have methyl groups on C-13 and an acetylene group on C-17.

20.D   Both groups are derived from a common precursor, PGH_2, which is catalyzed by the COX enzymes.

20.1   It is an ester of glycerol and contains a phosphate group; therefore it is a glycerophospholipid. Besides glycerol and phosphate it has a myristic acid and a linoleic acid component. The other alcohol is serine. Therefore, it belongs to the subgroup of cephalins.

20.3   Hydrophobic means water hating. It is important because if the body did not have such molecules there could be no structure since the water would dissolve everything.

*20.5   The melting point would increase. This happens because trans double bonds would fit more in the packing of the long hydrophobic tails, creating more order and therefore more interaction between chains. This would require more energy to disrupt, hence a higher melting point.

20.7   highest: A and B (no double bonds); lowest C and D (two double bonds in each diglyceride)

20.9   (b) because its molecular weight is higher

20.11   lowest (c); then (b); highest (a)

20.13   the more long chain groups the lower the solubility; lowest (a); then (b); highest (c)

20.15   glycerol, sodium palmitate, sodium stearate, and sodium linolenate

20.17   (a) They are found around cells and around small structures inside cells. (b) They separate cells from the external environment and allow selective passage of nutrients and waste products into and out of cells.

20.19   complex lipids and cholesterol

*20.21   Phosphatidyl inositol; the inositol has five —OH groups, which can form H-bonds with water.

*20.23   (c) because it has three charges, while the others have only two

20.25   No. For example, in red blood cells lecithins are on the outside, facing the plasma; cephalins are on the inside.

20.27   (a) yes   (b) because high serum cholesterol has been correlated with such diseases as atherosclerosis

*20.29   (a) Carbon stereocenters are circled.

(b) $2^8 = 256$ (c) only 1

20.31   It contains polar groups of phospholipids, cholesterol, and proteins.

20.33   It increases the glucose and glycogen concentrations in the body. It is also an anti-inflammatory agent.

20.35   Loss of a methyl group at the junction of rings A and B, together with a hydrogen at position 1, produces a double bond. This causes tautomerization of the keto group at C-3, resulting in an aromatic ring A, with a phenolic group at the C-3 position.

20.37   (a) They both have a steroid ring structure.
(b) RU486 has a *para*-aminophenyl group on ring C and a triple-bond group on ring D.

20.39   (a) PGE_2 has a ring, a ketone group, and two OH groups.   (b) The C=O group on PGE_2 is an OH group on $PGF_{2\alpha}$.

20.41   Prostaglandins contain a 5-membered ring; leukotrienes have no ring.

20.43   (a) oxidation of the double bond to aldehydes and other compounds
(b) exclude oxygen and sunlight; keep refrigerated

20.45   Soaps are salts of carboxylic acids; detergents are salts of sulfonic acids.

20.47   (a) Sphingomyelin acts as an insulator.
(b) The insulator is degraded, impairing nerve conduction.

20.49   It blocks cholesterol synthesis, thus helping to remove cholesterol from the blood.

20.51   They prevent ovulation.

20.53   It inhibits prostaglandin formation by preventing ring closure.

20.55   NSAIDS inhibit cyclooxygenases (COX enzymes) that are needed for ring closure. Leukotrienes have no ring in their structure; therefore they are not affected by COX inhibitors.

20.57   (See Figure 20.2.) Polar molecules cannot penetrate the bilayer. They are insoluble in lipids. Nonpolar molecules can interact with the interior of the bilayer (like dissolves like).

20.59   The various esters of cholesterol have different fatty acids (saturated, unsaturated, etc.). This prevents them from packing into a crystal lattice. Cholesterol, being uniform, does form a crystal lattice.

20.61   They are both salts of sulfonic acids.

20.63   (a) aldehyde, ketone, alcohol, C=C double bond
(b) Cortisone has all these groups except the aldehyde

20.65   Celebrex inhibits only COX-2 enzymes that cause inflammation, but not COX-1, which is needed to manufacture prostaglandins needed for normal physiological functions.

Chapter 21 Proteins

21.A Glycine does not have a chiral carbon. It has two identical substituents, H, on the α-carbon.

21.B $NH_3^+-CH-CH_2-CH_2-\overset{O}{\underset{}{C}}-NH-CH-\overset{O}{\underset{}{C}}-NH-CH_2-COO^-$

with COO^- on the first CH, and CH_2-S branch on the middle CH

$NH_3^+-CH-CH_2-CH_2-\overset{O}{\underset{}{C}}-NH-CH-\overset{O}{\underset{}{C}}-NH-CH_2-COO^-$

with COO^- on the first CH, and S on the middle CH

21.C To transform an α-helix into a β-pleated sheet, H-bonds between the backbone must be broken. After being in the random coil form, new H-bonds form either intramolecularly or intermolecularly.

21.D The heme and the polypeptide chain form the quaternary structure of cytochrome c. This is a conjugated protein.

21.1 $NH_2-CH-\overset{O}{\underset{}{C}}-OH$ with $CH-CH_3$ and CH_3 $NH_2-CH-\overset{O}{\underset{}{C}}-OH$ with CH_2 and phenyl

$\downarrow$

$NH_2-CH-\overset{O}{\underset{}{C}}-NH-CH-\overset{O}{\underset{}{C}}-OH$ with $CH-CH_3$, CH_3 on first; CH_2, phenyl on second

21.2 salt bridge

21.3 Most proteins are highly specific in their actions, so a great many are needed to do all the tasks required.

21.5 Tyrosine has an extra —OH group in the side chain. The terminal group of the side chain of tyrosine is a phenol; in phenylalanine, it is a benzene.

21.7 arginine

21.9

pyrrolidine ring with COOH, N—H

pyrrolidines (saturated pyrroles)

21.11 They supply most of the amino acids we need in our bodies.

21.13 Their structures are the same, except that a hydrogen of alanine is replaced by a phenyl group in phenylalanine.

21.15 $CH_3-CH-\overset{H}{\underset{NH_2}{C}}-COOH$ with CH_3

L-Valine

$HOOC-\overset{H}{\underset{NH_2}{C}}-CH-CH_3$ with CH_3

D-Valine

21.17 $CH_3-CH-COO^- + H_3O^+ \longrightarrow$ with NH_3^+

$CH_3-CH-COOH + H_2O$ with NH_3^+

$CH_3-CH-COO^- + OH^- \longrightarrow$ with NH_3^+

$CH_3-CH-COO^- + H_2O$ with NH_2

21.19 at pH 1 $CH_3-CH-CH-COOH$ with CH_3 and NH_3^+

at pH 12 $CH_3-CH-CH-COO^-$ with CH_3 and NH_2

*21.21 There are six. One of them is

$NH_2-CH-\overset{O}{\underset{}{C}}-NH-CH-\overset{O}{\underset{}{C}}-NH-CH-COOH$

first CH: CH_3-CH, OH; middle CH: CH_2, CH_2, CH_2, NH, $C=NH$, NH_2; last CH: CH_2, CH_2, S, CH_3

21.23 $CH_3-CH-CH_2-CH-\overset{}{\underset{O}{C}}-N$ (pyrrolidine ring with COOH) with CH_3 and NH_2

21.25

$$NH_2-\underset{\underset{CH_3}{|}}{CH}-\overset{\overset{O}{\|}}{C}-OH + NH_2-\underset{\underset{\underset{\underset{NH_2}{|}}{\underset{C=O}{|}}}{\underset{\underset{CH_2}{|}}{CH_2}}}{CH}-\overset{\overset{O}{\|}}{C}-OH \longrightarrow$$

$$NH_2-\underset{\underset{CH_3}{|}}{CH}-\overset{\overset{O}{\|}}{C}-NH-\underset{\underset{\underset{\underset{NH_2}{|}}{\underset{C=O}{|}}}{\underset{\underset{CH_2}{|}}{CH_2}}}{CH}-\overset{\overset{O}{\|}}{C}-OH \quad \text{or}$$

$$NH_2-\underset{\underset{\underset{\underset{NH_2}{|}}{\underset{C=O}{|}}}{\underset{\underset{CH_2}{|}}{CH_2}}}{CH}-\overset{\overset{O}{\|}}{C}-NH-\underset{\underset{CH_3}{|}}{CH}-\overset{\overset{O}{\|}}{C}-OH$$

21.27 at pH 2.0

$$\overset{+}{N}H_3-\underset{\underset{\underset{S-CH_3}{|}}{\underset{CH_2}{|}}}{\underset{CH_2}{|}}CH-\overset{O}{\underset{\|}{C}}-NH-\underset{\underset{\underset{OH}{|}}{CH_2}}{CH}-\overset{O}{\underset{\|}{C}}-NH-\underset{\underset{\underset{SH}{|}}{CH_2}}{CH}-COOH$$

at pH 7.0

$$\overset{+}{N}H_3-\underset{\underset{\underset{S-CH_3}{|}}{\underset{CH_2}{|}}}{\underset{CH_2}{|}}CH-\overset{O}{\underset{\|}{C}}-NH-\underset{\underset{\underset{OH}{|}}{CH_2}}{CH}-\overset{O}{\underset{\|}{C}}-NH-\underset{\underset{\underset{SH}{|}}{CH_2}}{CH}-COO^-$$

at pH 10.0

$$NH_2-\underset{\underset{\underset{S-CH_3}{|}}{\underset{CH_2}{|}}}{\underset{CH_2}{|}}CH-\overset{O}{\underset{\|}{C}}-NH-\underset{\underset{\underset{OH}{|}}{CH_2}}{CH}-\overset{O}{\underset{\|}{C}}-NH-\underset{\underset{\underset{SH}{|}}{CH_2}}{CH}-COO^-$$

21.29 It would acquire a net positive charge and become more water-soluble.

21.31 (a) 256   (b) $20^4 = 160\,000$

21.33 valine or isoleucine

21.35 (a) secondary   (b) tertiary and quaternary
(c) quaternary   (d) primary

21.37 Above pH 6.0 the COOH groups are converted to COO$^-$ groups. The negative charges repel each other, disrupting the compact α-helix and converting it to a random coil.

21.39 (1) C-terminal end   (2) N-terminal end
(3) pleated sheet   (4) random coil
(5) hydrophobic interaction
(6) disulfide bridge
(7) α-helix   (8) salt bridge
(9) hydrogen bonds

21.41 valine, phenylalanine, alanine, leucine, isoleucine

21.43 a protein that contains a non-amino acid portion called a prosthetic group

*21.45

21.47 It breaks the disulfide bridges of keratin and allows the hair to be set into a desired shape.

21.49 to destroy bacteria on the surface of the skin

21.51 In younger people, the AGE products are degraded; in older people, the metabolism slows down and the AGE products accumulate.

21.53 It provided some immunity against malaria.

21.55 Hydroxyurea therapy promotes the manufacture of fetal hemoglobin, which does not carry beta chains and, therefore, the sickle cell mutations. Cells with fetal hemoglobin do not sickle.

21.57 calcium hydroxyapatite

21.59 Photofrin is a photosensitizer. When it is absorbed by a tumor, a laser beam is directed toward the tumor. The energy of the laser transforms the Photofrin to an active compound that destroys the tumor.

21.61 more soluble because carbohydrates have many OH groups that form hydrogen bonds with water

21.63 (a) $4^2 = 16$ (b) $20^2 = 400$

21.65 The protein-digesting enzymes in the stomach and intestines would hydrolyze it before it could reach the blood.

21.67 It can stabilize 3-dimensional structures if it is oxidized and forms S—S bridges.

21.69 (a) hydrophobic   (b) salt bridge
(c) hydrogen bond   (d) hydrophobic

21.71 glycine

21.73 one positive charge: NH_3^+

Chapter 22 Enzymes

22.A If the number of active sites are greater than the maximum substrate concentration, the rate would linearly increase with concentration. No saturation point is reached because the number of active sites was not exhausted.

22.B (a) less active at normal body temperature
(b) The activity decreases.

22.C (A) Large circle = enzyme
(B) hexagon = active site
(C) triangle = glucose
(D) larger triangle = fluoroglucose
(E) small square = Mg^{2+}
(F) half hexagon = ATP
(G) black circle = Cd^{2+}
(a) assemble ABCEF
(b) ABDEF
(c) ABCEFH

22.D Phosphorylase b Phosphorylase a

22.1 A catalyst is a substance that speeds up a reaction without being used up itself. An enzyme is a catalyst that is a protein or a nucleic acid.

22.3 more than 3000

22.5 because enzymes are highly specific, and there are a great many different reactions that must be catalyzed

22.7 Trypsin catalyzes the hydrolysis of peptides only at specific points (for example, at a lysine), while lipases catalyze the hydrolysis of all kinds of triglycerides.

22.9 (a) isomerase   (b) hydrolase
(c) oxidoreductase   (d) lyase

22.11 A cofactor is a nonprotein portion of an enzyme. A coenzyme is an organic cofactor.

22.13 See Section 22.3.

22.15 No, at high substrate concentration the enzyme surface is saturated, and doubling of the substrate concentration will produce only a slight increase in the rate of the reaction or no increase at all.

22.17 the enzymes of thermophile bacteria that live in the ocean floor near vents.

22.19 It is too bulky to fit into the cavity where the active site is located.

22.21 In competitive inhibition the same maximum rate can be obtained as in the reaction without inhibitor, although at a higher substrate concentration. In noncompetitive inhibition the maximum rate is always lower than that without inhibitor.

22.23 an allosteric mechanism

22.25 There is no difference. They are the same.

22.27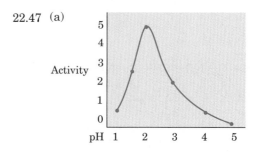

22.29 The enzyme is now named for the compound (alanine) that donates the amino group. The former name refers to the products.

22.31 H_4

22.33 digestive enzymes such as Pancreatin or Acro-lase

22.35 Succinylcholine is a competitive inhibitor of acetylcholinesterase and prevents acetylcholine from triggering muscle contractions.

22.37 Trypsin. Both trypsin and papain convert proteins to peptides. Carboxypeptidase yields amino acids by cleaving peptide bonds starting at the C-terminal end.

22.39 by administering antibiotics

*22.41 Alanyl and threonyl residues both provide the same surface environment. In both, the terminal group of the side chain is a nonpolar methyl group. The nature of interaction is hydrophobic interaction. Glycyl residue also contributes to the hydrophobic interaction.

*22.43 Humans do not synthesize their folic acid; they get it in their diets.

22.45 (a) pasteurized milk, canned tomatoes
(b) yogurt, pickles

22.47 (a)

(b) 2   (c) zero activity

22.49 cofactors

22.51 ALT

22.53 No, the nerve gas binds irreversibly by a covalent bond, so it cannot be removed simply by adding substrate.

22.55 a transferase

22.57 They catalyze the self-splicing of introns.

Chapter 23 Chemical Communication: Neurotransmitters and Hormones

23.A   A chemical messenger operates between cells; secondary messengers signal inside a cell in the cytoplasm.

23.B   An agonist stimulates the receptor; therefore, more glucose would enter the cells, and its concentration in the serum would decrease.

23.C   The phosphorylation by this enzyme can
(a) activate an enzyme (phosphorylase),
(b) inhibit an enzyme (glycogen synthase),
(c) open ion gates.

23.D   The luteinizing hormone is a peptide. It acts through the G-protein–adenylate cyclase cascade, producing secondary messengers. Progesterone is a steroid hormone; it acts in the nucleus of cells, influencing protein synthesis.

23.1   They are mostly glycoproteins.

23.3   The presynaptic end stores vesicles containing neurotransmitters; the postsynaptic end contains receptors.

23.5   The concentration of Ca^{2+} in neurons controls the process. When it reaches $10^{-4}\ M$ the vesicles release the neurotransmitters into the synapse.

23.7   anterior pituitary gland

23.9   Upon binding of acetylcholine, the conformation of the proteins in the receptor changes and the central core of ion channel opens.

23.11   Succinylcholine is a competitive inhibitor of acetylcholinesterase.

23.13   Taurine is a β-amino acid; its acidic group is $-SO_2OH$ instead of $-COOH$.

23.15   The amino group in GABA is in the gamma position; proteins contain only alpha amino acids.

23.17   (a) norepinephrine and histamine
(b) They activate a secondary messenger, cAMP, inside the cell.
(c) amphetamines and histidine

*23.19   It is phosphorylated by an ATP molecule.

23.21

23.23   (a) Amphetamines increase and (b) reserpine decreases the concentration of the adrenergic neurotransmitter.

23.25   the corresponding aldehyde

23.27   (a) the ion-translocating protein
(b) It gets phosphorylated and changes its shape.
(c) It activates the protein kinase that does the phosphorylation of the ion-translocating protein.

23.29   They are pentapeptides.

23.31   Calcium sparks are localized at specific parts of the cell; global calcium waves travel to different parts of the cell.

23.33   (a) They stop nerve transmission.
(b) They bind to acetylcholinesterase.

23.35   Facial tics are caused by uncontrolled release of acetylcholine, which, when absorbed by muscle cells, causes uncontrolled contractions. The botulinum prevents release of the acetylcholine, thus preventing the contractions.

23.37   Acetylcholinesterase inhibitors, such as Cognex, inhibit the enzyme that decomposes the neurotransmitter.

23.39   by binding to the dopamine transporter

23.41   in the conversion of arginine to citrulline

23.43

23.45   Tamoxifen binds to the estrogen receptor. Therefore, estrogen cannot bind to its receptor and cannot reach the DNA in the nucleus to influence protein synthesis.

23.47   It opens the ion gates and allows the translocation of ions.

23.49   Yes. It fits into the same receptor site as enkephalins.

23.51   It is associated with the receptor.

23.53   (a) Both deliver a message to a receptor.
(b) Neurotransmitters act rapidly over short distances. Hormones act more slowly and over longer distances. They are carried by the blood from the source to the target cell.

23.55   Cholera toxin affects G-protein.

23.57   Cobra toxin is an antagonist of the nicotinic cholinergic receptor.

Chapter 24 Nucleotides, Nucleic Acids, and Heredity

24.A   glycosidic bond between the sugar and the base, and the phosphate ester bond between the sugar and the phosphate

24.B   The phosphate groups carry H ions, which make the DNA and RNA acidic.

24.C   DNA; it contains up to millions of base pairs.

24.D   H-bonds, no covalent bond is broken; hence, the primary structure does not change during replication.

24.1   They are made of DNA and basic proteins called histones.

24.3   hemophilia, sickle-cell anemia, etc.

24.5   (a) base, sugar, phosphate
(b) base, sugar

24.7 Thymine has a methyl group in the 5 position; uracil has a hydrogen there.

24.9 (a)

(b) the same, but without the —OH in the 2 position.

24.11 Ribose has an —OH in the 2 position; deoxyribose has an H there.

24.13 C-3 and C-5 to the phosphate group; C-1 to the base

24.15 (a)

(b)

24.17 (a) the 3'—OH end and the 5'—OH end
(b) A is the 5'—OH end. C is the 3'—OH end.

*24.19 two

24.21 RNA

24.23 amino and carbonyl groups

24.25

24.27 at either end of the chain or in the middle

24.29 polymerases

24.31 The train model assumes that polymerases move along the DNA just like a locomotive does along a train track. The model of "polymerase factory" assumes that the number of polymerase proteins participating in the replication are immobilized as a factory, and the DNA winds its way through them.

24.33 rRNA

24.35 mRNA

24.37 It is an electrostatic attraction.

24.39 mRNA contains exons and introns when it is first synthesized, but before it actually functions the introns are cut out.

24.41 eight

24.43

24.45 TTAGGG

24.47 Telomerase resynthesizes the shortened DNA end in telomers; thus, the cell has no signal telling it when to stop dividing and therefore never dies.

24.49 The DNA fragments of the child, mother, and alleged father are run on the same gel, and their patterns are compared.

*24.51 Human DNA contains introns and other noncoding sequences. Bacterial DNA, like that of all prokaryotes, does not have such sequences, and so the total genome is much smaller.

24.53 The DNA fragments in multiple of 180 base pairs.

24.55 because every time an error is introduced it may cause a mutation that can be harmful or even detrimental to the propagation of the species

*24.57 the nucleosomes

Chapter 25 Gene Expression and Protein Synthesis

25.A (a) In transcription they are the polymerases and transcription factors
(b) In translation they are ribonucleic acid and proteins of the ribosomes and proteins of the elongation factors.

25.B Since all the diverse life forms on Earth share the same genetic code, it must mean that they evolved from one common ancestry.

25.C The codons are on the mRNA, and the anticodons on the tRNA.

25.D No. It would produce too many small fragments and no sticky ends.

25.1 When transcription factors are phosphorylated in the nucleus of a cell, a transcription cascade is initiated.

25.3 the 3′ end of the DNA template

25.5 (a) the larger (60S) portion　(b) See Section 25.5.

25.7 (a) CGA　(b) alanine

25.9 Yes they are both AUG; however, two different tRNAs perform the two different functions.

25.11 The A site is where the incoming tRNA carrying the amino acid binds. The P site is where the growing peptide chain binds, and the E site is where the exiting tRNA binds before being recycled.

25.13 They enable the tRNA to bind to the A site of the rRNA.

25.15 operator and promoter

25.17 In prokaryotes the RNA polymerase binds directly to the promoter. In eukaryotes the RNA polymerase has little affinity for the DNA; it needs transcription factors for binding.

25.19 (a) UAU and UAC. Both code for tyrosine.
(b) If UAU is changed to UAA, then instead of tyrosine, the chain will be terminated. This might well be fatal.

25.21 Yes; if the mutation occurred on a recessive gene.

25.23 Restriction endonucleases cleave DNA at specific sites.

25.25 In genetic engineering a new gene is inserted into the genome. However, in the corn that acquired insect resistance by natural selection, mutations occurred in the native genes.

25.27 They do not have the molecules necessary to reproduce themselves and must rely on their hosts.

25.29 T (lymphocyte) cells

*25.31 The promoter enabled the HIV protease to be expressed in the lens of the mouse. Thus expressed, the lens became cataractous. In this manner, it could serve as a bioessay to prove the efficacy of new HIV protease–inhibitor drugs.

25.33 A change from guanine to thymine in the gene results in a valine in place of a glycine.

25.35 When p53 binds to DNA, it arrests the cell cycle; thus, it allows more time for the cell defenses to repair damaged DNA.

25.37 (a) The 5′ end and the section just before the 3′ end must be base paired.　(b) There must be no base pairing in the anticodon loop.

25.39 For each amino acid, there is a specific enzyme that attaches it to its own specific tRNA.

25.41 because for most amino acids there is more than one codon

*25.43 When the combination of the first and second bases are AU, UA, UG, AG, and GG, the third base of the codon is relevant. In the eleven other cases, the third base is irrelevant.

Chapter 26 Bioenergetics. How the Body Converts Food to Energy

26.A (a) NADH is oxidized to NAD^+ at complex I of the electron transport chain. NAD^+ is reduced to NADH in the citric acid cycle, for example, at the conversion of malate to oxaloacetate.

26.B Succinate is oxidized by FAD, and the oxidation product is fumarate.

26.C Cytochrome C and CoQ.

26.D The energy of motion appears first in the ion channel where, upon passage of H^+, the proteins making this channel rotate.

26.1 ATP

26.3 (a) 2　(b) the outer membrane

26.5 the matrix

26.7 2

26.9 Neither; they yield the same energy.

26.11 phosphate ester

26.13 the 2 N atoms that are part of C=N bonds

26.15 (a) adenosine　(b) NAD^+ and FAD
(c) acetyl groups

26.17 amide

*26.19 No. The pantothenic acid portion is not the active part.

26.21 acetyl coenzyme A

26.23 α-ketoglutarate

26.25 oxidative decarboxylation

26.27 It removes two hydrogens from succinate to produce fumarate.

26.29 No, but GTP is produced in step ⑤.

26.31 It allows the energy to be released in small packets.

26.33 cis-aconitate and fumarate

26.35 to NAD^+, which becomes NADH + H^+

26.37 (a) 3　(b) 2

26.39 in the inner membranes of the mitochondria

26.41 (a) 0.5　(b) 12

26.43 the proton translocating ATPase

26.45 1.5 kcal

26.47 (a) by sliding the thick filaments (myosin) and the thin filaments (actin) past each other
(b) from the hydrolysis of ATP

26.49 ATP transfers a phosphate group to the serine residue at the active site of the enzyme.

*26.51 No. It would harm humans because they would not synthesize enough ATP molecules.

26.53 (a) hydroxylation　(b) the air we breathe in

*26.55 60 g or 1 mole of CH_3COOH

26.57 They are both hydroxy acids.

26.59 myosin

26.61 In NADH the reduction occurred at a carbon atom; in $FADH_2$ it occurred on two nitrogen atoms.

26.63 F_0; it is made of 12 subunits.

26.65 No, it largely comes from the chemical energy as a result of the breaking of bonds in the O_2 molecule.

Chapter 27 Specific Catabolic Pathways. Carbohydrate, Lipid and Protein Metabolism

27.A Glyceraldehyde 3-phosphate is oxidized first. The energy production begins with the creation of NADH, which can enter oxidative phosphorylation.

27.B No, there is no discrepancy. Table 27.1 lists the ATP yield per glucose molecule. Glucose yields 2 glyceraldehyde-3-phosphate molecules when C_6 becomes 2 C_3. Thus, the yield is one NADH per glyceraldehyde 3-phosphate but 2 NADH per glucose.

27.C No, the same amount of ATP/C atoms is produced in both.

27.D It is energy consuming. Three ATP molecules are hydrolyzed per cycle.

27.1 glycogen

27.3 They serve as building blocks for the synthesis of proteins.

27.5 The two C_3 fragments are in equilibrium, and as the glyceraldehyde phosphate is used up, the equilibrium shifts and converts the other C_3 fragment (dihydroxyacetone phosphate) to glyceraldehyde phosphate.

27.7 (a) steps ① and ③   (b) steps ⑥ and ⑨

27.9 step ⑤

27.11 NADPH

27.13 6 moles

27.15 12

*27.17 Two net ATP molecules are produced in both cases.

*27.19 glycerol kinase

*27.21 (a) thiokinase and thiolase   (b) —SH
        (c) Both enzymes insert a CoA—SH into a compound.

27.23 $CH_3(CH_2)_4CO—CoA$, three acetyl CoA, three $FADH_2$, three NADH + H^+

27.25 112

27.27 carbohydrates

27.29 See Section 27.7.

27.31 It enters the citric acid cycle.

27.33 $CH_3—\overset{\underset{|}{NH_3^+}}{CH}—COO^- + NAD^+ + H_2O \longrightarrow$

$CH_3—\overset{\underset{\parallel}{O}}{C}—COO^- + NADH + H^+ + NH_4^+$

27.35 because pyruvate can be converted to glucose when the body needs it

27.37 fumarate

27.39 ketone bodies

27.41 It is stored in ferritin and reused.

27.43 lactic acid accumulation

27.45 the bicarbonate/carbonic acid buffer

27.47 $Na^{+ \, -}OOC—CH_2CH_2—\overset{\underset{|}{NH_3^+}}{CH}—COO^-$

27.49 Glycine, the C-terminal amino acid of ubiquitin, forms an amide linkage with a lysine residue in the target protein.

*27.51

Phenylalanine   α-Ketoglutarate

Phenylpyruvate   Glutamate

27.53 diabetes

*27.55 (a) in steps ⑤ and ⑫
        (b) in steps ⑪ and ⑩
        (c) If enough O_2 is present in muscles, a net increase of NADH + H^+ occurs. If not enough O_2 is present (as in yeast), there is no net increase of either NAD^+ or NADH + H^+.

*27.57 Yes, they are glucogenic and can be converted to glucose.

27.59 $CH_3—\overset{\underset{\parallel}{O}}{C}—COO^- \; + \; \overset{\underset{\displaystyle |}{CH_2}}{\underset{\displaystyle COO^-}{\overset{\displaystyle COO^-}{\overset{\displaystyle |}{CH}}—NH_3^+}}$

*27.61 in the carbon dioxide exhaled by the animal

27.63 bilirubin

Chapter 28 Biosynthetic Pathways

28.A The statement is true. Carbohydrate is synthesized by reducing CO_2.

28.B Gluconeogenesis, because the other pathways metabolize glucose, and only gluconeogenesis manufactures it.

28.C Exactly the same way. The only difference is that for the membrane more unsaturated fatty acids are needed than in the fat depot. Therefore, the oxidation of the synthesized saturated fatty acids is more frequent when the membrane's needs are served.

28.D   When we are exposed to cold, energy is needed in the form of heat for survival. α-Ketoglutarate is part of the citric acid cycle, which produces energy. Thus α-ketoglutarate is used up and the equilibrium is shifted to the left, deamidating glutamic acid. This produces α-ketoglutaric acid, replacing that used up in the citric acid cycle.

28.1   for flexibility and to overcome unfavorable equilibria

28.3   because the presence of a large inorganic phosphate pool would shift the reaction to the degradation process so that no substantial amount of glycogen would be synthesized

28.5   Photosynthesis is the reverse of respiration.

28.7   (a) pyruvate   (b) oxaloacetate   (c) alanine

28.9   conversions of (1) pyruvate to oxaloacetate
(2) oxaloacetate to phosphoenol pyruvate
(3) fructose 1,6-bisphosphate to fructose 6-phosphate
(4) glucose 6-phosphate to glucose

28.11   the conversion of oxaloacetate to malate

28.13   uracil, ribose, and three phosphates

28.15   (a) the cytosol   (b) no

28.17   malonyl ACP

28.19   malonyl ACP

*28.21   It is an oxidation step because hydrogen is removed from the substrate. The oxidizing agent is O_2. NADPH is also oxidized during this step.

*28.23   NADPH is bulkier than NADH; it also has two more negative charges.

*28.25   No, the body makes other unsaturated fatty acids such as oleic and arachidonic.

*28.27   palmitoyl CoA, serine, acyl CoA, UDP-glucose

*28.29 CH_2—OH
       |
       C=O
       |
       COO$^-$

*28.31   aspartic acid

28.33   valine and α-ketoglutarate

28.35   NADPH

28.37   by oxidation of a saturated fatty acid by certain enzymes and NADPH + H$^+$

28.39   malonyl ACP

28.41   No, it can add C_2 fragments to a growing chain up to C_{16}. After that another enzyme system takes over.

28.43   No, cholesterol is constantly synthesized in the liver.

Chapter 29   Nutrition and Digestion

29.A   Yes; minerals and water go through the body essentially unchanged.

29.B   No; neither can synthesize the enzyme cellulase needed to digest wood. Termites eat wood because they have bacteria that do manufacture cellulase.

29.1   No, they differ according to body weight, age, occupation, and sex.

29.3   Only the calcium propionate. When this is metabolized it yields energy.

29.5   the section at the bottom that gives daily caloric needs

29.7   It is necessary for proper operation of the digestive system, and a lack of it may lead to colon cancer.

29.9   for daily activities involving motion

29.11   1830 Cal/day

29.13   Yes, but one must eat a wide range of vegetables and cereals.

29.15   linoleic acid

29.17   dietary deficiency diseases, since rice cannot supply the essential amino acids lysine or threonine or essential fatty acids

29.19   to prevent scurvy

29.21   It assists blood clotting.

29.23   scurvy

29.25   biotin, thiamine

29.27   hydrolysis of acetal, ester, and amide linkages

29.29   neither

29.31   yes, randomly

29.33   If glucose supplied all caloric needs, it would cause deterioration of the patient.

29.35   No, these are two different dipeptides. In aspartame, aspartic acid is at the N-terminal; in phenylalanylaspartic acid, it is at the C-terminal.

29.37   Both are derivatives of disaccharides.

29.39   arginine

29.41   (a) linoleic acid, linolenic acid   (b) unsaturated

29.43   nicotinic acid (niacin)

29.45   cellulose

29.47   The body would treat it as a food protein and digest it (hydrolyze it) before it could get into the blood serum.

29.49   See Table 21.1.

29.51   No. Arsenic is not an essential nutrient.

Chapter 30   Immunochemistry

30.A   in the lymph nodes, which act as filtering junctions

30.B   Two monoclonal antibodies isolated from the same population of lymphocytes would differ in their V region. They would interact with different antigens. They probably would have the same C region (constant region).

30.C   The variable portions of the chain of TcR bind to the antigen presenting MHCs.

30.D   The variable region of the chain acts as a binding site, and the MHC presenting the epitope of the antigen binds to it.

30.E   Cytokines carry the signal from one cell to another to proliferate.

30.1 Skin and mucin (or tear)

30.3 The skin fights bacteria by secreting lactic and fatty acids and thereby lowering the pH.

30.5 Acquired immunity is selective, and it has memory.

30.7 T cells mature and differentiate in the thymus gland, B cells in the bone marrow.

30.9 B cells differentiate to plasma cells after encountering the antigen.

30.11 only peptide antigens

30.13 It binds peptide antigen inside an infected cell and it brings it to the surface of the cell to be presented to a T cell.

30.15 Both bind antigens, and both belong to the immunoglobulin superfamily.

30.17 IgA are in secretions. They are the largest, with molecular weights of 200 000 to 700 000. IgE are involved in allergic reactions, their molecular weight is 190 000. IgG are the most important antibodies in blood; their molecular weight is 150 000.

30.19

Red blood   Anti-A   Red blood
cell   antibody   cell

30.21 by disulfide bridges

30.23 Monoclonal antibodies are identical copies of a single antibody against a specific antigen.

30.25 B cells differentiate to plasma cells, which manufacture and secrete soluble immunoglobulins. They respond to all kinds of antigens. T cells respond to peptide antigens only. They differentiate either to killer T cells or memory cells. Killer T cells attack and kill invading pathogens by direct contact.

*30.27 TcR receptor complex contains at least two more proteins besides the TcR receptor—for example, the CD3 and CD4 molecules, which are also members of the immunoglobulin superfamily.

30.29 TcR and coreceptors such as CD3 and CD4 or CD8

30.31 CD4 adhesion molecule.

30.33 Cytokines are small glycoprotein molecules.

30.35 (a) tumor necrosis factor   (b) interleukin
(c) epidermal growth factor

30.37 They have four cysteine residues. Cysteine residue number 1 is linked to residue number 3 by disulfide linkage. Cysteine number 2 is linked to cysteine number 4.

30.39 Cysteine

*30.41 Antibody is manufactured and secreted by B cells. They are the only cells that manufacture soluble immunoglobulins.

*30.43 Monoclonal antibody treatment is only mildly toxic. It specifically attacks the lymphoma cells. Chemotherapy kills many other cells besides the quickly reproducing cancer cells. Therefore, it is highly toxic.

30.45 Plasma cells are differentiated B cells, and memory cells are differentiated T cells.

30.47 The chemokine, when it interacts with the surface of a leukocyte, leads to alternative polymerization and breakdown of actin inside the cells. This allows the cell to change its shape. It allows the leukocyte to extend leg-like extrusions for moving along the endothelium. It allows the leukocyte to flatten out and the endothelial cells to change their shape and to develop gaps among the tight junctions. All this enables the flattened cells to pass through the gaps.

30.49 in the thoracic duct

30.51 They have variable regions that react with antigens.

30.53 on many cells, not just tumor cells

Chapter 31   Body Fluids

31.A The pH of an active muscle cell is lower than the pH of blood because the presence of CO_2 which, when it interacts with water, forms carbonic acid. CO_2 is generated in the active muscle by the citric acid cycle part of the metabolism.

31.B The specific gravity of urine becomes lower because the urine is more diluted with water.

31.C The inhibition of ACE stops the production of the vasoconstriction.

31.1 water

31.3 (a) yes   (b) through semipermeable membranes

31.5 They carry oxygen to the cells and carbon dioxide away from the cells.

31.7 See Sections 31.1 and 31.2.

31.9 Proper osmotic pressure is not provided, and water from the blood oozes into the interstitial fluid and causes swelling of tissues (edema).

31.11 globulins

31.13 four

31.15 20 mm Hg

31.17 Each heme has a cooperative effect on the other hemes.

31.19 (a) at the terminal NH_2 groups of the four polypeptide chains
(b) carbaminohemoglobin

*31.21 $K = 10$ (Significant figures!)

31.23 in the proximal tubule

31.25 inorganic ions

31.27 the distal tubule and the collecting tubule

31.29 No glucose reaches the tissues or the cells. Because 12 mm Hg is lower than the osmotic pressure, nutrients flow from the interstitial fluid into the blood.

31.31 drowsiness

31.33 platelets

31.35 restricted flow of urine

31.37 By blocking calcium channels heart muscles contract less frequently, thereby reducing blood pressure.

31.39 The concentration is higher. Since fibrinogen is

removed when plasma becomes serum, the albumin is dissolved in a smaller amount of material, and its concentration is higher.

31.41 Less oxygen is carried; more is released to the tissues.

31.43 a drug that increases the volume of urine

31.45 vasopressin

*31.47 Where the two circulations meet, the oxygen pressure is the same. However, the saturation of fetal hemoglobin is lower than that of adult hemoglobin. This means that at the same oxygen pressure, fetal hemoglobin is less saturated than the adult kind, so fetal hemoglobin can accept oxygen from adult hemoglobin.

GLOSSARY

A site *(Section 25.5)* The site on the large ribosomal subunit where the incoming tRNA molecule binds.

Acetal *(Section 17.5)* A molecule containing two —OR groups bonded to the same carbon.

Acetyl coenzyme A *(Section 26.3)* A biomolecule in which an acetyl group is bonded to the —SH group of coenzyme A by a thioester bond.

Achiral *(Section 15.2)* An object that lacks chirality; an object that is superposable on its mirror image.

Acid *(Section 8.1)* An Arrhenius acid is a substance that ionizes in aqueous solution to give H^+. *(Section 8.3)* A Brønsted-Lowry acid is a substance that donates a proton.

Acid dissociation constant (K_a) *(Section 8.4)* The equilibrium ratio of the products to reactants for the ionization of an acid in aqueous solution.

Acid salt *(Section 8.9)* The salt of a strong acid and a weak base.

Acid strength *(Section 8.3)* The willingness of a compound to give up a proton.

Acidic solution *(Section 8.8)* An aqueous solution with a pH lower than 7.

Acidosis *(Box 8E)* Blood with a pH lower than the normal value of 7.35–7.45.

ACP *(Section 28.3)* Acyl Carrier Protein, a large molecule that activates a fatty acid for fatty acid synthesis. The fatty acid is bonded to ACP as a thioester.

Acquired immunity *(Section 30.1)* The second line of defense that vertebrates have against invading organisms. It is based on a response that is specific, has memory, and involves the use of antibodies and T cells.

Activation energy *(Section 7.2)* The minimum energy necessary to cause a chemical reaction.

Activation, of an amino acid *(Section 25.5)* The process by which an amino acid is bonded to an AMP molecule and then bonded to the 3′—OH of a tRNA molecule.

Activation, of enzymes *(Section 22.3)* Any process by which an inactive enzyme is transformed into an active enzyme.

Active site *(Section 22.3)* The part of the enzyme where the substrates bind and where the catalysis takes place.

Actual yield *(Section 4.6)* The mass a product formed in a chemical reaction.

Adrenergic neurotransmitter *(Section 23.4)* A monoamine neurotransmitter/hormone, the most common of which are epinephrine (adrenalin), serotonin, histamine, and dopamine.

Adrenocorticoid hormone *(Section 20.10)* A hormone produced by the adrenal glands; adrenocorticoid hormones regulate the concentrations of ions and carbohydrates.

Agonist *(Section 23.1)* A molecule that mimics the structure of a natural neurotransmitter/hormone, binds to the same receptor, and elicits the same response.

AIDS *(Section 30.5)* Acquired Immune Deficiency Syndrome. The disease caused by the human immunodeficiency virus, which attacks and depletes T cells.

Alcohol *(Section 10.4)* A compound containing an —OH (hydroxyl) group bonded to a tetrahedral carbon atom.

Aldehyde *(Section 10.4)* A compound containing a carbonyl group bonded to a hydrogen; a —CHO group.

Alditol *(Section 19.5)* The product formed when the CH=O group of a monosaccharide is reduced to a CH_2OH group.

Aldose *(Section 19.2)* A monosaccharide containing an aldehyde group.

Aliphatic amine *(Section 16.2)* An amine in which nitrogen is bonded only to alkyl groups.

Aliphatic hydrocarbon *(Section 11.2)* An alternative word to describe an alkane.

Alkali *(Box 8B)* A base.

Alkali metal *(Section 2.5)* Those elements in Group 1A of the Periodic Table.

Alkaloid *(Box 16B)* A basic nitrogen-containing compound of plant origin; many alkaloids have physiological activity when administered to humans.

Alkalosis *(Box 8E)* A condition in which blood has a pH higher than the normal value of 7.35–7.45.

Alkane *(Section 11.2)* A saturated hydrocarbon whose carbons are arranged in an open chain.

Alkene *(Section 12.2)* An unsaturated hydrocarbon that contains a carbon-carbon double bond.

Alkyl group *(Section 11.4)* A group derived by removing a hydrogen from an alkane; given the symbol R—.

Alkyne *(Section 12.2)* An unsaturated hydrocarbon that contains a carbon-carbon triple bond.

Allosterism *(Section 22.6)* A type of enzyme regulation based on an event occurring at a location on the enzyme other than the active site but that then makes a change in the active site to affect the enzyme rate. Allosteric enzymes often have multiple polypeptide chains with the possibility of chemical communication between the chains.

Alpha amino acids *(Section 21.2)* The common 20 amino acids where the carboxyl and the amino group are both attached to the alpha carbon.

Alpha helix *(Section 21.8)* A repeating secondary structure of a protein where the chain adopts a helical conformation and the structure is held together by hydrogen bonds from the peptide backbone N—H to the backbone C=O four amino acids farther up the chain.

Alpha particle *(Section 9.2)* A helium nucleus, $^4_2$He (α).

Amide *(Section 18.5)* A compound in which a carbonyl group is bonded to a nitrogen, $RCONR'_2$; one or both of the R′ groups may be H, alkyl, or aryl groups.

Amino acid *(Section 21.2)* An organic compound containing an amino group and a carboxyl group.

Amino acid neurotransmitter *(Section 23.5)* A class of neurotransmitter/hormones that are amino acids. Some are members of the 20 protein-derived amino acids; others, such as taurine and GABA, are not found in proteins.

Amino acid pool *(Section 27.1)* The pool of free amino acids that is maintained by the digestion of proteins and various amino acid-producing pathways.

Amino group *(Section 10.4)* An —NH_2 group.

Amino sugar *(Section 19.2)* A monosaccharide in which an —OH group is replaced by an —NH_2 group.

Amorphous solid *(Section 5.10)* A solid with random crystalline structure.

Amphiprotic *(Section 8.3)* A substance that can act as either an acid or a base.

Anabolism *(Section 26.1)* The pathways by which biomolecules are synthesized.

Anaerobic pathway *(Section 27.2)* The pathway that generates lactic acid from pyruvate when the oxygen supply is too low for the pyruvate to be oxidized aerobically through the common pathway.

Angiotensin *(Section 31.8)* A hormone that is a vasoconstrictor and raises the blood pressure.

Anhydride *(Section 18.3)* A compound in which two carbonyl groups are bonded to the same oxygen; RCO—O—COR'.

Anhydrous *(Section 6.6)* A term meaning without water.

Anion *(Section 3.2)* An ion with a negative electric charge.

Anode *(Section 6.6)* An electrode toward which anions move; the electrode at which oxidation occurs.

Anomeric carbon *(Section 19.3)* The hemiacetal carbon of the cyclic form of a monosaccharide.

Anomers *(19.3)* Monosaccharides that differ in configuration only at their anomeric carbons.

Antacid *(Box 8C)* A substance taken orally to neutralize stomach acidity.

Antagonist *(Section 23.1)* A molecule that binds to a neurotransmitter receptor but does not elicit the natural response.

Antibody *(Section 30.4)* A defense glycoprotein synthesized by the immune system of vertebrates, also called an immunoglobulin.

Anticoagulant *(Box 31B)* A drug given to prevent blood clot formation.

Anticodon *(Section 25.3)* A sequence of three bases on a tRNA molecule that is the complement of a three-base codon on an mRNA molecule.

Antidiuretic *(Section 31.7)* Any agent that reduces the volume of urine produced.

Antigen *(Section 30.3)* Any molecule, usually foreign, capable of eliciting synthesis of a specific antibody in vertebrates.

Antiseptic *(Box 4C)* A compound used to kill bacteria, fungi, and other pathogens on wounds and exposed surfaces.

Apoenzyme *(Section 22.3)* The protein portion of an enzyme that has cofactors or prosthetic groups.

Apoptosis *(Box 24E)* Planned cell death characterized by small numbers of cells dying at a time and being cleaved into apoptopic bodies containing organelles and large nuclear fragments.

Aqueous *(Section 4.7)* A solution in which water is the solvent.

Ar— *(Section 14.1)* The symbol used for an aryl group.

Arene *(Section 12.2)* A compound containing one or more benzene rings.

Aromatic amine *(Section 16.2)* An amine in which nitrogen is bonded to one or more aryl groups.

Aromatic compound *(Section 14.1)* A term used to classify benzene and its derivatives.

Aryl group *(Section 14.1)* A group derived from an aromatic compound.

Atom *(Section 2.1)* The basic unit of matter.

Atomic mass unit (amu) *(Section 2.4)* The mass equivalent to 1.661×10^{-24} g.

Atomic number *(Section 2.4)* The number of protons in the nucleus of an atom of an element.

Atomic weight *(Section 2.4)* The weighted average of the masses of the naturally occurring isotopes of an element.

ATP *(Section 26.1)* Adenosine triphosphate.

Avogadro's law *(Section 5.5)* Law stating that equal volumes of gases at the same temperature and pressure contain the same number of molecules.

Avogadro's number *(Section 4.3)* 6.02×10^{23}.

Axial position *(Section 11.7)* A position on a chair conformation of a cyclohexane ring that extends from the ring parallel to the imaginary axis of the ring.

Axon *(Section 23.2)* The long part of a nerve cell that comes out of the main cell body and eventually connects with another nerve cell or a tissue cell.

B cell *(Section 30.2)* A type of lymphocyte that is produced in and matures in the bone marrow. B cells produce antibody molecules.

Balanced equation *(Section 4.4)* A chemical equation in which the left and right sides contain the same numbers and kinds of atoms.

Barometer *(Section 5.3)* An instrument for measuring atmospheric pressure.

Baroreceptors *(Section 31.8)* The receptors in the neck that detect a drop in blood pressure and send a signal to the heart to pump harder and to the muscles around the blood vessels to contract to raise blood pressure.

Basal caloric requirement *(Section 29.3)* The caloric requirement for an individual at rest, usually given in Cal/day.

Base, in acid-base reactions *(Section 8.1)* An Arrhenius base is a substance that releases hydroxide ions in aqueous solution. *(Section 8.3)* A Brønsted-Lowry base is a substance that accepts a proton.

Base, in nucleic acids *(Section 24.2)* A heterocyclic aromatic amine; most commonly adenine, guanine, cytosine, thymine, and uracil.

Base dissociation constant (K_b) *(Section 8.4)* The equilibrium ratio of products to reactants for the ionization of base in aqueous solution.

Base strength *(Section 8.3)* The willingness of a compound to accept a proton.

Basic salt *(Section 8.9)* The salt of a weak acid and a strong base.

Basic solution *(Section 8.8)* An aqueous solution with a pH above 7.

Becquerel (Bq) *(Section 9.5)* One disintegration (count) per second.

Beta oxidation *(Section 27.5)* The spiral pathway where a fatty acid is oxidized to generate acetyl-CoA, also producing reduced electron carriers.

Beta particle *(Section 9.2)* An electron, e⁻ (β).

Beta sheet *(Section 21.8)* Also called the β-pleated sheet, this is another repeating secondary structure based on hydrogen bonding within and between peptide chains. The shape loops back on itself to give a flatter pleated appearance.

Bile salts *(Section 20.11)* Oxidation products of cholesterol that are emulsifying agents for fatty acids. They are produced in the liver and stored in the gallbladder.

Binary compound *(Section 3.8)* A compound containing only two elements.

Binary ionic compound *(Section 3.8)* A compound containing two elements, each present as ions.

Binary molecular compound *(Section 3.8)* A compound containing two elements joined by a covalent bond.

Biochemical pathway *(Section 26.1)* A series of consecutive biochemical reactions that converts one molecule into another molecule; the pathway may be anabolic or catabolic.

Biosynthetic pathway *(Section 28.1)* A pathway that leads to the production of a compound. These pathways often build larger molecules from smaller ones and consume energy.

Bleaching *(Section 4.8)* A process in which color is lost.

Blood cellular elements *(Section 31.2)* The cell components that are suspended in blood, namely the erythrocytes, leukocytes, and platelets.

Blood plasma *(Section 31.1)* The fluid portion of the blood.

Blood-brain barrier *(Section 31.1)* The limited exchange between the blood and the interstitial fluid of the brain.

Bohr effect *(Section 31.3)* The relationship between the oxygen-carrying capacity of hemoglobin and the level of H⁺ and CO_2.

Boiling point *(Section 5.1)* The temperature at which the vapor pressure of a liquid is equal to the atmospheric pressure.

Bond angle *(Section 3.5)* The angle between two atoms bonded to a central atom.

Bonding electrons *(Section 3.4)* Valence electrons involved in forming a covalent bond; that is, shared electrons.

Bowman's capsule *(Section 31.5)* A cup-shaped receptacle in the kidney that is the initial, expanded segment of the nephron where filtrate enters from the blood.

Boyle's law *(Section 5.4)* Law stating that the volume of a gas at constant temperature is inversely proportional to the applied pressure.

Brackish water *(Box 6E)* A mixture of sea water (salt water) and fresh water.

Brownian motion *(Section 6.7)* The random motion of colloidal-size particles.

Buckminsterfullerene *(Section 5.10)* The crystalline structure of carbon in which there are 60, 72, or 80 carbons arranged in pentagons and hexagons to form a sphere.

Buckyball *(Section 5.10)* Buckminsterfullerene.

Buffer capacity *(Section 8.10)* The extent to which a buffer can prevent a significant change in the pH of a solution on the addition of hydrogen or hydroxide ion.

Buffer solution *(Section 8.10)* A solution that does not change pH to the extent of added hydrogen or hydroxide ion; the solution of a weak acid and the salt of that acid; the solution of a weak base and the salt of that base.

C-terminus *(Section 21.5)* The amino acid at the end of a peptide chain that has a free carboxyl group.

Calorie *(Section 1.9)* The amount of heat necessary to raise 1 g of liquid water by 1°C.

Calvin cycle *(Box 28A)* The dark reactions of photosynthesis that produce glucose from CO_2 and H_2O by a cyclic process.

Carbocation *(Section 12.6)* A species containing a carbon with only three bonds to it and bearing a positive charge.

Carbohydrate *(Section 19.1)* A polyhydroxyaldehyde or polyhydroxyketone or a substance that gives these compounds upon hydrolysis.

Carbonyl group *(Section 10.4)* A C=O group.

Carboxyl group *(Section 10.4)* A —COOH group.

Carboxylic acid *(Section 10.4)* A compound containing —COOH group.

Carcinogen *(Section 25.7)* A chemical mutagen that can cause cancer.

Catabolism *(Section 26.1)* The biochemical pathways that are involved in generating energy by breaking down large nutrient molecules into smaller molecules with the concurrent production of ATP.

Catalyst *(Section 7.4)* A substance that increases the rate of a chemical reaction by providing an alternative pathway with a lower activation energy.

Cathode *(Section 6.6)* An electrode to which cations are attracted or move; the electrode at which reduction occurs.

Cation *(Section 3.2)* An ion with a positive electric charge.

Celsius scale (°C) *(Section 1.4)* A temperature scale based on 0° as the freezing point of water and 100° as the normal boiling point of water.

Central dogma *(Section 25.1)* Doctrine stating the basic directionality of heredity where DNA leads to RNA, which leads to protein. This is true in almost all life forms, except certain viruses.

Cephalin *(Section 20.6)* A glycerophospholipid similar to phosphatidylcholine with the exception that another molecule is attached to the phosphate instead of choline.

Ceramide *(Section 20.7)* The combination of a fatty acid linked to sphingosine by an amide bond.

Cerebroside *(Section 20.8)* A glycolipid based on the ceramide backbone where one or more carbohydrates are attached to the ceramide.

Chain reaction, nuclear *(Section 9.7)* A nuclear reaction that results from fission of a nucleus, which produces a particle that causes fission in another nucleus and continues with accelerated speed.

Chair conformation *(Section 11.7)* The most stable conformation of a cyclohexane ring; all bond angles are approximately 109.5°.

Chaperone *(Section 21.9)* A protein in cells that helps nascent proteins to fold correctly.

Charles's law *(Section 5.4)* Law stating that the volume of a gas at constant pressure is directly proportional to its temperature in degrees kelvin.

Chemical change *(Section 1.1)* Chemical reaction.

Chemical equation *(Section 4.4)* A representation in words and/or symbols of the process that occurs when starting materials are converted to the products.

Chemical kinetics *(Section 7.1)* The study of chemical reaction rates.

Chemical messenger *(Section 23.1)* Any chemical that is released from one location and travels to another location before acting. These may be hormones, neurotransmitters, or simple ions.

Chemical property *(Section 1.1)* The chemical reactions a substance undergoes.

Chemical reaction *(Section 1.1)* As substances are used up, others appear in their place.

Chemiosmotic hypothesis *(Section 26.6)* The hypothesis proposed by Peter Mitchell to explain how movement of electrons down the electron transport chain creates a proton gradient that, in turn, leads to the production of ATP.

Chemistry *(Section 1.1)* The science that deals with matter.

Chemokine *(Section 30.6)* A chemotactic cytokine that facilitates the migration of leukocytes from the blood vessels to the site of injury or inflammation.

Chiral *(Section 15.2)* From the Greek *cheir,* meaning hand; objects that are not superposable on their mirror images.

Chloroplast *(Box 22A)* The subcellular organelle in plants that contains chlorophyll and is responsible for photosynthesis.

Cholinergic neurotransmitter *(Section 23.3)* A neurotransmitter/hormone based on acetylcholine.

Chromosome *(Section 24.1)* Structures within the nucleus of eukaryotes that contain DNA and protein and that are replicated as units during mitosis. Each chromosome is made up of one long DNA molecule that contains many heritable genes.

Cis *(Section 11.8)* A prefix meaning on the same side.

Cis-trans isomers *(Section 11.8)* Isomers that have the same order of attachment of their atoms but a different arrangement of their atoms in space due to the presence of either a ring or a carbon-carbon double bond.

Citric acid cycle *(Section 26.4)* The central biochemical pathway in metabolism whereby an acetyl-CoA combines with oxaloacetate to form citrate, which then undergoes a series of reactions to regenerate oxaloacetate while producing energy in the form of GTP, NADH, and $FADH_2$.

Cloning *(Section 24.6)* A process whereby DNA is amplified by inserting it into a host and having the host replicate it along with the host's own DNA.

Cluster determinant *(Section 30.6)* A set of membrane proteins on T cells that help the binding of antigens to the T cell receptors.

Coated pit *(Section 20.9)* An area of a cell membrane that contains LDL receptors and is involved in bringing LDL into the cell.

Codon *(Section 25.4)* A three-base sequence on mRNA that codes for a particular amino acid.

Coefficient *(Section 4.4)* A number placed in front of a chemical entity in a balanced chemical equation.

Coenzyme *(Section 22.3)* An organic cofactor, such as NAD^+.

Coenzyme A *(Section 26.3)* An important activating group often attached to a fatty acid or an acetate group. Coenzyme A is based on a structure composed of an adenine-based nucleotide, pantothenic acid, and mercaptoethylamine. The reactive portion is the sulfhydryl at the end.

Coenzyme Q *(Section 26.5)* Also called ubiquinone. An electron carrier found in the inner mitochondrial membrane that acts as an electron receptor from a variety of sources during electron transport.

Cofactor *(Section 22.3)* A nonprotein portion of an enzyme that is involved in a chemical reaction.

Colligative property *(Section 6.8)* A physical property of a solvent that depends on the concentration of solute particles; for example, freezing point, boiling point, vapor pressure, and osmotic pressure.

Colloid *(Section 6.7)* A two-part homogeneous mixture in which the suspended solute particles are from 1 to 1000 nm in size.

Combined gas law *(Section 5.4)* $PV = nRT$.

Combustion *(Section 4.8)* Burning in air.

Common catabolic pathway *(Section 26.1)* A term used for the citric acid cycle and the electron transport chain, the two metabolic pathways responsible for the production of the majority of energy during catabolism.

Competitive inhibitor *(Section 22.3)* An inhibitor that binds to the active site of an enzyme and competes with the natural substrate.

Complete protein *(Section 29.4)* A protein source that contains sufficient quantities of all amino acids required for normal growth and development.

Complex lipids *(Section 20.4)* Lipids found in membranes; the two main classes are glycolipids and phospholipids.

Compound *(Section 2.2)* A pure substance composed of more than one type of atom with a fixed ratio of the component atoms.

Computer-assisted tomography (CAT or CT Scan) *(Box 9E)* Multiple x-rays used to image tissues on the basis of density.

Conformation *(Section 11.7)* Any three-dimensional arrangement of atoms in a molecule that results by rotation about a single bond.

Conjugate acid *(Section 8.3)* In the Brønsted-Lowry theory, the substance formed when a base accepts a proton.

Conjugate base *(Section 8.3)* In the Brønsted-Lowry theory the substance formed when an acid loses a proton.

Conjugated protein *(Section 21.9)* A protein that contains a nonprotein part, such as hemoglobin.

Constitutional isomers *(Section 11.3)* Compounds with the same molecular formula but a different order of attachment of their atoms.

Contraindication *(Box 5E)* The caution that a medication or process should not be used due to its danger or an increased danger when used with another medication or process at the same time.

Control sites *(Section 25.6)* A DNA sequence that is part of a prokaryotic operon. This sequence is upstream of the structural gene DNA and plays a role in controlling whether the structural gene is transcribed.

Conversion factor *(Section 1.5)* A ratio, including units, used as a multiplier to change from one system or unit to another.

Covalent bond *(Section 3.2)* The interatomic attraction resulting from the sharing of electrons between atoms.

Crenation *(Section 6.8)* The shriveling of red blood cells.

Cristae *(Section 26.2)* The inner mitochondrial membrane.

Crystal lattice *(Section 5.10)* The three-dimensional arrangement of molecules or ions in a solid.

Crystallization *(Section 5.10)* The formation of a solid from a liquid during which the molecules become organized in a nonrandom manner.

Curie (Ci) *(Section 9.5)* A measure of radioactive decay equal to 3.7×10^{10} disintegrations per second.

Cyclic ether *(Section 13.3)* An ether in which the ether oxygen is one of the atoms of a ring.

Cycloalkane *(Section 11.5)* A saturated hydrocarbon that contains carbon atoms bonded to form a ring.

Cyclooxygenase *(Section 20.12)* The enzyme that catalyzes the first step in the conversion of arachidonic acid to prostaglandins and thromboxanes.

Cystine *(Section 21.4)* A dimer of cysteine in which the two amino acids are covalently bonded by disulfide bond between their side chain —SH groups.

Cytochrome *(Section 26.5)* A class of compounds that act as electron carriers in the inner mitochondrial membrane during electron transport. They are based on an iron-containing porphyrin ring similar to the heme group found in hemoglobin.

Cytokines *(Section 30.6)* Small protein molecules that are produced in one cell type and control and coordinate the immune response in other cell types.

Cytosol *(Section 27.2)* The soluble portion of a cell that is inside the cell membrane but outside any of the subcellular organelles.

D-Monosaccharide *(Section 19.2)* A monosaccharide that, when written as a Fischer projection, has the hydroxyl on its penultimate carbon facing to the right.

Dalton's law *(Section 5.6)* Law stating that the pressure of a mixture of gases is equal to the sum of the partial pressure of each gas.

Debranching enzyme *(Section 29.7)* The enzyme that catalyzes the hydrolysis of the 1,6-glycosidic bonds in starch and glycogen.

Decay *(Section 9.3)* The change of the nucleus of a radioactive atom into a different nucleus.

Degenerate code *(Section 25.4)* The genetic code is said to be degenerate because more than one codon can code for the same amino acid.

Dehydration *(Section 13.5)* Elimination of a molecule of water from an alcohol; an OH is removed from one carbon, and an H is removed from an adjacent carbon.

Dehydrogenase *(Section 22.6)* A class of enzymes that catalyzes oxidation-reduction reactions, often using NAD^+ or FAD as cofactors.

Denaturation *(Section 21.11)* The process of destroying the native conformation of a protein via chemical or physical means. Some denaturations are reversible, whereas others permanently damage the protein.

Dendrite *(Section 23.2)* The hair-like projections that extend from the cell body of a nerve cell on the opposite side from the axon.

Density *(Section 1.7)* The ratio of mass to volume for a substance.

Deoxyribonucleic acid *(Section 24.2)* The macromolecule of heredity in eukaryotes and prokaryotes. It is composed of chains of nucleotide monomers of a nitrogenous base, deoxyribose, and phosphate.

Dextrorotatory *(Section 15.5)* Clockwise (to the right) rotation of the plane of polarized light in a polarimeter.

Dialysis *(Section 6.8)* The separation of larger molecules, dissolved substances, or colloidal particles from smaller molecules, substances, or particles by a semipermeable membrane.

Diastereomers *(Section 15.4)* Stereoisomers that are not mirror images of each other.

Diastolic blood pressure *(Box 5C)* Arterial blood pressure during the relaxation of the heart; the lower number in the fraction.

Diatomic element *(Section 2.3)* An element that occurs in nature as two atoms of the element joined by a covalent bond.

Diffusion *(Section 5.6)* The movement of molecules through a medium from a higher concentration to a lower concentration.

Digestion *(Section 29.6)* The process of hydrolysis of starches, fats, and proteins into smaller units.

Diglyceride *(Section 20.2)* An ester of glycerol and two fatty acids.

Dilute *(Section 6.5)* A qualitative term referring to a solution that contains less solute than does a concentrated solution.

Dilution *(Section 6.5)* The process of reducing the concentration of the solute in a solution.

Diol *(Section 13.3)* A compound containing two —OH (hydroxyl) groups.

Dipeptide *(Section 21.5)* A peptide with two amino acids.

Dipole-dipole attractions *(Section 5.7)* The attraction of a pole in one molecule to a pole of opposite charge in another molecule or in the same molecule if it is large enough, as a protein.

Disaccharide *(Section 19.7)* A carbohydrate containing two monosaccharide units joined by a glycosidic bond.

Discriminatory curtailment diet *(Section 29.2)* A diet where a person attempts to both eat less and control the type and quality of nutrient consumed.

Disinfectant *(Box 4C)* A compound used to kill bacteria, fungi, and other pathogens on inanimate objects and surfaces.

Dissociation *(Section 4.7)* The separation of a compound into its component ions.

Dissociation (ionization) constant *(Section 8.4)* The result of the dissociation expression; ratio of the products to the reactants at equilibrium; K_a for an acid, K_b for a base.

Dissociation expression *(Section 8.4)* The mathematical expression that defines the dissociation constant; the product of the products divided by the product of the reactions, all in molar concentrations; pure liquids and solids are omitted.

Distal tubule *(Section 31.5)* The part of the kidney tubule farthest from the Bowman's capsule.

Disulfide *(Section 13.7)* A compound containing an —S—S— group.

Diuresis *(Section 31.7)* The production of urine in the kidney.

Diuretic *(Box 31E)* A compound that increases the flow of urine from the kidney.

DNA *(Section 24.2)* Deoxyribonucleic acid.

Double bond *(Section 3.4)* A bond formed by sharing two pairs of electrons; represented by two lines between the two bonded atoms.

Double helix *(Section 24.3)* The structure to which most DNA conforms. It is based on two chains of DNA that wind around each other with well-characterized parameters of distances and angles.

Ductility *(Section 2.5)* Can be drawn into a wire.

Edema *(Section 31.2)* A swelling of tissues due to water seeping from the blood into the interstitial fluids.

Effective collision *(Section 7.2)* A collision between two molecules that results in chemical reaction.

EGF *(Section 30.6)* Epidermal Growth Factor, a cytokine that stimulates epidermal cells during healing of wounds.

Electrolyte *(Section 6.6)* A compound that, when dissolved in water, produces a solution that conducts electricity.

Electromagnetic spectrum *(Section 9.1)* The array of electromagnetic phenomena by wavelength.

Electron *(Section 2.4)* A particle that has a charge of −1 and a mass of approximately 1/1837 amu.

Electron configuration *(Section 2.7)* The arrangement of electrons in the orbitals about a nucleus.

Electron transport chain *(Section 26.6)* A series of electron carriers imbedded or attached to the inner mitochondrial membrane that are alternately reduced and then oxidized as electrons pass from an initial electron carrier (NADH or $FADH_2$) to oxygen.

Electronegativity *(Section 3.2)* A measure of an atom's attraction for the electrons it shares in a chemical bond with another atom.

Element *(Section 2.2)* A pure substance composed of one type of atom.

Elongation *(Section 25.5)* The phase of protein synthesis during which activated tRNA molecules deliver new amino acids to ribosomes where they are joined by peptide bonds to form a polypeptide.

Elongation factor *(Section 25.5)* Small protein molecules that are involved in the process of tRNA binding and movement of the ribosome on the mRNA during elongation.

Emulsion *(Box 6D)* A suspension produced from small globules of a liquid in another liquid.

Enantiomers *(Section 15.2)* Stereoisomers that are nonsuperposable mirror images; refers to a relationship between pairs of objects.

End point *(Section 8.11)* The point at which an acid/base reaction has reached neutrality.

Endergonic reaction *(Section 4.9)* A reaction that absorbs energy in any form.

Endocrine gland *(Section 23.2)* A gland, such as the pancreas, pituitary, and hypothalamus, that produces hormones involved in the control of the chemical reactions and metabolism.

Endothermic reaction *(Section 4.9)* A reaction that absorbs heat.

Energy *(Section 1.8)* The capacity to do work; the joule is the base SI unit.

Energy yield *(Section 26.7)* The calculation of how many ATP molecules are formed from electrons passing down the electron transport chain. It is usually calculated as a number of ATP per NADH used, $FADH_2$ used, or H^+ passing through the ATPase.

Enkephalins *(Section 23.6)* Pentapeptides found in nerve cells of the brain that act to control pain perception.

Enol *(Section 17.6)* A molecule containing an —OH group bonded to a carbon of a carbon-carbon double bond.

Entropy *(Box 5A)* The measure of the disorder (chaotic condition) of a system.

Enzyme *(Section 22.1)* A biological catalyst that increases the rate of a chemical reaction by providing an alternative pathway with a lower activation energy.

Enzyme activity *(Section 22.4)* The rate at which an enzyme-catalyzed reaction proceeds; commonly measured as amount of product produced per minute.

Enzyme-substrate complex *(Section 22.5)* A part of an enzyme reaction mechanism where the enzyme is bound to the substrate.

Epitope *(Section 30.3)* The specific portion of an antigen that binds to the antibody-binding site.

Equatorial position *(Section 11.7)* A position on a chair conformation of a cyclohexane ring that extends from the ring roughly perpendicular to the imaginary axis of the ring.

Equilibrium *(Section 5.9)* A condition in which two opposed physical forces are equal; a reversible chemical reaction in which the rate of the forward and reverse reactions are equal.

Equilibrium constant *(Section 7.6)* The concentrations of the products of a chemical equilibrium divided by the concentration of the reactants, each term raised to the power equal to its coefficient in the balanced chemical equation.

Equivalence point *(Section 8.11)* The point at which there is an equal amount of acid and base in a neutralization reaction.

Equivalent dose *(Section 9.5)* The sum of the doses of different forms of radiation.

Equivalent weight *(Section 8.11)* The formula weight divided by the number of hydrogen or hydroxide ions per molecule expressed in grams.

Erythrocyte *(Section 31.2)* A red blood cell.

Essential amino acid *(Section 28.5)* An amino acid that cannot be synthesized in the human body in sufficient quantities for normal growth and development.

Essential fatty acid *(Section 20.3)* A polyunsaturated fatty acid that cannot be synthesized in the human body in sufficient quantities for normal growth and development.

Ester *(Section 18.2)* A compound in which the H of a carboxyl group, RCOOH, is replaced by an —OR′ group; RCOOR′.

Ether *(Section 13.2)* A compound containing an oxygen bonded to two carbons.

Eukaryote *(Section 25.6)* An organism that has a true

nucleus and organelles. Eukaryotes include animals, plants, and fungi.

Excitatory neurotransmitter *(Section 23.4)* A neurotransmitter that increases the transmission of nerve impulses.

Excited state, of an atom *(Section 2.6)* An electron configuration other than the ground-state electron configuration.

Exergonic reaction *(Section 4.9)* A reaction that gives off energy in any form.

Exon *(Section 24.6)* A section of DNA that codes for protein or RNA when transcribed.

Exothermic *(Section 4.9)* Refers to reactions that give off heat.

Exponential notation *(Section 1.3)* The expression of a number in base 10.

External innate immunity *(Section 30.1)* The innate protection against foreign invaders characteristic of the skin barrier, tears, and mucous.

Extracellular fluids *(Section 31.1)* All the body fluids that are not inside the cell.

Fact *(Section 1.2)* A statement based on experience.

Factor-label method *(Section 1.5)* A method of solving numerical problems that requires manipulation of units.

FAD/FADH$_2$ *(Section 26.3)* Flavin Adenine Dinucleotide, a coenzyme involved in metabolic oxidation-reductions.

Fat *(Section 20.2)* A triester of fatty acids and glycerol.

Fat depot *(Section 27.1)* The supply of stored fat in the form of triacylglycerols, usually found in fat cells or distributed in other tissues.

Feedback control *(Section 22.6)* A type of enzyme regulation where the product of a series of reactions inhibits the enzyme that catalyzes the first reaction in the series.

Fiber *(Section 29.2)* Nondigestible components from plants, usually based on cellulose, and necessary for proper functioning of the digestive system.

Fibrous protein *(Section 21.1)* A protein used for structural purposes; fibrous proteins are insoluble in water and have a high percentage of secondary structures, such as alpha helices and/or beta-pleated sheets.

Fischer esterification *(Section 18.2)* The process of forming an ester by refluxing a carboxylic acid and an alcohol in the presence of an acid catalyst, commonly sulfuric acid.

Fischer projection *(Section 19.2)* A two-dimensional representation for showing the configuration of a stereoisomer; horizontal lines represent bonds projecting forward from the stereocenter, and vertical lines represent bonds projecting to the rear.

Fission, nuclear *(Section 9.8)* The fragmentation of a heavy nucleus into smaller nuclei.

Fluid mosaic model *(Section 20.5)* A model for the structure and function of biological membranes.

Formula weight (FW) *(Section 4.2)* The sum of the atomic weights of the components of a molecule or ion expressed in grams.

Free radical *(Box 9C)* A molecule or ion with one or more unpaired electrons.

Frequency *(Section 9.1)* The number of wave crests passing a point per minute.

Fullerene *(Section 5.10)* Buckminsterfullerene.

Functional group *(Section 10.4)* An atom or group of atoms within a molecule that shows a characteristic set of physical and chemical properties.

Furanose *(Section 19.3)* A five-membered cyclic hemiacetal form of a monosaccharide.

Fusion, nuclear *(Section 9.7)* A combining of two or more lighter nuclei to form a heavier nucleus.

G-protein *(Section 23.5)* A protein that is either stimulated or inhibited when a hormone binds to a receptor and that subsequently alters the activity of another protein, such as adenylate cyclase.

Gamma ray *(Section 9.1)* A form of electromagnetic radiation characterized by very short wave length and very high energy.

Gas *(Section 1.6)* A physical state of matter in which there is no definite volume or shape.

Gas constant *(Section 5.5)* The constant R in the equation $PV = nRT$; the value of R is 0.0821 L·atm/mol·K.

Gay-Lussac's law *(Section 5.4)* Law stating that the pressure of a gas at constant volume is directly proportional to its temperature in degrees K.

Geiger-Müller counter *(Section 9.4)* An instrument for measuring ionizing radiation.

Gene *(Sections 24.1, 25.1)* A segment of DNA that carries the base sequence that directs the synthesis of a particular protein, tRNA, or rRNA.

Gene expression *(Section 25.1)* The turning on or activation of a gene so that it gets transcribed and translated.

Gene regulation *(Section 25.6)* The various methods used by organisms to control which genes will be expressed and when.

Genetic code *(Section 25.4)* The sequence of three bases on mRNA that codes for a particular amino acid.

Genetic engineering *(Section 25.8)* The process by which genes are inserted into cells.

Genome *(Box 24D)* The complete DNA sequence of an organism.

Globular protein *(Section 21.1)* Protein that is used mainly for nonstructural purposes and is largely soluble in water.

Glomeruli *(Section 31.5)* Blood vessels that connect with the kidney Bowman's capsule.

Glucogenic amino acid *(Sections 27.9, 28.2)* An amino acid whose carbon skeleton can be used for the synthesis of glucose.

Gluconeogenesis *(Section 28.2)* The biosynthetic pathway that converts pyruvate, citric acid cycle intermediates, or glucogenic amino acids into glucose.

Glycerophospholipid *(Section 20.4)* A diester of glycerol and two fatty acids. The third —OH group of glycerol is esterified with phosphoric acid, which is, in turn, esterified with a low-molecular weight alcohol such as choline or ethanolamine.

Glycogenesis *(Section 28.2)* The synthesis of glycogen from glucose.

Glycogenolysis *(Section 27.3)* The hydrolysis of glycogen to glucose.

Glycol *(Section 13.3)* A compound with two hydroxyl (—OH) groups on adjacent carbons.

Glycolipid *(Section 20.4)* A complex lipid that contains at least one monosaccharide.

Glycolysis *(Section 27.1)* The pathway by which glucose is converted to pyruvate, generating two ATPs and two $NADH+H^+$ in the process.

Glycoprotein *(Section 21.10)* A protein to which one or more carbohydrate molecules are bonded.

Glycoside *(Section 19.5)* A carbohydrate in which the —OH on its anomeric carbon is replaced by —OR.

Glycosidic bond *(Section 19.5)* The bond from the anomeric carbon of a glycoside to an —OR group.

Gp120 *(Section 30.5)* A 120 000 molecular weight glycoprotein on the surface of the human immunodeficiency virus that binds strongly to the CD4 molecules on T cells.

Graham's law *(Section 5.6)* The rate of diffusion of a molecule is inversely proportional to the square root of its molecule weight or density.

Gray (Gy) *(Section 9.5)* The SI unit of measure of the rad; 100 rad.

Ground state, of an atom *(Section 2.6)* The electron configuration of lowest energy state.

Group of the Periodic Table *(Section 2.5)* A column of the elements in the Periodic Table.

Haber process *(Box 7E)* The industrial process by which N_2 and H_2 are converted to NH_3.

Half-life, of a radioisotope *(Section 9.3)* The time it takes for one half of a sample to decay.

Halogen *(Section 2.5)* Those elements in Group 7A of the Periodic Table.

Haworth projection *(Section 19.3)* A way to view furanose and pyranose forms of monosaccharides; the ring is drawn flat and viewed through its edge with the anomeric carbon on the right.

HDL *(Section 20.9)* High Density Lipoprotein, a lipoprotein with high protein content (approximately 50 percent).

Heat of combustion *(Section 4.9)* The heat given off during a combustion reaction.

Heat of fusion *(Section 5.11)* The heat necessary to melt 1 g of a solid.

Heat of reaction *(Section 4.9)* The heat gained or lost during a reaction.

Heat of vaporization *(Section 5.11)* The heat necessary to vaporize 1 g of a liquid.

Helicase *(Section 24.4)* An unwinding protein that acts at a replication fork to unwind DNA so that DNA polymerase can synthesize a new DNA strand.

Hemiacetal *(Section 17.5)* A molecule containing a carbon bonded to an —OH and an —OR group; the product of adding one molecule of alcohol to the carbonyl group of an aldehyde or ketone.

Hemodialysis *(Box 6F)* The dialysis of blood.

Hemolysis *(Section 6.8)* The bursting of red blood cells.

Henle's loop *(Section 31.5)* A U-shaped twist in a kidney tubule between the proximal tubule and the distal tubule.

Hess's law *(Box 4E)* Law stating that the energy of a reaction depends only on the relative energies of reactants and products and not on the pathway by which the reaction takes place.

Heterocyclic aliphatic amine *(Section 16.2)* A heterocyclic amine in which nitrogen is bonded only to alkyl groups.

Heterocyclic amine *(Section 16.2)* An amine in which nitrogen is one of the atoms of a ring.

Heterocyclic aromatic amine *(Section 16.2)* An amine in which nitrogen is one of the atoms of an aromatic ring.

Heterogeneous *(Section 5.11)* Of inconsistent composition.

Histone *(Section 24.1)* A basic protein that is found in complexes with DNA in eukaryotes.

HIV *(Section 30.5)* Human Immunodeficiency Virus.

Homeostasis *(Section 31.1)* The process of maintaining the proper levels of nutrients and temperature of the blood.

Homogeneous *(Section 5.11)* The same composition throughout.

Hormone *(Section 23.2)* A compound that is synthesized in one location, travels large distances, usually in the blood, and then acts at a remote location.

Hund's rule *(Section 2.6)* Rule stating that the filling of *p, d,* and *f* orbitals in sequence so that each degenerate orbital is half filled before any one orbital is completely filled.

Hydrate *(Section 6.6)* The water contained in a crystalline compound in a specific ratio; *with water*.

Hydration *(Section 12.6)* Addition of water.

Hydrocarbon *(Section 11.2)* A compound that contains only carbon and hydrogen.

Hydrogen bonding *(Sections 5.7, 13.4)* A noncovalent interaction between a partial positive charge on a hydrogen bonded to an atom of high electronegativity and a partial negative charge on a nearby oxygen, nitrogen, or fluorine.

Hydrolase *(Section 22.2)* An enzyme that catalyzes a hydrolysis reaction.

Hydrolysis *(Section 17.5)* A reaction of a compound with water (hydro-) in which one or more bonds are broken (-lysis) and the —H and —OH of water add to the ends of the bond or bonds broken.

Hydronium ion *(Section 8.1)* The hydrated hydrogen ion, H_3O^+.

Hydrophilic *(Section 18.2)* From the Greek meaning water-loving.

Hydrophobic *(Section 18.2)* From the Greek meaning water-hating.

Hydrophobic interaction *(Section 21.9)* Interaction by London dispersion forces between hydrophobic groups.

Hydroxyl group *(Section 10.4)* An —OH group bonded to a tetrahedral carbon atom.

Hygroscopic *(Section 6.6)* A substance that absorbs water from the air.

Hyperbaric medicine *(Box 5E)* The exposure of a patient to a high partial pressure of oxygen in order to promote the healing process and to reverse the effect of some toxic substances.

Hyperglycemia *(Box 29D)* A condition whereby the blood glucose level reaches 140 mg/dL or higher.

Hypertension *(Box 31E)* The clinical term for high blood pressure.

Hyperthermia *(Box 1B)* Having a body temperature higher than normal.

Hyperthermophile *(Section 22.4)* An organism that lives at extremely high temperatures.

Hypertonic solution *(Section 6.8)* A solution that is more concentrated than another.

Hypoglycemia *(Box 29D)* A condition whereby the blood glucose level drops below the normal 65–100 mg/dL.

Hypothermia *(Box 1B)* Having a body temperature lower than normal.

Hypothesis *(Section 1.2)* A statement advanced to explain a fact.

Hypotonic solution *(Section 6.8)* A solution that is less concentrated than another.

Ideal gas *(Section 5.2)* A gas that conforms exactly to the assumptions of the kinetic molecular theory.

Ideal gas law *(Section 5.5)* $PV = nRT$.

Immunogen *(Section 30.3)* Another term for antigen.

Immunoglobulin *(Section 30.4)* An antibody protein generated against and capable of binding specifically to an antigen.

Immunoglobulin superfamily *(Section 30.4)* A family of molecules based on a similar structure that includes the immunoglobulins, T cell receptors, and other membrane proteins that are involved in cell communications. All molecules in this class have a certain portion that can react with antigens.

Indicator *(Section 8.8)* A substance that changes color at a specific pH.

Induced-fit model *(Section 22.5)* One model for enzyme–substrate interactions based on the postulate that there is a conformational shift in the active site of an enzyme when it binds a substrate.

Inhibition, of enzyme activity *(Section 22.3)* Any reversible or irreversible process that makes an enzyme less active.

Inhibitory neurotransmitters *(Section 23.4)* Neurotransmitters that decrease the transmission of nerve impulses.

Initiation, of protein synthesis *(Section 25.5)* The first step in the process whereby the base sequence of an mRNA is translated into the primary structure of a polypeptide.

Initiation signal *(Section 25.2)* A sequence on DNA that identifies the location where transcription is to begin.

Innate immunity *(Section 30.1)* The first line of defense against foreign invaders, which includes skin resistance to penetration, tears, mucus, and nonspecific macrophages that engulf bacteria.

Inorganic phosphate *(Section 26.3)* PO_4^{3-}.

Interleukin *(Section 30.6)* A cytokine that controls and coordinates the action of leukocytes.

Internal innate immunity *(Section 30.1)* The type of innate immunity that is used once a pathogen has already penetrated a tissue.

International System of Units (SI) *(Section 1.4)* A system of units of measurement based in part on the metric system.

Interstitial fluid *(Section 30.2)* The fluid that surrounds and fills in spaces between cells.

Intron *(Section 24.6)* A section of DNA that does not code for anything functional.

Ion *(Section 2.4)* An atom or a group of atoms with an unequal number of protons and electrons.

Ion product of water (K_w) *(Section 8.7)* The concentration of H_3O^+ multiplied by the concentration of OH^-: $K_w = [H_3O^+][OH^-] = 1 \times 10^{14}$.

Ionic bond *(Section 3.2)* A chemical bond resulting from the electrostatic attraction between positive and negative ions.

Ionic compound *(Sections 3.3, 3.8)* A compound composed of two or more ions.

Ionization *(Section 6.6)* Dissociation of a compound of compounds into ions.

Ionization energy *(Section 2.7)* The energy required to remove the most loosely held electron from an atom.

Ionizing radiation *(Section 9.4)* Radiation that causes one or more electrons to be ejected from an atom or molecule thereby creating a positive ion.

Isoelectric point *(Section 21.3)* The pH at which a molecule has no net charge, abbreviated pI.

Isoenzyme *(Section 22.6)* An enzyme that can be found in multiple forms but with each catalyzing the same reaction; also called an isozyme.

Isomers *(Section 11.3)* Different compounds that have the same molecular formula.

Isomerase *(Section 22.2)* An enzyme that catalyzes an isomerization reaction.

Isotonic solutions *(Section 6.8)* Two solutions that are of equal concentration.

Isotopes *(Section 2.4)* Atoms with the same number of protons but different numbers of neutrons.

Joule (J) *(Section 1.9)* SI base unit for heat; 1 J is 4.184 cal.

Kelvin scale (K) *(Section 1.4)* The SI base unit of temperature.

Ketoacidosis *(Box 27C)* A biochemical condition caused by starvation or diabetes where ketone bodies build up in the blood and cause the pH to decrease.

Ketogenic amino acid *(Section 27.9)* An amino acid that, when catabolized, can only form acetyl-CoA or acetoacetyl-CoA and therefore cannot lead to glucose production.

Ketone *(Section 10.4)* A compound containing a carbonyl group bonded to two carbon groups.

Ketone body *(Section 27.7)* Acetoacetate, acetone, and β-hydroxybutyrate, which are produced by an imbalance between fat and carbohydrate utilization.

Ketose *(Section 19.2)* A monosaccharide containing a ketone group.

Killer T cell *(Section 30.2)* A T cell that kills invading foreign cells by cell-to-cell contact.

Kinase *(Section 22.6)* An enzyme that covalently modifies a protein with a phosphate group, usually through the —OH group on the side chain of a serine, threonine, or tyrosine.

Kinetic energy *(Section 1.8)* The energy of motion; energy that is in the process of doing work.

Kinetic molecular theory *(Section 5.2)* A theory advanced to explain the behavior of gases.

Krebs cycle *(Section 26.4)* Another name for the citric acid cycle, used to honor Hans Krebs, the discoverer of the pathway.

Kwashiorkor *(Section 29.4)* A disease caused by insufficient protein intake and characterized by a swollen stomach, skin discoloration, and retarded growth.

L-Monosaccharide *(Section 19.2)* A monosaccharide that, when written as a Fischer projection, has the —OH on its penultimate carbon to the left.

Lactam *(Section 18.5)* A cyclic amide.

Lactone *(Section 18.4)* A cyclic ester.

Lagging strand *(Section 24.4)* A discontinuously synthesized DNA that elongates in a direction away from the replication fork.

Law of conservation of energy *(Section 1.8)* Law stating that energy can be neither created nor destroyed.

Law of conservation of mass *(Section 2.3)* Law stating that matter can be neither created nor destroyed.

Law of constant composition *(Section 2.3)* Law stating that each compound is composed of elements in a specific ratio.

LDL *(Section 20.9)* Low-Density Lipoprotein, a circulating lipoprotein where the cholesterol content is very high (45 percent) and the protein content is 25 percent or less.

Le Châtelier's principle *(Section 7.8)* Principle stating that when a stress is applied to a chemical system in equilibrium, the position of the equilibrium shifts in the direction to relieve the applied stress.

Leading strand *(Section 24.4)* The continuously synthesized DNA strand that elongates toward the replication fork.

Lecithin *(Section 20.6)* The common name for a phosphatidylcholine.

Leukocyte *(Sections 30.2, 31.2)* White blood cells, which are the principal parts of the acquired immunity system and act via phagocytosis or antibody production.

Leukotriene *(Section 20.12)* A class of acyclic arachidonic acid derivatives that act as mediators of hormonal responses.

Levorotatory *(Section 15.5)* Counterclockwise rotation of the plane of polarized light in a polarimeter.

Lewis structure, of an atom *(Sections 2.7, 3.4)* The symbol of the element surrounded by a number of dots equal to the number of electrons in the valence shell of an atom of the element.

Ligase *(Section 22.2)* A class of enzyme that catalyzes a reaction joining two molecules. These are often called synthetases or synthases.

Limiting reagent *(Section 4.5)* The reactant that is consumed, leaving an excess of the other reagent or reagents.

Line-angle drawing *(Section 11.5)* An abbreviated way to draw structural formulas in which each angle and line terminus represents a carbon and each line represents a bond.

Lipase *(Section 29.8)* An enzyme that catalyzes the hydrolysis of an ester bond between a fatty acid and glycerol.

Lipid *(Section 20.1)* A class of compounds found in living organisms that is insoluble in water but soluble in nonpolar solvents.

Lipid bilayer *(Section 20.5)* A back-to-back arrangement of phospholipids, often forming a closed vesicle or membrane.

Lipoproteins *(Section 20.9)* Combinations of lipid and protein, which are the circulating form of lipids in the blood.

Liquid state *(Section 1.6)* A physical state of matter in which there is a definite volume but no specific shape.

Lock and key model *(Section 22.5)* A model for enzyme substrate interaction based on the postulate that the active site of an enzyme is a perfect fit for the substrate.

London dispersion forces *(Section 5.7)* Extremely weak attractive forces between atoms and molecules due to electrostatic attraction between temporary positive and negative dipoles.

Lyase *(Section 22.2)* An enzyme that catalyzes the addition of two atoms or groups of atoms to a double bond or their removal to form a double bond.

Lymphocyte *(Section 30.2)* A white blood cell that spends most of its time in the lymphatic tissues. Those that mature in the bone marrow are B cells. Those that mature in the thymus are T cells.

Lymphoid organs *(Section 30.2)* The main organs of the immune system, such as the lymph nodes, spleen, and thymus, which are connected together by lymphatic capillary vessels.

Lysosome *(Section 26.2)* A subcellular organelle that cleanses a cell of damaged cellular components and some foreign material.

Macrophage *(Section 30.2)* An amoeboid white blood cell that moves through tissue fibers engulfing dead cells and bacteria by phagocytosis and then displays some of the engulfed antigens on its surface.

Magnetic resonance imaging (MRI) *(Box 9E)* The detection of the inversion of spin (flip) of hydrogen nuclei caused by exposure to radiofrequency radiation.

Major histocompatibility complex (MHC) *(Section 30.3)* A large set of cell surface proteins that aid in the binding of antigen epitopes to T cell receptors.

Malleability *(Section 2.5)* The ability to be pounded into a thin sheet.

Manometer *(Section 5.3)* An instrument for measuring the pressure of the atmosphere in a container.

Marasmus *(Section 29.3)* Another term for chronic starvation whereby the individual does not have adequate caloric intake. It is characterized by arrested growth, muscle wasting, anemia, and general weakness.

Markovnikov's rule *(Section 12.6)* Rule stating that, in the addition of HX or H_2O to an alkene, hydrogen adds to the carbon of the double bond having the greater number of hydrogens.

Mass *(Section 1.4)* The quantity of matter in an object; the SI base unit is the kilogram; often referred to as weight.

Mass number *(Section 2.4)* The sum of the number of protons and neutrons in an atom.

Matter *(Section 1.1)* Anything that has mass and takes up space.

Melting point *(Section 5.1)* The temperature at which a substance changes from the solid state to the liquid state.

Membrane *(Section 20.5)* A structure composed of a lipid bilayer, protein, and often carbohydrate that separates one cell compartment from another or one cell from another.

Memory *(Section 30.1)* The characteristic of the acquired immunity seen as a more rapid and vigorous response to a pathogen the second time it is encountered.

Memory cells *(Section 30.2)* A type of T cell that stays in the blood after an infection is over and acts as a quick line of defense if the same antigen is encountered again.

Mercaptan *(Section 13.3)* A common name for any molecule containing an —SH group.

Messenger RNA (mRNA) *(Sections 24.5, 25.1)* RNA that is transcribed from DNA that codes for proteins. It is characterized by being relatively long, unstable, and with no secondary structure.

Meta (*m*) *(Section 14.3)* Refers to groups occupying the 1 and 3 positions on a benzene ring.

Metabolism *(Section 26.1)* The sum of all the chemical reactions involved in maintaining the dynamic state of a cell or organism.

Metal-binding finger *(Section 25.6)* A type of transcription factor containing heavy metal ions, such as Zn^{2+}, that is involved in helping RNA polymerase bind to the DNA to be transcribed.

Metalloid *(Section 2.5)* An element that displays some characteristics of both metals and nonmetals.

Metal *(Section 2.5)* An element that is generally shiny, malleable, ductile and conductive electrically and thermally and that tends to give up electrons.

Metastasis *(Box 25E)* The process whereby cancerous cells proliferate and spread to other tissues.

Meter *(Section 1.4)* The SI base unit of length.

Metric system *(Section 1.4)* A system in which measurements of parameter are related by powers of ten.

Mineral in diet *(Section 29.5)* An inorganic ion that is required for structure and metabolism, such as Ca^{2+}, Fe^{2+}, and Mg^{2+}.

Mirror image *(Section 15.1)* The reflection of an object in a mirror.

Mitochondrion *(Section 26.2)* A subcellular organelle characterized by an outer membrane and a highly folded inner membrane. Mitochondria are responsible for the generation of most of the energy for cells.

Mixture *(Section 2.2)* A sample of matter composed of two or more substances in a variable ratio that can be separated by normal physical means; each component maintains its own identity and properties.

Molarity (*M*) *(Section 6.5)* Concentration expressed as the number of moles of a solute per liter of solution.

Mole *(Section 4.3)* The amount of any substance that contains as many entities as the number of atoms in exactly 12 g of carbon-12 (6.02×10^{23} atoms of carbon).

Molecular compound *(Section 3.8)* A compound composed for two or more elements joined by covalent bonds.

Molecular weight (MW) *(Section 4.2)* The sum of the atomic weights expressed in grams of the components of a molecular compound.

Molecule *(Section 2.3)* A compound in which all bonds are covalent.

Monoatomic element *(Section 2.3)* An element that exists in nature as single atoms.

Monoclonal antibody *(Section 30.5)* An antibody that is produced by fusing one B cell with a myeloma cell to create a hybridoma cell. The hybridoma cell will be immortal and will produce high quantities of only one specific antibody molecule.

Monoglyceride *(Section 20.2)* An ester of glycerol and a single fatty acid.

Monomer *(Section 12.7)* From the Greek *mono,* single, and *meros,* part; the simplest nonredundant unit from which a polymer is synthesized.

Monosaccharide *(Section 19.2)* A carbohydrate of the general formula $C_nH_{2n}O_n$, where *n* varies from 1 to 9.

Mutagen *(Section 25.7)* A chemical substance that induces a base change, or mutation, in DNA.

Mutarotation *(Section 19.3)* The change in optical activity that occurs when an α- or β- form of a carbohydrate is converted to an equilibrium mixture of the two forms.

Mutation *(Section 9.5)* A heritable change in the base sequence of DNA.

N-Linked saccharide *(Section 21.10)* A carbohydrate that is bonded by a N-glycosidic bond to a protein, most commonly to the side-chain nitrogen of asparagine.

N-terminus *(Section 21.5)* The amino acid at the end of a peptide chain that has a free amino group.

NAD⁺/NADH *(Section 26.3)* Nicotinamide adenine dinucleotide, a coenzyme participating in many metabolic oxidation-reduction reactions.

Nanotubes *(Section 5.10)* A classification of fullerenes.

Native structure *(Section 21.10)* The natural or normal conformation of a protein in the biological setting.

Negative modulation *(Section 22.6)* The process whereby an allosteric regulator inhibits enzyme action.

Nephron *(Section 31.5)* The tubular excretory unit of the kidney that filters the blood to remove waste products and to reabsorb nutrients.

Net ionic equation *(Section 4.7)* A chemical equation that does not contain spectator ions.

Network solid *(Section 5.10)* Extremely large crystal composed of one molecule.

Neuron *(Section 23.1)* Another name for a nerve cell.

Neuropeptide Y *(Section 23.6)* A brain peptide that affects the hypothalamus and is an appetite-stimulating agent.

Neurotransmitter *(Section 23.1)* Any compound that is involved in the communication between neurons or between neurons and target tissues.

Neutral salt *(Section 8.9)* The salt of a strong acid and a strong base; the salt of a weak acid and a weak base.

Neutral solution *(Section 8.8)* An aqueous solution with a pH of 7.

Neutralization *(Section 8.5)* The process by which acids and bases react with each other.

Neutron *(Section 2.4)* A nucleon with no charge and a mass of 1 amu.

Noble gas *(Section 2.5)* An element in Group 8A of the Periodic Table.

Nonbonding electrons *(Section 3.4)* Valence electrons not involved in forming covalent bonds; that is, unshared electrons.

Noncompetitive inhibitor *(Section 22.3)* An inhibitor that binds to an enzyme at a site other than the active site but that causes a shift in the tertiary structure of the enzyme such that its activity is reduced.

Nonmetal *(Section 2.5)* An element that does not have the characteristics of metals and tends to accept electrons.

Nonpolar covalent bond *(Section 3.4)* A covalent bond between atoms whose difference in electronegativity is less than 0.5.

Normal boiling point *(Section 5.9)* The temperature at which a substance boils when the pressure is 1 atm.

Normality (N) *(Section 8.11)* The number of equivalent weights per liter of solution.

Nuclear fission *(Section 9.8)* The process of splitting a nucleus.

Nuclear fusion *(Section 9.7)* The process of the joining of nuclei.

Nucleic Acid *(Section 24.1)* A polymer consisting of many nucleotide monomers; DNA and RNA.

Nucleon *(Section 2.4)* A subatomic particle found in the nucleus.

Nucleoside *(Section 24.2)* The combination of a heterocyclic aromatic amine attached via a β-glycosidic bond to either D-ribose or 2-deoxy-D-ribose.

Nucleosome *(Section 24.6)* Combinations of DNA and histone proteins.

Nucleotide *(Section 24.2)* A phosphoric ester of a nucleoside.

Nucleus *(Section 2.4)* The geographic center of an atom; the location of the nucleons. *(Section 26.2)* A subcellular organelle found in eukaryotes where the DNA is stored and replicated.

Nutrient *(Section 29.2)* Components of food and drink that provide energy, replacement, and growth.

O-Linked saccharide *(Section 21.10)* A carbohydrate that is bonded by a glycosidic bond to a protein, most commonly to the side chain OH of a serine, threonine, or hydroxylysine.

Obesity *(Section 29.3)* Accumulation of body fat beyond the norm.

Octet rule *(Section 3.2)* Rule stating that, when undergoing chemical reaction, atoms of Group 1A–7A elements gain, lose, or share electrons to achieve a configuration having eight valence electrons.

Okazaki fragment *(Section 24.4)* A piece of DNA that is synthesized discontinuously as the lagging strand as synthesis proceeds away from the replication fork.

Oligosaccharide *(Section 19.6)* A carbohydrate containing from four to ten monosaccharide units, each joined to the next by a glycosidic bond.

Oncogene *(Box 25E)* A gene that in some way participates in the development of cancer.

Operator site *(Section 25.6)* An upstream portion of an operon that controls the binding of RNA polymerase. In many cases, a repressor protein binds to the operator and impedes the binding of the polymerase.

Operon *(Section 25.6)* A system of gene regulation in prokaryotes. A series of DNA segments that make up

the regulatory genes, promoter and operator control sites, and structural genes that code for proteins.

Optically active *(Section 15.5)* Showing that a compound rotates the plane of polarized light.

Orbital *(Section 2.6)* A region in space about a nucleus where electrons can be found.

Order of precedence of functional groups *(Section 17.3)* A system for ranking functional groups in order of priority for the purposes of IUPAC nomenclature.

Organelle *(Section 26.2)* A special structure within a cell that carries out a particular part or parts of metabolism.

Organic chemistry *(Section 10.1)* The study of the compounds of carbon.

Ortho (o) *(Section 14.3)* Refers to groups occupying the 1 and 2 positions on a benzene ring.

Osmolarity *(Section 6.8)* The molarity of a solution multiplied by the number of particles produced by the solute molecules.

Outer transition element *(Section 2.5)* An element with two incomplete inner orbitals.

Oxidation *(Section 4.8)* The chemical process resulting in the loss of electrons.

Oxidation-reduction reaction *(Section 4.8)* A chemical reaction in which one or more electrons are transferred from one species to another.

Oxidative deamination *(Section 27.1)* An oxidation reaction in which a $CHNH_2$ group is converted to a $C{=}O$ group and NH_4^+.

Oxidative phosphorylation *(Section 26.5)* The process of generating ATPs by producing a hydrogen ion gradient during electron transport and then harnessing the energy of the gradient by letting the ions flow through the ATPase.

Oxidizing agent *(Section 4.8)* An entity that accepts the electrons in an oxidation-reduction reaction.

Oxidoreductase *(Section 22.2)* An enzyme that catalyzes an oxidation-reduction reaction.

Oxonium ion *(Section 12.6)* An ion in which oxygen is bonded to three other atoms and bears a positive charge.

P site *(Section 25.5)* The site on the large ribosomal subunit where the current peptide is bound before peptidyl transferase links it to the amino acid attached at the A site during elongation.

Para (p) *(Section 14.3)* Refers to groups occupying the 1 and 4 positions on a benzene ring.

Parenteral nutrition *(Box 29A)* The technical term for intravenous feeding.

Partial pressure *(Section 5.6)* The pressure of one gas in a mixture of gases.

Pauli exclusion principle *(Section 2.6)* A statement that any single orbital can hold a maximum of two electrons that have opposite spin (paired electrons).

Pentose phosphate pathway *(Section 27.2)* A pathway, often called the hexose monophosphate shunt, in which glucose 6-phosphate is decarboxylated to yield ribulose 5-phosphate. When necessary, this pathway can produce ribose for nucleotide synthesis and NADPH, which is required for many synthetic pathways.

Penultimate carbon *(Section 19.2)* The stereocenter of a monosaccharide farthest from the carbonyl group, as for example carbon 5 of glucose.

Peptide *(Section 21.5)* A short chain of amino acids linked via peptide bonds.

Peptide backbone *(Section 21.6)* The repeating pattern of peptide bonds in a polypeptide or protein.

Peptide bond *(Section 21.5)* The name given to an amide bond in a polypeptide or protein; also called a peptide linkage.

Peptidergic neurotransmitter *(Section 23.6)* A type of neurotransmitter/hormone that is based on a peptide, such as glucagon, insulin, and the enkephalins.

Percent composition *(Section 6.5)* A concentration expression in which the amount of a substance in a solution is compared to the total amounts of the solution components (weight/volume, weight/weight, volume/volume).

Percent yield *(Section 4.6)* The actual yield divided by the theoretical yield times 100.

Perforin *(Section 30.2)* A protein produced by killer T cells that punches holes in the membrane of target cells.

Period *(Section 2.5)* A row in the Periodic Table.

Periodic Table of the elements *(Section 2.5)* An organized array of the known elements.

pH *(Section 8.8)* The negative log of the hydrogen ion concentration; $pH = -\log[H^+]$.

Phagocytosis *(Section 30.6)* The process by which large particulates, including bacteria, are pulled inside a white cell called a phagocyte.

Phase change *(Section 5.11)* A change of physical state.

Phase diagram *(Section 5.11)* A graph of temperature versus pressure that illustrates the phase changes of a substance.

Phenol *(Section 14.5)* A compound that contains an —OH bonded to a benzene ring.

Phenyl group *(Section 14.3)* The C_6H_5— group.

Pheromone *(Box 12A)* A chemical secreted by an insect to influence the behavior of another member of the same species.

Phosphatidylcholine *(Section 20.6)* A glycerophospholipid where the substituent attached via a phosphoric ester bond is choline.

Phosphatidylinositol *(Section 20.6)* A glycerophospholipid where the substituent attached via a phosphoric ester bond is inositol.

Phospholipid *(Section 20.4)* A complex lipid comprising an alcohol, such as glycerol or sphingosine, a phosphate group, and fatty acids.

Photon *(Section 9.1)* The smallest unit of electromagnetic radiation.

Photosynthesis *(Sections 19.1, 28.2)* The process in plants whereby the sun's energy is used to turn CO_2 and H_2O into glucose and oxygen.

Physical change *(Section 1.1)* A change in matter in which it does not lose its identity.

Physical property *(Section 1.1)* Characteristics of a substance that are not chemical properties; those properties that are not a result of a chemical change.

pI *(Section 21.3)* Isoelectric point of a peptide or protein with ionizable acidic and basic groups. The pH at which such a molecule has no net charge.

PIP$_2$ *(Section 20.6)* Phosphatidyl inositol 3,4-bisphosphate.

pK *(Section 8.8)* The negative log of an equilibrium constant

Plane of symmetry *(Section 15.2)* An imaginary plane passing through an object, dividing it such that one half is the mirror image of the other half.

Plane polarized light *(Section 15.5)* Light vibrating in only parallel planes.

Plasma cell *(Section 30.2)* A cell derived from a B cell that has been exposed to an antigen.

Plasmid *(Section 25.8)* A small, circular, double-stranded DNA molecule of bacterial origin. Plasmids were found to naturally confer antibiotic resistance for certain bacteria. Now they are used for genetic engineering purposes.

Plastic *(Box 12C)* A polymer that can be molded when hot and that retains its shape when cooled.

Platelets *(Section 31.2)* Also called thrombocytes, these are special cells in the blood that control bleeding when there is an injury.

pOH *(Section 8.8)* The negative log of the hydroxide ion concentration; $pOH = -\log[OH^-]$.

Polar covalent bond *(Section 3.4)* A covalent bond between atoms whose difference in electronegativity is between 0.5 and 1.9.

Polarimeter *(Section 15.5)* An instrument for measuring the ability of a compound to rotate the plane of polarized light.

Polarity *(Section 3.2)* The result of having two centers of opposite charge; partial positive, δ^+, and partial negative, δ^-.

Polyamide *(Section 18.8)* A polymer in which each monomer unit is bonded to the next by an amide bond, as for example nylon 66.

Polyatomic element *(Section 2.3)* More than two atoms of an element per molecule.

Polyatomic ion *(Section 3.7)* An ion composed of two or more elements.

Polycarbonate *(Section 18.8)* A polyester in which the carboxyl groups are derived from carbonic acid.

Polyester *(Section 18.8)* A polymer in which each monomer unit is bonded to the next by an ester bond, as for example poly(ethylene terephthalate).

Polymer *(Section 12.7)* From the Greek *poly,* many, and *meros,* parts; any long-chain molecule synthesized by bonding together many single parts called monomers.

Polymerase chain reaction (PCR) *(Sections 22.4, 24.6)* An automated technique for amplifying DNA using a heat-stable DNA polymerase from thermophilic bacteria.

Polynuclear aromatic hydrocarbon *(Section 14.3)* A hydrocarbon containing two or more benzene rings, each of which shares two carbons with another benzene ring.

Polypeptide *(Section 21.5)* A long chain of amino acids bonded via peptide bonds.

Polysaccharide *(Section 19.7)* A carbohydrate containing a large number of monosaccharide units, each joined to the next by one or more glycosidic bonds.

Positive modulation *(Section 22.6)* The process whereby an allosteric regulator increases enzyme action.

Positron *(Section 9.3)* A particle with the mass of an electron and a positive charge, $_{+1}^{0}e$ (β^+).

Positron emission tomography (PET) *(Box 9E)* The detection of positrons emitted.

Postsynaptic membrane *(Section 23.2)* The membrane on the side of the synapse nearest the dendrite of the neuron receiving the transmission.

Potential energy *(Section 1.8)* Energy that is being stored; energy that is available for later use.

Presynaptic membrane *(Section 23.2)* The membrane on the side of the synapse nearest the axon of the neuron transmitting the signal.

Primary (1°) alcohol *(Section 10.4)* An alcohol in which the carbon bearing the —OH group is bonded to only one other carbon.

Primary (1°) amine *(Section 10.4)* An amine in which nitrogen is bonded to one carbon and two hydrogens.

Primary structure, of DNA *(Section 24.3)* The order of the bases in DNA.

Primary structure, of proteins *(Section 21.7)* The order of amino acids in a peptide, polypeptide, or protein.

Principal energy level *(Section 2.6)* The number that is used to designate an orbit or an orbital.

Proenzyme *(Section 22.6)* An inactive form of an enzyme that must have part of its polypeptide chain cleaved before it becomes active.

Prokaryote *(Section 25.6)* An organism that has no true nucleus or organelles.

Promoter *(Box 25C)* An upstream DNA sequence that is used for RNA polymerase recognition and binding to DNA.

Proportional counter *(Section 9.4)* A Geiger-Müller counter.

Prostaglandin *(Section 20.12)* A fatty acid-like substance derived from arachidonic acid that is involved in inflammation and smooth muscle metabolism.

Prosthetic group *(Section 21.9)* The non-amino-acid part of a conjugated protein.

Protein *(Section 21.1)* A long chain of amino acids linked via peptide bonds. There are usually 30 to 50 amino acids in a chain before it is considered a protein.

Protein modification *(Section 22.7)* The process of affecting the enzyme activity by covalently modifying the enzyme, such as phosphorylating a particular amino acid.

Proteoglycans *(Section 21.10)* A special class of glycoprotein where the carbohydrate part is a long chain of acidic polysaccharides called glycosaminoglycans.

Protic *(Section 8.3)* Refers to the number of protons (hydrogens) in an acid; monoprotic—one hydrogen, diprotic—two hydrogens, triprotic—three protons.

Proton *(Section 2.4)* A nucleon with a +1 charge and a mass of 1 amu.

Proton channel *(Section 26.6)* An opening in a membrane that will let hydrogen ions pass through it. The ATPase has such a channel that will allow hydrogen ions to pass back into the mitochondria while the energy is used to make ATP.

Proton gradient *(Section 26.6)* A difference in proton (hydrogen ion) concentration across a membrane. During the electron transport process, hydrogen ions are pumped out of the mitochondria, causing a gradient to form, with the outside having a higher $[H^+]$.

Proton translocating ATPase *(Section 26.6)* The ATPase of the inner mitochondrial membrane allows protons to pass through it and uses the energy to phosphorylate

ADP to ATP. The enzyme could also hydrolyze ATP and use this energy to move hydrogen ions out of the matrix.

Proximal tubule *(Section 31.5)* The part of the kidney tubule close to the Bowman's capsule.

Pyramidal shape *(Section 3.5)* A three-dimensional shape composed of three equal-sized and -shaped triangular faces and one triangular face of different dimensions (NH_3).

Pyranose *(Section 19.3)* A six-membered cyclic hemiacetal form of a monosaccharide.

Quantized energy of an electron *(Section 2.6)* Specific energy levels of an electron in an atom.

Quaternary structure *(Section 21.9)* The organization of a protein that has multiple polypeptide chains, or subunits. This refers principally to the way the multiple chains interact.

R *(Section 15.3)* From the Latin *rectus*, meaning straight, correct; used in the R,S system to show that, when the group of lowest priority is away from you, the order of priority of groups on a stereocenter is clockwise.

R,S system *(Section 15.3)* A set of rules for specifying configuration about a stereocenter.

R— *(Section 11.4)* A symbol used to represent an alkyl group.

Racemic mixture *(Section 15.2)* A mixture of equal amounts of two enantiomers.

Rad *(Section 9.5)* Radiation absorbed dose. The SI unit is the gray (Gy).

Radiation, nuclear *(Section 9.1)* Radiation emitted from a nucleus during nuclear decay; includes alpha particles, beta particles, gamma rays, and positrons.

Radioactive *(Section 9.1)* Refers to a substance that emits radiation during nuclear decay.

Radioactive dating *(Box 9B)* The process of establishing the age of a substance by means of the analysis of radio-isotope abundance as compared with the current amount.

Radioactivity *(Section 9.1)* The process by which radiation is emitted from a nucleus and/or the nuclear components change and/or the nucleus splits.

Random coils *(Section 21.8)* Proteins that do not exhibit any repeated pattern.

Rate constant *(Section 7.4)* A proportionality constant between the molar concentration of reactant(s) and the rate of reaction (rate = k[compound]).

RDA *(Section 29.2)* Recommended Dietary Allowance, an average daily requirement for nutrients published by the U.S. Food and Drug Administration.

Reaction mechanism *(Section 12.6)* A step-by-step description of how a chemical reaction occurs.

Real gas *(Section 5.2)* A gas that is not an ideal gas but generally conforms to the kinetic molecular theory.

Receptor *(Section 23.1)* A membrane protein that can bind a chemical messenger and then perform a function, such as synthesizing a second messenger or opening an ion channel. *(Section 30.1)* Antibody-like molecule that is attached to the surface of the membrane of a cell such as a T cell and that recognizes and binds to antigens.

Recognition site *(Section 25.3)* The area of the tRNA molecule that recognizes the mRNA codon.

Recombinant DNA *(Section 25.8)* DNA from two sources that have been combined into one molecule.

Redox *(Section 4.8)* An oxidation-reduction reaction.

Reducing agent *(Section 4.8)* An entity that donates electrons in an oxidation-reduction reaction.

Reducing sugar *(Section 19.5)* A carbohydrate that reacts with a mild oxidizing agent under basic conditions to give an aldonic acid; the carbohydrate reduces the oxidizing agent.

Reduction *(Section 4.8)* The chemical process in which electrons are gained.

Regioselective reaction *(Section 12.6)* A reaction in which one direction of bond forming or bond breaking occurs in preference to all other directions.

Regulator *(Section 22.6)* A molecule that binds to an allosteric enzyme and changes its activity. This change could be positive or negative.

Regulatory gene *(Section 25.6)* A gene that produces a protein that controls the transcription of a structural gene.

Regulatory site *(Section 22.6)* A site, other than the active site, where a regulator binds to an allosteric enzyme and affects the rate of reaction.

Rem (roentgen equivalent man) *(Section 9.5)* One rem is a measure of the effect of 1 roentgen on a human.

Replication *(Section 24.4)* The process whereby DNA is duplicated to form two exact replicas from the original DNA molecule.

Representative element *(Section 2.5)* An A-group element in the Periodic Table.

Residue *(Section 21.5)* Another term for an amino acid in a peptide chain.

Respiration *(Section 4.8)* Oxidation of substances occurring within living organisms.

Respiratory chain *(Section 26.3)* Another term for the electron transport chain.

Response element *(Section 25.6)* A sequence of DNA upstream from a promoter that interacts with a transcription factor to initiate transcription in eukaryotes.

Restriction endonuclease *(Section 25.8)* An enzyme, usually purified from bacteria, that cuts DNA at a specific base sequence. Restriction endonucleases are the tools that allow the creation of recombinant DNA.

Retrovirus *(Box 25B)* A virus, such as HIV, that has an RNA genome and converts it to DNA when it infects its host.

Reuptake *(Section 23.4)* The transport of a neurotransmitter from its receptor back through the presynaptic membrane into the neuron.

Ribonucleic acid *(Section 24.2)* A type of nucleic acid consisting of nucleotide monomers of a nitrogenous base, ribose, and phosphate.

Ribosomal RNA (rRNA) *(Section 24.5)* The type of RNA that is complexed with proteins and makes up the ribosomes used in translation of mRNA into protein.

Ribosome *(Section 24.5)* A cell organelle constructed in the nucleolus, functioning as the site of protein synthesis in the cytosol. It is composed of protein and rRNA and has many subunits.

Ribozyme *(Section 22.1, Section 24.5)* An enzyme that is made up of ribonucleic acid. The currently recognized ones catalyze cleavage of part of their own sequences in mRNA and tRNA.

RNA *(Section 24.2)* An abbreviation for ribonucleic acid.

Roentgen (R) *(Section 9.5)* The amount of radiation that produces ions having 2.58×10^{-4} coulombs per kilogram.

Rusting *(Section 4.8)* Oxidation of iron.

S *(Section 15.3)* From the Latin *sinister,* meaning left; used in the R,S system to show that, when the group of lowest priority is away from you, the order of priority of groups on a stereocenter is counterclockwise.

Saccharide *(Section 19.1)* A simpler member of the carbohydrate family, such as glucose.

Salt bridge *(Section 21.9)* A name for the electrostatic attraction that occurs between the charged side chains of acidic or basic amino acids in a polypeptide or protein.

Saponification *(Section 18.4)* Hydrolysis of an ester in aqueous NaOH or KOH to an alcohol and the sodium or potassium salts of a carboxylic acid.

Satellites *(Section 24.6)* Short sequences of DNA that are repeated hundreds or thousands of times but that do not code for any protein or RNA.

Saturated *(Section 6.4)* A solution in which the solvent has dissolved the maximum amount of solute at a specific temperature and pressure.

Saturated fatty acid *(Section 20.2)* A fatty acid that has no carbon-carbon double bonds.

Saturated hydrocarbon *(Section 11.2)* A hydrocarbon containing only carbon-carbon single bonds.

Saturation curve *(Section 22.4)* A graph of enzyme rate versus substrate concentration. At high levels of substrate, the enzyme becomes saturated, and the velocity does not increase linearly with increasing substrate.

Scientific method *(Section 1.2)* A method of acquiring knowledge by testing theories.

Scintillation counter *(Section 9.4)* A device containing a phosphor that emits light on exposure to ionizing radiation that can be sensed.

Secondary (2°) alcohol *(Section 10.4)* An alcohol in which the carbon bearing the —OH group is bonded to two other carbons.

Secondary (2°) amine *(Section 10.4)* An amine in which nitrogen is bonded to two carbons and one hydrogen.

Secondary messenger *(Section 23.1)* A molecule that is created/released due to the binding of a hormone or neurotransmitter, which then proceeds to carry and amplify the signal inside the cell.

Secondary structure in proteins *(Section 21.8)* Repeating structures within polypeptides that are based solely on interactions of the peptide backbone and that do not include interactions of the side chains. Examples are the alpha helix and the beta-pleated sheet.

Secondary structure, of DNA *(Section 24.3)* Specific forms taken by DNA due to pairing of complementary bases.

Self ionization *(Section 8.7)* The ionization of a solvent producing an equilibrium.

Semiconservative replication *(Section 24.4)* Replication of DNA strands whereby each daughter molecule has one parental strand and one newly synthesized strand.

Semipermeable membrane *(Section 6.8)* A membrane that allows only certain substances to pass through.

Serendipity *(Section 1.2)* A chance observation.

Serum *(Section 31.2)* A clear liquid that can be extracted from blood plasma. It has all the soluble components of plasma but lacks the fibrinogen that makes blood clots.

Shell *(Section 2.6)* All the orbitals in a principal energy level.

SI *(Section 1.4)* International System of Units.

Side chains *(Section 21.6)* The part of an amino acid that varies one from the other. The side chain is attached to the alpha carbon and the nature of the side chain determines the characteristics of the amino acid.

Sievert (Sv) *(Section 9.5)* One sievert is the value of 100 rem.

Signal transduction *(Section 23.5)* The process that occurs after a messenger binds to its receptor that passes and amplifies the signal.

Significant figures *(Appendix II)* The total number of digits in a number with the exception of place holders; the digits that express the certainty of a number.

Single bond *(Section 3.4)* A bond formed by sharing one pair of electrons; represented by a single line between two bonded atoms.

Solid *(Section 1.6)* A physical state of matter in which there is a definite volume and shape.

Solid state *(Section 5.1)* The physical state in which a substance is a solid.

Solubility *(Section 6.4)* The maximum amount of a substance that can be dissolved in a solute at a specific temperature and pressure.

Soluble immunoglobulins *(Section 30.1)* Another term for the antibodies secreted by B cells, which surround and immobilize antigens.

Solute *(Section 6.2)* The substance or substances that are dissolved in a solvent to produce a solution.

Solution *(Section 6.1)* A homogeneous mixture composed of a solvent and solute or solutes (solute particle size below 1 nm).

Solvent *(Section 6.2)* The fraction of a solution in which the other components are dissolved.

Specific gravity *(Section 1.7)* The density of a substance compared to water as a standard.

Specific heat *(Section 1.9)* The amount of heat (calories) necessary to raise the temperature of 1 g of a substance by 1°C.

Specific rotation *(Section 15.5)* The number of degrees by which an optically active compound rotates the plane of polarized light; it is given the symbol [α].

Specificity *(Section 30.1)* A characteristic of acquired immunity based on the fact that cells make specific antibodies to a wide range of pathogens.

Spectator ions *(Section 4.7)* An ion in a chemical reaction that is unchanged.

Sphingolipid *(Section 20.4)* A phospholipid with sphingosine as the alcohol backbone.

Sphygmomanometer *(Box 5C)* The device with which blood pressure is taken.

Standard conditions (STP) *(Section 5.5)* One atmosphere of pressure and 0°C (273 K).

Starting materials *(Section 4.1)* The chemicals that are being converted to other chemicals during a chemical reaction.

Step-growth polymerization *(Section 18.8)* A polymerization in which chain growth occurs in a stepwise manner between difunctional monomers, as for example between adipic acid and hexamethylenediamine to form nylon 66.

Stereocenter *(Section 15.2)* A tetrahedral carbon to which four different groups are bound.

Stereoisomers *(Section 11.8)* Isomers that have the same connectivity (the same order of attachment of their atoms) but a different orientation of their atoms in space.

Steroid *(Section 20.9)* A molecule that contains the characteristic four-ring steroid nucleus.

Stoichiometry *(Section 4.5)* The mass relationships in a chemical reaction.

STP *(Section 5.5)* Standard conditions; 0°C and 1 atm.

Structural genes *(Section 25.6)* Genes that code for the product proteins of an operon rather than for proteins that control the operon.

Sublimation *(Section 5.11)* A change in the physical state from a solid directly to a gas.

Subshell *(Section 2.6)* All the orbitals having the same principal quantum number and letter designation of the specific electron cloud (s, p, d, f).

Substance P *(Section 23.6)* An 11 amino acid peptidergic neurotransmitter involved in the transmission of pain signals.

Substrate *(Section 22.3)* The compound or compounds whose reaction an enzyme catalyzes.

Subunit *(Section 22.6)* An individual polypeptide chain of an enzyme that has multiple chains.

Supersaturated *(Section 6.4)* A solution in which the solvent has dissolved an amount of solute beyond the maximum amount at a specific temperature and pressure.

Surface presentation *(Section 30.1)* The process where a portion of an antigen from a foreign pathogen that infected a cell is brought to the surface of the cell.

Surface tension *(Section 5.8)* The layer on the surface of a liquid produced by uneven intermolecular attractions at its surface.

Suspension *(Section 6.7)* An unstable system in which the solute particles are larger than 1000 nm and that separates with time.

Synapse *(Section 23.2)* The space between the axon of one neuron and a dendrite of another.

Systolic blood pressure *(Box 5C)* The arterial blood pressure during the pumping stage of the heart.

T cell *(Section 30.2)* The type of lymphoid cell that matures in the thymus and reacts with antigens via bound receptors on its cell surface. T cells can differentiate into memory T cells or killer T cells.

T cell receptor complex *(Section 30.5)* The combination of T cell receptors, antigens, and cluster determinants (CD) that are all involved in the T cell's ability to bind antigen.

Tautomers *(Section 17.6)* Constitutional isomers that differ in the location of hydrogen and a double bond relative to an O or N.

Temperature *(Section 1.9)* The measure of the amount of heat in a substance; measurement is made in a temperature scale (°F, °C, or K); K is the SI base unit.

Termination *(Section 25.5)* The final stage of translation where a termination sequence on mRNA tells the ribosomes to dissociate and release the newly synthesized peptide.

Termination sequence *(Section 25.2)* A sequence of DNA that tells the RNA polymerase to terminate synthesis.

Terpene *(Section 12.5)* A compound whose carbon skeleton can be divided into two or more units identical to the carbon skeleton of isoprene.

Tertiary (3°) alcohol *(Section 10.4)* An alcohol in which the carbon bearing the —OH group is bonded to three other carbons.

Tertiary (3°) amine *(Section 10.4)* An amine in which nitrogen is bonded to three carbons.

Tertiary structure *(Section 21.9)* The overall conformation of a polypeptide chain, including the interactions of the side chains and the position of every atom in the polypeptide.

Tetrahedral shape *(Section 3.5)* A three-dimensional figure having the shape of a triangular-based pyramid.

Theoretical yield *(Section 4.6)* The amount of product that should be received after a chemical reaction has been performed according to the stoichiometry of the reaction.

Theory *(Section 1.2)* A hypothesis that is supported by evidence; a hypothesis that has passed tests.

Thiol *(Section 13.2)* A compound containing an —SH (sulfhydryl) group bonded to a tetrahedral carbon.

Thrombocyte *(Section 31.2)* A cell whose function is to control bleeding after an injury; also called platelets.

Thrombosis *(Box 31B)* The process where a blood clot breaks loose from one part of the body and lodges in an artery in another part of the body.

Thromboxane *(Section 20.12)* An arachidonic acid derivative involved in blood clotting.

Titration *(Section 8.11)* The neutralization reaction brought about by the addition of acid to a base or base to an acid.

TNF *(Section 30.6)* Tumor Necrosis Factor, a type of cytokine produced by T cells and macrophages that has the ability to lyse susceptible tumor cells.

Trans *(Section 11.8)* A prefix meaning across from.

Transamination *(Section 27.8)* A reaction that involves the interchange of the alpha amino group from the amino acid and a keto group from an alpha ketoacid.

Transcription *(Section 25.1)* The process by which RNA polymerase synthesizes RNA from a DNA template.

Transcription factors *(Section 25.6)* Binding proteins that facilitate the binding of RNA polymerase to the DNA to be transcribed.

Transfection *(Box 25C)* The process wherein a virus infects another organism and the viral genome is then expressed in the host cells.

Transfer RNA (tRNA) *(Sections 24.5, 25.1)* A small RNA molecule that is attached to an amino acid and is used in translation to align an amino acid in its correct order.

Transferase *(Section 22.2)* A class of enzymes that catalyzes a reaction where a group of atoms, such as an acetyl group or amino group, is transferred from one molecule to another. *(Section 25.5)* An abbreviation for peptidyl transferase, the enzyme activity of the ribosomal complex that is responsible for formation of peptide bonds between the amino acids of the growing peptide.

Transgenic *(Box 25C)* An organism that expresses foreign proteins due to transfection.

Transition element *(Section 2.7)* An element having one or more incomplete inner shells; an element located in a group other than an A group of the Periodic Table.

Translation *(Section 25.1)* The process wherein a base sequence of mRNA is used to create a protein.

Translocation *(Section 25.5)* The part of translation where the ribosome moves down the mRNA a distance of three bases, so that the new codon is on the A site.

Transuranium element *(Section 9.7)* An element with an atomic number greater than 92.

Triacylglycerol *(Section 20.2)* The more scientific term for triglyceride.

Tricarboxylic acid cycle *(Section 26.4)* Another term for the citric acid cycle.

Triglyceride *(Section 20.2)* A lipid composed of glycerol esterified to a fatty acid at each of its three hydroxyls.

Triple bond *(Section 3.4)* A bond formed by sharing three pairs of electrons; represented by three lines between the two bonded atoms.

Triple helix *(Section 21.8)* The collagen triple helix is composed of three peptide chains. Each chain is itself a left-handed helix. These chains are twisted around each other in a right-handed helix.

Triple point *(Section 5.11)* The unique condition of temperature and pressure for any substance at which solid, liquid, and vapor phases coexist in equilibrium.

Trisaccharide *(Section 19.7)* A combination of three monosaccharides linked by glycosidic bonds.

Tumor suppression factor *(Box 25F)* A protein that controls replication of DNA so that cells do not divide constantly. Many cancers are caused by mutated tumor suppression factors.

Tyndall effect *(Section 6.7)* A characteristic of a colloid in which light passing through the colloid scatters.

Ubiquitinylation *(Box 27E)* The process of attaching one or more of the protein called ubiquitin to an aging protein that is ready for degradation.

Universal gas constant (R) *(Section 5.5)* 0.0821 L·atm/mol·K.

Unsaturated *(Section 6.4)* A solution in which the solvent has not dissolved the maximum amount of solute at a specific temperature and pressure.

Unsaturated fatty acid *(Section 20.2)* A fatty acid containing one or more carbon-carbon double bonds.

Unwinding protein *(Section 24.4)* Special proteins that help unwind DNA so that it can be replicated.

Urea cycle *(Section 27.1)* The pathway of nitrogen metabolism where excess nitrogen is converted to urea.

Uronic acid *(Section 19.5)* The oxidized form of a monosaccharide where the terminal CH_2OH group is oxidized to a COOH group.

Valence electron *(Section 2.7)* An electron in the outermost shell of an atom.

Valence shell *(Section 2.7)* The outermost shell of an atom.

Valence-shell electron-pair repulsion model (VSEPR) *(Section 3.5)* A model for predicting the bond angles about an atom within a molecule or ion based on the number of regions of electron density about that atom.

Vapor *(Section 5.9)* A gas.

Vapor pressure *(Section 5.9)* The pressure of a vapor above its liquid in a closed container.

Vaporize *(Section 5.9)* The process of converting a substance from the solid or liquid state to the gaseous state.

Vesicle, synaptic *(Section 23.2)* A compartment containing a neurotransmitter that fuses with a presynaptic membrane and releases its contents when a nerve impulse arrives.

Vitamin *(Section 29.5)* An organic substance required in small quantities in the diet of most species, which generally functions as a cofactor in important metabolic reactions.

Voltaic cell *(Box 4B)* A redox reaction used to generate an electric current between two solutions.

Volume *(Section 1.4)* The space a substance occupies; the cubic meter is the base SI unit.

VSEPR model *(Section 3.5)* Valence-shell electron-pair repulsion model.

Water of hydration *(Section 6.6)* The water molecules that are in a compound that has been hydrated.

Wavelength *(Section 9.1)* The distance from the top of one wave to the top of the next wave.

Weight *(Section 1.4)* The result of a mass acted upon by gravity; the base unit of measure is a gram (g).

X-ray *(Section 9.1)* A type of electromagnetic radiation with a wave length shorter than ultraviolet light and longer than gamma rays.

Zaitsev's rule *(Section 13.5)* Dehydration of an alcohol normally forms the alkene with the more substituted double bond.

Zwitterion *(Section 21.3)* A molecule that has equal numbers of positive and negative charges, giving it a net charge of zero.

Zymogen *(Section 22.6)* An inactive form of an enzyme that must have part of its polypeptide chain cleaved before it becomes active. It is a synonym for proenzyme.

INDEX

Key: **Boldface** indicates glossary entry; *italics* indicates figure; 30*t* indicates table.

Some Important Organic Functional Groups

Functional Group*	Example	IUPAC (Common) Name
Alcohol $\quad$ —$\ddot{\text{O}}$H	CH₃CH₂OH	Ethanol (Ethyl alcohol)
Aldehyde $\quad$ —C(=$\ddot{\text{O}}$)—H	CH₃CH (C=O)	Ethanal (Acetaldehyde)
Alkane	CH₃CH₃	Ethane
Alkene $\quad$ C=C	CH₂=CH₂	Ethene (Ethylene)
Alkyne $\quad$ —C≡C—	HC≡CH	Ethyne (Acetylene)
Amide $\quad$ —$\ddot{\text{N}}$—C(=$\ddot{\text{O}}$)—	CH₃CNH₂ (C=O)	Ethanamide (Acetamide)
Amine $\quad$ —$\ddot{\text{N}}$H₂	CH₃CH₂NH₂	Ethanamine (Ethylamine)
Anhydride $\quad$ —C(=$\ddot{\text{O}}$)—$\ddot{\text{O}}$—C(=$\ddot{\text{O}}$)—	CH₃COOCCH₃ (C=O, C=O)	Ethanoic anhdride (Acetic anhydride)
Arene	benzene	Benzene
Carboxylic acid $\quad$ —C(=$\ddot{\text{O}}$)—$\ddot{\text{O}}$H	CH₃COH (C=O)	Ethanoic acid (Acetic acid)
Disulfide $\quad$ —$\ddot{\text{S}}$—$\ddot{\text{S}}$—	CH₃SSCH₃	Dimethyl disulfide
Ester $\quad$ —C(=$\ddot{\text{O}}$)—$\ddot{\text{O}}$—C—	CH₃COCH₃ (C=O)	Methyl ethanoate (Methyl acetate)
Ether $\quad$ —$\ddot{\text{O}}$—	CH₃CH₂OCH₂CH₃	Diethyl ether
Haloalkane (Alkyl halide) $\quad$ —$\ddot{\text{X}}$: $\quad$ X = F, Cl, Br, I	CH₃CH₂Cl	Chloroethane (Ethyl chloride)
Ketone $\quad$ —C(=$\ddot{\text{O}}$)—	CH₃CCH₃ (C=O)	Propanone (Acetone)
Phenol $\quad$ benzene—$\ddot{\text{O}}$H	benzene—OH	Phenol
Sulfide $\quad$ —$\ddot{\text{S}}$—	CH₃SCH₃	Dimethyl sulfide
Thiol $\quad$ —$\ddot{\text{S}}$H	CH₃CH₂SH	Ethanethiol (Ethyl mercaptan)

* Where bonds to an atom are not specified, the atom is assumed to be bonded to one or more carbon or hydrogen atoms in the rest of the molecule.